Mathématiques et Applications

Volume 90

Le but de cette collection, créée par la Société de Mathématiques Appliquées et Industrielles (SMAI), est d'éditer des cours avancés de Master et d'école doctorale ou de dernière année d'école d'ingénieurs. Les lecteurs concernés sont donc des étudiants, mais également des chercheurs et ingénieurs qui veulent s'initier aux méthodes et aux résultats des mathématiques appliquées. Certains ouvrages auront ainsi une vocation purement pédagogique alors que d'autres pourront constituer des textes de référence. La principale source des manuscrits réside dans les très nombreux cours qui sont enseignés en France, compte tenu de la variété des diplômes de fin d'études ou des options de mathématiques appliquées dans les écoles d'ingénieurs. Mais ce n'est pas l'unique source: certains textes pourront avoir une autre origine.

This series was founded by the "Société de Mathématiques Appliquées et Industrielles" (SMAI) with the purpose of publishing graduate-level textbooks in applied mathematics. It is mainly addressed to graduate students, but researchers and engineers will often find here advanced introductions to current research and to recent results in various branches of applied mathematics. The books arise, in the main, from the numerous graduate courses given in French universities and engineering schools ("grandes écoles d'ingénieurs"). While some are simple textbooks, others can also serve as references.

Thierry Gallouët • Raphaèle Herbin

Weak Solutions of Partial Differential Equations

 Springer

Thierry Gallouët
Saint François Longchamp
France

Raphaèle Herbin
Institut de Mathématiques de Marseille
Aix-Marseille University
Marseille, France

ISSN 1154-483X   ISSN 2198-3275   (electronic)
Mathématiques et Applications
ISBN 978-3-031-98981-0   ISBN 978-3-031-98982-7   (eBook)
https://doi.org/10.1007/978-3-031-98982-7

Mathematics Subject Classification (2020): 35-01

This Springer imprint is published by the registered company Springer Nature Switzerland AG
The registered company address is: Gewerbestrasse 11, 6330 Cham, Switzerland

If disposing of this product, please recycle the paper.

Preface

This book describes some tools for the mathematical study of partial differential equations (PDEs). A partial differential equation is a mathematical equation whose unknown is a function of several variables that involves the partial derivatives of this function with respect to these variables. For example, one might want to calculate the temperature at a point in space and over time. The unknown function is then the temperature, and the PDE involves the partial derivatives of this function with respect to time and space variables. Partial differential equations are therefore involved in many models of physics, engineering and biology, such as the propagation of heat or sound, fluid flow, electrodynamics, the spread of epidemics. They are also used for weather forecasting models and climate models. PDEs came into existence after the work of Leibniz and Newton on infinitesimals at the beginning of the 18th century; it was Euler who formulated in 1757 the first system of PDEs (known as Euler's equations) to describe the motion of an incompressible fluid. PDEs can be linear, as is the case with the classical heat equation, for example, or non-linear, like the Euler or Navier–Stokes equations that describe the motion of a fluid. Several examples of non-linear problems are studied, with recent tools allowing the existence and possibly the uniqueness of the solution to be shown.

Over the course of the book, elliptic, parabolic and hyperbolic PDEs will be discussed. This historical denomination refers to a classification of second order linear PDEs with constant coefficients, which is written in the form $a\partial^2_{xx}u + 2b\partial^2_{xy}u + c\partial^2_{yy}u + d\partial^2_{xx}u + e\partial_y u + f = 0$, where $u : \mathbb{R}^2 \to \mathbb{R}$ is a real-valued function of two real variables x and y usually representing the space variables in a bi-dimensional domain, or the time variable and the space variable of a one-dimensional domain; the symbols ∂_α and $\partial^2_{\alpha\beta}u$ designate the partial derivatives of order 1 with respect to α and of order 2 with respect to α and β, for α and $\beta = x$ or y. This PDE is qualified as elliptic, (resp. parabolic, hyperbolic) if the coefficients a, b, c, d, e, f satisfy the condition $b^2 - ac > 0$ (resp. $b^2 - ac = 0$, $b^2 - ac < 0$), in reference to the planar curves of the same name, because the assumption on the coefficients is the same as the condition for the analytic equation $ax^2 + 2bxy + cy^2 + dx + ey + f = 0$ for these curves. The typical example of an elliptic PDE is the equation of heat conduction in steady state, while the typical example of a parabolic PDE is the equation of heat conduction in transient regime. The wave propagation equation is on the other hand a hyperbolic PDE. Each type of PDE has certain characteristics, in particular regarding the regularity of solutions, the speed of propagation of information, and the effect of initial conditions and boundary conditions. In general, the solutions of elliptic or parabolic PDEs are more regular than the initial conditions and boundary conditions. Furthermore, the boundary conditions affect the solution at every point in the domain. As for linear hyperbolic PDEs, the regularity of the solution depends on the regularity of the initial and boundary conditions; for example, if there is a jump in the initial data, this jump will propagate as a discontinuity in the solution. If the hyperbolic PDE is non-linear, discontinuous solutions can appear even if the initial conditions and boundary conditions are regular. For a system modelled by a hyperbolic PDE, information moves at a finite speed, called the wave speed.

The tools we develop in this book aim to obtain existence (and often uniqueness) results for some examples of PDE problems of various natures. It is generally not possible to obtain explicit solutions of a PDE; the PDEs and models mentioned above are solved by numerical methods that provide approximate solutions. The development of these methods and the increase in the power of computers have allowed a constant improvement in the accuracy of approximate solutions in recent decades. In fact, some of the tools presented are also useful for the development and analysis of numerical methods for approximate resolution.

The content of this book comes from several graduate courses taught at the University of Aix-Marseille, at the ENS Lyon and at the University of Savoie; we would like to thank our students and colleagues for their interest and their comments, and especially Marc Durand, Robert Eymard and Arax Leroy for their careful proofreading.

We also sincerely thank the reviewers for their insightful comments, which have significantly improved the clarity and quality of our writing.

We do not claim here to cover the entire field of PDEs; our goal is rather to transmit a certain culture of PDEs that we had the chance to acquire through our years of study and research, and which is certainly very much influenced by our knowledge of numerical methods for their resolution. Even though we recall most of the advanced notions that we use, it goes without saying that a profitable reading of this book requires good knowledge of real analysis, Lebesgue integration theory and functional analysis. We recommend on this subject the works [26] and [13] (or [12] for a French version, or [25] also in French).

The book contains a very large number of solved problems. Some of these problems are continuations of results covered in the course, while others explore relatively recent results. The problems are "starred", the number of stars (from 1 to 4) giving an indication of their difficulty.

In Chapter 1 we give the definition of the Sobolev functional spaces: these spaces are fundamental for the obtention of the so-called weak formulation of a given problem. We start with the notion of *weak derivative*, introduced by Jean Leray in 1934 [35], and its generalisation that we call *derivative by transposition*. This latter allows us to introduce all the necessary tools without the complicated notion of distribution, which is commonly used in most textbooks, but is not necessary within the scope of our study. Some functional analysis results are then stated, as well as classical density, trace and compactness theorems, and the famous Sobolev embeddings.

Chapter 2 is dedicated to linear elliptic equations; a classical example is the steady state heat equation, which involves the second order partial derivatives of the unknown function (the temperature) with respect to the space variables. We study their weak or variational formulation, as well as the regularity of their solutions. The problems in this chapter introduce the notion of weighted Sobolev spaces, explore some deeper spectral analysis results, and allow the reader to prove the existence of solutions to the Stokes problem and Schrödinger's equations.

Nonlinear elliptic equations are studied in Chapter 3. We distinguish three types of methods to demonstrate the existence of solutions: compactness methods, either by fixed point, or by topological degree; monotonicity methods; and methods by minimisation of functionals. We give examples of the application of these methods on several nonlinear problems, both in the main text and in the problems. In particular, by Schauder's fixed point for a semi-linear diffusion problem, by topological degree for a nonlinear diffusion convection reaction problem, and by monotonicity for Leray–Lions type equations.

Chapter 4 is dedicated to parabolic type equations; a typical example of a parabolic equation is the heat equation in transient regime, which involves the first order partial derivative of the unknown function with respect to time and the second order partial derivatives with respect to the space variables. We start with an overview of some methods used for their theoretical study. The resolution of the weak formulations of these equations requires integration with values in a Banach space; some results on this integration theory are recalled in the second section, as well as the extension of the notions of weak derivative and derivative by transposition in these spaces. The existence, uniqueness and regularity of the solutions of the heat equation is then addressed. We also give the proof of the equivalence of two weak formulations commonly used in analysis and numerical analysis. Fixed point methods, topological degree and monotonicity are applied to several examples of nonlinear parabolic equations in the main text (diffusion, convection-diffusion-reaction, Leray–Lions equations) and in the problems (the Stefan problem). Aubin–Simon type compactness results are generalised to sequences of spaces, to allow their use in the context of the study of the convergence of numerical schemes.

Finally, the last chapter focuses on hyperbolic conservation laws (equations and systems), which involve the first order derivatives of the unknown function. The solutions of these equations have a very different behaviour from that of the solutions of parabolic equations, because, even starting from regular initial data, the solutions can become discontinuous and therefore not have a classical sense. We first study the scalar case, with a one-dimensional Cauchy problem, that is, a hyperbolic equation involving an unknown function of one space variable and the time variable, with an initial condition and posed on all $\mathbb{R}$ (thus without boundary conditions). We introduce the notion of entropic solution, which allows us to determine a unique solution. The case of boundary conditions is briefly presented and revisited in more detail in the multidimensional framework. We then study the case of hyperbolic systems in the one-dimensional case with the spatial variable in the whole of $\mathbb{R}$. The case of hyperbolic systems with boundary conditions and the multidimensional case are not covered in this course, the mathematical theory of hyperbolic systems still being very incomplete. Hyperbolic systems often occur in fluid mechanics, and their numerical resolution involves the Riemann problem (the Cauchy problem with constant initial data on either side of the origin). We provide a thorough study of the Riemann problem, with, in particular, the complete solution of the Riemann problem for the Barré de Saint-Venant equations.

At the end of this book, we provide a bibliography of the key articles related to the results presented, as well as a few reference works. It goes without saying that this bibliography is far from exhaustive.

Marseille,   *Thierry Gallouët and Raphaèle Herbin*
March 2025

Contents

Chapter 1
Sobolev Spaces

Sobolev[1] spaces are functional spaces, that is to say, vector spaces whose elements are functions, more precisely "classes of functions", from an open subset of $\mathbb{R}^N$ ($N \geq 1$) with values in $\mathbb{R}$ (or $\mathbb{C}$ but, unless otherwise indicated, we consider that the target space is $\mathbb{R}$), and these functions are such that their powers and the powers of their derivatives (in the sense of transposition, or in the weak sense, which we specify in the sequel) are Lebesgue-integrable.[2] Just like Lebesgue spaces, these spaces are Banach[3] spaces.[4] The fact that Sobolev spaces are complete spaces is very important for proving the existence of solutions to many partial differential equations; completeness is the reason why classes of functions rather than functions are considered.

The results concerning Sobolev spaces as well as the reminders of functional analysis and integration theory are, for the most part, only stated in this chapter. Proofs and further information on the subject can be found in the works [1], [13] and [26], for example.

---

[1] Sergueï Lvovitch Sobolev (1908–1989), Russian mathematician and physicist.

[2] Henri-Léon Lebesgue (1875–1941), French mathematician, recognised for his theory of integration.

[3] Stefan Banach (1892–1945), Polish mathematician, founder of modern functional analysis.

[4] A Banach space is a complete normed vector space (n.v.s.) (that is to say, every Cauchy sequence for the norm on the n.v.s. converges for this norm).

© The Author(s), under exclusive license to Springer Nature Switzerland AG 2025
T. Gallouët, R. Herbin, *Weak Solutions of Partial Differential Equations*,
Mathématiques et Applications 90, https://doi.org/10.1007/978-3-031-98982-7_1

1.1 Weak Derivative, Derivative by Transposition

The concept of *weak derivative* already appears in a famous article by Jean Leray[5] published in 1934, on the Navier[6]–Stokes[7] equations; in this article, it appears under the name of "quasi-derivative" ([35] page 205). This concept is fundamental for the study of the existence of solutions to partial differential equations. In this work, we use the concept of *derivative by transposition*, a simplified form of the theory of distributions formalised by Laurent Schwartz,[8] see for instance [46], which extends the notion of weak derivative.

In the following, Ω denotes an open subset of $\mathbb{R}^N$, $N \geq 1$, and $\mathcal{D}(\Omega)$ denotes the set of C^∞ real-valued functions with compact support in Ω, that is to say

$$\mathcal{D}(\Omega) = \{u \in C^\infty(\Omega) \, ; \, \exists K \subset \Omega, \, K \text{ compact} \, ; \, u = 0 \text{ on } K^c\}. \tag{1.1}$$

Furthermore, the open subset Ω will always be equipped with the Borel σ-algebra), denoted by $\mathcal{B}(\Omega)$, and the Lebesgue measure, denoted by λ if $N = 1$ and λ_N if $N > 1$ ($\lambda_N(\Omega)$ is therefore the area of Ω if $N = 2$ and its volume if $N = 3$). Unless otherwise indicated, the integrals are with respect to this Lebesgue measure. The Lebesgue spaces obtained with the measurable space $(\Omega, \mathcal{B}(\Omega), \lambda_N)$ (or λ if $N = 1$) are denoted by $L^p(\Omega)$ ($p \in [1, +\infty]$). The reader is assumed to be acquainted with the Lebesgue integration theory (in particular, recall that an element of $L^p(\Omega)$ is systematically conflated with one of its representatives).

Definition 1.1 (Space L^p_{loc}). Let Ω be an open subset of $\mathbb{R}^N$ and $p \in [1, +\infty[$; a function f from Ω to $\mathbb{R}$ belongs to $L^p_{\text{loc}}(\Omega)$ if, for every compact subset K of Ω, the restriction of f to K extended by 0 belongs to $L^p(\Omega)$.

Lemma 1.2 is fundamental, as it allows us to conflate a locally integrable function $f \in L^1_{\text{loc}}(\Omega)$ with the linear mapping $T_f : \mathcal{D}(\Omega) \to \mathbb{R}$ defined for $\varphi \in \mathcal{D}(\Omega)$ by $T_f(\varphi) = \int_\Omega f(x)\varphi(x) \, dx$. Recall that two functions are equal almost everywhere (or a.e. for short) for the Lebesgue measure if they are equal except on a set of measure zero.

Let $\mathcal{D}^\star(\Omega)$ be the set of linear forms on $\mathcal{D}(\Omega)$, that is, the algebraic dual space of $\mathcal{D}(\Omega)$. If $T \in \mathcal{D}^\star(\Omega)$ and $\varphi \in \mathcal{D}(\Omega)$, the action of T on φ is defined as the real number $T(\varphi)$ (which is therefore the image of φ by T and is also denoted by $\langle T, \varphi \rangle_{\mathcal{D}^\star(\Omega), \mathcal{D}(\Omega)}$). Let $f \in L^1_{\text{loc}}$, then the mapping T_f defined by

---

[5] Jean Leray (1906–1998), French mathematician; he worked on partial differential equations and on algebraic topology.

[6] Claude Louis Marie Henri Navier (1785–1836), French engineer, mathematician and economist who established the equations describing the movement of fluids.

[7] George Gabriel Stokes (1819–1903), Irish physicist and mathematician, professor at the University of Cambridge.

[8] Laurent Schwartz (1915–2002), French mathematician; he was awarded the Fields Medal in 1950 for his work on the theory of distributions.

$$\langle T_f, \varphi \rangle_{\mathcal{D}^\star(\Omega), \mathcal{D}(\Omega)} = \int_\Omega f\,\varphi\,\mathrm{d}x$$

is a linear form on $\mathcal{D}(\Omega)$.

Lemma 1.2 (Fundamental Lemma: almost everywhere equality and equality in $\mathcal{D}^\star$). *Let Ω be an open subset of $\mathbb{R}^N$, $N \geq 1$, and let f and $g \in L^1_{\mathrm{loc}}(\Omega)$. Then:*

$$\left[\forall \varphi \in \mathcal{D}(\Omega), \int_\Omega f(x)\varphi(x)\,\mathrm{d}x = \int_\Omega g(x)\varphi(x)\,\mathrm{d}x \right] \iff [f = g \ \ a.e.]\,.$$

Proof: The proof of this lemma uses the regularisation of an integrable function by convolution with a sequence of regularising kernels: see [26, Problem 8.7 page 480]. ∎

If f and $g \in L^1_{\mathrm{loc}}$, then, thanks to Lemma 1.2, $f = g$ a.e. if and only if $T_f = T_g$. Hence any function $f \in L^1_{\mathrm{loc}}$ may be identified with the associated linear form T_f; this identification is systematic in the sequel. We may thus define the derivative by transposition of a function L^1_{loc} in the following way:

Definition 1.3 (Derivative by transposition, weak derivative). Let Ω be an open subset of $\mathbb{R}^N$, $N \geq 1$;

- Let $f \in L^1_{\mathrm{loc}}(\Omega)$. We call the derivative by transposition of f with respect to its i-th variable the linear form $D_i f$ on $\mathcal{D}(\Omega)$ defined by:

$$\langle D_i f, \varphi \rangle_{\mathcal{D}^\star(\Omega), \mathcal{D}(\Omega)} = -\int_\Omega f\,\partial_i \varphi\,\mathrm{d}x,$$

  where $\partial_i \varphi$ denotes the classical partial derivative of φ with respect to its i-th variable. Therefore $D_i f$ belongs to $\mathcal{D}^\star(\Omega)$. Note that if $f \in C^1(\Omega)$, then $D_i f$ is nothing but $\partial_i f$ because we conflate $\partial_i f$ and $T_{\partial_i f}$ (which is the element of $\mathcal{D}^\star(\Omega)$ induced by $\partial_i f$). It is therefore indeed a generalisation of the notion of derivative. If the linear form $D_i f$ can be conflated with a locally integrable function, then thanks to Lemma 1.2, this function is unique up to a set of null measure, and we say that this function is the weak derivative of f in the direction i. If $D_i f$ can be conflated with a function of $L^p(\Omega)$, we say that f admits a weak derivative in $L^p(\Omega)$.
  In the case $N = 1$, $D_1 f$ is denoted by Df.
- Let $T \in \mathcal{D}^\star(\Omega)$; we define the derivative by transposition $D_i T$ of T by:

$$\langle D_i T, \varphi \rangle_{\mathcal{D}^\star(\Omega), \mathcal{D}(\Omega)} = -\langle T, \partial_i \varphi \rangle_{\mathcal{D}^\star(\Omega), \mathcal{D}(\Omega)}, \ \ \forall \varphi \in \mathcal{D}(\Omega).$$

In the case $N = 1$, $D_1 T$ is again denoted by DT.

The concept of derivative by transposition is close to that of derivative in the sense of distributions, but the latter requires the definition of a topology on $\mathcal{D}(\Omega)$; the topological dual space of $\mathcal{D}(\Omega)$ is the subspace of $\mathcal{D}^\star(\Omega)$ consisting of continuous linear forms for this topology, and is denoted by $\mathcal{D}'(\Omega)$. Here and throughout the rest of this course, we do not use distributions, and therefore we do not equip $\mathcal{D}(\Omega)$ with a topology. However, note that when $\mathcal{D}(\Omega)$ is equipped with this topology, the two definitions of derivative coincide.

Here is an example of a derivative by transposition. The Heaviside[9] function, defined by

$$H(x) = \begin{cases} 1 \text{ if } x \geq 0, \\ 0 \text{ if } x < 0, \end{cases}$$

is locally integrable. Therefore, it admits a derivative by transposition, denoted by DH; we note that for $\varphi \in \mathcal{D}(\mathbb{R})$, we have:

$$-\int_{\mathbb{R}} H(x)\varphi'(x)\,\mathrm{d}x = -\int_0^{+\infty} \varphi'(x)\,\mathrm{d}x = \varphi(0),$$

and so DH is the linear form $\varphi \mapsto \varphi(0)$, also called the "Dirac measure at 0": $DH = \delta_0$. However, this derivative is not a weak derivative, because δ_0 cannot be assimilated to a function of L^1_{loc}, in the sense that there does not exist a function $g \in L^1_{\mathrm{loc}}(\mathbb{R})$ such that $\delta_0(\varphi) = \int_{\mathbb{R}} g\varphi\,\mathrm{d}x$ for all $\varphi \in \mathcal{D}(\mathbb{R})$ (see [26, Problem 5.1] and Problem 1.1 below).

Definition 1.3 allows us to define derivatives by transposition of a function L^1_{loc} (or of an element of $\mathcal{D}^\star(\Omega)$) of all orders. By identifying a function L^1_{loc} with the element of $\mathcal{D}^\star(\Omega)$ it represents, we can also define the notion of weak derivative of all orders.

Definition 1.4 (Higher order derivative). For $\alpha = (\alpha_1, \ldots \alpha_N) \in \mathbb{N}^N$, and $u \in L^1_{\mathrm{loc}}(\mathbb{R}^N)$, let $|\alpha| = \alpha_1 + \ldots + \alpha_N$ and let the weak derivative $D^\alpha u = D_1^{\alpha_1} \ldots D_N^{\alpha_N} u \in L^1_{\mathrm{loc}}(\mathbb{R}^N)$ (of order $|\alpha|$), if it exists, be defined by

$$\int_{\mathbb{R}^N} D^\alpha u(x)\,\varphi(x)\,\mathrm{d}x = \int_{\mathbb{R}^N} D_1^{\alpha_1} \ldots D_N^{\alpha_N} u(x)\,\varphi(x)\,\mathrm{d}x$$

$$= (-1)^{|\alpha|} \int_{\mathbb{R}^N} u(x)\,\partial_1^{\alpha_1} \ldots \partial_N^{\alpha_N} \varphi(x)\,\mathrm{d}x, \forall \varphi \in \mathcal{D}(\mathbb{R}^N),$$

where $\partial_i^{\alpha_i}\varphi$ denotes the partial (classical) derivative of order α_i with respect to the i-th variable.

Definition 1.5 (Convergence in $\mathcal{D}^\star(\Omega)$, positivity of an element of $\mathcal{D}^\star(\Omega)$). Let Ω be an open subset of $\mathbb{R}^N$ ($N \geq 1$).

1. Let $(T_n)_{n \in \mathbb{N}}$ be a sequence of elements of $\mathcal{D}^\star(\Omega)$ and $T \in \mathcal{D}^\star(\Omega)$. We say that $T_n \to T$ in $\mathcal{D}^\star(\Omega)$, as $n \to +\infty$, if

$$\langle T_n, \varphi \rangle_{\mathcal{D}^\star(\Omega), \mathcal{D}(\Omega)} \to \langle T, \varphi \rangle_{\mathcal{D}^\star(\Omega), \mathcal{D}(\Omega)} \text{ for all } \varphi \in \mathcal{D}(\Omega). \tag{1.2}$$

---

[9] Oliver Heaviside (1850–1925), self-taught British physicist.

2. Let $T \in \mathcal{D}^{\star}(\Omega)$, we say that $T \geq 0$ if for all $\varphi \in \mathcal{D}(\Omega)$ $\varphi \geq 0$ (that is to say $\varphi(x) \geq 0$ for all $x \in \Omega$) implies $\langle T, \varphi \rangle_{\mathcal{D}^{\star}(\Omega), \mathcal{D}(\Omega)} \geq 0$. (Of course, $T \leq 0$ if $-T \geq 0$.)

Remark 1.6 (What about distributions?). We have just seen that convergence in $\mathcal{D}^{\star}(\Omega)$ is simple convergence in the set of mappings from $\mathcal{D}(\Omega)$ to $\mathbb{R}$. When we are interested in distributions, we add a topological structure to the space $\mathcal{D}(\Omega)$ (not described in this book) and, instead of working with $\mathcal{D}^{\star}(\Omega)$, we work with the strictly smaller space of continuous linear mappings from $\mathcal{D}(\Omega)$ to $\mathbb{R}$, a space denoted $\mathcal{D}'(\Omega)$. However, even when we work with the space $\mathcal{D}'(\Omega)$, the notion of convergence is still given by (1.2). Moreover, all the elements of $\mathcal{D}^{\star}(\Omega)$ that we consider in this book are in fact elements of $\mathcal{D}'(\Omega)$, but this information is not of any use in the present framework.

Remark 1.7 (Product $\mathcal{D}^{\star}(\Omega)$ with $\mathcal{D}(\Omega)$). Let Ω be an open subset of $\mathbb{R}^N$ ($N \geq 1$). Let $T \in \mathcal{D}^{\star}(\Omega)$ and $\psi \in \mathcal{D}(\Omega)$, it is natural to define ψT as the element of $\mathcal{D}^{\star}(\Omega)$ defined by

$$\langle \psi T, \varphi \rangle_{\mathcal{D}^{\star}(\Omega), \mathcal{D}(\Omega)} = \langle T, \psi \varphi \rangle_{\mathcal{D}^{\star}(\Omega), \mathcal{D}(\Omega)}, \text{ for all } \varphi \in \mathcal{D}(\Omega).$$

This definition is compatible with the product of functions: if $T \in \mathcal{D}^{\star}(\Omega)$ is given by a function $f \in L^1_{\mathrm{loc}}(\Omega)$, then ψT is indeed given by the function ψf. On the other hand, we note that $D_i(\psi T) = \psi D_i T + (\partial_i \psi) T$ for all $i = \{1, \ldots, N\}$.

1.2 Definition and Properties

We provide below the precise definition and properties of Sobolev spaces. We refer to the works [1, 51] for the proofs that are not detailed here. Schematically, a Sobolev space is a vector space of (classes of) functions equipped with a norm that is a combination of the L^p norms of the functions and their derivatives (in the weak sense or in the sense of the transposition that we have just introduced) up to a given order. Sobolev spaces are complete, so they are Banach spaces. Their interest comes from the fact that weak solutions (which are introduced in later chapters) of certain PDEs exist in appropriate Sobolev spaces, even though there may not exist solutions in spaces of continuous functions with derivatives in the classical sense. The higher the order of the Sobolev space, the more regular are the solutions of the PDEs sought in this space.

Definition 1.8 (Sobolev Spaces). Let Ω be an open subset of $\mathbb{R}^N$, $N \geq 1$.

1. The space $H^1(\Omega)$ is defined by:

$$H^1(\Omega) = \{u \in L^2(\Omega) \;;\; D_i u \in L^2(\Omega), \; \forall i = 1, \ldots, N\}.$$

Recall that $D_i u \in L^2(\Omega)$ means (see Definition 1.3) that

$$\exists g \in L^2(\Omega) \quad \langle D_i f, \varphi \rangle_{\mathcal{D}^\star(\Omega), \mathcal{D}(\Omega)} = \int_\Omega g\varphi \, dx, \quad \forall \varphi \in \mathcal{D}(\Omega).$$

2. The space $H^m(\Omega)$ is defined for $m \in \mathbb{N}$ by:

$$H^m(\Omega) = \{u \in L^2(\Omega) \quad D^\alpha u \in L^2(\Omega), \forall \alpha \in \mathbb{N}^N \quad |\alpha| \leq m\}.$$

3. The space $W^{m,p}(\Omega)$ is defined for $1 \leqslant p \leqslant \infty$ (note that p is not necessarily an integer) and $m \in \mathbb{N}$, by

$$W^{m,p}(\Omega) = \{u \in L^p(\Omega); D^\alpha u \in L^p(\Omega), \forall \alpha \in \mathbb{N}^N ; |\alpha| \leq m.\}$$

4. Note that for $m = 0$, the space $W^{m,p}(\Omega)$ is the Lebesgue space $L^p(\Omega)$.

Let $(\cdot|\cdot)_{L^2}$ be the inner product in $L^2(\Omega)$, i.e.

$$(u|v)_{L^2} = \int_\Omega uv \, dx,$$

and $\|\cdot\|_{L^p}$ the norm in $L^p(\Omega)$, i.e.

$$\|u\|_{L^p} = \left(\int_\Omega |u|^p \, dx \right)^{\frac{1}{p}} .$$

Here we consider (as in most of this book) functions with values in $\mathbb{R}$. The generalisation to functions with values in $\mathbb{C}$ requires a few modifications. One of these modifications consists of replacing $\int_\Omega uv \, dx$ by $\int_\Omega u\bar{v} \, dx$ in the definition of the inner product in $L^2(\Omega)$.

Proposition 1.9 (Norms and inner products on Sobolev spaces). *The spaces $H^m(\Omega)$ are Hilbert spaces[10] [11] when they are equipped with the inner product*

$$(u|v)_{H^m} = \sum_{|\alpha| \leq m} (D^\alpha u | D^\alpha v)_{L^2}.$$

Note that $W^{m,2}(\Omega) = H^m(\Omega)$. A natural norm on $W^{m,p}(\Omega)$ is defined by:

---

[10] David Hilbert (1862–1943), German mathematician, known for his work in many branches of mathematics. In 1900 he stated 23 problems, for which he had an intuition of their importance, some of which are still not solved.

[11] Recall that a Hilbert space is a complete normed vector space (n.v.s.) whose norm is induced by an inner product.

$$\|u\|_{W^{m,p}} = \begin{cases} \left(\displaystyle\sum_{0\leqslant|\alpha|\leqslant m} \|D^{\alpha}u\|_{L^p}^p\right)^{1/p}, & \text{if } 1 \leqslant p < +\infty \\[2em] \displaystyle\max_{0\leqslant|\alpha|\leqslant m} \|D^{\alpha}u\|_{L^\infty}, & \text{if } p = +\infty. \end{cases} \tag{1.3}$$

Equipped with this norm $W^{m,p}(\Omega)$ is a Banach space. *We can show that the norm:*

$$\|u\|_{m,p} = \begin{cases} \displaystyle\sum_{0\leqslant|\alpha|\leqslant m} \|D^{\alpha}u\|_{L^p}, & \text{if } 1 \leqslant p < +\infty \\[1.5em] \displaystyle\sum_{0\leqslant|\alpha|\leqslant m} \|D^{\alpha}u\|_{L^\infty}, & \text{if } p = +\infty. \end{cases} \tag{1.4}$$

is a norm equivalent to the norm defined by (1.3): this is a consequence of the equivalence between the norms in $\mathbb{R}^q$ where $q = card(\{\alpha \in \mathbb{N}^N \; |\alpha| \leq m\})$. Both norms are indifferently denoted $\|\cdot\|_{m,p}$ or $\|\cdot\|_{W^{m,p}}$. The main interest of the norm (1.3) is that in the case $p = 2$ it confers on H^m a Hilbertian structure, which is not the case with the norm defined by (1.4)

Remark 1.10 (Sobolev spaces and continuity). In one space dimension ($N = 1$), with $a, b \in \mathbb{R}, a < b, 1 \leq p \leq +\infty$, any element of $W^{1,p}(]a, b[)$ (which is therefore a class of functions) can be assimilated to a continuous function, in the sense that there exists a representative of the class which is continuous (this continuous representative is unique, see in this regard Problem 1.3). This is due to the fact that in dimension 1, any (class of) function(s) from $W^{1,p}(]a, b[)$ can be written a.e. as the integral of its derivative.

$$u \in W^{1,p}(]a, b[) \Longleftrightarrow \{\exists \tilde{u} \in C([a, b]) \text{ and } \exists v \in L^p(]a, b[) \; : \; u = \tilde{u} \text{ a.e. and}$$

$$\forall x \in]a, b[, \; \tilde{u}(x) = \tilde{u}(a) + \int_a^x v(s)\,\mathrm{d}s\Big\}.$$

In dimension greater than 1, this is false. In particular $H^1(\Omega) \not\subset C(\overline{\Omega})$, as can be seen through the following example: let $\Omega = \{x = (x_1, x_2)^t \in \mathbb{R}^2, |x_i| < \frac{1}{2}, i = 1, 2\}$, and u the function defined on Ω by $u(x) = (-\ln(|x|))^\gamma$, with $\gamma \in]0, \frac{1}{2}[$, where $|\cdot|$ denotes the Euclidean norm. Then $u \in H^1(\Omega)$ but $u \notin L^\infty(\Omega)$ (see Problem 1.5), and therefore, in particular, $u \notin C(\overline{\Omega})$.

Proposition 1.11 (Separability). *Let Ω be an open subset of $\mathbb{R}^N$ ($N \geq 1$), $m \in \mathbb{N}$ and $1 \leq p < +\infty$; the space $W^{m,p}(\Omega)$ is a **separable space** (that is to say, a normed vector space which contains a countable dense subset).*

The proof of this proposition is the subject of Problem 1.11, where it is also shown by a counterexample that the separability result is not true for $p = +\infty$.

The notion of separability is important, because it allows any element of the space to be approximated as close as one wants by an element of a countable family: in the context of a Hilbert space, it can be shown that this property is equivalent to the existence of a countable Hilbert basis (see for example [26, Proposition 6.62]).

Another important notion is the reflexivity of a space.

Definition 1.12 (Reflexive space). Let E be a real normed vector space, and let E' be its topological dual space, that is to say the set of linear continuous forms from E to $\mathbb{R}$ equipped with its natural norm (note that E' is a Banach space). For all $x \in E$, a mapping J_x from E' to $\mathbb{R}$ is defined by $J_x(T) = T(x)$ for all $T \in E'$. We have

$$|J_x(T)| = |T(x)| \leq \|T\|_{E'}\|x\|_E$$

and therefore J_x is a continuous linear form on E', which also reads $J_x \in E''$ where E'' is the bidual space of E, that is to say the topological dual space of E'. We can show by the Hahn[12]–Banach theorem, which we recall in the following section, that $\|J_x\|_{E''} = \|x\|_E$ (see item 3 of Problem 1.8). The mapping J defined from E to E'' by $J(x) = J_x$, for all $x \in E$, is therefore a linear isometry of E onto its image, denoted $\mathrm{Im}(J)$, and we obviously have $\mathrm{Im}(J) \subset E''$. We say that E is a **reflexive** space if $\mathrm{Im}(J) = E''$, which is equivalent to saying that J is surjective. Note that every reflexive space E is necessarily complete since the dual space of any normed vector space is always complete.

Proposition 1.13 (Reflexivity). *Let Ω be an open subset of $\mathbb{R}^N$, $N \geq 1$, and $m \in \mathbb{N}$. For all p such that $1 < p < +\infty$, the space $W^{m,p}(\Omega)$ is a **reflexive Banach space**.*

For a proof of this result, see for example [26, Proposition 6.73].

1.3 Functional Analysis in a Nutshell

We recall here some results of functional analysis that are particularly important in our study of PDEs. Let us start with a fundamental theorem (see for example [13]):

Theorem 1.14 (Hahn–Banach). *Let E be a vector space over $\mathbb{R}$ and p a mapping from E to $\mathbb{R}$ such that*

1. $p(u + v) \leq p(u) + p(v)$, *for all $u, v \in E$,*
2. $p(\lambda u) = \lambda p(u)$, *for all $u \in E$ and $\lambda > 0$.*

Let F be a subspace of E, and T a linear form on F that satisfies $T(x) \leq p(x)$ for all $x \in F$. There then exists a linear form from E to $\mathbb{R}$, still denoted T, which is equal to T on F and still satisfies the condition: $T(x) \leq p(x)$ for all $x \in E$.

The following corollary is essential:

Corollary 1.15 (Extension of a linear mapping). *Let E be a normed vector space over $\mathbb{R}$, F a vector subspace of E not reduced to $\{0\}$ and T a continuous linear form on F. We can then extend T to a continuous linear form defined on E, with the same norm as T.*

---

[12] Hans Hahn (1879–1934), Austrian mathematician and philosopher, known for his contributions to functional analysis, topology, set theory and real analysis.

The following theorem on finite-dimensional spaces is well known:

Theorem 1.16 (Necessary and sufficient condition on dimension). *A Banach space E is of finite dimension if and only if its closed unit ball is compact.*

The concepts of weak convergence and weak-$\star$ convergence are fundamental.

Definition 1.17 (Weak and weak-$\star$ convergences). Let E be a Banach space.

1. **Weak convergence** Let $(u_n)_{n\in\mathbb{N}} \subset E$ and $u \in E$. We say that $u_n \to u$ weakly in E when $n \to +\infty$ if $T(u_n) \to T(u)$ for all $T \in E'$.
2. **Weak-$\star$ convergence** Let $(T_n)_{n\in\mathbb{N}} \subset E'$ and $T \in E'$. We say that $T_n \to T$ $\star$-weakly in E' when $n \to +\infty$ if $T_n(u) \to T(u)$ for all $u \in E$.

Remark 1.18 (On convergence in a normed vector space). Throughout the rest of this book, if E is a normed vector space, $(u_n)_{n\in\mathbb{N}}$ a sequence of elements of E and $u \in E$, we say that the sequence $(u_n)_{n\in\mathbb{N}}$ converges to u in E as $n \to +\infty$ if $\|u_n - u\|_E \to 0$ as $n \to +\infty$. This classical notion of convergence is sometimes called "strong convergence" in the literature, as opposed to the concept of weak convergence.

Example 1.19 (Weak convergence in L^1 and weak-$\star$ in L^∞). Weak-$\star$ convergence is interesting in the case of non-reflexive spaces. For example, the spaces $L^1(\mathbb{R})$ of (classes of) integrable functions on $\mathbb{R}$ for the Lebesgue measure and the space $L^\infty(\mathbb{R})$ of (classes of) essentially bounded functions on $\mathbb{R}$ for the Lebesgue measure are not reflexive. Classically, the dual space of $L^1(\mathbb{R})$ can be identified with $L^\infty(\mathbb{R})$ but the dual space of $L^\infty(\mathbb{R})$ is not identifiable with $L^1(\mathbb{R})$, see for example [26, Remark 6.71]. Given this identification of $(L^1(\mathbb{R}))'$ with $L^\infty(\mathbb{R})$, a sequence $(u_n)_{n\in\mathbb{N}} \subset L^1(\mathbb{R})$ converges weakly to $u \in L^1(\mathbb{R})$ if for all $v \in L^\infty(\mathbb{R})$,

$$\int u_n(x)v(x)\,\mathrm{d}x \to \int u(x)v(x)\,\mathrm{d}x \text{ when } n \to +\infty.$$

A sequence $(v_n)_{n\in\mathbb{N}} \subset L^\infty(\mathbb{R})$ converges $\star$-weakly towards $v \in L^\infty(\mathbb{R})$ if for all $u \in L^1(\mathbb{R})$,

$$\int v_n(x)u(x)\,\mathrm{d}x \to \int v(x)u(x)\,\mathrm{d}x \text{ when } n \to +\infty.$$

Remark 1.20 (Weak convergence and limit inf). Let E be a Banach space. It is interesting to note that if a sequence $(u_n)_{n\in\mathbb{N}} \subset E$ converges weakly in E to $u \in E$, then

$$\|u\|_E \leq \liminf_{n\to+\infty} \|u_n\|_E.$$

The proof of this result is the subject of question 4b of Problem 1.8.

The following theorem is a sequential version of the Banach–Alaoglu[13] theorem (see [13]), sufficient in the context of this course. Here and throughout the rest of this book $\mathbb{R}_+$ denotes the set of non-negative real numbers and $\mathbb{R}_+^{\star}$ denotes the set of positive real numbers.

Theorem 1.21 (Weak-$\star$ compactness of bounded subsets of the dual space of a separable space, Sequential Banach–Alaoglu). *Let E be a separable Banach space, and let $(T_n)_{n \in \mathbb{N}}$ be a bounded sequence in E', that is, such that there exists a $C \in \mathbb{R}_+$ such that $\|T_n\|_{E'} \leq C$ for all $n \in \mathbb{N}$. Then there exists a subsequence, still denoted $(T_n)_{n \in \mathbb{N}}$, and $T \in E'$ such that $T_n \to T$ $\star$-weakly in E' (in the sense of Definition 1.17).*

We refer to the functional analysis course notes [25] for the proof of this theorem and for its important corollaries, such as the result of sequential weak compactness of bounded subsets of a reflexive Banach space (Theorem 1.22).

A remarkable application of Theorem 1.21 is the following: if Ω is an open subset of $\mathbb{R}^N$ and $(u_n)_{n \in \mathbb{N}}$ is a bounded sequence in $L^{\infty}(\Omega)$, then there exists a subsequence, still denoted $(u_n)_{n \in \mathbb{N}}$, and $u \in L^{\infty}(\Omega)$ such that $\int_{\Omega} u_n \varphi \, dx \to \int_{\Omega} u \varphi \, dx$ for all $\varphi \in L^1(\Omega)$. This follows from the fact that there exists a natural isometry between $L^{\infty}(\Omega)$ and the dual space of $L^1(\Omega)$ and that $L^1(\Omega)$ is separable.

Theorem 1.22 (Weak compactness of bounded subsets of a reflexive space). *Let E be a reflexive Banach space, and let $(u_n)_{n \in \mathbb{N}}$ be a bounded sequence in E, that is to say, such that there exists a $C \in \mathbb{R}_+$ such that $\|u_n\|_E \leq C$ for all $n \in \mathbb{N}$. Then there exists a subsequence, still denoted $(u_n)_{n \in \mathbb{N}}$, and $u \in E$ such that $u_n \to u$ weakly in E.*

Note that a Hilbert space is always a reflexive Banach space.

1.4 Density Theorems

It is often useful to know how to approximate a function by a more regular function. This is the subject of the density theorems of functional spaces that we present here. These theorems require a certain regularity of the boundary of the domain, namely that it is Lipschitz.[14]

Definition 1.23 (Lipschitz boundary). A bounded open subset Ω of $\mathbb{R}^N$ is said to have a Lipschitz boundary if there exist $n \in \mathbb{N}$, some open subsets $(\Omega_0, \Omega_1, \ldots, \Omega_n)$ of $\mathbb{R}^N$ and some mappings $(\phi_0, \phi_1, \ldots, \phi_n)$ such that

1. $\overline{\Omega} \subset \bigcup_{i=0}^{n} \Omega_i$ and $\Omega_0 \subset \Omega$.
2. $\phi_0 : \Omega_0 \to B_{1,N} = \{x \in \mathbb{R}^N \; ; \; \|x\| < 1\}$ is bijective and ϕ_0 and ϕ_0^{-1} are Lipschitz-continuous.

---

[13] Leonidas Alaoglu (1914–1981), Canadian mathematician, famous for his functional analysis result called Alaoglu's theorem (also called the Banach–Alaoglu theorem).

[14] Rudolf Otto Sigismund Lipschitz (1832–1903) was a German mathematician.

3. For all $i \geq 1$, $\phi_i : \Omega_i \to B_{1,N}$ is bijective, ϕ_i and ϕ_i^{-1} are Lipschitz-continuous, $\phi_i(\Omega_i \cap \Omega) = B_{1,N} \cap \mathbb{R}_+^N$ and $\phi_i(\Omega_i \cap \partial\Omega) = B_{1,N} \cap \{(0, y), y \in \mathbb{R}^{N-1}\}$ (where $\mathbb{R}_+^N = \{(x, y) \in \mathbb{R}^N \ x \in \mathbb{R}, x > 0 \text{ and } y \in \mathbb{R}^{N-1}\}$).

Moreover, if the mappings $\phi_0, \phi_1, \ldots, \phi_n$ are of class C^∞, the open subset Ω is said to be of class C^∞.

Remark 1.24 (Strongly Lipschitz boundary). A bounded open subset Ω of $\mathbb{R}^N$ is said to have a strongly Lipschitz boundary if the boundary of Ω is locally the graph of a Lipschitz-continuous function and that Ω is (locally) on one side of this graph. An open (bounded) set with a strongly Lipschitz boundary is an open subset with a Lipschitz boundary but the converse is false, as shown by Problem 1.16, see also [20], which gives a lot of details about this section and the following ones.

Theorem 1.25 (Density and extension). *Let $N \geq 1$, let Ω be $\mathbb{R}^N$, $\Omega = \mathbb{R}_+^N$ or an open bounded subset of $\mathbb{R}^N$ with Lipschitz boundary, and let $1 \leq p \leq +\infty$. Let $C_c^\infty(\overline{\Omega})$ be the set of restrictions to Ω of functions belonging to $\mathcal{D}(\mathbb{R}^N)$. Then*

1. *If $p < +\infty$, $C_c^\infty(\overline{\Omega})$ is dense in $W^{1,p}(\Omega)$.*
2. *There exists a continuous linear mapping $P : W^{1,p}(\Omega) \to W^{1,p}(\mathbb{R}^N)$ such that*

$$\forall u \in W^{1,p}(\Omega), P(u) = u \text{ a.e. in } \Omega.$$

Similar results are true with $W^{m,p}(\Omega)$ $(m > 1)$ instead of $W^{1,p}(\Omega)$ but require more regularity on Ω (see [1]).

Brief proof of Theorem 1.25.

• The density property is first proved in the case $\Omega = \mathbb{R}^N$ in two steps:

Step 1, truncation – We prove in this step the density in $W^{1,p}(\mathbb{R}^N)$ of elements of $W^{1,p}(\mathbb{R}^N)$ with compact support (the element u of $W^{1,p}(\mathbb{R}^N)$ has compact support if there exists a compact set K such that $u = 0$ a.e. on K^c).

Let $\psi \in \mathcal{D}(\mathbb{R}^N)$ such that $0 \leq \psi(x) \leq 1$ for all x, $\psi(x) = 1$ if $|x| \leq 1$, $\psi(x) = 0$ if $|x| \geq 2$ (recall that $|x|$ denotes the Euclidean norm of $x \in \mathbb{R}^N$). Define u_n for $n \in \mathbb{N}^\star$ by $u_n(x) = u(x)\psi(x/n)$. Then, one can prove that $u_n \to u$ in $W^{1,p}(\mathbb{R}^N)$ as $n \to +\infty$.

Step 2, regularisation – Let $\rho \in \mathcal{D}(\mathbb{R}^N)$ such that $\rho(x) = 0$ if $|x| \geq 1$ and $\int_{\mathbb{R}^N} \rho(x) = 1$. For $n \in \mathbb{N}^\star$, let ρ_n be the function defined by $\rho_n(x) = n^N \rho(nx)$. Let $u \in W^{1,p}(\mathbb{R}^N)$, u with compact support and set $u_n = u \star \rho_n$. Then one can prove that $u_n \in \mathcal{D}(\mathbb{R}^N)$ and $u_n \to u$ in $W^{1,p}(\mathbb{R}^N)$ as $n \to +\infty$.

We have thus proved that $C_c^\infty(\overline{\Omega})$ is dense in $W^{1,p}(\Omega)$. (Note that this result is false if $p = +\infty$.)

In the case $\Omega = \mathbb{R}_+^N$, the first step is identical to that of the case $\Omega = \mathbb{R}^N$ but the second step needs to be slightly modified. We start by extending u by 0 outside of $\mathbb{R}_+^N$. Then, we still choose a function $\rho \in \mathcal{D}(\mathbb{R}^N)$ such that $\rho(x) = 0$ if $|x| \geq 1$ and $\int_{\mathbb{R}^N} \rho(x) = 1$ but we add that $\rho(x_1, y) = 0$ if $x_1 \geq 0$ (and $y \in \mathbb{R}^{N-1}$). Set $u_n = u \star \rho_n$ (still with $\rho_n = n^N \rho(n\cdot)$). We then prove that $u_n \in \mathcal{D}(\mathbb{R}^N)$ and $u_n \to u$ in $W^{1,p}(\mathbb{R}_+^N)$ as $n \to +\infty$ (more precisely, the restriction of u_n to $\mathbb{R}_+^N$).

To show that $u_n \to u$ in $W^{1,p}(\mathbb{R}_+^N)$ it is important to note that $\partial_i u_n = \widetilde{D_i u} \star \rho_n$ on $\mathbb{R}_+^N$, where $= \widetilde{D_i u}$ is equal to $D_i u$ extended by 0 outside of $\mathbb{R}_+^N$, and therefore $\partial_i u_n \to D_i u$ in $L^p(\mathbb{R}_+^N)$ as $n \to +\infty$.

Finally, in the case where Ω is an open bounded subset (of $\mathbb{R}^N$) with a Lipschitz boundary, we reduce to the case $\Omega = \mathbb{R}_+^N$ by using the functions ϕ_i from Definition 1.23 and a partition of unity (Problem 1.25).

• We now indicate how we prove the second property of Theorem 1.25 in the case $p < +\infty$, thanks to the density. The proof in the case $p = +\infty$ differs from this one, because the density property of Theorem 1.25 is then false; however, in this case we can directly construct the extension by noticing that $W^{1,\infty}(\Omega)$ is the space of Lipschitz-continuous functions on $\bar{\Omega}$, see Problem 1.15.

Let us therefore consider the case $p < +\infty$, and start with $\Omega = \mathbb{R}_+^N$. If $u \in C_c^\infty(\overline{\mathbb{R}}_+^N)$ (that is to say, u is the restriction to $\mathbb{R}_+^N$ of an element of $\mathcal{D}(\mathbb{R}^N)$ still denoted u), we define $\tilde{u}$ by $\tilde{u}(x) = u(x)$ if $x = (x_1, y)^t$ with $x_1 \geq 0$ and $\tilde{u}(x) = u(-x_1, y)$ if $x = (x_1, y)^t$ with $x_1 < 0$. The function $\tilde{u}$ is not of class C^∞, nor even of class C^1, but it is continuous and we show (using classical integration by parts) that the derivatives by transposition of $\tilde{u}$ are represented by the classical derivatives of $\tilde{u}$. Hence $\tilde{u} \in W^{1,p}(\mathbb{R}^N)$ and $\|\tilde{u}\|_{W^{1,p}(\mathbb{R}^N)} = \sqrt{2}\,\|u\|_{W^{1,p}(\mathbb{R}_+^N)}$. The mapping $u \mapsto \tilde{u}$ is therefore a continuous linear operator from $C_c^\infty(\overline{\mathbb{R}}_+^N) \subset W^{1,p}(\mathbb{R}_+^N)$ to $W^{1,p}(\mathbb{R}^N)$ with norm $\sqrt{2}$. By density of $C_c^\infty(\overline{\mathbb{R}}_+^N)$ in $W^{1,p}(\mathbb{R}_+^N)$, it therefore extends (in a unique way) to an operator P that is linear continuous from $W^{1,p}(\mathbb{R}_+^N)$ to $W^{1,p}(\mathbb{R}^N)$ and whose norm equals $\sqrt{2}$. We have, of course, $P(u) = u$ a.e. in $\mathbb{R}_+^N$.

Finally, as in the proof of the density, in the case where Ω is an open bounded subset (of $\mathbb{R}^N$) with a Lipschitz boundary, we reduce to the case $\Omega = \mathbb{R}_+^N$.   ∎

It can also be shown that $C_c^\infty(\overline{\Omega})$ is dense in $W^{m,p}(\Omega)$ if $N \geq 1$, $m \in \mathbb{N}$ and $1 \leq p < +\infty$, with Ω equal to $\mathbb{R}^N$, $\mathbb{R}_+^N$ or a bounded open subset of $\mathbb{R}^N$ with a "sufficiently regular" boundary (for example if the functions ϕ_i of the Definition 1.23 are of class C^∞, see, for example, [1]).

Remark 1.26 (Mapping and operator). In the proof of the previous theorem and throughout the rest of the work, the term mapping or operator is used interchangeably in the case of a mapping of an n.v.s. into an n.v.s.

Extension operators can also be constructed as in the second property of Theorem 1.25. A particular case is the subject of Problem 1.21. Finally, it is important to note that if, for example, Ω is an open bounded subset, the space $\mathcal{D}(\Omega)$ is not dense in $H^1(\Omega)$. Its closure is a strict subspace of $H^1(\Omega)$, denoted $H_0^1(\Omega)$.

Definition 1.27 (The spaces $W_0^{m,p}(\Omega)$ and their dual spaces). Let Ω be an open subset of $\mathbb{R}^N$, $N \geq 1$.

1. Let $H_0^1(\Omega)$ be the closure of $\mathcal{D}(\Omega)$ in $H^1(\Omega)$:

$$H_0^1(\Omega) = \overline{\mathcal{D}(\Omega)}^{H^1(\Omega)}.$$

2. For $m > 0$ and $1 \leq p < +\infty$, we define the subspace $W_0^{m,p}(\Omega)$ of $W^{m,p}(\Omega)$ as the closure of $\mathcal{D}(\Omega)$ in $W^{m,p}(\Omega)$:

$$W_0^{m,p}(\Omega) = \overline{\mathcal{D}(\Omega)}^{W^{m,p}(\Omega)}.$$

3. For $1 \leq p < +\infty$, and $q = \frac{p}{p-1}$ if $p > 1$, $q = +\infty$ if $p = 1$; the dual space of $W_0^{1,p}(\Omega)$ is denoted $W^{-1,q}(\Omega)$.

4. for $q = 2$, the space $W^{-1,2}(\Omega)$ is also denoted $H^{-1}(\Omega)$.

As previously stated, if $\Omega = \mathbb{R}^N$ we have $H_0^1(\Omega) = H^1(\Omega)$, whereas the inclusion is strict if Ω is an open bounded subset.

The concept of differentiability is only defined on open subsets. This difficulty is circumvented by introducing the following concept on the closure of an open subset of $\mathbb{R}^N$, which we need later.

Definition 1.28 (Space $C^k(\bar{\Omega})$). Let Ω be an open subset of $\mathbb{R}^N$, $N \geq 1$, let $k \in \mathbb{N}^\star \cup \{+\infty\}$ and let φ be a function from Ω to $\mathbb{R}$. We say that $\varphi \in C^k(\bar{\Omega})$ if there exists a function ψ from $\mathbb{R}^N$ to $\mathbb{R}$, of class C^k, such that $\psi = \varphi$ in Ω. If Ω is bounded, it is of course possible to require that the function ψ has compact support in $\mathbb{R}^N$, as in Theorem 1.25 (and thus $C^k(\bar{\Omega}) = C_c^k(\bar{\Omega})$, where $C_c^k(\bar{\Omega})$ is defined as $C^k(\bar{\Omega})$ with $\mathcal{D}(\mathbb{R}^N)$ instead of $C^k(\mathbb{R}^N)$).

It is interesting to note that it is possible to take the same definition for $k = 0$, since it is compatible with the usual notion of continuity on $\bar{\Omega}$: indeed, if φ is continuous from $\bar{\Omega}$ to $\mathbb{R}$, then there exists a function ψ that is continuous from $\mathbb{R}^N$ to $\mathbb{R}$ such that $\psi = \varphi$ in $\bar{\Omega}$, see Problem 1.17.

1.5 Trace Theorems

We state here some fundamental results on the trace operator, first in the half-space and then for an open bounded subset Ω with a Lipschitz boundary. We refer to the reference works [13] and [22] as well as to the monograph (in French) [20, chapter 4] for a clear and precise presentation on these issues, with proofs.

Definition 1.29 (Half-space). The half-space $\mathbb{R}_+^N$ of $\mathbb{R}^N$, $N \geq 1$, is the set $\{(x, y) \in \mathbb{R}^N : x \in \mathbb{R}, \ x > 0 \text{ and } y \in \mathbb{R}^{N-1}\}$.

Theorem 1.30 (Trace, half-space). *Let $\Omega = \mathbb{R}_+^N$. For all p such that $1 \leq p < +\infty$, there exists a unique continuous linear mapping γ from $W^{1,p}(\Omega)$ to $L^p(\mathbb{R}^{N-1})$ such that $\gamma u = u(0, \cdot)$ a.e. on $\mathbb{R}^{N-1}$ if $u \in C_c^\infty(\overline{\mathbb{R}_+^N})$. Note that the equality $\gamma u = u(0, \cdot)$ a.e. on $\mathbb{R}^{N-1}$ is to be taken in the sense of the $(N - 1)$-dimensional Lebesgue measure.*

Remark 1.31 (Link with the classical trace). Assume that $\Omega = \mathbb{R}^N_+$. Then:

1. If $u \in H^1(\Omega) \cap C(\overline{\Omega})$, then $\gamma u = u$ a.e. on $\partial\Omega$ (in the sense of the Lebesgue $N - 1$-dimensional measure).
2. $\mathrm{Ker}\gamma = W^{1,p}_0(\mathbb{R}^N_+)$.

See in this regard Problem 1.20.

Theorem 1.32 (Trace, bounded open subset). *Let Ω be an open bounded subset with a Lipschitz boundary and $1 \le p < +\infty$. Then, there exists a unique (linear continuous) mapping γ defined from $W^{1,p}(\Omega)$ to $L^p(\partial\Omega)$ such that*

$$\gamma u = u \ a.e. \ on \ \partial\Omega \ if \ u \in W^{1,p}(\Omega) \cap C(\overline{\Omega}).$$

Here again, a.e. is to be understood in the sense of the $(N-1)$-dimensional Lebesgue measure on $\partial\Omega$.

Furthermore $\mathrm{Ker}\gamma = W^{1,p}_0(\Omega)$.

Note that if $p > N$, we can show (see Theorem 1.41) that $W^{1,p}(\Omega) \subset C(\overline{\Omega})$ and γu is then the value of u on the boundary in the classical sense.

The following theorem generalises the property of integration by parts of regular functions. The result is obtained (in both cases of Theorem 1.33) by density of $C^\infty_c(\overline{\Omega})$ in $H^1(\Omega)$. Recall that $C^\infty_c(\overline{\Omega})$ denotes the set of restrictions to Ω of elements of $\mathcal{D}(\mathbb{R}^N)$.

Theorem 1.33 (Integration by parts).

- *If $\Omega = \mathbb{R}^N_+$, then*

$$\int_\Omega u\, D_i v\, \mathrm{d}x = \begin{cases} -\int_\Omega D_i u\, v\, \mathrm{d}x, \ \forall (u,v) \in (H^1(\Omega))^2, & if\, 2 \le i \le N, \\ -\int_\Omega D_1 u\, v\, \mathrm{d}x + \int_{\partial\Omega} \gamma u(y)\, \gamma v(y)\, \mathrm{d}\gamma(y), & if\, i = 1, \\ \qquad \forall (u,v) \in (H^1(\Omega))^2, \end{cases}$$

- *if Ω is an open bounded subset with a Lipschitz boundary, then, for all $i = 1, \ldots, N$,*

$$\int_\Omega u\, D_i v\, \mathrm{d}x = -\int_\Omega D_i u\, v\, \mathrm{d}x + \int_{\partial\Omega} \gamma u(y)\, \gamma v(y) n_i(y)\, \mathrm{d}\gamma(y), \forall (u,v) \in (H^1(\Omega))^2,$$

where γu denotes the trace of u on the boundary $\partial\Omega$ and $\mathrm{d}\gamma(y)$ denotes integration with respect to the appropriate measure on $\partial\Omega$ (which can be seen as an $(N-1)$-dimensional Lebesgue measure, see the following remark), and $n = (n_1, \ldots, n_N)^t$ is the outward normal unit vector to $\partial\Omega$.

Remark 1.34 (Measure on $\partial\Omega$). The measure used on $\partial\Omega$ is defined with all the technical details in J. Droniou's monograph [20, section 2.2.1]. The proof of the previous trace theorem may also be found there [ibid., theorem 4.2.1].

1.6 Compactness Theorems

The proofs of existence of solutions of PDEs sometimes use compactness arguments: For example, once estimates on the approximate solutions of a given problem (for example by reducing to problems in finite dimension) are obtained, the compactness theorems presented below allow us to extract convergent subsequences of some sequence of these approximate solutions. It then remains to show that the limit of such a subsequence is indeed a solution to the problem under study.

The Rellich compactness theorem[15] and its generalisations are a consequence of the Kolmogorov compactness theorem[16] (also sometimes called the Fréchet–Kolmogorov theorem), see [26, chapter 8]. Kolmogorov's theorem is adapted in the case of space-time in Chapter 4 (see Theorems 4.43 and 4.44). It is itself a consequence of Ascoli's[17] theorem, which is used several times in the rest of the book.

Theorem 1.35 (Ascoli (or Arzelà–Ascoli)). *Let (K, d) be a compact metric space and (E, d') a complete metric space. The space $C(K, E)$ of continuous functions from K to E, equipped with the uniform distance, is a complete metric space. Recall that the uniform distance is defined by*

$$d(f, g) = \sup_{x \in K} d'(f(x), g(x)).$$

A subset A of $C(K, E)$ is relatively compact (that is, included in a compact set) if and only if, for every point x of K:

- *A is equicontinuous at x, that is, for every $\varepsilon > 0$, there exists a δ such that for every $f \in A$ and for every y such that $d(x, y) < \delta$, $d'(f(x), f(y)) < \varepsilon$,*
- *the set $\{f(x) | f \in A\}$ is relatively compact.*

Theorem 1.36 (Rellich). *Let Ω be an open bounded subset of $\mathbb{R}^N$ ($N \geq 1$) and $1 \leq p < +\infty$. Every bounded subset of $W_0^{1,p}(\Omega)$ is relatively compact in $L^p(\Omega)$. This is equivalent to saying that from any bounded sequence in $W_0^{1,p}(\Omega)$, one can extract a subsequence that converges in $L^p(\Omega)$.*

The previous theorem remains true for $W^{1,p}(\Omega)$ provided the boundary is Lipschitz.

Theorem 1.37 (Compactness of the bounded subsets of $W^{1,p}(\Omega)$ in $L^p(\Omega)$). *Let Ω be an open bounded subset of $\mathbb{R}^N$ ($N \geq 1$), with Lipschitz boundary, and $1 \leq p < +\infty$. Any bounded subset of $W^{1,p}(\Omega)$ is relatively compact in $L^p(\Omega)$. This is equivalent to saying that from any bounded sequence in $W^{1,p}(\Omega)$, we can extract a subsequence that converges in $L^p(\Omega)$.*

---

[15] Franz Rellich (1906–1955), Austrian-German mathematician, specialist in mathematical physics and partial differential equations.

[16] Andreï Nikolaïevitch Kolmogorov (1903–1987), Soviet mathematician known for his contributions in probability, topology, logic, turbulence, mechanics, information theory and complexity.

[17] Giulio Ascoli (1843–1896), Italian mathematician.

We also need a version of Theorem 1.36 in the dual spaces of L^p and $W_0^{1,p}$. Recall that for $p < +\infty$, the dual space of L^p is identified with the space L^q, where $q = \frac{p}{p-1}$, and that the dual space of $W_0^{1,p}$ is denoted $W^{-1,q}$.

Theorem 1.38 (Compactness of the bounded subsets of $L^q(\Omega)$ in $W^{-1,q}(\Omega)$). *Let Ω be an open bounded subset of $\mathbb{R}^N$ ($N \geq 1$) and $1 < q < +\infty$. Any bounded subset of $L^q(\Omega)$ is relatively compact in $W^{-1,q}(\Omega)$. In particular, for $q = 2$, the space $W^{-1,2}(\Omega)$ is also denoted $H^{-1}(\Omega)$. Any bounded subset of $L^2(\Omega)$ is therefore relatively compact in $H^{-1}(\Omega)$.*

Proof This result is a consequence of Rellich's theorem (Theorem 1.36), see Problem 1.26. ∎

Remark 1.39 (On the compatibility of identifications). Let Ω be an open subset of $\mathbb{R}^N$, $N \geq 1$. If $f \in L^1_{\text{loc}}(\Omega)$, we identify f with the element of $\mathcal{D}^\star(\Omega)$ it represents. Let $1 \leq p \leq +\infty$. If $f \in L^p(\Omega)$ (and therefore $f \in L^1_{\text{loc}}(\Omega)$), it is usual to identify f with the element of $(L^q(\Omega))'$ it represents, where $q = \frac{p}{p-1} \in [1, +\infty]$. These two identifications are compatible in the sense that if $f \in L^p(\Omega)$ and $\varphi \in \mathcal{D}(\Omega)$ (and therefore $\varphi \in L^q(\Omega)$),

$$\langle f, \varphi \rangle_{\mathcal{D}^\star(\Omega), \mathcal{D}(\Omega)} = \langle f, \varphi \rangle_{L^q(\Omega)', L^q(\Omega)} = \int_\Omega f(x)\varphi(x)\, \mathrm{d}x,$$

where $\mathrm{d}x$ is the integration sign with respect to the Lebesgue measure.

Now consider $f \in H^1(\Omega)(\subset L^1_{\text{loc}}(\Omega))$. Since there is a natural isomorphism between a Hilbert space and its dual space, one might be tempted to identify f with the element of $H^1(\Omega)'$ given by this isomorphism. But this identification is incompatible with the identification of f with the element of $\mathcal{D}^\star(\Omega)$ it represents. Indeed, these two identifications are based on the equalities

$$\langle f, \varphi \rangle_{\mathcal{D}^\star(\Omega), \mathcal{D}(\Omega)} = \int_\Omega f(x)\varphi(x)\, \mathrm{d}x,$$

$$\langle f, \varphi \rangle_{H', H} = \int_\Omega \nabla f(x) \cdot \nabla \varphi(x)\, \mathrm{d}x + \int_\Omega f(x)\varphi(x)\, \mathrm{d}x,$$

and so, $\langle f, \varphi \rangle_{\mathcal{D}^\star(\Omega), \mathcal{D}(\Omega)} \neq \langle f, \varphi \rangle_{H', H}$ (except in very specific cases).

Similarly, if Ω is an open bounded subset of $\mathbb{R}^d$ and $f \in H_0^1(\Omega) \subset L^2(\Omega)$, and we identify f with

$$T_f \in (L^2(\Omega))' : v \mapsto (f|v)_{L^2(\Omega)} = \int_\Omega f(x)v(x)\, \mathrm{d}x,$$

we can no longer identify f with

$$\tilde{T}_f \in (H_0^1(\Omega))' : v \mapsto (f|v)_{H_0^1(\Omega)} = \int_\Omega \nabla f(x) \cdot \nabla v(x)\, \mathrm{d}x.$$

Here we have taken in $H_0^1(\Omega)$ the inner product $(u \mid v)_{H_0^1(\Omega)} = \int_\Omega \nabla u(x) \cdot \nabla v(x)\, dx$, which is equivalent, in $H_0^1(\Omega)$, to the inner product of $H^1(\Omega)$.

1.7 Sobolev Embeddings

Sobolev embeddings[18] are very useful tools for the analysis of PDEs. They establish the fact that a function u such that its power u^p (for some fixed p) and its derivative are integrable (that is to say, $u \in W^{1,p}$) is in fact in a "better" space in terms of integration or regularity. We distinguish three different cases, depending on whether the power is less than, equal to, or greater than the dimension of the space N. The third case involves Hölder functions, whose definition is as follows.

Definition 1.40 (Hölder[19] functions). For $\alpha > 0$ and $K \subset \mathbb{R}^N$, the set of functions from K to $\mathbb{R}$ that are Hölder with exponent α, denoted $C^{0,\alpha}(K)$, is defined by

$$C^{0,\alpha}(K) = \{u \in C(K, \mathbb{R}) \mid \exists k \in \mathbb{R} \ \ |u(x) - u(y)| \leq k\|x - y\|^\alpha, \forall (x, y) \in K^2\}.$$

Theorem 1.41 (Sobolev embeddings). *Let Ω be an open subset of $\mathbb{R}^N$, $N \geq 1$, which is either bounded with Lipschitz boundary, or equal to $\mathbb{R}^N$.*

1. *If $1 \leq p < N$, then $W^{1,p}(\Omega) \subset L^{p^\star}(\Omega)$, with $p^\star = \frac{Np}{N-p}$. The inclusion map, or canonical injection[20] from $W^{1,p}(\Omega)$ to $L^{1,p^\star}(\Omega)$ is defined as the mapping $u \in W^{1,p}(\Omega) \mapsto u \in L^{p^\star}(\Omega)$. This injective mapping is continuous, that is, there exists a $C \in \mathbb{R}_+$ (depending only on p, N and Ω) such that the following inequality (known as Sobolev's inequality) is satisfied:*

$$\forall u \in W^{1,p}(\Omega), \ \|u\|_{L^{p^\star}} \leq C\|u\|_{W^{1,p}}.$$

   We also say that $W^{1,p}(\Omega)$ is continuously embedded in $L^{p^\star}(\Omega)$. In particular, $W^{1,1}(\Omega)$ is continuously embedded in $L^{\frac{N}{N-1}}(\Omega)$.

2. *In the case $N = 1$, the choice $p = N$ is allowed, that is to say, for $N = 1$, $W^{1,1}(\Omega)$ is continuously embedded in $L^\infty(\Omega)$.*

3. *If $p > N$, then we write, with a certain abuse of notation,*

$$W^{1,p}(\Omega) \subset C^{0,1-\frac{N}{p}}(\bar{\Omega})$$

   in the sense that for any class of functions $u \in W^{1,p}(\Omega)$, there exists a function $v \in C^{0,1-\frac{N}{p}}(\bar{\Omega})$ belonging to the class u. In practice, the class u is then conflated with the function v which is the unique continuous function belonging to the

---

[18] Recall that an embedding is an injective mapping from one set to another.

[19] Otto Ludwig Hölder (1859–1937), German mathematician.

[20] More generally, if two sets E and F are such that $E \subset F$, the injective map $x \in E \mapsto x \in F$ is said to be canonical.

class u. We can in fact show that the embedding of $W^{1,p}(\Omega)$ to $C^{0,1-\frac{N}{p}}(\bar{\Omega})$ is continuous, for a norm to be defined, see Problem 1.18.

4. *In the case where Ω is bounded with a Lipschitz boundary, the space $W^{1,N}(\Omega)$ is continuously embedded in the space $L^q(\Omega)$, for any q such that $1 \leq q < +\infty$ (and the case $q = +\infty$ is allowed if $N = 1$). This result is false in the case where $\Omega = \mathbb{R}^N$, see the counterexample in Problem 1.5.*

If Ω is an open bounded subset without any assumption of regularity on the boundary, the four previous assertions remain true if we replace the space $W^{1,p}(\Omega)$ by the space $W_0^{1,p}(\Omega)$.

Remark 1.42 (Sobolev embedding for dual spaces). Let E, F be two (real) Banach spaces and let $T \in \mathcal{L}(E, F)$, where $\mathcal{L}(E, F)$ is the set of continuous linear mappings from E to F. For $g \in F'$ we define $T^t g \in E'$ by

$$\langle T^t g, u \rangle_{E',E} = \langle g, Tu \rangle_{F',F}.$$

It is clear that $T^t g$ is indeed an element of E' for all $g \in F'$ and it can be shown that $T^t \in \mathcal{L}(F', E')$ (see Problem 1.26). A consequence of this result is that F' is continuously embedded in E' if E is continuously embedded in F. As an example, if Ω is an open bounded subset of $\mathbb{R}^N$ with Lipschitz boundary and $1 \leq p < N$, the space $L^{p^\star}(\Omega)'$ is continuously embedded in the space $W^{1,p}(\Omega)'$; therefore, since $L^{p^\star}(\Omega)'$ is generally identified with $L^{(p^\star)'}(\Omega)$ (where $(p^\star)'$ is the conjugate exponent of $p^\star$), the space $L^{(p^\star)'}(\Omega)$ is continuously embedded in $W^{1,p}(\Omega)'$.

Remark 1.43 (Compact embedding of $W_0^{1,p}(\Omega)$ into $L^q(\Omega)$). Let Ω be an open bounded subset of $\mathbb{R}^N$. A consequence of the Sobolev embedding theorem (Theorem 1.41) and the Rellich compactness theorem (Theorem 1.36) is that the embedding is compact from $W_0^{1,p}(\Omega)$ to $L^q(\Omega)$ if $1 \leq p \leq N$ and $q < p^\star = \frac{pN}{N-p}$, i.e. from a bounded sequence in $W_0^{1,p}(\Omega)$ we can extract a subsequence that converges in $L^q(\Omega)$. To prove this property, we can limit ourselves to the case $p < N$ (thanks to the embedding of $W^{1,N}(\Omega)$ into $W^{1,p}(\Omega)$ for all $p < N$). It is then sufficient to note that convergence in L^p gives convergence in L^1 (because Ω is bounded) and use Hölder's inequality, which gives (for $1 < q < p^\star$ and $u \in L^{p^\star}(\Omega)$) $\|u\|_{L^q} \leq \|u\|_{L^1}^\theta \|u\|_{L^{p^\star}}^{1-\theta}$ (with $\theta = (p^\star - q)/(p^\star - 1)$), from which we get that $u_n \to u$ in L^1 and $(u_n)_{n \in \mathbb{N}}$ bounded in $L^{p^\star}$ implies $u_n \to u$ in L^q as $n \to +\infty$.

If $p > N$, a consequence of the Sobolev embedding theorem (Theorem 1.41) and Ascoli's theorem (Theorem 1.35) is that the mapping $u \mapsto u$ is compact from $W^{1,p}(\Omega)$ into $C(\bar{\Omega})$, i.e. from a bounded sequence in $W_0^{1,p}(\Omega)$ we can extract a subsequence uniformly convergent in $\bar{\Omega}$. Of course, it is clear here that each element of $W^{1,p}(\Omega)$ is conflated with its continuous representative.

1.8 Problem Set

Problem 1.1 (Example of derivative ($\star\star\star$)). *Solution on page 31.* Let $N \geq 1$, $\Omega = \{x = (x_1,\ldots,x_N)^t \in \mathbb{R}^N, |x_i| < 1, i = 1,\ldots,N\}$ and $u : \mathbb{R}^N \to \mathbb{R}$ be defined by $u(x) = 1$ if $x \in \Omega$ and $u(x) = 0$ if $x \notin \Omega$.

1. For $i = \{1,\ldots,N\}$ and $\varphi \in \mathcal{D}(\mathbb{R}^N)$, show that $\int_{\mathbb{R}^N} u(x)\partial_i \varphi(x)\,dx$ only depends on the values taken by φ on the boundary of Ω.
2. Show that $u \notin W^{1,1}(\mathbb{R}^N)$.

Problem 1.2 (A function with zero derivative is constant a.e. ($\star\star$)). *Solution on page 32.* Let $u \in L^1_{\mathrm{loc}}(]0,1[)$ such that $Du = 0$. Show that

$$\exists a \in \mathbb{R} \quad u = a \text{ a.e..}$$

Problem 1.3 (Sobolev spaces in one space dimension ($\star\star\star$)). *Solution on page 33.* Let $1 \leq p \leq \infty$.

1. Let $u \in W^{1,p}(]0,1[)$.

   a. Show that there exists a $C \in \mathbb{R}$ such that $u(x) = C + \int_0^x Du(t)\,dt$, for almost all $x \in]0,1[$. Deduce that $u \in C([0,1],\mathbb{R})$ (in the sense that there exists a $v \in C([0,1],\mathbb{R})$ such that $u = v$ a.e. on $]0,1[$; by identifying u and v, we can therefore say that $W^{1,p}(]0,1[) \subset C([0,1],\mathbb{R})$).
   b. Show that $\|u\|_\infty \leq \|u\|_{W^{1,p}(]0,1[)}$.
   c. If $p > 1$, show that u is a Hölder function with exponent $1 - \dfrac{1}{p}$.

2. Let $u \in C([0,1],\mathbb{R})$. Assume that there exists a function $w \in L^p(]0,1[)$ such that $u(x) = u(0) + \int_0^x w(t)\,dt$, for all $x \in]0,1[$. Show that $u \in W^{1,p}(]0,1[)$ and $Du = w$.

Problem 1.4 (A function with zero gradient is constant a.e. ($\star\star\star\star$)). *Solution on page 35.* Let $N \geq 1$, $B = \{x \in \mathbb{R}^N, |x| < 1\}$, $|x|$ denoting the Euclidean norm of x, and $u \in L^1_{\mathrm{loc}}(B)$.

1. Assuming that $D_i u = 0$ for all $i \in \{1,\ldots,N\}$, show that there exists an $a \in \mathbb{R}$ such that $u = a$ a.e. (u is therefore the constant function equal to a.) [One could, for example, reason as follows: Let $\varepsilon \in]0, \frac{1}{2}[$ and $(\rho_n)_{n\in\mathbb{N}^\star}$ be a sequence of regularising kernels, that is to say:

$$\rho \in \mathcal{D}(\mathbb{R}^N), \quad \int_{\mathbb{R}^N} \rho\,dx = 1, \ \rho \geq 0, \ \rho(x) = 0 \text{ if } |x| \geq 1,$$

and, for $n \in \mathbb{N}^\star$, $x \in \mathbb{R}^N$, $\rho_n(x) = n^N \rho(nx)$.

Set $u_\varepsilon(x) = u$ if $|x| \leq 1 - \varepsilon$ and $u_\varepsilon = 0$ otherwise, and $u_{\varepsilon,n} = u_\varepsilon \star \rho_n$.

Show that $u_{\varepsilon,n} \in \mathcal{D}(\mathbb{R}^N)$ and that, if $\frac{1}{n} < \varepsilon$, $u_{\varepsilon,n}$ is constant on the ball centred at 0 and of radius $1 - 2\varepsilon$. Then, conclude. . . .]

2. Assuming now that for all $i \in \{1, \ldots, N\}$, $D_i u$ is a continuous function, show that $u \in C^1(B, \mathbb{R})$ (in the sense "there exists a $v \in C^1(B, \mathbb{R})$ such that $u = v$ a.e."). [One could, for example, take up the hint in question 1 above and reason as follows: Show that for all $x, y \in \mathbb{R}^N$ we have

$$u_{\varepsilon,n}(y) - u_{\varepsilon,n}(x) = \int_0^1 \nabla u_{\varepsilon,n}(ty + (1-t)x) \cdot (y-x) \, dt,$$

and that for z in the ball centred at 0 and of radius $1 - 2\varepsilon$ and $i \in \{1, \ldots, N\}$ we have

$$\partial_i u_{\varepsilon,n}(z) = \int_B D_i u(\bar{z}) \rho_n(z - \bar{z}) \, d\bar{z}.$$

Recall that $\partial_i u$ denotes the partial derivative of u with respect to the i-th variable. Deduce that for almost all $x, y \in B$, we have, with $Du = \{D_1 u, \ldots, D_N u\}^t$,

$$u(y) - u(x) = \int_0^1 Du(ty + (1-t)x) \cdot (y-x) \, dt.$$

Then show that u is continuous and that the previous formula holds for all $x, y \in B$. Finally conclude that $u \in C^1(B, \mathbb{R})$.]
3. Assume now that B is any open subset of $\mathbb{R}^N$ and that, as in the first question, $D_i u = 0$ for all $i \in \{1, \ldots, N\}$. Show that u is constant on each connected component of B. (As usual, saying that u is constant means that there exists an $a \in \mathbb{R}$ such that $u = a$ a.e.)

Problem 1.5 (An H^1 function is not necessarily continuous if $N > 1$ ($\star\star$)). *Solution on page 37.* Let $\Omega = \{x = (x_1, x_2)^t \in \mathbb{R}^2, |x_i| < \frac{1}{2}, i = 1, 2\}$, $\gamma \in]0, \frac{1}{2}[$, and let $u : \Omega \to \mathbb{R}$ be defined by $u(x) = (-\ln(|x|))^\gamma$, $|x|$ denoting the Euclidean norm of x. Show that $u \in H^1(\Omega)$. Deduce that $H^1(\Omega) \not\subset C(\overline{\Omega})$.

Problem 1.6 (Laplacian of an element of $H_0^1(\Omega)$ ($\star$)). *Solution on page 38.* Let Ω be an open bounded subset of $\mathbb{R}^N$ ($N \geq 1$) and $u \in H_0^1(\Omega)$. For $u \in H^1(\Omega)$, we define $\Delta u = \sum_{i=1}^N D_i^2 u$, where $D_i^2 u$ is the second order derivative by transposition of u with respect to the variable x_i (see Definition 1.4).

1. Show that, for all $\varphi \in \mathcal{D}(\Omega)$, we have

$$\langle \Delta u, \varphi \rangle_{\mathcal{D}^\star(\Omega), \mathcal{D}(\Omega)} = \int_\Omega u(x) \Delta \varphi(x) \, dx = - \int_\Omega \nabla u(x) \cdot \nabla \varphi(x) \, dx.$$

2. Recall that $H_0^1(\Omega)$ is a closed subspace of $H^1(\Omega)$. Equipped with the norm on $H^1(\Omega)$, the space $H_0^1(\Omega)$ is therefore a Hilbert space. Let $H^{-1}(\Omega)$ be the (topological) dual space of $H_0^1(\Omega)$. Deduce from the previous question that $\Delta u \in H^{-1}(\Omega)$ (that is, the element of $\mathcal{D}^\star(\Omega)$, denoted Δu, extends in a unique way to an element of $H^{-1}(\Omega)$, still denoted Δu) and that

$$\|\Delta u\|_{H^{-1}(\Omega)} \leq \| |\nabla u| \|_{L^2(\Omega)}.$$

Problem 1.7 (Point singularity ($\star\star\star$)). *Solution on page 40.* For $x \in \mathbb{R}^2 \setminus \{0\}$, set $G(x) = \ln(|x|)$.

1. Show that $G \in C^\infty(\mathbb{R}^2 \setminus \{0\})$ and $\Delta G = 0$ (in the classical sense) in $\mathbb{R}^2 \setminus \{0\}$. Deduce that $\Delta G = 0$ in $\mathcal{D}^\star(\mathbb{R}^2 \setminus \{0\})$. What is the value of ΔG in $\mathcal{D}^\star(\mathbb{R}^2)$?
2. Show that $G \in L^p_{\text{loc}}(\mathbb{R}^2)$ for all $1 \leq p < +\infty$ and $\nabla G \in L^p_{\text{loc}}(\mathbb{R}^2)$ for $1 \leq p < 2$.
3. In this question, we take $\Omega =]0, 1[^2$. Show that $u \in L^2(\Omega)$, $\Delta u = 0 \Rightarrow u \in H^1(\Omega)$. Recall that for any vector function

$$v : \begin{bmatrix} x_1 \\ x_2 \end{bmatrix} \in \mathbb{R}^2 \mapsto \begin{bmatrix} v_1(x_1, x_2) \\ v_2(x_1, x_2)) \end{bmatrix} \in \mathbb{R}^2,$$

the divergence of v and its curl are functions of $\mathbb{R}^2$ to $\mathbb{R}$ defined respectively by $\operatorname{div} v = \partial_1 v_1 + \partial_2 v_2$ and $\operatorname{curl} v = \partial_1 v_2 - \partial_2 v_1$.
Show that the fact that $v \in (H^1(\Omega)')^2$, $\operatorname{div} v = 0$ in $\mathcal{D}^\star(\Omega)$ and $\operatorname{curl} v = 0$ in $\mathcal{D}^\star(\Omega)$ does not imply that $v \in (L^2(\Omega))^2$.
4. (Removable singularity) Let Ω be an open subset of $\mathbb{R}^2$ containing 0. Assume that $u \in H^1(\Omega)$ and that $\Delta u = 0$ in $\mathcal{D}^\star(\Omega \setminus \{0\})$. Show that $\Delta u = 0$ in $\mathcal{D}^\star(\Omega)$.

Problem 1.8 (Some consequences of the Hahn–Banach theorem ($\star\star$)). *Solution on page 42.* Let E be a real Banach space.

1. Let $x \in E$, $x \neq 0$. Show that there exists a $T \in E'$ such that $T(x) = \|x\|_E$ and $\|T\|_{E'} = 1$. Deduce that

$$\|x\|_E = \max\{S(x), \ S \in E' \ \ \|S\|_{E'} = 1\}.$$

2. Let F be a subspace of E and $x \in E$. Show that $x \notin \bar{F}$ if and only if there exists a $T \in E'$ such that $T(x) \neq 0$ and $T(y) = 0$ for all $y \in F$.
3. For $x \in E$, we define $J(x)$ from E' to $\mathbb{R}$ by $J(x)(T) = T(x)$ for all $T \in E'$. Show that $J(x) \in E''$ for all $x \in E$ and that the mapping $J : x \mapsto J(x)$ is an isometry of E onto $\operatorname{Im}(J) \subset E''$.
4. Let E be a Banach space.

   a. Let $(T_n)_{n \in \mathbb{N}}$ be a sequence in E' and $T \in E'$ such that $T_n \to T$ $\star$-weakly in E' as $n \to +\infty$. Show that $\|T\|_{E'} \leq \liminf_{n \to +\infty} \|T_n\|_{E'}$.
   b. Let $(x_n)_{n \in \mathbb{N}}$ be a sequence in E and $x \in E$ such that $x_n \to x$ weakly in E as $n \to +\infty$. Show that $\|x\|_E \leq \liminf_{n \to +\infty} \|x_n\|_E$. This proves the result stated in Remark 1.20.

Problem 1.9 (Comparison $\ell^p(\mathbb{N})$-$\ell^q(\mathbb{N})$ ($\star$)). *Solution on page 43.* For $1 \leq p < +\infty$, we define the space

$$\ell^p = \ell^p(\mathbb{N}) = \{x = (x_n)_{n \in \mathbb{N}}; \ x_n \in \mathbb{R} \text{ for all } n \in \mathbb{N}, \ \sum_{n=0}^{\infty} |x_n|^p < +\infty\}$$

which we equip with a norm defined by $\|x\|_p = (\sum_{n=0}^{+\infty} |x_n|^p)^{\frac{1}{p}}$ for $x = (x_n)_{n \in \mathbb{N}} \in \ell^p$.

We also define, for $p = +\infty$, the space

$$\ell^\infty = \ell^\infty(\mathbb{N}) = \{x = (x_n)_{n\in\mathbb{N}}; \ x_n \in \mathbb{R} \text{ for all } n \in \mathbb{N}, \ \sup\{|x_n|, \ n \in \mathbb{N}\} < +\infty\},$$

which we equip with the norm defined by $\|x\|_\infty = \sup_{n\in\mathbb{N}}|x_n|$ for $x = (x_n)_{n\in\mathbb{N}} \in \ell^\infty$. With this norm, ℓ^p is a Banach space for all $1 \le p \le +\infty$ (see for example [26, Problem 6.27]).

1. Let $1 \le p < q \le +\infty$. Show that $\ell^p \subset \ell^q$ and that, for all $x \in \ell^p$,

$$\|x\|_q \le \|x\|_p .$$

2. Let $1 \le p \le q < +\infty$. Show that ℓ^p is dense in ℓ^q.
3. Let $1 \le p < +\infty$. Show that the closure of ℓ^p in ℓ^∞ is the set A defined by

$$A = \{x = (x_n)_{n\in\mathbb{N}}; \ \lim_{n\to+\infty} x_n = 0\}.$$

Deduce that ℓ^p is not dense in ℓ^∞.

N.B. When $(X, \mathcal{T}, m)$ is a finite measure space, we still have the reverse inequality, that is $L^p(X, \mathcal{T}, m) \subset L^q(X, \mathcal{T}, m)$ if $1 \le q \le p \le +\infty$.

Problem 1.10 (Characterisation of the density of a v.s.s. of a Banach space $(\star)$). *Solution on page 44.* Let E be a real Banach space and G a v.s.s. of E.

1. Show that $\bar{G} = E$ if and only if

$$(f \in E', \langle f, u\rangle_{E',E} = 0 \text{ for all } u \in G) \Rightarrow f = 0. \tag{1.5}$$

   [Use Problem 1.8.]
2. We now assume that $E = F'$ (where F is a real Banach space).

   a. Assuming F is reflexive, show that $\bar{G} = F'$ if and only if

   $$(v \in F, \langle g, v\rangle_{F',F} = 0 \text{ for all } g \in G) \Rightarrow v = 0. \tag{1.6}$$

   b. We no longer assume that F is reflexive. Give an example where $\bar{G} \ne F'$ and yet (1.6) is satisfied, that is

   $$(v \in F, \langle g, v\rangle_{F',F} = 0 \text{ for all } g \in G) \Rightarrow v = 0.$$

   [Take $F = \ell^1$ and identify $(\ell^1)'$ with ℓ^∞, $G = \ell^1$.]

Problem 1.11 (Separability of L^p $(\star\star\star)$). *Solution on page 44.* We denote by L^p the space $L^p(\mathbb{R}, \mathcal{B}(\mathbb{R}), \lambda)$.

1. Let $1 \le p < +\infty$; show that L^p is separable.
2. Show that $L^\infty(\mathbb{R})$ is not separable.

Problem 1.12 (Separability of a subset of a separable space ($\star$)). *Solution on page 45.* Let E be a separable normed vector space, A a countable dense subset of E and F a subset of E. Let $n \in \mathbb{N}^{\star}$. For all $x \in A$, if there exists at least one point y from F such that $\|x - y\|_E \leq \frac{1}{n}$, we choose such a point and denote it $a_{x,n}$ (so we have $a_{x,n} \in F$ and $\|x - a_{x,n}\|_E \leq \frac{1}{n}$). Let B_n be the set of points obtained in this way (note that $B_n \subset F$).

1. Let $n \in \mathbb{N}^{\star}$. Construct an injective mapping from B_n to A (which proves that B_n is countable).
2. With the help of the sets B_n ($n \in \mathbb{N}^{\star}$), construct a subset of F that is countable and dense in F (which proves that F is separable).

Problem 1.13 (Closed vector subspace of a reflexive Banach space ($\star\star$)). *Solution on page 46.* Let E be a reflexive (real) Banach space and F a closed vector subspace of E. The space F (equipped with the norm on E) is therefore also a Banach space. For $G = E$ or $G = F$ let J_G be the injective mapping from G to G'' (see Problem 1.8) and let $u \in F''$.

1. If $f \in E'$, let $f_{|F}$ be the restriction of f to F (and therefore $f_{|F} \in F'$). Show that the mapping $v : f \mapsto \langle u, f_{|F} \rangle_{F'',F'}$ is a continuous linear mapping from E' to $\mathbb{R}$, and thus belongs to E'' (note that $\langle v, f \rangle_{E'',E'} = \langle u, f_{|F} \rangle_{F'',F'}$). Show that $\|v\|_{E''} \leq \|u\|_{F''}$.
2. (Lone question) Do we have $\|v\|_{E''} = \|u\|_{F''}$?
3. Show that there exists an $x \in E$ such that $\langle v, f \rangle_{E'',E'} = \langle f, x \rangle_{E',E}$ for all $f \in E'$. [Use the fact that E is reflexive.]
4. Consider the element x of E found in question 3.

   a. Show that $x \in F$. [Use a consequence of the Hahn–Banach theorem, Problem 1.8.]
   b. Show that $J_F(x) = u$.

5. Deduce from the previous questions that F is reflexive.

Problem 1.14 (Continuity of a mapping from L^p to L^q ($\star\star\star$)). *Solution on page 47.* Let (E, T, m) be a finite measure space, $p, q \in [1, \infty[$ and g a continuous function from $\mathbb{R}$ to $\mathbb{R}$ such that:

$$\exists C \in \mathbb{R}_+^{\star} \quad |g(s)| \leq C|s|^{\frac{p}{q}} + C, \ \forall s \in \mathbb{R}. \tag{1.7}$$

1. Let $u \in \mathcal{L}_{\mathbb{R}}^p(E, T, m)$. Show that $g \circ u \in \mathcal{L}_{\mathbb{R}}^q(E, T, m)$.

Setting $L^r = L_{\mathbb{R}}^r(E, T, m)$, for $r = p$ and $r = q$, we define, for $u \in L^p$, $G(u) = \{h \in \mathcal{L}_{\mathbb{R}}^q(E, T, m); h = g \circ v \text{ a.e.}\}$, with $v \in u$, i.e. $v \in \mathcal{L}_{\mathbb{R}}^p(E, T, m)$ and v is one of the functions belonging to the equivalence class u. In this problem, we make a clear distinction between L^p and $\mathcal{L}^p$, a distinction that, in general, is not made, including in this book. Note that $G(u) \in L^q$. In practice, and this is what is done in the remainder of this book, this element $G(u)$ of L^q is denoted, incorrectly, $g(u)$.

2. Show that the previous definition makes sense, i.e. that $G(u)$ does not depend on the choice of v in u.

3. Let $(u_n)_{n\in\mathbb{N}} \subset L^p$ and assume that $u_n \to u$ a.e., as $n \to +\infty$, and there exists an $F \in L^p$ such that $|u_n| \leq F$ a.e., for all $n \in \mathbb{N}$. Show that $G(u_n) \to G(u)$ in L^q.
4. Show that G is continuous from L^p to L^q.
5. Here we consider $(E, T, m) = ([0, 1], \mathcal{B}([0, 1]), \lambda)$ and we take $p = q = 1$. We assume that g does not satisfy (1.7). We construct $u \in L^1$ such that $G(u) \notin L^1$.

   a. Let $n \in \mathbb{N}^\star$. Show that there exists an $\alpha_n \in \mathbb{R}$ such that: $|g(\alpha_n)| \geq n|\alpha_n|$ and $|\alpha_n| \geq n$.
   b. Let $(\alpha_n)_{n\in\mathbb{N}^\star}$ be a sequence satisfying the conditions given in question 5a above. Show that there exists an $\alpha > 0$ s.t.

   $$\sum_{n=1}^{+\infty} \frac{\alpha}{|\alpha_n|n^2} = 1.$$

   c. Let $(a_n)_{n\in\mathbb{N}^\star}$ be a sequence defined by: $a_1 = 1$ and $a_{n+1} = a_n - \dfrac{\alpha}{|\alpha_n|n^2}$ (where α_n and α are defined in questions 5a and 5b above). Define $u = \sum_{n=1}^{+\infty} \alpha_n \mathbb{1}_{[a_{n+1}, a_n[}$, where $\mathbb{1}_A$ denotes the characteristic function of a set A, that is $\mathbb{1}_A(x) = 1$ if $x \in A$, 0 otherwise. Show that $u \in L^1$ and $G(u) \notin L^1$.

Problem 1.15 (Lipschitz-continuous functions ($\star\star\star$)). *Solution on page 48.* Let Ω be an open subset of $\mathbb{R}^N$.

1. Let $u : \Omega \to \mathbb{R}$ be a bounded Lipschitz-continuous function. Show that $u \in W^{1,\infty}(\Omega)$. (Note that if Ω is bounded and u is Lipschitz-continuous from $\Omega \to \mathbb{R}$, then u is continuous on Ω and extends to a continuous function on $\bar{\Omega}$, so the function u is bounded.)
2. Let $u \in W^{1,\infty}(\mathbb{R}^N)$. Show that u is Lipschitz-continuous (in the sense: there exists a Lipschitz-continuous function $v: \mathbb{R}^N \to \mathbb{R}$ such that $u = v$ a.e.).

The result of question 2 allows us to show that if Ω is an open bounded subset with a Lipschitz boundary and u is a function from Ω to $\mathbb{R}$, then $u \in W^{1,\infty}(\Omega)$ if and only if u is Lipschitz-continuous (in the sense: there exists a Lipschitz-continuous function $v : \Omega \to \mathbb{R}$ such that $u = v$ a.e. Note that v then extends to a Lipschitz-continuous function on $\bar{\Omega}$.)

Problem 1.16 (A Lipschitz open subset that is not strongly Lipschitz ($\star\star\star$)). *Solution on page 50.* A Lipschitz open subset is defined as an open subset with a Lipschitz boundary (Definition 1.23) and a strongly Lipschitz open subset is one with a strongly Lipschitz boundary (Remark 1.24). To construct an example of a Lipschitz open subset that is not strongly Lipschitz, the idea is to construct an open subset that does not satisfy the segment property.

Definition 1.44 (Segment property). An open subset Ω in $\mathbb{R}^N$ is said to satisfy the segment property if for every $z \in \partial\Omega$, there exist a neighbourhood V of z, $d \in \mathbb{R}^N$, $d \neq 0$ and $t \in \mathbb{R}_+^\star$ such that, for any $\bar{z} \in V \cap \bar{\Omega}$, $\{\bar{z} + sd, s \in]0, t]\} \subset \Omega$.

A strongly Lipschitz open subset in $\mathbb{R}^N$ satisfies the segment property: it suffices to consider (without loss of generality) that the boundary is locally aligned with the x_1 axis in Cartesian coordinates and to choose $d = (0, \ldots, 0, 1)$. To construct an open subset that does not satisfy the segment property, the idea used here (apparently due to Zerner, see [30]) is to take as an open subset a sort of "road" leading to the point $(0, 0)$ with an infinity of turns, without changing the radius of curvature of the turns (which means that the open subset is Lipschitz), but (of course) making sure that the width of the turns tends towards 0 when nearing $(0, 0)$. Because of the turns, the open subset does not satisfy the segment property and therefore is not strongly Lipschitz.

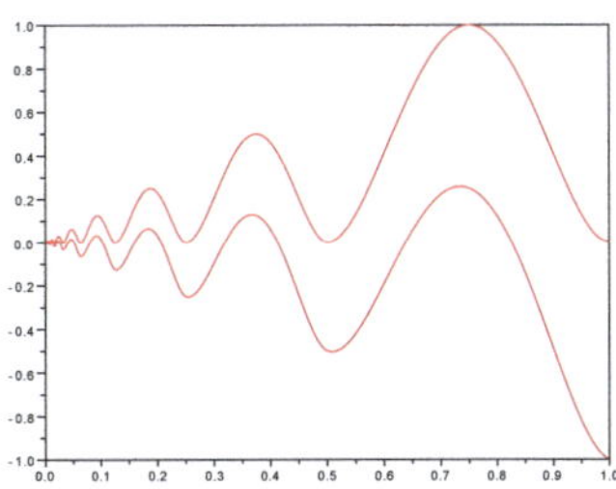

Fig. 1.1 **The open subset** Ω

Construction of the open subset Ω. Let $\bar{\varphi}$ be a continuous function from $[0, 1]$ to $\mathbb{R}$, null at 0 and 1. Assume that

$$\bar{\varphi}\left(\frac{1}{4}\right) \geq 1. \tag{1.8}$$

The function $\bar{\varphi}$ can be, for example, a "hat" function or a C^∞ class function with compact support in $]0, 1[$. Set $a_n = \frac{1}{2^n}$ for $n \in \mathbb{N}$ and define φ from $[0, 1]$ to $\mathbb{R}$ by

$$\varphi(x) = a_{n-1}\bar{\varphi}\left(\frac{x - a_n}{a_{n-1}}\right) \text{ if } x \in]a_n, a_{n-1}] \text{ and } n \geq 1.$$

The function φ is therefore Lipschitz-continuous from $[0, 1]$ to $\mathbb{R}$ (by adding $\varphi(0) = 0$). Also note that, for all $n \in \mathbb{N}$, $\varphi(a_n) = 0$. Let us then set $\Omega = \{(x, y)^t \in \mathbb{R}^2; x \in]0, 1[, \varphi(x) - x < y < \varphi(x)\}$.

1. **The open subset** Ω **is not strongly Lipschitz.** Show that the segment property is not satisfied for Ω and $z = 0$ and deduce that Ω is not strongly Lipschitz.
2. **The open subset** Ω **is weakly Lipschitz.** Set $T = \{(x, y)^t, x \in]0, 1[, -x < y < x\}$ and define a bijection ψ from Ω to the triangle T by setting

$$\psi(x, y) = (x, x + 2(y - \varphi(x))).$$

Show that ψ is Lipschitz-continuous as well as its inverse and deduce that Ω is a Lipschitz open subset.

Problem 1.17 (Extension of a continuous function $(\star)$). *Solution on page 51.* Let Ω be an open subset of $\mathbb{R}^N$, $N \geq 1$, and let $f \in C(\bar{\Omega}, \mathbb{R})$. The aim of the problem is to show that there exists a $g \in C(\mathbb{R}^N, \mathbb{R})$ such that $g = f$ in $\bar{\Omega}$ (see Definition 1.28).

If $x \in \bar{\Omega}$, set $g(x) = f(x)$.

If $x \notin \bar{\Omega}$, set $d_x = \inf\{|x - y|, y \in \Omega\}$ (so we have $d_x > 0$), $B_x = \{z \in \mathbb{R}^N, |x - z| < 2d_x\}$ and

$$g(x) = \frac{\int_{\Omega \cap B_x} f(z)\, dz}{\int_{\Omega \cap B_x} dz}.$$

Show that g is well defined and is continuous from $\mathbb{R}^N$ to $\mathbb{R}$.

Problem 1.18 (Sobolev inequalities for $p > N$ $(\star\star\star\star)$). *Solution on page 52.* The purpose here is to prove the Sobolev embedding for $p > N$.

Let $x = (x_1, \bar{x}) \in \mathbb{R}^N$ $(N \geq 1)$, with $x_1 \in \mathbb{R}$ and $\bar{x} \in \mathbb{R}^{N-1}$. We define $H = \{(t, (1 - |t|)a), t \in]-1, 1[, a \in B_{N-1}\}$, where $B_{N-1} = \{x \in \mathbb{R}^{N-1}, |x| < 1\}$ (recall that $|\cdot|$ always designates the Euclidean norm.)

Let $N < p < +\infty$.

1. Let $u \in C^1(\mathbb{R}^N, \mathbb{R})$. Show that there exists a $C_1 \in \mathbb{R}$, depending only on N and p, such that

$$|u(1, 0) - u(-1, 0)| \leq C_1 \, \|(|\nabla u|)\|_{L^p(H)} . \tag{1.9}$$

[One may start by writing $u(1, 0) - u(0, a)$ as an integral using $\nabla u(t, (1 - t)a)$ for $t \in]0, 1[$, and integrate for $a \in B_{N-1}$ to compare $u(1, 0)$ and its mean value on B_{N-1}. Dealing with the case $N = 2$ avoids unnecessary complications.]

2. Let $u \in C_c^1(\mathbb{R}^N, \mathbb{R})$. Show that there exists a $C_2 \in \mathbb{R}$, not depending only on N and p, such that

$$|u(x) - u(y)| \leq C_2 \, \|(|\nabla u|)\|_{L^p(\mathbb{R}^N)} \, |x - y|^{1 - \frac{N}{p}} . \tag{1.10}$$

[Notice that it is possible to get back to (1.9) through a rotation, a translation and a homothety.]

For $\alpha \in]0, 1]$ and K closed subset of $\mathbb{R}^N$, we define

$$C^{0,\alpha}(K) = \{u \in C(K, \mathbb{R}), \ \|u\|_{L^\infty(K)} < +\infty \ \text{and} \ \sup_{x,y \in K,\, x \neq y} \frac{|u(x) - u(y)|}{|x - y|^\alpha} < +\infty\},$$

and, if $u \in C^{0,\alpha}(K)$,

$$\|u\|_{0,\alpha} = \|u\|_{L^\infty(K)} + \sup_{x,y \in K,\, x \neq y} \frac{|u(x) - u(y)|}{|x - y|^\alpha} .$$

Note that $C^{0,\alpha}(K)$, equipped with this norm, is a Banach space.

3. Let $u \in C_c^1(\mathbb{R}^N, \mathbb{R})$. Show that there exists a $C_3 \in \mathbb{R}$, only depending on N and p, such that

$$\|u\|_{L^\infty(\mathbb{R}^N)} \leq C_3 \|u\|_{W^{1,p}(\mathbb{R}^N)} . \qquad (1.11)$$

[This question is more delicate... Use (1.10) and the fact that $u \in L^p(\mathbb{R}^N)$.]

4. (Sobolev embedding in $\mathbb{R}^N$.) Show that $W^{1,p}(\mathbb{R}^N) \subset C^{0,\alpha}(\mathbb{R}^N)$, with $\alpha = 1 - \frac{N}{p}$, and that there exists a $C_4 \in \mathbb{R}$, only depending on N and p, such that

$$\|u\|_{C^{0,\alpha}(\mathbb{R}^N)} \leq C_4 \|u\|_{W^{1,p}(\mathbb{R}^N)} . \qquad (1.12)$$

5. (Sobolev embedding in Ω.) Let Ω be an open bounded subset of $\mathbb{R}^N$ ($N \geq 1$), with a Lipschitz boundary. Show that $W^{1,p}(\Omega) \subset C^{0,\alpha}(\bar{\Omega})$, with $\alpha = 1 - \frac{N}{p}$, and that there exists a $C_5 \in \mathbb{R}$, only depending on Ω, N and p, such that

$$\|u\|_{C^{0,\alpha}(\bar{\Omega})} \leq C_5 \|u\|_{W^{1,p}(\Omega)} .$$

Problem 1.19 (Sobolev embeddings for $p \leq N$ (★★★★)). *Solution on page 54.* The purpose here is to prove the Sobolev embedding for $1 \leq p \leq N$.

The proof proposed here is due to L. Nirenberg.[21] It consists of first considering the case $p = 1$, then deducing the case $1 < p < N$. Note that in fact, the case $1 < p < N$ was proved before the case $p = 1$ (and the case $p = 1$ remained an open problem for a long time).

1. Let $u \in C_c^1(\mathbb{R}^N)$.

   a. Assume here $N = 1$. Show that $\|u\|_\infty \leq \|u'\|_1$. (Recall that u' denotes the classical derivative of u).
   b. Show by induction on N that $\|u\|_{\frac{N}{N-1}} \leq \|\partial_1 u\|_1^{\frac{1}{N}} \ldots \|\partial_N u\|_1^{\frac{1}{N}}$.
   c. Show that $\|u\|_{\frac{N}{N-1}} \leq \| |\nabla u| \|_1$.
   d. Let $1 \leq p < N$. Show that there exists a $C_{N,p}$ depending only on N and p such that $\|u\|_{p^\star} \leq C_{N,p} \| |\nabla u| \|_p$, with $p^\star = (Np)/(N-p)$.

2. Let $1 \leq p < N$. Show that $\|u\|_{p^\star} \leq C_{N,p} \| |\nabla u| \|_p$, for all $u \in W^{1,p}(\mathbb{R}^N)$ ($C_{N,p}$ and $p^\star$ are given in the previous question). Deduce that the embedding from $W^{1,p}(\mathbb{R}^N)$ to $L^q(\mathbb{R}^N)$ is continuous for all $q \in [p, p^\star]$.

3. Let $p = N$. Show that the embedding from $W^{1,N}(\mathbb{R}^N)$ to $L^q(\mathbb{R}^N)$ is continuous for all $q \in [N, \infty[$. (For $N = 1$, the case $q = +\infty$ is allowed.)

4. We now assume that Ω is an open bounded subset with Lipschitz boundary. For $1 \leq p < N$, show that the embedding from $W^{1,p}(\Omega)$ to $L^q(\Omega)$ is continuous for all $q \in [p, p^\star]$ ($p^\star = (Np)/(N-p)$). Show that the embedding from $W^{1,N}(\Omega)$ to $L^q(\Omega)$ is continuous for all $q \in [N, \infty[$. (For $N = 1$, the case $q = +\infty$ is allowed.)

---

[21] Louis Nirenberg (1925–2020), Canadian mathematician, specialist in the analysis of partial differential equations.

Problem 1.20 (Kernel of the trace operator ($\star\star\star\star$)).   *Solution on page 58.* Let $\Omega = \mathbb{R}_+^N$, $N > 1$, $1 \leq p < +\infty$ and $\gamma : W^{1,p}(\Omega) \to L^p(\partial\Omega)$ be the trace operator defined in Theorem 1.30.

1. Show that $\mathrm{Ker}\gamma = W_0^{1,p}(\Omega)$.
2. Let $u \in W^{1,p}(\Omega) \cap C(\overline{\Omega})$. Show that $\gamma u = u$ a.e. (for the $N-1$-dimensional Lebesgue measure on $\partial\Omega$).

Problem 1.21 (Extension of a function belonging to H^2 ($\star\star$)).   *Solution on page 59.* Let $N \geq 1$, $\Omega = \mathbb{R}_+^N$ and $p \in [1, \infty[$.

1. Show that $C_c^\infty(\overline{\Omega})$ is dense in $W^{2,p}(\Omega)$. [Suggestion: follow the ideas of the proof of the density of $C_c^\infty(\overline{\Omega})$ in $W^{1,p}(\Omega)$, Theorem 1.25].
2. Show that there exists a continuous linear operator $P : W^{2,p}(\Omega) \to W^{2,p}(\mathbb{R}^N)$ such that $Pu = u$ a.e. in Ω, for all $u \in W^{2,p}(\Omega)$. [Suggestion: for $u \in C_c^\infty(\overline{\Omega})$, look for P of the form $Pu(x_1, y) = \alpha u(-x_1, y) + \beta u(-2x_1, y)$, for $x_1 \in \mathbb{R}_-$ and $y \in \mathbb{R}^{N-1}$, with α and β well chosen in $\mathbb{R}$.]
3. We now take $p = +\infty$. Is $C_c^\infty(\overline{\Omega})$ dense in $W^{2,\infty}(\Omega)$? Is there a continuous linear operator $P : W^{2,\infty}(\Omega) \to W^{2,\infty}(\mathbb{R}^N)$ such that $Pu = u$ a.e. in Ω, for all $u \in W^{2,\infty}(\Omega)$?

Problem 1.22 (Weak convergence and continuous operator ($\star$)).   *Solution on page 60.* Let E and F be two Banach spaces.

1. Let $T \in \mathcal{L}(E, F)$, that is, T is a continuous linear mapping from E to F. Let $(u_n)_{n\in\mathbb{N}}$ be a sequence of elements of E which converges weakly to u in E as $n \to +\infty$. Show that $T(u_n) \to T(u)$ weakly in F.
2. Assume that E is continuously embedded in F, that is, that $E \subset F$ and that the mapping $u \mapsto u$ is continuous from E to F. Let $(u_n)_{n\in\mathbb{N}}$ be a sequence of elements of E converging weakly to u in E as $n \to +\infty$. Show that $u_n \to u$ weakly in F.
3. Let Ω be an open bounded subset of $\mathbb{R}^N$, $N \geq 1$, and $(u_n)_{n\in\mathbb{N}}$ be a sequence of elements of $H_0^1(\Omega)$ converging weakly to u in $H_0^1(\Omega)$ as $n \to +\infty$. Let $i \in \{1, \ldots, N\}$. Show that $D_i u_n \to D_i u$ weakly in $L^2(\Omega)$ as $n \to +\infty$.

Problem 1.23 (Non-continuous function belonging to $H^1(\mathbb{R}^2) \cap L^\infty(\mathbb{R}^2)$($\star\star\star$)). *Solution on page 60.* In this problem, we construct a function $v \in H^1(\mathbb{R}^2) \cap L^\infty(\mathbb{R}^2)$ such that $v \notin C(\mathbb{R}^2, \mathbb{R})$ (that is, there does not exist $w \in C(\mathbb{R}^2, \mathbb{R})$ such that $v = w$ a.e.). For this purpose, we reuse the function from Problem 1.5.

Let $\gamma \in]0, \frac{1}{2}[$ and define u by

$$u(x) = \begin{cases} (-\ln(|x|))^\gamma & \text{if } 0 < |x| < 1, \\ 0 & \text{if } |x| \geq 1, \end{cases}$$

and, for example, $u(x) = 0$. For each $n \in \mathbb{N}^\star$, we define the function u_n by $u_n(x) = 1 - (1 - \bar{u}_n(x))^+\}$ with $\bar{u}_n(x) = (u(x) - n)^+$ (which can also be written $u_n(x) = u(x) - n$ if $n \leq u(x) \leq n + 1$, $u_n(x) = 0$ if $u(x) < n$ and $u_n(x) = 1$ if $u(x) < n + 1$).

1. Show that $u_n \in H^1(\mathbb{R}^2)$ for all $n \in \mathbb{N}^\star$ and that

$$\sum_{n=1}^{\infty} \| |\nabla u_n| \|_{L^2(\mathbb{R}^2)}^2 < +\infty.$$

[Use Problem 1.5 and the generalisation given in Remark 2.27 of Lemma 2.26.]

2. Let $n \in \mathbb{N}^\star$. Show that u_n takes its values between 0 and 1 and that the support of u_n is a ball centred at 0 and whose radius tends towards 0 as $n \to +\infty$.

For $n \in \mathbb{N}^\star$, set $x_n = (\frac{1}{n}, 0)^t \in \mathbb{R}^2$ and choose m_n such that the support of u_{m_n} is a ball centred at 0 with radius, denoted r_n, smaller than $(\frac{1}{2})(\frac{1}{n} - \frac{1}{n+1})$ and such that the sequence $(m_n)_{n \in \mathbb{N}^\star}$ is increasing. Then for $x \in \mathbb{R}^2$, let $v_n(x) = u_{m_n}(x - x_n)$.

3. Show that all the functions v_n have disjoint supports.
4. Show that the series $\sum_{n \in \mathbb{N}^\star} v_n$ is convergent in $H^1(\mathbb{R}^2)$. Let v be the sum of this series. Show that v belongs to $H^1(\mathbb{R}^2) \cap L^\infty(\mathbb{R}^2)$ but that $v \notin C(\mathbb{R}^2, \mathbb{R})$, that is, there does not exist $w \in C(\mathbb{R}^2, \mathbb{R})$ such that $v = w$ a.e.

Problem 1.24 (On the embedding from $W^{1,1}$ to $L^{1^\star}(\star\star)$). *Solution on page 62.* Let Ω be an open bounded connected set of $\mathbb{R}^N$ ($N \geq 1$) with a Lipschitz boundary and let ω be a Borel subset of Ω with positive Lebesgue measure, that is, $\lambda_N(\omega) > 0$, where λ_N denotes the Lebesgue measure on the Borel sets of $\mathbb{R}^N$. Define the set W_ω by

$$W_\omega = \{u \in W^{1,1}(\Omega) \text{ such that } u = 0 \text{ a.e. in } \omega\}.$$

The aim of this problem is to show, by two different methods, that there exists a $C \in \mathbb{R}_+$, depending only on Ω and ω, such that

$$\forall u \in W_\omega, \ \forall 1 \leq p \leq \frac{N}{N-1}, \ \|u\|_{L^p(\Omega)} \leq C \| |\nabla u| \|_{L^1(\Omega)}. \qquad (1.13)$$

I- First method (direct method)

1. We suppose that there exists a sequence $(u_n)_{n \in \mathbb{N}^\star}$ of elements of W_ω such that $\|u_n\|_{L^1(\Omega)} = 1$ for all $n \in \mathbb{N}^\star$ and

$$\|u_n\|_{L^1(\Omega)} > n \| |\nabla u_n| \|_{L^1(\Omega)} \text{ for all } n \in \mathbb{N}^\star.$$

Using a compactness theorem, show that we can assume, up to a subsequence, that $u_n \to u$ in $L^1(\Omega)$ as $n \to +\infty$. Then show that $u = 0$ a.e. and that $\|u\|_{L^1(\Omega)} = 1$ (which is impossible...).

2. Deduce from the previous question that there exists a C_1, depending only on Ω and ω, such that

$$\|u\|_{L^1(\Omega)} \leq C_1 \| |\nabla u| \|_{L^1(\Omega)} \text{ for all } u \in W_\omega.$$

3. We recall (Theorem 1.41) that there exists C_2, depending only on Ω, such that, by setting $1^\star = \frac{N}{N-1}$, we have $\|u\|_{L^{1^\star}(\Omega)} \le C_2 \|u\|_{W^{1,1}(\Omega)}$ for all $u \in W^{1,1}(\Omega)$. Using the result of the previous question, prove that there exists $C \in \mathbb{R}_+$, depending only on Ω and ω, satisfying (1.13).

II- Second method (using the mean value of u)

1. Let $H = \{u \in W^{1,1}(\Omega) \quad \int_\Omega u(x)\,\mathrm{d}x = 0\}$. Show that there exists a C_3 depending only on Ω such that $\|u\|_{L^1(\Omega)} \le C_3 \|\,|\nabla u|\,\|_{L^1(\Omega)}$ for all $u \in H$.

   Using the hint from question 3 of the first part, deduce that there exists a C_4 depending only on Ω such that

$$\|u - m\|_{L^{1^\star}(\Omega)} \le C_4 \|\,|\nabla u|\,\|_{L^1(\Omega)} \quad \text{for all } u \in W^{1,1}(\Omega), \tag{1.14}$$

   with $m\lambda_N(\Omega) = \int_\Omega u(x)\,\mathrm{d}x$. [Note that $u - m \in H$.]

2. Let $u \in W_\omega$ and m such that $m\lambda_N(\Omega) = \int_\Omega u(x)\,\mathrm{d}x$. Show that

$$|m| \le \frac{C_4}{\lambda_N(\omega)^{\frac{1}{1^\star}}} \|\,|\nabla u|\,\|_{L^1(\Omega)}$$

   and deduce that

$$\|u\|_{L^{1^\star}(\Omega)} \le C_4 \left(1 + \left(\frac{\lambda_N(\Omega)}{\lambda_N(\omega)}\right)^{\frac{1}{1^\star}}\right) \|\,|\nabla u|\,\|_{L^1(\Omega)}.$$

Problem 1.25 (Partition of unity ($\star\star$)). *Solution on page 64.* Let K be a compact subset of $\mathbb{R}^N$ ($N \ge 1$) and $\Omega_1, \ldots, \Omega_n$, $n \in \mathbb{N}^\star$, a finite family of open subsets such that $K \subset \cup_{i=1}^n \Omega_i$. We show here that there exist functions $\varphi_1, \ldots, \varphi_n$ such that

(p1) For all $i \in \{1, \ldots, n\}$, $\varphi_i \in C_c^\infty(\mathbb{R}^N, \mathbb{R}_+)$, $\bar{S}_i \subset \Omega_i$ with $S_i = \{x \in \mathbb{R}^N, \varphi_i(x) \ne 0\}$,

(p2) $\sum_{i=1}^n \varphi_i = 1$ on K.

For $\varepsilon > 0$ and $i \in \{1, \ldots, n\}$, set $\Omega_{i,\varepsilon} = \{x \in \Omega_i, d(x, \Omega_i^c) > \varepsilon\}$, where $d(x, \Omega_i^c) = \inf\{|x - y|, y \notin \Omega_i\}$.

1. Show that there exists an $\varepsilon > 0$ such that $K \subset \cup_{i=1}^n \Omega_{i,\varepsilon}$.

Let ε now be given by question 1.

2. Show that there exist n functions $f_1, \ldots, f_n$ such that, for all i, $f_i = 0$ on $\Omega_{i,\varepsilon}^c$ and $\sum_{i=1}^n f_i = 1$ on $\cup_{i=1}^n \Omega_{i,\varepsilon}$.
3. By a regularisation method, show the existence of n functions $\varphi_1, \ldots, \varphi_n$ satisfying (p1) and (p2).

Problem 1.26 (Transposed operator, continuity and compactness ($\star\star\star\star$)). *Solution on page 64.* Let E, F be two (real) Banach spaces and let $\mathcal{L}(E, F)$ be the set of continuous linear mappings from E to F. Let $T \in \mathcal{L}(E, F)$, and let T^t be the transpose of the operator T defined by (see Definition 2.13)

$$g \in F' \mapsto T^t g \in E' \quad \text{with} \quad \langle T^t g, u\rangle_{E',E} = \langle g, Tu\rangle_{F',F}.$$

1. Verify that $T^t g$ is indeed an element of E' for all $g \in F'$ and that $T^t \in \mathcal{L}(F', E')$.
2. Show that $\|T^t\|_{\mathcal{L}(F', E')} = \|T\|_{\mathcal{L}(E, F)}$.

We now assume that T is a compact operator, i.e. that from any bounded sequence in E we can extract a subsequence whose image under T converges in F. Let $B_E = \{u \in E, \|u\|_E \leq 1\}$ denote the unit ball.

3. Show that the set $T(B_E) = \{T(u), u \in B_E\}$ is precompact, i.e. that for all $\varepsilon > 0$, there exists an $I \subset B_E$ such that $\mathrm{card}(I) < +\infty$ and

$$T(B_E) \subset \cup_{u \in I} B_F(Tu, \varepsilon),$$

where $B_F(Tu, \varepsilon) = \{v \in F, \|v - Tu\|_F < \varepsilon\}$.

For $p \in \mathbb{N}^\star$, choose I_p according to question 3 with $\varepsilon = \frac{1}{p}$ and set $I = \cup_{p \in \mathbb{N}^\star} I_p$ (so that I is countable). Let $(g_n)_{n \in \mathbb{N}}$ be a bounded sequence in F'.

4. Show that there exists a subsequence of the sequence $(g_n)_{n \in \mathbb{N}}$ such that for this subsequence, still denoted $(g_n)_{n \in \mathbb{N}}$, the sequence $(\langle T^t g_n, u \rangle_{E', E})_{n \in \mathbb{N}}$ converges for all $u \in I$. [*Use the diagonal process, see for example [26, Step 2 of proposition 8.19].*]
   For the next two questions, consider this subsequence.
5. Show that the sequence $(\langle T^t g_n, u \rangle_{E', E})_{n \in \mathbb{N}}$ converges for all $u \in E$.
6. Show that there exists an $f \in E'$ such that $T^t g_n \to f$ in E'.
7. Deduce from the previous questions that T^t is a compact operator.
8. Prove the compactness of the bounded subsets of $L^q(\Omega)$ in $W^{-1,q}(\Omega)$ (result given in Theorem 1.38).

Problem 1.27 (Transpose of a continuous embedding $(\star)$). *Solution on page 66.* Let E and F be two Banach spaces such that E is continuously embedded in F, i.e. $E \subset F$ and the mapping $u \mapsto u$ is continuous from E to F. Show that F' is continuously embedded in E'.

1.9 Problem Set Solutions

Problem 1.1 (Example of derivative)

1. We take, for example, $i = 1$ (the other values of i are treated similarly). For all $\varphi \in \mathcal{D}(\mathbb{R}^N)$ we have

$$\int_{\mathbb{R}^N} u(x) \partial_1 \varphi(x) \, dx = \int_{]-1,1[^N} \partial_1 \varphi(x) \, dx = \int_{]-1,1[^{N-1}} \int_{-1}^{1} \partial_1 \varphi(x_1, y) \, dx_1 \, dy$$

and therefore

$$\int_{\mathbb{R}^N} u(x) \partial_1 \varphi(x) \, dx = \int_{]-1,1[^{N-1}} \varphi(1, y) \, dy - \int_{]-1,1[^{N-1}} \varphi(-1, y) \, dy.$$

This clearly shows that $\int_{\mathbb{R}^N} u(x)\partial_1\varphi(x)\,dx$ only depends on the values taken by φ on the boundary of Ω.

2. We reason by contradiction: we assume that $u \in W^{1,1}(\mathbb{R}^N)$. There then exists (in particular) a $g \in L^1(\mathbb{R}^N)$ such that

$$\int_{\mathbb{R}^N} u(x)\partial_1\varphi(x)\,dx = \int_{\mathbb{R}^N} g(x)\varphi(x)\,dx \text{ for all } \varphi \in \mathcal{D}(\mathbb{R}^N).$$

For $n \in \mathbb{N}^\star$, set $A_n =]1 - \frac{1}{n}, 1 + \frac{1}{n}[\times] - 1, 1[^{N-1}$.

Let $\varphi \in \mathcal{D}(\mathbb{R}^N)$ be a function such that $0 \le \varphi(x) \le 1$ for all $x \in \mathbb{R}^N$, $\varphi(x) = 0$ if $x \notin A_1$ and $\varphi(x) = 1$ if $x = (1, y)$ with $y \in] - \frac{1}{2}, \frac{1}{2}[^{N-1}$ (such a function φ exists). For $n \in \mathbb{N}^\star$, define φ_n by $\varphi_n(1 + x_1, y) = \varphi(1 + nx_1, y)$ for all $x_1 \in \mathbb{R}$ and $y \in \mathbb{R}^{N-1}$ (so that $\varphi_n = 0$ outside of A_n).

For $n \in \mathbb{N}^\star$, we indeed have $\varphi_n \in \mathcal{D}(\mathbb{R}^N)$ and the choice of φ_n gives

$$\int_{\mathbb{R}^N} u(x)\partial_1\varphi_n(x)\,dx = \int_{]-1,1[^{N-1}} \varphi_n(1, y)\,dy - \int_{]-1,1[^{N-1}} \varphi_n(-1, y)\,dy \ge 1$$

and

$$\left| \int_{\mathbb{R}^N} g(x)\varphi_n(x)\,dx \right| \le \int_{A_n} |g(x)|\,dx.$$

Hence $\int_{A_n} |g(x)|\,dx \ge 1$ for all $n \in \mathbb{N}^\star$, which is impossible because the (N-dimensional) Lebesgue measure of A_n tends towards 0 as $n \to +\infty$, and therefore $\int_{A_n} |g(x)|\,dx \to 0$ as $n \to +\infty$ (see for example [26, proposition 4.50]).

Problem 1.2 (A function with a null derivative is constant a.e.)

Let $\varphi_0 \in \mathcal{D}(]0, 1[)$ such that $\int_0^1 \varphi_0(x)\,dx = 1$, and for $\psi \in \mathcal{D}(]0, 1[)$, let φ be the function defined by

$$\varphi(x) = \int_0^x \psi(t)\,dt - \left(\int_0^1 \psi(t)\,dt \right) \int_0^x \varphi_0(t)\,dt \text{ for } x \in]0, 1[.$$

With this choice of φ one has $\varphi \in \mathcal{D}(]0, 1[)$ and therefore, since $Du = 0$,

$$0 = \langle Du, \varphi \rangle_{\mathcal{D}^\star, \mathcal{D}} = - \int_0^1 u(x)\varphi'(x)\,dx.$$

Since $\varphi' = \psi - (\int_0^1 \psi(t)\,dt)\varphi_0$, we therefore have

$$\int_0^1 u(x)\psi(x)\,dx - \left(\int_0^1 \psi(t)\,dt \right)\left(\int_0^1 u(x)\varphi_0(x)\,dx \right) = 0.$$

Set $a = \int_0^1 u(x)\varphi_0(x)\,\mathrm{d}x$, we thus have

$$\int_0^1 u(x)\psi(x)\,\mathrm{d}x = \int_0^1 a\psi(x)\,\mathrm{d}x \quad \text{for all } \psi \in \mathcal{D}(]0,1[).$$

Lemma 1.2 then gives $u = a$ a.e.

Another method consists in first considering the case $u \in L^1(]0,1[)$ and proceeding, for example, by density. The function u can be approximated by convolution with regularising kernels ρ_n supported in $]-\frac{1}{n},\frac{1}{n}[$. Extending u by 0 outside of $[0,1]$, let $u_n = u \star \rho_n$, so that $u_n' = u \star \rho_n'$. It may then be shown that $u_n'(x) = -\langle Du, \rho_n(x-\cdot)\rangle_{\mathcal{D}\star,\mathcal{D}}$ for all $x \in]\frac{1}{n}, 1-\frac{1}{n}[$, so that $u_n'(x) = 0$ for all $x \in]\frac{1}{n}, 1-\frac{1}{n}[$. The proof is then concluded thanks to the fact that $u_n \mathbb{1}_{]\frac{1}{n},1-\frac{1}{n}[}$ tends towards u in L^1 ($\mathbb{1}_A$ denotes the characteristic function of a set A, that is $\mathbb{1}_A(x) = 1$ if $x \in A$, 0 otherwise).

In the case $u \in L^1_{\mathrm{loc}}(]0,1[)$ we first consider the function $u_\varepsilon = u\mathbb{1}_{[\varepsilon,1-\varepsilon]}$ with $\varepsilon > 0$.

The interest of this second method is that it generalises to the multidimensional case (see Problem 1.4).

Problem 1.3 (Sobolev space in one dimension)

1. a. For $x \in [0,1]$, set $F(x) = \int_0^x Du(t)\,\mathrm{d}t$. Since $Du \in L^1(]0,1[)$, we have $F \in C([0,1],\mathbb{R})$. We can also show that F is differentiable a.e. and that $F' = Du$ a.e. but this is unnecessary here. We are more interested in the derivative by transposition of F, that is to say DF, and we show that $DF = Du$. Let $\varphi \in \mathcal{D}(]0,1[)$, we have

$$\langle DF, \varphi\rangle_{\mathcal{D}\star,\mathcal{D}} = -\int_0^1 F(x)\varphi'(x)\,\mathrm{d}x$$

$$= -\int_0^1 \left(\int_0^1 \mathbb{1}_{]0,x[}(t)Du(t)\,\mathrm{d}t\right)\varphi'(x)\,\mathrm{d}x.$$

   Noting that $\mathbb{1}_{]0,x[}(t) = \mathbb{1}_{]t,1[}(x)$ for all $t,x \in]0,1[$ and using Fubini's theorem, we therefore have

$$\langle DF, \varphi\rangle_{\mathcal{D}\star,\mathcal{D}} = -\int_0^1 \left(\int_0^1 \mathbb{1}_{]t,1[}(x)\varphi'(t)\,\mathrm{d}x\right)Du(t)\,\mathrm{d}t$$

$$= \int_0^1 \varphi(t)Du(t)\,\mathrm{d}t,$$

which proves that $DF = Du$. Hence $D(u - F) = 0$ and owing to the result of Problem 1.2 there exists a $C \in \mathbb{R}$ such that $u - F = C$ a.e., that is

$$u(x) = C + \int_0^x Du(t)\, dt \text{ for almost all } x \in\,]0, 1[.$$

b. We now choose for u (which is a class of functions) its continuous representative; we then have for all $x \in [0, 1]$

$$u(x) = u(0) + \int_0^x Du(t)\, dt.$$

We then also have for all $x, y \in [0, 1]$, $u(x) = u(y) + \int_y^x Du(t)\, dt$, and we get that

$$|u(x)| \leq |u(y)| + \int_0^1 |Du(t)|\, dt.$$

Integrating this inequality over $[0, 1]$ (with respect to y), we obtain for all $x \in [0, 1]$

$$|u(x)| \leq \|u\|_{L^1} + \|Du\|_{L^1} = \|u\|_{W^{1,1}}$$

and therefore, by taking the max over x and using Hölder's inequality,

$$\|u\|_{L^\infty} \leq \|u\|_{L^1} + \|Du\|_{L^1} \leq \|u\|_{L^p} + \|Du\|_{L^p} = \|u\|_{W^{1,p}}.$$

c. We still choose for u its continuous representative. Let $x, y \in [0, 1]$, $y > x$. We have

$$u(y) - u(x) = \int_x^y Du(t)\, dt$$

and therefore, using Hölder's inequality,

$$|u(y) - u(x)| \leq \left(\int_x^y |Du(t)|^p dt \right)^{\frac{1}{p}} |y - x|^{1 - \frac{1}{p}} \leq \|u\|_{W^{1,p}} |y - x|^{1 - \frac{1}{p}}.$$

2. It is clear that $u \in L^p(]0, 1[)$. To show that $u \in W^{1,p}(]0, 1[)$ it is therefore sufficient to show that $Du = w$, that is $\langle Du, \varphi \rangle_{\mathcal{D}^\star, \mathcal{D}} = \int_0^1 w(t)\varphi(t)\, dt$ for all $\varphi \in \mathcal{D}(]0, 1[)$. Let $\varphi \in \mathcal{D}(]0, 1[)$. We have

$$\langle Du, \varphi \rangle_{\mathcal{D}^\star, \mathcal{D}} = -\int_0^1 u(t)\varphi'(t)\, dt = -\int_0^1 \left(\int_0^t w(x)\, dx \right)\varphi'(t)\, dt$$

$$= -\int_0^1 \left(\int_0^1 \mathbb{1}_{]0,t[}(x) w(x)\, dx \right)\varphi'(t)\, dt.$$

Again owing to Fubini's theorem and since $\mathbb{1}_{]0,t[}(x) = \mathbb{1}_{]x,1[}(t)$ (for all $x, t \in\,]0, 1[$), we obtain

$$\langle Du, \varphi \rangle_{\mathcal{D}^\star, \mathcal{D}} = - \int_0^1 \left(\int_0^1 \mathbb{1}_{]x,1[}(t)\varphi'(t)\,\mathrm{d}t \right) w(x)\,\mathrm{d}x$$

$$= - \int_0^1 \left(\int_x^1 \varphi'(t)\,\mathrm{d}t \right) w(x)\,\mathrm{d}x = \int_0^1 \varphi(x)w(x)\,\mathrm{d}x,$$

which indeed gives $Du = w$.

Problem 1.4 (A function with a null gradient is a.e. constant)

1. We have $u_\varepsilon \in L^1(\mathbb{R}^N)$. For all $n \in \mathbb{N}^\star$, the function $u_{\varepsilon,n}$ is therefore well defined on all $\mathbb{R}^N$. The fact that $u_{\varepsilon,n}$ is of class C^∞ is classical and the derivatives of $u_{\varepsilon,n}$ are equal to the convolution of u_ε with the derivatives of ρ_n. We also note that $u_{\varepsilon,n}$ is a function with compact support because u_ε and ρ_n are functions with compact support.

Let B_r be the ball of centre 0 and radius r. We now show that for all i, the function $\partial_i u_{\varepsilon,n}$ is null on $B_{1-2\varepsilon}$ if $\frac{1}{n} < \varepsilon$.

Let $i \in \{1, \ldots, N\}$ and $x \in \mathbb{R}^N$. We have

$$\partial_i u_{\varepsilon,n}(x) = \left(u_\varepsilon \star \partial_i \rho_n \right)(x) = \int_{\mathbb{R}^N} u_\varepsilon(y)\partial_i \rho_n(x - y)\,\mathrm{d}y.$$

If $\frac{1}{n} < \varepsilon$ and $x \in B_{1-2\varepsilon}$, the function $\rho_n(x - \cdot)$ belongs to $\mathcal{D}(B)$ and is null outside of $B_{1-\varepsilon}$. Let τ be the function defined by $\tau(y) = \rho_n(x - y)$ (the variable x is here fixed), so that

$$\partial_i \rho_n(x - y) = -\partial_i \tau(y) \text{ for all } y \in \mathbb{R}^N,$$

(recall that ∂_i designates the derivative with respect to the i-th argument) and therefore

$$\partial_i u_{\varepsilon,n}(x) = \int_B u(y)\partial_i \rho_n(x - y)\,\mathrm{d}y = - \int_B u(y)\partial_i \tau(y)\,\mathrm{d}y$$

$$= \langle D_i u, \tau \rangle_{\mathcal{D}^\star(B), \mathcal{D}(B)} = \langle D_i u, \rho_n(x - \cdot) \rangle_{\mathcal{D}^\star(B), \mathcal{D}(B)} = 0.$$

We have thus shown that for $\frac{1}{n} < \varepsilon$, the function $\partial_i u_{\varepsilon,n}$ is, for all i, null on $B_{1-2\varepsilon}$. Hence the function $u_{\varepsilon,n}$ is constant on $B_{1-2\varepsilon}$. Indeed, it is enough to note that for all $x \in B_{1-2\varepsilon}$ we have

$$u_{\varepsilon,n}(x) - u_{\varepsilon,n}(0) = \int_0^1 \nabla u_{\varepsilon,n}(tx) \cdot x\,\mathrm{d}t = 0.$$

Since $u_\varepsilon \in L^1(\mathbb{R}^N)$, the sequence $(u_{\varepsilon,n})_{n\in\mathbb{N}}$ converges in $L^1(\mathbb{R}^N)$ towards u_ε. Considering the restrictions of these functions to the ball $B_{1-2\varepsilon}$ (on which $u_\varepsilon = u$), the sequence $(u_{\varepsilon,n})_{n\in\mathbb{N}}$ converges in $L^1(B_{1-2\varepsilon})$ towards u. Since $u_{\varepsilon,n}$ is a constant function on $B_{1-2\varepsilon}$ (for $\frac{1}{n} < \varepsilon$), its limit (in L^1) is therefore also a constant function. This shows that the function u is constant on $B_{1-2\varepsilon}$, that is, there exists an $a_\varepsilon \in \mathbb{R}$ such that $u = a_\varepsilon$ a.e. on $B_{1-2\varepsilon}$. Since $\varepsilon > 0$ is arbitrary, it follows that a_ε does not depend on ε and that u is constant on B.

2. Let $\varepsilon > 0$. The function $u_{\varepsilon,n}$ is of class C^∞. Therefore, for all $x, y \in \mathbb{R}^N$,

$$u_{\varepsilon,n}(y) - u_{\varepsilon,n}(x) = \int_0^1 \nabla u_{\varepsilon,n}(ty + (1-t)x) \cdot (y - x)\, dt. \tag{1.15}$$

In this formula $\nabla u_{\varepsilon,n}$ denotes the vector function defined by the classical derivatives of $u_{\varepsilon,n}$. For $z \in \mathbb{R}^N$ and $i \in \{1, \ldots, N\}$ we have

$$\partial_i u_{\varepsilon,n}(z) = \int_{\mathbb{R}^N} u_\varepsilon(\bar{z}) \partial_i \rho_n(z - \bar{z})\, d\bar{z}.$$

If $z \in B_{1-2\varepsilon}$ and $\frac{1}{n} < \varepsilon$, the function $\rho_n(z - \cdot)$ belongs to $\mathcal{D}(B)$ and is null outside of $B_{1-\varepsilon}$ (and on $B_{1-\varepsilon}$ we have $u_\varepsilon = u$). It follows that

$$\partial_i u_{\varepsilon,n}(z) = \langle D_i u, \rho_n(z - \cdot)\rangle_{\mathcal{D}^\star(B), \mathcal{D}(B)} = \int_B D_i u(\bar{z}) \rho_n(z - \bar{z})\, d\bar{z}.$$

Since $D_i u$ is uniformly continuous on $B_{1-\varepsilon}$, it follows from the previous formula that $\partial_i u_{\varepsilon,n}$ converges towards $D_i u$ uniformly on $B_{1-2\varepsilon}$. Therefore, for all $x, y \in B_{1-2\varepsilon}$

$$\lim_{n \to +\infty} \int_0^1 \nabla u_{\varepsilon,n}(ty + (1-t)x) \cdot (y - x)\, dt. = \int_0^1 Du(ty + (1-t)x) \cdot (y - x)\, dt.$$

The sequence $(u_{\varepsilon,n})_{n\in\mathbb{N}^\star}$ converges in $L^1(\mathbb{R}^N)$ towards u_ε. After possibly extracting a subsequence, we can therefore assume that this sequence converges a.e. towards u_ε and therefore a.e. towards u on the ball $B_{1-\varepsilon}$. Passing to the limit as $n \to +\infty$ in the equality (1.15), we obtain for almost all x, y in $B_{1-2\varepsilon}$

$$u(y) - u(x) = \int_0^1 Du(ty + (1-t)x) \cdot (y - x)\, dt. \tag{1.16}$$

Since $\varepsilon > 0$ is arbitrary, formula (1.16) is valid for almost all $x, y \in B$.

To conclude, fix a point $x \in B$ for which (1.16) is true for almost all $y \in B$ and set

$$v(y) = u(x) + \int_0^1 Du(ty + (1-t)x) \cdot (y - x)\, dt \text{ for all } y \in B.$$

The function v is of class C^1 (because Du is a continuous function and therefore v is differentiable over all B and $\nabla v = Du$). Since $u = v$ a.e., this concludes the question.

3. Let Ω the open subset replacing B. The previous reasoning shows that on any ball included in Ω, u is a.e. equal to a constant. For the equality to be true on the whole ball (and not only a.e.), it is enough to define v on Ω by

$$v(x) = \lim_{h \to 0,\, h>0} \frac{1}{\lambda_N(B_{x,h})} \int_{B(x,h)} u(y)\, dy \quad \text{for all } x \in \Omega,$$

where $B(x, h)$ denotes the ball of centre x and radius h and $\lambda_N(B(x, h))$ the N-dimensional Lebesgue measure of this ball. We then have $u = v$ a.e. (v is therefore a representative of the class u) and on any ball included in Ω, v is equal to a constant.

The function v is therefore locally constant. Hence v is constant on each connected component of Ω. Indeed, let $x \in \Omega$ and U be the connected component of Ω containing x. Let $a = v(x)$. We have just shown that the set $\{y \in U;\, v(y) = a\}$ is a non-empty open subset of U and the set $\{y \in U;\, v(y) \neq a\}$ is also an open subset of U disjoint from the previous one. Since U is connected, this latter set is therefore empty, which proves that $v = a$ over all U.

Problem 1.5 (An H^1 function is not necessarily continuous if $N > 1$)

The function u is of class C^∞ on $\bar{\Omega} \setminus \{0\}$ (noting that $|x| \leq \sqrt{2}/2 < 1$ for all $x \in \bar{\Omega}$). The classical derivatives of u are for $x = (x_1, x_2)^t \neq 0$

$$\partial_i u(x) = -\gamma(-\ln(|x|))^{\gamma-1} \frac{x_i}{|x|^2}, \quad i = 1, 2.$$

We note that $u \in L^2(\Omega)$ (and even $u \in L^p(\Omega)$ for all $1 \leq p < +\infty$). Since $\gamma < \frac{1}{2}$, we can also show that the classical derivatives of u belong to $L^2(\Omega)$. To do this, it is enough to notice that, for $a > 0$,

$$\int_0^a \frac{1}{r|\ln(r)|^{2(1-\gamma)}}\, dr < +\infty.$$

To show that $u \in H^1(\Omega)$, it is therefore enough to show that the derivatives by transposition of u are represented by the classical derivatives, i.e. that for all $\varphi \in \mathcal{D}(\Omega)$ and for $i = 1, 2$, we have

$$\int_\Omega u(x)\partial_i \varphi(x)\, dx = - \int_\Omega \partial_i u(x)\varphi(x)\, dx. \tag{1.17}$$

We now show (1.17) for $i = 1$ (the case $i = 2$ is treated similarly). Let $\varphi \in \mathcal{D}(\Omega)$ and $0 < \varepsilon < \frac{1}{2}$. Set $L_\varepsilon = [-\varepsilon, \varepsilon] \times [-\frac{1}{2}, \frac{1}{2}]$. Integrating by parts and noting that $u(\varepsilon, x_2) = u(-\varepsilon, x_2)$, we have

$$\int_{\Omega \setminus L_\varepsilon} u(x)\partial_1 \varphi(x)\, dx = - \int_{\Omega \setminus L_\varepsilon} \partial_1 u(x)\varphi(x)\, dx$$
$$- \int_{-\frac{1}{2}}^{\frac{1}{2}} u(\varepsilon, x_2)(\varphi(\varepsilon, x_2) - \varphi(-\varepsilon, x_2))\, dx_2. \tag{1.18}$$

By the dominated convergence theorem,

$$\lim_{\varepsilon \to 0} \int_{\Omega \setminus L_\varepsilon} u(x) \partial_1 \varphi(x) \, dx = \int_\Omega u(x) \partial_1 \varphi(x) \, dx \quad \text{and}$$

$$\lim_{\varepsilon \to 0} \int_{\Omega \setminus L_\varepsilon} \partial_1 u(x) \varphi(x) \, dx = \int_\Omega \partial_1 u(x) \varphi(x) \, dx.$$

It remains to show that the second term on the right-hand side of (1.18) tends to 0. Observing that the function φ is regular, there exists a $C \in \mathbb{R}_+$ depending only on φ such that

$$\left| \int_{-\frac{1}{2}}^{\frac{1}{2}} u(\varepsilon, x_2)(\varphi(\varepsilon, x_2) - \varphi(-\varepsilon, x_2)) \, dx_2 \right| \le |\ln(\varepsilon)|^\gamma C \varepsilon.$$

Therefore

$$\lim_{\varepsilon \to 0} \int_{-\frac{1}{2}}^{\frac{1}{2}} u(\varepsilon, x_2)(\varphi(\varepsilon, x_2) - \varphi(-\varepsilon, x_2)) \, dx_2 = 0,$$

which completes the proof of (1.17) for $i = 1$. Finally, we have thus shown that the derivatives by transposition of u are represented by the classical derivatives and that $u \in H^1(\Omega)$.

There is no continuous function in the class u. The quickest way to see this is to note that $u \notin L^\infty(\Omega)$ because for all C, $\lambda_2(\{u \ge C\}) > 0$.

Problem 1.6 (Laplacian of an element of $H_0^1(\Omega)$)

1. By definition of the derivative by transposition, we have

$$\langle \Delta u, \varphi \rangle_{\mathcal{D}^\star(\Omega), \mathcal{D}(\Omega)} = \sum_{i=1}^N \langle D_i D_i u, \varphi \rangle_{\mathcal{D}^\star(\Omega), \mathcal{D}(\Omega)} = \sum_{i=1}^N \int_\Omega u(x) \partial_i^2 \varphi(x) \, dx.$$

Since $u \in H_0^1(\Omega)$, the linear form (on $\mathcal{D}(\Omega)$) $D_i u$ is represented by an element of $L^2(\Omega)$, still denoted $D_i u$, and we have, for all $i \in \{1, \ldots, N\}$,

$$\int_\Omega u(x) \partial_i^2 \varphi(x) \, dx = -\langle D_i u, \partial_i \varphi \rangle_{\mathcal{D}^\star(\Omega), \mathcal{D}(\Omega)} = - \int_\Omega D_i u(x) \partial_i \varphi(x) \, dx.$$

Since ∇u is the element of $L^2(\Omega)^N$ whose components are $D_i u$, we obtain

$$\langle \Delta u, \varphi \rangle_{\mathcal{D}^\star(\Omega), \mathcal{D}(\Omega)} = - \int_\Omega \nabla u(x) \cdot \nabla \varphi(x) \, dx.$$

2. For all $\varphi \in \mathcal{D}(\Omega)$ we have, using the Cauchy[22]–Schwarz[23] inequality,

$$|\langle \Delta u, \varphi \rangle_{\mathcal{D}^\star(\Omega),\mathcal{D}(\Omega)}| \le \int_\Omega |\nabla u(x)||\nabla \varphi(x)|\,dx$$
$$\le \||\nabla u|\|_{L^2(\Omega)}\,\||\nabla \varphi|\|_{L^2(\Omega)} \le \|u\|_{H^1(\Omega)}\,\|\varphi\|_{H^1(\Omega)}\,.$$

This shows that the mapping $\varphi \mapsto \langle \Delta u, \varphi \rangle_{\mathcal{D}^\star(\Omega),\mathcal{D}(\Omega)}$ is a continuous linear mapping from $\mathcal{D}(\Omega)$, equipped with the norm $H^1(\Omega)$, into $\mathbb{R}$. Since $\mathcal{D}(\Omega)$ is dense in $H^1_0(\Omega)$ this mapping therefore extends, by density, in a unique way to a continuous linear mapping from $H^1_0(\Omega)$ to $\mathbb{R}$, that is, to an element of $H^{-1}(\Omega)$; this element of $H^{-1}(\Omega)$ is still denoted Δu and the extension by density gives, for all $v \in H^1_0(\Omega)$,

$$\langle \Delta u, v \rangle_{H^{-1}(\Omega),H^1_0(\Omega)} = -\int_\Omega \nabla u(x) \cdot \nabla v(x)\,dx.$$

Hence, for all $v \in H^1_0(\Omega)$,

$$|\langle \Delta u, v \rangle_{H^{-1}(\Omega),H^1_0(\Omega)}| \le \int_\Omega |\nabla u(x)||\nabla v(x)|\,dx \le \|u\|_{H^1(\Omega)}\,\|v\|_{H^1(\Omega)}\,. \tag{1.19}$$

Equipping $H^1_0(\Omega)$ with the norm $H^1(\Omega)$ yields

$$\|\Delta u\|_{H^{-1}(\Omega)} = \sup_{v \in H^1_0(\Omega),\,v \ne 0} \frac{|\langle \Delta u, v \rangle_{H^{-1}(\Omega),H^1_0(\Omega)}|}{\|v\|_{H^1(\Omega)}}, \tag{1.20}$$

so that owing to (1.19),

$$\|\Delta u\|_{H^{-1}(\Omega)} \le \|u\|_{H^1(\Omega)}\,.$$

N.B. In fact, it is shown in Chapter 2 that in $H^1_0(\Omega)$ the norm $H^1(\Omega)$ is equivalent to the norm denoted $\|\cdot\|_{H^1_0(\Omega)}$ defined by $\|u\|_{H^1_0(\Omega)} = \||\nabla u|\|_{L^2(\Omega)}$. The choice of the norm $\|\cdot\|_{H^1_0(\Omega)}$ in $H^1_0(\Omega)$ also modifies the norm in $H^{-1}(\Omega)$, then defined by (1.20) with $\|\cdot\|_{H^1_0(\Omega)}$ instead of $\|\cdot\|_{H^1(\Omega)}$, and we obtain $\|\Delta u\|_{H^{-1}(\Omega)} = \|u\|_{H^1_0(\Omega)}$.

---

[22] Augustin Louis, Baron Cauchy (1789–1857), French mathematician who is particularly known for numerous results in analysis.

[23] Hermann Schwarz (1843–1921), German mathematician, known for his work in complex analysis. (Not to be confused with Laurent Schwartz, French mathematician of the 20th century.)

Problem 1.7 (Point singularity)

1. Recall that, for $x = (x_1, x_2)^t$, $|x|$ is defined by $|x|^2 = x_1^2 + x_2^2$. The mapping $x \mapsto |x|$ is of class C^∞ on $\mathbb{R}^2 \setminus \{0\}$ and has positive values. By composition with the logarithm function, we obtain $G \in C^\infty(\mathbb{R}^2 \setminus \{0\})$. We now calculate $\Delta G(x)$ for $x \in \mathbb{R}^2 \setminus \{0\}$.

$$\partial_i G(x) = \frac{x_i}{|x|^2}, \quad \partial_i^2 G(x) = \frac{1}{|x|^2} - 2\frac{x_i^2}{|x|^4} \text{ for } i = 1, 2,$$

and therefore $\Delta G(x) = \partial_1^2 G(x) + \partial_2^2 G(x) = 0$ for all $x \in \mathbb{R}^2 \setminus \{0\}$.

Since $G \in C^\infty(\mathbb{R}^2 \setminus \{0\})$, the derivatives by transposition of G in $\mathbb{R}^2 \setminus \{0\}$ are given (at all orders) by the classical derivatives and therefore $\Delta G = 0$ in $\mathcal{D}^\star(\mathbb{R}^2 \setminus \{0\})$.

For $\varepsilon > 0$, let $C_\varepsilon = \{x \in \mathbb{R}^2, |x| = \varepsilon\}$ and $A_\varepsilon = \{x \in \mathbb{R}^2, |x| > \varepsilon\}$. Let $\varphi \in \mathcal{D}(\mathbb{R}^2)$; since $G \in L^p_{\text{loc}}(\mathbb{R}^2)$,

$$\langle \Delta G, \varphi \rangle_{\mathcal{D}^\star(\mathbb{R}^2), \mathcal{D}(\mathbb{R}^2)} = \int_{\mathbb{R}^2} G(x)\Delta\varphi(x)\, dx = \lim_{\varepsilon \to 0} \int_{A_\varepsilon} G(x)\Delta\varphi(x)\, dx.$$

For $\varepsilon > 0$, integrating by parts[24] gives

$$\int_{A_\varepsilon} G(x)\Delta\varphi(x)\, dx = \int_{A_\varepsilon} \Delta G(x)\varphi(x)\, dx$$

$$+ \int_{C_\varepsilon} (\nabla\varphi(x) \cdot n(x)G(x) - \nabla G(x) \cdot n(x)\varphi(x))\, d\gamma(x)$$

$$= \int_{C_\varepsilon} \nabla\varphi(x) \cdot n(x)G(x)\, d\gamma(x) - \int_{C_\varepsilon} \nabla G(x) \cdot n(x)\varphi(x))\, d\gamma(x).$$

where $n(x)$ is the normal vector to C_ε at point x, outward to A_ε. The measure γ on C_ε is of total mass $2\pi\varepsilon$ and therefore

$$\left| \int_{C_\varepsilon} \nabla\varphi(x) \cdot n(x)G(x)\, d\gamma(x) \right| \leq \sup_{x \in \mathbb{R}^2} |\nabla\varphi(x)| \ln\varepsilon\, 2\pi\varepsilon \to 0 \text{ when } \varepsilon \to 0.$$

For $x \in C_\varepsilon$, $\nabla G(x) \cdot n(x) = -\frac{1}{\varepsilon}$, because $\nabla G(x) = \frac{x}{\varepsilon^2}$ and $n(x) = -\frac{x}{\varepsilon}$ and therefore

$$\int_{C_\varepsilon} \nabla G(x) \cdot n(x)\varphi(x)\, d\gamma(x) = -\frac{1}{\varepsilon} \int_{C_\varepsilon} \varphi(x)\, d\gamma(x)$$

$$\to -2\pi\varphi(0) \text{ when } \varepsilon \to 0.$$

---

[24] See Theorem 1.33, but here in a simpler case because the functions G and φ are of class C^∞ on $\mathbb{R}^2 \setminus \{0\}$.

Hence $\langle \Delta G, \varphi \rangle_{\mathcal{D}^\star(\mathbb{R}^2), \mathcal{D}(\mathbb{R}^2)} = 2\pi\varphi(0)$ and therefore $\Delta G = 2\pi\delta_0$.

2. Let $1 \le p < +\infty$. To show that $G \in L^p_{\mathrm{loc}}(\mathbb{R}^2)$, it is enough to notice that the mapping $x \mapsto |x|(-\ln(|x|)^p$ is bounded on $\{x \in \mathbb{R}^2, |x| < 1\}$.

Let $1 \le p < 2$. For $x \in \mathbb{R}^2$, $|\nabla G(x)|^2 = \frac{1}{|x|^2}$. This gives $|\nabla G(x)|^p = \frac{1}{|x|^p}$. Hence $\nabla G \in L^p_{\mathrm{loc}}(\mathbb{R}^2)$. However, $\nabla G \notin L^2_{\mathrm{loc}}(\mathbb{R}^2)$. Indeed, noting that the Jacobian of the change to polar coordinates is $|x|$, see for example [26, section 7.6], we therefore have

$$\int_{|x|<1} |\nabla G(x)|^2 \, dx = \int_0^{2\pi} \int_0^1 r\frac{1}{r^2} \, dr = +\infty.$$

3. It is enough to take $u = G$. Indeed, we have $u \in L^2(\Omega)$, $\Delta u = 0$ and $u \notin H^1(\Omega)$ (because $\int_\Omega |\nabla u(x)|^2 \, dx = \int_0^{\pi/2} \int_0^1 \frac{1}{r} \, dr = +\infty$).

We now take $v = \nabla G$. Question 2 gives $v \in (L^p(\Omega))^2$ for all $p < 2$, $i = 1$, 2. However, the theorem on Sobolev embeddings (Theorem 1.41) states that for $1 \le q < +\infty$, $H^1(\Omega) = W^{1,2}(\Omega)$ is continuously embedded in $L^q(\Omega)$; by duality (see Remark 1.42), for $r > 1$, $L^r(\Omega)$ is continuously embedded in $H^1(\Omega)'$. Hence $v \in (H^1(\Omega)')^2$. Let us then note that $\operatorname{div} v = \Delta G = 0$ in $\mathcal{D}^\star(\Omega)$ (and even in the classical sense) and $\operatorname{curl} v = 0$ in Ω in the sense of classical derivatives (because v is the gradient of a class C^2 function) and therefore $\operatorname{curl} v = 0$ in $\mathcal{D}^\star(\Omega)$. Finally, we showed in question 2 that $v \notin (L^2(\Omega))^2$.

4. We choose a function $\psi_0 \in \mathcal{D}(\mathbb{R}^2)$ such that $\psi_0(x) = 1$ if $|x| \le 1$ and $\psi_0(x) = 0$ if $|x| \ge 2$. For $n \in \mathbb{N}^\star$ and $x \in \mathbb{R}^2$, set $\psi_n(x) = \psi_0(nx)$. Since $0 \in \Omega$ and Ω is open, there exists an n_0 such that $\{x \in \mathbb{R}^2, |x| < 2/n_0\} \subset \Omega\}$.

Let $\varphi \in \mathcal{D}(\Omega)$. The objective is to show that $\langle \Delta u, \varphi \rangle_{\mathcal{D}^\star(\Omega), \mathcal{D}(\Omega)} = 0$. We use for this ψ_n with $n \ge n_0$ and the fact that $(1 - \psi_n)\varphi \in \mathcal{D}(\Omega \setminus \{0\})$,

$$\begin{aligned}
\langle \Delta u, \varphi \rangle_{\mathcal{D}^\star(\Omega), \mathcal{D}(\Omega)} &= \langle \Delta u, \varphi\psi_n \rangle_{\mathcal{D}^\star(\Omega), \mathcal{D}(\Omega)} + \langle \Delta u, \varphi(1 - \psi_n) \rangle_{\mathcal{D}^\star(\Omega), \mathcal{D}(\Omega)} \\
&= \langle \Delta u, \varphi\psi_n \rangle_{\mathcal{D}^\star(\Omega), \mathcal{D}(\Omega)} + \langle \Delta u, \varphi(1 - \psi_n) \rangle_{\mathcal{D}^\star(\Omega\setminus\{0\}), \mathcal{D}(\Omega\setminus\{0\})} \\
&\hspace{4cm} = \langle \Delta u, \varphi\psi_n \rangle_{\mathcal{D}^\star(\Omega), \mathcal{D}(\Omega)}.
\end{aligned}$$

We now use the fact that $u \in H^1(\Omega)$,

$$\begin{aligned}
\langle \Delta u, \varphi\psi_n \rangle_{\mathcal{D}^\star(\Omega), \mathcal{D}(\Omega)} &= \int_\Omega \nabla u(x) \cdot \nabla(\varphi\psi_n)(x) \, dx \\
&= \int_\Omega (\nabla u(x) \cdot \nabla\varphi(x))\psi_n(x) \, dx \\
&\quad + \int_\Omega (\nabla u(x) \cdot \nabla\psi_n(x))\varphi(x) \, dx.
\end{aligned}$$

The fact that $\lim_{n\to+\infty} \int_\Omega (\nabla u(x) \cdot \nabla\varphi(x))\psi_n(x) \, dx = 0$ is given by the dominated convergence theorem because $|\nabla u| \in L^1(\{x, |x| < 2/n_0\})$. For the last term, we note that $\nabla\psi_n(x) = n\nabla\psi_0(nx)$ and therefore by the Cauchy–Schwarz inequality, there exists a $C \in \mathbb{R}_+$ depending only on ψ_0 and φ such that

$$\left| \int_\Omega \nabla u(x) \cdot \nabla \varphi_n(x)\varphi(x)\,\mathrm{d}x \right| \le Cn \int_{\{x,\,|x|<2/n\}} |\nabla u(x)|\,\mathrm{d}x$$

$$\le Cn\,(\lambda_2(\{x,\,|x| < 2/n\}))^{\frac{1}{2}} \left(\int_{\{x,\,|x|<2/n\}} |\nabla u(x)|^2\,\mathrm{d}x \right)^{\frac{1}{2}},$$

where λ_2 denotes the Lebesgue measure on $\mathbb{R}^2$. Since $n(\lambda_2(\{x,\,|x| < 2/n\}))^{\frac{1}{2}}$ is independent of n and $|\nabla u| \in L^2(\Omega)$, it follows that

$$\lim_{n \to +\infty} \int_\Omega (\nabla u(x) \cdot \nabla \varphi_n(x))\varphi(x)\,\mathrm{d}x = 0$$

and therefore finally that $\langle \Delta u, \varphi \rangle_{\mathcal{D}^\star(\Omega), \mathcal{D}(\Omega)} = 0$. We have indeed shown that $\Delta u = 0$ in $\mathcal{D}^\star(\Omega)$.

Problem 1.8 (Some consequences of the Hahn–Banach theorem)

1. Set $F = \mathrm{Vect}\{x\}$ and define the linear continuous mapping T_1 from F to $\mathbb{R}$ by

$$T_1(\alpha x) = \alpha\,\|x\|_E \quad \text{for all } \alpha \in \mathbb{R}.$$

The mapping T_1 belongs to F' (F being equipped with the norm on E) and $\|T_1\|_{F'} = 1$. By the Corollary 1.15, T_1 extends to $T \in E'$ with $\|T\|_{E'} = \|T_1\|_{F'}$. The mapping T satisfies the requested conditions.
Then, since $|S(x)| \le \|x\|_E$ for all $S \in E'$ such that $\|S\|_{E'} = 1$, we get that

$$\|x\|_E = \max\{S(x),\ S \in E' \quad \|S\|_{E'} = 1\}.$$

2. If $x \in \bar{F}$, $T(y) = 0$ for all $y \in F$ implies $T(x) = 0$ (it is enough to consider a sequence $(x_n)_{n \in \mathbb{N}}$ of elements of T such that $\lim_{n \to +\infty} x_n = x$).
If $x \notin \bar{F}$, there exists an $\varepsilon > 0$ such that $\|x - y\|_E \ge \varepsilon$ for all $y \in F$. Let us then consider $G = F \oplus \mathrm{Vect}\{x\}$ and define the linear mapping S from G to $\mathbb{R}$ by

$$S(y + \alpha x) = \alpha \quad \text{for all } \alpha \in \mathbb{R} \text{ and for all } y \in F.$$

The mapping S belongs to G' because for all $\alpha \in \mathbb{R}^\star$ and for all $y \in F$,

$$\|y + \alpha x\|_E = |\alpha|\,\left\| x - \frac{y}{-\alpha} \right\|_E \ge |\alpha|\varepsilon \quad \text{because } \frac{y}{-\alpha} \in F,$$

and therefore

$$|S(y + \alpha x)| = |\alpha| \le \frac{1}{\varepsilon}\,\|y + \alpha x\|_E .$$

This gives $\|S\|_{G'} \le \frac{1}{\varepsilon}$. By Corollary 1.15, S extends to $T \in E'$ and we have indeed $T(y) = 0$ for all $y \in F$ and $T(x) \ne 0$.

3. Let $x \in E$; the mapping $J(x)$ is indeed linear from E' to $\mathbb{R}$. It is also continuous because $|J(x)(T)| \le \|T\|_{E'} \|x\|_E$ and therefore $J(x) \in E''$ and $\|J(x)\|_{E''} \le \|x\|_E$. On the other hand, question 1 shows that there is a linear form $T \in E'$ such that $T(x) = \|x\|_E$ and $\|T\|_{E'} = 1$. Hence $\|J(x)\|_{E''} = \|x\|_E$ and therefore the mapping J (which is, of course, linear) is an isometry of E onto its image.

4. a. Let $x \in E$; for all $n \in \mathbb{N}^\star$, $T_n(x) \le \|T_n\|_{E'} \|x\|_E$. Hence

$$T(x) = \lim_{n \to +\infty} T_n(x) = \liminf_{n \to +\infty} T_n(x) \le \liminf_{n \to +\infty} \|T_n\|_{E'} \|x\|_E$$
$$= (\liminf_{n \to +\infty} \|T_n\|_{E'}) \|x\|_E$$

and therefore $\|T\|_{E'} \le \liminf_{n \to +\infty} \|T_n\|_{E'}$.

b. Using the notation from question 3, $x_n \to x$ weakly in E means $J(x_n) \to J(x)$ $\star$-weakly in E''. Question 4a then gives

$$\|J(x)\|_{E''} \le \liminf_{n \to +\infty} \|J(x_n)\|_{E''}.$$

Since J is an isometry from E onto its image, this gives

$$\|x\|_E \le \liminf_{n \to +\infty} \|x_n\|_E.$$

Problem 1.9 (Comparison $\ell^p(\mathbb{N})$-$\ell^q(\mathbb{N})$)

1. We first consider the case $q < +\infty$.

   Let $x = (x_n)_{n \in \mathbb{N}} \in \ell^p$. Assuming $\|x\|_p = 1$, we then have, for all $n \in \mathbb{N}$, $|x_n| \le 1$ and therefore $|x_n|^q \le |x_n|^p$. Summing over n yields

   $$\sum_{n \in \mathbb{N}} |x_n|^q \le \sum_{n \in \mathbb{N}} |x_n|^p = 1,$$

   and therefore $x \in \ell^q$ and $\|x\|_q \le 1$.

   Now let $x \in \ell^p$, $x \ne 0$. Since $\left\|\frac{x}{\|x\|_p}\right\|_p = 1$, we get from the previous result that $\frac{x}{\|x\|_p} \in \ell^q$ and $\left\|\frac{x}{\|x\|_p}\right\|_q \le 1$. This gives $x \in \ell^q$ and $\|x\|_q \le \|x\|_p$.

   We now consider the case $q = +\infty$.

   Let $x = (x_n)_{n \in \mathbb{N}} \in \ell^p$. For all $m \in \mathbb{N}$, $|x_m| \le (\sum_{n \in \mathbb{N}} |x_n|^p)^{\frac{1}{p}} = \|x\|_p$ and therefore $x \in \ell^\infty$ and $\|x\|_\infty \le \|x\|_p$.

2. Let B be the set of real sequences having only a finite number of non-zero terms. It is then sufficient to note that $B \subset \ell^p$ and that B is dense in ℓ^q (because if $x = (x_n)_{n \in \mathbb{N}} \in \ell^q$, $\lim_{n \to +\infty} \sum_{i > n} |x_i|^q = 0$).

3. The set A is a closed subset of ℓ^∞ and it is the closure in ℓ^∞ of the set B introduced in question 2. Since $B \subset \ell^p \subset A$, the set A is also the closure of ℓ^p in ℓ^∞.

   Since $A \ne \ell^\infty$, ℓ^p is not dense in ℓ^∞.

Problem 1.10 (Characterisation of the density of a v.s.s. of a Banach space)

1. The condition (1.5) can be written as "$G \subset \mathrm{Ker}(f) \Rightarrow \mathrm{Ker}(f) = E$".

   Assume $\bar{G} = E$. Let $f \in E'$ such that $G \subset \mathrm{Ker}(f)$. Since $\mathrm{Ker}(f)$ is closed, we indeed have $\mathrm{Ker}(f) = E$.

   Conversely, we assume that, for all $f \in E'$, "$G \subset \mathrm{Ker}(f) \Rightarrow \mathrm{Ker}(f) = E$". We reason by contradiction. If $\bar{G} \neq E$, there exists a $u \notin \bar{G}$. Problem 1.8 (second point) then shows that there exists an $f \in E'$ such that $\langle f, u \rangle_{E',E} \neq 0$ and $G \subset \mathrm{Ker}(f)$, in contradiction with the hypothesis (because $u \notin \mathrm{Ker}(f)$).

2. a. The previous result gives that $\bar{G} = F'$ if and only if

$$\left(f \in F'', \langle f, u \rangle_{F'',F'} = 0 \text{ for all } u \in G \right) \Rightarrow f = 0. \tag{1.21}$$

   Let $J : F \to F''$ be the mapping defined in Definition 1.12.
   For all $x \in F$ and all $u \in F'$, $\langle J(x), u \rangle_{F'',F'} = \langle u, x \rangle_{F',F}$. Since F is assumed to be reflexive, J is an isometry between F and F''. The condition (1.21) is therefore equivalent to

$$\left(x \in F, \langle u, x \rangle_{F',F} = 0 \text{ for all } u \in G \right) \Rightarrow x = 0,$$

   which gives the desired result.

   b. Let $F = \ell^1(\mathbb{N})$ (see Problem 1.9 for its definition) and identify F' with $\ell^\infty(\mathbb{N})$ (see for instance [26, Exercise 6.61]). Let $G = \ell^1(\mathbb{N})$ (recall that $\ell^1(\mathbb{N}) \subset \ell^\infty(\mathbb{N})$).
   Let $v \in \ell^1(\mathbb{N})$ such that $\langle g, v \rangle_{\ell^\infty(\mathbb{N}),\ell^1(\mathbb{N})} = 0$ for all $g \in \ell^1(\mathbb{N})$. Taking $g = v$ yields $v = 0$ so that condition (1.6) is satisfied. However $\bar{G} \neq \ell^\infty(\mathbb{N})$; indeed, the sequence $y = (y_n)_{n \in \mathbb{N}}$, with $y_n = 1$ for all n, belongs to $\ell^\infty(\mathbb{N})$ but not to $\bar{G}$ since for all $x \in G$, $\|x - y\|_{\ell^\infty(\mathbb{N})} \geq 1$.

Problem 1.11 (Separability of L^p)

1. We break the proof down into two steps.

   Step 1. Density of $C_c(\mathbb{R}, \mathbb{R})$ in L^p.
   Let $A \in \mathcal{B}(\mathbb{R})$ and $\lambda(A) < +\infty$; the key point of this step is to approximate as closely as we want (in the L^p norm) the function $f = \mathbb{1}_A$ by a continuous function with compact support. Let us first observe that it is sufficient to consider the case where A is bounded. For a bounded set A, the regularity of the Lebesgue measure gives for all $\varepsilon > 0$ the existence of an open set O and a compact set K such that $K \subset A \subset O$ and $\lambda(O \setminus K) \leq \varepsilon$. To approximate the characteristic function of A, denoted $\mathbb{1}_A$, we then construct an element $\varphi \in C_c(\mathbb{R}, [0, 1])$ such that $\varphi = 1$ on K and $\varphi = 0$ on $O^c = \mathbb{R} \setminus O$.
   Consider the case where f is a non-negative step function and in L^p: it is then a finite linear combination of functions of the type $\mathbb{1}_A$; then consider the case where f is a non-negative measurable function and in L^p: it is a non-decreasing

sequence of non-negative step functions. Consider finally the case $f \in L^p$ by decomposing $f = f^+ - f^-$.

Step 2. For $N \in \mathbb{N}$ and $n \in \mathbb{N}$, denote by $A_{N,n}$ the set of functions f that only take rational values, that are null outside of $] - N, N[$ and constant on each interval $]i/n, (i + 1)n[$, $i \in \mathbb{Z}$, and set $B_N = \cup_{n \in \mathbb{N}} A_{N,n}$ and $A = \cup_{N \in \mathbb{N}} B_N$. For all $N \in \mathbb{N}$ and $n \in \mathbb{N}$, the set $A_{N,n}$ is countable (because it is in bijection with $\mathbb{Q}^{2nN}$). The set A is therefore also countable. The fact that we can approximate $f \in C_c(\mathbb{R}, \mathbb{R})$ in the L^p norm as close as we want by an element of A then follows from the uniform continuity of f and the density of $\mathbb{Q}$ in $\mathbb{R}$.

2. Let B be the set of functions that take (a.e.) only the values 1 or 0 and which are (a.e.) constant on each interval $]n, n + 1[$, $n \in \mathbb{N}$. The set B is a subset of L^∞. The proof that $L^\infty(\mathbb{R})$ is not separable is performed in two steps.

Step 1 We show in this step that B is uncountable.

Let $f \in B$; we define $\psi(f) = \{n \in \mathbb{N}; f = 1 \text{ a.e. in }]n, n + 1[\}$. The mapping ψ is a bijection between B and $\mathcal{P}(\mathbb{N})$. Since $\mathrm{card}(\mathcal{P}(\mathbb{N})) > \mathrm{card}(\mathbb{N})$, the set B is uncountable.

Step 2 We now show that L^∞ is not separable.

Let $A \subset L^\infty$, A dense in L^∞. To show that A is uncountable, which implies that L^∞ is not separable, we construct an injection ψ from B to A: this gives $\mathrm{card}(A) \geq \mathrm{card}(B)$ and therefore A is uncountable.

Since A is dense in L^∞, we can choose, for every $f \in B$, an element φ_f of A such that $\left\| f - \varphi_f \right\|_{L^\infty} < \frac{1}{2}$. Let us then define $\phi : B \to A$ by setting $\phi(f) = \varphi_f$ and show that ϕ is injective.

Let $f, g \in B$ such that $\varphi_f = \phi(f) = \phi(g) = \varphi_g$. Then

$$\| f - g \|_{L^\infty} \leq \left\| f - \varphi_f \right\|_{L^\infty} + \left\| \varphi_f - \varphi_g \right\|_{L^\infty} + \left\| \varphi_g - g \right\|_{L^\infty} < \frac{1}{2} + \frac{1}{2} = 1. \quad (1.22)$$

Now, for all $f, g \in B$, $f \neq g$ (in L^∞) implies $\| f - g \|_{L^\infty} = 1$ (because there exists an $n \in \mathbb{N}$ such that $|f - g| = 1$ a.e. in $]n, n + 1[$). The inequality (1.22) therefore gives $f = g$. The mapping ϕ is injective.

Finally $\mathrm{card}(A) \geq \mathrm{card}(B) > \mathrm{card}(N)$ and therefore any set A dense in L^∞ is uncountable. The space L^∞ is non-separable.

Problem 1.12 (Separability of a subset of a separable space)

1. For all $y \in B_n$, choose a point $x_y \in A$ such that $y = a_{x_y,n}$; note that such a point exists but is not necessarily unique. The mapping $y \mapsto x_y$ from B_n to A is injective, because if $x_y = x_z$, with $y, z \in B_n$, one has $y = a_{x_y,n} = a_{x_z,n} = z$.
2. Consider $B = \cup_{n \in \mathbb{N}^\star} B_n$. Then $B \subset F$ and B is countable since it is a countable union of countable sets. Let us show that B is dense in F.

Let $y \in F$ and $n \in \mathbf{N}^{\star}$; since A is dense in E, there exists an $x \in A$ such that $\|x - y\|_E \leq \frac{1}{n}$. Observe then that

$$\left\|y - a_{x,n}\right\|_E \leq \|y - x\|_E + \left\|x - a_{x,n}\right\|_E \leq \frac{2}{n}.$$

Since $a_{x,n} \in B$, this proves the density of B in F.

Problem 1.13 (Closed vector subspace of a reflexive Banach space)

1. The mapping v is indeed linear. It is continuous because, for all $f \in E'$, $\left\|f_{|F}\right\|_{F'} \leq \|f\|_{E'}$ and therefore

$$\left|\langle u, f_{|F}\rangle_{F'',F'}\right| \leq \|u\|_{F''} \left\|f_{|F}\right\|_{F'} \leq \|u\|_{F''} \|f\|_{E'} .$$

   This shows that $v \in E''$ and $\|v\|_{E''} \leq \|u\|_{F''}$.
2. The answer is "yes". Indeed, let $\varepsilon > 0$. There exists a $g \in F'$ such that $\|g\|_{F'} = 1$ and $\langle u, g\rangle_{F'',F'} \geq \|u\|_{F''} - \varepsilon$. The Hahn–Banach theorem gives the existence of $f \in E'$ such that $f_{|F} = g$ and $\|f\|_{E'} = \|g\|_{F'}$. Hence

$$\|v\|_{E''} \geq \langle v, f\rangle_{E'',E'} = \langle u, g\rangle_{F'',F'} \geq \|u\|_{F''} - \varepsilon.$$

   Since ε is arbitrary, this gives $\|v\|_{E''} \geq \|u\|_{F''}$ and therefore finally $\|v\|_{E''} = \|u\|_{F''}$.
3. Since E is reflexive, $\mathrm{Im}(J_E) = E''$ and therefore there exists an $x \in E$ such that $v = J_E(x)$ and therefore

$$\langle v, f\rangle_{E'',E'} = \langle J_E(x), f\rangle_{E'',E'} = \langle f, x\rangle_{E',E} \text{ for all } f \in E'.$$

4. Consider the element x of E found in question 3.

   a. If $x \notin F$, the Hahn–Banach theorem gives that there exists an $f \in E'$ such that $f_{|F} = 0$ and $\langle f, x\rangle_{E',E} \neq 0$, which is impossible because

   $$0 \neq \langle f, x\rangle_{E',E} = \langle v, f\rangle_{E'',E'} = \langle u, f_{|F}\rangle_{F'',F'} = 0.$$

   b. Let $g \in F'$. By the Hahn–Banach theorem, there exists an $f \in E'$ such that $f_{|F} = g$, and since $x \in F$,

   $$\langle J_F(x), g\rangle_{F'',F'} = \langle g, x\rangle_{F',F} = \langle f, x\rangle_{E',E} = \langle J_E(x), f\rangle_{E'',E'}$$
   $$= \langle v, f\rangle_{E'',E'} = \langle u, f_{|F}\rangle_{F'',F'} = \langle u, g\rangle_{F'',F'}.$$

   Therefore $J_F(x) = u$.

5. The previous questions show that $\mathrm{Im}(J_F) = F''$, so that F is reflexive.

Problem 1.14 (Continuity of a mapping from L^p to L^q)

1. The function u is measurable from E equipped with the σ-algebra T to $\mathbb{R}$ equipped with the σ-algebra $\mathcal{B}(\mathbb{R})$ and g is a Borel-measurable function, that is to say measurable from $\mathbb{R}$ to $\mathbb{R}$, equipped with the σ-algebra $\mathcal{B}(\mathbb{R})$. By composition, $g \circ u$ is measurable from E to $\mathbb{R}$.

   For $s \in [-1, 1]$, we have $|g(s)| \leq 2C$ and therefore $|g(s)|^q \leq 2^q C^q$. For $s \in \mathbb{R} \setminus [-1, 1]$, we have $|g(s)| \leq 2C|s|^{\frac{p}{q}}$ and therefore $|g(s)|^q \leq 2^q C^q |s|^p$. So we have, for all $s \in \mathbb{R}$, $|g(s)|^q \leq 2^q C^q + 2^q C^q |s|^p$. It follows that, for all $x \in E$, $|g \circ u(x)|^q = |g(u(x))|^q \leq 2^q C^q + 2^q C^q |u(x)|^p$, and therefore:

$$\int |g \circ u|^q \, dm \leq 2^q C^q \, \|u\|_p^p + 2^q C^q m(E),$$

   which gives $g \circ u \in \mathcal{L}_{\mathbb{R}}^q (E, T, m)$.
2. Let $v, w \in u$. There exists an $A \in \mathcal{T}$ such that $m(A) = 0$ and $v = w$ on A^c. We also have $g \circ v = g \circ w$ on A^c and therefore $g \circ v = g \circ w$ a.e. It follows that

$$\{h \in \mathcal{L}_{\mathbb{R}}^q (E, T, m); h = g \circ v \ a.e.\} = \{h \in \mathcal{L}_{\mathbb{R}}^q (E, T, m); h = g \circ w \ a.e.\}.$$

   Therefore $G(u)$ does not depend on the choice of v in u.
3. For all $n \in \mathbb{N}$, choose a representative of u_n, still denoted u_n, and choose representatives of u and F, still denoted u and F. Since $u_n \to u$ a.e. as $n \to +\infty$ and since g is continuous, $g \circ u_n \to g \circ u$ a.e., so that $G(u_n) \to G(u)$ a.e.

   Observe then that

$$|g \circ u_n| \leq C|u_n|^{\frac{p}{q}} + C \leq C|F|^{\frac{p}{q}} + C \ a.e.,$$

   and therefore $|G(u_n)| \leq C|F|^{\frac{p}{q}} + C$ a.e., for all $n \in \mathbb{N}$.

   Since $F \in L^p$, $|F|^{\frac{p}{q}} \in L^q$, and since $m(E) < +\infty$, the constant functions also belong to L^q. Hence $C|F|^{\frac{p}{q}} + C \in L^q$. The dominated convergence theorem in L^q gives that $G(u_n) \to G(u)$ in L^q as $n \to +\infty$.
4. By contradiction, suppose that G is not continuous from L^p to L^q; then there exists a $u \in L^p$ and $(u_n)_{n \in \mathbb{N}} \subset L^p$ such that $u_n \to u$ in L^p and $G(u_n) \not\to G(u)$ in L^q as $n \to +\infty$.

   Since $G(u_n) \not\to G(u)$, there exists an $\varepsilon > 0$ and $\varphi : \mathbb{N} \to \mathbb{N}$ such that $\varphi(n) \to \infty$ when $n \to \infty$ and:

$$\left\| G(u_{\varphi(n)}) - G(u) \right\|_q \geq \varepsilon \text{ for all } n \in \mathbb{N}. \tag{1.23}$$

   (The sequence $(G(u_{\varphi(n)}))_{n \in \mathbb{N}}$ is a subsequence of the sequence $(G(u_n))_{n \in \mathbb{N}}$.)

   Since $u_{\varphi(n)} \to u$ in L^p, there exists a function $\psi : \mathbb{N} \to \mathbb{N}$ and $F \in L^p$ such that $\psi(n) \to \infty$ when $n \to \infty$, $u_{\varphi \circ \psi(n)} \to u$ a.e. and $|u_{\varphi \circ \psi(n)}| \leq F$ a.e., for all $n \in \mathbb{N}$ (see [26, Theorem 6.11]). The sequence $(u_{\varphi \circ \psi(n)})_{n \in \mathbb{N}}$ is a subsequence of the sequence $(u_{\varphi(n)})_{n \in \mathbb{N}}$.

Apply now the result of question 2 to the sequence $(u_{\varphi\circ\psi(n)})_{n\in\mathbb{N}}$. It gives that $G(u_{\varphi\circ\psi(n)}) \to G(u)$ in L^q as $n \to +\infty$, which contradicts (1.23).

5. a. By contradiction, assume that $|g(s)| < n|s|$ for all s such that $|s| \geq n$ and set $M = \max\{|g(s)|, s \in [-n, n]\}$. Since g is continuous on the compact set $[-n, n]$ (note that n is fixed), $M < +\infty$. Setting $C = \max\{n, M\}$ yields:

$$|g(s)| \leq C|s| + C, \text{ for all } s \in \mathbb{R},$$

which contradicts the assumption that g does not satisfy (1.7).

Therefore, there exists an s such that $|s| \geq n$ and $|g(s)| \geq n|s|$, which proves the existence of α_n.

b. Since $\alpha_n \geq n$, we have $\frac{1}{|\alpha_n|n^2} \leq \frac{1}{n^3}$ and therefore:

$$0 < \beta = \sum_{n\in\mathbb{N}^\star} \frac{1}{|\alpha_n|n^2} < +\infty.$$

Let us then choose $\alpha = \frac{1}{\beta}$.

c. For $n \geq 2$, we have $a_n = 1 - \sum_{p=1}^{n-1} \frac{\alpha}{|\alpha_p|p^2}$. Thanks to the choice of α, we therefore have $a_n > 0$ for all $n \in \mathbb{N}^\star$, and $a_n \downarrow 0$ as $n \to +\infty$.

The function u is indeed measurable and, by the monotone convergence theorem, we obtain:

$$\int |u|\,\mathrm{d}\lambda = \sum_{n\in\mathbb{N}^\star} |\alpha_n|(a_n - a_{n+1}) = \sum_{n\in\mathbb{N}^\star} \frac{\alpha}{n^2} < +\infty.$$

So, $u \in \mathcal{L}^1$ and also $u \in L^1$ by conflating, as usual, u with its class.

Let us then note that $g \circ u = \sum_{n=1}^{+\infty} g(\alpha_n) \mathbb{1}_{[a_{n+1}, a_n[}$. Hence:

$$\int |g \circ u|\,\mathrm{d}\lambda = \sum_{n\in\mathbb{N}^\star} |g(\alpha_n)|(a_n - a_{n+1}) \geq \sum_{n\in\mathbb{N}^\star} \frac{\alpha}{n} = +\infty.$$

This shows that $g \circ u \notin \mathcal{L}^1$ and therefore $G(u) \notin L^1$.

Problem 1.15 (Lipschitz-continuous functions)

Let Ω be an open subset of $\mathbb{R}^N$.

1. The function u is bounded and therefore $u \in L^\infty(\Omega)$. Let L be a Lipschitz constant of u, that is, such that $|u(x) - u(y)| \leq L|x - y|$ for all $x, y \in \Omega$. We show that $D_1 u \in L^\infty(\Omega)$ and $\|D_1 u\|_{L^\infty(\Omega)} \leq L$. A similar reasoning gives $D_i u \in L^\infty(\Omega)$ and $\|D_i u\|_{L^\infty(\Omega)} \leq L$ for $i > 1$.

Let $\varphi \in \mathcal{D}(\Omega)$ and extend it by 0 outside of Ω. For $n \in \mathbb{N}^{\star}$, we define the function φ_n by $\varphi_n(x) = \varphi(x_1 + \frac{1}{n}, y)$, where $x = (x_1, y)^t$, $x_1 \in \mathbb{R}$, $y \in \mathbb{R}^{n-1}$. Since φ has compact support in Ω, there exists an n_0 such that $\varphi_n \in \mathcal{D}(\Omega)$ if $n \geq n_0$. For $n \geq n_0$, a change of variables in $\int_{\Omega} u(x)\varphi_n(x)\,dx$ gives

$$
\left| \int_{\Omega} u(x)(\varphi_n(x) - \varphi(x))\,dx \right| = \left| \int_{\Omega} u(x)\varphi_n(x)\,dx - \int_{\Omega} u(x)\varphi(x)\,dx \right|
$$

$$
= \left| \int_{\Omega} (u(x_1 - \frac{1}{n}, y) - u(x_1, y))\varphi(x)\,dx \right|
$$

$$
\leq \int_{\Omega} \left| u(x_1 - \frac{1}{n}, y) - u(x_1, y) \right| |\varphi(x)|\,dx
$$

$$
\leq \frac{L}{n} \|\varphi\|_{L^1(\Omega)} .
$$

Since $n(\varphi_n - \varphi) \to \partial_1 \varphi$ uniformly as $n \to +\infty$ and $u \in L^1_{\mathrm{loc}}(\Omega)$,

$$
\left| \langle D_1 u, \varphi \rangle_{\mathcal{D}^{\star}(\Omega), \mathcal{D}(\Omega)} \right| = \left| \int_{\Omega} u(x)\partial_1\varphi(x)\,dx \right| \leq L \|\varphi\|_{L^1(\Omega)} . \tag{1.24}
$$

By density of $\mathcal{D}(\Omega)$ in $L^1(\Omega)$, the mapping $u \mapsto \langle D_i u, \varphi \rangle_{\mathcal{D}^{\star}(\Omega), \mathcal{D}(\Omega)}$ therefore extends to an element of $L^1(\Omega)'$. Hence there exists a $v \in L^{\infty}(\Omega)$ such that

$$
\langle D_1 u, \varphi \rangle_{\mathcal{D}^{\star}(\Omega), \mathcal{D}(\Omega)} = - \int u(x)v(x)\,dx,
$$

that is to say, $D_1 u$ is represented by the function v, and we write $D_1 u = v$. This shows that $u \in W^{1,\infty}(\Omega)$. The inequality (1.24) also gives $\|v\|_{L^{\infty}} \leq L$.

2. For any bounded open subset Ω and $1 \leq p < +\infty$, the function u, or rather its restriction to Ω, belongs to $W^{1,p}(\Omega)$. Taking $p > N$, the Sobolev embedding theorem (Theorem 1.41) then gives $u \in C(\bar{\Omega}, \mathbb{R})$. Since Ω is arbitrary, we get that $u \in C(\mathbb{R}^N, \mathbb{R})$.

Let us show that u is Lipschitz-continuous with Lipschitz constant $L = \sum_{i=1}^{N} \|D_i u\|_{L^{\infty}(\mathbb{R}^N)}$.

Choose a function $\varphi_0 \in \mathcal{D}(\mathbb{R}^N)$ such that $\varphi_0(x) \geq 0$ for all x, $\varphi_0(x) = 0$ if $|x| \geq 1$ and $\int_{\mathbb{R}^N} \varphi_0(x)\,dx = 1$.

For $n \in \mathbb{N}^{\star}$ and $x \in \mathbb{R}^N$, set $\varphi_n(x) = n^N \varphi_0(nx)$ so that $\varphi_n(x) = 0$ if $|x| \geq \frac{1}{n}$ and $\int_{\mathbb{R}^N} \varphi_n(x)\,dx = 1$. For $n \in \mathbb{N}^{\star}$ and $x \in \mathbb{R}^N$, set $u_n(x) = \int_{\mathbb{R}^N} u(x + z)\varphi_n(z)\,dz$. Since u is continuous $\lim_{n \to +\infty} u_n(x) = u(x)$ (for all $x \in \mathbb{R}^N$). Let $x, y \in \mathbb{R}^N$. A simple change of variables gives

$$
u_n(x) - u_n(y) = \int_{\mathbb{R}^N} u(x + z)\varphi_n(z)\,dz - \int_{\mathbb{R}^N} u(y + z)\varphi_n(z)\,dz
$$

$$
= \int_{\mathbb{R}^N} u(z)\varphi_n(z - x)\,dz - \int_{\mathbb{R}^N} u(z)\varphi_n(z - y)\,dz.
$$

Since $\varphi_n(z-x) - \varphi_n(z-y) = \int_0^1 \nabla\varphi_n(z-y+t(y-x)) \cdot (y-x)\,\mathrm{d}t$, we have, thanks to Fubini's theorem,

$$u_n(x) - u_n(y) = \int_{\mathbb{R}^N} u(z)\left(\int_0^1 \nabla\varphi_n(z-y+t(y-x)) \cdot (y-x)\,\mathrm{d}t\right)\mathrm{d}z$$

$$= \int_0^1 \left(\int_{\mathbb{R}^N} u(z)\nabla\varphi_n(z-y+t(y-x)) \cdot (y-x)\,\mathrm{d}z\right)\mathrm{d}t. \quad (1.25)$$

Denoting by $(y-x)_i$ the components of $y-x$, since the function $z \mapsto \varphi_n(z-y+t(y-x))$ belongs to $\mathcal{D}(\mathbb{R}^N)$,

$$\int_{\mathbb{R}^N} u(z)\nabla\varphi_n(z-y+t(y-x)) \cdot (y-x)\,\mathrm{d}z$$

$$= \sum_{i=1}^{N}\left(\int_{\mathbb{R}^N} u(z)\partial_i\varphi_n(z-y+t(y-x))\,\mathrm{d}z\right)(y-x)_i$$

$$= -\sum_{i=1}^{N}\left(\int_{\mathbb{R}^N} D_i u(z)\varphi_n(z-y+t(y-x))\right)(y-x)_i.$$

Since $\|\varphi_n\|_{L^1(\mathbb{R}^N)} = 1$, we get

$$\left|\int_{\mathbb{R}^N} u(z)\nabla\varphi_n(z-y+t(y-x)) \cdot (y-x)\,\mathrm{d}z\right| \leq \sum_{i=1}^{N} \|D_i u\|_{L^\infty(\mathbb{R}^N)}\,|y-x|.$$

This gives, returning to (1.25), $|u_n(x)-u_n(y)| \leq \sum_{i=1}^{N} \|D_i u\|_{L^\infty(\mathbb{R}^N)}\,|y-x|$. When $n \to +\infty$, we obtain $|u(x) - u(y)| \leq L|y-x|$, with $L = \sum_{i=1}^{N} \|D_i u\|_{L^\infty(\mathbb{R}^N)}$.

Problem 1.16 (A Lipschitz open subset that is not strongly Lipschitz)

1. Let $d = (d_1, d_2)^t \in \mathbb{R}^2$, $d \neq 0$, $t \in \mathbb{R}_+^\star$ and $S = \{sd, s \in]0, t]\}$. Let us show that $S \cap \Omega^c \neq \emptyset$. Since $\Omega \subset \mathbb{R}_+^\star \times \mathbb{R}$, we have, of course, $S \subset \Omega^c$ if $d_1 \leq 0$. We therefore assume $d_1 > 0$ and (since t is arbitrary) it is sufficient to consider the case $d_1 = 1$.

   If $d_2 \geq 0$, we have $a_n d \in S$ for n such that $a_n \leq t$ and $a_n d \notin \Omega$ because $a_n d_2 \geq \varphi(a_n) = 0$. Therefore, $S \cap \Omega^c \neq \emptyset$.

   If $d_2 < 0$, we have $(a_n + \frac{a_n}{2})d \in S$ for n such that $a_n + \frac{a_n}{2} \leq t$. But using (1.8), we have

$$\varphi\left(a_n + \frac{a_n}{2}\right) = a_{n-1}\bar{\varphi}\left(\frac{1}{4}\right) \geq a_{n-1} \geq a_n + \frac{a_n}{2}.$$

Hence $\varphi(a_n + \frac{a_n}{2}) - (a_n + \frac{a_n}{2}) \geq 0$ and, since $(a_n + \frac{a_n}{2})d_2 < 0$, this shows that $(a_n + \frac{a_n}{2})d \notin \Omega$. We have thus shown that $S \cap \Omega^c \neq \emptyset$.

Finally, we have indeed shown that Ω is not strongly Lipschitz.

2. The function ψ is Lipschitz-continuous (because so is φ) and its inverse is also Lipschitz-continuous because its inverse is the mapping $\bar{\psi}$ defined by

$$\bar{\psi}(x, y) = (x, \varphi(x) + \frac{1}{2}(y - \varphi(x))).$$

Since the triangle T is strongly Lipschitz, we get (this point is left to the reader) that Ω is Lipschitz.

Problem 1.17 (Extension of a continuous function)

The function g is well defined because if $x \notin \bar{\Omega}$, since Ω is open, B_x contains an open subset included in Ω and therefore $\int_{\Omega \cap B_x} dz > 0$. On the other hand, f is integrable over $\Omega \cap B_x$ and therefore $\int_{\Omega \cap B_x} f(z)\, dz \in \mathbb{R}$. This indeed gives $g(x) \in \mathbb{R}$.

Of course, g is continuous at every point of Ω because Ω is open and $f = g$ in Ω. The aim is to show that g is continuous if $x \notin \Omega$. We distinguish the cases $x \in \bar{\Omega} \setminus \Omega$ and $x \notin \bar{\Omega}$.

First case Assume that $x \in \bar{\Omega} \setminus \Omega$ and let $(x_n)_{n \in \mathbb{N}}$ be a sequence in $\mathbb{R}$ such that $x_n \to x$ as $n \to +\infty$. Let $n \in \mathbb{N}$, if $x_n \notin \bar{\Omega}$, since $d_{x_n} \leq |x - x_n|$ (because $x \in \bar{\Omega}$), $z \in B_{x_n}$ implies $|z - x| < 2d_{x_n} + |x - x_n| \leq 3|x - x_n|$ and therefore

$$|f(x) - g(x_n)| \leq \max_{z \in C_n} |f(x) - f(z)| \text{ with } C_n = \Omega \cap \{z;\ |z - x| \leq 3|x - x_n|\}.$$

This inequality is also true if $x_n \in \bar{\Omega}$ (because then $g(x_n) = f(x_n)$). Since f is continuous at point x and $g(x) = f(x)$ (because $x \in \bar{\Omega}$), we get that $\lim_{n \to +\infty} g(x_n) = f(x) = g(x)$. Hence g is continuous at point x.

Second case Assume that $x \notin \bar{\Omega}$. Let $(x_n)_{n \in \mathbb{N}}$ be a sequence in $\mathbb{R}$ such that $x_n \to x$ as $n \to +\infty$; since $x \notin \bar{\Omega} \setminus \Omega$, we can assume (after removing some initial terms) that $x_n \notin \bar{\Omega}$ for all $n \in \mathbb{N}$. It is now enough to notice that $\mathbb{1}_{B_{x_n}} \to \mathbb{1}_{B_x}$ a.e. as $n \to +\infty$ (indeed $\mathbb{1}_{B_{x_n}}(z) \to \mathbb{1}_{B_x}(z)$ when $z \in B_x$ and $z \notin \bar{B}_x$ and the Lebesgue N-dimensional measure of $\bar{B}_x \setminus B_x$ is null). Since f is integrable on every compact subset of $\bar{\Omega}$, the dominated convergence theorem yields that

$$\int_\Omega f(z)\mathbb{1}_{B_{x_n}}(z)\, dz \to \int_\Omega f(z)\mathbb{1}_{B_x}(z)\, dz \text{ and}$$

$$\int_\Omega \mathbb{1}_{B_{x_n}}(z)\, dz \to \int_\Omega \mathbb{1}_{B_x}(z)\, dz \text{ as } n \to +\infty.$$

Hence $\lim_{n \to +\infty} g(x_n) = g(x)$ and therefore g is continuous at point x.

Problem 1.18 (Sobolev inequalities for $p > N$)

1. Let us consider here the case $N \geq 2$ (the case $N = 1$ is simpler, see Problem 1.3).
   Let $a \in B_{N-1}$ and let $\varphi \in C^1(\mathbb{R}, \mathbb{R})$ be defined by $\varphi(t) = u(t, (1 - t)a)$ so that

$$u(1,0) - u(0,a) = \varphi(1) - \varphi(0) = \int_0^1 \varphi'(t)\, dt = \int_0^1 \nabla u(t, (1 - t)a) \cdot (1, -a)^t\, dt.$$

Integrating this equality over the ball B_{N-1} and denoting by m_N the $(N - 1)$-dimensional measure of B_{N-1} and m the mean value of u on B_{N-1} yields

$$m_N(u(1,0) - m) = \int_{B_{N-1}} \left(\int_0^1 \nabla u(t, (1 - t)a) \cdot (1, -a)^t\, dt \right) da.$$

A similar calculation with $\varphi(t) = u(t, (1 + t)a)$ gives

$$m_N(u(-1,0) - m) = \int_{B_{N-1}} \left(\int_{-1}^0 (\nabla u(t, (1 + t)a) \cdot (1, a)^t\, dt \right) da.$$

Taking the absolute value of the difference of these two equations yields

$$m_N|u(1,0) - u(-1,0)| \leq 2 \int_{B_{N-1}} \left(\int_{-1}^1 |\nabla u(t, (1 - t)a)|\, dt \right) da$$

$$= 2 \int_{-1}^1 \left(\int_{B_{N-1}} |\nabla u(t, (1 - |t|)a)|\, da \right) dt.$$

thanks to the Fubini–Tonelli theorem for the last equality. In the integral with respect to a (t is fixed), the change of variables $(1 - |t|)a = \bar{x}$ gives, with $A_t = \{(1 - |t|)a, a \in B_{N-1}\}$,

$$m_N|u(1,0) - u(-1,0)| \leq 2 \int_{-1}^1 \left(\int_{A_t} |\nabla u(t, \bar{x})|\, d\bar{x} \right) \frac{1}{(1 - |t|)^{N-1}}\, dt$$

$$= 2 \int_H |\nabla u(t, \bar{x})| \frac{1}{(1 - |t|)^{N-1}}\, d(t, \bar{x}).$$

Owing to Hölder's inequality with p and $\frac{p}{p-1}$,

$$m_N|u(1,0) - u(-1,0)| \leq$$

$$\left(\int_H |\nabla u(t, \bar{x})|^p\, d(t, \bar{x}) \right)^{\frac{1}{p}} \left(\int_H (1 - |t|)^{\frac{p(1-N)}{p-1}}\, d(t, \bar{x}) \right)^{1 - \frac{1}{p}}.$$

Finally, noting that the $(N-1)$-dimensional measure of A_t is $(1-t)^{N-1}m_N$,

$$\int_H (1-|t|)^{\frac{p(1-N)}{p-1}}\, d(t,\bar{x}) = 2\int_0^1 (1-t)^{\frac{p(1-N)}{p-1}}(1-t)^{N-1}m_N\, dt$$

$$= 2\int_0^1 (1-t)^{\frac{1-N}{p-1}}m_N\, dt.$$

This last integral is finite if and only if $\frac{N-1}{p-1} < 1$, that is, if $N < p$. For $p > N$, the requested inequality is obtained with C_1 depending only on N and p through m_N and $\int_0^1 (1-t)^{\frac{1-N}{p-1}}\, dt$.

2. Since both the Lebesgue measure and the Euclidean norm of the gradient of a function are invariant under translation and rotation, it is sufficient to consider the case $x = (b,0)$ and $y = (-b,0)$ with $b > 0$.

Let v be the function defined by $v(x) = u(bx)$. The inequality (1.9) gives

$$|u(b,0) - u(-b,0)| \le C_1\, \||\nabla v|\|_{L^p(\mathbb{R}^N)}\, .$$

Since $\nabla v(x) = b\nabla u(bx)$,

$$\int_{\mathbb{R}^N} |\nabla v(x)|^p\, dx = b^p\int_{\mathbb{R}^N} |\nabla u(bx)|^p\, dx = b^p\int_{\mathbb{R}^N} |\nabla u(x)|^p b^{-N}\, dx,$$

which gives

$$|u(b,0) - u(-b,0)| \le C_1 b^{1-\frac{N}{p}}\, \||\nabla u|\|_{L^p(\mathbb{R}^N)}\, .$$

We conclude by noting that $|(b,0)^t - (-b,0)^t| = 2b$. We can therefore choose $C_2 = C_1$.

3. Let $u \in W^{1,p}(\mathbb{R}^N)$ such that $\|u\|_{W^{1,p}}(\mathbb{R}^N) = 1$. Given $x \in \mathbb{R}^N$, let us show that $|u(x)| \le C_3 = NC_2 + \frac{1}{M_N}$, where M_N^p is the measure of the unit ball of $\mathbb{R}^N$. Indeed, suppose that $|u(x)| > C_3$. Then, thanks to (1.10) and as

$$\|(|\nabla u|)\|_{L^p(\mathbb{R}^N)} \le N\|u\|_{W^{1,p}}(\mathbb{R}^N) = N$$

with the $W^{1,p}$ norm of Proposition 1.9, it follows that $|u(y)| \ge C_3 - NC_2$ for all $y \in B(x,1)$, where $B(x,1)$ denotes the ball of centre x and radius 1. Hence

$$\|u\|_{L^p(\mathbb{R}^N)}^p = \int_{\mathbb{R}^N} |u(y)|^p\, dy \ge \int_{B(x,1)} |u(y)|^p\, dy \ge (C_3 - NC_2)^p M_N^p,$$

and therefore $\|u\|_{L^p(\mathbb{R}^N)} \ge (C_3 - NC_2)M_N$. However,

$$\|u\|_{L^p(\mathbb{R}^N)} \le \|u\|_{W^{1,p}}(\mathbb{R}^N) = 1.$$

Therefore $(C_3 - NC_2)M_N \leq 1$, that is to say $C_3 < NC_2 + \frac{1}{M_N}$, in contradiction with the choice of C_3. Hence (1.11) is satisfied if $\|u\|_{W^{1,p}(\mathbb{R}^N)} = 1$.

If now $u \in W^{1,p}(\mathbb{R}^N)$, u not null, we obtain (1.11) by reducing to the previous case with $v = u/\|u\|_{W^{1,p}(\mathbb{R}^N)}$.

4. (Sobolev embedding into $\mathbb{R}^N$.)

   Thanks to the two previous questions, we obtain (1.12) with $C_4 = C_3 + NC_2$.

5. (Sobolev embedding into Ω.)

   It is sufficient to apply the extension theorem (Theorem 1.25). Let C_P be the norm of the continuous linear operator (from $W^{1,p}(\Omega)$ to $W^{1,p}(\mathbb{R}^N)$) P given in the extension theorem (Theorem 1.25); then

   $$\|u\|_{C^{0,\alpha}(\bar{\Omega})} \leq \|Pu\|_{C^{0,\alpha}(\mathbb{R}^N)} \leq C_4 \|Pu\|_{W^{1,p}(\mathbb{R}^N)} \leq C_4 C_P \|u\|_{W^{1,p}(\Omega)} .$$

   It is therefore sufficient to take $C_5 = C_4 C_P$.

Problem 1.19 (Sobolev embeddings for $p \leq N$)

1. a. Since $u \in C_c^1(\mathbb{R})$ we have, for $x \in \mathbb{R}$, $u(x) = \int_{-\infty}^x u'(t)\, dt$ and therefore

   $$|u(x)| \leq \int_{-\infty}^x |u'(t)|\, dt \leq \|u'\|_1 .$$

   Hence $\|u\|_\infty \leq \|u'\|_1$.

   b. The previous question allows us to initialize a proof by induction: we have for all functions u belonging to $C_c^1(\mathbb{R})$, $\|u\|_\infty \leq \|u'\|_1$.

   Now let $N \geq 1$ and assume that any function $u \in C_c^1(\mathbb{R}^N)$ satisfies

   $$\|u\|_{L^{\frac{N}{N-1}}(\mathbb{R}^N)} \leq \|\partial_1 u\|_{L^1(\mathbb{R}^N)}^{\frac{1}{N}} \cdots \|\partial_N u\|_{L^1(\mathbb{R}^N)}^{\frac{1}{N}} ,$$

   where ∂_i denotes the derivative with respect to the i-th variable.

   Let $u \in C_c^1(\mathbb{R}^{N+1})$, and for $x \in \mathbb{R}^{N+1}$ let $x = (x_1, y)^t$ with $x_1 \in \mathbb{R}$ and $y \in \mathbb{R}^N$. For $x_1 \in \mathbb{R}$, the Hölder inequality gives

   $$\begin{aligned}
   \int_{\mathbb{R}^N} |u(x_1, y)|^{\frac{N+1}{N}}\, dy &= \int_{\mathbb{R}^N} |u(x_1, y)| |u(x_1, y)|^{\frac{1}{N}}\, dy \\
   &\leq \left(\int_{\mathbb{R}^N} |u(x_1, y)|^{\frac{N}{N-1}}\, dy \right)^{\frac{N-1}{N}} \left(\int_{\mathbb{R}^N} |u(x_1, y)|\, dy \right)^{\frac{1}{N}} .
   \end{aligned} \tag{1.26}$$

   Applying the induction hypothesis to the function $y \mapsto u(x_1, y)$ (which indeed belongs to $C_c^1(\mathbb{R}^N)$) yields

   $$\|u(x_1, \cdot)\|_{L^{\frac{N}{N-1}}(\mathbb{R}^N)} \leq \|\partial_2 u(x_1, \cdot)\|_{L^1(\mathbb{R}^N)}^{\frac{1}{N}} \cdots \|\partial_{N+1} u(x_1, \cdot)\|_{L^1(\mathbb{R}^N)}^{\frac{1}{N}} .$$

Next, applying the case $N = 1$ (proved in question (a)) to the function $z \mapsto u(z, y)$ (which belongs to $C_c^1(\mathbb{R})$) gives that for all $y \in \mathbb{R}^N$ and all $x_1 \in \mathbb{R}$

$$|u(x_1, y)| \le \|\partial_1 u(\cdot, y)\|_{L^1(\mathbb{R})} \, .$$

Therefore, integrating with respect to y,

$$\int_{\mathbb{R}^N} |u(x_1, y)| \, dy \le \|\partial_1 u\|_{L^1(\mathbb{R}^{N+1})} \, .$$

Injecting these upper bounds into (1.26) we obtain that for all $x_1 \in \mathbb{R}$

$$\int_{\mathbb{R}^N} |u(x_1, y)|^{\frac{N+1}{N}} \, dy$$

$$\le \|\partial_2 u(x_1, \cdot)\|_{L^1(\mathbb{R}^N)}^{\frac{1}{N}} \cdots \|\partial_{N+1} u(x_1, \cdot)\|_{L^1(\mathbb{R}^N)}^{\frac{1}{N}} \, \|\partial_1 u\|_{L^1(\mathbb{R}^{N+1})}^{\frac{1}{N}} \, .$$

Integrating this inequality with respect to x_1 and using Hölder's inequality once again (with the product of N functions in L^N) yields the desired inequality, that is

$$\|u\|_{L^{\frac{N+1}{N}}(\mathbb{R}^{N+1})}^{\frac{N}{N+1}} \le \|\partial_1 u\|_{L^1(\mathbb{R}^{N+1})}^{\frac{1}{N}} \cdots \|\partial_{N+1} u\|_{L^1(\mathbb{R}^{N+1})}^{\frac{1}{N}} \, ,$$

or equivalently (the norms below being in $\mathbb{R}^{N+1}$)

$$\|u\|_{\frac{N+1}{N}} \le \|\partial_1 u\|_1^{\frac{1}{N+1}} \cdots \|\partial u_{N+1}\|_1^{\frac{1}{N+1}} \, ,$$

which concludes the proof by induction.

c. The geometric mean of N non-negative numbers is smaller than the arithmetic mean of these same numbers.[25] Hence

$$\|u\|_{\frac{N}{N-1}} \le \|\partial_1 u\|_1^{\frac{1}{N}} \cdots \|\partial_N u\|_1^{\frac{1}{N}} \le \frac{1}{N} \sum_{i=1}^{N} \|\partial_i u\|_1 \, .$$

Since $\|\partial_i u\|_1 \le \| \, |\nabla u| \, \|_1$ for all i, we indeed have

$$\|u\|_{\frac{N}{N-1}} \le \| \, |\nabla u| \, \|_1 \, .$$

d. For $p = 1$, we have seen that $C_{N,p} = 1$ is suitable. We now assume $1 < p < N$ and set $\alpha = \dfrac{p(N-1)}{N-p}$ (so that $\alpha \frac{N}{N-1} = p^\star$) and $v = |u|^{\alpha-1} u$. Since $\alpha > 1$ and $u \in C_c^1(\mathbb{R}^N)$, we also have $v \in C_c^1(\mathbb{R}^N)$. We can therefore apply the result of question (c) to the function v. We obtain

---

[25] This can be proved using, for example, the convexity of the exponential function.

$$\left(\int_{\mathbb{R}^N} |u(x)|^{p^\star} \right)^{\frac{N-1}{N}} = \left(\int_{\mathbb{R}^N} |u(x)|^{\alpha \frac{N}{N-1}} \right)^{\frac{N-1}{N}} \leq \| \, |\nabla v| \, \|_1 .$$

Since $|\nabla v| = \alpha |u|^{\alpha-1} |\nabla u|$, Hölder's inequality (with p and $q = \frac{p}{p-1}$) gives

$$\| \, |\nabla v| \, \|_1 = \alpha \left\| |u|^{\alpha-1} |\nabla u| \right\|_1 \leq \alpha \left\| |u|^{\alpha-1} \right\|_q \| \, |\nabla u| \, \|_p .$$

Since $(\alpha - 1)q = (\alpha - 1)\frac{p}{p-1} = p^\star$, we therefore have

$$\|u\|_{p^\star}^{\frac{p^\star (N-1)}{N}} = \left(\int_{\mathbb{R}^N} |u(x)|^{p^\star} \right)^{\frac{N-1}{N}} \leq \alpha \|u\|_{p^\star}^{\frac{p^\star (p-1)}{p}} \| \, |\nabla u| \, \|_p ,$$

which gives, with $C_{N,p} = \alpha = \dfrac{p(N-1)}{N-p}$,

$$\|u\|_{p^\star} \leq C_{N,p} \| \, |\nabla u| \, \|_p .$$

2. Let $u \in W^{1,p}(\mathbb{R}^N)$. There exists a sequence $(u_n)_{n \in \mathbb{N}}$ of functions belonging to $C_c^1(\mathbb{R}^N)$ such that $u_n \to u$ in $W^{1,p}(\mathbb{R}^N)$ as $n \to +\infty$. By the previous question, $(u_n)_{n \in \mathbb{N}}$ is a Cauchy sequence in $L^{p^\star}$. By uniqueness of the limit (for example in $L_{\text{loc}}^1(\mathbb{R}^N)$) this limit is necessarily equal to u. Passing to the limit as $n \to +\infty$ in the inequality $\|u_n\|_{p^\star} \leq C_{N,p} \| \, |\nabla u_n| \, \|_p$ yields

$$\|u\|_{p^\star} \leq C_{N,p} \| \, |\nabla u| \, \|_p \quad \text{for all } u \in W^{1,p}(\mathbb{R}^N),$$

which proves the continuous embedding from $W^{1,p}(\mathbb{R}^N)$ to $L^{p^\star}(\mathbb{R}^N)$.

The continuous embedding from $W^{1,p}(\mathbb{R}^N)$ to $L^p(\mathbb{R}^N)$ is immediate because $\|u\|_p \leq \|u\|_{W^{1,p}}$ for all $u \in W^{1,p}(\mathbb{R}^N)$.

Now let $q \in]p, p^\star[$. To show that $W^{1,p}(\mathbb{R}^N)$ is continuously embedded in $L^q(\mathbb{R}^N)$ it is sufficient to use the following classical inequality (which is proved using Hölder's inequality) with $p < q < r = p^\star$.

$$\|u\|_q \leq \|u\|_p^\theta \|u\|_r^{1-\theta} , \tag{1.27}$$

with $\theta = \frac{p(r-q)}{q(r-p)} \in]0, 1[$.

3. Let us start with the case $N = 1$. Question 2 gives the continuous embedding from $W^{1,1}(\mathbb{R})$ to $L^\infty(\mathbb{R})$. Since $W^{1,1}(\mathbb{R})$ also is continuously embedded in $L^1(\mathbb{R})$, we also obtain a continuous embedding from $W^{1,1}(\mathbb{R})$ to $L^q(\mathbb{R})$ for all $q \in]1, +\infty[$ (using (1.27) with $r = +\infty$, $p = 1$ and $\theta = \frac{1}{q}$).

We now assume $N > 1$. We have a continuous embedding from $W^{1,N}(\mathbb{R})$ to $L^N(\mathbb{R})$. The only case to consider is therefore $N < q < +\infty$. We give below a proof that uses the homogeneous character of the norm, which is an interesting tool *per se*. We choose on $W^{1,N}(\mathbb{R}^N)$ the norm $\|u\|_{W^{1,N}} = \|u\|_N + \| \, |\nabla u| \, \|_N$ (which is equivalent to the norm defined in Proposition 1.9).

Let $N < q < +\infty$. There then exists a $p \in]1, N[$ such that $p^\star = Np/(N-p) = q$. We use question 1 with this value of p.

Define $\varphi \in C^1(\mathbb{R}, \mathbb{R})$ by:

$$\varphi(s) = \begin{cases} 0 & \text{if } |s| \le 1, \\ \frac{1}{2}(|s|-1)^2 & \text{if } 1 < |s| \le 2, \\ |s| - \frac{3}{2} & \text{if } 2 < |s|. \end{cases}$$

We have $|\varphi(s)| \le |s|$ and $|\varphi'(s)| \le 1$ for all $s \in \mathbb{R}$. Let $u \in C_c^1(\mathbb{R}^N)$ such that $\|u\|_{W^{1,N}} = 1$. We have $\varphi(u) \in C_c^1(\mathbb{R}^N)$ (we denote by $\varphi(u)$, somewhat incorrectly, the function $\varphi \circ u$). By question 1 (and the definition of φ) we obtain

$$\|\varphi(u)\|_q \le C_{N,p} \, \| \, |\nabla \varphi(u)| \, \|_p$$

$$\le C_{N,p} \Big(\int_{\{|u| \ge 1\}} |\nabla u(x)|^p \, dx \Big)^{\frac{1}{p}}$$

$$\le C_{N,p} \Big(\int_{\mathbb{R}^N} |\nabla u(x)|^N \, dx \Big)^{\frac{1}{N}} \lambda_N(\{|u| \ge 1\})^{\frac{1}{p} - \frac{1}{N}}.$$

We have $\lambda_N(\{|u| \ge 1\}) \le \int_{\mathbb{R}^N} |u(x)|^N \, dx \le 1$ and $\int_{\mathbb{R}^N} |\nabla u(x)|^N \, dx \le 1$ and we get that $\|\varphi(u)\|_q \le C_{N,p}$. Since $|u|^q \le 2^q \varphi(u)^q + 4^q$, we therefore have

$$\int_{\mathbb{R}^N} |u(x)|^q \, dx = \int_{\{|u| \le 1\}} |u(x)|^q \, dx + \int_{\{|u| > 1\}} |u(x)|^q \, dx$$

$$\le \int_{\mathbb{R}^N} |u(x)|^N \, dx + 2^q \int_{\mathbb{R}^N} |\varphi(u(x))|^q \, dx + 4^q \lambda_N(\{|u| > 1\})$$

$$\le 1 + 2^q C_{N,p}^q + 4^q. \tag{1.28}$$

Set $D_{N,q} = (1 + 2^q C_{N,p}^q + 4^q)^{\frac{1}{q}}$, which depends only on N and q, and note that $\|u\|_q \le D_{N,q}$ if $u \in C_c^1(\mathbb{R}^N)$ and $\|u\|_{W^{1,N}} = 1$. Thanks to the homogeneous nature of the norm, $\|u\|_q \le D_{N,q} \|u\|_{W^{1,N}(\mathbb{R}^N)}$ for all $u \in C_c^1(\mathbb{R}^N)$. Finally, by density of $C_c^1(\mathbb{R}^N)$ in $W^{1,N}(\mathbb{R}^N)$, $W^{1,N}(\mathbb{R}^N) \subset L^q(\mathbb{R}^N)$ and

$$\|u\|_q \le D_{N,q} \|u\|_{W^{1,N}(\mathbb{R}^N)} \quad \text{for all } u \in W^{1,N}(\mathbb{R}^N),$$

which clearly shows that there is a continuous embedding from $W^{1,N}(\mathbb{R}^N)$ to $L^q(\mathbb{R}^N)$.

4. This question consists only in using the extension theorem (Theorem 1.25) which gives the existence of a continuous linear operator P from $W^{1,p}(\Omega)$ to $W^{1,p}(\mathbb{R}^N)$ such that $Pu = u$ a.e. in Ω. Indeed, thanks to this operator, question 4 is a consequence of the previous questions (and of the inequality (1.27)).

Problem 1.20 (Kernel of the trace operator)

1. Let us first show that $W_0^{1,p}(\Omega) \subset \mathrm{Ker}\gamma$.

   Let $u \in W_0^{1,p}(\Omega)$. By Definition 1.27 of $W_0^{1,p}(\Omega)$, there exists a sequence $(u_n)_{n \in \mathbb{N}} \subset \mathcal{D}(\Omega)$ such that $u_n \to u$ in $W^{1,p}(\Omega)$ as $n \to +\infty$. Owing to Theorem 1.30, $\gamma(u_n) \to \gamma(u)$ in $L^p(\partial\Omega)$ as $n \to +\infty$, where the functions u_n are extended by 0 outside of Ω. Since, for all $n \in \mathbb{N}$, $\gamma(u_n) = 0$ a.e. on $\partial\Omega$, we get that $\gamma(u) = 0$ a.e. on $\partial\Omega$ (for the $(N-1)$–dimensional measure), that is, $u \in \mathrm{Ker}\gamma$.

   Let us now show that $\mathrm{Ker}\gamma \subset W_0^{1,p}(\Omega)$.

   Let $u \in \mathrm{Ker}\gamma$; the proof that $u \in W_0^{1,p}(\Omega)$ is performed in two steps. Denoting by $\tilde{u}$ the function u extended by 0 outside of Ω, we show $\tilde{u} \in W^{1,p}(\mathbb{R}^N)$ in a first step, and we get $u \in W_0^{1,p}(\Omega)$ in a second step.

   Step 1 Obviously, $\tilde{u} \in L^p(\mathbb{R}^N)$ and $\|\tilde{u}\|_{L^p(\mathbb{R}^N)} = \|u\|_{L^p(\Omega)}$. It is now a matter of showing the same equality with $D_i\tilde{u}$ and $D_i u$ (instead of $\tilde{u}$ and u).

   There exists a sequence $(u_n)_{n \in \mathbb{N}} \subset \mathcal{D}(\mathbb{R}^N)$ such that $u_n \to u$ in $W^{1,p}(\Omega)$ as $n \to +\infty$ (u_n is here rather the restriction of u_n to Ω). Let $\varphi \in \mathcal{D}(\mathbb{R}^N)$ and $i \in \{1, \ldots, N\}$,

   $$\langle D_i\tilde{u}, \varphi\rangle_{\mathcal{D}^\star(\mathbb{R}^N), \mathcal{D}(\mathbb{R}^N)} = -\int_{\mathbb{R}^N} \tilde{u}(x)\partial_i\varphi(x)\,\mathrm{d}x = -\int_\Omega u(x)\partial_i\varphi(x)\,\mathrm{d}x.$$

   But, since u_n and φ belong to $\mathcal{D}(\mathbb{R}^N)$,

   $$-\int_\Omega u_n(x)\partial_i\varphi(x)\,\mathrm{d}x = \int_\Omega \partial_i u_n(x)\varphi(x)\,\mathrm{d}x + \int_{\mathbb{R}^{N-1}} u_n(0, y)\varphi(0, y)n_i\,\mathrm{d}y,$$

   with $n_1 = 1$ and $n_i = 0$ for $i > 1$. By passing to the limit in this equality as $n \to +\infty$, since $u_n \to u$ in $W^{1,p}(\Omega)$, $u_n \to \gamma u$ in $L^p(\mathbb{R}^{N-1})$, φ with compact support and $\gamma u = 0$ a.e. on $\mathbb{R}^{N-1}$ (because $u \in \mathrm{Ker}\gamma$) we obtain

   $$\langle D_i\tilde{u}, \varphi\rangle_{\mathcal{D}^\star(\mathbb{R}^N), \mathcal{D}(\mathbb{R}^N)} = \int_\Omega D_i u(x)\varphi(x)\,\mathrm{d}x, \tag{1.29}$$

   i.e. that $D_i\tilde{u}$ is the function equal to $D_i u$ on Ω and 0 outside of Ω. We have finally shown that $\tilde{u} \in W^{1,p}(\mathbb{R}^N)$ and $\|\tilde{u}\|_{W^{1,p}(\mathbb{R}^N)} = \|u\|_{W^{1,p}(\Omega)}$.

   Step 2 Let $\varphi_0 \in \mathcal{D}(\mathbb{R}^N)$ be such that $\varphi_0(x) \geq 0$ for all x, $\varphi_0(x) = 0$ if $|x| \geq 2$, $\varphi_0(x_1, y) = 0$ if $x_1 \leq 1$ (and $y \in \mathbb{R}^{N-1}$) and $\int_{\mathbb{R}^N} \varphi(x)\,\mathrm{d}x = 1$. For $n \in \mathbb{N}^\star$, we define φ_n by $\varphi_n(x) = n^N \varphi_0(nx)$. Set $u_n = \tilde{u} \star \varphi_n$ so that $u_n \in \mathcal{D}(\mathbb{R}^N)$ and the restriction of u_n to Ω, still denoted u_n, belongs to $\mathcal{D}(\Omega)$ (because $u_n(x_1, y) = 0$ if $x_1 < \frac{1}{n}$). Finally, as seen in the proof of Theorem 1.25, $u_n \to \tilde{u}$ in $W^{1,p}(\mathbb{R}^N)$ as $n \to +\infty$, and therefore, by taking the restrictions to Ω, $u_n \to \tilde{u}$ in $W^{1,p}(\Omega)$, which proves that $u \in W_0^{1,p}(\Omega)$.

2. Let us use the method used to prove the first property of Theorem 1.25. With the notations of the proof of the first property of Theorem 1.25, $u_n = u \star \varphi$ (the function u has been extended by 0 outside of Ω and we have seen that $u_n \to u$

in $W^{1,p}(\Omega)$ and therefore also $u_n \to \gamma(u)$ in $L^p(\partial\Omega)$ as $n \to +\infty$. But, since $u \in C(\overline{\Omega})$, we note that $u_n \to u$ uniformly on $\overline{\Omega}$ (this is due to the fact that $\varphi_0(x_1, y) = 0$ for $x_1 \geq 0$). Hence $\gamma(u) = u$ a.e. on $\partial\Omega$.

Problem 1.21 (Extension of a function belonging to H^2)

1. Let us revisit the proof of the first property of Theorem 1.25.
    In a first step, we prove the density in $W^{2,p}(\mathbb{R}_+^N)$ of the set of compactly supported functions belonging to $W^{2,p}(\mathbb{R}_+^N)$. To this end, choose $\psi \in \mathcal{D}(\mathbb{R}^N)$ such that $0 \leq \psi(x) \leq 1$ for all x, $\psi(x) = 1$ if $|x| \leq 1$, $\psi(x) = 0$ if $|x| \geq 2$ and define u_n for $n \in \mathbb{N}^\star$ by $u_n(x) = u(x)\psi(x/n)$. Then $u_n \to u$ in $W^{2,p}(\mathbb{R}^N +)$ as $n \to +\infty$.

    In a second step, consider $u \in W^{2,p}(\mathbb{R}_+^N)$ with compact support and extend u by 0 outside of $\mathbb{R}_+^N$. Then, choose (as in Theorem 1.25 for the case $\mathbb{R}_+^N$) a function $\rho \in \mathcal{D}(\mathbb{R}^N)$ such that $\rho(x) = 0$ if $|x| \geq 1$, $\int_{\mathbb{R}^N} \rho(x) = 1$ and $\rho(x_1, y) = 0$ if $x_1 \geq 0$ (and $y \in \mathbb{R}^{N-1}$), and set $u_n = u \star \rho_n$. Then $u_n \in \mathcal{D}(\mathbb{R}^N)$ and, thanks to the fact that $\rho(x_1, y) = 0$ if $x_1 \geq 0$, as in Theorem 1.25 for the case $\mathbb{R}_+^N$, $u_n \to u$ in $W^{2,p}(\mathbb{R}_+^N)$ as $n \to +\infty$ (more precisely it is the restriction to $\mathbb{R}_+^N$ of u_n).
2. Take α and β such that $\alpha + \beta = 1$ and $-\alpha - 2\beta = 1$, that is to say, $\beta = -2$ and $\alpha = 3$.
    If $u \in C_c^\infty(\overline{\Omega})$ (that is to say, u is the restriction to Ω of an element of $\mathcal{D}(\mathbb{R}^N)$ still denoted by u), we define Pu by

$$Pu(x) = \begin{cases} u(x) & \text{if } x = (x_1, y)^t \text{ with } x_1 \geq 0, \\ \alpha u(-x_1, y) + \beta u(-2x_1, y) & \text{if } x = (x_1, y)^t \text{ with } x_1 < 0. \end{cases}$$

The function Pu is of class C^1 and its first and second derivatives are represented by the classical derivatives. Hence $Pu \in W^{2,p}(\mathbb{R}^N)$ and $\|\tilde{u}\|_{W^{2,p}(\mathbb{R}^N)} = C \|u\|_{W^{1,p}(\Omega)}$; note that C can be calculated explicitly with respect to α and β (but this calculation is unnecessary here). The operator P is therefore a continuous linear operator from $C_c^\infty(\overline{\Omega}) \subset W^{2,p}(\overline{\Omega})$ to $W^{2,p}(\mathbb{R}^N)$ and its norm is equal to C. By density of $C_c^\infty(\overline{\Omega})$ in $W^{2,p}(\overline{\Omega})$, it therefore extends (in a unique way) to a continuous linear operator P from $W^{2,p}(\Omega)$ to $W^{2,p}(\mathbb{R}^N)$ and of norm equal to C. Of course, $P(u) = u$ a.e. in Ω.

3. The set $C_c^\infty(\overline{\Omega})$ is not dense in $W^{2,\infty}(\Omega)$. A limit in $L^\infty(\Omega)$ of continuous functions on $\overline{\Omega}$ is necessarily continuous. To show that $C_c^\infty(\overline{\Omega})$ is not dense in $W^{2,\infty}(\Omega)$, it is therefore sufficient to find an element of $W^{2,\infty}(\Omega)$ for which (at least) one second derivative is not continuous. A possible example is to take, for instance, $u(x) = \varphi(x_1)\psi(x)$ for $x = (x_1, y)^t$, $x_1 > 0$ and $y \in \mathbb{R}$ with $\varphi(x_1) = ((1 - x_1)^+)^2$ and $\psi \in \mathcal{D}(\mathbb{R}^N)$, $\psi(x) = 1$ if $|x| < 1$. Indeed, $u \in W^{2,\infty}(\Omega)$ but the function $D_1 D_1 u$ is discontinuous, which is enough to affirm that u is not in the closure of $C_c^\infty(\overline{\Omega})$ in $W^{2,\infty}(\Omega)$.

    However, there does exist a continuous linear operator P from $W^{2,\infty}(\Omega)$ to $W^{2,\infty}(\mathbb{R}^N)$ such that $Pu = u$ a.e. in Ω for every $u \in W^{2,\infty}(\Omega)$. This is due to the fact that a function from $W^{2,\infty}(\Omega)$ belongs to $C^1(\overline{\Omega}, \mathbb{R})$ (see Problem

1.15). A possible choice of Pu (for $u \in W^{2,\infty}(\Omega)$) is then that of question 2 with the same values of α and β:

$$Pu(x) = \begin{cases} u(x) & \text{if } x = (x_1, y)^t \text{ with } x_1 \geq 0, \\ \alpha u(-x_1, y) + \beta u(-2x_1, y) & \text{if } x = (x_1, y)^t \text{ with } x_1 < 0. \end{cases}$$

Problem 1.22 (Weak convergence and continuous operator)

1. Let T^t denote the transpose of the operator T given by Definition 2.13): $\langle T^t g, u \rangle_{E',E} = \langle g, Tu \rangle_{F',F}$. Observe that $T^t \in \mathcal{L}(F', E')$ (see for example Problem 1.26).

   For all $f \in F'$ and all $n \in \mathbb{N}$, $\langle f, T(u_n) \rangle_{F',F} = \langle T^t f, u_n \rangle_{E',E}$. Since $T^t f \in E'$ and $u_n \to u$ weakly in E as $n \to +\infty$,

$$\lim_{n \to +\infty} \langle f, T(u_n) \rangle_{F',F} = \lim_{n \to +\infty} \langle T^t f, u_n \rangle_{E',E} = \langle T^t f, u \rangle_{E',E} = \langle f, Tu \rangle_{F',F}.$$

   This proves that $T(u_n) \to Tu$ weakly in F.
2. This is an immediate application of question 1 applied to $T \in \mathcal{L}(E, F)$ defined by $Tu = u$.
3. This is yet another immediate application of the question 1, taking $E = H_0^1(\Omega)$, $F = L^2(\Omega)$ and T defined by $T(u) = D_i u$.

Problem 1.23 (Non-continuous function belonging to $H^1(\mathbb{R}^2) \cap L^\infty(\mathbb{R}^2)$)

1. Let B be the ball of centre 0 and radius 1 (for the Euclidean norm). Problem 1.5 shows that the restriction of u to B belongs to $H^1(B)$. Then since u is continuous in the neighbourhood of the edge of B and $u = 0$ on the edge of B, $u \in H_0^1(B)$, see the proof of question 2 of Problem 1.20). Finally, the function u consists of extending by 0 its restriction to B, the proof of question 1 of Problem 1.20 then gives $u \in H^1(\mathbb{R}^2)$.

   Remark 2.27 of Lemma 2.26 then gives that the restriction of u_n to B belongs to $H_0^1(B)$ (and $u_n \subset H^1(\mathbb{R}^2)$) and that

$$\| |\nabla u_n| \|_{L^2(\mathbb{R}^2)}^2 = \int_{\{n < |u(x)| < n+1\}} |\nabla u(x)|^2 \, dx.$$

   Hence $\sum_{n=1}^{\infty} \| |\nabla u_n| \|_{L^2(\mathbb{R}^2)}^2 \leq \|u\|_{H^1(B)}^2 < +\infty$.
2. For $x \in \mathbb{R}^2$, $u_n(x) = 0$ or 1 or $u(x) - n$ if $n < u(x) < n + 1$, so that u_n takes its values between 0 and 1. Then, again for $x \in \mathbb{R}^2$, $u_n(x) \neq 0$ if and only if $u(x) > n$, that is $(-\ln |x|)^\gamma > n$ and therefore $|x| < \exp(-n^{1/\gamma})$. The support of u_n is therefore the (closed) ball of centre 0 and radius $\exp(-n^{1/\gamma})$. The radius of this ball does indeed tend towards 0 as $n \to +\infty$.

3. For all $n \in \mathbb{N}^\star$, the support of v_n is included in the ball $B_n = \{z \in \mathbb{R}^2, |z - x_n| \le r_n\}$.

   Let $p > n \ge 1$. Suppose there exists an $x \in \mathbb{R}^2$ such that $x \in B_n$ and $x \in B_p$. Since $|x - x_n| \le r_n$ and $|x - x_p| \le r_p$,

   $$\frac{1}{n} - \frac{1}{n+1} \le |x_n - x_p| \le r_n + r_p$$
   $$\le \frac{1}{2}\left(\frac{1}{n} - \frac{1}{n+1}\right) + \frac{1}{2}\left(\frac{1}{p} - \frac{1}{p+1}\right) < \frac{1}{n} - \frac{1}{n+1},$$

   which is impossible. The functions v_n therefore have disjoint supports.

4. For all $n \in \mathbb{N}^\star$, the restriction of v_n belongs to $H_0^1(B_n)$ (with B_n defined in the previous question). Hence v_n (which consists of extending by 0 its restriction to B_n) belongs to $H^1(\mathbb{R}^2)$ and, for $i \in \{1, 2\}$, $D_i v_n$ is equal to D_i of the restriction of v_n to B_n extended by 0 outside of B_n (see equality (1.29)). Therefore,

   $$\langle D_i v_n, \varphi \rangle_{\mathcal{D}^\star(\mathbb{R}^2), \mathcal{D}(\mathbb{R}^2)} = \int_{B_n} \partial_i v_n(x) \varphi(x) \, dx.$$

   Since the supports of the functions v_n are disjoint, we get that for $p > n \ge 1$,

   $$\langle D_i \sum_{q=n}^{p} v_q, \varphi \rangle_{\mathcal{D}^\star(\mathbb{R}^N), \mathcal{D}(\mathbb{R}^N)} = \sum_{q=n}^{p} \int_{B_q} \partial_i v_q(x) \varphi(x) \, dx,$$

   and therefore $D_i(\sum_{q=n}^{p} v_q) \in L^2(\mathbb{R}^2)$ and

   $$\left\| D_i\left(\sum_{q=n}^{p} v_q\right) \right\|_{L^2(\mathbb{R}^2)}^2 = \sum_{q=n}^{p} \int_{\{m_q < |u(x)| < m_q + 1\}} |\nabla u(x)|^2 \, dx \le \int_{\{m_n < |u(x)|\}} |\nabla u(x)|^2 \, dx.$$

   Moreover, denoting by λ_2 the Lebesgue measure on $\mathbb{R}^2$,

   $$\left\| \sum_{q=n}^{p} v_q \right\|_{L^2(\mathbb{R}^2)} \le \sum_{q=n}^{p} \lambda_2(B_q).$$

   Since $u \in H^1(\mathbb{R}^2)$ and all the balls B_n are disjoint and included in the ball B with centre 0 and radius 2, we get from these last two inequalities that the series $\sum_{n \in \mathbb{N}^\star} v_n$ is convergent in $H^1(\mathbb{R}^2)$. Let v be the sum of this series (and therefore $v \in H^1(\mathbb{R}^2)$).

   Now observe that for all $x \in \mathbb{R}^2$ the series $\sum_{n \in \mathbb{N}^\star} v_n(x)$ is convergent in $\mathbb{R}$ (there is at most one non-zero term in this series), let $\bar{v}(x)$ be its limit. Since a sequence that converges in $L^2(\mathbb{R}^2)$ admits a subsequence that is convergent a.e., we have $v = \bar{v}$ a.e. Since $0 \le \bar{v}(x) \le 1$ for all x, we therefore have $v \in L^\infty(\mathbb{R}^2)$.

Now assume that there exists a $w \in C(\mathbb{R}^2, \mathbb{R})$ such that $w = v$ a.e. Therefore, $w = \bar{v}$ a.e., so that $w(x) = \bar{v}(x)$ for all $x \neq 0$ (indeed, recall that two continuous functions on an open subset and equal a.e. are equal everywhere on this open subset). But this contradicts the continuity of w at 0; indeed, let $y_n = x_n + (r_n, 0)^t$, then

$$w(0) = \lim_{n \to +\infty} \bar{v}(x_n) = 1 \quad \text{and} \quad w(0) = \lim_{n \to +\infty} \bar{v}(y_n) = 0.$$

Problem 1.24 (On the embedding from $W^{1,1}$ to $L^{1^\star}$)

I- First method (direct method)

1. Since $\|u_n\|_{L^1(\Omega)} = 1$ and $\| |\nabla u_n| \|_{L^1(\Omega)} < \frac{1}{n}$, the sequence $(u_n)_{n \in \mathbb{N}^\star}$ is bounded in $W^{1,1}(\Omega)$. Theorem 1.37 then gives that we can assume, after extraction of a subsequence, that $u_n \to u$ in $L^1(\Omega)$ as $n \to +\infty$. Hence $\|u\|_{L^1(\Omega)} = \lim_{n \to +\infty} \|u_n\|_{L^1(\Omega)} = 1$.

   Then $\| |\nabla u_n| \|_{L^1(\Omega)} \to 0$ as $n \to +\infty$, and therefore $D_i u_n \to 0$ in $L^1(\Omega)$ and also in $\mathcal{D}^\star(\Omega)$. Since $D_i u_n \to D_i u$ in $\mathcal{D}^\star(\Omega)$ (because $u_n \to u$ in $L^1(\Omega)$ and therefore in $\mathcal{D}^\star(\Omega)$), we therefore have $D_i u = 0$ (in $\mathcal{D}^\star(\Omega)$). Question 3 of Problem 1.4 gives that there exists an $a \in \mathbb{R}$ such that $u = a$ a.e. Now observe that, if necessary up to the extraction of another subsequence, $u_n \to u$ a.e. as $n \to +\infty$ and therefore $u = 0$ a.e. on ω. Since $\lambda_N(\omega) > 0$ and $u = a$ a.e. on ω, we therefore have $a = 0$, which contradicts $\|u\|_{L^1(\Omega)} = 1$.
2. By reasoning by contradiction, we are reduced to the previous question. If C_1 does not exist, there then exists a sequence $(u_n)_{n \in \mathbb{N}^\star}$ of elements of W_ω such that, for all $n \in \mathbb{N}^\star$,

$$\|u_n\|_{L^1(\Omega)} > n \| |\nabla u_n| \|_{L^1(\Omega)} \quad \text{for all } n \in \mathbb{N}^\star.$$

By a homogeneity argument we can assume that $\|u_n\|_{L^1(\Omega)} = 1$ but the previous question showed that such a sequence does not exist.
3. First consider the case $p = 1^\star = \frac{N}{N-1}$ (and therefore $p = +\infty$ if $N = 1$). Let $u \in W_\omega$. Using the result of the previous question,

$$\|u\|_{L^{1^\star}(\Omega)} \le C_2 \|u\|_{W^{1,1}(\Omega)} = C_2 \Big(\|u\|_{L^1(\Omega)} + \sum_{i=1}^N \|D_i u\|_{L^1(\Omega)} \Big)$$

$$\le C_2 \big(C_1 \| |\nabla u| \|_{L^1(\Omega)} + N \| |\nabla u| \|_{L^1(\Omega)} \big) = C_2(C_1 + N) \| |\nabla u| \|_{L^1(\Omega)} .$$

Hence (1.13) holds for $p = 1$ and for $p = 1^\star$, taking $C = \max\{C_1, C_2(C_1 + N)\}$. Hölder's inequality then gives that inequality (1.13) is then true for $1 \le p \le 1^\star$.

II- Second method (using the mean value of u)

1. Let us proceed by contradiction. If C_3 does not exist, there exists a sequence $(u_n)_{n\in\mathbb{N}^\star}$ of elements of H such that, for all $n \in \mathbb{N}^\star$,

$$\|u_n\|_{L^1(\Omega)} > n\,\|\,|\nabla u_n|\,\|_{L^1(\Omega)} \quad \text{for all } n \in \mathbb{N}^\star.$$

By a homogeneity argument we can suppose that $\|u_n\|_{L^1(\Omega)} = 1$. The proof is then very similar to that of question 1 of the first method. The sequence $(u_n)_{n\in\mathbb{N}^\star}$ is bounded in $W^{1,1}(\Omega)$ and we can therefore suppose, after extraction of a subsequence, that $u_n \to u$ in $L^1(\Omega)$ as $n \to +\infty$. Hence $\|u\|_{L^1(\Omega)} = \lim_{n\to+\infty} \|u_n\|_{L^1(\Omega)} = 1$.

Then $\|\,|\nabla u_n|\,\|_{L^1(\Omega)} \to 0$ as $n \to +\infty$, and therefore $D_i u_n \to 0$ in $\mathcal{D}^\star(\Omega)$ (for all i). Hence $D_i u = 0$ (in $\mathcal{D}^\star(\Omega)$) and question 3 of Problem 1.4 shows that there exists an $a \in \mathbb{R}$ such that $u = a$ a.e.

Now observe that $\int_\Omega u(x)\,dx = \lim_{n\to+\infty} \int_\Omega u_n(x)\,dx = 0$ (because $u_n \to u$ in $L^1(\Omega)$ and $u_n \in H$). This shows that $a = 0$, which contradicts $\|u\|_{L^1(\Omega)} = 1$.

Using the hint from question 3 of the first part and the fact that $u - m \in H$ (and $\nabla(u - m) = \nabla u$ a.e.) yields

$$\|u - m\|_{L^{1^\star}(\Omega)} \le C_2\,\|u - m\|_{W^{1,1}(\Omega)}$$

$$= C_2\Big(\|u - m\|_{L^1(\Omega)} + \sum_{i=1}^{N} \|D_i u\|_{L^1(\Omega)}\Big)$$

$$\le C_2\big(C_3\,\|\,|\nabla u|\,\|_{L^1(\Omega)} + N\,\|\,|\nabla u|\,\|_{L^1(\Omega)}\big) = C_2(C_3 + N)\,\|\,|\nabla u|\,\|_{L^1(\Omega)}.$$

Thus (1.14) holds with $C_4 = C_2(C_3 + N)\}$.

2. Since $u = 0$ a.e. on ω and owing to the result of the previous question,

$$|m|\lambda_N(\omega)^{\frac{1}{1^\star}} = \Big(\int_\omega |u(x) - m|^{1^\star}\Big)^{\frac{1}{1^\star}} dx \le \|u - m\|_{L^{1^\star}(\Omega)} \le C_4\,\|\,|\nabla u|\,\|_{L^1(\Omega)},$$

and therefore

$$|m| \le \frac{C_4}{\lambda_N(\omega)^{\frac{1}{1^\star}}}\,\|\,|\nabla u|\,\|_{L^1(\Omega)}.$$

Then,

$$\|u\|_{L^{1^\star}(\Omega)} \le \|u - m\|_{L^{1^\star}(\Omega)} + |m|\lambda_N(\Omega)^{\frac{1}{1^\star}}$$

$$\le C_4\Big(1 + \Big(\frac{\lambda_N(\Omega)}{\lambda_N(\omega)}\Big)^{\frac{1}{1^\star}}\Big)\,\|\,|\nabla u|\,\|_{L^1(\Omega)}.$$

Problem 1.25 (Partition of unity)

1. For $i \in \{1, \ldots, n\}$, we have $\Omega_i = \cup_{\varepsilon>0}\Omega_{i,\varepsilon}$ and therefore $K \subset \cup_{i=1}^n \cup_{\varepsilon>0} \Omega_{i,\varepsilon}$. Since K is compact and $\Omega_{i,\varepsilon}$ is open (for all i and ε), there exist $\varepsilon_1, \ldots, \varepsilon_n$ such that $\varepsilon_i > 0$ for all i and $K \subset \cup_{i=1}^n \Omega_{i,\varepsilon_i}$. Taking $\varepsilon = \min\{\varepsilon_1, \ldots, \varepsilon_n\}$, we have $K \subset \cup_{i=1}^n \Omega_{i,\varepsilon}$.

2. For all $i \in \{1, \ldots, n\}$, set

$$f_i(x) = \begin{cases} 1 & \text{if } x \in (\Omega_{i,\varepsilon} \setminus \cup_{j<i}\Omega_{j,\varepsilon}), \\ 0 & \text{if } x \notin (\Omega_{i,\varepsilon} \setminus \cup_{j<i}\Omega_{j,\varepsilon}). \end{cases}$$

   With this definition, $f_i = 0$ on $\Omega_{i,\varepsilon}^c$. Moreover, if $x \in \cup_{i=1}^n \Omega_{i,\varepsilon}$ and $i = \min\{j; x \in \Omega_{j,\varepsilon}$, we have $f_i(x) = 1$ and $f_j(x) = 0$ if $j \neq i$. Therefore $\sum_{i=1}^n f_i(x) = 1$.

3. Since K is compact and $(\cup_{i=1}^n \Omega_{i,\varepsilon})^c$ is closed, we have $d(K, (\cup_{i=1}^n \Omega_{i,\varepsilon})^c) = \delta > 0$. Take η such that $0 < \eta < \min\{\delta, \varepsilon\}$ and let $\rho \in C_c^\infty(\mathbb{R}^n, \mathbb{R}_+)$ such that $\rho(x) = 0$ if $|x| \geq \eta$ and such that $\int_{\mathbb{R}^n} \rho(x)\, dx = 1$. For $i \in \{1, \ldots, n\}$ take $\varphi_i = f_i \star \rho$. Then $\varphi_i \in C_c^\infty(\mathbb{R}^n, \mathbb{R}_+)$ and since $f_i = 0$ on $\Omega_{i,\varepsilon}^c$ and $\eta < \varepsilon$, the function φ_i satisfies (p1). Since $\sum_{i=1}^n \varphi_i = (\sum_{i=1}^n f_i) \star \rho$ and $\eta < \delta$, the functions φ_i satisfy (p2).

Problem 1.26 (Transposed operator, continuity and compactness)

1. Let $g \in F'$, the mapping $u \mapsto \langle g, Tu \rangle_{F',F}$ is linear and

$$|\langle g, Tu \rangle_{F',F}| \leq \|T\|_{\mathcal{L}(E,F)} \|g\|_{F'} \|u\|_E .$$

   This proves that $T^t g$ belongs to E' and that $\|T^t g\|_{E'} \leq \|T\|_{\mathcal{L}(E,F)} \|g\|_{F'}$. Hence $T^t \in \mathcal{L}(F', E')$ and $\|T^t\|_{\mathcal{L}(F',E')} \leq \|T\|_{\mathcal{L}(E,F)}$.

2. Let $u \in E$. A consequence of the Hahn–Banach theorem (see Problem 1.8, question 1) gives the existence of $g \in F'$ such that $\|g\|_{F'} = 1$ and $\|Tu\|_F = \langle g, Tu \rangle_{F',F}$. Indeed, setting $v = Tu \in F$ and assuming $v \neq 0$ (the case $v = 0$ is trivial), we choose $\widetilde{g} : \mathbb{R}v \to \mathbb{R}$ linear and such that $\widetilde{g}(v) = \|v\|$, so that $\|\widetilde{g}\| = 1$. By the Hahn–Banach theorem, we can extend $\widetilde{g}$ to $g \in F'$ with $\|g\|_{F'} = 1$ and $\langle g, v \rangle_{F',F} = \|v\|_F$, which also reads $\|Tu\|_F = \langle g, Tu \rangle_{F',F}$. Hence $\|Tu\|_F \leq \|T^t\|_{\mathcal{L}(F',E')} \|u\|_E$. This proves that $\|T\|_{\mathcal{L}(E,F)} \leq \|T^t\|_{\mathcal{L}(F',E')}$ and finally $\|T\|_{\mathcal{L}(E,F)} = \|T^t\|_{\mathcal{L}(F',E')}$.

3. By hypothesis, the set $T(B_E)$ is relatively compact and therefore precompact. Indeed, let $\varepsilon > 0$, suppose by contradiction that there does not exist a finite covering of $T(B_E)$ by balls of the form $B_F(Tu, \varepsilon)$. Let $u_0 \in B_E$, so we have $T(B_E) \not\subset B_F(Tu_0, \varepsilon)$ Then, by induction, we suppose $u_0, \ldots, u_n$ have been chosen in B_E. Since $T(B_E) \not\subset \cup_{i=0}^n B_F(Tu_i, \varepsilon)$, we choose $u_{n+1} \in B_E$ such that $T(u_{n+1}) \notin \cup_{i=0}^n B_F(Tu_i, \varepsilon)$. Hence the constructed sequence $(u_n)_{n\in\mathbb{N}} \subset B_E$ is such that the sequence $(T(u_n))_{n\in\mathbb{N}}$ admits no convergent subsequence (because

$\|T(u_n) - T(u_m)\|_F \geq \varepsilon$ if $n \neq m$), which contradicts the compactness assumption of T.

4. For all $u \in I$, the sequence $(\langle T^t g_n, u \rangle_{E',E})_{n \in \mathbb{N}}$ is bounded in $\mathbb{R}$ and therefore it admits a convergent subsequence. Since I is countable, the diagonal process, described for example in the proof of proposition 8.19 of [26], allows us to extract a subsequence such that the sequence $(\langle T^t g_n, u \rangle_{E',E})_{n \in \mathbb{N}}$ is convergent for all $u \in I$. In the following let f_u denote this limit. Note that, for this question, it is sufficient that I_p is finite or countable.

5. Let $u \in B_E$. We then notice that the sequence $(\langle T^t g_n, u \rangle_{E',E})_{n \in \mathbb{N}}$ is a Cauchy sequence. Indeed, let $\varepsilon > 0$ and choose $p \in \mathbb{N}^\star$ such that $\frac{1}{p} \leq \varepsilon$. There exists a $v \in I_p$ such that $\|Tv - Tu\|_F \leq \varepsilon$. We then have, for all $n, m \in \mathbb{N}$, with $C = \sup_n \|g_n\|_{F'}$,

$$\begin{aligned}
|\langle T^t g_n, u \rangle_{E',E} - \langle T^t g_m, u \rangle_{E',E}| &\leq |\langle T^t g_n, v \rangle_{E',E} - \langle T^t g_m, v \rangle_{E',E}| \\
&\quad + |\langle g_n, Tv - Tu \rangle_{F',F}| + |\langle g_m, Tv - Tu \rangle_{F',F}| \\
&\leq |\langle T^t g_n, v \rangle_{E',E} - \langle T^t g_m, v \rangle_{E',E}| + 2C\varepsilon.
\end{aligned}$$

Then, since $v \in I_p \subset I$, there exists an n_0 such that

$$|\langle T^t g_n, v \rangle_{E',E} - \langle T^t g_m, v \rangle_{E',E}| \leq \varepsilon$$

for $n, m \geq n_0$. Hence, for $n, m \geq n_0$,

$$|\langle T^t g_n, u \rangle_{E',E} - \langle T^t g_m, u \rangle_{E',E}| \leq (2C + 1)\varepsilon. \tag{1.30}$$

This clearly shows that the sequence $(\langle T^t g_n, u \rangle_{E',E})_{n \in \mathbb{N}}$ is a Cauchy sequence. If $u \in E$, $u \neq 0$, we reduce to the previous case by dividing u by its norm. We thus obtain the convergence of the sequence $(\langle T^t g_n, u \rangle_{E',E})_{n \in \mathbb{N}}$ for all $u \in E$ and we still denote this limit by f_u. The mapping $u \mapsto f_u$ is trivially linear (from E to $\mathbb{R}$) because it is the limit of linear mappings. But it is also continuous because $|f_u| \leq C \|T\|_{\mathcal{L}(E,F)} \|u\|_E$. Therefore, there exists an $f \in E'$ such that $f_u = \langle f, u \rangle_{E',E}$ for all $u \in E$.

This question shows that $T^t g_n \to f$ $\star$-weakly in E' as $n \to +\infty$.

6. It is for this question that we use that I_p is finite. We use the method from the previous question.

Let $\varepsilon > 0$ and choose $p \in \mathbb{N}^\star$ such that $\frac{1}{p} \leq \varepsilon$. Let $u \in B_E$. There exists a $v \in I_p$ such that $\|Tv - Tu\|_F \leq \varepsilon$. The inequality (1.30) then gives

$$|\langle T^t g_n, u \rangle_{E',E} - \langle T^t g_m, u \rangle_{E',E}| \leq (2C + 1)\varepsilon, \tag{1.31}$$

for $n, m \geq n_0$. But, as I_p is finite, n_0 can be chosen independently of v (and therefore of u). We thus obtain, when $m \to +\infty$ in (1.31), for all $n \geq n_0$ and all $u \in B_E$,

$$|\langle T^t g_n, u \rangle_{E',E} - \langle f, u \rangle_{E',E}| \leq (2C + 1)\varepsilon,$$

and therefore, for all $n \geq n_0$, $\|T^t g_n - f\|_{E'} \leq (2C + 1)\varepsilon$. We have indeed shown that $T^t g_n \to f$ in E'.

7. We showed that from any bounded sequence in F' we can extract a subsequence whose image under T^t converges in E'. This indeed shows that T^t is a compact operator.

8. Let Ω be an open bounded subset of $\mathbb{R}^N$ ($N \geq 1$) and $1 < q < +\infty$. Theorem 1.36 shows that the mapping $T : u \mapsto u$ from $W_0^{1,p}(\Omega)$ to $L^p(\Omega)$ is compact. The operator T^t from $L^p(\Omega)'$ to $W_0^{1,p}(\Omega)'$ is therefore compact. The space $L^p(\Omega)'$ is identified with $L^q(\Omega)$, where $q = \frac{p}{p-1}$, and $W_0^{1,p}(\Omega)'$ is denoted $W^{-1,q}(\Omega)$. For $u \in (L^p)'(\Omega)$ and $v \in W_0^{1,p}(\Omega)$, as $L^p(\Omega)'$ is identified with $L^q(\Omega)$,

$$\langle T^t u, v \rangle_{W^{-1,q}, W_0^{1,p}} = \langle u, Tv \rangle_{(L^p)', L^p} = \int_\Omega u(x) v(x) \, \mathrm{d}x,$$

i.e. $T^t u$ is identified with u. The mapping $u \mapsto u$ is therefore compact from $L^q(\Omega)$ to $W^{-1,q}(\Omega)$.

Problem 1.27 (Transpose of a continuous embedding)

Let T be the operator from E to F defined by $T(u) = u$. The transposed operator T^t (defined in Problem 1.26) belongs to $\mathcal{L}(F', E')$. For all $f \in F'$, $T^t f$ is defined by

$$\langle T^t f, u \rangle_{E', E} = \langle f, Tu \rangle_{F', F} = \langle f, u \rangle_{F', F} \text{ for all } u \in E.$$

The element $T^t f$ of E' is therefore simply the restriction of f to E. The operator T^t is therefore the operator which to f (element of F') associates its restriction to E (element of E'). We still denote by f the restriction of f to E. The mapping $f \mapsto f$ is therefore continuous from F' to E', i.e. F' is continuously embedded in E'.

Chapter 2
Linear Elliptic Problems

Elliptic PDEs are used to describe a wide variety of phenomena, in particular the steady-state regimes of physical problems such as heat conduction. The basic example of an elliptic PDE is the steady-state heat equation in two space dimensions, $-\Delta u := -\partial^2_{xx} u(x, y) - \partial^2_{yy} u(x, y) = f(x, y)$, where $u(x, y)$ is the temperature at position (x, y) of a given two-dimensional domain and f is a source term (here we have chosen the thermal diffusivity equal to 1). The operator Δ is called the *Laplacian*. The theory of elliptic equations is now well developed. In particular, the question of the regularity of their solutions has been a major issue in research on elliptic PDEs since the mid 20th century, and we refer to [22] for a thorough study of this question. In this chapter, we mainly focus on results concerning weak solutions of linear elliptic PDEs. One of the advantages of considering weak solutions rather than classical solutions (i.e., solutions of the PDE considering derivatives in the classical sense) is that the proof of existence is greatly facilitated by the fact that the notion of solution and the notion of derivative are weakened.

The outline of this chapter is the following: the *weak formulation* of a linear elliptic equation is first addressed, as well as the result of existence and uniqueness of the weak solution in the case of homogeneous Dirichlet conditions. The spectral analysis of elliptic equations is then introduced. Regularity results of weak solutions are then presented in Section 2.3. An important point is the positivity of the solutions of the Dirichlet problem which is the subject of Section 2.4. The study of linear elliptic equations with non-homogeneous Dirichlet conditions concludes this chapter.

2.1 Homogeneous Dirichlet Boundary Conditions

At first glance, the solutions of the one-dimensional steady-state heat equation $-u'' = f$, with f continuous, should be at least twice continuously differentiable. This is what is called a *strong solution* or classical solution of the equation, which is itself often referred to as a *strong formulation*. Unfortunately, PDEs do not always have strong solutions. The good news is that they often have solutions referred

© The Author(s), under exclusive license to Springer Nature Switzerland AG 2025
T. Gallouët, R. Herbin, *Weak Solutions of Partial Differential Equations*,
Mathématiques et Applications 90, https://doi.org/10.1007/978-3-031-98982-7_2

to as *weak*. Indeed, the strong formulation can be weakened by multiplying the equation by a function, often called a *test function*, and by performing integrations by parts. Consider for example the heat equation $-u'' = f$, posed on $]0, 1[$, and impose $u(0) = u(1) = 0$. A weak formulation can be obtained by the following formal process (in the sense that the mathematical justifications are exposed later): multiplying the equation by a sufficiently regular function v, integrating between 0 and 1 and integrating by parts gives

$$\int_0^1 u'(x)v'(x)\,\mathrm{d}x = \int_0^1 f(x)v(x)\,\mathrm{d}x. \tag{2.1}$$

Equation (2.1) is a *weak formulation* in the sense that it only involves the first derivative of u, moreover in an integral form, so that it makes sense for functions u significantly less regular than the strong solution, the latter requiring a C^2 regularity. In the following, we rigorously define the concept of *weak solution* for linear elliptic problems. In the example considered here, a function u is a weak solution if it belongs to the Sobolev space $H_0^1(]0, 1[)$ and it satisfies (2.1) for any test function $v \in H_0^1(]0, 1[)$.

Let us now introduce the complete framework of the linear elliptic problems studied in this chapter, of which the heat equation is an example. Let Ω be an open bounded subset of $\mathbb{R}^N$ ($N \geq 1$) with boundary $\partial\Omega = \overline{\Omega} \setminus \Omega$. Let $a_{i,j} \in L^\infty(\Omega)$, for $i, j = 1, \ldots, N$ and assume that the functions $a_{i,j}$ satisfy the uniform ellipticity assumption, that is:

$$\exists \alpha > 0 \quad \forall \xi = (\xi_1, \ldots, \xi_N) \in \mathbb{R}^N, \ \sum_{i,j=1}^N a_{i,j}\xi_i\xi_j \geq \alpha|\xi|^2 \ \text{ a.e. in } \Omega. \tag{2.2}$$

For given functions $f \in L^2(\Omega)$ and $g : \partial\Omega \to \mathbb{R}$, we seek a solution to the problem:

$$-\sum_{i=1}^N \sum_{j=1}^N \partial_i(a_{i,j}(x)\partial_j u)(x) = f(x), \ x \in \Omega, \tag{2.3a}$$

$$u(x) = g(x), \ x \in \partial\Omega, \tag{2.3b}$$

where $\partial_i u$ denotes the partial derivative of u with respect to its i-th variable. The boundary condition (2.3b) is called the Dirichlet[1] boundary condition. It is said to be homogeneous if $g = 0$ and non-homogeneous otherwise.

Example 2.1 (The Laplacian). If we take $a_{i,j} = \delta_{i,j}$ (that is, 1 if $i = j$, 0 if $i \neq j$), then the problem (2.3) becomes

$$-\Delta u = f \text{ on } \Omega,$$

$$u = g \text{ on } \partial\Omega,$$

---

[1] Johann Peter Gustav Lejeune Dirichlet (1805–1859), Prussian mathematician specialising in particular in number theory and mathematical analysis.

where Δ is the Laplace[2] operator, also named the Laplacian, defined by

$$\Delta u = \sum_{i=1}^{N} \partial_i^2 u, \tag{2.4}$$

and $\partial_i^2 u$ denotes the second partial derivative of u with respect to the i-th space variable x_i.

Definition 2.2 (Classical Solution). Assume that $a_{i,j} \in C^1(\bar{\Omega})$ for all $i, j = 1, \ldots, N$ and assume that $f \in C(\bar{\Omega})$ and $g \in C(\partial\Omega)$. A classical solution to (2.3) is a function $u \in C^2(\bar{\Omega})$ satisfying (2.3).

Recall that for all $k \in \mathbb{N} \cup \{+\infty\}$, $C^k(\bar{\Omega})$ denotes the set of restrictions to Ω of the functions belonging to $C^k(\mathbb{R}^N)$ (see Definition 1.28).

There does not necessarily exist a classical solution to (2.3), but there exist solutions in a weaker sense that we define below. To understand the nature of weak solutions, let us first consider the case $g = 0$, with $a_{i,j} \in C^1(\bar{\Omega})$ and $f \in C(\bar{\Omega})$, and suppose that there exists a classical solution $u \in C^2(\bar{\Omega})$. It satisfies:

$$-\sum_{i=1}^{N} \sum_{j=1}^{N} \partial_i(a_{i,j}(x)\partial_j u)(x) = f(x), \ \forall x \in \Omega.$$

Let $\varphi \in \mathcal{D}(\Omega)$; multiply the previous equation by $\varphi(x)$ and integrate over Ω:

$$-\int_{\Omega} \left(\sum_{i=1}^{N} \sum_{j=1}^{N} \partial_i(a_{i,j}(x)\partial_j u)(x) \right) \varphi(x)\, \mathrm{d}x = \int_{\Omega} f(x)\varphi(x)\, \mathrm{d}x, \ \forall \varphi \in \mathcal{D}(\Omega).$$

An integration by parts then gives:

$$\int_{\Omega} \left(\sum_{i=1}^{N} \sum_{j=1}^{N} a_{i,j}(x)\partial_j u(x)\partial_i \varphi(x) \right) \mathrm{d}x = \int_{\Omega} f(x)\varphi(x)\, \mathrm{d}x, \ \forall \varphi \in \mathcal{D}(\Omega). \tag{2.5}$$

Since $u \in C^2(\bar{\Omega})$, we have $\partial_j u \in C^1(\overline{\Omega}) \subset C(\overline{\Omega}) \subset L^2(\Omega)$, and $D_j u = \partial_j u$ a.e (as seen in Chapter 1). Moreover $u \in C(\overline{\Omega}) \subset L^2(\Omega)$ and therefore $u \in H^1(\Omega)$. Finally, since $u = 0$ on $\partial\Omega$, one has $u \in H_0^1(\Omega)$, at least for Ω regular enough (see Problem 1.20).

Let $v \in H_0^1(\Omega)$, by density of $\mathcal{D}(\Omega)$ in $H_0^1(\Omega)$, there exists a sequence $(\varphi_n)_{n \in \mathbb{N}} \subset \mathcal{D}(\Omega)$ such that $\varphi_n \to v$ in $H^1(\Omega)$, that is to say $\varphi_n \to v$ in $L^2(\Omega)$ and $\partial_i \varphi_n \to D_i v$ in $L^2(\Omega)$ for $i = 1, \ldots, N$. By writing (2.5) with $\varphi = \varphi_n$, we obtain:

---

[2] Pierre-Simon de Laplace (1749–1827), mathematician, astronomer, physicist and French politician, very influential in science and politics during the Napoleonic era.

$$\int_{\Omega}\left(\sum_{i=1}^{N}\sum_{j=1}^{N} a_{i,j}(x)\partial_j u(x)\partial_i \varphi_n(x)\right) dx = \int_{\Omega} f(x)\varphi_n(x)\, dx;$$

passing to the limit as $n \to +\infty$, we find that u satisfies the *weak formulation* of the problem (2.3), which reads (recall that we are considering here the case $g = 0$):

$$u \in H_0^1(\Omega),$$
$$\int_{\Omega}\left(\sum_{i=1}^{N}\sum_{j=1}^{N} a_{i,j}(x)D_j u(x)D_i v(x)\right) dx = \int_{\Omega} f(x)v(x)\, dx,\ \forall v \in H_0^1(\Omega). \tag{2.6}$$

Therefore, *any classical solution to problem (2.3) (when $g = 0$) is a weak solution, that is to say, satisfies (2.6)*. The existence and uniqueness of the solution to problem (2.6) follows from the Lax[3]–Milgram[4] theorem.

Theorem 2.3 (Lax–Milgram). *Let H be a real Hilbert space equipped with the inner product denoted $(\cdot|\cdot)$, with associated norm denoted $\|\cdot\|$, and $a(\cdot,\cdot)$ a bilinear mapping from $H \times H$ to $\mathbb{R}$ which is*

- *continuous, which is equivalent to saying that there exists a $c > 0$ such that, for all $(u,v) \in H^2$, we have $|a(u,v)| \le c\|u\|\|v\|$,*
- *coercive on H (some authors prefer the term H-elliptic), that is to say there exists an $\alpha > 0$ such that, for all $u \in H$, $a(u,u) \ge \alpha\|u\|^2$,*

and let T be a continuous linear form on H. Then there exists a unique u in H such that the equation $a(u,v) = T(v)$ is satisfied for all v in H:

$$\exists!\, u \in H,\ \forall v \in H,\quad a(u,v) = T(v). \tag{2.7}$$

If in addition the bilinear form a is symmetric, then u is the unique element of H that minimises the functional $J : H \to \mathbb{R}$ defined by $J(v) = \frac{1}{2}a(v,v) - T(v)$ for all v in H, that is to say:

$$J(u) = \min_{v \in H} J(v)\ \text{ and }\ J(u) < J(v)\ \text{if}\ u \ne v.$$

See for instance [26, Problem 6.35] for the proof of existence and uniqueness of u in Theorem 2.3.

Remark 2.4 (Symmetric case). Under the assumptions of the Lax–Milgram theorem, if the bilinear form a is symmetric, it defines an inner product on H equivalent to the initial inner product (this is an immediate consequence of the continuity and coercivity of a). In this case, the existence and uniqueness of the solution u to

---

[3] Peter Lax, contemporary American mathematician of Hungarian origin, Abel prize laureate, who made several important contributions in PDEs and scientific computing.

[4] Arthur Norton Milgram (1912–1961), American mathematician, who worked in functional analysis, in combinatorics, in differential geometry, in general topology, in PDEs and in Galois theory.

(2.7) (given in the Lax–Milgram theorem) is a direct consequence of the Riesz[5] representation theorem[6] in a Hilbert space.

To apply the Lax–Milgram theorem to problem (2.6), we need the Poincaré[7] inequality.

Lemma 2.5 (Poincaré's Inequality). *Let Ω be an open bounded subset of $\mathbb{R}^N$, $N \geq 1$.[8] Then there exists a C_Ω depending only on Ω such that*

$$\|u\|_{L^2(\Omega)} \leq C_\Omega \| \, |\nabla u| \, \|_{L^2(\Omega)}, \ \forall u \in H_0^1(\Omega). \tag{2.8}$$

This inequality implies that the mapping

$$u \mapsto \|u\|_{H_0^1(\Omega)} =: \| \, |\nabla u| \, \|_{L^2(\Omega)},$$

where $|\cdot|$ denotes the Euclidean norm in $\mathbb{R}^N$, is a norm on $H_0^1(\Omega)$, equivalent to the norm $\|\cdot\|_{H^1(\Omega)}$. Be careful, this can be false if Ω is unbounded.

In Lemma 2.5 note that $\| \, |\nabla u| \, \|_{L^2(\Omega)}^2 = \int_\Omega |\nabla u(x)|^2 \, \mathrm{d}x = \sum_{i=1}^N \int_\Omega D_i u(x)^2 \, \mathrm{d}x$.

Proof By hypothesis on Ω, there exists an $a > 0$ such that $\Omega \subset]-a, a[\times \mathbb{R}^{N-1}$. Let $u \in \mathcal{D}(\Omega)$ and extend u by 0 outside of Ω, then:

$$u \in \mathcal{D}(\mathbb{R}^N), u = 0 \text{ on } \Omega^c.$$

Let $x = (x_1, \ldots, x_N)^t = (x_1, y)^t \in \Omega$, with $x_1 \in]-a, a[$ and $y = (x_2, \ldots, x_N) \in \mathbb{R}^{N-1}$. We have:

$$u(x_1, y) = \int_{-a}^{x_1} \partial_1 u(t, y) \, \mathrm{d}t,$$

and therefore, by the Cauchy–Schwarz inequality,

$$|u(x_1, y)|^2 \leq \left(\int_{-a}^{a} |\partial_1 u(t, y)| \, \mathrm{d}t \right)^2 \leq 2a \int_{-a}^{a} (\partial_1 u(t, y))^2 \, \mathrm{d}t.$$

Integrating between $-a$ and a yields:

$$\int_{-a}^{a} |u(x_1, y)|^2 \, \mathrm{d}x_1 \leq 4a^2 \int_{-a}^{a} (\partial_1 u(t, y))^2 \, \mathrm{d}t,$$

---

[5] Frigyes Riesz (1880–1956), Hungarian mathematician. He is one of the founders of functional analysis.

[6] Riesz representation theorem: Let H be a Hilbert space equipped with its inner product denoted $(\cdot|\cdot)$ and $T \in H'$ a continuous linear form on H. Then there exists a unique $y \in H$ such that for all $x \in H$ we have $\langle T, x \rangle_{H', H} = (x|y)$.

[7] Henri Poincaré (1854–1912), French mathematician, theoretical physicist and philosopher of science, author of important results in optics and infinitesimal calculus, and precursor of the theory of dynamical systems.

[8] As seen in the proof, it is sufficient that there exists an $a > 0$ such that $\Omega \subset]-a, a[\times \mathbb{R}^{N-1}$, i.e. that Ω is bounded in one direction.

and therefore, integrating with respect to y,

$$\int_\Omega |u(x)|^2 \, dx \le 4a^2 \int_\Omega (\partial_1 u(x))^2 \, dx, \quad \forall u \in \mathcal{D}(\Omega). \tag{2.9}$$

Let us then proceed by density; for $u \in H_0^1(\Omega)$, there exists a sequence $(u_n)_{n \in \mathbb{N}} \subset \mathcal{D}(\Omega)$ such that $u_n \to u$ in $H_0^1(\Omega)$. Therefore, we have $u_n \to u$ in $L^2(\Omega)$ and $\partial_i u \to D_i u$ in $L^2(\Omega)$. Writing (2.9) for u_n and taking the limit as $n \to +\infty$ leads to:

$$\|u\|_{L^2(\Omega)}^2 \le 4a^2 \|D_1 u\|_{L^2(\Omega)}^2 \le 4a^2 \sum_{i=1}^N \|D_i u\|_{L^2(\Omega)}^2 = 4a^2 \| |\nabla u| \|_{L^2(\Omega)}^2 .$$

$\blacksquare$

Theorem 2.6 (Existence and uniqueness of the solution to (2.6)). *Let Ω be an open bounded subset of $\mathbb{R}^N$, $f \in L^2(\Omega)$, and let $(a_{i,j})_{i,j=1,\ldots,N} \subset L^\infty(\Omega)$ and $\alpha > 0$ such that (2.2) is satisfied. Then there exists a unique solution to (2.6).*

Proof The problem (2.6) reads:

$$u \in H,$$
$$a(u, v) = T(v) \text{ for all } v \in H,$$

with $H = H_0^1(\Omega)$ (which is indeed a Hilbert space, equipped with the norm defined by $\|u\|_{H^1(\Omega)} = (\|u\|_{L^2(\Omega)}^2 + \| |\nabla u| \|_{L^2(\Omega)}^2)^{\frac{1}{2}}$), and with a and T defined by

$$a(u, v) = \int_\Omega \left(\sum_{i=1}^N \sum_{j=1}^N a_{i,j}(x) D_j u(x) D_i v(x) \right) dx \text{ and } T(v) = \int_\Omega f(x) v(x) \, dx.$$

Observe that the linear form T is continuous. Indeed,

$$|T(v)| \le \|v\|_{L^2(\Omega)} \|f\|_{L^2(\Omega)} \le \|v\|_{H^1(\Omega)} \|f\|_{L^2(\Omega)} .$$

The form a is also obviously bilinear, and satisfies:

$$|a(u, v)| \le \sum_{i,j=1}^N \|a_{i,j}\|_{L^\infty(\Omega)} \|D_i u\|_{L^2(\Omega)} \|D_j v\|_{L^2(\Omega)}$$
$$\le C \|u\|_{H^1(\Omega)} \|v\|_{H^1(\Omega)},$$

with $C = \sum_{i,j=1}^N \|a_{i,j}\|_{L^\infty(\Omega)}$. It is therefore continuous.

Let us check that a is coercive: we need to show that there exists a $\beta \in \mathbb{R}_+$ such that $a(u, u) \ge \beta \|u\|_{H^1(\Omega)}^2$, for all $u \in H_0^1(\Omega)$. By hypothesis on a, we have:

$$a(u, u) = \int_{\Omega} \left(\sum_{i=1}^{N} \sum_{j=1}^{N} a_{i,j}(x) D_j u(x) D_i u(x) \right) dx$$

$$\geq \alpha \int_{\Omega} \left(\sum_{i=1}^{N} |D_i u(x)|^2 \right) dx = \alpha \int_{\Omega} |\nabla u(x)|^2 \, dx.$$

We then apply Poincaré's inequality (2.8):

$$\|u\|_{H^1(\Omega)}^2 = \|u\|_{L^2(\Omega)}^2 + \sum_{i=1}^{N} \|D_i u\|_{L^2(\Omega)}^2 \leq (C_{\Omega}^2 + 1) \sum_{i=1}^{N} \|D_i u\|_{L^2(\Omega)}^2,$$

so:

$$\sum_{i=1}^{N} \|D_i u\|_{L^2(\Omega)}^2 \geq \frac{1}{C_{\Omega}^2 + 1} \|u\|_{H^1(\Omega)}^2,$$

and therefore

$$a(u, u) \geq \alpha \sum_{i=1}^{N} \|D_i u\|_{L^2(\Omega)}^2 \geq \frac{\alpha}{C_{\Omega}^2 + 1} \|u\|_{H^1(\Omega)}^2,$$

which proves the coercivity of a. The Lax–Milgram theorem (Theorem 2.3) then gives the existence and uniqueness of the solution to problem (2.6). ∎

In Theorem 2.6, if $a_{i,j} = a_{j,i}$ a.e. for $i \neq j$, it follows from the Lax–Milgram theorem (since the bilinear form a is symmetric) that u is a solution to (2.6) if and only if u is a solution of the following *variational formulation*:

$$u \in H_0^1(\Omega),$$
$$J(u) \leq J(v), \forall v \in H_0^1(\Omega), \tag{2.10}$$

where the functional J is defined by:

$$J(v) = \frac{1}{2} \int_{\Omega} \left(\sum_{i=1}^{N} \sum_{j=1}^{N} a_{i,j} D_i v D_j v \right) dx - \int_{\Omega} f(x) v(x) \, dx.$$

Remark 2.7. Lemma 2.5 is still true with $1 \leq p \leq +\infty$ instead of $p = 2$. If Ω is an open bounded subset of $\mathbb{R}^N$, $N \geq 1$, and $1 \leq p \leq +\infty$, there exists a $C_{p,\Omega}$ depending only on p and Ω such that

$$\|u\|_{L^p(\Omega)} \leq C_{p,\Omega} \| \, |\nabla u| \, \|_{L^p(\Omega)}, \ \forall u \in W_0^{1,p}(\Omega).$$

This allows us to define a norm on $W_0^{1,p}(\Omega)$ equivalent to the norm $W^{1,p}(\Omega)$, see Definition 2.8. In the case $p = 2$, this equivalence of norms follows from the proof of Theorem 2.6.

Definition 2.8 (Norm on $W_0^{1,p}(\Omega)$). Let Ω be an open bounded subset of $\mathbb{R}^N$ ($N \geq 1$) and $1 \leq p < +\infty$. For $u \in W_0^{1,p}(\Omega)$, set

$$\|u\|_{W_0^{1,p}(\Omega)} = \Big(\int_\Omega |\nabla u(x)|^p \, \mathrm{d}x \Big)^{\frac{1}{p}}.$$

According to Remark 2.7, this is a norm on $W_0^{1,p}(\Omega)$ that is equivalent to the norm on $W^{1,p}(\Omega)$. For $p = 2$, the space $W_0^{1,2}(\Omega)$ is also denoted $H_0^1(\Omega)$ and the norm $\|\cdot\|_{W_0^{1,2}(\Omega)}$ is the norm $\|\cdot\|_{H_0^1(\Omega)}$.

If the right-hand side of (2.6) belongs to the dual space $H^{-1}(\Omega)$ of $H_0^1(\Omega)$ (that is to say the set of continuous linear forms on $H_0^1(\Omega)$) the existence and uniqueness of the solution again follow by the Lax–Milgram theorem.

Theorem 2.9 (Existence and uniqueness, $T \in H^{-1}$). *Let Ω be an open bounded subset of $\mathbb{R}^N$, $(a_{i,j})_{i,j=1,\ldots,N} \subset L^\infty(\Omega)$ and $\alpha > 0$ such that (2.2) is satisfied. Let $T \in H^{-1}(\Omega)$, there then exists a unique solution u of:*

$$u \in H_0^1(\Omega),$$

$$\int_\Omega \left(\sum_{i=1}^N \sum_{j=1}^N a_{i,j}(x) D_j u(x) D_i v(x) \right) \mathrm{d}x = T(v), \forall v \in H_0^1(\Omega). \tag{2.11}$$

Right-hand sides in L^q may also be considered. For instance, let Ω be an open bounded subset of $\mathbb{R}^N$ and $f \in L^q(\Omega)$. If the mapping $\varphi \mapsto \int_\Omega f(x)\varphi(x)\,\mathrm{d}x$ (defined, for example, for $\varphi \in \mathcal{D}(\Omega)$) extends to an element of $H^{-1}(\Omega)$ (and in this case, the extension is unique by density of $\mathcal{D}(\Omega)$ in $H_0^1(\Omega)$), then existence and uniqueness of the solution to problem (2.6) can again be obtained by Theorem 2.9. If $N = 1$, the hypothesis $f \in L^1(\Omega)$ is sufficient. If $N \geq 3$, the hypothesis $f \in L^q(\Omega)$, with $q = \frac{2N}{N+2}$, is sufficient. If $N = 2$, the hypothesis $f \in L^q(\Omega)$, with $q > 1$, is sufficient. A more precise result (for $N = 2$) is given in Problem 2.14.

We have only dealt here with the case of homogeneous Dirichlet boundary conditions (that is, $g = 0$ in problem (2.3)). The case of non-homogeneous Dirichlet boundary conditions is treated in Section 2.5.

The existence and uniqueness of weak solutions is possible with other boundary conditions. Problem 2.8 deals with the case of Neumann[9] conditions and Problem 2.12 with so-called Fourier[10] (or Robin[11], according to the authors) conditions. The resolution of the Neumann problem also leads to a useful decomposition of an

---

[9] Carl Gottfried Neumann (1832–1925), German mathematician who worked notably on the Dirichlet principle and on the theory of integral equations.

[10] Jean-Baptiste Joseph Fourier (1768–1830), French mathematician and physicist, known in particular for having determined, by calculation, the diffusion of heat using the decomposition of any function into a convergent trigonometric series (Fourier series).

[11] Victor Gustave Robin (1855–1897), French mathematician, specialist in thermodynamics and potential theory.

element of $L^2(\Omega)^N$, called the Hodge[12] decomposition, see Problem 2.15. Problem 2.11 concerns boundary conditions appearing in solid mechanics. It is also possible to couple an elliptic problem on an open subset Ω of $\mathbb{R}^2$ with a one-dimensional elliptic problem on the boundary of Ω, which is the subject of Problem 2.16.

Problems 2.17 and 2.13 prove the existence (and uniqueness or a "partial uniqueness") for elliptic systems (Stokes problem and the Schrödinger[13] equation).

Finally, it is possible to deal with elliptic problems with unbounded coefficients $a_{i,j}$. We then introduce the so-called *weighted* Sobolev spaces, see Problem 2.6.

2.2 Spectral Theory

Spectral theory is a generalisation of the theory of eigenvalues and eigenvectors (reduction of endomorphisms) to infinite-dimensional linear equations. It allows the resolution of certain linear PDEs arising from physics. Here we focus on the study of self-adjoint equations of a real Hilbert space, which are the infinite-dimensional counterpart of symmetric endomorphisms on $\mathbb{R}^n$ (represented by real symmetric $n \times n$ matrices) in finite dimension. An application of this theory allows us in particular to identify a *Hilbert basis* of $L^2(\Omega)$ consisting of the eigenfunctions of the Laplacian.

Definition 2.10 (Regular values, spectral values, eigenvalues). Let E be a real Banach space, and T a continuous linear mapping from E to E and consider the following sets:

- The set $\rho(T) = \{\lambda \in \mathbb{R} \mid T - \lambda \operatorname{Id}$ is bijective$\}$ is the set of *regular values* of T.
- The set $\sigma(T) = \{\lambda \in \mathbb{R} \mid T - \lambda \operatorname{Id}$ is non-bijective$\} = \mathbb{R} \setminus \sigma(T)$ is the set of *non-regular values* (or *spectral values*) of T. This set is also called the *spectrum* of the operator.
- $\mathcal{VP}(T) = \{\lambda \in \mathbb{R} \mid T - \lambda \operatorname{Id}$ is non-injective$\}$ is the set of *eigenvalues* of T.

Here and throughout the rest of this book, Id denotes the identity mapping of a set into itself.

Remark 2.11 (Finite dimension). Note that if E is of finite dimension, a consequence of the rank theorem is that the notions of non-regular value and eigenvalue coincide. Hence in this case $\sigma(T) = \mathcal{VP}(T)$.

Note that if $\lambda \in \rho(T)$, the mapping $T - \lambda \operatorname{Id}$ is continuous; this follows from Banach's theorem.

Theorem 2.12 (Banach). *Let E and F be Banach spaces and T a continuous linear mapping from E to F. If T is bijective and continuous, then T^{-1} is continuous.*

---

[12] William Vallance Douglas Hodge (1903–1975), British mathematician who specialised in geometry.

[13] Erwin Rudolf Josef Alexander Schrödinger (1887–1961), Austro-Irish physicist who is credited with many results in quantum mechanics.

It is often important to generalise the definitions of $\rho(T)$, $\sigma(T)$ and $\mathcal{VP}(T)$ by taking $\lambda \in \mathbb{C}$ instead of $\lambda \in \mathbb{R}$. However it is unnecessary in this section since only self-adjoint operators in real Hilbert spaces are considered.

Definition 2.13 (Transposed operator, adjoint and self-adjoint, compact operator). Let E and F be Banach spaces and T a continuous linear mapping from E to F. The *transpose* T^t of the operator T, defined by $\langle T^t g, u \rangle_{E',E} = \langle g, Tu \rangle_{F',F}$ is a continuous linear mapping from F' to E' (see Problem 1.26). We say that the mapping T is *compact* if from any bounded sequence in E we can extract a subsequence whose image under T converges in F.

In the case where $E = F$ is a Hilbert space, equipped with the inner product $(\cdot|\cdot)_E$, the space E' can be identified with E by the Riesz representation theorem, see footnote 6 page 71. The transpose operator is then an operator from E to E, it is called the adjoint operator of T and is denoted T^*; it is therefore defined by $(T^* x|y)_E = (x|Ty)_E$ (see, for example, [25] definition 5.2). We say that the operator T is self-adjoint if it is identical to its adjoint T^*, and we then have $(Tx|y)_E = (x|Ty)_E$. In the case of a real Hilbert space E, we also say that the operator is symmetric.

If E is a finite-dimensional Hilbert space and T a self-adjoint linear operator from E to E, then we have $\mathcal{VP}(T) = \sigma(T)$. We have a similar result in infinite dimension, provided that the operator T is compact linear and self-adjoint. More precisely, in this case we have: $\mathcal{VP}(T) \setminus \{0\} = \sigma(T) \setminus \{0\}$. A consequence of this result is Proposition 2.14. It gives a spectral decomposition property for separable Hilbert spaces and for a self-adjoint operator.

Proposition 2.14 (Compact self-adjoint continuous linear operator). *Let E be a separable Hilbert space equipped with the inner product $(\cdot, \cdot)_E$, and let T be a compact self-adjoint continuous linear operator. Then*

1. $\mathcal{VP}(T) \setminus \{0\} = \sigma(T) \setminus \{0\}$.
2. *If $\mathcal{VP}(T) \setminus \{0\}$ has infinite cardinality, $\mathcal{VP}(T) \setminus \{0\} = \{\lambda_n, n \in \mathbb{N}^\star\}$ with $\lambda_n \to 0$ when $n \to +\infty$.*
3. *There exists a Hilbert basis of E formed of eigenvectors of T, that is to say elements of E, denoted e_n, $n \in \mathbb{N}$, such that $T(e_n)$ is collinear to e_n, $(e_n | e_m)_E = \delta_{n,m}$ and if $u \in E$, then u can be written as $u = \sum_{n \in \mathbb{N}} (u | e_n)_E e_n$, this series being convergent in E.*

In the previous proposition, the eigenspaces associated with non-zero eigenvalues are all of finite dimension. In the case of a non-separable Hilbert space and a compact self-adjoint continuous linear operator T, the kernel of T is of infinite dimension and non-separable.

We now consider, for simplicity, the case of the Laplacian. Let Ω be an open bounded subset of $\mathbb{R}^N$. Recall that for a sufficiently regular function u, its Laplacian is defined by $\Delta u = \sum_{i=1}^N \partial_i^2 u$. To extend this definition to only locally integrable functions, set $\Delta u = \sum_{i=1}^N D_i^2 u$ for $u \in L^1_{\text{loc}}(\Omega)$. We now define an operator $\mathcal{A}$ from a subset of $L^2(\Omega)$ to $L^2(\Omega)$ by first defining its *domain* $D(\mathcal{A})$:

$$D(\mathcal{A}) = \{u \in H_0^1(\Omega) \, ; \, \Delta u \in L^2(\Omega)\}, \qquad (2.12a)$$

$$\text{and for } u \in D(\mathcal{A}), \text{ set } \mathcal{A}u = -\Delta u. \qquad (2.12b)$$

A linear operator $\mathcal{A} : D(\mathcal{A}) \subset L^2(\Omega) \to L^2(\Omega)$ is thus defined.

By Theorem 2.6, if $f \in L^2(\Omega)$, there is a unique solution to problem (2.6) which reads in the case of the Laplacian (that is to say, with the values $a_{i,j} = \delta_{i,j}$, $i, j = 1, \ldots, N$)

$$u \in H_0^1(\Omega),$$

$$\int_\Omega \nabla u(x) \nabla v(x) \, dx = \int_\Omega f(x)v(x) \, dx, \forall v \in H_0^1(\Omega). \qquad (2.13)$$

Thanks to the density of $\mathcal{D}(\Omega)$ in $H_0^1(\Omega)$, the function u is a solution to (2.13) if and only if $u \in D(\mathcal{A})$ and $-\Delta u = f$ a.e. (i.e., $-\Delta u = f$ in $L^2(\Omega)$). The operator $\mathcal{A}$ is therefore invertible. Let $T = \mathcal{A}^{-1}$ be its inverse, which is defined from $L^2(\Omega)$ to $L^2(\Omega)$ by $Tf = u$ where u is a solution to (2.13). This operator is injective but not surjective. Both operators are linear. The proof that there exists a Hilbert basis formed by the eigenvectors of $T = \mathcal{A}^{-1}$ is based on the following theorem:

Theorem 2.15 (Inverse of the Laplace operator). *Let Ω be an open bounded subset of $\mathbb{R}^N$. Let $\mathcal{A}$ be the Laplace operator[14] (with Dirichlet condition) defined by (2.12) The operator $T = \mathcal{A}^{-1}$ is a continuous, compact, self-adjoint linear operator from $L^2(\Omega)$ to $L^2(\Omega)$. Moreover, $\ker(T) = \{f \in L^2(\Omega), Tf = 0 \text{ a.e.}\} = \{0\}$.*

Proof It is immediate to see that T is linear. First note that $\ker(T) = \{f \in E, Tf = 0$ a.e.$\} = \{0\}$. Indeed, let $f \in L^2(\Omega)$ such that $Tf = 0$ a.e. Hence, according to (2.13),

$$\int_\Omega fv \, dx = 0 \text{ for all } v \in H_0^1(\Omega).$$

By density of $H_0^1(\Omega)$ in $L^2(\Omega)$ (in fact, $\mathcal{D}(\Omega)$ is dense in $L^2(\Omega)$), $f = 0$ a.e.

Let us now prove that T is continuous. Let $f \in L^2(\Omega)$ and $u = Tf$; taking $v = u$ in (2.13) yields

$$\|u\|^2_{H_0^1(\Omega)} = \int_\Omega \nabla u \cdot \nabla u \, dx = \int_\Omega fu \, dx \leq \|f\|_{L^2(\Omega)} \|u\|_{L^2(\Omega)}.$$

By the Poincaré inequality, there exists a $C_\Omega \in \mathbb{R}_+$ depending only on Ω such that $\|u\|_{L^2(\Omega)} \leq C_\Omega \|u\|_{H_0^1(\Omega)}$, and therefore:

$$\|u\|^2_{H_0^1(\Omega)} = \sum_{i=1}^{N} \|D_i u\|^2_{L^2(\Omega)} \leq \|f\|_{L^2(\Omega)} \|u\|_{L^2(\Omega)} \leq C_\Omega \|f\|_{L^2(\Omega)} \|u\|_{H_0^1(\Omega)}.$$

---

[14] Note here the abuse of language, as $\mathcal{A}$ is actually the "minus Laplacian" operator.

Hence $\|u\|_{H_0^1(\Omega)} \leq C_\Omega \|f\|_{L^2(\Omega)}$ and so:

$$\|Tf\|_{L^2(\Omega)} = \|u\|_{L^2(\Omega)} \leq C_\Omega \|u\|_{H_0^1(\Omega)} \leq C_\Omega^2 \|f\|_{L^2(\Omega)},$$

which proves the continuity of T.

Let us now show that the operator T is compact, i.e. that the image under T of a bounded subset B of $L^2(\Omega)$ is relatively compact in $L^2(\Omega)$. The operator T may be written $T = I \circ T_0$, where I is the canonical embedding of $H_0^1(\Omega)$ to $L^2(\Omega)$, defined by $I : v \in H_0^1(\Omega) \mapsto v \in L^2(\Omega)$, and T_0 is the mapping which to $f \in L^2(\Omega)$ associates $u = Tf \in H_0^1(\Omega)$. Since $\|Tf\|_{H_0^1(\Omega)} \leq C_\Omega \|f\|_{L^2(\Omega)}$, the mapping T_0 is continuous from $L^2(\Omega)$ to $H_0^1(\Omega)$ and the mapping I is compact by Rellich's theorem (Theorem 1.36 page 15); the operator T is therefore compact.

Let us now show that the operator T is self-adjoint, i.e. that

$$(Tf \mid g)_{L^2(\Omega)} = (f \mid Tg)_{L^2(\Omega)}, \quad \forall f, g \in L^2(\Omega). \tag{2.14}$$

Let f and $g \in L^2(\Omega)$, let u be the unique solution to (2.13) and let v the unique solution to (2.13) where f is replaced by g in the right-hand side. Since v is the solution to (2.13) with f replaced by g,

$$(Tf \mid g)_{L^2(\Omega)} = \int_\Omega Tf \, g \, dx = \int_\Omega u \, g \, dx = \int_\Omega \nabla u \cdot \nabla v \, dx.$$

Similarly, $(f \mid Tg)_{L^2(\Omega)} = \int_\Omega \nabla u \cdot \nabla v \, dx$, which proves (2.14). $\blacksquare$

The following is the consequence of Theorem 2.15 and of Proposition 2.14 for the Laplace operator with homogeneous Dirichlet conditions.

Theorem 2.16 (Hilbert basis of $L^2(\Omega)$ of eigenfunctions of $-\Delta$). *Let Ω be an open bounded subset of $\mathbb{R}^N$ and let $\mathcal{A}$ be the Laplace operator (with homogeneous Dirichlet condition) defined by (2.12). There then exists a (countable) Hilbert basis of $L^2(\Omega)$ consisting of eigenfunctions of $\mathcal{A}$ associated with the eigenvalues $(\mu_n)_{n \in \mathbb{N}^\star}$ which are in ascending order, i.e. $\mu_n \leq \mu_{n+1}$ for all $n \in \mathbb{N}^\star$; moreover, $\mu_1 > 0$ and $\lim_{n \to +\infty} \mu_n = +\infty$.*

Proof For $f \in L^2(\Omega)$, let Tf be the unique solution to (2.13). According to Theorem 2.15 and Proposition 2.14, there exists a Hilbert basis $(e_n)_{n \in \mathbb{N}^\star}$ of $L^2(\Omega)$ consisting of eigenfunctions of T. The associated eigenvalues are all positive. Indeed, if $f \in L^2(\Omega)$ and $f \neq 0$, then $u = Tf \neq 0$ and

$$(Tf \mid f)_{L^2(\Omega)} = (u \mid f)_{L^2(\Omega)} = \int_\Omega \nabla u \cdot \nabla u \, dx = \|u\|_{H_0^1(\Omega)}^2 > 0.$$

If λ_n is an eigenvalue of T associated with the eigenvector $e_n \neq 0$, we have $Te_n = \lambda_n e_n$, and since $e_n \neq 0$,

$$\lambda_n (e_n \mid e_n)_{L^2(\Omega)} = (\lambda_n e_n \mid e_n)_{L^2(\Omega)} = (Te_n \mid e_n)_{L^2(\Omega)} > 0.$$

The sequence $(\lambda_n)_{n\in\mathbb{N}}$ is therefore made up of positive numbers. Proposition 2.14 gives that $\lim_{n\to+\infty}\lambda_n = 0$. This sequence may be assumed to be non-increasing, if necessary by changing its order. Note that the eigenvalues of $\mathcal{A}$ are therefore the values $\mu_n = \frac{1}{\lambda_n}$ for all $n\in\mathbb{N}^\star$, with $\mu_n > 0$ for all $n\in\mathbb{N}^\star$ and $\mu_n \to +\infty$ when $n\to+\infty$. $\blacksquare$

Using the notation of Theorem 2.16, if $\mathcal{A}$ denotes the Laplace operator (in the sense $-\Delta$) with homogeneous Dirichlet conditions defined by (2.12), its domain $D(\mathcal{A})$ can be characterized as follows: Let $u\in L^2(\Omega)$, then $u\in D(\mathcal{A})$ if and only if $\sum_{n\in\mathbb{N}}\mu_n^2(u\,|\,e_n)^2_{L^2(\Omega)} < +\infty$. Moreover, if $u\in D(\mathcal{A})$, then $\mathcal{A}u = \sum_{n=1}^{+\infty}\mu_n(u\,|\,e_n)_{L^2(\Omega)}e_n$. We can thus define the powers of the operator $\mathcal{A}$:

Definition 2.17 (Power of the Laplace operator). Let Ω be an open bounded subset of $\mathbb{R}^N$, and $\mathcal{A}$ the Laplace operator defined by (2.12). Let $(e_n)_{n\in\mathbb{N}^\star}$ be a Hilbert basis of $L^2(\Omega)$ made up of eigenfunctions of $\mathcal{A}$, associated with the eigenvalues $(\mu_n)_{n\in\mathbb{N}^\star}$. Let $s \geq 0$; we define

$$D(\mathcal{A}^s) = \{u\in L^2(\Omega) : \sum_{n=1}^{+\infty}\mu_n^{2s}(u\,|\,e_n)^2_{L^2(\Omega)} < +\infty\}.$$

For $u\in D(\mathcal{A}^s)$, we can then define $\mathcal{A}^s u$ by:

$$\mathcal{A}^s u = \sum_{n=1}^{+\infty}\mu_n^s(u\,|\,e_n)_{L^2(\Omega)}e_n,$$

noting that the series on the right-hand side of this equality is convergent in $L^2(\Omega)$.

If $s = 0$ then $D(\mathcal{A}^0) = L^2(\Omega)$ and $\mathcal{A}^0 u = u$, so $\mathcal{A}^0$ is the identity operator.
If $s = 1$ then $\mathcal{A}^1 = \mathcal{A}$.
If $s = \frac{1}{2}$ then $D(\mathcal{A}^{\frac{1}{2}}) = \{u\in L^2(\Omega) \ \sum_{n=1}^{+\infty}\mu_n(u\,|\,e_n)^2_{L^2(\Omega)} < +\infty\}$.
It may be shown that $D(\mathcal{A}^{\frac{1}{2}}) = H_0^1(\Omega)$, and that $\mathcal{A}^{\frac{1}{2}}u = \sum_{n=1}^{+\infty}\sqrt{\mu_n}(u\,|\,e_n)_{L^2(\Omega)}e_n$.

The spectral decomposition theorem (Theorem 2.16) is detailed in Problem 2.3 in the one-dimensional setting.

2.3 Regularity of Weak Solutions

Let Ω be an open bounded subset of $\mathbb{R}^N$ and $f\in L^2(\Omega)$; under the assumptions (2.2), Theorem 2.6 gives that there exists a unique solution to problem (2.6); the question is now the regularity of this solution, depending on the data of the problem. The problem is quite simple in dimension $N = 1$, see Problem 2.2, but much more difficult in dimension $N > 1$. The answer depends on the regularity of the coefficients of the operator and the regularity of the boundary of the open subset; the boundary of Ω is of class C^k if it is locally the graph of a function of class C^k.

Let us start with the case $\Omega = \mathbb{R}_+^N = \{(x_1, y), y \in \mathbb{R}^{N-1}, x_1 > 0\}$, which is an open subset of $\mathbb{R}^N$ but is not bounded. Theorem 2.18 present the main idea of this regularity result in the case of the Dirichlet problem. See e.g. [22, chapter 6] for more general cases.

Theorem 2.18 (Nirenberg [41]). *Let* $\Omega = \mathbb{R}_+^N = \{(x_1, y), y \in \mathbb{R}^{N-1}, x_1 > 0\}$, $N > 1$, *and* $f \in L^2(\Omega)$. *Let* $u \in H_0^1(\Omega)$ *be a solution of the following problem:*

$$u \in H_0^1(\Omega),$$

$$\int_\Omega \nabla u \cdot \nabla v \, dx = \int_\Omega f v \, dx, \forall v \in H_0^1(\Omega). \tag{2.15}$$

Then $u \in H^2(\mathbb{R}_+^N)$.

Proof We only give the proof in the case $N = 2$. The generalisation to the case $N > 2$ does not bring any additional difficulties.

Let $u \in H_0^1(\Omega)$ be a solution to (2.15), which also reads:

$$\int_\Omega \nabla u \cdot \nabla v \, dx + \int_\Omega u \, v \, dx = \int_\Omega g v \, dx, \forall v \in H_0^1(\Omega), \text{ where } g = u + f \in L^2(\Omega).$$

As usual, the (class of) function(s) g is identified with the mapping $T_g \in H^{-1}(\Omega)$ defined by $T_g(v) = \int g v \, dx$.

Since, by definition,

$$\|g\|_{H^{-1}(\Omega)} = \sup\{\int_{\mathbb{R}_+^2} g \, v \, dx, \ v \in H_0^1(\Omega), \|v\|_{H^1(\Omega)} \le 1\},$$

the following equality holds

$$(u \mid v)_{H^1(\Omega)} = \int_{\mathbb{R}_+^2} g \, v \, dx \le \|g\|_{H^{-1}(\Omega)} \|v\|_{H^1(\Omega)}. \tag{2.16}$$

Now taking $v = u$ in (2.16) yields $\|u\|_{H^1(\Omega)} \le \|g\|_{H^{-1}(\Omega)}$.

The regularity on $D_2 u$ is proven by introducing, for $h > 0$, the function Ψ_h defined by $\Psi_h u = \frac{1}{h}(u_h - u)$ with $u_h \in H_0^1(\Omega)$ defined by $u_h(x) = u(x_1, x_2 + h)$. Since u satisfies (2.15), u_h satisfies

$$\int_\Omega \nabla u_h \cdot \nabla v \, dx = \int_\Omega f_h v \, dx \text{ with } f_h(x) = f(x_1, x_2 + h),$$

and therefore $\Psi_h u$ belongs to $H_0^1(\Omega)$ and satisfies

$$\int_\Omega \nabla(\Psi_h u) \cdot \nabla v \, dx = \int_\Omega (\Psi_h f) v \, dx \text{ for all } v \in H_0^1(\Omega).$$

Hence $(\Psi_h u \mid v)_{H^1(\Omega)} = \int_\Omega (\Psi_h g) v \, dx$, and therefore $\|\Psi_h u\|_{H^1(\Omega)} \leq \|\Psi_h g\|_{H^{-1}(\Omega)}$. Since $g \in L^2(\Omega)$, Lemma 2.21 below yields that

$$\|\Psi_h u\|_{H^1(\Omega)} \leq \|g\|_{L^2(\Omega)}.$$

Now take $h = \frac{1}{n}$ and let n tend to $+\infty$. Owing to the latter inequality, the sequence $(\Psi_{\frac{1}{n}} u)_{n \in \mathbb{N}^\star}$ is bounded in $H_0^1(\Omega)$, so there exists a subsequence, still denoted $(\Psi_{\frac{1}{n}} u)_{n \in \mathbb{N}^\star}$, and $w \in H_0^1(\Omega)$ such that $\Psi_{\frac{1}{n}} u \to w$ weakly in $H_0^1(\Omega)$, i.e. $S(\Psi_{\frac{1}{n}} u) \to S(w)$ for all $S \in H^{-1}(\Omega)$. Therefore $\Psi_{\frac{1}{n}} u \to w$ in $\mathcal{D}^\star(\Omega)$. But by Lemma 2.22 given below, $\Psi_{\frac{1}{n}} u \to D_2 u$ in $\mathcal{D}^\star(\Omega)$. By uniqueness of the limit, $D_2 u = w \in H_0^1(\Omega)$, and consequently, $D_1 D_2 u \in L^2(\Omega)$ and $D_2 D_2 u \in L^2(\Omega)$.

To conclude, all that remains is to show that $D_1 D_1 u \in L^2(\Omega)$. With this aim, we use the equation satisfied by u. Indeed, as u is a weak solution to (2.6), we have $-\Delta u = f$ in $\mathcal{D}^\star(\Omega)$, and therefore $D_1 D_1 u = -f - D_2 D_2 u$, which proves that $D_1 D_1 u \in L^2(\Omega)$. This concludes the proof. ∎

The regularity of the solutions to problem (2.6) for an open bounded subset Ω with C^k boundary is stated in the next theorem; it relies on the so-called local maps technique: a local parameterisation of the boundary $\partial\Omega$, see e.g. [11, Appendix C1]) or [20, Paragraph 2.1.1], is used to return back to the case $\Omega = \mathbb{R}_+^N = \{(x_1, y), y \in \mathbb{R}^{N-1}, x_1 > 0\}$ and a solution u of an elliptic problem for which the regularity is proven with a proof similar to that of Theorem 2.18, see [22, chapter 6].

Theorem 2.19 (Regularity of the solution of the Dirichlet problem). *Let Ω be an open bounded subset of $\mathbb{R}^N$, $N > 1$, and $f \in L^2(\Omega)$; under the assumptions (2.2), let $u \in H_0^1(\Omega)$ be the solution to (2.6).*

1. *If $a_{i,j} \in C^1(\overline{\Omega})$ for $i, j = 1, \ldots, N$ and Ω has a C^2 boundary, then $u \in H^2(\Omega)$.*
2. *If $a_{i,j} \in C^\infty(\overline{\Omega})$ for $i, j = 1, \ldots, N$, if Ω has a C^∞ boundary and if $f \in H^m(\Omega)$ with $m \geq 0$, then $u \in H^{m+2}(\Omega)$. Consequently, if $f \in C^\infty(\overline{\Omega})$, then $u \in C^\infty(\overline{\Omega})$ and therefore u is a classical solution. Similarly, if $f \in H^m(\Omega)$ with $m > \frac{N}{2}$, then $u \in C^2(\overline{\Omega})$ and therefore u is still a classical solution.*

Remark 2.20 (Optimality of assumptions). Note that item 1 of the previous theorem is false without the assumptions $a_{i,j} \in C^1(\overline{\Omega})$ and Ω has a C^2 boundary. However, in the case of the Laplacian, that is $a_{i,j} = \delta_{i,j}$, if Ω is convex,[15] then $u \in H^2(\Omega)$ as soon as $f \in L^2(\Omega)$.

The proof of Theorem 2.18 required the following two technical lemmas, which we state for $N = 2$, for simplicity:

Lemma 2.21. *Let $\Omega = \mathbb{R}_+ \times \mathbb{R} = \{(x_1, y), x_1 > 0, y \in \mathbb{R}\}$. Let $g \in L^2(\Omega)$ and, for $h > 0$, $\Psi_h g$ defined by: $\Psi_h g = \frac{1}{h}(g_h - g)$, where $g_h \in L^2(\Omega)$ is defined by $g_h(x) = g(x_1, x_2 + h)$. Then $\|\Psi_h g\|_{H^{-1}(\Omega)} \leq \|g\|_{L^2(\Omega)}$.*

---

[15] Recall that a set E is convex if for all $(x, y) \in E^2$, the set $\{tx + (1-t)y, t \in [0, 1]\}$ is included in E.

Proof Let $g \in L^2(\Omega)$, by definition,

$$\|\Psi_h g\|_{H^{-1}(\Omega)} = \sup\{\int_\Omega \Psi_h g \, v \, dx, \ v \in H_0^1(\Omega), \ \|v\|_{H^1(\Omega)} \leq 1\},$$

and therefore, by density of $\mathcal{D}(\Omega)$ in $H_0^1(\Omega)$,

$$\|\Psi_h g\|_{H^{-1}(\Omega)} = \sup\{\int_\Omega \Psi_h g \, v \, dx, \ v \in \mathcal{D}(\Omega), \ \|v\|_{H^1(\Omega)} \leq 1\}.$$

Let $v \in \mathcal{D}(\Omega)$ such that $\|v\|_{H^1(\Omega)} \leq 1$.

$$\begin{aligned}
\int_\Omega \Psi_h g \, v \, dx &= \frac{1}{h} \int_{\mathbb{R}_+} \int_{\mathbb{R}} (g(x_1, x_2 + h) - g(x_1, x_2)) \, v(x_1, x_2) \, dx_1 \, dx_2 \\
&= \frac{1}{h} \int_{\mathbb{R}_+} \int_{\mathbb{R}} g(x_1, \tilde{x}_2) v(x_1, \tilde{x}_2 - h) \, dx_1 \, d\tilde{x}_2 \\
&\qquad\qquad - \frac{1}{h} \int_{\mathbb{R}_+} \int_{\mathbb{R}} g(x_1, x_2) v(x_1, x_2) \, dx_1 \, dx_2 \\
&= - \int_{\mathbb{R}_+} \int_{\mathbb{R}} g(x_1, x_2) \frac{v(x_1, x_2 - h) - v(x_1, x_2)}{-h} \, dx_1 \, dx_2.
\end{aligned}$$

By the Cauchy–Schwarz inequality, we obtain that

$$\left| \int_\Omega \Psi_h g \, v \, dx \right| \leq \|g\|_{L^2(\Omega)} \left\| \frac{v - v(\cdot, \cdot - h)}{h} \right\|_{L^2(\Omega)}.$$

Writing that

$$v(x_1, x_2) - v(x_1, x_2 - h) = \int_{x_2 - h}^{x_2} \partial_2 v(x_1, s) \, ds$$

we have

$$\left\| \frac{v - v(\cdot, \cdot - h)}{h} \right\|_{L^2(\Omega)}^2 = \frac{1}{h^2} \int_{\mathbb{R}_+} \int_{\mathbb{R}} \left(\int_{\mathbb{R}} \mathbb{1}_{[x_2 - h, x_2]}(s) \partial_2 v(x_1, s) \, ds \right)^2 dx_2 \, dx_1.$$

By applying the Cauchy–Schwarz inequality once more and then the Fubini–Tonelli theorem, we obtain, noting that $\mathbb{1}_{[x_2 - h, x_2]}(s) = \mathbb{1}_{[s, s+h]}(x_2)$,

$$\begin{aligned}
\left\| \frac{v - v(\cdot, \cdot - h)}{h} \right\|_{L^2(\Omega)}^2 &\leq \frac{1}{h} \int_{\mathbb{R}_+} \int_{\mathbb{R}} \left(\int_{\mathbb{R}} \mathbb{1}_{[x_2 - h, x_2]}(s) \partial_2 v(x_1, s)^2 \, ds \right) dx_2 \, dx_1. \\
&\leq \frac{1}{h} \int_{\mathbb{R}_+} \int_{\mathbb{R}} \left(\int_{\mathbb{R}} \mathbb{1}_{[s, s+h]}(x_2) \, dx_2 \right) \partial_2 v(x_1, s)^2 \, ds \, dx_1. \\
&\leq \|\partial_2 v\|_{L^2(\Omega)}^2 \leq \|v\|_{H^1(\Omega)}^2.
\end{aligned}$$

Hence

$$\left| \int_\Omega \Psi_h g \, v \, dx \right| \le \|g\|_{L^2(\Omega)} \left\| \frac{v - v(\cdot, \cdot - h)}{h} \right\|_{L^2(\Omega)} \le \|g\|_{L^2(\Omega)} \|v\|_{H^1(\Omega)} \le \|g\|_{L^2(\Omega)}.$$

Finally, we indeed have $\|\Psi_h g\|_{H^{-1}(\Omega)} \le \|g\|_{L^2(\Omega)}$. ∎

Lemma 2.22. *Under the assumptions of Lemma 2.21, let $u \in L^1_{\mathrm{loc}}(\Omega)$, then $\Psi_h u \to D_2 u$ in $\mathcal{D}^\star(\Omega)$ when $h \to 0$.*

Proof Let $\varphi \in \mathcal{D}(\Omega)$; we want to show that

$$\int_\Omega \Psi_h u \, \varphi \, dx \to - \int_\Omega u \partial_2 \varphi \, dx = \langle D_2 u, \varphi \rangle_{\mathcal{D}^\star(\Omega), \mathcal{D}(\Omega)} \quad \text{when } h \to 0.$$

However

$$\int_\Omega \Psi_h u \varphi \, dx = \int_{\mathbb{R}_+} \int_{\mathbb{R}} \frac{u(x_1, x_2 + h) - u(x_1, x_2)}{h} \varphi(x_1, x_2) \, dx_1 \, dx_2$$

$$= - \int_{\mathbb{R}_+} \int_{\mathbb{R}} u(x_1, x_2) \frac{\varphi(x_1, x_2 - h) - \varphi(x_1, x_2)}{-h} \, dx_1 \, dx_2.$$

But $\frac{\varphi(x_1, x_2 - h) - \varphi(x_1, x_2)}{-h} \to \partial_2 \varphi(x_1, x_2)$ uniformly when $h \to 0$, and the support of this function is included in a compact subset K of Ω, independent of h if $|h| < 1$. Therefore $\lim_{h \to 0} \int_\Omega \Psi_h u \, \varphi \, dx = - \int_\Omega u \partial_2 \varphi \, dx$. ∎

Remark 2.23 (Regularity if $\Omega = \mathbb{R}^N$). Let $\Omega = \mathbb{R}^N$ ($N \ge 1$), $f \in L^2(\Omega)$ and $u \in H^1_0(\Omega)$ be a solution to (2.15) (note that (2.15) is equivalent to saying $u \in H^1_0(\Omega)$ and $-\Delta u = f$ in $\mathcal{D}^\star(\Omega)$). The proof of Theorem 2.18 directly gives $u \in H^2(\mathbb{R}^N)$; in fact it is easier since there is no difficulty in showing that $D_1 D_1 u \in L^2(\Omega)$).

Remark 2.24 (More regularity...). Let Ω be an open bounded subset of $\mathbb{R}^N$, $N > 1$, and $f \in L^2(\Omega)$. Under the assumptions (2.2), let $u \in H^1_0(\Omega)$ be the solution to (2.6).

1. Suppose that $a_{i,j} \in C^1(\overline{\Omega})$ and that Ω has a C^2 boundary. We have already seen that if $f \in L^2(\Omega)$ then $u \in H^2(\Omega)$. We can show that if $f \in L^p(\Omega)$ then $u \in W^{2,p}(\Omega)$ ($2 \le p < +\infty$).
2. Now suppose that we only have $a_{i,j} \in L^\infty(\Omega)$. We can show [40] that there exists a $p^\star > 2$ such that if $f \in L^p(\Omega)$ with $2 \le p \le p^\star$, then $u \in W^{1,p}_0(\Omega)$.
3. Still in the case $a_{i,j} \in L^\infty(\Omega)$, we can show (this result is due to Stampacchia[16]) that if $f \in L^p(\Omega)$, with $p > \frac{N}{2}$, then $u \in L^\infty(\Omega)$.
4. It is also possible to prove regularity results for other boundary conditions. Problem 2.8 provides an example with Neumann conditions, Problem 2.12 an example

---

[16] Guido Stampacchia, Italian mathematician (1922–1978), specialist in the calculus of variations and partial differential equations, among other fields.

with Fourier conditions and Problem 2.13 deals with the example of the elliptic system induced by the Schrödinger equation (which is usually presented as an equation whose unknown takes its values in $\mathbb{C}$).

2.4 Non-Negativity of the Solution

The purpose of this section is to prove that if the right-hand side of (2.6) is non-negative, then the corresponding weak solution is non-negative. More precisely, assume that Ω is an open bounded subset of $\mathbb{R}^N$, $N \geq 1$, $a_{i,j} \in L^\infty(\Omega)$, for $i, j = 1, \ldots, N$, that the functions $a_{i,j}$ satisfy (2.2) and that $f \in L^2(\Omega)$ is such that $f \geq 0$ a.e. The question is to show that under these conditions, the solution u of (2.6) satisfies $u \geq 0$ a.e.

2.4.1 The Case of Classical Solutions

Let Ω be an open bounded subset of $\mathbb{R}^N$ ($N \geq 1$). Assume that $u \in C^2(\bar{\Omega})$, $-\Delta u = f$ in Ω and $u = 0$ on the boundary of Ω: the function u is therefore a classical solution with $a_{i,j} = 0$ if $i \neq j$ and $a_{i,j} = 1$ if $i = j$. Assuming that $f > 0$ in Ω, we now prove by contradiction that $u \geq 0$ in Ω. Assume that there exists an $a \in \Omega$ such that $u(a) < 0$, and choose $\bar{x} \in \Omega$ such that $u(\bar{x}) = \min\{u(x), x \in \bar{\Omega}\}$ (such an $\bar{x}$ exists because $\bar{\Omega}$ is compact, u is continuous, $u(a) < 0$ and $u = 0$ on the boundary of Ω). Then

$$\partial_i u(\bar{x}) = 0 \text{ and } \partial_i^2 u(\bar{x}) \geq 0 \text{ for all } i \in \{1, \ldots, N\}.$$

This gives $\Delta u(\bar{x}) \geq 0$, which contradicts $\Delta u(\bar{x}) = -f(\bar{x}) < 0$. Hence $u(x) \geq 0$ for all $x \in \Omega$.

An additional argument allows us to replace the assumption $f > 0$ by $f \geq 0$. Indeed, let us only assume that $f \geq 0$. For $\varepsilon > 0$, set $u_\varepsilon(x) = u(x) - \varepsilon x_1^2$ so that $-\Delta u_\varepsilon = f + 2\varepsilon > 0$ in Ω. Let $\bar{x}_\varepsilon \in \bar{\Omega}$ such that $u_\varepsilon(\bar{x}_\varepsilon) = \min\{u_\varepsilon(x), x \in \bar{\Omega}\}$. If $\bar{x}_\varepsilon \in \Omega$, the previous argument shows that $\Delta u_\varepsilon(\bar{x}_\varepsilon) \geq 0$, which contradicts the fact that $-\Delta u_\varepsilon(\bar{x}_\varepsilon) = f(\bar{x}_\varepsilon) + 2\varepsilon > 0$. Hence $\bar{x}_\varepsilon \in \partial\Omega$ and

$$u_\varepsilon(y) \geq u_\varepsilon(\bar{x}_\varepsilon) \geq -\varepsilon \max_{x \in \partial\Omega} x_1^2 \text{ for all } y \in \bar{\Omega}.$$

Letting ε tend to 0 yields the desired result, that is, $u \geq 0$ in $\bar{\Omega}$.

The objective is now to extend this positivity property to weak solutions.

2.4.2 The Case of Weak Solutions

The non-negativity of the weak solutions is based on the following two lemmas due to G. Stampacchia.

Lemma 2.25. *Let Ω be an open bounded subset of $\mathbb{R}^N$ ($N \geq 1$) and $\varphi \in C^1(\mathbb{R}, \mathbb{R})$. Assume that φ' is bounded and $\varphi(0) = 0$, and let $u \in H_0^1(\Omega)$, then $\varphi(u) \in H_0^1(\Omega)$ and $D_i\varphi(u) = \varphi'(u)D_iu$ a.e. (for all $i \in \{1, \ldots, N\}$). The notation $\varphi(u)$ denotes the function $\varphi \circ u$. (This notation $\varphi(u)$ instead of $\varphi \circ u$ is usual and is used systematically in the rest of this book.)*

Proof There exists a sequence $(u_n)_{n \in \mathbb{N}}$ of functions belonging to $\mathcal{D}(\Omega)$ such that $u_n \to u$ in $H_0^1(\Omega)$ as $n \to +\infty$, that is,

$$u_n \to u \text{ in } L^2(\Omega),$$
$$D_iu_n \to D_iu \text{ in } L^2(\Omega), \text{ for all } i \in \{1, \ldots, N\}.$$

After possibly extracting a subsequence, we can even assume that there exists an $F \in L^2(\Omega)$ and, for all $i \in \{1, \ldots, N\}$, $F_i \in L^2(\Omega)$ s.t.

$$u_n \to u \text{ a.e. and } |u_n| \leq F \text{ a.e. and for all } n \in \mathbb{N},$$
$$D_iu_n \to D_iu \text{ a.e. and } |D_iu_n| \leq F_i \text{ a.e. and for all } n \in \mathbb{N}, \ i \in \{1, \ldots, N\}.$$

We then have $\varphi(u_n) \in C_c^1(\Omega)$ and for all $n \in \mathbb{N}$ and all $i \in \{1, \ldots, N\}$,

$$D_i\varphi(u_n) = \partial_i\varphi(u_n) = \varphi'(u_n)\partial_iu_n.$$

Set $M = \sup\{|\varphi'(s)|, s \in \mathbb{R}\}$, so that $|\varphi(s)| \leq M|s|$, for all $s \in \mathbb{R}$. Hence

$$\varphi(u_n) \to \varphi(u) \text{ a.e. and } |\varphi(u_n)| \leq MF \text{ a.e. and for all } n \in \mathbb{N}.$$

Since $MF \in L^2(\Omega)$, the dominated convergence theorem (in $L^2(\Omega)$) gives $\varphi(u_n) \to \varphi(u)$ in $L^2(\Omega)$. Hence $D_i\varphi(u_n) \to D_i\varphi(u)$ in $\mathcal{D}^\star(\Omega)$. We now recall that $D_i\varphi(u_n) = \varphi'(u_n)\partial_iu_n$. Since

$$\varphi'(u_n) \to \varphi'(u) \text{ a.e.},$$
$$\partial_iu_n \to D_iu \text{ a.e.},$$
$$|\varphi'(u_n)\partial_iu_n| \leq MF_i \text{ a.e. and for all } n \in \mathbb{N},$$

the dominated convergence theorem gives $\varphi'(u_n)\partial_iu_n \to \varphi'(u)D_iu$ in $L^2(\Omega)$ and therefore also in $\mathcal{D}^\star(\Omega)$. By uniqueness of the limit in $\mathcal{D}^\star(\Omega)$ we therefore have $D_i\varphi(u) = \varphi'(u)D_iu$ a.e. for $i \in [\![1, N]\!]$. Hence $\varphi(u_n) \to \varphi(u)$ in $H^1(\Omega)$ as $n \to +\infty$, and since $\varphi(u_n) \in C_c^1(\Omega)$, we get that $\varphi(u) \in H_0^1(\Omega)$ and $D_i\varphi(u) = \varphi'(u)D_iu$ a.e., for all i. $\blacksquare$

Lemma 2.26. *Let Ω be an open bounded subset of $\mathbb{R}^N$ ($N \geq 1$). Let $u \in H_0^1(\Omega)$; we define u^+ by $u^+(x) = \max\{u(x), 0\}$, for $x \in \Omega$. Then, $u^+ \in H_0^1(\Omega)$ and $D_i u^+ = \mathbb{1}_{u \geq 0} D_i u = \mathbb{1}_{u > 0} D_i u$ a.e. (for all $i \in \{1, \ldots, N\}$). In particular, we have $D_i u = 0$ a.e. (for all i) on the set $\{u = 0\}$.*

Proof For $n \in \mathbb{N}^\star$, we define $\varphi_n \in C^1(\mathbb{R}, \mathbb{R})$ by

$$
\varphi_n(s) = \begin{cases} 0 & \text{if } s \leq 0, \\ \frac{n}{2} s^2 & \text{if } 0 < s < \frac{1}{n}, \\ s - \frac{1}{2n} & \text{if } \frac{1}{n} \leq s. \end{cases}
$$

Hence $\varphi_n(s) \to s^+$ for all $s \in \mathbb{R}$ as $n \to +\infty$ and $|\varphi_n'(s)| \leq 1$ for all s and for all $n \in \mathbb{N}^\star$. Lemma 2.25 gives $\varphi_n(u) \in H_0^1(\Omega)$ and $D_i(\varphi_n(u)) = \varphi_n'(u) D_i u$ a.e. (and for all $i \in \{1, \ldots, N\}$). On the other hand, we have

$$
\varphi_n(u) \to u^+ \text{ a.e., } |\varphi_n(u)| \leq |u| \text{ a.e. and for all } n \in \mathbb{N}.
$$

The dominated convergence theorem therefore gives $\varphi_n(u) \to u^+$ in $L^2(\Omega)$ (and therefore that $D_i \varphi_n(u) \to D_i u^+$ in $\mathcal{D}^\star(\Omega)$). Then, we note that $\varphi_n'(u) \to \mathbb{1}_{\{u > 0\}}$ a.e. and therefore

$$
\varphi_n'(u) D_i u \to \mathbb{1}_{\{u > 0\}} D_i u \text{ a.e., } |\varphi_n'(u) D_i u| \leq |D_i u| \text{ a.e. and for all } n \in \mathbb{N},
$$

which, again by the dominated convergence theorem, gives $\varphi_n'(u) D_i u \to \mathbb{1}_{\{u > 0\}} D_i u$ in $L^2(\Omega)$ and therefore in $\mathcal{D}^\star(\Omega)$. Since $D_i(\varphi_n(u)) = \varphi_n'(u) D_i u$, the uniqueness of the limit in $\mathcal{D}^\star(\Omega)$ yields that $D_i u^+ = \mathbb{1}_{\{u > 0\}} D_i u$ a.e. The sequence $(\varphi_n(u))_{n \in \mathbb{N}}$ is therefore a sequence in $H_0^1(\Omega)$, it converges in $H^1(\Omega)$ towards u^+. Finally, $u^+ \in H_0^1(\Omega)$ and $D_i u^+ = \mathbb{1}_{\{u > 0\}} D_i u$ a.e. (and for all i).

Considering the sequence $(\psi_n(u))_{n \in \mathbb{N}}$ with ψ_n defined by $\psi_n(s) = \varphi(s + \frac{1}{n}) - 1/(2n)$, similar arguments may be used to show that $D_i u^+ = \mathbb{1}_{\{u \geq 0\}} D_i u$ a.e.; the essential difference between φ_n and ψ_n is that $\varphi_n'(0) = 0$ while $\psi_n'(0) = 1$.

In particular, we obtain $\mathbb{1}_{\{u \geq 0\}} D_i u = \mathbb{1}_{\{u > 0\}} D_i u$ a.e. and therefore $\mathbb{1}_{\{u = 0\}} D_i u = 0$ a.e., so $D_i u = 0$ a.e. on the set $\{u = 0\}$. $\blacksquare$

Remark 2.27 (Generalisation of Lemma 2.26). Lemma 2.26 can be generalised by replacing the function $u \mapsto u^+$ with a Lipschitz-continuous function that vanishes at 0, thus obtaining the following result:

Let Ω be an open bounded subset of $\mathbb{R}^N$ ($N \geq 1$) and φ a Lipschitz-continuous function from $\mathbb{R}$ to $\mathbb{R}$, vanishing at 0. Let $u \in H_0^1(\Omega)$; we then have $\varphi(u) \in H_0^1(\Omega)$ and $D_i \varphi(u) = \varphi'(u) D_i u$ a.e. (for all $i \in \{1, \ldots, N\}$).

This generalisation is rather easy if φ' has only a finite number of points of discontinuity. It is significantly more difficult in the general case of a Lipschitz-continuous function. An important example is to take $\varphi(s) = (s - k)^+$ for all $s \in \mathbb{R}$, with k given in $\mathbb{R}_+$. We thus obtain, for $u \in H_0^1(\Omega)$, $(u - k)^+ \in H_0^1(\Omega)$ and $D_i \varphi(u) = \mathbb{1}_{\{u > k\}} D_i u = \mathbb{1}_{\{u \geq k\}} D_i u$ a.e.

Thanks to these lemmas, we are in position to prove the non-negativity of the weak solutions.

Theorem 2.28 (Non-negativity of the weak solution). *Let Ω be an open bounded subset of $\mathbb{R}^N$, $N \geq 1$, $a_{i,j} \in L^\infty(\Omega)$, for $i, j = 1, \ldots, N$. Under the assumption (2.2), let $f \in L^2(\Omega)$ and let u be the solution to (2.6). If $f \geq 0$ a.e. then $u \geq 0$ a.e.*

Proof Assuming that $f \leq 0$ a.e., we are going to show that $u \leq 0$ a.e.; changing f to $-f$ and u to $-u$ yields the desired result. Since u is a solution to (2.6),

$$\int_\Omega \sum_{i,j=1}^n a_{i,j}(x) D_j u(x) D_i v(x) \, dx = \int_\Omega f(x) v(x) \, dx \text{ for all } v \in H_0^1(\Omega).$$

Take $v = u^+$ in this equality:

$$\alpha \int_\Omega |\nabla u(x)|^2 \mathbb{1}_{\{u \geq 0\}}(x) \, dx \leq \int_\Omega \sum_{i,j=1}^n a_{i,j}(x) D_j u(x) D_i u^+(x) \, dx$$

$$= \int_\Omega f(x) u^+(x) \, dx \leq 0.$$

Hence $\alpha \|u^+\|_{H_0^1(\Omega)}^2 = \alpha \int_\Omega |\nabla u^+(x)|^2 \, dx \leq \int_\Omega f(x) u^+(x) \, dx \leq 0$ and therefore $u^+ = 0$ a.e., that is, $u \leq 0$ a.e. ∎

2.5 Non-Homogeneous Dirichlet Conditions

So far only homogeneous boundary conditions (i.e. the solution is zero at the boundary of the domain) have been considered. We now wish to replace the condition "$u = 0$" on the boundary of Ω with "$u = g$" on the boundary of Ω. If Ω is sufficiently regular, the trace operator γ defined in Theorem 1.32 is well defined. If furthermore g belongs to the image of γ (i.e., $g = \gamma(G)$ with $G \in H^1(\Omega)$) the problem may be reduced to the Dirichlet problem with homogeneous boundary conditions:

More precisely, let Ω be an open bounded subset of $\mathbb{R}^N$ ($N \geq 1$) with a Lipschitz boundary $\partial\Omega$. Let $a_{i,j} \in L^\infty(\Omega)$, for $i, j = 1, \ldots, N$, satisfying the uniform ellipticity assumption (2.2). Let f be a function from Ω to $\mathbb{R}$ and g a function from $\partial\Omega$ to $\mathbb{R}$. Let us study problem (2.3). Theorem 2.6 allows us to prove the following theorem, where (2.17) is the weak formulation of problem (2.3).

Theorem 2.29 (Non-homogeneous Dirichlet condition (1)). *Let Ω be an open bounded subset of $\mathbb{R}^N$ ($N \geq 1$) with a Lipschitz boundary, $f \in L^2(\Omega)$, let γ denote the trace operator from $H^1(\Omega)$ to $L^2(\Omega)$ introduced in Theorem 1.32 and let $g \in \mathrm{Im}(\gamma)$. Let $(a_{i,j})_{i,j=1,\ldots,N} \subset L^\infty(\Omega)$ satisfying (2.2), then there exists a unique solution to the following problem:*

$$u \in H^1(\Omega), \gamma(u) = g,$$
$$\int_\Omega \left(\sum_{i=1}^N \sum_{j=1}^N a_{i,j}(x) D_j u(x) D_i v(x) \right) dx = \int_\Omega f(x) v(x) \, dx, \forall v \in H_0^1(\Omega). \tag{2.17}$$

Proof The proof is part of Problem 2.24. It consists in considering $G \in H^1(\Omega)$ such that $\gamma(G) = g$ and seeking $u - G$ as a weak solution to an elliptic problem posed in $H_0^1(\Omega)$ with a right-hand side in $H^{-1}(\Omega)$. ∎

Note that it is possible to replace the right-hand side of (2.17) by $T(v)$, where $T \in H^{-1}(\Omega)$, as stated in the following Theorem 2.30.

Theorem 2.30 (Non-homogeneous Dirichlet condition (2)). *Let Ω be an open bounded subset of $\mathbb{R}^N$ ($N \geq 1$) with a Lipschitz boundary, $T \in H^{-1}(\Omega)$, $g \in \text{Im}(\gamma)$ (where γ denotes the trace operator of $H^1(\Omega)$ in $L^2(\Omega)$ seen in Theorem 1.32). Let $(a_{i,j})_{i,j=1,\dots,N} \subset L^\infty(\Omega)$ and $\alpha > 0$ such that (2.2) is satisfied. Then there exists a unique solution to (2.18).*

$$u \in H^1(\Omega), \gamma(u) = g,$$
$$\int_\Omega \left(\sum_{i=1}^N \sum_{j=1}^N a_{i,j}(x) D_j u(x) D_i v(x) \right) dx = T(v), \forall v \in H_0^1(\Omega). \tag{2.18}$$

Proof The proof is similar to that of Theorem 2.29 (see Problem 2.24). ∎

Remark 2.31 (Maximum principle). Under the hypotheses of Theorem 2.29, a method similar to that given in Theorem 2.28 gives that if $f = 0$ and $A \leq g \leq B$ a.e., with $A, B \in \mathbb{R}$ (a.e. is to be understood here in the sense of the $(N - 1)$-dimensional Lebesgue measure on $\partial\Omega$), we then have $A \leq u \leq B$ a.e., where u is the solution to (2.17). This result is called the "maximum principle".

The rest of this section provides some additional information on the image of the trace operator γ defined on $H^1(\Omega)$ when Ω is an open bounded subset of $\mathbb{R}^N$ with a Lipschitz boundary.

Definition 2.32 ($H^{\frac{1}{2}}(\partial\Omega)$ space). Let Ω be an open bounded subset of $\mathbb{R}^N$ ($N \geq 1$) with a Lipschitz boundary. Let $H^{\frac{1}{2}}(\partial\Omega)$ be the set of traces of the functions $H^1(\Omega)$, that is to say, $H^{\frac{1}{2}}(\partial\Omega) = \text{Im}(\gamma)$, where γ is the trace operator of $H^1(\Omega)$ in $L^2(\partial\Omega)$ seen in Theorem 1.32. A norm on $H^{\frac{1}{2}}(\partial\Omega)$ is defined by setting

$$\|u\|_{H^{\frac{1}{2}}(\partial\Omega)} = \inf\{\|\overline{u}\|_{H^1(\Omega)}, \ \gamma(\overline{u}) = u\}.$$

Proposition 2.34 shows that $H^{\frac{1}{2}}(\partial\Omega)$ is then a Hilbert space and that the mapping $u \mapsto u$ is continuous from $H^{\frac{1}{2}}(\partial\Omega)$ to $L^2(\partial\Omega)$. (It is then said that $H^{\frac{1}{2}}(\partial\Omega)$ is continuously embedded in $L^2(\partial\Omega)$.) Let $H^{-\frac{1}{2}}(\partial\Omega)$ be the dual space of $H^{\frac{1}{2}}(\partial\Omega)$ (it is therefore also a Hilbert space).

Remark 2.33 (Compactness of $H^{\frac{1}{2}}(\partial\Omega)$ in $L^2(\partial\Omega)$). Under the hypotheses of Definition 2.32, the mapping $u \mapsto u$ from $H^{\frac{1}{2}}(\partial\Omega)$ to $L^2(\partial\Omega)$ may be shown to be compact, see e.g. [18, chapter 3].

Proposition 2.34 (Properties of the space $H^{\frac{1}{2}}(\partial\Omega)$). *Let Ω be an open bounded subset of $\mathbb{R}^N$ ($N \geq 1$) with a Lipschitz boundary. Let γ be the trace operator defined on $H^1(\Omega)$.*

1. *Let $u \in H^{\frac{1}{2}}(\partial\Omega)$. Then $\|u\|_{H^{\frac{1}{2}}(\partial\Omega)} = \|\bar{u}\|_{H^1(\Omega)}$, where $\bar{u}$ is the unique weak solution of $-\Delta\bar{u} + \bar{u} = 0$ in Ω with $\gamma(\bar{u}) = u$, that is, the unique solution of*

$$\bar{u} \in H^1(\Omega), \gamma(\bar{u}) = u,$$

$$\int_\Omega (\nabla\bar{u}(x)\nabla v(x) + \bar{u}(x)v(x))\, dx = 0, \forall v \in H_0^1(\Omega).$$

2. *The space $H^{\frac{1}{2}}(\partial\Omega)$ is a Hilbert space.*
3. *The space $H^{\frac{1}{2}}(\partial\Omega)$ is continuously embedded in $L^2(\partial\Omega)$.*

The proof of this proposition is the subject of Problem 2.25.

Let Ω be an open bounded subset of $\mathbb{R}^N$ ($N \geq 1$) with a Lipschitz boundary; owing to Definition 2.32 and Proposition 2.34, the trace operator defined on $H^1(\Omega)$ is a continuous linear operator from $H^1(\Omega)$ to $H^{\frac{1}{2}}(\partial\Omega)$ (and its norm is equal to 1). If now $u \in H^1(\Omega)^N$, the trace of u, still denoted $\gamma(u)$, may be defined by taking the trace of each of the components of u. So we have $\gamma(u) \in H^{\frac{1}{2}}(\partial\Omega)^N \subset L^2(\partial\Omega)^N$. Let $n(x)$ be the outward normal unit vector to $\partial\Omega$. Since Ω has a Lipschitz boundary, the vector $n(x)$ is defined a.e. at $x \in \partial\Omega$ (a.e. here means, as usual, a.e. for the $(N-1)$–dimensional Lebesgue measure on $\partial\Omega$) and the function $x \mapsto n(x)$ defines an element of $L^\infty(\partial\Omega)$. We thus obtain $\gamma(u) \cdot n \in L^2(\partial\Omega)$. This (class of) function(s) $\gamma(u) \cdot n$ is called the "normal trace of u on $\partial\Omega$". Problem 2.26 shows that $\gamma(u) \cdot n$ may be defined as an element of the dual space of $H^{\frac{1}{2}}(\Omega)$, let $H^{-\frac{1}{2}}(\Omega)$ be this dual space, under the assumption $u \in L^2(\Omega)^N$ with $\operatorname{div} u \in L^2(\Omega)$. Recall that for a function that is differentiable in the classical sense, $\operatorname{div} u$ is the function from $\mathbb{R}^N$ to $\mathbb{R}$ defined by $x = (x_1, \ldots, x_N) \mapsto \partial_1 u_1(x) + \cdots + \partial_N u_N(x)$.

Definition 2.35 (The space $H_{\operatorname{div}}(\Omega)$). Let Ω be an open subset of $\mathbb{R}^N$ with a Lipschitz boundary and $u \in L^2(\Omega)^N$; we define the divergence of u by $\operatorname{div} u = D_1 u_1 + \cdots + D_N u_N$. The space of functions $u \in L^2(\Omega)^N$ satisfying $\operatorname{div} u \in L^2(\Omega)$ is denoted $H_{\operatorname{div}}(\Omega)$, that is,

$$H_{\operatorname{div}}(\Omega) = \{u \in L^2(\Omega)^N ;\ \operatorname{div} u \in L^2(\Omega)\}.$$

The assumption $u \in H_{\operatorname{div}}(\Omega)$ is weaker than $u \in H^1(\Omega)^N$. In particular, we can have $\operatorname{div} u \in L^2(\Omega)$ without $D_i u_i \in L^2(\Omega)$ for all $i = 1, \ldots, N$.

Remark 2.36 (Normal trace on a part of the boundary). It is interesting to note that, if Ω is an open subset of $\mathbb{R}^N$ with a Lipschitz boundary and $u \in H_{\text{div}}(\Omega)$, its normal trace $\gamma(u) \cdot n$, which is therefore an element of $H^{-\frac{1}{2}}(\Omega)$, is not always represented by a function on $\partial\Omega$, and this induces a difficulty when we wish to consider the restriction of $\gamma(u) \cdot n$ to a part of the boundary of Ω, see in this regard Problem 2.27. We also recommend the works [39] and [30] for further study on this subject.

2.6 Problem Set

Problem 2.1 (A generalisation of the Lax–Milgram theorem ($\star$)). *Solution on page 115*. The purpose here is to prove the following generalisation of the Lax–Milgram theorem.

Theorem 2.37. *Let H be a real Hilbert space equipped with the inner product denoted $(\cdot|\cdot)$, with associated norm denoted $\|\cdot\|$, and $A \in \mathcal{L}(H)$ a continuous linear mapping from H to H. Let $A^\star$ be the adjoint operator of A (see Definition 2.13). Under the following assumptions:*

- *A and $A^\star$ are injective,*
- *if $(w_n)_{n\in\mathbb{N}} \subset H$ is a bounded sequence such that Aw_n converges to 0, then w_n converges to 0 (in H),*

the operator A is bijective.

Let us then consider the assumptions of Theorem 2.37.

1. Show that $\overline{\text{Im}(A)} = H$.
2. The purpose of this question is to show that $\text{Im}(A)$ is closed (and therefore with question 1 that A is surjective and therefore bijective because A is injective by assumption). Let $(w_n)_{n\in\mathbb{N}} \subset H$; set $f_n = Aw_n$ and assume that $f_n \to f \in H$ as $n \to +\infty$.

   a. Show that the sequence $(w_n)_{n\in\mathbb{N}}$ is bounded.
   b. Show that there exists a $w \in H$ such that $Aw = f$.

Problem 2.2 (Regularity in dimension 1 ($\star$)). *Solution on page 115*. Let $f \in L^2(]0, 1[)$; recall (see Theorem 2.6) that there exists a unique solution u of

$$u \in H_0^1(]0, 1[),$$
$$\int_0^1 Du(t)Dv(t)\,\mathrm{d}t = \int_0^1 f(t)v(t)\,\mathrm{d}t, \ \forall v \in H_0^1(]0, 1[). \tag{2.19}$$

We now assume that $f \in C([0, 1], \mathbb{R})(\subset L^2(]0, 1[)$. Set $F(x) = \int_0^x f(t)\,\mathrm{d}t$, for all $x \in [0, 1]$. Let u be the solution to (2.19); show that, for all $\varphi \in C([0, 1], \mathbb{R})$,

$$\int_0^1 (Du(t) + F(t))\varphi(t)\,\mathrm{d}t = \int_0^1 c\varphi(t)\,\mathrm{d}t$$

with $c \in \mathbb{R}$ suitably chosen (and independent of φ). Deduce that $Du = -F + c$ a.e., then that u is twice continuously differentiable on $]0, 1[$ and $-u''(x) = f(x)$ for all $x \in]0, 1[$ (and that $u(0) = u(1) = 0$).

Problem 2.3 (Spectral decomposition in dimension 1 ($\star\star$)). *Answer on page 117.* Set $E = L^2(]0, 1[)$ (equipped with the norm $\| \cdot \|_2$). For $f \in E$, recall that there exists a unique solution u of (2.19) (Theorem 2.6). Let T be the mapping from E to E that associates f with u (solution to (2.19), note that $H_0^1(]0, 1[) \subset E$). Recall that T is a compact self-adjoint linear operator from E to E.

1. Let $\lambda \in \mathcal{VP}(T)$. Show that there exists a $u \in C([0, 1], \mathbb{R}) \cap C^2(]0, 1[, \mathbb{R})$, $u \neq 0$, such that $-\lambda u'' = u$, on $]0, 1[$ and $u(0) = u(1) = 0$.
2. Show that $\mathcal{VP}(T) = \{\frac{1}{k^2 \pi^2}, \ k \in \mathbb{N}^\star\}$ and $\sigma(T) = \mathcal{VP}(T) \cup \{0\}$.
3. Let $f \in E$. For $n \in \mathbb{N}^\star$, setting $c_n = 2 \int_0^1 f(t) \sin(n\pi t) \, dt$, show that:

$$\|f - \sum_{p=1}^{n} c_p \sin(p\pi\cdot)\|_2 \to 0, \ \text{when } n \to +\infty.$$

(Compare with Fourier series.)
4. Let $\mu \in \mathbb{R}^\star$. Using the fact that T is compact, provide a necessary and sufficient condition on $f \in E$ for the following problem to have a solution:

$$u \in H_0^1(]0, 1[),$$
$$\int_0^1 Du(t) Dv(t) \, dt + \mu \int_0^1 u(t)v(t) \, dt = \int_0^1 f(t)v(t) \, dt, \ \forall v \in H_0^1(]0, 1[).$$

Problem 2.4 (First eigenvalue of $-\Delta$ ($\star\star\star$)). *Answer on page 119.* Let Ω be an open bounded subset of $\mathbb{R}^N$ ($N \geq 1$). For $u \in H_0^1(\Omega) \setminus \{0\}$, set

$$Q(u) = \frac{\int_\Omega |\nabla u(x)|^2 \, dx}{\int_\Omega u^2(x) \, dx}.$$

Set $\mu = \inf_{v \in H_0^1(\Omega) \setminus \{0\}} Q(v)$.

1. Show that $\mu > 0$ and that there exists a $u \in H_0^1(\Omega) \setminus \{0\}$ such that $Q(u) = \mu$. [Hint: consider a minimising sequence, i.e. a sequence $(u_n)_{n \in \mathbb{N}} \subset H_0^1(\Omega) \setminus \{0\}$ such that $\lim_{n \to +\infty} Q(u_n) = \mu$ and use Rellich's theorem.]
2. Let $u \in H_0^1(\Omega) \setminus \{0\}$ such that $Q(u) = \mu$, and let $\mathcal{A}$ be the Laplace operator defined by (2.12); show that $u \in D(\mathcal{A})$ and $\mathcal{A}u = \mu u$ a.e. Deduce that μ is the smallest eigenvalue of $\mathcal{A}$.
3. Let $u \in H_0^1(\Omega) \setminus \{0\}$ such that $Q(u) = \mu$, show that $u^+, u^- \in D(\mathcal{A})$ and $\mathcal{A}u^\pm = \mu u^\pm$ a.e.). [Recall that if $u \in H_0^1(\Omega)$ we also have $u^+, u^- \in H_0^1(\Omega)$, Lemma 2.26. We can then compare $Q(u)$ with $Q(u^+)$ and $Q(u^-)$ if u^+ and u^- are non-zero functions.]

Problem 2.5 (Mean Poincaré inequality on the boundary ($\star\star$)). *Solution on page 120.* Let Ω be an open bounded subset of $\mathbb{R}^N$, which we assume to be connected and with a Lipschitz boundary, and let $A \subset \partial\Omega$ be of non-zero measure in the sense of the $(N-1)$-dimensional Lebesgue measure on $\partial\Omega$. Assume that $u \in H^1(\Omega)$ and that $\int_A u(x)\,\mathrm{d}\gamma(x) = 0$. Show that there exists a $C \in \mathbb{R}_+$ depending only on Ω and A such that $\|u\|_{L^2(\Omega)} \le C\|\nabla u\|_{L^2(\Omega)}$.

Problem 2.6 (Elliptic problem with unbounded coefficients ($\star\star$)). *Solution on page 121.* Let Ω be an open bounded subset of $\mathbb{R}^N$, $N \ge 1$, and $p : \Omega \to \mathbb{R}$ a measurable function such that $\inf\{p(x), x \in \Omega\} = \alpha > 0$. Let $H^1(p, \Omega) = \{u \in L^2(\Omega) \text{ such that } D_i u \in L^1_{\mathrm{loc}}(\Omega) \text{ and } p\,D_i u \in L^2(\Omega) \text{ for all } i \in \{1, \ldots, N\}\}$. Recall that $D_i u$ denotes the derivative, in the sense of derivatives by transposition, of u with respect to the i-th component of $x = (x_1, \ldots, x_N)^t$. For $u \in H^1(p, \Omega)$, define $\|u\|$ by

$$\|u\|^2 = \|u\|_2^2 + \sum_{i=1}^{N} \|p\,D_i u\|_2^2, \text{ with } \|\cdot\|_2 = \|\cdot\|_{L^2(\Omega)}.$$

1. (Study of the functional space.)

   a. Show that $H^1(p, \Omega) \subset H^1(\Omega)$.
   b. Show that $H^1(p, \Omega)$, equipped with the norm $\|\cdot\|$, is a Hilbert space. [Notice that a Cauchy sequence in $H^1(p, \Omega)$ is also a Cauchy sequence in $H^1(\Omega)$.]

Let $H_0^1(p, \Omega) = H^1(p, \Omega) \cap H_0^1(\Omega)$.

2. (Functional space, continued.) Show that $H_0^1(p, \Omega)$ is a closed subspace of $H^1(p, \Omega)$.
3. (Weak solution.) Let $h \in L^2(\Omega)$, show that there exists a unique u such that

$$\begin{cases} u \in H_0^1(p, \Omega), \\ \displaystyle\int_\Omega p^2(x)\nabla u(x) \cdot \nabla v(x)\,\mathrm{d}x = \int_\Omega h(x)v(x)\,\mathrm{d}x, \ \forall v \in H_0^1(p, \Omega). \end{cases} \tag{2.20}$$

4. (More properties)

   a. Assuming that $p^2 \in L^1_{\mathrm{loc}}(\Omega)$. Show that $C_c^\infty(\Omega) \subset H_0^1(p, \Omega)$.
   b. Now take $N = 1$ and $\Omega =]0, 1[$. Give an example of a function p (with $p : \Omega \to \mathbb{R}$ measurable and such that $\inf\{p(x), x \in \Omega\} > 0$) for which $C_c^\infty(\Omega) \cap H_0^1(p, \Omega) = \{0\}$ (this question is more difficult).

Problem 2.7 (Two nested elliptic problems ($\star\star\star\star$)). *Solution on page 122.* Let Ω be an open bounded subset of $\mathbb{R}^d$, $d \ge 1$, and M and N two $d \times d$ matrices with coefficients in $L^\infty(\Omega)$. Assume that there exists an $\alpha > 0$ such that for almost all $x \in \Omega$ and for all $\xi \in \mathbb{R}^d$, the following inequality holds:

$$M(x)\xi \cdot \xi \ge \alpha|\xi|^2 \text{ and } N(x)\xi \cdot \xi \ge \alpha|\xi|^2.$$

1. Let $f \in L^2(\Omega)$. Show that there exists a unique u such that

$$
\begin{cases}
u \in H^1_0(\Omega), \\
\displaystyle\int_\Omega N(x)\nabla u(x) \cdot \nabla v(x)\,\mathrm{d}x = \int_\Omega (M(x) + N(x))\nabla w(x) \cdot \nabla v(x)\,\mathrm{d}x, \qquad (2.21) \\
\hspace{6cm} \forall v \in H^1_0(\Omega),
\end{cases}
$$

with solution w of

$$
\begin{cases}
w \in H^1_0(\Omega), \\
\displaystyle\int_\Omega M(x)\nabla w(x) \cdot \nabla v(x)\,\mathrm{d}x = \int_\Omega f(x)v(x)\,\mathrm{d}x, \ \forall v \in H^1_0(\Omega).
\end{cases}
\qquad (2.22)
$$

For the following questions, the unique solution to (2.21) with solution w to (2.22) is denoted by $T(f)$.

2. Show that T is a compact linear mapping[17] from $L^2(\Omega)$ to $L^2(\Omega)$ (that is, T is linear, continuous and maps bounded subsets of $L^2(\Omega)$ to relatively compact subsets of $L^2(\Omega)$).

3. Assume in this question (and only in this question) that there exists a $\lambda \in \mathbb{R}$ such that $M = \lambda N$. Show that there exists a matrix A, depending only on M and λ, such that, if $u = T(f)$,

$$
\int_\Omega A(x)\nabla u(x) \cdot \nabla v(x)\,\mathrm{d}x = \int_\Omega f(x)v(x)\,\mathrm{d}x \text{ for all } v \in H^1_0(\Omega). \qquad (2.23)
$$

Give the expression of A in terms of M and λ.

4. Assume now that $d = 2$ and $1 < p \le +\infty$. Show that for all $f \in L^p(\Omega)$ there exists a unique solution u of (2.21) with solution w to (2.22).

   Show that the mapping that associates u (solution to (2.21) with solution w to (2.22)) to f is compact from $L^p(\Omega)$ to $L^q(\Omega)$ for $1 \le q < +\infty$

5. Assume in this question that $d = 3$ and $p = 6/5$. Show that for all $f \in L^p(\Omega)$ there exists a unique solution u of (2.21) with solution w to (2.22).

   Show that the mapping that associates u (solution to (2.21) with solution w to (2.22)) to f is continuous from $L^p(\Omega)$ to $L^6(\Omega)$ and compact from $L^p(\Omega)$ to $L^q(\Omega)$ for $1 \le q < 6$.

Problem 2.8 (Neumann Problem (★★★★)). *Solution on page 124.* Let Ω be a bounded connected (non-empty) open subset of $\mathbb{R}^N$ ($N \ge 1$), with a Lipschitz boundary; set $H = \{u \in H^1(\Omega), \int_\Omega u(x)\,\mathrm{d}x = 0\}$. Recall that on such an open subset, a L^1_{loc} function whose derivatives (in the sense of derivatives by transposition) are null is necessarily constant (that is, there exists a $C \in \mathbb{R}$ such that this function is equal to C a.e.), see Problem 1.4.

---

[17] Recall that every compact linear mapping is continuous; this is false in the non-linear case.

1. *Mean Poincaré inequality.* Show that H is a closed subspace of $H^1(\Omega)$ and that, on H, the H^1 norm is equivalent to the $\|\cdot\|_m$ norm defined by $\|u\|_m = \|(|\nabla u|)\|_{L^2(\Omega)}$. [Suggestion: show, by contradiction, that there exists a sequence of elements of H, $(u_n)_{n\in\mathbb{N}}$ such that $\|u_n\|_{L^2(\Omega)} > n \|u_n\|_m$ for all $n \in \mathbb{N}$.]

2. *Characterisation of $(H^1(\Omega))'$.* Let $T \in (H^1(\Omega))'$, show that there exists an $a \in \mathbb{R}$ and $F \in (L^2(\Omega))^N$ such that

$$\langle T, u \rangle_{(H^1(\Omega))', H^1(\Omega)} = a \int_\Omega u(x)\,\mathrm{d}x + \int_\Omega F(x) \cdot \nabla u(x)\,\mathrm{d}x, \ \forall u \in H^1(\Omega). \quad (2.24)$$

[Suggestion: consider $T_{|H}$ and use a suitable injective mapping from H to $L^2(\Omega)^N$.]

For every $x \in \Omega$, let $A(x)$ be an $N \times N$ matrix, whose coefficients are denoted by $a_{i,j}(x)$, $i, j = 1, \ldots, N$. Assume that $a_{i,j} \in L^\infty(\Omega)$ for all $i, j = 1, \ldots, N$ and that there exists an $\alpha > 0$ such that $A(x)\xi \cdot \xi \geq \alpha|\xi|^2$, for all $\xi \in \mathbb{R}^N$ and a.e. in $x \in \Omega$. Let $a \in \mathbb{R}$ and $F \in (L^2(\Omega))^N$; we seek a solution u of

$$u \in H^1(\Omega), \ : \ \forall v \in H^1(\Omega),$$
$$\int_\Omega A(x)\nabla u(x) \cdot \nabla v(x)\,\mathrm{d}x = a \int_\Omega v(x)\,\mathrm{d}x + \int_\Omega F(x) \cdot \nabla v(x)\,\mathrm{d}x. \quad (2.25)$$

3. *Existence and uniqueness.*

   a. If $a \neq 0$, show that (2.25) has no solution.
   b. If $a = 0$, show that (2.25) has a solution and that this solution is unique if it is required to belong to H.
   c. In this question, assume that $a = 0$, $a_{i,j} \in C^\infty(\bar{\Omega}, \mathbb{R})$ for all $i, j = 1, \ldots, N$, $F \in C^\infty(\bar{\Omega}, \mathbb{R}^N)$, and that Ω is of class C^∞ in the sense of Definition 1.23; assume further that the solution (belonging to H) of (2.25) is also in $C^\infty(\bar{\Omega}, \mathbb{R})$. Show that $-\mathrm{div}(A\nabla u) = -\mathrm{div}\,F$ in Ω, and that $A\nabla u \cdot n = F \cdot n$ on $\partial\Omega$, where n is the outward normal unit vector to $\partial\Omega$. This condition is called a Neumann boundary condition.

4. *Dependence on parameters.* Assume that $a = 0$ and let u be the solution (belonging to H) of (2.25). Assume furthermore that, for all $n \in \mathbb{N}$, $u_n \in H$ is the solution to (2.25) with A_n instead of A and F_n instead of F (and $a = 0$); assume also that

   - $A_n = (a_{i,j}^{(n)})_{i,j=1,\ldots,N}$ satisfies, for all n, the same assumptions as A with α independent of n,
   - $(a_{i,j}^{(n)})_{n\in\mathbb{N}}$ is bounded in $L^\infty(\Omega)$, for all $i, j = 1, \ldots, N$,
   - $a_{i,j}^{(n)} \to a_{i,j}$ a.e., when $n \to \infty$, for all $i, j = 1, \ldots, N$,
   - $F_n \to F$ in $L^2(\Omega)^N$, when $n \to \infty$.

   Under these assumptions, show that $(u_n)_{n\in\mathbb{N}}$ is bounded in H, then that $u_n \to u$ weakly in $H^1(\Omega)$ (when $n \to \infty$) and finally that $u_n \to u$ in $H^1(\Omega)$.

5. (H^2 regularity by the reflection technique, question independent of the previous one.) Assume that $a = 0$ and that there exists an $f \in L^2(\Omega)$ such that $\int_\Omega F(x) \cdot \nabla v(x)\,dx = \int_\Omega f(x)v(x)\,dx$, for all $v \in H^1(\Omega)$. Let u be the solution (belonging to H) of (2.25) and assume that $N = 2$ and that $\Omega =]0,1[\times]0,1[$. Let $\Omega_s =]-1,1[\times]0,1[$ and define A, f and u on Ω_s by setting

$$\text{if } (x_1, x_2) \in]-1,0[\times]0,1[, \begin{cases} a_{i,i}(x_1, x_2) = a_{i,i}(-x_1, x_2) \\ a_{i,j}(x_1, x_2) = -a_{i,j}(-x_1, x_2) \text{ if } i \neq j, \\ f(x_1, x_2) = f(-x_1, x_2), \\ u(x_1, x_2) = u(-x_1, x_2) \end{cases}$$

Show that u is a solution to (2.25) with Ω_s instead of Ω. Using several reflections, show that $u \in H^2(\Omega)$ in the case $A(x) = \mathrm{Id}$ for all $x \in \Omega$.

Problem 2.9 (An example in $H^1(\mathbb{R}^N)$ ($\star$)). *Solution on page 129.* Let $f \in L^2(\mathbb{R}^N)$, $N \geq 1$.

1. Let $u \in H^1(\mathbb{R}^N)$ and $i \in \{1, \ldots, N\}$. Show that $\Delta u - u = D_i f$ in $\mathcal{D}^\star(\mathbb{R}^N)$ if and only if u satisfies

$$\int \nabla u(x) \cdot \nabla v(x)\,dx + \int u(x)v(x)\,dx$$
$$= \int f(x) D_i v(x)\,dx, \ \forall v \in H^1(\mathbb{R}^N). \quad (2.26)$$

2. Show that there exists a unique solution $u \in H^1(\mathbb{R}^N)$ to (2.26) and that $\|u\|_{H^1(\mathbb{R}^N)} \leq \|f\|_{L^2(\mathbb{R}^N)}$.

Problem 2.10 (H^2 norm on $\mathbb{R}^N$ ($\star\star$)). *Solution on page 130.* Let $N \geq 1$; the aim here is to show that the H^2 norm is equivalent to the sum of the L^2 norm of the function and the L^2 norm of its Laplacian in $\mathbb{R}^N$. This equivalence is useful for instance in the study of a problem involving the biharmonic operator.

1. Let $u \in H^2(\mathbb{R}^N)$. Show that there exist positive C_1 and C_2, depending (possibly) only on N, such that

$$C_1(\|u\|_{L^2}^2 + \|\Delta u\|_{L^2}^2) \leq \|u\|_{H^2}^2 \leq C_2(\|u\|_{L^2}^2 + \|\Delta u\|_{L^2}^2).$$

(Of course, L^2 denotes $L^2(\mathbb{R}^N)$ and H^2 denotes $H^2(\mathbb{R}^N)$.)

2. Let $H^{-2}(\mathbb{R}^N)$ be the (topological) dual space of $H^2(\mathbb{R}^N)$. Let $f \in H^{-2}(\mathbb{R}^N)$ and $\lambda > 0$.

a. Let $u \in H^2(\mathbb{R}^N)$. Show that $\Delta(\Delta u) + \lambda u = f$ in $\mathcal{D}^\star(\mathbb{R}^N)$ if and only if u satisfies

$$\int \Delta u(x) \Delta v(x)\,dx + \lambda \int u(x)v(x)\,dx = \langle f, v \rangle_{H^{-2}, H^2}. \quad (2.27)$$

b. Show that there exists one and only one solution $u \in H^2(\mathbb{R}^N)$ to (2.27).

Problem 2.11 (Modelling of a contact problem ($\star\star\star$)). *Solution on page 132.* Set $B = \{x \in \mathbb{R}^2, |x| < 2\}$, $I =] - 1, 1[$, and $\Omega = B \setminus ([-1, 1] \times \{0\})$ (Ω is therefore an open connected set of $\mathbb{R}^2$). Let $\partial B = \overline{B} - B$; recall that $|x|$ denotes the Euclidean norm of $x \in \mathbb{R}^2$ and $x \cdot y$ the corresponding inner product of x and y ($\in \mathbb{R}^2$). Let $f \in L^2(\Omega)$ and $g \in L^\infty(I)$ such that $g \geq 0$ a.e. (on I). Consider the following problem.

$$-\Delta u(x) = f(x), \ x \in \Omega, \tag{2.28}$$

$$u(x) = 0, \ x \in \partial B, \tag{2.29}$$

$$\partial_y u(x, 0^+) = \partial_y u(x, 0^-), \ x \in I, \tag{2.30}$$

$$\partial_y u(x, 0^+) = g(x)(u(x, 0^+) - u(x, 0^-)), \ x \in I. \tag{2.31}$$

1. (Search for a weak formulation) In this question, we assume that f is a continuous function on $\overline{\Omega}$ and g a continuous function on $\overline{I}$. Let $\Omega_+ = \Omega \cap \{(x, y), \ y > 0\}$ and $\Omega_- = \Omega \cap \{(x, y), \ y < 0\}$. Let $u \in C^2(\Omega, \mathbb{R})$ such that $u_{|\Omega_+} \in C^2(\overline{\Omega_+})$, i.e. $u_{|\Omega_+}$ is the trace on Ω_+ of an element of $C^2(\mathbb{R}^2)$, and $u_{|\Omega_-} \in C^2(\overline{\Omega_-})$. Note then that all the expressions in (2.28)–(2.31) make sense. We have, for example, $u(x, 0^+) = \lim_{y \to 0, \, y>0} u(x, y)$. Show that u is a "classical" solution to (2.28)–(2.31) (that is to say, it satisfies (2.28) for all $x \in \Omega$, (2.29) for all $x \in \partial B$ and (2.30),(2.31) for all $x \in I$ if and only if u satisfies:

$$u(x) = 0, \ \forall x \in \partial B,$$

$$\int_\Omega \nabla u(x) \cdot \nabla v(x) \, dx + \int_I g(x)(u(x, 0^+) - u(x, 0^-))(v(x, 0^+) - v(x, 0^-)) \, dx$$

$$= \int_\Omega f(x)v(x) \, dx, \tag{2.32}$$

for any function $v \in C^2(\Omega, \mathbb{R})$ such that $v_{|\Omega_+} \in C^2(\overline{\Omega_+})$, $v_{|\Omega_-} \in C^2(\overline{\Omega_-})$ and $v(x) = 0$ for all $x \in \partial B$.

2. (Traces and functional space) Assuming the existence of the trace operator (Theorem 1.32), show that there exists a continuous linear operator γ_0 from $H^1(\Omega)$ to $L^2(\partial B)$ such that $\gamma_0(u)(x) = u(x)$ a.e. (for the Lebesgue 1-dimensional measure on ∂B) if $u \in H^1(\Omega)$ and u is continuous on $\overline{B} \setminus [-1, 1] \times \{0\}$. Also show that there exists a γ_+ [resp. γ_-] continuous linear from $H^1(\Omega)$ to $L^2(I)$ such that $\gamma_+(u)(x) = u(x, 0+)$ [resp. $\gamma_-(u)(x) = u(x, 0-)$] a.e. for $x \in I$ if $u \in H^1(\Omega)$ and $u_{|\Omega_+}$ is continuous on $\overline{\Omega_+}$ [resp. $u_{|\Omega_-}$ is continuous on $\overline{\Omega_-}$].

In the following, consider the space $H = \mathrm{Ker}\gamma_0$ (where γ_0 is defined in question 2). The space H is therefore a closed vector subspace of $H^1(\Omega)$.

3. (Coercivity) Show that there exists a $C \in \mathbb{R}_+$ such that $\|u\|_{L^2(\Omega)} \leq C \, \||\nabla u|\|_{L^2(\Omega)}$ for all $u \in H$. [One could, for example, note that $u_{|\Omega_+} \in H^1(\Omega_+)$ and $u_{|\Omega_-} \in H^1(\Omega_-)$.]

4. (Existence and uniqueness of weak solutions) Show that there exists one and only one solution u of (2.33).

$$\begin{cases} u \in H, \\ \displaystyle\int_\Omega \nabla u(x) \cdot \nabla v(x)\,\mathrm{d}x + \int_I g(x)(\gamma_+ u(x) - \gamma_- u(x))(\gamma_+ v(x) - \gamma_- v(x))\,\mathrm{d}x \\ \displaystyle= \int_\Omega f(x)v(x)\,\mathrm{d}x, \ \forall v \in H. \end{cases} \tag{2.33}$$

5. For $n \in \mathbb{N}$, let u_n be the solution to (2.33) with g such that $g(x) = n$, for all $x \in I$. Show that $u_n \to u$ (in a sense to be specified) when $n \to \infty$, where u is the (unique (weak) solution of $-\Delta u = f$ in B, $u = 0$ on ∂B.

Problem 2.12 (From Fourier to Dirichlet... ($\star\star\star$)). *Solution on page 134.* Let $\sigma \geq 0$, $f \in L^2(\mathbb{R}^N_+)$ and $g \in L^2(\mathbb{R}^{N-1})$ and consider the following problem:

$$\begin{aligned} -\Delta u(x) + u(x) &= f(x), \ x \in \mathbb{R}^N_+, \\ -\partial_1 u(0, y) + \sigma u(0, y) &= g(y), \ y \in \mathbb{R}^{N-1}, \end{aligned} \tag{2.34}$$

where $\partial_1 u$ denotes the partial derivative of u with respect to its first argument. The second equation of this system is a boundary condition on the hyperplane $x_1 = 0$ which is called a Fourier or Robin boundary condition.

1. Give a definition of a classical solution to (2.34) and a weak solution to (2.34).
2. Show the existence and uniqueness of the weak solution to (2.34).

In the following questions, assume that $g = 0$ a.e. (for the Lebesgue 1-dimensional measure).

3. Show that the weak solution to (2.34) (found in the previous question) belongs to $H^2(\mathbb{R}^N_+)$. [*Consider the case $N = 2$; as in Theorem 2.18, the generalisation to any $N \geq 2$ does not bring any additional difficulties.*]
4. For $n \in \mathbb{N}$, let u_n be the (weak) solution to (2.34) corresponding to $\sigma = n$. Show that $u_n \to u$ in $H^1(\mathbb{R}^N_+)$, where u is the weak solution of:

$$-\Delta u(x) + u(x) = f(x), \ x \in \mathbb{R}^N_+, \tag{2.35a}$$

$$u(0, y) = 0, \ y \in \mathbb{R}^{N-1}. \tag{2.35b}$$

Problem 2.13 (Schrödinger ($\star\star\star$)). *Solution on page 136.* Let $N \geq 1$ and let Ω be the unit ball of $\mathbb{R}^N$ (in fact, the results of this problem remain true for sufficiently regular bounded open subsets of $\mathbb{R}^N$, but this generalisation is not detailed here).

For $f_1, f_2 \in L^2(\Omega)$, consider the system:

$$-\Delta u_1 + u_2 = f_1 \ \text{in } \Omega, \tag{2.36a}$$

$$-\Delta u_2 - u_1 = f_2 \ \text{in } \Omega, \tag{2.36b}$$

with various boundary conditions given later.

1. *Dirichlet conditions* - Consider in this first question the boundary conditions:

$$u_1 = 0, \; u_2 = 0 \text{ on } \partial\Omega. \tag{2.37}$$

Let $f_1, f_2 \in L^2(\Omega)$. The pair of functions (u_1, u_2) is said to be a weak solution to the problem (2.36)–(2.37) if

$$u_1 \in H_0^1(\Omega), \; u_2 \in H_0^1(\Omega) \text{ and } \forall \varphi \in H_0^1(\Omega),$$

$$\int_\Omega \nabla u_1(x) \cdot \nabla \varphi(x) \, \mathrm{d}x + \int_\Omega u_2(x) \varphi(x) \, \mathrm{d}x = \int_\Omega f_1(x) \varphi(x) \, \mathrm{d}x, \tag{2.38}$$

$$\int_\Omega \nabla u_2(x) \cdot \nabla \varphi(x) \, \mathrm{d}x - \int_\Omega u_1(x) \varphi(x) \, \mathrm{d}x = \int_\Omega f_2(x) \varphi(x) \, \mathrm{d}x.$$

a. Show that problem (2.38) has one and only one solution. [Use the space $V = H_0^1(\Omega) \times H_0^1(\Omega)$.]
b. Show that problem (2.36)–(2.37) admits one and only one solution in the following sense: $u_1 \in H^2(\Omega) \cap H_0^1(\Omega), u_2 \in H^2(\Omega) \cap H_0^1(\Omega)$ and the equations (2.36) are satisfied a.e. on Ω. [Use, in particular, the previous question and the regularity theorem (Theorem 2.19).]
c. Assume in this question that $f_1, f_2 \in C^\infty(\bar\Omega)$; show that the solution to (2.38), denoted (u_1, u_2), belongs to $H^m(\Omega)(= W^{m,2}(\Omega))$ for all $m \in \mathbb{N}$ (thanks to the Sobolev embedding theorems, this gives $u_1, u_2 \in C^\infty(\bar\Omega)$).
d. For $f = (f_1, f_2) \in L^2(\Omega) \times L^2(\Omega)$, let $u = (u_1, u_2)$ be the solution to (2.38), and define a mapping Φ by $u = \Phi(f)$. Show that the operator $\Phi : f \mapsto u$ is a continuous and compact linear operator from $L^2(\Omega) \times L^2(\Omega)$ to itself.

2. *Neumann boundary conditions* - Consider in this second question the boundary conditions:

$$\partial_n u_1 = 0, \; \partial_n u_2 = 0 \text{ on } \partial\Omega, \tag{2.39}$$

where n denotes the outward normal unit vector to $\partial\Omega$. To solve problem (2.36)–(2.39), we introduce a parameter, $k \in \mathbb{N}^\star$, intended to tend towards infinity.

Let $f_1, f_2 \in L^2(\Omega)$. For $k \in \mathbb{N}^\star$, consider the system:

$$
\begin{aligned}
-\Delta u_1 + u_2 + \frac{1}{k} u_1 &= f_1 \text{ in } \Omega, \\
-\Delta u_2 - u_1 + \frac{1}{k} u_2 &= f_2 \text{ in } \Omega,
\end{aligned}
\tag{2.40}
$$

with the boundary conditions (2.39). We say that (u_1, u_2) is a weak solution to problem (2.39)–(2.40) if

$$u_1 \in H^1(\Omega), \ u_2 \in H^1(\Omega), \quad \forall \varphi \in H^1(\Omega),$$

$$\int_\Omega \nabla u_1(x) \cdot \nabla \varphi(x) \, dx + \int_\Omega (u_2(x) + \frac{1}{k} u_1(x)) \varphi(x) \, dx = \int_\Omega f_1(x) \varphi(x) \, dx,$$

$$(2.41a)$$

$$\int_\Omega \nabla u_2(x) \cdot \nabla \varphi(x) \, dx + \int_\Omega (\frac{1}{k} u_2(x) - u_1(x)) \varphi(x) \, dx = \int_\Omega f_2(x) \varphi(x) \, dx.$$

$$(2.41b)$$

Also note that (u_1, u_2) is a weak solution to problem (2.36)–(2.39) if (u_1, u_2) is a solution to (2.41) by replacing $\frac{1}{k}$ with 0.

a. Let $k \in \mathbb{N}^\star$; show that the problem (2.41) has one and only one solution, denoted by $(u_1^{(k)}, u_2^{(k)})$ in the following.
b. Show that

$$\|u_1^{(k)}\|_{L^2(\Omega)}^2 + \|u_2^{(k)}\|_{L^2(\Omega)}^2 \leq \|f_1\|_{L^2(\Omega)}^2 + \|f_2\|_{L^2(\Omega)}^2.$$

Deduce that the sequences $(u_1^{(k)})_{n \in \mathbb{N}^\star}$ and $(u_2^{(k)})_{n \in \mathbb{N}^\star}$ are bounded in $H^1(\Omega)$.
c. Show that there is one and only one solution to problem (2.41) obtained by replacing $\frac{1}{n}$ with 0, that is, one and only one weak solution to problem (2.36)–(2.39). [For the proof of existence, use the sequences $(u_1^{(k)})_{k \in \mathbb{N}^\star}$, and $(u_2^{(k)})_{k \in \mathbb{N}^\star}$ from the previous question and let n tend to $+\infty$; then show uniqueness.]
d. Show that problem (2.36)–(2.39) has one and only one solution in the following sense: $u_1 \in H^2(\Omega)$, $u_2 \in H^2(\Omega)$, the equations (2.36) are satisfied a.e. on Ω and the equations (2.39) are satisfied a.e. (for the $N - 1$-dimensional Lebesgue measure) on $\partial\Omega$ using the trace operator of $H^1(\Omega)$ in $L^2(\partial\Omega)$ to give meaning to $\partial_n u_1$ and $\partial_n u_2$. [*We assume here that the first point of the regularity theorem (Theorem 2.19) is still valid if u is a solution to (2.6) with $H^1(\Omega)$ instead of $H_0^1(\Omega)$.*]
e. For $f = (f_1, f_2) \in L^2(\Omega) \times L^2(\Omega)$, let $u = (u_1, u_2)$ be the weak solution to (2.36)–(2.39) and define a mapping Φ by $u = \Phi(f)$. Show that $\Phi : f \mapsto u$ is a linear continuous and compact mapping from $L^2(\Omega) \times L^2(\Omega)$ to itself.

3. *Mixed boundary conditions* - Similarly, briefly indicate how to solve problem (2.36) with the boundary conditions:

$$u_1 = 0, \ \partial_n u_2 = 0 \ \text{on} \ \partial\Omega.$$

Problem 2.14 (Trudinger[18]–Moser[19] inequality and $L^1(\sqrt{\ln(L^1)}) \subset H^{-1}$, $N = 2$ ($\star\star\star\star$)). *Solution on page 140.*

Part I, decomposition in $H_0^1(\Omega)$

Let Ω be an open subset of $\mathbb{R}^N$, $N \geq 1$, and define φ from $\mathbb{R}$ to $\mathbb{R}$ by

$$\varphi(s) = \begin{cases} s, & \text{for } 0 \leq s \leq 1, \\ -\frac{s^2}{2} + 2s - \frac{1}{2}, & \text{for } 1 < s \leq 2, \\ \frac{3}{2}, & \text{for } 2 < s, \\ -\varphi(-s), & \text{for } s < 0. \end{cases}$$

For $k \in \mathbb{N}^\star$, define φ_k from $\mathbb{R}$ to $\mathbb{R}$ as $\varphi_k(s) = k\varphi(\frac{s}{k})$ for $s \in \mathbb{R}$.

1. Show that, for all $s \in \mathbb{R}$, $\varphi_k(s) \to s$ and $\varphi'_k(s) \to 1$ when $k \to \infty$ and that $|\varphi_k(s)| \leq |s|$, $\varphi'_k(s) \leq 1$.
2. Let $u \in H_0^1(\Omega)$; show that $\varphi_k(u) \in H_0^1(\Omega)$, for all $k \in N^\star$, and that $\varphi_k(u) \to u$ in $H_0^1(\Omega)$ when $k \to \infty$. [Use Lemma 2.25.]
3. From this, deduce that, for all $u \in H_0^1(\Omega)$ and for all $\varepsilon > 0$, there exist $u_1 \in L^\infty(\Omega)$ and $u_2 \in H_0^1(\Omega)$ such that $u = u_1 + u_2$ and $\|u_2\|_{H_0^1} \leq \varepsilon$.

Part II, Trudinger–Moser inequality

1. Show that there exists a $D > 0$ such that

$$\|u\|_{L^q(\mathbb{R}^2)} \leq Dq \|u\|_{H_0^1(\mathbb{R}^2)}, \quad \forall u \in H_0^1(\mathbb{R}^2), \ \forall q \in [2, \infty[. \qquad (2.42)$$

[It is suggested to specify the value of $D_{N,q}$ given by (1.28) in the solution of the third question of Problem 1.9.]

In the remainder of this problem, we admit that in question 1 it is possible to replace q with $\sqrt{q}$ in (2.42), that is to say, there exists a $D > 0$ such that

$$\|u\|_{L^q(\mathbb{R}^2)} \leq D\sqrt{q} \|u\|_{H_0^1(\mathbb{R}^2)}, \quad \forall u \in H_0^1(\mathbb{R}^2), \ \forall q \in [2, \infty[. \qquad (2.43)$$

Let Ω be an open bounded subset of $\mathbb{R}^2$.

2. Show that there exists a $C > 0$, only depending on Ω, such that

$$\|u\|_{L^q(\Omega)} \leq C\sqrt{q} \|u\|_{H_0^1(\Omega)}, \quad \forall u \in H_0^1(\Omega), \ \forall q \in [1, \infty[. \qquad (2.44)$$

3. Let $u \in H_0^1(\Omega)$, $u \neq 0$, such that $\|u\|_{H_0^1(\Omega)} \leq 1$. Show that there exist $\sigma > 0$ and $a > 0$, only depending on Ω, such that $e^{\sigma u^2} \in L^1(\Omega)$ and $\left\|e^{\sigma u^2}\right\|_{L^1(\Omega)} \leq a$. [Suggestion: Expand e^s in powers of s.]

---

[18] Neil Trudinger (1942–), Australian mathematician, specialist in non-linear elliptic PDEs.

[19] Jürgen Moser (1928–1999), German-American mathematician, specialist in Hamiltonian systems and PDEs.

4. Using part I (and question 3) show that $e^{\sigma u^2} \in L^p(\Omega)$ for all $u \in H_0^1(\Omega)$, all $\sigma > 0$ and all $p \in [1, \infty[$.

Part III, on the resolution of the Dirichlet problem

Let Ω be an open bounded subset of $\mathbb{R}^2$. Let $f \in L^1(\Omega)$ such that $f\sqrt{|\ln(|f|)|} \in L^1(\Omega)$.

1. (Preliminary.) Let $\sigma > 0$. Show that there exist $\beta, \gamma \in \mathbb{R}_+^\star$, only depending on σ, such that
$$st \le e^{\sigma s^2} + \beta t\sqrt{|\ln t|} + \gamma t, \ \forall s, t \in \mathbb{R}_+^\star.$$

   [Suggestion: notice that $st \le \max\{\beta t\sqrt{|\ln t|}, se^{\frac{s^2}{\beta^2}}\}$ for all $\beta > 0$ (and all $s, t > 0$), then choose β (depending on σ) and conclude.]
2. Show that $fu \in L^1(\Omega)$ for all $u \in H_0^1(\Omega)$ and that the mapping $T : u \mapsto \int_\Omega f(x)u(x)\,dx$ belongs to $H^{-1}(\Omega)$.
3. Show that there exists one and only one $u \in H_0^1(\Omega)$ such that $-\Delta u = f$ in $\mathcal{D}^\star(\Omega)$.

Part IV, counterexample

Let Ω be an open bounded subset of $\mathbb{R}^2$ and $\theta \in]0, \frac{1}{2}[$. Assuming that $0 \in \Omega$, let $\delta \in]0, \frac{1}{2}[$ be such that $B_{2\delta} = \{x \in \mathbb{R}^2, |x| < 2\delta\} \subset \Omega$.

1. Let $\gamma \in]0, \frac{1}{2}[$. Show that there exists a function $u \in H_0^1(\Omega)$ such that $u(x) = (-\ln|x|)^\gamma$ a.e. on B_δ. [Suggestion: consider the function $v \in H^1(B_{2\delta})$ defined by $v(x) = (-\ln(|x|))^\gamma$, see Problem 1.5].
2. Show that there exists a function $f \in L^1(\Omega)$ such that $f(|\ln|f||)^\theta \in L^1(\Omega)$ and $fu \notin L^1(\Omega)$ for some $u \in H_0^1(\Omega)$.
3. Show that there exists a function $f \in L^1(\Omega)$ such that $f(\ln|f|)^\theta \in L^1(\Omega)$ and such that there does not exist a $u \in H_0^1(\Omega)$ satisfying $-\Delta u = f$ in $\mathcal{D}^\star(\Omega)$.

Problem 2.15 (Hodge Decomposition ($\star$)). *Solution on page 144.* Let Ω be an open bounded subset connected to the Lipschitz boundary of $\mathbb{R}^N$ ($N \ge 1$) and $f \in (L^2(\Omega))^N$.

Show that there exists a function $u \in H^1(\Omega)$ such that
$$\int_\Omega \nabla u(x) \cdot \nabla\varphi(x)\,dx = \int_\Omega f(x) \cdot \nabla\varphi(x)\,dx, \ \ \forall\varphi \in H^1(\Omega).$$

[Suggestion: use Problem 2.8.] Deduce that there exist $u \in H^1(\Omega)$ and $g \in (L^2(\Omega))^N$ such that $f = \nabla u + g$, a.e. in Ω and $\int_\Omega g(x) \cdot \nabla\varphi(x)\,dx = 0$ for all $\varphi \in H^1(\Omega)$ (which is equivalent to saying that div $g = 0$, see Definition 2.35).

Assume now that $g \in C^1(\bar{\Omega})$ and that $\Omega = (]0, 1[)^N$. Show that div $g = 0$ on Ω and that $g \cdot = 0$ a.e. (for the $(N-1)$-dimensional Lebesgue measure) on $\partial\Omega$, where n is a normal vector to $\partial\Omega$.

Problem 2.16 (Wentzel[20] boundary conditions ($\star\star\star\star$)). *Solution on page 144.*
Let $H_p^1(0, 2\pi) = \{u \in H^1(]0, 2\pi[); u(0) = u(2\pi)\}$; recall that, if $u \in H^1(]0, 2\pi[)$,
u always admits a continuous representative on $[0, 2\pi]$ and is then identified with
this continuous representative.

Let $B = \{(x, y)^t \in \mathbb{R}^2, x^2 + y^2 < 1\}$. Recall that there exists a linear continuous
mapping $\gamma : H^1(B) \to L^2(\partial B)$ such that $\gamma(u) = u$ a.e. on ∂B if $u \in H^1(B) \cap C(\bar{B}, \mathbb{R})$.

If $w \in L^2(\partial B)$, define $j(w) \in L^2(]0, 2\pi[)$ by $j(w)(\theta) = w(\cos\theta, \sin\theta)$, for
$\theta \in [0, 2\pi[$. The mapping j is an isometry of $L^2(\partial B)$ onto $L^2(]0, 2\pi[)$, so that
$\bar{\gamma} = j \circ \gamma$ is a linear continuous mapping from $H^1(B)$ to $L^2(]0, 2\pi[)$. Set $H = \{u \in H^1(B) : \bar{\gamma}(u) \in H_p^1(0, 2\pi)\}$ and equip H with the following inner product

$$(u \,|\, v)_H = (u \,|\, v)_{H^1(B)} + (\bar{\gamma}(u) \,|\, \bar{\gamma}(v))_{H_p^1(0,2\pi)}.$$

Part I (Some functional analysis preliminaries)

1. Show that $H_p^1(0, 2\pi)$ is a Hilbert space.
2. Show that H is a Hilbert space.

Part II (Wentzel boundary conditions) For $(x, y) \in \mathbb{R}^2$, $(x, y) \neq (0, 0)$, define r
and θ by $r = (x^2 + y^2)^{\frac{1}{2}}$ and $\theta \in [0, 2\pi[$ such that $x = r\cos\theta$ and $y = r\sin\theta$. For
$u \in C^1(\mathbb{R}^N \setminus (0, 0), \mathbb{R})$, set $\bar{u}(r, \theta) = u(r\cos\theta, r\sin\theta)$ so that

$$\partial_r\bar{u}(r, \theta) = \cos\theta\partial_x u(r\cos\theta, r\sin\theta) + \sin\theta\partial_y u(r\cos\theta, r\sin\theta)$$
$$\partial_\theta\bar{u}(r, \theta) = -r\sin\theta\partial_x u(x, y) + r\cos\theta\partial_y u(x, y).$$

For given f and g, consider the following problem:

$$- \Delta u(x, y) + u(x, y) = f(x, y), \quad (x, y) \in B, \tag{2.45a}$$
$$\partial_r\bar{u}(1, \theta) - \partial_\theta^2\bar{u}(1, \theta) + \bar{u}(1, \theta) = g(\cos\theta, \sin\theta), \quad \theta \in [0, 2\pi[. \tag{2.45b}$$

Note that the boundary condition (2.45b) is indeed written for $(x, y) \in \partial B$. Let
$f \in L^2(B)$ and $g \in L^2(\partial B)$, and define a weak solution to (2.45) as a solution of the
following problem:

$$\begin{cases} u \in H, \\[2mm] \displaystyle\int_B \left(\sum_i D_i u(z)D_i v(z) + u(z)v(z)\right) \mathrm{d}z \\[4mm] \qquad + \displaystyle\int_0^{2\pi} (D\bar{\gamma}(u)(\theta)D\bar{\gamma}(v)(\theta) + \bar{\gamma}(u)(\theta)\bar{\gamma}(v)(\theta)) \,\mathrm{d}\theta \\[4mm] \qquad = \displaystyle\int_B f(z)v(z) \,\mathrm{d}z + \int_0^{2\pi} j(g)(\theta)j(\gamma(v))(\theta) \,\mathrm{d}\theta, \quad \forall v \in H. \end{cases} \tag{2.46}$$

---

[20] Gregor Wentzel (1898–1978), German physicist known for his development of quantum mechanics.

1. Let $f \in L^2(B)$ and $g \in L^2(\partial B)$; show that there exists one and only one solution to (2.46).
2. (More difficult question.) We remove in this question the term "uv" in the first integral of (2.46). Let $f \in L^2(B)$ and $g \in L^2(\partial B)$. Show that there still exists one and only one solution to (2.46).
3. Let $f \in C(\overline{B}, \mathbb{R})$, $g \in C(\partial B, \mathbb{R})$ and let $u \in C^2(\overline{B}, \mathbb{R})$. Show that u is a classical solution to (2.45) (i.e. satisfies (2.45a) for all $(x, y) \in \mathcal{B}$ and (2.45b) for all $(x, y) \in \partial B$) if and only if u is a weak solution to (2.45). [Use the fact that $C^2(\overline{B}, \mathbb{R})$ is dense in H.]
4. For $f \in L^2(B)$ and $g \in L^2(\partial B)$, define $T(f, g) = (u, \gamma(u)) \in L^2(B) \times L^2(\partial B)$, where u is the unique weak solution to (2.45a)–(2.45b). Define the inner product in $L^2(B) \times L^2(\partial B)$ by

$$((f, g) \mid (\varphi, \psi))_{L^2(B) \times L^2(\partial B)} = \int_B f(x)\varphi(x)\,dx + \int_0^{2\pi} j(g)(\theta) j(\psi)(\theta)\,d\theta.$$

Show that T is a compact self-adjoint linear operator of $L^2(B) \times L^2(\partial B)$ to itself.

Problem 2.17 (Stokes problem, velocity and pressure ($\star\star$)). *Solution on page 148.*
Let Ω be a bounded, connected open subset of $\mathbb{R}^N$ ($N \geq 1$) with a Lipschitz boundary and $f = (f_1, \ldots, f_N)^t \in (L^2(\Omega))^N$. The strong formulation of the (incompressible) Stokes problem is to find a solution $u = (u_1, \ldots, u_N)^t : \Omega \to \mathbb{R}^N$ and $p : \Omega \to \mathbb{R}$ to

$$\begin{aligned} -\Delta u + \nabla p &= f \text{ in } \Omega, \\ \operatorname{div} u &= 0 \text{ in } \Omega, \\ u &= 0 \text{ on } \partial\Omega. \end{aligned} \tag{2.47}$$

Note that the first equation of (2.47) is a vector equation.

Set $H = H_0^1(\Omega)^N$ and $V = \{u \in H : \operatorname{div} u = 0 \text{ a.e. in } \Omega\}$. A weak solution to (2.47) is defined as a pair of functions (u, p) which is a solution to

$$u = (u_1, \ldots, u_N)^t \in V, \ p \in L^2(\Omega) \ : \ \forall v = (v_1, \ldots, v_N)^t \in H,$$

$$\sum_{i=1}^N \int_\Omega \nabla u_i(x) \cdot \nabla v_i(x)\,dx - \int_\Omega p(x)\operatorname{div} v(x)\,dx = \int_\Omega f(x) \cdot v(x)\,dx. \tag{2.48}$$

Observe that a classical solution (u, p) of (2.47) is a solution to (2.48).

Part I, existence and uniqueness of u

1. Show that if (u, p) is a classical solution to (2.47), u is then a solution of

$$u = (u_1, \ldots, u_N)^t \in V \ : \ \forall v = (v_1, \ldots, v_N)^t \in V,$$

$$\sum_{i=1}^N \int_\Omega \nabla u_i(x) \cdot \nabla v_i(x)\,dx = \int_\Omega f(x) \cdot v(x)\,dx. \tag{2.49}$$

The rest of this first part is devoted to showing that (2.49) has one and only one solution and that if (u, p) is a solution to (2.48), then u is the unique solution to (2.49).

1. Show that V is a closed subspace of H.
2. Show that (2.49) admits one and only one solution. [Use the Lax–Milgram theorem.]
3. Let (u, p) be a solution to (2.48); show that u is the unique solution to (2.49).

Let u be the solution to (2.49). The rest of the problem aims at finding p such that (u, p) is a solution to (2.48).

Part II, functional analysis preliminaries

Let E and F be two (real) Hilbert spaces and let $(\cdot|\cdot)_E$ (resp. $(\cdot|\cdot)_F$) be the inner product in E (resp. F). Let A be a continuous linear operator from E to F and $A^\star$ be the adjoint operator of A; the operator $A^\star$ is a continuous linear operator from F to E. By Definition 2.13, for all $g \in F$, $A^\star g$ is the unique element of E defined by

$$(A^\star g|u)_E = (g|Au)_F \text{ for all } u \in E.$$

Note that the existence and uniqueness of $A^\star g$ is given by the Riesz representation theorem, see footnote 6 on page 71.

1. Show that $\mathrm{Ker}A = (\mathrm{Im}A^\star)^\perp$. (Recall that if $G \subset E$, $G^\perp = \{u \in E, (u|v)_E = 0$ for all $v \in G\}$.)
2. Show that $(\mathrm{Ker}A)^\perp = \overline{\mathrm{Im}A^\star}$.

Part III, Partial existence and uniqueness of p

In this part, we use the following lemma (often attributed to J. Nečas,[21], 1965), which we admit.

Lemma 2.38. *Let Ω be a bounded connected open subset of $\mathbb{R}^N$ ($N \geq 1$) with a Lipschitz boundary and $q \in L^2(\Omega)$ such that $\int_\Omega q(x)\,dx = 0$. Then there exists a $v \in (H_0^1(\Omega))^N$ such that $\mathrm{div}\, v = q$ a.e. in Ω and*

$$\|v\|_{H_0^1(\Omega)^N} \leq C \|q\|_{L^2(\Omega)},$$

where $C \geq 0$ depends only on Ω.

Let $F = L^2(\Omega)$, and set, for $u \in H$, $Au = \mathrm{div}\, u$, so that A is a continuous linear operator from H to F.

1. Let $(p_n)_{n\in\mathbb{N}}$ be a sequence in F and $v \in H$ such that $A^\star p_n \to v$ in H as $n \to +\infty$. For $n \in \mathbb{N}$, set $q_n = p_n - a_n$, where a_n is the mean value of p_n in Ω.

   a. Show that $A^\star p_n = A^\star q_n$.
   b. Show that the sequence $(q_n)_{n\in\mathbb{N}}$ is bounded in F. [Use Lemma 2.38.]
   c. Show that $v \in \mathrm{Im}A^\star$.

---

[21] Jindřich Nečas (1929–2002), Czech mathematician, specialist in PDEs.

2. Show that $(\mathrm{Ker} A)^{\perp} = \mathrm{Im} A^{\star}$ and that $\mathrm{Ker} A = V$.
3. The inner product in H is defined by

$$(u \mid v)_H = \sum_{i=1}^{N} \int_{\Omega} \nabla u_i(x) \cdot \nabla v_i(x) \, dx.$$

Thanks to the Riesz representation theorem, see footnote 6 on page 71, the element $T_f \in H$ is defined by $(T_f \mid v)_H = \int_{\Omega} f(x) \cdot v(x) \, dx$ for all $v \in H$.
Recall that u is the solution to (2.49).

a. Show that $u - T_f \in V^{\perp}$ and deduce that $u - T_f \in \mathrm{Im} A^{\star}$.
b. Show that there exists a function $p \in F$ such that (u, p) is a solution to (2.48).

4. Let (u_1, p_1) and (u_2, p_2) be two solutions of (2.48). Show that $u_1 = u_2 = u$ (where u is the unique solution to (2.49)) and that there exists an $a \in \mathbb{R}$ such that $p_1 - p_2 = a$ a.e.

Problem 2.18 (Stokes problem, penalisation ($\star$)). *Solution on page 150.* We use here the same notations and assumptions as in the previous problem, where it was shown that if a pair of functions (u, p) is a weak solution to the Stokes problem (2.47) (that is to say, a solution of (2.48)) then u is the unique solution to (2.49).

The aim here is to show that this solution can be obtained by a penalisation method. Let $n \in \mathbb{N}^{\star}$. Consider the following problem:

$$u = (u_1, \ldots, u_N)^t \in H \; : \; \forall i \in \{1, \ldots, N\},$$
$$\int_{\Omega} (\nabla u_i(x) \cdot \nabla v(x) + n \mathrm{div}\, u(x) D_i v(x)) \, dx = \int_{\Omega} f_i(x) v(x) \, dx, \qquad (2.50)$$
$$\forall v \in H_0^1(\Omega).$$

1. Let $n \in \mathbb{N}^{\star}$. Show that there exists a unique solution $u^{(n)}$ to (2.50). [Suggestion: use the Riesz representation theorem or the Lax–Milgram theorem on $(H_0^1(\Omega))^N$.]
2. Show that the sequence $(u^{(n)})_{n \in \mathbb{N}}$ is bounded in $(H_0^1(\Omega))^N$ and that the sequence $(\sqrt{n}\, \mathrm{div}\, u^{(n)})_{n \in \mathbb{N}}$ is bounded in $L^2(\Omega)$.
3. Show that $u^{(n)} \to u$ weakly in $(H_0^1(\Omega))^N$ as $n \to +\infty$, where u is the solution to (2.49).

Problem 2.19 (Sequential continuity of L^2-weak in H_0^1 ($\star$)). *Solution on page 151.* Let Ω be an open bounded subset of $\mathbb{R}^N$ ($N > 1$). For every $x \in \Omega$, let $A(x)$ be a given matrix whose coefficients are denoted $a_{i,j}(x)$, $i, j = 1, \ldots, N$; assume that $a_{i,j} \in L^{\infty}(\Omega)$ for all $i, j = 1, \ldots, N$ and that there exists an $\alpha > 0$ such that $A(x)\xi \cdot \xi \geq \alpha |\xi|^2$, for all $\xi \in \mathbb{R}^N$ and a.e. in $x \in \Omega$.

For $f \in L^2(\Omega)$, there exists a unique solution to the following problem:

$$\begin{cases} u \in H_0^1(\Omega), \\ \int_{\Omega} A(x)\nabla u(x) \cdot \nabla v(x) \, dx = \int_{\Omega} f(x) v(x) \, dx, \quad \text{for all } v \in H_0^1(\Omega). \end{cases} \qquad (2.51)$$

Let $(f_n)_{n\in\mathbb{N}}$ be a bounded sequence in $L^2(\Omega)$ and $f \in L^2(\Omega)$. Let u be the solution to (2.51) and, for $n \in \mathbb{N}$, let u_n be the solution to (2.51) with f_n instead of f. Assume that $f_n \to f$ weakly in $L^2(\Omega)$.

1. Show that the sequence $(u_n)_{n\in\mathbb{N}}$ is bounded in $H_0^1(\Omega)$.
2. Show that $u_n \to u$ weakly in $H_0^1(\Omega)$ and that $u_n \to u$ in $L^2(\Omega)$ as $n \to +\infty$.
3. Show that, as $n \to +\infty$,

$$\int_\Omega A(x)\nabla u_n(x) \cdot \nabla u_n(x)\, dx \to \int_\Omega A(x)\nabla u(x) \cdot \nabla u(x)\, dx.$$

[Use the fact that $\int_\Omega A(x)\nabla u_n(x) \cdot \nabla u_n(x)\, dx = \int_\Omega f_n(x)u_n(x)\, dx$ and pass to the limit in the right term of this equality.]

4. Show that $u_n \to u$ in $H_0^1(\Omega)$. [One may consider $\int_\Omega A(x)\nabla(u_n - u)(x) \cdot \nabla(u_n - u)(x)\, dx$.]

Problem 2.20 (Preliminary to Problem 2.21). Let φ be a non-increasing function from $\mathbb{R}_+$ to $\mathbb{R}_+$ and assume that there exist $C > 0$ and $\beta > 1$ such that

$$0 \le x < y \Rightarrow \varphi(y) \le C\frac{\varphi(x)^\beta}{y - x}. \tag{2.52}$$

Show that there exists an $a \in \mathbb{R}_+$ such that $\varphi(a) = 0$. [One may show the existence of an increasing sequence $(a_k)_{k\in\mathbb{N}^\star}$ such that $\varphi(a_k) \le \frac{1}{2^k}$ for all $k \in \mathbb{N}^\star$ and $\lim_{k\to\infty} a_k < +\infty$. This result can be obtained by showing that there exists an a_0 such that $\varphi(a_0) \le 1$; then, with a proof by induction, define a_{k+1} by $\frac{C}{a_{k+1}-a_k}\frac{1}{2^{k\beta}} = \frac{1}{2^{k+1}}$.]

Problem 2.21 (Bounded solutions of an elliptic problem ($\star\star\star$)). *Solution on page 153*. Let Ω be an open bounded subset of $\mathbb{R}^N$ ($N > 1$). For every $x \in \Omega$, let $A(x)$ be a given matrix whose coefficients are denoted $a_{i,j}(x)$, $i, j = 1, \ldots, N$. Assume that $a_{i,j} \in L^\infty(\Omega)$ for all $i, j = 1, \ldots, N$ and that there exists an $\alpha > 0$ such that $A(x)\xi \cdot \xi \ge \alpha|\xi|^2$, for all $\xi \in \mathbb{R}^N$ and a.e. in $x \in \Omega$.

If B is a Borel subset of $\mathbb{R}^N$, let $\lambda_N(B)$ denote the N-dimensional Lebesgue measure of B.

1. Let $F \in L^2(\Omega)^N$. Show that there exists a unique solution u of

$$\begin{cases} u \in H_0^1(\Omega), \\ \displaystyle\int_\Omega A(x)\nabla u(x) \cdot \nabla v(x)\, dx = \int_\Omega F(x) \cdot \nabla v(x)\, dx, \forall v \in H_0^1(\Omega). \end{cases} \tag{2.53}$$

Let $p > N$: assume for the rest of the problem that $F \in L^p(\Omega)^N$ (recall that $L^p(\Omega)^N \subset L^2(\Omega)^N$ because $p > 2$) and denote by u the unique solution to (2.53). For $k \in \mathbb{R}_+$, define the function S_k from $\mathbb{R}$ to $\mathbb{R}$ by

$$S_k(s) = \begin{cases} 0 & \text{if } -k \le s \le k, \\ s - k & \text{if } s > k, \\ s + k & \text{if } s < -k. \end{cases}$$

Recall that if $v \in H_0^1(\Omega)$ we have $S_k(v) \in H_0^1(\Omega)$ and $\nabla S_k(v) = \mathbb{1}_{A_k} \nabla v$ a.e., with $A_k = \{|v| > k\}$ (see Remark 2.27).

1. Let $k \in \mathbb{R}_+$, Show that

$$\alpha \, \|\|\nabla S_k(u)\|\|_{L^2(\Omega)} = \alpha \left(\int_{A_k} \nabla u(x) \cdot \nabla u(x) \, dx \right)^{\frac{1}{2}} \leq \lambda_N(A_k)^{\frac{1}{2} - \frac{1}{p}} \, \|\|F\|\|_{L^p(\Omega)} .$$

[Suggestion: take $v = S_k(u)$ in (2.53) and use Hölder's inequality.]
2. Set $1^\star = \frac{N}{N-1}$. Recall that there exists a C_1 depending only on N such that

$$\|w\|_{L^{1^\star}(\Omega)} \leq C_1 \, \|w\|_{W_0^{1,1}(\Omega)} = C_1 \, \|\|\nabla w\|\|_{L^1(\Omega)} \text{ for all } w \in W_0^{1,1}(\Omega).$$

Let $k, h \in \mathbb{R}_+$ such that $k < h$. Show that

$$(h - k)\lambda_N(A_h)^{\frac{N-1}{N}} \leq \left(\int_{A_h} |S_k(u(x)|^{1^\star} \, dx \right)^{\frac{1}{1^\star}} \leq C_1 \, \|\|\nabla S_k(u)\|\|_{L^1(\Omega)}$$

$$\leq C_1 \, \|\|\nabla S_k(u)\|\|_{L^2(\Omega)} \, \lambda_N(A_k)^{\frac{1}{2}}.$$

Deduce that there exists a C_2 depending only on C_1, α, F and p such that

$$(h - k)\lambda_N(A_h)^{\frac{N-1}{N}} \leq C_2 \lambda_N(A_k)^{1 - \frac{1}{p}} .$$

3. Show that $u \in L^\infty(\Omega)$ (that is, there exists an $a \in \mathbb{R}_+$ such that $\lambda_N(A_a) = 0$).
   [Suggestion: set $\varphi(k) = \lambda_N(A_k)^{\frac{N-1}{N}}$ and use Problem 2.20.]
4. Show that there exists a C_3 depending only on Ω, α and p such that

$$\|u\|_{L^\infty(\Omega)} \leq C_3 \, \|\|F\|\|_{L^p(\Omega)} .$$

Problem 2.22 (Bounded solutions of an elliptic problem, continuation ($\star\star\star$)).
Solution on page 155. Let Ω be an open bounded subset of $\mathbb{R}^N$ ($N > 1$); for all $x \in \Omega$, let $A(x)$ be a given matrix whose coefficients are denoted $a_{i,j}(x)$, $i, j = 1, \ldots, N$, and assume that $a_{i,j} \in L^\infty(\Omega)$ for all $i, j = 1, \ldots, N$ and that there exists an $\alpha > 0$ such that $A(x)\xi \cdot \xi \geq \alpha|\xi|^2$, for all $\xi \in \mathbb{R}^N$ and a.e. in $x \in \Omega$.

1. Let $f \in L^p(\Omega)$ with $p > 1$ if $N = 2$ and $p = \frac{2N}{N+2}$ if $N \geq 3$. Show that there exists a unique solution u of

$$\begin{cases} u \in H_0^1(\Omega), \\ \displaystyle\int_\Omega A(x)\nabla u(x) \cdot \nabla v(x) \, dx = \int_\Omega f(x)v(x) \, dx, \; \forall v \in H_0^1(\Omega). \end{cases} \qquad (2.54)$$

2. Let $N > 2$, $p > N/2$ and $f \in L^p(\Omega)$; show that there exists a unique solution u of (2.54). [Reduce to the previous question.]

Show that $u \in L^\infty(\Omega)$ and that there exists a $C \in \mathbb{R}_+$ depending only on Ω, α and p such that

$$\|u\|_{L^\infty(\Omega)} \leq C \|f\|_{L^p(\Omega)} .$$

[Refer back to Problem 2.21.]

Problem 2.23 (Evanescent diffusion and convection ($\star\star\star$)). *Solution on page 156.*

Part I Let Ω be an open bounded subset of $\mathbb{R}^N$ ($N \geq 1$) and $w = (w_1, \ldots, w_N)^t \in (L^\infty(\Omega))^N$ a vector function such that div $w = 0$ in $\mathcal{D}^\star(\Omega)$ (here, the operator div is taken in the sense of Definition 2.35). Let $u \in H_0^1(\Omega)$.

1. Show that $u^2 \in W_0^{1,1}(\Omega)$ and that $D_i(u^2) = 2uD_iu$, for all $i \in 1, \ldots, N$. [Use the density of $\mathcal{D}(\Omega)$ in $H_0^1(\Omega)$.]

2. Show that $\int_\Omega w(x) \cdot \nabla\varphi(x)\,\mathrm{d}x = 0$, for all $\varphi \in W_0^{1,1}(\Omega)$. [Use the density of $\mathcal{D}(\Omega)$ in $W_0^{1,1}(\Omega)$.]

3. Show that $\int_\Omega w(x) \cdot \nabla(u^2)(x)\,\mathrm{d}x = 2 \int_\Omega u(x)w(x) \cdot \nabla u(x)\,\mathrm{d}x = 0$ (recall that $w \cdot \nabla(u^2) = \sum_{i=1}^N w_iD_i(u^2)$).

Part II Let Ω be an open bounded subset of $\mathbb{R}^N$, with a Lipschitz boundary recall that under this assumption, Theorem 1.32 gives the existence of the trace operator, denoted γ, continuous linear from $H^1(\Omega)$ to $L^2(\partial\Omega)$ and such that $\gamma(u) = u$ on $\partial\Omega$ if $u \in C(\bar{\Omega}) \cap H^1(\Omega)$ and $\ker(\gamma) = H_0^1(\Omega)$). Let $a \in \mathbb{R}_+^\star$ and $w \in (L^\infty(\Omega))^N$ such that div $w = 0$ in $\mathcal{D}^\star(\Omega)$. Let $f \in L^2(\Omega)$ and $g \in \mathrm{Im}\,\gamma$. We seek a solution u of the following problem:

$$u \in H^1(\Omega),\ \gamma(u) = g\ (\text{in } L^2(\partial\Omega))\ :\ \forall v \in H_0^1(\Omega)$$

$$\int_\Omega a\nabla u(x) \cdot \nabla v(x)\,\mathrm{d}x + \int_\Omega u(x)w(x) \cdot \nabla v(x)\,\mathrm{d}x = \int_\Omega f(x)v(x)\,\mathrm{d}x. \quad (2.55)$$

1. Let $G \in H^1(\Omega)$ such that $\gamma(G) = g$ (in $L^2(\partial\Omega)$). Show that u is a solution to (2.55) if and only if $u = G + \bar{u}$ with $\bar{u}$ solution to the following problem.

$$\bar{u} \in H_0^1(\Omega)\ :\ \forall v \in H_0^1(\Omega),$$

$$\int_\Omega a\nabla\bar{u}(x) \cdot \nabla v(x)\,\mathrm{d}x + \int_\Omega \bar{u}(x)w(x) \cdot \nabla v(x)\,\mathrm{d}x = \int_\Omega f(x)v(x)\,\mathrm{d}x$$
$$- \int_\Omega a\nabla G(x) \cdot \nabla v(x)\,\mathrm{d}x - \int_\Omega G(x)w(x) \cdot \nabla v(x)\,\mathrm{d}x. \quad (2.56)$$

2. Show that (2.55) admits one and only one solution.

We denote this solution by u in the remainder of part II.

3. Assume in this question that $g = 0$ (so that $u \in H_0^1(\Omega)$). Show that

$$a \|u\|_{H_0^1}^2 \leq \int_\Omega f(x)u(x)\,\mathrm{d}x.$$

4. Let $b \in \mathbb{R}$. Assume, in this question, that $f \le 0$ a.e. in Ω and that $g \le b$ a.e. on $\partial\Omega$ (for the $N - 1$-dimensional measure on $\partial\Omega$). Show that $u \le b$ a.e. in Ω. [We may admit that $(u - b)^+ \in H_0^1(\Omega)$ and that $\nabla(u - b)^+ = \mathbb{1}_{u>b}\nabla u$ a.e. (this result is similar to that of Lemma 2.26), use (2.55) and part I.]

Part III In this part, set $N = 2$, $\Omega =]0, 1[^2$, $w = (-1, 0)$ and $g = 0$, assume that $f \in L^\infty(\Omega)$ and that $f \ge 0$ a.e. on Ω. Let u_n be the solution to (2.55) for $a = \frac{1}{n}$, $n \in \mathbb{N}^\star$ and let us study the limit of u_n as $n \to +\infty$.

1. Let $n \in \mathbb{N}^\star$; show that $u_n \ge 0$ a.e. [Use Part II, question 4.]
2. Let $n \in \mathbb{N}^\star$; show that there exists a C_1, depending only on f, such that $\|u_n\|_{L^\infty(\Omega)} \le C_1$. [Suggestion: find a problem of the type (2.55) whose solution is the function $u_n + \beta\psi$, with $\psi(x) = x_1$ and β suitably chosen, and use Part II, question 4.]
3. Let $n \in \mathbb{N}^\star$; show that there exists a C_2, depending only on f, such that $\|u_n\|_{H_0^1(\Omega)} \le C_2\sqrt{n}$.
4. Using Remark 2.20, show that $u_n \in H^2(\Omega)$ for all $n \in \mathbb{N}^\star$.
5. Let $n \in \mathbb{N}^\star$. If $u_n \in C^1(\overline{\Omega})$, deduce from question 1 of part III that $\partial_1 u_n(0, x_2) \ge 0$ and $\partial_1 u_n(1, x_2) \le 0$ for all $x_2 \in]0, 1[$ (similarly, $\partial_2 u_n(x_1, 0) \ge 0$ and $\partial_2 u_n(x_1, 1) \le 0$ for all $x_1 \in]0, 1[$). We admit, in the following, that this result is still true, with only $u_n \in H^2(\Omega)$, in the sense that $\gamma(D_1 u_n)(0, x_2) \ge 0$ and $\gamma(D_1 u_n)(1, x_2) \le 0$ a.e. in $x_2 \in]0, 1[$ (similarly $\gamma(D_2 u_n)(x_1, 0) \ge 0$ and $\gamma(D_2 u_n)(x_1, 1) \le 0$ a.e. in $x_1 \in]0, 1[$).
6. Using question 2 of part III, show that we can assume (up to a subsequence) that $u_n \to u$ $\star$-weakly in $L^\infty(\Omega)$ when $n \to +\infty$, that is to say:

$$\int_\Omega u_n(x)\varphi(x)\,\mathrm{d}x \to \int_\Omega u(x)\varphi(x)\,\mathrm{d}x, \quad \text{for all } \varphi \in L^1(\Omega).$$

Show that $u \ge 0$ a.e.

In the following, we seek the equation and the boundary conditions satisfied by u.

7. Show that $D_1 u = f$ in $\mathcal{D}^\star(\Omega)$.
8. Let $n \in \mathbb{N}^\star$ and $\varphi \in C^1(\overline{\Omega})$, show that

$$\frac{1}{n}\int_\Omega \nabla u_n(x)\nabla\varphi(x)\,\mathrm{d}x + \frac{1}{n}\int_0^1 \gamma(D_1 u_n)(0, x_2)\varphi(0, x_2)\,\mathrm{d}x_2$$
$$-\frac{1}{n}\int_0^1 \gamma(D_1 u_n)(1, x_2)\varphi(1, x_2)\,\mathrm{d}x_2 + \frac{1}{n}\int_0^1 \gamma(D_2 u_n)(x_1, 0)\varphi(x_1, 0)\,\mathrm{d}x_1$$
$$-\frac{1}{n}\int_0^1 \gamma(D_2 u_n)(x_1, 1)\varphi(x_1, 1)\,\mathrm{d}x_1 - \int_\Omega u_n(x)\partial_1\varphi(x)\,\mathrm{d}x$$
$$= \int_\Omega f(x)\varphi(x)\,\mathrm{d}x.$$

9. Let $\varphi \in C^1(\overline{\Omega})$ such that $\varphi \ge 0$ on $\partial\Omega$. Show that

$$-\int_\Omega u(x)\partial_1\varphi(x)\,\mathrm{d}x \le \int_\Omega f(x)\varphi(x)\,\mathrm{d}x. \tag{2.57}$$

10. Assume, in this question, that $u \in C^1(\overline{\Omega})$ and that $f \in C(\overline{\Omega})$. Show that $\partial_1 u = f$ everywhere in Ω and that $u(0, x_2) = 0$ for all $x_2 \in]0, 1[$.

    Is the function u then entirely determined by f?

11. Replacing $w = (-1, 0)$ by $w \in \mathbb{R}^2 \setminus \{0\}$, find the problem depending on w whose solution is u [distinguish the signs of the 2 components of w].

Problem 2.24 (Non-homogeneous Dirichlet condition $(\star)$). *Solution on page 163.* Let Ω be an open bounded subset of $\mathbb{R}^N$ with Lipschitz boundary and $g \in \mathrm{Im}(\gamma)$ (where γ denotes the trace operator seen in Theorem 1.32). Let $(a_{i,j})_{i,j=1,\dots,N} \subset L^\infty(\Omega)$ and $\alpha > 0$ such that (2.2) is satisfied.

1. Let $f \in L^2(\Omega)$; show that problem (2.17) has a unique solution.
2. Let $T \in H^{-1}(\Omega)$; show that problem (2.18) has a unique solution.
3. Assume in this question that $N = 2$ and $1 < p \le +\infty$. Show that for all $f \in L^p(\Omega)$ there exists a unique solution to problem (2.17).
4. Assume in this question that $N \ge 3$ and $p = \frac{2N}{N+2}$. Show that for all $f \in L^p(\Omega)$ there exists a unique solution to problem (2.17).

Problem 2.25 ($H^{\frac{1}{2}}(\partial\Omega)$ space $(\star\star)$). *Solution on page 164.* The aim of this problem is to prove Proposition 2.34. Let Ω be an open bounded subset of $\mathbb{R}^N$ ($N \ge 1$) with Lipschitz boundary. Let γ be the trace operator defined on $H^1(\Omega)$. We recall (see Section 1.5) that $H^{\frac{1}{2}}(\partial\Omega) = \mathrm{Im}(\gamma)$ and that $\|u\|_{H^{\frac{1}{2}}(\partial\Omega)} = \inf\{\|v\|_{H^1(\Omega)}, \ v \in H^1(\Omega), \gamma(v) = u\}$.

1. Let $u \in H^{\frac{1}{2}}(\partial\Omega)$. Show that $\|u\|_{H^{\frac{1}{2}}(\partial\Omega)} = \|\overline{u}\|_{H^1(\Omega)}$, where $\overline{u}$ is the unique weak solution to $-\Delta\overline{u} + \overline{u} = 0$ in Ω with $\gamma(\overline{u}) = u$, that is, the unique solution to

$$\overline{u} \in H^1(\Omega), \gamma(\overline{u}) = u, \tag{2.58}$$

$$\int_\Omega (\nabla\overline{u}(x) \cdot \nabla v(x) + \overline{u}(x)v(x)) \, dx = 0, \ \forall v \in H_0^1(\Omega). \tag{2.59}$$

2. Show that the space $H^{\frac{1}{2}}(\partial\Omega)$ is a Hilbert space.
3. Show that the space $H^{\frac{1}{2}}(\partial\Omega)$ is continuously embedded in $L^2(\partial\Omega)$.

Problem 2.26 (Normal trace of an element of H_{div} $(\star)$). *Solution on page 166.* Let Ω be an open bounded subset of $\mathbb{R}^2$ with Lipschitz boundary. We recall (see Definition 2.35) that $H_{\mathrm{div}}(\Omega) = \{v = (v_1, v_2) \in L^2(\Omega)^2 \text{ such that } \mathrm{div}\, v \in L^2(\Omega)\}$ and, for $v \in H_{\mathrm{div}}(\Omega)$,

$$\|v\|_{H_{\mathrm{div}}(\Omega)} = (\||v|\|_{L^2(\Omega)}^2 + \|\mathrm{div}\, v\|_{L^2(\Omega)}^2)^{\frac{1}{2}}. \tag{2.60}$$

1. Show that $H_{\mathrm{div}}(\Omega)$, equipped with the norm defined by (2.60), is a Hilbert space.
2. Let $v \in H_{\mathrm{div}}(\Omega)$.

   a. Show that for any function $\varphi \in \mathcal{D}(\Omega)$,

$$\int_\Omega \nabla\varphi \cdot v \, dx + \int_\Omega \varphi \, \mathrm{div}\, v \, dx = 0.$$

Deduce that this relation is still true for any function $\varphi \in H_0^1(\Omega)$.

b. Let $u_1, u_2 \in H^1(\Omega)$ such that $\gamma(u_1) = \gamma(u_2)$ (where γ is the trace operator defined on $H^1(\Omega)$). Show that

$$\int_\Omega \nabla u_1 \cdot v \, dx + \int_\Omega u_1 \, \mathrm{div}\, v \, dx = \int_\Omega \nabla u_2 \cdot v \, dx + \int_\Omega u_2 \, \mathrm{div}\, v \, dx.$$

Recall that $H^{\frac{1}{2}}(\partial\Omega) = \mathrm{Im}(\gamma)$ and that $H^{\frac{1}{2}}(\partial\Omega)$ is a Hilbert space with the norm defined in Problem 2.25. Let $H^{-\frac{1}{2}}(\partial\Omega)$ be the dual space of $H^{\frac{1}{2}}(\partial\Omega)$ (that is to say, the set of continuous linear mappings from $H^{\frac{1}{2}}(\partial\Omega)$ to $\mathbb{R}$).

3. Let $v \in H_{\mathrm{div}}(\Omega)$. Show that one can define an element of $H^{-\frac{1}{2}}(\partial\Omega)$, denoted $T(v)$, by setting, for $u \in H^{\frac{1}{2}}(\Omega)$,

$$\langle T(v), u \rangle_{H^{-\frac{1}{2}}(\partial\Omega), H^{\frac{1}{2}}(\partial\Omega)} = \int_\Omega \nabla \overline{u} \cdot v \, dx + \int_\Omega \overline{u} \, \mathrm{div}\, v \, dx, \tag{2.61}$$

with $\overline{u} \in H^1(\Omega)$ such that $\gamma(\overline{u}) = u$. (In particular, the term on the right of (2.61) is well defined and does not depend on $\overline{u}$ if $\overline{u} \in H^1(\Omega)$ and $\gamma(\overline{u}) = u$.)

We have thus defined a mapping T from $H_{\mathrm{div}}(\Omega)$ to $H^{-\frac{1}{2}}(\partial\Omega)$.

4. Show that the mapping T is linear and continuous from $H_{\mathrm{div}}(\Omega)$ to $H^{-\frac{1}{2}}(\partial\Omega)$.
5. Assume in this question that $v \in H^1(\Omega)^2$ (we therefore also have $v \in H_{\mathrm{div}}(\Omega)$). Let $\gamma(v)$ be the function obtained on $\partial\Omega$ by taking the trace of each of the components of v. Then $\gamma(v) \in H^{\frac{1}{2}}(\partial\Omega)^2 \subset L^2(\partial\Omega)^2$. Let $n(x)$ be the outward normal unit vector to $\partial\Omega$. Since Ω has a Lipschitz boundary, the vector $n(x)$ is defined a.e. at $x \in \partial\Omega$ (a.e. here means, as usual, a.e. for the Lebesgue 1-dimensional measure on $\partial\Omega$) and the function $x \mapsto n(x)$ defines an element of $L^\infty(\partial\Omega)$, see [20, Section 4.2]). We thus obtain $\gamma(v) \cdot n \in L^2(\partial\Omega)$. This (class of) function(s) $\gamma(v) \cdot n$ is called the "normal trace of v on $\partial\Omega$".

Show that, denoting by $d\lambda$ the integration with respect to the Lebesgue 1-dimensional measure on $\partial\Omega$,

$$\langle T(v), u \rangle_{H^{-\frac{1}{2}}(\partial\Omega), H^{\frac{1}{2}}(\partial\Omega)} = \int_{\partial\Omega} u \, \gamma(v) \cdot n \, d\lambda \text{ for all } u \in H^{\frac{1}{2}}(\partial\Omega). \tag{2.62}$$

N.B. This question explains why the mapping $T(v)$ is often denoted $v \cdot n$ even if $v \in H_{\mathrm{div}}(\Omega)$ (and not in $H^1(\Omega)^2$). It can also be shown that $H^{\frac{1}{2}}(\partial\Omega)$ is dense in $L^2(\partial\Omega)$. Hence if $v \in H^1(\Omega)^2$, $\gamma(v) \cdot n$ is the unique element of $L^2(\partial\Omega)$ satisfying (2.62).

Problem 2.27 (Normal trace on a part of the boundary ($\star\star$)). *Solution on page 167.* Using here the notation of Problem 2.26, we detail here Remark 2.36. In Problem 2.26, we constructed for every $v \in H_{\mathrm{div}}(\Omega)$ an element $T(v)$ of $H^{-\frac{1}{2}}(\partial\Omega)$ that is a continuous linear mapping from $H^{\frac{1}{2}}(\partial\Omega)$ to $\mathbb{R}$. The mapping T satisfies the following two properties:

- (Question 5 of Problem 2.26, T generalises the "classical" notion of normal trace) If $v \in C^1(\bar{\Omega})$,

$$\langle T(v), u\rangle_{H^{-\frac{1}{2}}(\partial\Omega), H^{\frac{1}{2}}(\partial\Omega)} = \int_{\partial\Omega} u\, v \cdot n\, d\lambda \text{ for all } u \in H^{\frac{1}{2}}(\partial\Omega).$$

- (Question 4 of Problem 2.26, continuity of the normal trace) T is continuous from $H_{\mathrm{div}}(\Omega)$ to $H^{-\frac{1}{2}}(\partial\Omega)$ and therefore, in particular,

$$\langle T(v_n), u\rangle_{H^{-\frac{1}{2}}(\partial\Omega), H^{\frac{1}{2}}(\partial\Omega)} \to \langle T(v), u\rangle_{H^{-\frac{1}{2}}(\partial\Omega), H^{\frac{1}{2}}(\partial\Omega)}$$

$$\text{if } v_n \to v \text{ in } H_{\mathrm{div}}(\Omega) \text{ and } u \in H^{\frac{1}{2}}(\partial\Omega).$$

Now let I be a part of the boundary of Ω, and suppose that the Lebesgue 1-dimensional measure of I is non-zero. It seems natural to denote by $H^{\frac{1}{2}}(I)$ the set of restrictions to I of the elements of $H^{\frac{1}{2}}(\Omega)$ (recall that $H^{\frac{1}{2}}(\Omega) \subset L^2(\partial\Omega)$), which is equivalent to writing

$$H^{\frac{1}{2}}(I) = \{u \text{ such that } u = \gamma(\overline{u}) \text{ a.e. on } I \text{ with } \overline{u} \in H^1(\Omega)\} \qquad (2.63)$$

(where a.e. means a.e. for the 1-dimensional Lebesgue measure on I).

The question is now: is it possible to construct for every $v \in H_{\mathrm{div}}(\Omega)$ a linear mapping S_v from $H^{\frac{1}{2}}(I)$ to $\mathbb{R}$ such that the mapping $v \mapsto S_v$ (from $H_{\mathrm{div}}(\Omega)$ in the algebraic dual space of $H^{\frac{1}{2}}(I)$) satisfies an analogue of the above-mentioned two properties of T? More precisely, construct S_v such that:

- *S generalises the classical notion of normal trace*

$$\text{If } v \in C^1(\bar{\Omega}),\ S_v(u) = \int_I u\, v \cdot n\, d\lambda \text{ for all } u \in H^{\frac{1}{2}}(I). \qquad (2.64)$$

- *Simple continuity of the normal trace*

$$S_{v_n}(u) \to S_v(u) \text{ if } v_n \to v \text{ in } H_{\mathrm{div}}(\Omega) \text{ and } u \in H^{\frac{1}{2}}(I). \qquad (2.65)$$

(For the sake of simplicity, $S_v(u)$ denotes the quantity $\langle S_v, u\rangle_{(H^{\frac{1}{2}}(I))^\star, H^{\frac{1}{2}}(I)}$, this latter notation being however more in line with the usual notations of this book.)

The answer is no. Let us give a simple example for which it is impossible to construct such a mapping S. We take $\Omega =]0, a[^2$, with $a > 0$ such that $a\sqrt{2} < 1$, and $I =]0, a[\times\{0\}$. The aim is to show that there is no mapping S (from $H_{\mathrm{div}}(\Omega)$ in the algebraic dual space of $H^{\frac{1}{2}}(I)$) satisfying (2.64)–(2.65).

We reason by contradiction: suppose that there exists an S satisfying (2.64)–(2.65). Let $0 < \beta < \frac{1}{2}$. For $x \in]0, \sqrt{2}[^2$, set

$$u(x) = (-\ln(|x|))^\beta.$$

We have $u \in C^\infty(\Omega)$ (more precisely, the restriction of u to Ω belongs to $C^\infty(\Omega)$) and we know that $u \in H^1(\Omega)$ (see Problem 1.5). The trace of u on I is equal (a.e. for λ) to the classical trace. Let x_1, x_2 denote the components of $x \in \mathbb{R}^2$ and set $v = (v_1, v_2)$ with

$$v_1 = -\partial_2 u, \quad v_2 = \partial_1 u.$$

1. Show that $\operatorname{div} v = \partial_1 v_1 + \partial_2 v_2 = 0$ and therefore that $v \in H_{\mathrm{div}}(\Omega)$.

For n such that $(a + \frac{1}{n})\sqrt{2} < 1$, define $v^{(n)}$ by

$$v^{(n)}(x_1, x_2) = v(x_1 + \frac{1}{n}, x_2).$$

2. Show that $v^{(n)} \in C^\infty(\bar{\Omega})$ and $v^{(n)} \to v$ in $H_{\mathrm{div}}(\Omega)$ as $n \to +\infty$. Infer that

$$S_{v^{(n)}}(\mathbb{1}_{\partial\Omega}) \left(= \int_I \mathbb{1}_{\partial\Omega} \, v^{(n)} \cdot n \, d\lambda(x)\right) \to S_v(\mathbb{1}_{\partial\Omega}) \text{ as } n \to +\infty,$$

where $\mathbb{1}_{\partial\Omega}$ is the function identically equal to 1 on $\partial\Omega$; note that $\mathbb{1}_{\partial\Omega} \in H^{\frac{1}{2}}(I)$ because it is the trace of the function equal to 1 on the whole set Ω.

3. Let $\psi_n = \int_0^a \beta \frac{(-\ln(x_1 + \frac{1}{n}))^{\beta-1}}{x_1 + \frac{1}{n}} \, dx_1$. Show that $\psi_n \to +\infty$ (recall that $\beta > 0$).

4. Noting that $v^{(n)} \cdot n = -v_2^{(n)}$ on I, show that $S_{v^{(n)}}(\mathbb{1}_{\partial\Omega}) = \psi_n$ and conclude the non-existence of S.

5. Show that $\langle T(v), \mathbb{1}_{\partial\Omega} \rangle_{H^{-\frac{1}{2}}(\partial\Omega), H^{\frac{1}{2}}(\partial\Omega)} = 0$, where T is the mapping of Problem 2.26. Show that $\lim_{n \to +\infty} \int_{\partial\Omega} \mathbb{1}_{\partial\Omega} \, v^{(n)} \cdot n \, d\lambda = 0$.

Problem 2.28 (A slight generalisation of the Liouville[22] theorem($\star\star\star\star$)). *Solution on page 168.* Liouville's theorem is stated as follows:

Theorem 2.39 (Liouville). *If f is a function defined and holomorphic[23] over the entire complex plane, then f is constant as long as it is bounded.*

The proof of this theorem is usually based on the Cauchy estimates of a holomorphic function defined on a neighbourhood of a closed disc, which provide bounds for each of the derivatives of this function at the centre of the disc.

Recall that if $f = \mathcal{R}e(f) + i\mathcal{I}m(f)$ is a holomorphic function from $\mathbb{C}$ to $\mathbb{R}$, then its real part $\mathcal{R}e(f)$ and imaginary part $\mathcal{I}m(f)$ are harmonic. We show here the following result, which generalises this theorem to locally integrable functions and functions from $\mathbb{R}^d$ to $\mathbb{R}$ that are bounded below: f is constant as soon as it is bounded.

---

[22] Joseph Liouville (1809–1882), French mathematician, known for his work in number theory and complex analysis, and founder of the Journal of Pure and Applied Mathematics.

[23] A holomorphic function is a function of a complex variable with complex values, defined and differentiable at every point of an open subset of the complex plane $\mathbb{C}$.

Theorem 2.40 (Generalised Liouville). *Let $d \geq 1$ and $u \in L^1_{\text{loc}}(\mathbb{R}^d)$ be a harmonic function, that is to say, such that $\Delta u = 0$ in $\mathcal{D}^\star(\mathbb{R}^d)$, and bounded below, that is to say, such that there exists a $c \in \mathbb{R}$ such that $u \geq c$ a.e., then u is constant, in the sense that there exists a $C \in \mathbb{R}$ such that $u = C$ a.e.*

1. Show that it is sufficient to prove the theorem with $c = 0$. Then, by regularising u with a sequence of regularising kernels, show that it is sufficient to prove the theorem in the case $u \in C^\infty(\mathbb{R}^d)$.

We now therefore assume that $u \in C^\infty(\mathbb{R}^d)$ and $u \geq 0$. For $r > 0$, let $B_r = \{x \in \mathbb{R}^d; |x| < r\}$ and $C_r = \{x \in \mathbb{R}^d; |x| = r\}$ and for all $a \in \mathbb{R}^d$, let $B_{a,r} = \{x \in \mathbb{R}^d; |x - a| < r\}$.

2. Let $r > 0$. Show that the integration of Δu over B_r gives

$$\int_{C_r} \nabla u(x) \cdot n(x) \, d\gamma(x) = 0,$$

where $n(x)$ is the outward normal unit vector to B_r and γ the $(d-1)$-dimensional Lebesgue measure on C_r (see Remark 1.34, the notation is slightly incorrect because this measure depends on r). By reducing to C_1 and differentiating under the sign $\int$, deduce that the quantity

$$\frac{1}{r^{d-1}} \int_{C_r} u(x) \, d\gamma(x)$$

is independent of r.

3. Let $r > 0$. With the change of variables $x \mapsto (r, y)$ with $r = |x|$ and $y \in C_1$, show that the quantity

$$\frac{1}{r^d} \int_{B_r} u(x) \, dx$$

is independent of r. Deduce that, for all $r > 0$,

$$\frac{1}{|B_r|} \int_{B_r} u(x) \, dx = u(0).$$

Similarly, show that, for all $a \in \mathbb{R}^d$ and all $r > 0$,

$$\frac{1}{|B_r|} \int_{B_{a,r}} u(x) \, dx = u(a).$$

Note that until now, only the fact that $u \in L^1_{\text{loc}}(\mathbb{R}^d)$ has been used. The fact that u is bounded below is only useful for the last question.

4. Let $a \in \mathbb{R}^d$. Since $u \geq 0$, the following inequality holds for all $r > \alpha = |a|$,

$$\int_{B_{r-\alpha}} u(x)\,dx \leq \int_{B_{a,r}} u(x)\,dx \leq \int_{B_{r+\alpha}} u(x)\,dx.$$

Deduce from this that $u(a) = u(0)$ and therefore that u is constant.

2.7 Problem Set Solutions

Problem 2.1 (A generalisation of the Lax–Milgram theorem)

1. If F is a closed subspace of a Hilbert space H, we always have $H = F \oplus F^{\perp}$. On the other hand, if $G \subset H$, we have $G^{\perp} = \bar{G}^{\perp}$. Taking $F = \overline{\mathrm{Im}(A)}$, we therefore have $H = \overline{\mathrm{Im}(A)} \oplus \mathrm{Im}(A)^{\perp}$. We now note that $\mathrm{Im}(A)^{\perp} \subset \mathrm{Ker}(A^{\star})$. Indeed, let $u \in (\mathrm{Im}(A))^{\perp}$; we then have, by setting $f = A^{\star}u$, $(A^{\star}u \,|\, A^{\star}u)_H = (f \,|\, A^{\star}u)_H = (Af \,|\, u)_H = 0$, because $Af \in \mathrm{Im}(A)$. Therefore, $A^{\star}u = 0$, that is $u \in \mathrm{Ker}A^{\star}$. Since $A^{\star}$ is an injective mapping, we get that $\mathrm{Im}(A)^{\perp} = \{0\}$ and therefore $\overline{\mathrm{Im}(A)} = H$.

   N.B. In fact, it is shown in Problem 2.17 that the equality $\overline{\mathrm{Im}(A)} = \mathrm{Ker}(A^{\star})^{\perp}$ still holds for $A \in \mathcal{L}(H)$ with H a Hilbert space.

2. a. Let us reason by contradiction: assuming (if necessary by extracting a subsequence) that $\lim_{n \to +\infty} \|w_n\|_H = +\infty$, set $\overline{w_n} = w_n / \|w_n\|_H$ so that $\|\overline{w_n}\|_H = 1$. The sequence $(\overline{w_n})_{n \in \mathbb{N}}$ is therefore bounded and since $A\overline{w_n} = \dfrac{f_n}{\|w_n\|_H} \to 0$, the second hypothesis of the theorem gives that $\overline{w_n} \to 0$, which is impossible because $\|\overline{w_n}\|_H = 1$ for all $n \in \mathbb{N}$.

   b. Since the sequence $(w_n)_{n \in \mathbb{N}}$ is bounded, we can assume (still if necessary by extracting a subsequence) that $w_n \to w$ weakly in H as $n \to +\infty$. We then have $Aw_n \to Aw$ weakly in H. Indeed, it is enough to note that, for all $v \in H$,

   $$(Aw_n \,|\, v)_H = (w_n \,|\, A^{\star}v)_H \to (w \,|\, A^{\star}v)_H = (Aw \,|\, v)_H.$$

   Since $Aw_n = f_n \to f$ as $n \to +\infty$, we therefore have $Aw = f$.

Problem 2.2 (Regularity in dimension 1)

Let $\varphi \in C([0, 1], \mathbb{R})$. For $x \in [0, 1]$, set

$$\psi(x) = \int_0^x \varphi(t)\,dt - x \int_0^1 \varphi(t)\,dt.$$

Then $\psi \in C^1([0,1])$, $\psi(0) = \psi(1) = 0$ and the weak derivative of ψ is a.e. equal to its classical derivative (see Definition 1.3), that is,

$$D\psi(x) = \psi'(x) = \varphi(x) - \int_0^1 \varphi(s)\,\mathrm{d}s \text{ for almost all } x \in]0,1[.$$

Hence $\psi \in L^2(\Omega)$ and $D\psi \in L^2(\Omega)$, which proves that $\psi \in H^1(]0,1[)$. Since $\psi(0) = \psi(1) = 0$, we even have $\psi \in H_0^1(\Omega)$ (see Section 1.5). Therefore, taking $v = \psi$ in (2.19) yields that

$$\int_0^1 Du(t)\varphi(t)\,\mathrm{d}t - \int_0^1 \varphi(t)\,\mathrm{d}t \int_0^1 Du(t)\,\mathrm{d}t = \int_0^1 f(x)\psi(x)\,\mathrm{d}x.$$

Since F is of class C^1 and $F' = f$, we have (also using $\psi(0) = \psi(1) = 0$)

$$\int_0^1 f(x)\psi(x)\,\mathrm{d}x = \int_0^1 F'(x)\psi(x)\,\mathrm{d}x$$

$$= -\int_0^1 F(x)\psi'(x)\,\mathrm{d}x$$

$$= -\int_0^1 F(x)\varphi(x)\,\mathrm{d}x + \int_0^1 F(x)\,\mathrm{d}x \int_0^1 \varphi(t)\,\mathrm{d}t.$$

Setting $c = \int_0^1 Du(t)\,\mathrm{d}t + \int_0^1 F(t)\,\mathrm{d}t$, we therefore have

$$\int_0^1 (Du(t) + F(t))\varphi(t)\,\mathrm{d}t = c \int_0^1 \varphi(t)\,\mathrm{d}t \text{ for all } \varphi \in C([0,1]).$$

Since $Du + F - c \in L^2(]0,1[)$ and $C([0,1])$ is dense in $L^2(]0,1[)$, we get

$$Du = -F + c \text{ a.e. in }]0,1[.$$

We now set

$$w(x) = \int_0^x (-F(t) + c)\,\mathrm{d}t \text{ for } x \in [0,1].$$

Since w is of class C^1 (the function w is even of class C^2) the derivative by transposition of w is a weak derivative and is a.e. equal to the classical derivative of w. Hence $Dw = w' = -F + c$ a.e., and therefore $Dw = Du$ a.e. so that $w - u$ is a function that is a.e. equal to a constant (see Problem 1.2). Identifying the (class of) function(s) u to its continuous representative, we therefore have u of class C^2, $u' = -F + c$ and $u'' = -F' = f$. We also get that $u(0) = u(1)$ (because $u \in H_0^1(]0,1[)$ and therefore the continuous representative of u satisfies $u(0) = u(1) = 0$).

Problem 2.3 (Spectral decomposition in dimension 1)

1. Owing to Theorem 2.15, $\ker(T) = \{f \in E, Tf = 0 \text{ a.e.}\} = \{0\}$, the eigenvalues of T are all positive and there exists a Hilbert basis of $L^2(]0, 1[)$ composed of eigenfunctions of T (Theorem 2.16). Let us seek here such a Hilbert basis. With this aim, let us first find the eigenvalues of T.

   Recall that, for $f \in E, Tf \in H_0^1(]0, 1[)$ and, setting $u = Tf$,

   $$\int_0^1 Du(t)Dv(t)\,\mathrm{d}t = \int_0^1 f(t)v(t)\,\mathrm{d}t \text{ for all } v \in H_0^1(]0, 1[).$$

   Let λ be an eigenvalue of T; we already know that $\lambda > 0$ and that there exists an $f \in E, f \neq 0$, such that $Tf = \lambda f$. Setting $u = Tf$, we get that $u \in H_0^1(]0, 1[)$, $u \neq 0$ and $f = u/\lambda$, which gives

   $$\int_0^1 Du(t)Dv(t)\,\mathrm{d}t = \frac{1}{\lambda}\int_0^1 u(t)v(t)\,\mathrm{d}t \text{ for all } v \in H_0^1(]0, 1[).$$

   Since $u \in H_0^1(]0, 1[)$, u is continuous on $[0, 1]$ (more specifically, u has a continuous representative and we identify u with this representative) and $u(0) = u(1) = 0$. Problem 2.2 then shows that u is of class C^2 and that

   $$-\lambda u''(x) = u(x) \text{ for all } x \in]0, 1[. \tag{2.66}$$

2. By the previous question, finding the eigenvalues amounts to solving a classical linear differential equation. It is well known (this is, for example, a consequence of the existence and uniqueness result known as the Cauchy–Lipschitz[24] theorem, see Theorem 5.3 in Chapter 5 for its full statement) that the set of solutions of (2.66) is a two-dimensional vector space generated by the functions $x \mapsto \sin(x/\sqrt{\lambda})$ and $x \mapsto \cos(x/\sqrt{\lambda})$.

   Hence if λ is an eigenvalue of T, there exists (by the previous question) $u \neq 0$ such that $Tu = \lambda u$, u of class C^2, with u continuous on $[0, 1]$, $u(0) = u(1) = 0$ and u a solution to (2.66). Thus, there exist $A, B \in \mathbb{R}$ such that

   $$u(x) = A\sin\left(\frac{x}{\sqrt{\lambda}}\right) + B\cos\left(\frac{x}{\sqrt{\lambda}}\right) \text{ for all } x \in [0, 1].$$

   Since $u(0) = 0$, then necessarily $B = 0$. Thus, since $u \neq 0$, then necessarily $A \neq 0$. Finally, since $u(1) = 0$, then necessarily $\sin(1/\sqrt{\lambda}) = 0$, which gives the existence of $k \in \mathbb{Z}$ such that $1/\sqrt{\lambda} = k\pi$. Since $\lambda > 0$, we therefore have $k \in \mathbb{N}^\star$, $1/\lambda = k^2\pi^2$ and $u(x) = A\sin(k\pi x)$ for all $x \in [0, 1]$ with $A \neq 0$; the function u satisfies $Tu = \lambda u$; indeed, by density of $\mathcal{D}(]0, 1[)$ in $H_0^1(]0, 1[)$, the weak formulation remains valid when considering test functions v in $\mathcal{D}(]0, 1[)$.

---

[24] Rudolph Otto Sigismund Lipschitz (1832–1903), German mathematician known in particular for his work in analysis, differential equations and number theory.

We have thus found all the eigenvalues of T, $\mathcal{VP}(T) = \{\frac{1}{k^2\pi^2}, \ k \in \mathbb{N}^\star\}$. Section 2.2 then gives that $\sigma(T) \setminus \{0\} = \mathcal{VP}(T) \setminus \{0\}$. Finally, since T is not surjective (which is always the case for a compact linear operator in infinite dimension), we have $0 \in \sigma(T)$ and therefore $\sigma(T) = \mathcal{VP}(T) \cup \{0\}$.

3. The previous question gave us the eigenvalues of T but also the corresponding eigenspaces. The answer to the question is then an immediate application of the results of Section 2.2. For $n \in \mathbb{N}^\star$, set $e_n(x) = \sqrt{2}\sin(p\pi x)$ for all $x \in [0, 1]$. The family $\{e_n, n \in \mathbb{N}^\star\}$ is a Hilbert basis of $L^2(]0, 1[)$ so that, for all $f \in L^2(]0, 1[)$,

$$\left\| f - \sum_{p=1}^{n} c_p \sin(p\pi\cdot) \right\|_2 \to 0, \ \text{when } n \to \infty,$$

that is to say, $f = \sum_{p=1}^{\infty} c_p \sin(p\pi\cdot)$, the convergence of the series being taken in the space $L^2(]0, 1[)$.

This series is not the Fourier series of f, since the Fourier series of f is obtained with the functions $\sin(2p\pi\cdot)$ and $\cos(2p\pi\cdot)$ $(p \in \mathbb{Z})$. The decomposition of f into a Fourier series also corresponds to the operator $u \mapsto u''$, but with periodic conditions $(u(0) = u(1)$ and $u'(0) = u'(1))$ instead of Dirichlet conditions $(u(0) = u(1) = 0)$.

4. Let $f \in E$. The function u is a solution to problem (2.67) if and only if $T(f - \mu u) = u$, that is to say,

$$T(u) + \frac{1}{\mu}u = \frac{T(f)}{\mu}. \tag{2.67}$$

Since T is compact, this problem has a solution if and only if f is orthogonal (in E) to the eigenspace of T associated with $(-1/\mu)$.

This can be proved from the previous questions. Indeed, we set $b_n = (f \mid e_n)_E$ (the family $\{e_n, n \in \mathbb{N}^\star\}$ being the Hilbert basis of E given in question 4), so that $f = \sum_{p=1}^{\infty} b_n e_n$; this series is convergent in E. Thus

$$T(f) = \sum_{n=1}^{+\infty} \frac{b_n}{n^2\pi^2} e_n,$$

and this series is also convergent in E.

Let $u \in E$ and set $a_n = (u \mid e_n)_E$, so that

$$T(u) + \frac{1}{\mu}u = \sum_{n=1}^{+\infty} a_n \frac{\mu + n^2\pi^2}{\mu n^2\pi^2} e_n,$$

and this series is again convergent in E. The function u is therefore a solution to (2.67) if and only if

$$a_n(\mu + n^2\pi^2) = \mu b_n \ \text{for all } n \in \mathbb{N}^\star.$$

If $\mu \neq -n^2\pi^2$ for all $n \in \mathbb{N}^\star$, there exists a unique solution to (2.67).

If there exists a $p \in \mathbb{N}^\star$ such that $\mu = p^2 \pi^2$, the equation (2.67) has a solution if and only if $b_p = 0$, that is to say if and only if f is orthogonal (in E) to e_p, which is equivalent to saying that f is orthogonal to the eigenspace of T associated with the eigenvalue $(-1/\mu)$.

Problem 2.4 (First eigenvalue of $-\Delta$)

1. Let $(u_n)_{n \in \mathbb{N}}$ be a sequence in $H_0^1(\Omega) \setminus \{0\}$ such that $\lim_{n \to +\infty} Q(u_n) = \mu$. By a homogeneity argument, that is, by replacing u_n by $\frac{u_n}{\|u_n\|_{L^2(\Omega)}}$, we can assume that $\|u_n\|_{L^2(\Omega)} = 1$. The sequence $(u_n)_{n \in \mathbb{N}}$ is therefore bounded in $H_0^1(\Omega)$ and we can assume (if necessary by extracting a subsequence) that it converges weakly in $H_0^1(\Omega)$ to a certain function u. By Rellich's theorem (Theorem 1.36) the sequence $(u_n)_{n \to +\infty}$ converges towards u in $L^2(\Omega)$ and therefore $\|u\|_{L^2(\Omega)} = 1$. Since $u \neq 0$, $Q(u) > 0$ (since $\nabla u = 0$ a.e. implies $u = 0$ a.e. because $u \in H_0^1(\Omega)$). Moreover, thanks to the weak convergence in $H_0^1(\Omega)$ of u_n towards u and the Cauchy–Schwarz inequality,

$$Q(u) = \lim_{n \to +\infty} \int_\Omega \nabla u_n(x) \cdot \nabla u(x) \, dx \leq \lim_{n \to +\infty} \sqrt{Q(u_n)}\sqrt{Q(u)} = \sqrt{\mu}\sqrt{Q(u)}.$$

It follows that $0 < Q(u) \leq \mu$ and therefore by definition of $\mu = \inf\{Q(v), v \in H_0^1(\Omega) \setminus \{0\}\}$, we obtain that $Q(u) = \mu$.

2. Let $\varphi \in \mathcal{D}(\Omega)$, $\varphi \neq 0$. For $0 < t < \dfrac{\|u\|_{H_0^1(\Omega)}}{\|\varphi\|_{H_0^1(\Omega)}}$, $u + t\varphi \neq 0$ and therefore $Q(u + t\varphi) \geq Q(u) = \mu$. It follows that

$$Q(u) + 2t \int_\Omega \nabla u(x) \cdot \nabla \varphi(x) \, dx + t^2 \int_\Omega \nabla \varphi(x) \cdot \nabla \varphi(x) \, dx$$
$$\leq \mu \left(1 + 2t \int_\Omega u(x)\varphi(x) \, dx + t^2 \int_\Omega \varphi(x)^2 \, dx\right),$$

and therefore, since $Q(u) = \mu$, by dividing by $2t$ and letting $t \to 0$,

$$\int_\Omega \nabla u(x) \cdot \nabla \varphi(x) \, dx \leq \mu \int_\Omega u(x)\varphi(x) \, dx.$$

By changing φ to $-\varphi$,

$$\langle -\Delta u, \varphi \rangle_{\mathcal{D}^\star(0), \mathcal{D}(\Omega)} = \int_\Omega \nabla u(x) \cdot \nabla \varphi(x) \, dx = \mu \int_\Omega u(x)\varphi(x) \, dx.$$

This means that $-\Delta u$ (an element of $\mathcal{D}^\star(\Omega)$) is represented by the function μu (an element of $L^2(\Omega)$) and therefore identified with μu. We have indeed shown that $u \in D(\mathcal{A})$ and $\mathcal{A}u = \mu u$ (in $L^2(\Omega)$, which we write as $\mathcal{A}u = \mu u$ a.e.).

Let us show that μ is indeed the smallest eigenvalue of $\mathcal{A}$. Indeed, let v be an eigenvalue of $\mathcal{A}$. There exists then $v \in H_0^1(\Omega)$, $v \neq 0$, such that $\mathcal{A}v = vv$. Hence, for all $w \in \mathcal{D}(\Omega)$ and therefore also (by density) for all $w \in H_0^1(\Omega)$,

$$\int_\Omega \nabla v(x) \cdot \nabla w(x)\,\mathrm{d}x = v \int_\Omega v(x)w(x)\,\mathrm{d}x.$$

Taking $w = v$ yields $Q(v) = v$ and therefore $v \geq \mu$.

3. If u is of constant sign, that is, $u \geq 0$ a.e. or $u \leq 0$ a.e., the result is immediate. We therefore assume that the sign of u is not constant. We then use the following lemma, whose proof is left to the reader:

Lemma 2.41. *Let a, b, c, $d \in \mathbb{R}_+^\star$. Then*

$$\min\{\frac{a}{b}, \frac{c}{d}\} \leq \frac{a+c}{b+d} \leq \max\{\frac{a}{b}, \frac{c}{d}\}.$$

Moreover, the inequalities are strict unless $\min\{\frac{a}{b}, \frac{c}{d}\} = \max\{\frac{a}{b}, \frac{c}{d}\}$.

By applying this lemma with $a = \int_\Omega |\nabla u^+(x)|^2\,\mathrm{d}x$, $b = \int_\Omega u^+(x)^2\,\mathrm{d}x$ and the equivalent for c and d with $-$ instead of $+$, we obtain that $Q(u)$ is between $Q(u^+)$ and $Q(u^-)$. Since $Q(u^\pm) \leq Q(u)$, we get $Q(u^+) = Q(u^-) = Q(u) = \mu$. We also have $Q(|u|) = \mu$.

Problem 2.5 (Mean Poincaré inequality on the boundary)

The easiest way is probably to reason by contradiction. If C does not exist, there is a sequence $(u_n)_{n\in\mathbb{N}}$ of $H^1(\Omega)$ such that, for all $n \in \mathbb{N}$

$$\|u_n\|_{L^2(\Omega)} \geq n \, \||\nabla u_n|\|_{L^2(\Omega)}.$$

By a homogeneity argument, we can assume $\|u_n\|_{L^2(\Omega)} = 1$. The sequence $(u_n)_{n\in\mathbb{N}}$ is then bounded in $H^1(\Omega)$. It therefore converges (after extracting a subsequence) weakly in $H^1(\Omega)$ towards a limit u. By Theorem 1.37, $u_n \to u$ in $L^2(\Omega)$ and therefore $\|u\|_{L^2(\Omega)} = 1$. On the other hand $\nabla u_n \to \nabla u$ weakly in $L^2(\Omega)^N$ and since $\nabla u_n \to 0$ in $L^2(\Omega)^N$ we therefore have $\nabla u = 0$ (in $L^2(\Omega)^N$). Since Ω is connected, this proves that u is constant, that is, there exists an $a \in \mathbb{R}$ such that $u = a$ a.e. in Ω (Problem 1.4). Since the boundary of Ω is assumed to be Lipschitz, Theorem 1.32 gives the existence of the trace of u on $\partial\Omega$, which in this case is also equal to a a.e. (for the $(N-1)$-dimensional Lebesgue measure on $\partial\Omega$); we therefore have $0 = \int_A u(x)\,\mathrm{d}\gamma(x) = \int_A a\,\mathrm{d}\gamma(x)$. This implies that $a = 0$, which contradicts $\|u\|_{L^2(\Omega)} = 1$.

Problem 2.6 (Elliptic problem with unbounded coefficients)

1. (Study of the functional space.)

   a. Let $u \in H^1(p, \Omega)$; since $p|D_i u| \geq \alpha|D_i u|$ a.e., $D_i u \in L^2(\Omega)$ for all i and therefore $u \in H^1(\Omega)$. We also note that

   $$\|u\|^2_{H^1(\Omega)} = \|u\|^2_2 + \sum_{i=1}^{N} \|D_i u\|^2_2 \leq \max\{1, \frac{1}{\alpha^2}\} \|u\|^2_{H^1(p,\Omega)} . \tag{2.68}$$

   b. It is clear that $H^1(p, \Omega)$ is a normed vector space and its norm is induced by an inner product. We now need to show that $H^1(p, \Omega)$ is complete.

   Let $(u_n)_{n\in\mathbb{N}}$ be a Cauchy sequence in $H^1(p, \Omega)$. The inequality (2.68) shows that the sequence $(u_n)_{n\in\mathbb{N}}$ is a Cauchy sequence in $H^1(\Omega)$. Therefore, there exists a $u \in H^1(\Omega)$ such that $u_n \to u$ and $D_i u_n \to D_i u$ (for all i) in $L^2(\Omega)$ as $n \to +\infty$. For $i \in \{1, \ldots, N\}$, the sequence $(pD_i u_n)_{n\in\mathbb{N}}$ is a Cauchy sequence in $L^2(\Omega)$. Therefore, there exists an $\xi_i \in L^2(\Omega)$ such that $pD_i u_n \to \xi_i$ in $L^2(\Omega)$ as $n \to +\infty$. But, up to the extraction of a subsequence, we also have $D_i u_n \to D_i u$ a.e. and $pD_i u_n \to \xi_i$ a.e., which proves that $\xi_i = pD_i u$. Finally, we therefore have $u \in H^1(p, \Omega)$ and $u_n \to u$ in $H^1(p, \Omega)$ as $n \to +\infty$, which proves that $H^1(p, \Omega)$ is a Hilbert space.

2. Let $(u_n)_{n\in\mathbb{N}}$ be a Cauchy sequence in $H^1_0(p, \Omega)$ converging in $H^1(p, \Omega)$. The inequality (2.68) shows that the sequence $(u_n)_{n\in\mathbb{N}}$ also converges in $H^1(\Omega)$. Since $H^1_0(\Omega)$ is closed in $H^1(\Omega)$, we therefore have $u \in H^1_0(\Omega)$ and therefore $u \in H^1_0(p, \Omega)$. We have indeed shown that $H^1_0(p, \Omega)$ is a closed linear subspace of $H^1(p, \Omega)$.

3. The space $H^1_0(p, \Omega)$ is a Hilbert space. The existence and uniqueness of a solution u to (2.20) is then a consequence of the Lax–Milgram theorem (Theorem 2.3). Indeed, we define the bilinear form a and the linear form T on $H^1_0(p, \Omega)$ by

$$a(u, v) = \int_{\Omega} p(x)\nabla u(x) \cdot \nabla v(x) \, dx,$$

$$T(v) = \int_{\Omega} h(x)v(x) \, dx,$$

Owing to the Cauchy–Schwarz inequality, $a(u, v) \leq \|u\|_{H^1_0(p,\Omega)} \|v\|_{H^1_0(p,\Omega)}$, and $T(v) \leq \|h\|_{L^2(\Omega)} \|v\|_{H^1_0(p,\Omega)}$. Hence a and T are continuous. The coercivity of a is a consequence of $p \geq \alpha$ a.e. and of the Poincaré inequality,

$$a(u, u) = \int_{\Omega} p^2(x)\nabla u(x) \cdot \nabla u(x) \, dx \geq \frac{\alpha^2}{\alpha^2 + C^2_{\Omega}} \|u\|_{H^1_0(p,\Omega)} ,$$

where C_{Ω} is given in Lemma 2.5.

4. (Clarifications...)

  a. Let $\varphi \in C_c^\infty(\Omega)$ and K be a compact subset of Ω such that $\varphi = 0$ in K^c. We then have $pD_i\varphi \in L^2(\Omega)$ because $D_i\varphi \in L^\infty(\Omega)$ and the restriction of p^2 to K is integrable. Hence $\varphi \in H_0^1(p,\Omega)$.

  b. We construct p from a function ψ from $]0,1[$ to $[1,+\infty[$ that is measurable (and even continuous) integrable on $]0,1[$ with integral 1 but whose square is not integrable on $]0,\varepsilon[$ for any $\varepsilon > 0$ (for example, consider $\psi(x) = 1/\sqrt{x}$) and a countable part A that is dense in $[0,1]$ (for instance, take $A = \mathbb{Q} \cap [0,1]$). We index the part A with $\mathbb{N}^\star$, that is to say $A = \{q_n,\ n \in \mathbb{N}^\star\}$, and we can add that $q_1 = 0$. We then define the function $\bar{p}$ by $\bar{p}(x) = \sum_{n\in\mathbb{N}^\star}(\frac{1}{n}^2)\psi(x - q_n)$. The series $\sum_{n\in\mathbb{N}^\star}(\frac{1}{n}^2)\psi(\cdot - q_n)$ is convergent in $\overline{\mathbb{R}}_+$ at every point and is absolutely convergent and therefore convergent in $L^1(]0,1[)$. So we have $\bar{p} < +\infty$ a.e. Taking $p(x) = 1$ if $\bar{p}(x) = +\infty$ and $p(x) = \bar{p}(x)$ otherwise, we thus obtain a measurable function p, bounded below by 1 and equal a.e. to $\bar{p}$. Let $\varphi \in C_c^\infty(\Omega)$, $\varphi \neq 0$. We now show that $p\varphi' \notin L^2(]0,1[)$ (and therefore $\varphi \notin H_0^1(p,\Omega)$). Since $\varphi \neq 0$, there exists an $a \in]0,1[$ such that $\varphi'(a) \neq 0$. By continuity of φ' there exists an $\varepsilon > 0$ and $\eta > 0$ such that $|\varphi'(x)| \geq \eta$ for all $x \in [a, a + 2\varepsilon[$. We then choose $n \in \mathbb{N}^\star$ such that $q_n \in [a, a + \varepsilon]$ and we note that $\bar{p}(x)^2 \varphi'(x)^2 \geq (\eta^2/n^4)\psi^2(x - q_n)$ for $x \in [a, a + 2\varepsilon$ and therefore $p\varphi' \notin L^2(]0,1[)$ because

$$\int_a^{a+2\varepsilon} \psi^2(x - q_n)\,\mathrm{d}x \geq \int_0^\varepsilon \psi^2(x)\,\mathrm{d}x = +\infty.$$

Problem 2.7 (Two nested elliptic problems)

1. Theorem 2.6 provides the existence and uniqueness of w as a solution to (2.22). For $v \in H_0^1(\Omega)$, we then set

$$S(v) = \int_\Omega (M(x) + N(x))\nabla w(x) \cdot \nabla v(x)\,\mathrm{d}x.$$

The mapping S is a continuous linear operator from $H_0^1(\Omega)$ to $\mathbb{R}$, thus it belongs to $H^{-1}(\Omega)$. Theorem 2.9 then provides the existence and uniqueness of u as a solution to (2.21), which is indeed the result requested.

2. The solution w to (2.22) depends linearly on f. Then, the solution u to (2.21) depends linearly on w. Hence u depends linearly on f and therefore the mapping T is linear from $L^2(\Omega)$ to $H_0^1(\Omega)$ and also linear from $L^2(\Omega)$ to $L^2(\Omega)$.

If w is the solution to (2.22), we have, by taking $v = w$ in (2.22),

$$\alpha \|w\|_{H_0^1(\Omega)}^2 \leq \|f\|_{L^2(\Omega)} \|w\|_{L^2(\Omega)}.$$

Owing to the Poincaré inequality (Lemma 2.5), there exists a C_Ω, depending only on Ω, such that $\|w\|_{L^2(\Omega)} \le C_\Omega \|w\|_{H_0^1(\Omega)}$. Hence, with $C_1 = \frac{C_\Omega}{\alpha}$,

$$\|w\|_{H_0^1(\Omega)} \le C_1 \|f\|_{L^2(\Omega)} . \tag{2.69}$$

Since M and N have coefficients in $L^\infty(\Omega)$, there exists a $\beta \in \mathbb{R}_+$ (depending only on M and N) such that, for all $\xi \in \mathbb{R}^d$,

$$|(M + N)\xi| \le \beta|\xi| \text{ a.e.}$$

Hence, for all $v \in H_0^1(\Omega)$ and S defined in the first question,

$$|S(v)| \le \beta \|w\|_{H_0^1(\Omega)} \|v\|_{H_0^1(\Omega)} .$$

If $u = T(f)$, taking $v = u$ in (2.21) yields that

$$\alpha \|u\|_{H_0^1(\Omega)}^2 \le \beta \|w\|_{H_0^1(\Omega)} \|u\|_{H_0^1(\Omega)} ,$$

and therefore, with (2.69) and $C_2 = \beta C_1/\alpha$,

$$\|u\|_{H_0^1(\Omega)} \le C_2 \|f\|_{L^2(\Omega)} .$$

This proves that the mapping $f \mapsto u$ is linear and continuous from $L^2(\Omega)$ to $H_0^1(\Omega)$. Since the mapping $u \mapsto u$ is compact from $H_0^1(\Omega)$ to $L^2(\Omega)$ (Theorem 1.36), it follows that Φ is a compact linear mapping from $L^2(\Omega)$ to $L^2(\Omega)$.

3. We start by noting that the hypotheses on M and N impose $\lambda > 0$. Then, if $u = T(f)$, (2.21) and (2.22) give, for all $v \in H_0^1(\Omega)$,

$$\int_\Omega M(x)\nabla u(x) \cdot \nabla v(x)\, dx = \int_\Omega (\lambda + 1)M(x)\nabla w(x) \cdot \nabla v(x)\, dx$$
$$= (\lambda + 1) \int_\Omega f(x)v(x)\, dx,$$

which indeed shows that u is a solution to (2.23) with $A = M/(\lambda + 1)$.

4. Let $f \in L^p(\Omega)$. Let p' be the conjugate exponent of p, that is $p' = \frac{p}{p-1}$. The Sobolev embedding theorem (Theorem 1.41) gives the existence of C_p (in fact only depending on p) such that, for all $v \in H_0^1(\Omega)$, we have $v \in L^{p'}(\Omega)$ and

$$\|v\|_{L^{p'}(\Omega)} \le C_p \|v\|_{H_0^1(\Omega)} .$$

With Hölder's inequality, it follows that the mapping $v \mapsto \int_\Omega f(x)v(x)\, dx$ belongs to $H^{-1}(\Omega)$ and that

$$\left| \int_\Omega f(x)v(x)\, dx \right| \le C_p \|f\|_{L^p(\Omega)} \|v\|_{H_0^1(\Omega)} .$$

We can then take up (slightly adapting them) the proofs of the first two questions. Theorem 2.9 gives the existence and uniqueness of a solution w to (2.22) and we have $\alpha \|w\|_{H_0^1(\Omega)} \leq C_p \|f\|_{L^p(\Omega)}$. Then, Theorem 2.9 gives the existence and uniqueness of a solution u to (2.21) and, with β defined in question 2, we obtain

$$\|u\|_{H_0^1(\Omega)} \leq \frac{\beta C_p}{\alpha^2} \|f\|_{L^p(\Omega)} .$$

This shows that the mapping $f \mapsto u$ is a continuous linear mapping from $L^p(\Omega)$ to $H_0^1(\Omega)$. Then, since the mapping $u \mapsto u$ is compact from $H_0^1(\Omega)$ to $L^q(\Omega)$ for $1 \leq q < +\infty$ (see Remark 1.43), it follows that the mapping $f \mapsto u$ is a compact linear mapping from $L^p(\Omega)$ to $L^q(\Omega)$ for $1 \leq q < +\infty$.

5. The proof here is very similar to the previous one. We have here $p = 6/5$ and thus the conjugate of p is $p' = 6 = 2^\star$. Let $f \in L^{6/5}(\Omega)$. The Sobolev embedding theorem (Theorem 1.41) provides the existence of $C \geq 0$ (depending on nothing) such that, for all $v \in H_0^1(\Omega)$, we have $v \in L^6(\Omega)$ and

$$\|v\|_{L^6(\Omega)} \leq C \|v\|_{H_0^1(\Omega)} .$$

With Hölder's inequality, we get that the mapping $v \mapsto \int_\Omega f(x)v(x)\,\mathrm{d}x$ belongs to $H^{-1}(\Omega)$ and that

$$\left| \int_\Omega f(x)v(x)\,\mathrm{d}x \right| \leq C \|f\|_{L^{6/5}(\Omega)} \|v\|_{H_0^1(\Omega)} .$$

Theorem 2.9 provides the existence and uniqueness of a solution w to (2.22) and we have $\alpha \|w\|_{H_0^1(\Omega)} \leq C \|f\|_{L^{6/5}(\Omega)}$. Then, Theorem 2.9 provides the existence and uniqueness of a solution u to (2.21) and, with β defined in question 2, we obtain

$$\|u\|_{H_0^1(\Omega)} \leq \frac{\beta C}{\alpha^2} \|f\|_{L^{6/5}(\Omega)} .$$

This shows that the mapping $f \mapsto u$ is a continuous linear mapping from $L^{6/5}(\Omega)$ to $H_0^1(\Omega)$. Then, since the mapping $u \mapsto u$ is continuous from $H_0^1(\Omega)$ to $L^6(\Omega)$ (Theorem 1.41) and is compact from $H_0^1(\Omega)$ to $L^q(\Omega)$ for $1 \leq q < 6 = 2^\star$ (see Remark 1.43), we get that the mapping $f \mapsto u$ is a continuous linear mapping from $L^{6/5}(\Omega)$ to $L^6(\Omega)$ and a compact linear mapping from $L^{6/5}(\Omega)$ to $L^q(\Omega)$ for $1 \leq q < 6$.

Problem 2.8 (Neumann problem)

1. *"Average Poincaré" inequality.* For $u \in H^1(\Omega)$, we set $S(u) = \int_\Omega u(x)\,\mathrm{d}x$. The mapping S is well defined on $H^1(\Omega)$ (because $H^1(\Omega) \subset L^2(\Omega) \subset L^1(\Omega)$). It is linear. Finally, it is continuous because

$$S(u) \le \|u\|_{L^1(\Omega)} \le \|u\|_{L^2(\Omega)}\, \lambda_N(\Omega)^{\frac{1}{2}} \le \|u\|_{H^1(\Omega)}\, \lambda_N(\Omega)^{\frac{1}{2}},$$

where $\lambda_N(\Omega)$ is the (N-dimensional) Lebesgue measure on Ω. Since $H = \mathrm{Ker}(S)$, we get that H is a closed subspace of $H^1(\Omega)$.

For all $u \in H^1(\Omega)$, we have $\| \, |\nabla u| \, \|^2_{L^2(\Omega)} \le \| \, |\nabla u| \, \|^2_{L^2(\Omega)} + \|u\|^2_{L^2(\Omega)} = \|u\|^2_{H^1(\Omega)}$. Therefore, we have $\|u\|_m \le \|u\|_{H^1(\Omega)}$ for all $u \in H$. To show that $\|\cdot\|_m$ is equivalent in H to $\|\cdot\|_{H^1(\Omega)}$, it is therefore sufficient to show that there exists a $C > 0$ (depending only on Ω) such that

$$\|u\|_{L^2(\Omega)} \le C\,\|u\|_m \quad \text{for all } u \in H. \tag{2.70}$$

(We then have $\|u\|^2_{H^1(\Omega)} \le (C^2 + 1)\,\|u\|^2_m$ for all $u \in H$.)

To prove (2.70), we reason by contradiction. We suppose that there exists a sequence of elements of H, $(u_n)_{n\in\mathbb{N}}$ such that

$$\|u_n\|_{L^2(\Omega)} > n\,\|u_n\|_m \quad \text{for all } n \in \mathbb{N}.$$

By a homogeneity argument, we can suppose $\|u_n\|_{L^2(\Omega)} = 1$. We then also have $\|u_n\|_m \le \frac{1}{n}$, which proves that the sequence $(u_n)_{n\in\mathbb{N}}$ is bounded in $H^1(\Omega)$. By the compactness theorems seen in Chapter 1 (Section 1.6), we get that the sequence $(u_n)_{n\in\mathbb{N}}$ is relatively compact in $L^2(\Omega)$. We can suppose (after extracting a subsequence) that there exists a $u \in L^2(\Omega)$ such that $u_n \to u$ in $L^2(\Omega)$, as $n \to +\infty$. Since $\|u_n\|_{L^2(\Omega)} = 1$ for all $n \in \mathbb{N}$, we also have $\|u\|_{L^2(\Omega)} = 1$. We also note that the derivatives (by transposition) of u_n converge towards the derivatives of u in $\mathcal{D}^\star(\Omega)$. However, from $\|u_n\|_m \le \frac{1}{n}$ we get $\nabla u_n \to 0$ in $L^2(\Omega)^N$. Since the L^2 convergence implies the convergence in $\mathcal{D}^\star(\Omega)$, we therefore have $\nabla u = 0$. This shows that u is constant on Ω (Problem 1.4). Since $u_n \to u$ in $H^1(\Omega)$ and since $u_n \in H$ for all $n \in \mathbb{N}$, we also have $u \in H$ and therefore $\int_\Omega u(x)\,\mathrm{d}x = 0$. Hence $u = 0$ a.e., which is impossible because $\|u\|_{L^2(\Omega)} = 1$.

2. *Characterisation of* $(H^1(\Omega))'$. For $v = (v_1, \ldots, v_N)^t \in L^2(\Omega)^N$, we set $\|v\|_{L^2(\Omega)^N} = \int_\Omega |v(x)|^2\,\mathrm{d}x$, so that $L^2(\Omega)^N$ equipped with this norm is a Hilbert space. For $u \in H$, we set $J(u) = \nabla u = (D_1 u, \ldots, D_N u)^t$. The mapping J is then an isometry of H (equipped with the norm $\|\cdot\|_m$) into a subset of $L^2(\Omega)^N$, denoted $\mathrm{Im}(J)$.

Let $v \in \mathrm{Im}(J)$. There exists a unique $u \in H$ such that $v = J(u)$. Set $S(v) = \langle T, u\rangle_{(H^1(\Omega))', H^1(\Omega)}$. Since J is an isometry and the norm $\|\cdot\|_{H^1(\Omega)}$ is equivalent in H to the norm $\|\cdot\|_m$, the mapping S is a continuous linear mapping from the subspace $\mathrm{Im}(J)$ of $L^2(\Omega)^N$ into $\mathbb{R}$. By the Hahn–Banach theorem, we can therefore extend S to $\tilde{S}$, an element of the topological dual space of $L^2(\Omega)^N$. Then, owing to the Riesz representation theorem in Hilbert spaces, there exists an $F \in L^2(\Omega)^N$ such that

$$\tilde{S}(v) = \int_\Omega F(x) \cdot v(x)\,\mathrm{d}x.$$

Hence, for all $u \in H$,

$$\langle T, u \rangle_{(H^1(\Omega))', H^1(\Omega)} = \int_\Omega F(x) \cdot \nabla u(x) \, dx.$$

Now set

$$a = \frac{1}{\lambda_N(\Omega)} \langle T, 1_\Omega \rangle_{(H^1(\Omega))', H^1(\Omega)},$$

(where 1_Ω denotes the constant function equal to 1 in Ω).

For $u \in H^1(\Omega)$, we have $u = u - m + m$ (or, more rigorously, $u = u - m 1_\Omega + m 1_\Omega$ a.e.) with

$$m = \frac{1}{\lambda_N(\Omega)} \int_\Omega u(x) \, dx.$$

Since $u - m \in H$ and $\nabla(u - m) = \nabla u$ a.e. we have $\langle T, u \rangle_{(H^1(\Omega))', H^1(\Omega)} = \int_\Omega F(x) \cdot \nabla u(x) \, dx$ and therefore

$$\langle T, u \rangle_{(H^1(\Omega))', H^1(\Omega)} = \langle T, u - m \rangle_{(H^1(\Omega))', H^1(\Omega)} + m \langle T, 1_\Omega \rangle_{(H^1(\Omega))', H^1(\Omega)}$$

$$= \int_\Omega F(x) \cdot \nabla u(x) \, dx + a \int_\Omega u(x) \, dx.$$

3. (Existence and uniqueness.)

    a. Assume that u is a solution to (2.25); taking $v = 1_\Omega$ in (2.25), we then have

$$0 = a \lambda_N(\Omega) + 0,$$

which proves that $a = 0$. We apply the Lax–Milgram theorem (Theorem 2.3) in the Hilbert space H (equipped with the norm $\|\cdot\|_m$) with

$$a(u, v) = \int_\Omega A(x) \nabla u(x) \cdot \nabla v(x) \, dx,$$

and

$$T(v) = \int_\Omega F(x) \cdot \nabla v(x) \, dx.$$

The continuity of a comes from the fact that $a_{i,j} \in L^\infty(\Omega)$ for all i, j. The coercivity of a comes from the existence of $\alpha > 0$ given in the assumptions on A. Finally, the continuity of T comes from the fact that $F \in L^2(\Omega)^N$.

We thus obtain a unique $u \in H$ such that (2.25) is true for all $v \in H$. Since (2.25) is also true if v is a constant function, we also obtain the existence and uniqueness of $u \in H$ such that (2.25) is true for all $v \in H^1(\Omega)$.

    b. We apply the Lax–Milgram theorem (Theorem 2.3) in the Hilbert space H (equipped with the norm $\|\cdot\|_m$) with

$$a(u, v) = \int_\Omega A(x) \nabla u(x) \cdot \nabla v(x) \, dx,$$

and

$$T(v) = \int_\Omega F(x) \cdot \nabla v(x)\,dx.$$

The continuity of a comes from the fact that $a_{i,j} \in L^\infty(\Omega)$ for all i, j. The coercivity of a comes from the existence of $\alpha > 0$ given in the assumptions on A. Finally, the continuity of T comes from the fact that $F \in L^2(\Omega)^N$.

We thus obtain a unique $u \in H$ such that (2.25) is true for all $v \in H$. Since (2.25) is also true if v is a constant function, we also obtain the existence and uniqueness of $u \in H$ such that (2.25) is true for all $v \in H^1(\Omega)$.

c. We first take $v \in \mathcal{D}(\Omega)$ in (2.25) (with $a = 0$). The regularity of A, F, u and v allows us to integrate by parts (the regularity of Ω is not used for this step). We obtain

$$\int_\Omega (-\mathrm{div}(A(x)\nabla u(x)) + \mathrm{div}(F(x)))v(x)\,dx = 0 \text{ for all } v \in \mathcal{D}(\Omega).$$

Hence owing to the fundamental lemma (Lemma 1.2), $-\mathrm{div}(A(x)\nabla u(x)) + \mathrm{div}(F(x)) = 0$ a.e. Then, by continuity of the function $-\mathrm{div}(A\nabla u) + \mathrm{div}\,F$, $-\mathrm{div}(A(x)\nabla u(x)) + \mathrm{div}(F(x)) = 0$ for all $x \in \Omega$.

We now take functions $v \in C^\infty(\bar{\Omega})$ in (2.25). We can also integrate by parts (here we use the regularity of Ω) and obtain

$$\int_{\partial\Omega} (A(x)\nabla u(x) - F(x)) \cdot n(x)v(x)\,d\gamma(x) = 0 \text{ for all } v \in C^\infty(\bar{\Omega}),$$

where $\partial\Omega$ is the boundary of Ω and $d\gamma(x)$ denotes integration with respect to the $(N-1)$-dimensional measure on $\partial\Omega$.

By the *local maps* technique (see for example [20, Section 2.1.1]), we can reduce to the case of the fundamental lemma (Lemma 1.2) to deduce that $(A\nabla u - F) \cdot n = 0$ a.e. on $\partial\Omega$ then everywhere on $\partial\Omega$. But it is quicker to see that it is possible to choose v such that $v = (A\nabla u - F) \cdot n$ on $\partial\Omega$. We thus directly obtain $(A\nabla u - F) \cdot n = 0$ on $\partial\Omega$.

4. *Dependence on parameters.* We take $v = u_n$ in (2.25) with A_n instead of A and F_n instead of F (and $a = 0$). We obtain, with the Cauchy–Schwarz inequality

$$\alpha \|u_n\|_m^2 \le \| |F_n| \|_{L^2(\Omega)} \|u_n\|_m .$$

Hence the sequence $(u_n)_{n\in\mathbb{N}}$ is bounded in H (and therefore in $H^1(\Omega)$). We can therefore assume, after extracting a subsequence, that $u_n \to w$ weakly in $H^1(\Omega)$ as $n \to +\infty$. Since $\int_\Omega u_n(x)\,dx \to \int_\Omega w(x)\,dx$, we get that $w \in H$.

Then, passing to the limit as $n \to +\infty$ in (2.25) (with u_n instead of u, A_n instead of A and F_n instead of F, recall that $a = 0$), we find that w is a solution to (2.25) and therefore $w = u$. To justify the passage to the limit as $n \to +\infty$ in (2.25), we use for the right-hand side term that $F_n \to F$ in $L^2(\Omega)^N$ and that $\nabla u_n \to \nabla w$ weakly in $L^2(\Omega)^N$. For the left-hand side term, we use that for all i,

$j \in \{1, \ldots, N\}$, $a_{i,j}^{(n)} D_j v \to a_{i,j} D_j v$ in $L^2(\Omega)$ (by the dominated convergence theorem) and that $D_i u_n \to D_i w$ weakly in $L^2(\Omega)$. Since the limit of the extracted sequence is still the same function u, we can also affirm that $u_n \to u$ weakly in $H^1(\Omega)$ as $n \to +\infty$ without extraction of a subsequence.

It remains to show that $u_n \to u$ in $H^1(\Omega)$. With this aim, note that since $F_n - A_n \nabla u \to F - A \nabla u$ in $L^2(\Omega)^N$ and $\nabla(u_n(x) - u(x)) \to 0$ weakly in $L^2(\Omega)^N$,

$$\alpha \|u_n - u\|_m^2 \le \int_\Omega A_n(x) \nabla(u_n(x) - u(x)) \cdot \nabla(u_n(x) - u(x)) \, dx$$

$$= \int_\Omega (F_n(x) - A_n(x) \nabla u(x)) \cdot \nabla(u_n(x) - u(x)) \, dx \to 0 \text{ as } n \to +\infty,$$

and therefore $u_n \to u$ in $H^1(\Omega)$.

5. (Regularity H^2 by the reflection technique.). Let $\widetilde{\Omega} =]-1, 0[\times]0, 1[$. We first show that $u \in H^1(\Omega_s)$. It is clear that $u \in L^2(\Omega_s)$. We now show that $D_1 u \in L^2(\Omega_s)$ (the proof of $D_2 u \in L^2(\Omega_s)$ is rather easier). Let $\varphi \in \mathcal{D}(\Omega_s)$

$$\int_{\Omega_s} D_1 u(x) \varphi(x) \, dx = - \int_{\Omega_s} u(x) \partial_1 \varphi(x) \, dx$$

$$= - \int_\Omega u(x) \partial_1 \varphi(x) \, dx - \int_{\widetilde{\Omega}} u(x) \partial_i \varphi(x) \, dx.$$

The trace of u in $L^2(]0, 1[)$ on the side $x_1 > 0$ is the same as the trace of u in $L^2(]0, 1[)$ on the side $x_1 < 0$ because $u(x_1, x_2) = u(-x_1, x_2)$ (the equality of the traces is immediate if $u \in C_c^\infty(\overline{\Omega})$ and therefore true by density if $u \in H^1(\Omega)$). The integration by parts formula (Theorem 1.33) then gives

$$\int_{\Omega_s} D_1 u(x) \varphi(x) \, dx = \int_\Omega D_1 u(x) \varphi(x) \, dx + \int_{\widetilde{\Omega}} D_1 u(x) \varphi(x) \, dx.$$

This proves that $D_1 u \in L^2(\Omega_s)$ (note that $D_1 u(x) = -D_1 u(\widetilde{x})$ for almost all $x \in \widetilde{\Omega}$ with $\widetilde{x} = (-x_1, x_2)^t$ if $x = (x_1, x_2)^t \in \Omega$).

We now show that u is a solution to (2.25) with Ω_s instead of Ω. For $v \in H^1(\Omega_s)$, we define $w \in H^1(\Omega)$ by $w(x_1, x_2) = v(-x_1, x_2)$ $(x_1, x_2 \in]0, 1[)$. We then have

$$\int_{\widetilde{\Omega}} A(x) \nabla u(x) \cdot \nabla v(x) \, dx$$

$$= \sum_{i,j=1}^2 \int_0^1 \int_{-1}^0 a_{i,j}(x_1, x_2) D_i u(x_1, x_2) D_j v(x_1, x_2) \, dx_1 \, dx_2$$

$$= \sum_{i,j=1}^2 \int_0^1 \int_{-1}^0 a_{i,j}(-x_1, x_2) D_i u(-x_1, x_2) D_j w(-x_1, x_2) \, dx_1 \, dx_2$$

$$= \sum_{i,j=1}^{2} \int_0^1 \int_0^1 a_{i,j}(x_1,x_2) D_i u(x_1,x_2) D_j w(x_1,x_2)\, dx_1\, dx_2$$

$$= \int_{\widetilde{\Omega}} A(x)\nabla u(x) \cdot \nabla w(x)\, dx.$$

Hence

$$\int_{\Omega_s} A(x)\nabla u(x) \cdot \nabla v(x)\, dx = \int_{\Omega} A(x)\nabla u(x) \cdot \nabla v(x)\, dx$$
$$+ \int_{\Omega} A(x)\nabla u(x) \cdot \nabla w(x)\, dx.$$

Similarly,

$$\int_{\widetilde{\Omega}} f(x)v(x)\, dx = \int_0^1 \int_{-1}^0 f(x_1,x_2)v(x_1,x_2)\, dx_1\, dx_2$$
$$= \int_0^1 \int_{-1}^0 f(-x_1,x_2)w(-x_1,x_2)\, dx_1\, dx_2$$
$$= \int_0^1 \int_0^1 f(x_1,x_2)w(x_1,x_2)\, dx_1\, dx_2 = \int_{\Omega} f(x)w(x)\, dx.$$

Hence

$$\int_{\Omega_s} f(x)v(x)\, dx = \int_{\Omega} f(x)v(x)\, dx + \int_{\Omega} f(x)w(x)\, dx.$$

Since u is a solution to (2.25) (on Ω) with v and w, we get that u is a solution to (2.25) with Ω_s instead of Ω.

Set $\Omega_e =]-1,2[\times]-1,2[$. With one more reflection in the x_1 direction and then two reflections in the x_2 direction, we thus construct $u \in H^1(\Omega_e)$ and $f \in L^2(\Omega_e)$ such that u is a solution to (2.25) with Ω_e instead of Ω. Assume that $A(x) = \mathrm{Id}$ for all $x \in \Omega$ (and therefore all $x \in \Omega_e$). Now let $\varphi \in \mathcal{D}(\Omega_e)$, such that $\varphi = 1$ in Ω. The function φu belongs to $H^1(\mathbb{R}^2)$ and $\Delta(\varphi u) \in L^2(\mathbb{R}^2)$. Remark 2.23 then gives that φu belongs to $H^2(\mathbb{R}^2)$, which proves that $u \in H^2(\Omega)$.

Problem 2.9 (An example in $H^1(\mathbb{R}^N)$)

1. Since $u, f \in L^1_{\mathrm{loc}}(\mathbb{R}^N)$, $\Delta u - u = D_i f$ in $\mathcal{D}^\star(\mathbb{R}^N)$ means

$$\forall v \in \mathcal{D}(\mathbb{R}^N), \quad \int u(x)\Delta v(x)\, dx - \int u(x)v(x)\, dx = -\int f(x)\partial_i v(x)\, dx.$$

(All integrals are over $\mathbb{R}^N$.)

Since $u \in H^1(\mathbb{R}^N)$, $\int u(x)\Delta v(x)\,dx = -\int \nabla u(x) \cdot \nabla v(x)\,dx$ for all $v \in \mathcal{D}(\mathbb{R}^N)$. Hence $\Delta u - u = D_i f$ in $\mathcal{D}^\star(\mathbb{R}^N)$ if and only if

$$\forall v \in \mathcal{D}(\mathbb{R}^N), \quad \int \nabla u(x) \cdot \nabla v(x)\,dx + \int u(x)v(x)\,dx = \int f(x)\partial_i v(x)(x)\,dx.$$

Since $f \in L^2(\mathbb{R}^N)$ and $\mathcal{D}(\mathbb{R}^N)$ is dense in $H^1(\mathbb{R}^N)$, this is equivalent to (2.26).

2. Define the mapping T from $H^1(\mathbb{R}^N)$ to $\mathbb{R}$ by $T(v) = \int f(x)D_i v(x)\,dx$. The Cauchy–Schwarz inequality shows that T is well defined, belongs to $H^{-1}(\mathbb{R}^N)$ (the topological dual space of $H^1(\mathbb{R}^N)$) and that $\|T\|_{H^{-1}(\mathbb{R}^N)} \leq \|f\|_{L^2(\mathbb{R}^N)}$. The equality (2.26) reads $(u\,|\,v)_{H^1(\mathbb{R}^N)} = T(v)$. The existence and uniqueness of a solution u to (2.26) is then a consequence of the Riesz representation theorem in a Hilbert space. Taking $v = u$ in (2.26) we obtain $\|u\|_{H^1(\mathbb{R}^N)} \leq \|f\|_{L^2(\mathbb{R}^N)}$.

Problem 2.10 (H^2 norm on $\mathbb{R}^N$)

1. Recall that

$$\|u\|_{H^2}^2 = \|u\|_{L^2}^2 + \sum_{i=1}^{N} \|D_i u\|_{L^2}^2 + \sum_{i=1}^{N}\sum_{j=1}^{N} \|D_j D_i u\|_{L^2}^2 .$$

However

$$\|\Delta u\|_{L^2}^2 = \int_\Omega \left(\sum_{i=1}^{N} D_i D_i u \right)^2 dx \leq N^2 \max_{i=1,\ldots,N} \|D_i D_i u\|_{L^2}^2$$

$$\leq N^2 \sum_{i=1}^{N} \|D_i D_i u\|_{L^2}^2 ,$$

and therefore

$$\|u\|_{L^2}^2 + \|\Delta u\|_{L^2}^2 \leq \|u\|_{L^2}^2 + N^2 \sum_{i=1}^{N} \|D_i D_i u\|_{L^2}^2 \leq N^2 \|u\|_{H^2}^2 .$$

We can therefore take $C_1 = \frac{1}{N}^2$.

To show the existence of C_2, we use integration by parts (with regular functions) and the density of $\mathcal{D}(\mathbb{R}^N)$ in $H^2(\mathbb{R}^N)$. For $i, j \in \{1,\ldots,N\}$ and $\varphi \in \mathcal{D}(\mathbb{R}^N)$, denoting by ∂_i the partial derivative in the direction i,

$$\int \partial_i \partial_j \varphi(x)\partial_i \partial_j \varphi(x)\,dx = \int \partial_i \partial_i \varphi(x)\partial_j \partial_j \varphi(x)\,dx,$$

and therefore

$$\sum_{i=1}^{N}\sum_{j=1}^{N}\int (\partial_i\partial_j\varphi(x))^2\,\mathrm{d}x = \int \Delta\varphi(x)^2\,\mathrm{d}x.$$

By density of $\mathcal{D}(\mathbb{R}^N)$ in $H^2(\mathbb{R}^N)$ we get that for all $u \in H^2(\mathbb{R}^N)$,

$$\sum_{i=1}^{N}\sum_{j=1}^{N}\left\|D_jD_iu\right\|_{L^2}^2 = \|\Delta u\|_{L^2}^2\,. \tag{2.71}$$

For $i \in \{1,\dots,N\}$ and $\varphi \in \mathcal{D}(\mathbb{R}^N)$ we also have

$$\int \partial_i\varphi(x)\partial_i\varphi(x)\,\mathrm{d}x = \int \partial_i\partial_i\varphi(x)\varphi(x)\,\mathrm{d}x,$$

and therefore

$$\sum_{i=1}^{N}\int (\partial_i\varphi(x))^2\,\mathrm{d}x = \int \Delta\varphi(x)\varphi(x)\,\mathrm{d}x \le \int \Delta\varphi(x)^2\,\mathrm{d}x + \int \varphi(x)^2\,\mathrm{d}x.$$

By density of $\mathcal{D}(\mathbb{R}^N)$ in $H^2(\mathbb{R}^N)$ we get that for all $u \in H^2(\mathbb{R}^N)$,

$$\sum_{i=1}^{N}\|D_iu\|_{L^2}^2 \le \|\Delta u\|_{L^2}^2 + \|u\|_{L^2}^2\,. \tag{2.72}$$

With (2.71) and (2.72), we obtain that $C_2 = 2$ is suitable.

2. a. Since $u \in L^1_{\mathrm{loc}}(\mathbb{R}^N)$, $\Delta\Delta u + \lambda u = f$ in $\mathcal{D}^\star(\mathbb{R}^N)$ means

$$\forall v \in \mathcal{D}(\mathbb{R}^N),$$
$$\int u(x)\Delta\Delta v(x)\,\mathrm{d}x + \lambda\int u(x)v(x)\,\mathrm{d}x = \langle f,v\rangle_{\mathcal{D}^\star(\mathbb{R}^N),\mathcal{D}(\mathbb{R}^N)}.$$

(The integrals are over $\mathbb{R}^N$.) Since $u \in H^2(\mathbb{R}^N)$, $\int u(x)\Delta\Delta v(x)\,\mathrm{d}x = \int \Delta u(x)\Delta v(x)\,\mathrm{d}x$ for all $v \in \mathcal{D}(\mathbb{R}^N)$. Hence $\Delta\Delta u + \lambda u = f$ in $\mathcal{D}^\star(\mathbb{R}^N)$ if and only if

$$\forall v \in \mathcal{D}(\mathbb{R}^N),$$
$$\int \Delta u(x)\Delta v(x)\,\mathrm{d}x + \lambda\int u(x)v(x)\,\mathrm{d}x = \langle f,v\rangle_{\mathcal{D}^\star(\mathbb{R}^N),\mathcal{D}(\mathbb{R}^N)}.$$

Since $f \in H^{-2}(\mathbb{R}^N)$ and $\mathcal{D}(\mathbb{R}^N)$ is dense in $H^2(\mathbb{R}^N)$, this is equivalent to (2.27).

b. Define the inner product $(\cdot\,|\,\cdot)_\lambda$ on $H^2(\mathbb{R}^N)$ by

$$(u\,|\,v)_\lambda = \int \Delta u(x)\Delta v(x)\,\mathrm{d}x + \lambda\int u(x)v(x)\,\mathrm{d}x.$$

Question 1 shows that this inner product is equivalent to the usual inner product of $H^2(\mathbb{R}^N)$. The existence and uniqueness of a solution u to (2.27) is then a consequence of Riesz's representation theorem in a Hilbert space.

Problem 2.11 (Modelling of a contact problem)

1. (Search for a weak formulation) Assume that u is a classical solution to (2.28)–(2.31). Of course, we have $u(x) = 0$ for all $x \in \partial B$. Let $v \in C^2(\Omega, \mathbb{R})$ such that $v_{|\Omega_+} \in C^2(\overline{\Omega_+})$, $v_{|\Omega_-} \in C^2(\overline{\Omega_-})$ and $v(x) = 0$ for all $x \in \partial B$. Multiplying (2.28) by $v(x)$ and integrating over Ω, integration by parts gives

$$\int_\Omega \nabla u(x) \cdot \nabla v(x)\, dx + \int_I \partial_y u(x, 0^+) v(x, 0^+)\, dx - \int_I \partial_y u(x, 0^-) v(x, 0^-)\, dx$$
$$= \int_\Omega f(x) v(x)\, dx.$$

Using (2.30), (2.31), we indeed find that u satisfies (2.32).

Conversely, we now assume that u satisfies (2.32). Taking $v \in C^\infty(\Omega_+)$ and extending v by 0 outside of Ω_+), integrating by parts yields

$$-\int_{\Omega_+} \Delta u(x) v(x)\, dx = \int_{\Omega_+} f(x) v(x)\, dx$$

and therefore, since v is arbitrary, $-\Delta u(x) = f(x)$ for all $x \in \Omega_+$. A similar reasoning gives $-\Delta u(x) = f(x)$ for all $x \in \Omega_-$ and therefore (by continuity) $-\Delta u(x) = f(x)$ for all $x \in \Omega$. We now take $w \in C_c^2(I)$, which we extend by 0 outside of I and we define v on Ω by setting

$$v(x, y) = \begin{cases} w(x)((1 - 2y)^+)^3 & \text{if } y \geq 0, \\ 0 & \text{if } y < 0. \end{cases}$$

This function v is acceptable in (2.32). It gives, integrating by parts, since $-\Delta u(x) = f(x)$ for all $x \in \Omega$,

$$-\int_I \partial_y u(x, 0^+) w(x)\, dx + \int_I g(x)(u(x, 0^+) - u(x, 0^-)) w(x)\, dx.$$

Since w is arbitrary in $C_c^2(I)$, this gives $\partial_y u(x, 0^+) = g(x)(u(x, 0^+) - u(x, 0^-))$ for all $x \in I$. An analogous reasoning gives $\partial_y u(x, 0^-) = g(x)(u(x, 0^+) - u(x, 0^-))$ for all $x \in I$. We have indeed shown that u is a classical solution to (2.28)–(2.31).
2. (Traces and functional space) Let $D = \{x \in B,\ 1 < |x| < 2\}$, so that $D \subset \Omega$. Since D has a Lipschitz boundary, Theorem 1.32 provides the existence of an operator γ that is linear continuous from $H^1(D)$ to $L^2(\partial D)$, extending the notion of classical trace. The operator $\tilde\gamma$ which associates to $u \in H^1(D)$ the restriction of

$\gamma(u)$ to ∂B is therefore linear continuous from $H^1(D)$ to $L^2(\partial B)$. For $u \in H^1(\Omega)$ we define $\gamma_0(u)$ as the image by $\tilde{\gamma}$ of the restriction of u to D, which is indeed an element of $H^1(D)$.

Since Ω_+ has a Lipschitz boundary, Theorem 1.32 provides the existence of a linear continuous operator γ from $H^1(\Omega_+)$ to $L^2(\partial\Omega_+)$ extending the notion of classical trace. For $u \in H^1(\Omega)$, define $\gamma_+(u)$ as the restriction to I of $\gamma(u_{|\Omega_+})$ and define γ_- in an analogous manner.

3. (Coercivity) It is sufficient to show that there exists a $C \in \mathbb{R}_+$ such that $\|u\|_{L^2(\Omega_\pm)} \le C \||\nabla u|\|_{L^2(\Omega_\pm)}$ for all $u \in H$. The quickest proof is probably by contradiction, using Theorem 1.37 for $H^1(\Omega_+)$ (and $H^1(\Omega_-)$). Following this idea, assume that there exists a sequence $(u_n)_{n\in\mathbb{N}}$ of H such that

$$\|u_n\|_{L^2(\Omega_+)} > n \||\nabla u_n|\|_{L^2(\Omega_+)} .$$

By a homogeneity argument, we can assume that $\|u_n\|_{L^2(\Omega_+)} = 1$ and therefore

$$\lim_{n\to+\infty} \||\nabla u_n|\|_{L^2(\Omega_+)} = 0.$$

Hence the sequence $(u_n)_{n\in\mathbb{N}}$ is bounded in $H^1(\Omega_+)$; more precisely, it is the sequence of restrictions to Ω_+ of the sequence $(u_n)_{n\in\mathbb{N}}$. We can assume, if necessary by extracting a subsequence, that $u_n \to u$ weakly in $H^1(\Omega_+)$ as $n \to +\infty$ and Theorem 1.37 gives $u_n \to u$ in $L^2(\Omega_+)$ and therefore $\|u\|_{L^2(\Omega_+)} = 1$. Since $\nabla u_n \to 0$ in $L^2(\Omega_+)^N$, $\nabla u = 0$ a.e. and the function u is therefore constant (Problem 1.4) and $u_n \to u$ in $H^1(\Omega_+)$. However, the trace of u_n on the edge of Ω_+ is zero (because u belongs to H); the trace operator being continuous from $H^1(\Omega_+)$ to $L^2(\partial\Omega_+)$ we get that the trace of u on the edge of Ω_+ is zero and therefore $u = 0$ a.e., in contradiction with $\|u\|_{L^2(\Omega_+)} = 1$. Of course, a similar reasoning can be done with Ω_-.

4. (Existence and uniqueness of weak solutions) The space H is a Hilbert space (with the inner product of $H^1(\Omega)$). Define a from $H \times H$ to $\mathbb{R}$ by

$$a(u,v) = \int_\Omega \nabla u(x) \cdot \nabla v(x)\,\mathrm{d}x + \int_I g(x)(\gamma_+ u(x) - \gamma_- u(x))(\gamma_+ v(x) - \gamma_- v(x))\,\mathrm{d}x.$$

The form a is a continuous symmetric bilinear form (thanks to the continuity of the operators γ_+ and γ_- from H to $L^2(I)$). Question 3 shows that it defines an inner product on H equivalent to the inner product of $H^1(\Omega)$ because $a(u,u) \ge \int_\Omega \nabla u(x) \cdot \nabla u(x)\,\mathrm{d}x$. On the other hand, the mapping $v \mapsto \int_\Omega f(x)v(x)\,\mathrm{d}x$ belongs to H' (because $f \in L^2(\Omega)$). The Riesz representation theorem in a Hilbert space (see for example [26], theorem 6.56) then gives the existence and uniqueness of u solution to (2.33).

5. Taking $v = u_n$ in (2.33), noting that $\int_\Omega f(x)u_n(x)\,\mathrm{d}x \le \|f\|_{L^2(\Omega)} \|u\|_{L^2(\Omega)}$ and using question 3, we show that the sequence $(u_n)_{n\in\mathbb{N}}$ is bounded in $H^1(\Omega)$ and that there exists a $C \in \mathbb{R}_+$ such that, for all $n \in \mathbb{N}$,

$$\int_\Omega n(\gamma_+ u_n(x) - \gamma_- u_n(x))^2 \, dx \le C. \tag{2.73}$$

We can therefore assume that, up to a subsequence, $u_n \to u$ weakly in $H^1(\Omega)$ and therefore also weakly in $H^1(\Omega_+)$ and $H^1(\Omega_-)$ (more precisely it is about the convergence of the restrictions of the sequence $(u_n)_{n\in\mathbb{N}}$ to Ω_+ and Ω_-. The operators $\gamma_\pm$ being continuous from $H^1(\Omega_\pm)$ to $L^2(I)$, we also have $\gamma_\pm u_n \omega \gamma_\pm u$ weakly in $L^2(I)$ because a continuous operator between two Banach spaces maps a weakly convergent sequence to a weakly convergent sequence, see on this subject Problem 1.22. In fact, here, we could even show the convergence of the sequence $(\gamma_\pm u_n)_{n\in\mathbb{N}}$ in $L^2(I)$. The inequality (2.73) gives that $\gamma_+ u_n(x) - \gamma_- u_n(x) \to 0$ in $L^2(I)$ and therefore $\gamma_+ u = \gamma_- u$ a.e. on I. Taking $\varphi \in \mathcal{D}(\Omega)$, an integration by parts (Theorem 1.33) on Ω_+ and Ω_- gives $u \in H^1(B)$ and $D_i u = D_i u_{|\Omega_\pm}$ a.e. on $\Omega_\pm$. Since $\gamma_0 u = 0$ a.e. on ∂B, finally $u \in H_0^1(B)$. Now taking $v \in H_0^1(B)$ in (2.33), we obtain that u is the solution to the problem

$$u \in H_0^1(B),$$

$$\int_\Omega \nabla u(x) \cdot \nabla v(x) \, dx = \int f(x) v(x) \, dx \text{ for all } v \in H_0^1(B).$$

Owing to Theorem (2.6), the solution to this problem is unique and a classical proof by contradiction gives that the whole sequence $(u_n)_{n\in\mathbb{N}}$ converges to u weakly in $H^1(\Omega)$.

Problem 2.12 (From Fourier to Dirichlet)

1. Here are possible definitions. The function u is a classical solution to (2.34) if $u \in C^2(\overline{\mathbb{R}_+^N}, \mathbb{R})$ and satisfies $-\Delta u(x) + u(x) = f(x)$ for all $x \in \mathbb{R}_+^N$ and $-\partial_1 u(0, y) + \sigma u(0, y) = g(y)$ for all $y \in \mathbb{R}^{N-1}$. The function u is a "weak" solution to (2.34) if $u \in H^1(\mathbb{R}_+^N)$ and satisfies, for all $v \in H^1(\mathbb{R}_+^N)$,

$$\int_{\mathbb{R}_+^N} (\nabla u(x) \cdot \nabla v(x) + u(x)v(x)) \, dx + \int_{\mathbb{R}^{N-1}} \sigma \gamma u(y) \gamma v(y) \, dy$$

$$= \int_{\mathbb{R}_+^N} f(x) v(x) \, dx + \int_{\mathbb{R}^{N-1}} g(y) \gamma v(y) \, dy,$$

where γ is the trace operator, continuous linear from $H^1(\mathbb{R}_+^N)$ to $L^2(\mathbb{R}^{N-1})$, whose existence is given by Theorem 1.30.

2. Define a from $H^1(\mathbb{R}_+^N)^2$ to $\mathbb{R}$ and T from $H^1(\mathbb{R}_+^N)$ to $\mathbb{R}$ by

$$a(u, v) = \int_{\mathbb{R}_+^N} (\nabla u(x) \cdot \nabla v(x) + u(x)v(x)) \, dx + \int_{\mathbb{R}^{N-1}} \sigma \gamma u(y) \gamma v(y) \, dy,$$

$$T(v) = \int_{\mathbb{R}_+^N} f(x) v(x) \, dx + \int_{\mathbb{R}^{N-1}} g(y) \gamma v(y) \, dy.$$

The form a defines an inner product on $H^1(\mathbb{R}^N_+)$ that is equivalent to the usual inner product because $\sigma \geq 0$ and γ is continuous from $H^1(\mathbb{R}^N_+)$ to $L^2(\mathbb{R}^{N-1})$. The mapping T belongs to $H^1(\mathbb{R}^N_+)'$, because $f \in L^2(\mathbb{R}^N_+)$, $g \in L^2(\mathbb{R}^{N-1})$ and γ is continuous from $H^1(\mathbb{R}^N_+)$ to $L^2(\mathbb{R}^{N-1})$. The Riesz representation theorem in a Hilbert space (see for example [26], theorem 6.56) then gives the existence and uniqueness of u weak solution to (2.34). Since $a(u,u) \geq \|u\|_{H^1(\mathbb{R}^N_+)}$, it is useful for the following to note that $\|u\|_{H^1(\mathbb{R}^N_+)} \leq \|T\|_{H^1(\mathbb{R}^N_+)'}$.

3. Let u be the weak solution to (2.34), that is to say, a solution u to

$$u \in H^1(\mathbb{R}^N_+),$$

$$\int_{\mathbb{R}^N_+} (\nabla u(x) \cdot \nabla v(x) + u(x)v(x))\, \mathrm{d}x + \int_{\mathbb{R}^{N-1}} \sigma \gamma u(y) \gamma v(y)\, \mathrm{d}y$$

$$= \int_{\mathbb{R}^N_+} f(x)v(x)\, \mathrm{d}x. \qquad (2.74)$$

To show that $H^2(\mathbb{R}^N_+)$, we use the method given in the proof of Theorem 2.18. As shown in the previous question, for all $T \in H^1(\mathbb{R}^N_+)'$, there exists one and only one solution w_T of

$$w_T \in H^1(\mathbb{R}^N_+)\ :\ \forall v \in H^1(\mathbb{R}^N_+),$$

$$\int_{\mathbb{R}^N_+} (\nabla w_T(x) \cdot \nabla v(x) + w_T(x)v(x))\, \mathrm{d}x \qquad (2.75)$$

$$+ \int_{\mathbb{R}^{N-1}} \sigma \gamma w_T(y) \gamma v(y)\, \mathrm{d}y = T(v),$$

and $\|w_T\|_{H^1(\mathbb{R}^N_+)} \leq \|T\|_{H^1(\mathbb{R}^N_+)'}$. Let $n \in \mathbb{N}^\star$, we take for T the mapping $v \mapsto \int_{\mathbb{R}^N_+} n(f(x_1, x_2 + \frac{1}{n}) - f(x))\, \mathrm{d}x$ (where x_1 and x_2 are the two components of x, it is only to avoid burdening the notation that we limit here to $N = 2$). The solution to (2.75) is then the function $w_T = n(u(x_1, x_2 + \frac{1}{n}) - u(x))$. An immediate adaptation of Lemma 2.21 (where $H^1_0(\mathbb{R}^N_+)$ can be replaced by $H^1(\mathbb{R}^N_+)$ and $\mathcal{D}(\mathbb{R}^N_+)$ by $C^\infty_c(\overline{\mathbb{R}^N_+})$, which is dense in $H^1(\mathbb{R}^N_+)$) shows that $\|T\|_{H^1(\mathbb{R}^N_+)'} \leq \|f\|_{L^2(\mathbb{R}^N_+)}$ and therefore, for all $n \in \mathbb{N}^\star$,

$$\left\| n\left(u\left(x_1, x_2 + \frac{1}{n}\right) - u(x)\right) \right\|_{H^1(\mathbb{R}^N_+)} \leq \|f\|_{L^2(\mathbb{R}^N_+)}.$$

Set $\psi_n = n(u(x_1, x_2 + \frac{1}{n}) - u(x))$. The sequence $(\psi_n)_{n \in \mathbb{N}^\star}$ is bounded in $H^1(\mathbb{R}^N_+)$. Therefore, we can assume that, up to a subsequence, it converges weakly in $H^1(\mathbb{R}^N_+)$. Let ψ be this weak limit. Moreover, the proof of Lemma 2.22 gives $\psi_n \to D_2 u$ in $\mathcal{D}^\star(\mathbb{R}^N_+)$ as $n \to +\infty$. Hence $D_2 u = \psi \in H^1(\mathbb{R}^N_+)$ and therefore $D_1 D_2 u \in L^2(\mathbb{R}^N_+)$ and $D_2 D_2 u \in L^2(\mathbb{R}^N_+)$. To conclude, all that remains is to show that $D_1 D_1 u \in L^2(\mathbb{R}^N_+)$. With this aim, we use the equation satisfied by u. Indeed, this gives $-\Delta u + u = f$ in $\mathcal{D}^\star(\mathbb{R}^N_+)$, and therefore $D_1 D_1 u =$

$u - f - D_2 D_2 u$, which proves that $D_1 D_1 u \in L^2(\mathbb{R}^N_+)$. Finally, we have indeed shown that $u \in H^2(\mathbb{R}^N_+)$.

4. Since $\|u_n\|_{H^1(\mathbb{R}^N_+)} \leq \|f\|_{L^2(\mathbb{R}^N_+)}$, the sequence $(u_n)_{n\in\mathbb{N}}$ is bounded in $H^1(\mathbb{R}^N_+)$. Therefore, we can assume that up to a subsequence, it converges weakly in $H^1(\mathbb{R}^N_+)$. Let u be this weak limit. Taking $v = u_n$ in (2.74) (with $u = u_n$ and $\sigma = n$), we notice that $\gamma u_n \to 0$ in $L^2(\mathbb{R}^{N-1})$ as $n \to +\infty$. But $\gamma u_n \to \gamma u$ at least weakly in $L^2(\mathbb{R}^{N-1})$ (a continuous operator between two Banach spaces maps a weakly convergent sequence to a weakly convergent sequence, see Problem 1.22). Hence $\gamma u = 0$, that is, $u \in \mathrm{Ker}\gamma = H^1_0(\mathbb{R}^N_+)$ (Remark 1.31). Now taking $v \in H^1_0(\mathbb{R}^N_+)$ in (2.74) (with $u = u_n$ and $\sigma = n$) and letting $n \to +\infty$, we obtain

$$(u \mid v)_{H^1_0(\mathbb{R}^N_+)} = \int_{\mathbb{R}^N_+} f(x)v(x)\,\mathrm{d}x.$$

This proves that u is a weak solution to (2.35). This solution is unique (for example because $f = 0$ a.e. implies $u = 0$ a.e.). This uniqueness result allows us (by contradiction) to assert that the whole sequence $(u_n)_{n\in\mathbb{N}}$ converges to u weakly in $H^1(\mathbb{R}^N_+)$ as $n \to +\infty$. It remains to prove that $u_n \to u$ in $H^1(\mathbb{R}^N_+)$, as $n \to +\infty$. The weak convergence in $H^1(\mathbb{R}^N_+)$ gives us

$$(u \mid u)_{H^1(\mathbb{R}^N_+)} \leq \liminf_{n\to+\infty} (u_n \mid u_n)_{H^1(\mathbb{R}^N_+)}$$

(see Remark 1.20). But, we also notice that, for all $n \in \mathbb{N}$,

$$(u_n \mid u_n)_{H^1(\mathbb{R}^N_+)} \leq (f \mid u_n)_{L^2(\mathbb{R}^N_+)}$$

and thus

$$\limsup_{n\to+\infty} (u_n \mid u_n)_{H^1(\mathbb{R}^N_+)} \leq \lim_{n\to+\infty} (f \mid u_n)_{L^2(\mathbb{R}^N_+)} = (f \mid u)_{L^2(\mathbb{R}^N_+)}$$

$$= (u \mid u)_{H^1(\mathbb{R}^N_+)}.$$

Hence $(u_n \mid u_n)_{H^1(\mathbb{R}^N_+)} \to (u \mid u)_{H^1(\mathbb{R}^N_+)}$ as $n \to +\infty$, which is sufficient to assert that $u_n \to u$ in $H^1(\mathbb{R}^N_+)$, as $n \to +\infty$, because, using again that $u_n \to u$ weakly in $H^1(\mathbb{R}^N_+)$,

$$(u_n - u \mid u_n - u)_{H^1(\mathbb{R}^N_+)} = (u_n \mid u_n)_{H^1(\mathbb{R}^N_+)} - 2(u_n \mid u)_{H^1(\mathbb{R}^N_+)} + (u \mid u)_{H^1(\mathbb{R}^N_+)}$$

$$\to 0 \text{ when } n \to +\infty.$$

Problem 2.13 (Schrödinger equation)

1. *Dirichlet conditions –*

   a. For $u, v \in V$, let u_1, u_2 be the components of u and v_1, v_2 the components of v, and define an inner product on V by

$$(u \mid v)_V = \int_\Omega \nabla u_1(x) \cdot \nabla v_1(x)\, \mathrm{d}x + \int_\Omega \nabla u_2(x) \cdot \nabla v_2(x)\, \mathrm{d}x.$$

Thanks to the Poincaré inequality (Lemma 2.5), the mapping $u, v \mapsto (u \mid v)_V$ is indeed an inner product on V and the space V equipped with this inner product is a Hilbert space. For $u, v \in V$, we define a and T by

$$a(u, v) = (u \mid v)_V + \int_\Omega u_2(x)v_1(x)\, \mathrm{d}x - \int_\Omega u_1(x)v_2(x)\, \mathrm{d}x,$$

$$T(v) = \int_\Omega f_1(x)v_1(x)\, \mathrm{d}x + \int_\Omega f_2(x)v_2(x)\, \mathrm{d}x.$$

Thanks again to the Poincaré inequality, the form a is bilinear continuous (from $V \times V$ to $\mathbb{R}$) and $T \in V'$. Moreover, a is coercive because $a(u, u) = (u \mid u)_V$ (note that a is not symmetric). We can therefore apply the Lax–Milgram theorem (Theorem 2.3), which gives the existence and uniqueness of $u \in V$ such that $a(u, v) = T(v)$ for all $v \in V$ and therefore the existence and the uniqueness of a solution u to the problem (2.38) (because $a(u, v) = T(v)$ for all $v \in V$ if and only if u is a solution to (2.38)).

b. Let u be the solution (2.38). Since f_1, f_2, u_1 and u_2 belong to $L^2(\Omega)$ and since the open subset considered is the unit ball and therefore of class C^∞, the regularity Theorem 2.19 gives $u_1, u_2 \in H^2(\Omega)$. Equation (2.35a) then gives, for all $\varphi \in \mathcal{D}(\Omega)$,

$$-\int_\Omega \Delta u_1(x)\varphi(x)\, \mathrm{d}x = -\langle \Delta u_1, \varphi \rangle_{\mathcal{D}^\star(\Omega), \mathcal{D}(\Omega)} = \int_\Omega \nabla u_1(x) \cdot \nabla \varphi(x)\, \mathrm{d}x$$

$$= -\int_\Omega u_2(x)\varphi(x)\, \mathrm{d}x + \int_\Omega f_1(x)\varphi(x)\, \mathrm{d}x.$$

Owing to Lemma 1.2, Equation (2.36a) is satisfied a.e. on Ω. A similar reasoning gives that (2.36b) is satisfied a.e. on Ω.

Conversely, if $u_1, u_2 \in H^2(\Omega)$ and the equations (2.36) are satisfied a.e. on Ω, the equations (2.38) are satisfied for all $\varphi \in \mathcal{D}(\Omega)$ and therefore, by density, for all $\varphi \in H_0^1(\Omega)$. If we add that $u_1, u_2 \in H_0^1(\Omega)$, we obtain that u is a solution to (2.38). Since the solution to (2.38) is unique, this ends the question.

c. Let us show by induction on m that $u_1, u_2 \in H^{2m}(\Omega)$. Indeed, the previous question shows that $u_1, u_2 \in H^2(\Omega)$. Then if $u_1, u_2 \in H^{2m}(\Omega)$, the regularity theorem (Theorem 2.19) gives $u_1, u_2 \in H^{2m+2}(\Omega) = H^{2(m+1)}(\Omega)$ (because $(-u_2 + f_1)$ and $(u_1 + f_2)$ belong to $H^{2m}(\Omega)$).

d. The mapping $f \mapsto u$, where u is the solution to (2.38), is continuous from $L^2(\Omega) \times L^2(\Omega)$ to V (noting that $\|u\|_V^2 = a(u, u) = T(u)$ with the notation of question 1a). Then the mapping $u \mapsto u$ from V to $L^2(\Omega) \times L^2(\Omega)$ is compact (by Theorem 1.37). By composition, we get that the mapping $\Phi : f \mapsto u$, where u is the solution to (2.38), is compact from $L^2(\Omega) \times L^2(\Omega)$ to $L^2(\Omega) \times L^2(\Omega)$.

2. *Neumann boundary conditions –*

   a. Consider the Hilbert space $W = H^1(\Omega) \times H^1(\Omega)$ equipped with the inner product induced by $H^1(\Omega)$, that is to say,

   $$(u \mid v)_W = (u_1 \mid v_1)_{H^1(\Omega)} + (u_2 \mid v_2)_{H^1(\Omega)}.$$

   For $u, v \in W$, define a and T by

   $$a(u,v) = \int_\Omega (\nabla u_1(x) \cdot \nabla v_1(x) + (u_2(x) + \frac{1}{k} u_1(x)) v_1(x)) \, dx$$
   $$+ \int_\Omega (\nabla u_2(x) \cdot \nabla v_2(x) + (\frac{1}{k} u_2(x) - u_1(x)) v_2(x)) \, dx,$$
   $$T(v) = \int_\Omega f_1(x) v_1(x) \, dx + \int_\Omega f_2(x) v_2(x) \, dx.$$

   Here again the form a is bilinear continuous (from $W \times W$ to $\mathbb{R}$) and $T \in W'$. Moreover, a is coercive because $a(u,u) \geq (\frac{1}{k})(u \mid u)_W$. We can therefore apply the Lax–Milgram theorem (Theorem 2.3), which gives the existence and uniqueness of $u \in W$ such that $a(u,v) = T(v)$ for all $v \in W$ and therefore the existence and uniqueness of a solution u to problem (2.41).

   b. Taking $\varphi = u_2^{(k)}$ in (2.41a), $\varphi = -u_1^{(k)}$ in (2.41b) and adding the obtained equations yields, thanks to the Cauchy–Schwarz inequality,

   $$\|u_1^{(k)}\|_{L^2(\Omega)}^2 + \|u_2^{(k)}\|_{L^2(\Omega)}^2 = \int_\Omega f_1(x) u_2^{(k)}(x) \, dx - \int_\Omega f_2(x) u_1^{(k)}(x) \, dx$$
   $$\leq \|f_1\|_{L^2(\Omega)} \left\|u_2^{(k)}\right\|_{L^2(\Omega)} + \|f_2\|_{L^2(\Omega)} \left\|u_1^{(k)}\right\|_{L^2(\Omega)}$$
   $$\leq \frac{1}{2}(\|f_1\|_{L^2(\Omega)}^2 + \left\|u_2^{(k)}\right\|_{L^2(\Omega)}^2 + \|f_2\|_{L^2(\Omega)}^2 + \left\|u_1^{(k)}\right\|_{L^2(\Omega)}^2),$$

   which proves the required inequality.

   The sequences $(u_1^{(k)})_{k \in \mathbb{N}^\star}$ and $(u_2^{(k)})_{k \in \mathbb{N}^\star}$ are therefore bounded in $L^2(\Omega)$. Now taking $\varphi = u_1^{(k)}$ in (2.41a) and $\varphi = u_2^{(k)}$ in (2.41b) yields a bound on the sequences $(|\nabla u_1^{(k)}|)_{k \in \mathbb{N}^\star}$ and $(|\nabla u_2^{(k)}|)_{k \in \mathbb{N}^\star}$ in $L^2(\Omega)$; therefore the sequences $(u_1^{(k)})_{k \in \mathbb{N}^\star}$ and $(u_2^{(k)})_{k \in \mathbb{N}^\star}$ are bounded in $H^1(\Omega)$. More precisely,

   $$\left\|u_1^{(k)}\right\|_{H^1(\Omega)}^2 + \left\|u_2^{(k)}\right\|_{H^1(\Omega)}^2 \leq 2(\|f_1\|_{L^2(\Omega)}^2 + \|f_2\|_{L^2(\Omega)}^2). \tag{2.76}$$

   c. The sequences $(u_1^{(k)})_{k \in \mathbb{N}^\star}$ and $(u_2^{(k)})_{k \in \mathbb{N}^\star}$ are bounded in $H^1(\Omega)$, so that up to a subsequence, they converge weakly in $H^1(\Omega)$ to some limits $u_1, u_2 \in H^1(\Omega)$. By passing to the limit as $k \to +\infty$ in the equations of (2.41) (written with $u_1^{(k)}$ and $u_2^{(k)}$) we obtain that u_1, u_2 are solutions of (2.41) with 0 instead of

$\frac{1}{k}$ (note for this that the sequences $(u_1^{(k)})_{k \in \mathbb{N}^\star}$ and $(u_2^{(k)})_{k \in \mathbb{N}^\star}$ are bounded in $L^2(\Omega)$).

Since we extracted a subsequence, the uniqueness of the solution to (2.41) with 0 instead of $\frac{1}{k}$ does not follow from the previous reasoning. But if there are two solutions of (2.41) with 0 instead of $\frac{1}{k}$ (and the same f_1 and f_2) their difference, still denoted u_1, u_2, is a solution to (2.41) with 0 instead of $\frac{1}{k}$ and $f_1 = f_2 = 0$ a.e. In the first equation of (2.41), take $\varphi = u_2$. In the second equation of (2.41), take $\varphi = -u_1$. Adding the resulting equations yields $u_1 = u_2 = 0$ a.e. This indeed proves the requested uniqueness.

Finally, we note that (2.76) gives, by passing to the lower limit as $k \to +\infty$,

$$\|u_1\|_{H^1(\Omega)}^2 + \|u_2\|_{H^1(\Omega)}^2 \leq 2(\|f_1\|_{L^2(\Omega)}^2 + \|f_2\|_{L^2(\Omega)}^2). \tag{2.77}$$

d. Let u be the solution (2.41) with 0 instead of $\frac{1}{k}$. Since u_2, $u_1 \in L^2(\Omega)$, the regularity theorem (Theorem 2.19), valid with $H^1(\Omega)$ instead of $H_0^1(\Omega)$, gives u_1, $u_2 \in H^2(\Omega)$. The reasoning of question 1b then gives that the equations (2.36) are satisfied a.e. on Ω. Now taking $\varphi \in H^1(\Omega)$ in (2.41), and integrating by parts thanks to Theorem 1.33 gives

$$\int_{\partial\Omega} \nabla u_i(x) \cdot n(x) v(x) \, dx = 0 \text{ for all } v \in H^{\frac{1}{2}}(\partial\Omega) \text{ and } i = 1, 2, \tag{2.78}$$

where n is the outward normal unit vector to $\partial\Omega$. For $i = 1$ or 2, the function $\nabla u_i(x) \cdot n(x)$ belongs to $H^{\frac{1}{2}}(\partial\Omega)$ because $u_i \in H^2(\Omega)$). We can therefore take $v = H^{\frac{1}{2}}(\partial\Omega)$ in (2.78), and we get $\nabla u_i(x) \cdot n(x) = 0$ a.e. on $\partial\Omega$ (for the $N - 1$-dimensional Lebesgue measure). The equations (2.39) are satisfied a.e. (for the measure Lebesgue $N - 1$-dimensional) on $\partial\Omega$.

Conversely let us assume that u_1, $u_2 \in H^2(\Omega)$, that the equations (2.36) are satisfied a.e. on Ω and that the equations (2.39) are satisfied a.e. (for the $N - 1$-dimensional Lebesgue measure) on $\partial\Omega$ (with the trace operator). It is then sufficient to use integration by parts thanks to Theorem 1.33 to show that $u = (u_1, u_2)$ is a solution (2.41) with 0 instead of $\frac{1}{k}$. Since this solution is unique, this ends the question.
Note that if f_1, $f_2 \in C^\infty(\bar\Omega)$, it can also be shown that u_1, $u_2 \in C^\infty(\bar\Omega)$.

e. The proof is identical to that of question 1d. Thanks to the estimate (2.77), the mapping $f \mapsto u$, where u is the solution to (2.41) (with 0 instead of $\frac{1}{k}$), is continuous from $L^2(\Omega) \times L^2(\Omega)$ to W. Then the mapping $u \mapsto u$ from W to $L^2(\Omega) \times L^2(\Omega)$ is compact (by Theorem 1.37). By composition, the mapping $f \mapsto u$, where u is the solution to (2.41) (with 0 instead of $\frac{1}{k}$), is compact from $L^2(\Omega) \times L^2(\Omega)$ to $L^2(\Omega) \times L^2(\Omega)$.

3. *Mixed boundary conditions* – The method suggested by the previous questions consists in introducing the following weak problem:

$$u_1 \in H_0^1(\Omega), \ u_2 \in H^1(\Omega) \ : \ \forall \varphi \in H^1(\Omega),$$

$$\int_\Omega \nabla u_1(x) \cdot \nabla \varphi(x) \, dx + \int_\Omega u_2(x)\varphi(x) \, dx = \int_\Omega f_1(x)\varphi(x) \, dx,$$

$$\int_\Omega \nabla u_2(x) \cdot \nabla \varphi(x) \, dx + \int_\Omega (\frac{1}{k}u_2(x) - u_1(x))\varphi(x) \, dx = \int_\Omega f_2(x)\varphi(x) \, dx.$$

One first shows that this weak problem has one and only one solution. Estimates on the solution (u_1, u_2) that are independent of n are then obtained in $H_0^1(\Omega) \times H^1(\Omega)$. These estimates allow us to pass to the limit as $k \to +\infty$ and obtain a solution of the desired problem. It is then possible to show the regularity of the obtained solution and finally the compactness of the operator obtained from $L^2(\Omega) \times L^2(\Omega)$ in itself.

Problem 2.14 (Trudinger–Moser inequality and $L^1(\sqrt{\ln(L^1)}) \subset H^{-1}$, $N = 2$)

Part I, decomposition in $H_0^1(\Omega)$

1. For $1 < s \le 2$, $\varphi'(s) = -s + 2$. We then have $\varphi \in C^1(\mathbb{R}, \mathbb{R})$ and therefore $\varphi_k \in C^1(\mathbb{R}, \mathbb{R})$ for $k \in \mathbb{N}^\star$. For $|s| \le k$, $\varphi_k(s) = s$ and $\varphi_k'(s) = 1$. Hence for all $s \in \mathbb{R}$, $\lim_{k\to\infty} \varphi_k(s) = s$ and $\lim_{k\to\infty} \varphi_k'(s) = 1$. For $k \le |s| \le 2k$, $0 \le \varphi_k'(s) = \varphi'(\frac{s}{k}) = -|\frac{s}{k}| + 2 \le 1$ and $|\varphi_k(s)| \le |s|$. For $|s| \ge 2k$, $\varphi_k'(s) = 0 \le 1$ and $|\varphi_k(s)| = \frac{3}{2} \le |s|$.

2. By Lemma 2.25, we have $\varphi_k(u) \in H_0^1(\Omega)$ for all $k \in \mathbb{N}^\star$. By the previous question, $|\varphi_k(u)| \le |u|$ a.e. for all $k \in \mathbb{N}^\star$ and $\varphi_k(u) \to u$ a.e. when $k \to \infty$. Moreover, for all $i \in \{1, \ldots, N\}$, $|D_i\varphi_k(u)| = |\varphi_k'(u)D_iu| \le |D_iu|$ a.e. and $D_i\varphi_k(u) \to D_iu$ a.e. when $k \to \infty$. The dominated convergence theorem then gives that $\varphi_k(u) \to u$ in $H_0^1(\Omega)$ when $k \to \infty$.

3. Let $u \in H_0^1(\Omega)$ and $\varepsilon > 0$. Question 2 gives the existence of $k > 0$ such that $\|u - \varphi_k(u)\|_{H_0^1(\Omega)} \le \varepsilon$. It is then sufficient to take $u_1 = \varphi_k(u)$ and $u_2 = u - \varphi_k(u)$. We have indeed $u_1 \in L^\infty(\Omega)$ because $|u_1| \le (3/2)k$ a.e.

Part II, Trudinger–Moser inequality

1. Recall that $H_0^1(\mathbb{R}^2) = H^1(\mathbb{R}^2) = W^{1,2}(\mathbb{R}^2)$ (see Theorem 1.25). The upper bound (1.28) of the problem solution 1.9 gives, for $N = 2$, $q \ge 2$ and $u \in H_0^1(\mathbb{R}^2)$,

$$\|u\|_{L^q(\mathbb{R}^2)} \le D_{2,q} \|u\|_{H_0^1(\mathbb{R}^2)} \text{ for all } u \in H_0^1(\mathbb{R}^2),$$

with $D_{2,q} = (1 + 2^q C_{2,p}^q + 4^q)^{\frac{1}{q}}$, $C_{2,p} = \frac{p}{2-p}$ and $q = 2\frac{p}{2-p}$. So we have $p = \frac{2q}{2+q}$, $C_{2,p} = \frac{q}{2}$ and therefore

$$D_{2,q} \le (3\max\{1, 2^q C_{2,p}^q, 4^q\})^{\frac{1}{q}} \le 3(1 + 2C_{2,p} + 4) = 3(5 + q).$$

Hence there exists a $D \in R_+$ satisfying (2.42).

2. By extending u by 0 outside of Ω, the inequality (2.43) gives (2.44) if $q \geq 2$. Then for $q < 2$, the generalized Hölder inequality gives that $\|u\|_{L^q(\mathbb{R}^2)} \leq \lambda_2(\Omega)^{\frac{1}{q} - \frac{1}{2}} \|u\|_{L^2(\mathbb{R}^2)}$ which yields the existence of C depending only on Ω satisfying (2.44).

3. For $x \in \Omega$,

$$e^{\sigma u(x)^2} = \sum_{n \in \mathbb{N}} \frac{\sigma^n u(x)^{2n}}{n!},$$

and therefore, by the monotone convergence theorem and question 2,

$$\left\| e^{\sigma u^2} \right\|_{L^1(\Omega)} = \lambda_2(\Omega) + \sum_{n \geq 1} \frac{\sigma^n}{n!} \|u\|_{L^{2n}(\Omega)}^{2n}$$

$$\leq \lambda_2(\Omega) + \sum_{n \geq 1} \frac{\sigma^n}{n!} (C\sqrt{2n})^{2n} \leq \lambda_2(\Omega) + \sum_{n \geq 1} \frac{\sigma^n}{n!} (2C^2)^n n^n.$$

Let $a_n = \frac{\sigma^n}{n!} (2C^2)^n n^n$ so that

$$\frac{a_{n+1}}{a_n} = \frac{\sigma}{n} 2C^2 \frac{(n+1)^{n+1}}{n^n} = \sigma 2C^2 \frac{n+1}{n} \left(\frac{n+1}{n}\right)^n.$$

Therefore $\lim_{n \to +\infty} \frac{a_{n+1}}{a_n} = \sigma 2eC^2$. Let us then choose σ_0 such that $0 < \sigma_0 < 1/(2eC^2)$ and set

$$a = \lambda_2(\Omega) + \sum_{n \geq 1} \frac{\sigma_0^n}{n!} (2C^2)^n n^n.$$

We then obtain $\left\| e^{\sigma_0 u^2} \right\|_{L^1(\Omega)} \leq a$.

4. We use for this question the numbers σ_0 and a found in question 3. Let $\sigma > 0$ and $1 \leq p < +\infty$. Choose $\varepsilon > 0$ such that $(\frac{2p\sigma}{\sigma_0})^{\frac{1}{2}} \varepsilon = 1$. Let $u \in H_0^1(\Omega)$, $u \neq 0$. Part I gives the existence of u_1 and u_2 such that $u_1 \in L^\infty(\Omega)$ and $\|u_2\|_{H_0^1(\Omega)} \leq \varepsilon$. Hence, since $u^2 \leq 2u_1^2 + 2u_2^2$ a.e., with $A > 0$ such that $e^{2p\sigma u_1^2} \leq A$ a.e.,

$$e^{p\sigma u^2} \leq e^{2p\sigma u_1^2} e^{2p\sigma u_2^2} \leq A e^{\sigma_0 \frac{2p\sigma}{\sigma_0} u_2^2} \text{ a.e.}$$

The choice of ε gives $\left\| (\frac{2p\sigma}{\sigma_0})^{\frac{1}{2}} u_2 \right\|_{H_0^1(\Omega)} = (\frac{2p\sigma}{\sigma_0})^{\frac{1}{2}} \|u_2\|_{H_0^1(\Omega)} \leq 1$. Hence $e^{\sigma u^2} \in L^p(\Omega)$ and

$$\left\| e^{\sigma u^2} \right\|_{L^p(\Omega)} \leq (Aa)^{\frac{1}{p}}.$$

Part III, on the resolution of the Dirichlet problem

1. Let $\beta \in \mathbb{R}_+^\star$ and let $s, t \in \mathbb{R}_+^\star$. If $st > \beta t \sqrt{|\ln t|}$, then we have $s > \beta \sqrt{|\ln t|}$, so that $s^2 > \beta^2 |\ln t|$ and therefore $e^{\frac{s^2}{\beta^2}} > e^{|\ln t|} > e^{\ln t} = t$. Hence $st < se^{\frac{s^2}{\beta^2}}$ and we

have indeed shown that

$$st \le \max\{\beta t\sqrt{|\ln t|}, se^{\frac{s^2}{\beta^2}}\},$$

and therefore

$$st \le \beta t\sqrt{|\ln t|} + se^{\frac{s^2}{\beta^2}}.$$

Now choose $\beta > 0$ such that $\sigma > 1/\beta^2$. Then $\frac{1}{s}e^{\sigma - \frac{s^2}{\beta^2}} \to +\infty$ as $s \to +\infty$, and there exists $\gamma > 0$ such that $e^{\sigma s^2} > se^{\frac{s^2}{\beta^2}}$ for $s > \gamma$. Hence, for all $s, t \in \mathbb{R}_+^\star$,

$$st \le \beta t\sqrt{|\ln t|} + se^{\sigma s^2} \le \begin{cases} \beta t\sqrt{|\ln t|} + e^{\sigma s^2} & \text{if } s > \gamma, \\ \gamma t & \text{if } s \le \gamma, \end{cases}$$

so that finally, for all $s, t \in \mathbb{R}_+^\star$,

$$st \le e^{\sigma s^2} + \gamma t + \beta t\sqrt{|\ln t|}.$$

2. Let $u \in H_0^1(\Omega)$ such that $\|u\|_{H_0^1(\Omega)} \le 1$. Choose $\sigma = \sigma_0 > 0$ found in question 3 of part II and $\beta > 0$, $\gamma > 0$ found in question 1. Question 1 gives

$$|fu| \le e^{\sigma_0 u^2} + \gamma|f| + \beta|f|\sqrt{|\ln|f||} \text{ a.e.}$$

Hence $fu \in L^1(\Omega)$ and

$$\|fu\|_{L^1(\Omega)} \le a + \gamma\|f\|_{L^1(\Omega)} + \beta\left\|f\sqrt{|\ln(|f|)|}\right\|_{L^1(\Omega)} = M,$$

with a found in question 3 of part II. By linearity, we therefore have, for all $u \in H_0^1(\Omega)$, $fu \in L^1(\Omega)$ and $\|fu\|_{L^1(\Omega)} \le M\|u\|_{H_0^1(\Omega)}$, which proves that the mapping $T : u \mapsto \int_\Omega f(x)u(x)\,dx$ belongs to $H^{-1}(\Omega)$.
3. Since $T \in H^{-1}(\Omega)$, Theorem 2.9 gives the existence and uniqueness of $u \in H_0^1(\Omega)$ such that

$$\int_\Omega \nabla u \cdot \nabla v\,dx = \langle T, u\rangle_{H^{-1}(\Omega), H_0^1(\Omega)} = \int_\Omega fu\,dx \text{ for all } v \in H_0^1(\Omega). \qquad (2.79)$$

Taking $v \in \mathcal{D}^\star(\Omega)$ in (2.79) this shows that $-\Delta u = f$ in $\mathcal{D}^\star(\Omega)$.
Conversely if $u \in H_0^1(\Omega)$ is such that $-\Delta u = f$ in $\mathcal{D}^\star(\Omega)$. We have (2.79) for all $v \in \mathcal{D}^\star(\Omega)$ and therefore by density u is the solution to (2.79).

Part IV, counterexample

1. Choose a non-negative function $\varphi \in C_c^\infty(B_{2\delta})$ such that $\varphi = 1$ over B_δ (the existence of such a function is a classical result by regularisation, see for example [26, section 8.1.2]) and extend it by 0 so that $\varphi \in C_c^\infty(\Omega)$. The function $u = v\varphi$ belongs to $H_0^1(\Omega)$.

2. Choose f in the form $f(x) = \frac{1}{|x|^2(-\ln(|x|))^\alpha}$ for $x \in B(0, \delta)$ and $f = 0$ outside of $B(0, \delta)$. For $f \in L^1(\Omega)$ it is sufficient that $\alpha > 1$. We have

$$f(x)(\ln|f(x)|)^\theta = \frac{1}{|x|^2(-\ln(|x|))^\alpha}(-2\ln|x| - \alpha\ln(-\ln(|x|)))^\theta.$$

Therefore $f(\ln|f|)^\theta \in L^1(\Omega)$ if $\alpha - \theta > 1$, that is $\alpha > 1 + \theta$. Finally, choosing u constructed in question 1 with $\gamma \in]0, \frac{1}{2}[$ (so that $u \in H_0^1(\Omega)$),

$$f(x)u(x) = \frac{1}{|x|^2(-\ln|x|)^{\alpha-\gamma}} \quad \text{for } x \in B_\delta,$$

and therefore $fu \notin L^1(\Omega)$ if $\alpha - \gamma \leq 1$. It is therefore sufficient to choose $\gamma = \alpha - 1$, which is possible by taking $1 + \theta < \alpha < 3/2$. We indeed have $f \in L^1(\Omega)$, $f(\ln|f|)^\theta \in L^1(\Omega)$ and $fu \notin L^1(\Omega)$ for some $u \in H_0^1(\Omega)$.

3. Let f be the function constructed in the previous question and assume that $-\Delta u = f$ in $\mathcal{D}^\star(\Omega)$ with $u \in H_0^1(\Omega)$. Since $u \in H_0^1(\Omega)$, we have $-\Delta u \in H^{-1}(\Omega)$, that is, the element of $\mathcal{D}^\star(\Omega)$ denoted $-\Delta u$ extends to a continuous linear mapping on $H_0^1(\Omega)$. This is impossible because we constructed in the previous question a function v in $H_0^1(\Omega)$ such that $fv \notin L^1(\Omega)$. A simple way to see this impossibility is to use the fact that the functions f and v are non-negative on $B(0, \delta)$. We then consider $v_n = \min\{v, n\}$ so that $v_n \in H_0^1(\Omega)$ and $\|v_n\|_{H_0^1(\Omega)} \leq \|v\|_{H_0^1(\Omega)}$. Since $fv_n \in L^1(\Omega)$, the fact that $-\Delta u \in H^{-1}(\Omega)$ (with $u \in H_0^1(\Omega)$) would give, assuming that

$$\int_\Omega fv_n\, dx = \langle -\Delta u, v_n \rangle_{H^{-1}(\Omega),H_0^1(\Omega)} \tag{2.80}$$

which we prove below, the existence of C such that

$$\int_\Omega fv_n\, dx \leq C\,\|v_n\|_{H_0^1(\Omega)} \leq C\,\|v\|_{H_0^1(\Omega)}\,.$$

By monotone convergence we would deduce $fv \in L^1(\Omega)$, which is false.

Let us now prove the equality (2.80). Since $v_n \in L^\infty(\Omega) \cap H_0^1(\Omega)$, there exists a sequence $(\varphi_p)_{p\in\mathbb{N}}$ bounded in $L^\infty(\Omega)$ of elements of $C_c^\infty(\Omega)$ such that $\varphi_p \to v_n$ in $H_0^1(\Omega)$ as $p \to +\infty$. We obtain $\int_\Omega fv_n\, dx = \langle -\Delta u, v_n \rangle_{H^{-1}(\Omega),H_0^1(\Omega)}$ by letting $p \to +\infty$ in the equality $\int_\Omega f\varphi_p\, dx = \langle -\Delta u, \varphi_p \rangle_{H^{-1}(\Omega),H_0^1(\Omega)}$ (with the dominated convergence theorem for the left-hand side and the convergence of φ_p to v_n in $H_0^1(\Omega)$ for the right-hand side).

Problem 2.15 (Hodge Decomposition)

Problem 2.8 provides the existence of a function $u \in H^1(\Omega)$ such that

$$\int_\Omega \nabla u(x) \cdot \nabla \varphi(x)\, dx = \int_\Omega f(x) \cdot \nabla \varphi(x)\, dx, \quad \forall \varphi \in H^1(\Omega).$$

We can also add the condition $\int_\Omega u(x)\, dx = 0$ and we then have existence and uniqueness of u (see Problem 2.8). Then, setting $g = f - \nabla u$, the functions u and g satisfy the requested conditions.

Assuming now that $g \in C^1(\bar\Omega)$ and that $\Omega = (]0, 1[)^N$, we have

$$\int_\Omega g(x) \cdot \nabla \varphi(x)\, dx = 0 \text{ for all } \varphi \in H^1(\Omega).$$

We reason as in Problem 2.8. Taking $\varphi \in C_c^\infty(\Omega)$, owing to the fundamental lemma (Lemma 1.2) div $g = 0$ everywhere in Ω. Then, taking $\varphi \in C^1(\bar\Omega)$, an integration by parts (rather easier than in Problem 2.8) gives

$$\int_{\partial\Omega} g(x) \cdot n(x)\varphi(x)\, d\gamma(x) = 0 \text{ for all } \varphi \in C^1(\bar\Omega).$$

From this equality, it may be shown that $g \cdot n = 0$ a.e. on $\partial\Omega$. If $N = 2$, the proof is as follows: let $d\gamma(x) = dx_1$ or dx_2, depending on the parts of $\partial\Omega$, with $x = (x_1, x_2)^t$. With $g = (g_1, g_2)^t$, $g_1(x) = 0$ everywhere on $\{0\} \times [0, 1] \cup \{1\} \times [0, 1]$ and $g_2(x) = 0$ everywhere on $[0, 1] \times \{0\} \cup [0, 1] \times \{1\}$, which indeed gives $g \cdot n = 0$ a.e. on $\partial\Omega$. The generalisation to the case $N \geq 1$ poses no difficulty.

Problem 2.16 (Wentzel boundary conditions)

Part I (Functional analysis preliminary)

1. The space $H_p^1(0, 2\pi)$ is closed in $H^1(]0, 2\pi[)$ (since the mapping $u \mapsto u$ is continuous from $H^1(]0, 2\pi[)$ to $C([0, 2\pi], \mathbb{R})$ equipped with the uniform norm). The space $H_p^1(0, 2\pi)$ equipped with the inner product of $H^1(]0, 2\pi[)$ is therefore a Hilbert space.
2. The mapping $(u, v) \mapsto (u \mid v)_H$ is clearly an inner product on H (it is symmetric bilinear and $a(u, u) > 0$ if u is non-zero). It remains to show that H is complete with this inner product.
   Let $(u_n)_{n \in \mathbb{N}}$ be a Cauchy sequence in H. The sequence $(u_n)_{n \in \mathbb{N}}$ is therefore a Cauchy sequence in $H^1(B)$ and the sequence $(\bar\gamma(u_n))n \in \mathbb{N}$ is a Cauchy sequence in $H_p^1(0, 2\pi)$. Hence there exist $u \in H^1(B)$ and $g \in H_p^1(0, 2\pi)$ such that $u_n \to u$ in $H^1(B)$ and $\bar\gamma(u_n) \to g$ in $H_p^1(0, 2\pi)$ as $n \to +\infty$. Since $u_n \to u$ in $H^1(B)$ we also have $\bar\gamma(u_n) \to \bar\gamma(u)$ in $L^2(]0, 2\pi[)$. Hence $\bar\gamma(u) = g$ a.e. and therefore

$u \in H$. Finally, since $u_n \to u$ in $H^1(B)$ and $\bar{\gamma}(u_n) \to \bar{\gamma}(u)$ as $n \to +\infty$, $u_n \to u$ in H. The space H is indeed complete.

Part II (Wentzel boundary conditions)

1. Let us set

$$T(v) = \int_B f(z)v(z)\,\mathrm{d}z + \int_0^{2\pi} j(g)(\theta)j(\gamma(v))(\theta)\,\mathrm{d}\theta.$$

The problem (2.46) is then written as

$$u \in H$$
$$(u \,|\, v)_H = T(v) \text{ for all } v \in H.$$

The mapping T belongs to H'. Indeed,

$$\left| \int_B f(z)v(z)\,\mathrm{d}z \right| \leq \|f\|_{L^2(B)} \|v\|_{H^1(B)} \text{ and}$$

$$\left| \int_0^{2\pi} j(g)(\theta)j(\gamma(v))(\theta)\,\mathrm{d}\theta \right| \leq \|g\|_{L^2(\partial B)} \|\gamma(v)\|_{L^2(\partial B)} \,.$$

Owing to the continuity of the operator γ from $H^1(B)$ to $L^2(\partial B)$, the operator T thus belongs to H'. The Riesz representation theorem (see for example [26] theorem 6.56) gives the existence and uniqueness of a solution u to (2.46).

It is interesting to note, for question 4 below, that the mapping $(f, g) \mapsto (u, \gamma(u))$ (where u is the solution to (2.46)) sends a bounded subset of $L^2(B) \times L^2(\partial B)$ into a bounded subset of $H^1(B) \times H_p^1(0, 2\pi)$.

2. For $u, v \in H$ we set $a(u, v) = \int_B \nabla u(x) \cdot \nabla v(x)\,\mathrm{d}x + (\bar{\gamma}(u) \,|\, \bar{\gamma}(v))_{H_p^1(0,2\pi)}$. The problem (2.46) (without "uv" in the first integral of (2.46)) is then written, with the same mapping T as in the previous question,

$$u \in H$$
$$a(u, v) = T(v) \text{ for all } v \in H.$$

To show the existence and uniqueness of the solution u to (2.46) (without "uv") it is therefore sufficient (by the Riesz representation theorem) to show that the bilinear form a defines on H an inner product equivalent to the inner product $(\cdot \,|\, \cdot)_H$. It is immediate that $a(u, u) \leq (u \,|\, u)_H$. It is therefore sufficient to show that there exists a $c > 0$ such that, for all $u \in H$, $a(u, u) \geq c(u \,|\, u)_H$.

Let us prove the existence of c by contradiction. If c does not exist, there exists, for all $n \in \mathbb{N}^\star$, $u_n \in H$ such that

$$a(u_n, u_n) < \frac{1}{n}(u_n \,|\, u_n)_H. \tag{2.81}$$

Note that $u_n \neq 0$ (since $a(u, u) \geq 0$ for all $u \in H$). By a homogeneity argument, we can assume that $\|u_n\|_H = 1$. The sequence $(u_n)_{n \in \mathbb{N}}$ is therefore bounded in $H^1(B)$ and the sequence $(\bar{\gamma}(u_n))_{n \in \mathbb{N}}$ is bounded in $H_p^1(0, 2\pi)$. We can therefore also assume that there exist $u \in H^1(B)$ and $h \in H_p^1(0, 2\pi)$ such that, up to a subsequence, $u_n \to u$ weakly in $H^1(B)$ and $\bar{\gamma}(u_n) \to h$ weakly in $H_p^1(0, 2\pi)$ as $n \to +\infty$.

Since γ is continuous from $H^1(B)$ to $L^2(\partial B)$, the weak convergence of u_n to u in $H^1(B)$ gives $\gamma(u_n) \to \gamma(u)$ weakly in $L^2(\partial B)$ (see for example Problem 1.22). Since j is an isometry, we also have as $n \to +\infty$ $\bar{\gamma}(u_n) \to \bar{\gamma}(u)$ weakly in $L^2(\partial B)$ and therefore $h = \bar{\gamma}u$ a.e. Finally, we also have (for example by the Problem 1.22) $D_i u_n \to D_i u$ weakly in $L^2(B)$ for $i = 1, 2$.

Now observe that $(u_n \mid u_n)_H = 1$; hence the inequality (2.81) gives $D_i u_n \to 0$ (for $i = 1, 2$) in $L^2(B)$ and $\bar{\gamma}(u_n) \to 0$ in $H_p^1(0, 2\pi)$ as $n \to +\infty$ (in particular $h = \bar{\gamma}(u) = 0$ a.e.). Hence $\nabla u = 0$ a.e. and there exists a $b \in \mathbb{R}$ such that $u = b$ a.e. (Problem 1.4). this gives $\gamma(u) = b$ and $\bar{\gamma}(u) = b$ a.e. and therefore $u = 0$ a.e. To conclude, owing to Theorem 1.37, the weak convergence of u_n to u in $H^1(B)$ gives convergence in $L^2(B)$ and therefore $u_n \to 0$ in $L^2(B)$. But, we already know that $D_i u_n \to 0$ (for $i = 1, 2$) in $L^2(B)$ and therefore $u_n \to 0$ in $H^1(B)$, which is impossible because $\|u_n\|_H = 1$ and

$$\|u_n\|_H^2 = \|u_n\|_{H^1(B)}^2 + \|\bar{\gamma}(u_n)\|_{H_p^1(0,2\pi)} \to 0 \text{ when } n \to +\infty.$$

Thus, the bilinear form a defines on H an inner product equivalent to the inner product $(\cdot \mid \cdot)_H$, which proves therefore the existence and uniqueness of the solution u to (2.46) (without the term "uv").

Here too we notice that the mapping $(f, g) \mapsto (u, \gamma(u))$ (where u is the solution to (2.46) without "uv") sends a bounded subset of $L^2(B) \times L^2(\partial B)$ into a bounded subset of $H^1(B) \times H_p^1(0, 2\pi)$.

3. Assume that u is a solution in the classical sense. Let $v \in C^2(\overline{B}, \mathbb{R})$. Set $\bar{v}(r, \theta) = v(r \cos \theta, r \sin \theta)$. We multiply (2.45a) by v and integrate by parts over B, we obtain

$$\int_B \nabla u(z) \cdot \nabla v(z) \, dz + \int_B u(z)v(z) \, dz$$
$$- \int_0^{2\pi} \partial_x u(\cos \theta, \sin \theta) \cos \theta + \partial_y u(\cos \theta, \sin \theta) \sin \theta)\bar{v}(1, \theta) \, d\theta$$
$$= \int_B f(z)v(z) \, dz,$$

that is to say,

$$\int_B \nabla u(z) \cdot \nabla v(z) \, dz + \int_B u(z)v(z) \, dz - \int_0^{2\pi} \partial_r \bar{u}(1, \theta)\bar{v}(1, \theta) \, d\theta$$
$$= \int_B f(z)v(z) \, dz.$$

Using (2.45b) and an integration by parts, we obtain

$$\int_{\mathcal{B}} \nabla u(z) \cdot \nabla v(z)\, dz + \int_{B} u(z)v(z)\, dz + \int_{0}^{2\pi} \partial_{\theta}\bar{u}(1,\theta)\partial_{\theta}\bar{v}(1,\theta)\, d\theta$$

$$+ \int_{0}^{2\pi} \bar{u}(1,\theta)\bar{v}(1,\theta)\, d\theta = \int_{B} f(z)v(z)\, dz + \int_{0}^{2\pi} g(\cos\theta, \sin\theta)\bar{v}(1,\theta)\, d\theta.$$

Since $\bar{u}(1,\theta) = \bar{\gamma}(u)(\theta)$ and $\bar{v}(1,\theta) = \bar{\gamma}(v)(\theta)$, this gives exactly the equation (2.46) for $v \in C^2(\bar{B}, \mathbb{R})$. The density of $C^2(\bar{B}, \mathbb{R})$ in H allows us to conclude that u is the unique solution to (2.46).

Conversely, we now assume that u is the unique solution to (2.46). Since (2.46) is satisfied for all $v \in \mathcal{D}(B)$ and since $u \in C^2(B, \mathbb{R}))$, u satisfies (2.45a) for all $(x, y) \in B$. Taking $v \in C^2(\bar{B}, \mathbb{R})$ in (2.46) and integrating by parts over B and over $]0, 2\pi[$ yields, since $u \in C^2(\bar{B}, \mathbb{R})$ and u satisfies (2.45a),

$$\int_{0}^{2\pi} (\partial_r \bar{u}(1,\theta)\bar{v}(1,\theta) - \partial_{\theta}^2 \bar{u}(1,\theta) + \bar{u}(1,\theta))\bar{v}(1,\theta)\, d\theta$$

$$= \int_{0}^{2\pi} g(\cos\theta, \sin\theta)\bar{v}(1,\theta)\, d\theta.$$

Introducing the function $\theta \mapsto \bar{v}(1,\theta)$ as the restriction to $[0, 2\pi]$ of an arbitrary 2π-periodic function of class C^2 from $\mathbb{R}$ to $\mathbb{R}$, we get that u satisfies (2.45b) for all $(x, y) \in \partial B$.

4. The linearity of T is immediate.

The mapping $(f, g) \mapsto (u, \gamma(u))$ (where u is the solution to (2.46)) sends a bounded subset of $L^2(B) \times L^2(\partial B)$ to a bounded subset of $H^1(B) \times H_p^1(0, 2\pi)$ and then the mapping $(u, v) \mapsto (u, v)$ sends (by Theorem 1.37) a bounded subset of $H^1(B) \times H_p^1(0, 2\pi)$ into a relatively compact subset of $L^2(B) \times L^2(\partial B)$; hence T is compact.

Finally, let u be the solution to (2.46) associated with (f, g) and let v be the solution to (2.46) associated with (φ, ψ) instead of (f, g). Observe that, since $T(f, g) = (u, \gamma(u))$, with u the solution to (2.46)

$$(u \mid v)_H = \int_{B} f(z)v(z)\, dz + \int_{0}^{2\pi} j(g)(\theta)j(\gamma(v))(\theta)\, d\theta$$

$$= ((f, g) \mid v, \gamma(v))_{L^2(B) \times L^2(\partial B)} = ((f, g) \mid T(\varphi, \psi))_{L^2(B) \times L^2(\partial B)}.$$

Now since $T(\varphi, \psi) = (v, \gamma(v))$, with v the solution to (2.46) with (φ, ψ) instead of (f, g),

$$(v \mid u)_H = \int_{B} \varphi(z)u(z)\, dz + \int_{0}^{2\pi} j(\psi)(\theta)j(\gamma(u))(\theta)\, d\theta$$

$$= ((\varphi, \psi) \mid u, \gamma(u))_{L^2(B) \times L^2(\partial B)} = ((\varphi, \psi) \mid T(f, g))_{L^2(B) \times L^2(\partial B)},$$

which clearly shows that $T = T^{\star}$.

Problem 2.17 (Stokes problem, velocity and pressure)

Part I, existence and uniqueness of u

1. Let (u, p) be a classical solution to (2.47); first note that $u \in V$. Then multiply the first equation of (2.47) by $v \in V$ and integrate over Ω. The functions u and v are sufficiently regular to integrate by parts and thus obtain that u is then a solution to (2.49).

2. The mapping div $: u \in H_0^1(\Omega)^N \mapsto \operatorname{div} u \in L^2(\Omega)$ is a linear continuous operator. Since $V = \operatorname{Ker} \operatorname{div}$, it follows that V is a closed subspace of $H_0^1(\Omega)^N$.

3. Here it is sufficient to apply the Lax–Milgram theorem, Theorem 2.3 (or the Riesz representation theorem in Hilbert spaces), noting that V is a Hilbert space equipped with the natural inner product on $H_0^1(\Omega)^N$, with a and T defined by

$$a(u, v) = \sum_{i=1}^{N} \int_\Omega \nabla u_i(x) \cdot \nabla v_i(x)\, dx, \ \text{ and } T(v) = \int_\Omega f(x) \cdot v(x)\, dx.$$

4. For $v \in V$, we have $\operatorname{div} v = 0$ a.e. in Ω and therefore $\int_\Omega p \operatorname{div} v\, dx = 0$. It follows that u is a solution to (2.49). By the previous question, the (vector) function u is therefore the unique solution to (2.49).

Part II, functional analysis preliminaries

1. Let $u \in \operatorname{Ker} A$ (so we have $Au = 0$); for $v \in \operatorname{Im}(A^*)$. There exists a $g \in F$ such that $v = A^* g$, so we have

$$(v|u)_E = (A^* g|u)_E = (g|Au)_F = 0,$$

which shows that $u \in (\operatorname{Im}(A^*))^\perp$. So we have $\operatorname{Ker} A \subset (\operatorname{Im}(A^*))^\perp$.

Conversely, let $u \in (\operatorname{Im}(A^*))^\perp$. Setting $f = Au$, since $A^* f \in \operatorname{Im}(A^*)$,

$$(Au|Au)_F = (f|Au)_F = (A^* f|u)_E = 0.$$

Therefore, $Au = 0$, that is $u \in \operatorname{Ker} A$, and $(\operatorname{Im}(A^*))^\perp \subset \operatorname{Ker} A$, which concludes the proof that $(\operatorname{Im}(A^*))^\perp = \operatorname{Ker} A$.

2. If G is a closed subspace of a Hilbert space E, we always have $E = G \oplus G^\perp$. Moreover, for any subspace G of E, $G^\perp = \bar{G}^\perp$.

Hence for any subspace of G of a Hilbert space E,

$$E = \bar{G} \oplus G^\perp \ \text{ and } \ E = G^\perp \oplus (G^\perp)^\perp,$$

so that $(G^\perp)^\perp = \bar{G}$.

Applying this result with $G = \operatorname{Im}(A^*)$ and using the result of the previous question yields

$$\overline{\operatorname{Im}(A^*)} = ((\operatorname{Im}(A^*))^\perp)^\perp = (\operatorname{Ker} A)^\perp.$$

Part III, Partial existence and uniqueness of p

1. a. Let $v \in H$. We have

$$(A^* p_n | v)_H = (p_n | Av)_F = \int_\Omega p_n \operatorname{div} v \, dx,$$

and

$$(A^* q_n | v)_H = (q_n | Av)_F = \int_\Omega q_n \operatorname{div} v \, dx = \int_\Omega p_n \operatorname{div} v \, dx - a_n \int_\Omega \operatorname{div} v \, dx.$$

Since $v \in H$, we have (integrating by parts) $\int_\Omega \operatorname{div} v \, dx = 0$ and therefore

$$(A^* p_n | v)_H = (A^* q_n | v)_H \text{ for all } v \in H.$$

This clearly shows that $A^* p_n = A^* q_n$.
   b. By Lemma 2.38, there exists a $v_n \in H$ such that $\operatorname{div} v_n = q_n$ a.e. in Ω and $\|v_n\|_{H_0^1(\Omega)^N} \leq C \|q_n\|_{L^2(\Omega)}$. We then obtain

$$(A^* q_n | v_n)_H = \int_\Omega q_n \operatorname{div} v_n \, dx = \int_\Omega q_n^2 \, dx = \|q_n\|_F^2 .$$

The previous question gives $A^* p_n = A^* q_n$. Hence

$$\|q_n\|_F^2 = (A^* p_n | v_n)_H \leq \|A^* p_n\|_H \, \|v_n\|_H \leq C \|A^* p_n\|_H \, \|q_n\|_F ,$$

and therefore
$$\|q_n\|_F \leq C \|A^* p_n\|_H .$$

The convergence assumption on $A^* p_n$ shows that the sequence $(A^* p_n)_{n \in \mathbb{N}}$ is bounded in H. Hence the sequence $(q_n)_{n \in \mathbb{N}}$ is bounded in F.
   c. Since the sequence $(q_n)_{n \in \mathbb{N}}$ is bounded in the Hilbert space F, up to a subsequence, it converges weakly in F to some limit $q \in F$ as $n \to +\infty$; let us show that $v = A^* q$.
Let $w \in H$, we have $\lim_{n \to +\infty} (q_n | Aw)_F = (q | Aw)_F$. But

$$(q_n | Aw)_F = (A^* q_n | w)_H = (A^* p_n | w)_H.$$

Since $A^* p_n \to v$ in H, $\lim_{n \to +\infty} (q_n | Aw)_F = (v | w)_H$, and so

$$(q | Aw)_F = (v | w)_H \text{ for all } w \in H.$$

This gives $(A^* q | w)_H = (v | w)_H$ for all $w \in H$, and therefore $v = A^* q$. We have indeed shown that $v \in \operatorname{Im}(A^*)$.
2. The previous question shows that $\operatorname{Im}(A^*)$ is closed in H. Owing to part II, we therefore have $(\operatorname{Ker} A)^\perp = \operatorname{Im}(A^*)$. We have already seen that $\operatorname{Ker} A = V$, hence $V^\perp = \operatorname{Im}(A^*)$.

3. a. We have $(u|v)_H = \int_\Omega fv\,dx = (T_f|v)_H$ for all $v \in V$. This indeed means that $u - T_f \in V^\perp$ and therefore that $u - T_f \in \mathrm{Im}(A^*)$.
   b. Since $u - T_f \in \mathrm{Im}(A^*)$, there exists a $p \in F = L^2(\Omega)$ such that $u - T_f = A^* p$. Hence for all $v \in H$,

$$(u|v)_H - \int_\Omega fv\,dx = (u - T_f|v)_H = (A^* p|v)_H = (p|Av)_F = \int_\Omega p\,\mathrm{div}\,v\,dx,$$

   which indeed means that (u, p) is a solution to (2.48).
4. We have already shown in question 3 of part I that $u_1 = u_2 = u$ where u is the unique solution to (2.49).

   Now $\int_\Omega p_1\mathrm{div}\,v\,dx = \int_\Omega p_2\mathrm{div}\,v\,dx$ for all $v \in H$. Taking $v = (v_1, \ldots, v_N)^t$ with $v_1 \in \mathcal{D}(\Omega)$ and $v_i = 0$ for $i \geq 2$ yields that $D_1(p_1 - p_2) = 0$ (in $\mathcal{D}^\star(\Omega)$). Similarly, we have $D_i(p_1 - p_2) = 0$ for all $i \in \{1, \ldots, N\}$, hence there exists an $a \in \mathbb{R}$ such that $p_1 - p_2 = a$ a.e. (see Problem 1.4).

Problem 2.18 (Stokes problem, penalisation)

1. With the same notation as that of the previous problem, we first note that the problem (2.50) is equivalent to

$$u \in H,$$
$$(u\,|\,v)_H + n\int_\Omega \mathrm{div}\,u(x)\mathrm{div}\,v(x)\,dx = \int_\Omega f(x)\cdot v(x)\,dx \text{ for all } v \in H.$$

   Let the bilinear form a on H be defined by

$$a(u, v) = (u\,|\,v)_H + n\int_\Omega \mathrm{div}\,u(x)\mathrm{div}\,v(x)\,dx.$$

   Observe that a is an inner product on H equivalent to the natural inner product $(\cdot\,|\,\cdot)_H$. Applying once again the Riesz representation theorem gives the existence and uniqueness of a solution $u \in H$ to (2.50).
2. Take $v = u^{(n)}$ in (2.50) and use the Cauchy–Schwarz inequality to obtain

$$(u^{(n)}\,|\,u^{(n)})_H \leq (u^{(n)}\,|\,u^{(n)})_H + n\int_\Omega (\mathrm{div}\,u^{(n)}(x))^2\,dx$$
$$= \int_\Omega f(x)\cdot u^{(n)}(x)\,dx \leq \sum_{i=1}^n \|f_i\|_{L^2(\Omega)}\left\|u_i^{(n)}\right\|_{L^2(\Omega)}.$$

   Owing to the Poincaré inequality (Lemma 2.5), the sequence $(u^{(n)})_{n\in\mathbb{N}}$ is bounded in H, and the sequence $(\sqrt{n}\,\mathrm{div}\,u^{(n)})_{n\in\mathbb{N}}$ is bounded in $L^2(\Omega)$.

3. The sequence $(u^{(n)})_{n\in\mathbb{N}}$ is bounded in H and therefore, up to a subsequence, converges weakly in H to some limit $u \in H$. Taking $v \in V$, so that $\operatorname{div} v = 0$ a.e., yields

$$(u^{(n)} \mid v)_H = \int_\Omega f(x) \cdot v(x)\, dx.$$

Passing to the limit as $n \to +\infty$, we obtain

$$(u \mid v)_H = \int_\Omega f(x) \cdot v(x)\, dx.$$

The function u belongs to H and satisfies the requested equation in (2.49). It remains to verify that $u \in V$; indeed, since $\operatorname{div} u^{(n)} \to \operatorname{div} u$ at least weakly in $L^2(\Omega)$ and $\operatorname{div} u^{(n)} \to 0$ in $L^2(\Omega)$ (because the sequence $(\sqrt{n}\,\operatorname{div} u^{(n)})_{n\in\mathbb{N}}$ is bounded in $L^2(\Omega)$), we get that $\operatorname{div} u = 0$ a.e., which proves that $u \in V$. The function u is therefore a solution to (2.49). The uniqueness of the solution to (2.49) then allows us to affirm that the whole sequence $(u^{(n)})_{n\in\mathbb{N}}$ converges to u weakly in V as $n \to +\infty$.

Problem 2.19 (Sequential continuity of L^2-weak in H_0^1) ($\star$)

1. Taking $v = u_n$ in (2.51) (with f_n and u_n instead of f and u), we obtain

$$\alpha \|u_n\|^2_{H_0^1(\Omega)} \le \|f_n\|_{L^2(\Omega)} \|u_n\|_{L^2(\Omega)} \le \|f_n\|_{L^2(\Omega)} \, C_\Omega \|u_n\|_{H_0^1(\Omega)},$$

where C_Ω is given by Poincaré's inequality. Therefore, for all $n \in \mathbb{N}$,

$$\|u_n\|_{H_0^1(\Omega)} \le \frac{C_\Omega}{\alpha} \sup_{p\in\mathbb{N}} (\|f_p\|_{L^2(\Omega)}) = M < +\infty,$$

the sequence $(f_n)_{n\in\mathbb{N}}$ being bounded in $L^2(\Omega)$.

2. If $u_n \nrightarrow u$ weakly in $H_0^1(\Omega)$, there exist $\varepsilon > 0$, $\psi \in H^{-1}(\Omega)$ and a subsequence, still denoted $(u_n)_{n\in\mathbb{N}}$, such that

$$|\langle \psi, u_n - u \rangle_{H^{-1}(\Omega), H_0^1(\Omega)}| \ge \varepsilon \text{ for all } n \in \mathbb{N}. \tag{2.82}$$

After a possible further extraction, we can assume that $u_n \to \bar{u}$ weakly in $H_0^1(\Omega)$. Let $v \in H_0^1(\Omega)$; we have

$$\int_\Omega A\nabla u_n \cdot \nabla v\, dx = \int_\Omega f_n v\, dx,$$

and therefore, when $n \to \infty$,

$$\int_\Omega A\nabla \bar{u} \cdot \nabla v\, dx = \int_\Omega f v\, dx.$$

Hence $\bar{u} = u$, which contradicts (2.82). Note that it is necessary to reason by contradiction to show that the entire sequence $(f_n)_{n \in \mathbb{N}}$ converges. We have therefore shown that $u_n \to u$ weakly in $H_0^1(\Omega)$ and, by Rellich's theorem, that $u_n \to u$ in $L^2(\Omega)$.

3.

$$\int_\Omega A\nabla u_n \cdot \nabla u_n \, dx = \int_\Omega f_n u_n \, dx \to \int_\Omega f u \, dx = \int A\nabla u \cdot \nabla u \, dx,$$

because $u_n \to u$ in $L^2(\Omega)$ and $f_n \to f$ weakly in $L^2(\Omega)$.

4. We have

$$\int_\Omega A(\nabla u_n - \nabla u) \cdot (\nabla u_n - \nabla u) \, dx = \int_\Omega A\nabla u_n \cdot \nabla u_n \, dx$$

$$- \int_\Omega A\nabla u_n \cdot \nabla u - \int_\Omega A\nabla u \cdot \nabla u_n + \int_\Omega A\nabla u \cdot \nabla u \, dx.$$

The four terms on the right of this equality tend towards $\int_\Omega \nabla u \cdot \nabla u \, dx$ as $n \to +\infty$. The term on the left (which is non-negative) therefore tends towards 0. This gives $\alpha \|u_n - u\|^2_{H_0^1(\Omega)} \to 0$ as $n \to +\infty$ and therefore, as $n \to +\infty$,

$$u_n \to u \quad \text{in } H_0^1(\Omega).$$

Remark on weak topology: The mapping $f \mapsto u$ (where u is the solution to (2.51)) is therefore sequentially continuous from $L^2(\Omega)$-weak to $H_0^1(\Omega)$, that is, it maps the weakly convergent sequences of $L^2(\Omega)$ into convergent sequences of $H_0^1(\Omega)$. It is therefore also sequentially continuous from $L^2(\Omega)$-weak to $L^2(\Omega)$. After defining the weak topology of $L^2(\Omega)$ (which is not done in this book), we can however note that this mapping is not continuous from $L^2(\Omega)$-weak (that is, $L^2(\Omega)$ equipped with the weak topology) in $L^2(\Omega)$ (that is, $L^2(\Omega)$ equipped with the topology associated with its norm).

Problem 2.20 (Preliminary to Problem 2.21)

Taking $x = 0$ in (2.52) yields $\lim\limits_{y \to \infty} \varphi(y) = 0$. Therefore, there exists an a_0 such that $\varphi(a_0) \leq 1$.

Let us proceed by induction and define a sequence $(a_k)_{k \in \mathbb{N}}$ by

$$\frac{C}{a_{k+1} - a_k} \left(\frac{1}{2^k}\right)^\beta = \frac{1}{2^{k+1}}.$$

For $k = 0$ we have $\varphi(a_0) \le 1$. Then, for $k \ge 0$, if $\varphi(a_k) \le \dfrac{1}{2^k}$, we have

$$\varphi(a_{k+1}) \le \frac{C}{a_{k+1} - a_k}\,\varphi(a_k)^\beta \le \frac{C}{a_{k+1} - a_k}\,\frac{1}{2^{k\beta}} = \frac{1}{2^{k+1}}.$$

Hence $\varphi(a_k) \le \dfrac{1}{2^k}$ for $k \in \mathbb{N}$. In order to show that $\lim_{k\to\infty} a_k < +\infty$, let us note that

$$a_{k+1} - a_k = 2C\frac{2^k}{2^{k\beta}} = 2C\frac{1}{2^{k(\beta-1)}} = 2Cb^k \ \text{ with } \ b = \frac{1}{2^{\beta-1}}.$$

Since $\beta > 1$, $b = \frac{1}{2^{\beta-1}} < 1$ and therefore

$$a_k = a_0 + \sum_{p=0}^{k-1} 2Cb^p \le a_0 + 2C\sum_{p=0}^{\infty} b^p = a_0 + \frac{2C}{1-b}.$$

Taking $a = a_0 + \dfrac{2C}{1-b}$, since φ is decreasing,

$$0 \le \varphi(a) \le \varphi(a_k) \ \text{ for all } k \in \mathbb{N},$$

and therefore

$$0 \le \varphi(a) \le \frac{1}{2^k} \ \text{ for all } k \in \mathbb{N},$$

which gives $\varphi(a) = 0$.

Problem 2.21 (Bounded solutions of an elliptic problem)

1. For $u, v \in H_0^1(\Omega)$ we set

$$a(u,v) = \int_\Omega A\nabla u \cdot \nabla v \, dx \ \text{ and } \ T(v) = \int_\Omega F \cdot \nabla v \, dx.$$

As seen in this chapter, the form a is a continuous coercive bilinear form on $H_0^1(\Omega)$. Then, for $v \in H_0^1(\Omega)$, we have

$$T(v) \le \int_\Omega |F \cdot \nabla v| \, dx \le \||F|\|_{L^2(\Omega)}\||\nabla v|\|_{L^2(\Omega)} = \||F|\|_{L^2(\Omega)}\|v\|_{H_0^1(\Omega)}.$$

Hence $T \in (H_0^1(\Omega))'$ and therefore there exists a unique solution u of (2.53).

2. Taking $v = S_k(u)$ in (2.53) and noting that

$$\int_{A_k} |F|^2 dx \le \left(\int_\Omega |F|^p \, dx\right)^{\frac{2}{p}} \lambda_N(A_k)^{1-\frac{2}{p}},$$

it follows that

$$\alpha \|\|\nabla(S_k(u))\|\|^2_{L^2(\Omega)} = \alpha \int_{A_k} \nabla u \cdot \nabla u \, dx = \int_{A_k} F \cdot \nabla u \, dx$$

$$\leq \|\|F\|\|_{L^2(A_k)} \|\|\nabla(S_k(u))\|\|_{L^2(\Omega)}$$

$$\leq \|\|F\|\|_{L^p(\Omega)} (\lambda_N(A_k))^{\frac{1}{2}-\frac{1}{p}} \|\|\nabla(S_k(u))\|\|_{L^2(\Omega)}.$$

We thus obtain

$$\alpha \|\|\nabla(S_k(u))\|\|_{L^2(\Omega)} \leq \|\|F\|\|_{L^p(\Omega)} \lambda_N(A_k)^{\frac{1}{2}-\frac{1}{p}}.$$

3. For $h > k$, we have $|S_k(u)| \geq (h-k)$ on A_h. Hence

$$(h-k)(\lambda_N(A_h))^{\frac{1}{1^\star}} \leq \left(\int_\Omega |S_k(u)|^{1^\star} dx \right)^{\frac{1}{1^\star}} \leq C_1 \|\|\nabla S_k(u)\|\|_{L^1(\Omega)}$$

$$\leq C_1 \int_{A_k} |\nabla S_k(u)| \, dx \leq C_1 \|\|\nabla S_k(u)\|\|_{L^2(\Omega)} \lambda_N(A_k)^{\frac{1}{2}}.$$

By question 1, we obtain

$$(h-k)(\lambda_N(A_h))^{\frac{1}{1^\star}} \leq \frac{C_1}{\alpha} \|\|F\|\|_{L^p} \lambda_N(A_k)^{1-\frac{1}{p}},$$

and therefore, with $C_2 = \frac{C_1}{\alpha} \|\|F\|\|_{L^p(\Omega)}$,

$$(h-k)\lambda_N(A_h)^{\frac{N-1}{N}} \leq C_2 \lambda_N(A_k)^{1-\frac{1}{p}}.$$

4. For $k \in \mathbb{R}_+$, we set $\varphi(k) = (\lambda_N(A_k))^{\frac{N-1}{N}}$. We then have, for $h \geq k \geq 0$,

$$(h-k)\varphi(h) \leq C_2 \varphi(k)^{\frac{N}{N-1} \frac{p-1}{p}}.$$

Set $\beta = \dfrac{N}{N-1}\dfrac{p-1}{p}$. Since $p > N$, we get that $\beta > 1$; indeed $\dfrac{N}{N-1}\dfrac{p-1}{p} > 1 \Leftrightarrow Np - N > Np - p$. Owing to the results of Problem 2.20, there exists an $a \in \mathbb{R}_+$ such that $\varphi(a) = 0$ and therefore $\|u\|_{L^\infty(\Omega)} \leq a$.

5. First assume that $\|\|F\|\|_{L^p(\Omega)} = 1$, which gives, with the notations of the previous questions, $C_2 = \dfrac{C_1}{\alpha}$. Then, recalling the solution of Problem 2.20, the choice of a_0 is such that $\varphi(a_0) \leq 1$. Since

$$\varphi(0) \leq \lambda_N(\Omega)^{\frac{N-1}{N}}, \ \beta\frac{N-1}{N} = \frac{p-1}{p} \text{ and } C_2 = C_1/\alpha,$$

it is therefore sufficient to take a_0 such that

$$\frac{C_1 \lambda_N(\Omega)^{\frac{p-1}{p}}}{\alpha a_0} \leq 1.$$

We can therefore choose $a_0 = \dfrac{C_1 \lambda_N(\Omega)^{\frac{p-1}{p}}}{\alpha}$. We then have $\|u\|_{L^\infty(\Omega)} \le a$ with

$$a = a_0 + \frac{2C_2}{1-b} = a_0 + \frac{C_1}{\alpha} \frac{2}{1 - \frac{1}{2^{\beta-1}}}.$$

Hence $\|u\|_{L^\infty(\Omega)} \le C_3$, with $C_3 = a_0 + \dfrac{C_1}{\alpha} \dfrac{2}{1 - \frac{1}{2^{\beta-1}}}$. It is clear that C_3 depends only on Ω, α and p (note that N is implicitly in Ω).

We can now assume that F is arbitrary in $L^p(\Omega)^N$ (the function u is still the solution to (2.53)). For $\gamma > 0$ the function u/γ is then a solution to (2.53) with F/γ instead of F. If $\|\|F\|\|_{L^p(\Omega)} > 0$, by choosing $\gamma = \|\|F\|\|_{L^p(\Omega)}$ (so that $\|\|F/\gamma\|\|_{L^p(\Omega)} = 1$) we therefore have $\|u/\gamma\|_{\le} C_3$, which gives

$$\|u\|_{L^\infty(\Omega)} \le C_3 \|\|F\|\|_{L^p(\Omega)}.$$

(Also note that the inequality is obvious if $\|\|F\|\|_{L^p(\Omega)} = 0$.)

Problem 2.22 (Bounded solutions of an elliptic problem, continuation)

1. The Sobolev embedding theorem (Theorem 1.41) provides the existence of C, depending only on Ω (and q, $q < +\infty$, if $N = 2$) such that, for all $u \in H^1_0(\Omega)$,

$$\text{if } N = 2, \ \|u\|_{L^q(\Omega)} \le C \, \|u\|_{H^1_0(\Omega)},$$

$$\text{if } N > 2, \ \|u\|_{L^{2^\star}(\Omega)} \le C \, \|u\|_{H^1_0(\Omega)}, \ 2^\star = \frac{2N}{N-2}.$$

Taking $q = \frac{p}{p-1}$, that is $q = \frac{2N}{N-2}$ if $N > 2$, Hölder's inequality yields that for all $u \in H^1_0(\Omega)$,

$$\int_\Omega f(x)u(x)\,\mathrm{d}x \le \|f\|_{L^p(\Omega)} \|u\|_{L^q(\Omega)} \le C \|f\|_{L^p(\Omega)} \|u\|_{H^1_0(\Omega)}.$$

The mapping $f \mapsto \int_\Omega f(x)u(x)\,\mathrm{d}x$ therefore belongs to $H^{-1}(\Omega)$ and so, by Theorem 2.9, there exists a unique solution u of (2.54).

2. For $N > 2$, we have $N/2 > \frac{2N}{N+2}$ and therefore $f \in L^{\frac{2N}{N+2}}(\Omega)$. The previous question also shows that there exists a unique solution u of (2.54). Let $q = \frac{p}{p-1}$ so that $q < N/(N-2)$ ($q < +\infty$ if $N = 2$), and let r be such that $r^\star (= \frac{Nr}{N-r}) = q$ so that $r < \frac{N}{N-1}$; owing to Hölder's inequality with q and the Sobolev embedding of $W^{1,r}_0$ in $L^q(\Omega)$, we get

$$\int_\Omega f(x)u(x)\,\mathrm{d}x \le \|f\|_{L^p(\Omega)} \|u\|_{L^q(\Omega)} \le C \|f\|_{L^p(\Omega)} \|u\|_{W^{1,r}_0(\Omega)}.$$

Let $G = \{\nabla v,\, v \in W_0^{1,r}(\Omega)\}$ and consider the mapping $\nabla v \mapsto \int_\Omega f(x)v(x)\,dx$. This mapping is well defined because an element of $W_0^{1,r}(\Omega)$ is entirely determined by its gradient (see Remark 2.7); it is also linear and continuous with respect to the norm on $L^r(\Omega)^N$. Owing to the Hahn–Banach theorem, it is thus extendable to a continuous linear function denoted by T on all $L^r(\Omega)^N$. By the natural isomorphism between $L^r(\Omega)'$ and $L^{r'}(\Omega)$, with $r' = r/(r-1) > N$, there exists an $F \in (L^{r'})^N$ such that

$$\int_\Omega f(x)v(x)\,dx = \int_\Omega F(x) \cdot \nabla v(x)\,dx,$$

for all $v \in W_0^{1,r}(\Omega)$ and therefore also for all $v \in W_0^{1,2}(\Omega)$ because $r < 2$. We are thus reduced to Problem 2.21.

N.B. Additional remarks on the existence of F in the above arguments:

1. To find for example the first component of F, we consider the continuous linear mapping from $L^r(\Omega)$ to $\mathbb{R}$ defined by $w \mapsto (w, 0, \dots, 0) \mapsto T(w, 0, \dots, 0)$ and we use the isomorphism between $L^r(\Omega)'$ and $L^{r'}(\Omega)$.
2. The existence of F can be shown (in a similar way) for a more general class of continuous linear operators: let Ω be an open bounded subset of $\mathbb{R}^N$ ($N > 1$) and $q \in [1, +\infty[$; a linear mapping T from $W_0^{1,q}(\Omega)$ to $\mathbb{R}$ is continuous and thus belongs to $W_0^{1,q}(\Omega)')$ if and only if there exists an $F \in (L^p)^N$, $p = q(q-1)$ ($p = +\infty$ if $q = 1$), such that

$$T(g) = \int F(x) \cdot \nabla g(x)\,dx \text{ for all } g \in W_0^{1,q}(\Omega).$$

(Note that this is equivalent to saying that $T = -\operatorname{div} F$ in $\mathcal{D}^\star(\Omega)$.)

Problem 2.23 (Evanescent diffusion and convection)

Part I

1. Let $u_n \in \mathcal{D}(\Omega)$ such that $u_n \to u$ in $H_0^1(\Omega)$. Hence, as $n \to +\infty$, $u_n \to u$ in $L^2(\Omega)$ and $\partial_i u_n \to D_i u$ in $L^2(\Omega)$ for all $i \in \{1, \dots, N\}$. (Recall that $\partial_i u_n$ denotes the classical partial derivative of u_n with respect to its i-th variable.) We can also assume that, up to a subsequence, $u_n \to u$ a.e. and that there exists an $F \in L^2(\Omega)$ such that $|u_n| \le F$ for all $n \in \mathbb{N}$. Thus, by the dominated convergence theorem, $u_n^2 \to u^2$ in $L^1(\Omega)$ as $n \to +\infty$. Let $\varphi \in \mathcal{D}(\Omega)$ and $i \in \{1, \dots, N\}$, we then have

$$\langle D_i(u^2), \varphi \rangle_{\mathcal{D}^\star(\Omega), \mathcal{D}(\Omega)} = -\int_\Omega u^2 \partial_i \varphi \,dx$$

$$= -\lim_{n \to +\infty} \int_\Omega u_n^2 \partial_i \varphi \,dx. \tag{2.83}$$

Since u_n and φ belong to $\mathcal{D}(\Omega)$, we have, integrating by parts

$$\int_\Omega u_n^2 \partial_i \varphi \, dx = -2 \int_\Omega \varphi u_n \partial_i u_n \, dx.$$

Since $u_n \to u$ in $L^2(\Omega)$ and $\partial_i u_n \to D_i u$ in $L^2(\Omega)$, we have $u_n \partial_i u_n \to u D_i u$ in $L^1(\Omega)$ and since $\varphi \in L^\infty(\Omega)$,

$$\lim_{n \to +\infty} \int_\Omega \varphi u_n \partial_i u_n \, dx = \int_\Omega \varphi u \partial_i u \, dx.$$

Returning to (2.83), we get that

$$\langle D_i(u^2), \varphi \rangle_{\mathcal{D}^\star(\Omega), \mathcal{D}(\Omega)} = - \int_\Omega u^2 \partial_i \varphi \, dx = 2 \int_\Omega \varphi u D_i u \, dx,$$

which indeed proves that $D_i(u^2) = 2u D_i u$ a.e. (the derivatives by transposition of u and u^2 are in fact weak derivatives and therefore identified as functions). The previous proof also gives that $u_n^2 \to u^2$ in $L^1(\Omega)$ and $\partial_i(u_n^2) = 2u_n \partial_i u_n \to 2u D_i u = D_i(u^2)$ in $L^1(\Omega)$. Hence $u_n^2 \to u^2$ in $W^{1,1}(\Omega)$, which gives, since $u_n \in \mathcal{D}(\Omega)$, that $u^2 \in W_0^{1,1}(\Omega)$.

2. Let $\varphi \in W_0^{1,1}(\Omega)$. There exists a sequence $(\varphi_n)_{n \in \mathbb{N}}$ of elements of $\mathcal{D}(\Omega)$ such that $\varphi_n \to \varphi$ in $W^{1,1}(\Omega)$. Since $\operatorname{div} w = 0$ in $\mathcal{D}^\star(\Omega)$, we have, for all $n \in \mathbb{N}$

$$0 = \langle \operatorname{div} w, \varphi_n \rangle_{\mathcal{D}^\star(\Omega), \mathcal{D}(\Omega)} = - \sum_{i=1}^N \int_\Omega w_i \partial_i \varphi_n \, dx.$$

Since $\partial_i \varphi_n \to D_i \varphi$ in $L^1(\Omega)$ (and $w \in L^\infty(\Omega)^N$), passing to the limit as $n \to +\infty$ yields that

$$\sum_{i=1}^N \int_\Omega w_i D_i \varphi \, dx = 0,$$

that is to say, $\int_\Omega w(x) \cdot \nabla\varphi(x) \, dx = 0$.

3. Using the result of the previous question with $\varphi = u^2$ and the fact that $D_i(u^2) = 2u D_i u$, we obtain

$$0 = \int_\Omega w \cdot \nabla(u^2) \, dx = \sum_{i=1}^N \int_\Omega 2w_i u D_i u \, dx = 2 \int_\Omega uw \cdot \nabla u \, dx.$$

Part II

1. Assume that u is a solution to (2.55) and set $\bar{u} = u - G$. Then $\bar{u} \in H^1(\Omega)$ and, since γ is a linear operator, $\gamma(\bar{u}) = \gamma(u) - \gamma(G) = g - g = 0$ (in $L^2(\partial\Omega)$), which proves that $\bar{u} \in H_0^1(\Omega)$. If $v \in H_0^1(\Omega)$, replacing u with $\bar{u} + G$ in (2.55), we show that $\bar{u}$ is indeed a solution of (2.56).

Conversely, assume that $\overline{u}$ is a solution to (2.56) and set $u = \overline{u} + G$. We have $u \in H^1(\Omega)$ and $\gamma(u) = \gamma(\overline{u})+\gamma(G) = 0+g = g$ (in $L^2(\partial\Omega)$). Then, if $v \in H_0^1(\Omega)$, replacing $\overline{u}$ with $u - G$ in (2.56), we show that u is indeed a solution of (2.55).

We thus proved the desired equivalence.

2. Let us show that thanks to the Lax–Milgram theorem (Theorem 2.3), (2.56) admits one and only one solution and thanks to the previous question, (2.55) admits one and only one solution.

Problem (2.56) can be written

$$\overline{u} \in H, \tag{2.84a}$$

$$\bar{a}(\overline{u}, v) = T(v), \ \forall v \in H, \tag{2.84b}$$

with $H = H_0^1(\Omega)$,

$$\bar{a}(\overline{u}, v) = \int_\Omega a\nabla\overline{u}(x) \cdot \nabla v(x)\,\mathrm{d}x + \int_\Omega \overline{u}(x)w(x) \cdot \nabla v(x)\,\mathrm{d}x$$

and

$$T(v) = \int_\Omega f(x)v(x)\,\mathrm{d}x - \int_\Omega a\nabla G(x) \cdot \nabla v(x)\,\mathrm{d}x - \int_\Omega G(x)w(x) \cdot \nabla v(x)\,\mathrm{d}x.$$

The space H is indeed a Hilbert space (with its natural norm). The mapping T is indeed linear from H to $\mathbb{R}$ and, using Hölder's inequality, we see that T is continuous (we use here the fact that $f \in L^2(\Omega)$, $G \in H^1(\Omega)$ and $w \in L^\infty(\Omega)^N$). The mapping $\bar{a}$ is indeed bilinear from $H \times H$ to $\mathbb{R}$ and continuous (thanks again to the fact that $w \in L^\infty(\Omega)^N$).

To prove the coercivity of $\bar{a}$, we use question 3 from part I, which gives, for all $u \in H$,

$$\begin{aligned}
\bar{a}(u, u) &= \int_\Omega a\nabla u(x) \cdot \nabla u(x)\,\mathrm{d}x + \int_\Omega u(x)w(x) \cdot \nabla u(x)\,\mathrm{d}x \\
&= \int_\Omega a\nabla u(x) \cdot \nabla u(x)\,\mathrm{d}x \\
&= a\,\|u\|_{H_0^1(\Omega)}.
\end{aligned}$$

Since $a > 0$, we get that $\bar{a}$ is coercive. We can therefore apply Theorem 2.3, which gives the existence and uniqueness of the solution $\overline{u}$ to (2.56). Thanks to the previous question, we get the existence and uniqueness of the solution u to (2.55).

Note that the non-uniqueness of G is not a problem for the uniqueness of the solution to (2.55). Indeed, we fix G and denote by $\overline{u}$ the unique solution to (2.56). If u is a solution to (2.55), $u - G$ is a solution to (2.56) and therefore $u = \overline{u} + G$. The function $\overline{u} + G$ is therefore the unique solution to (2.55). In fact, the function $\overline{u}$ depends on the choice of G but the function $\overline{u} + G$ does not depend on G.

3. It is sufficient to take $v = u$ in (2.55). With question 3 from part I we get

$$a \|u\|^2_{H_0^1(\Omega)} = \int_\Omega f(x)u(x)\, dx.$$

4. Since $(u - b)^+ \in H_0^1(\Omega)$, we can take $v = (u - b)^+$ in (2.55). We get, using $\nabla(u - b)^+ = \mathbb{1}_{u>b}\nabla u$ a.e. and $f \le 0$ a.e.,

$$\int_{u>b} a\nabla u(x) \cdot \nabla u(x)\, dx + \int_\Omega u(x)w(x) \cdot \nabla(u - b)^+(x)\, dx$$

$$= \int_\Omega f(x)(u - b)^+(x)\, dx \le 0. \quad (2.85)$$

We now note that

$$\int_{u>b} a\nabla u(x) \cdot \nabla u(x)\, dx = \int_\Omega a\nabla(u - b)^+ \cdot \nabla(u - b)^+ = a \left\|(u - b)^+\right\|_{H_0^1(\Omega)}$$

and that

$$\int_\Omega u(x)w(x) \cdot \nabla(u - b)^+(x)\, dx = \int_\Omega (u(x) - b)w(x) \cdot \nabla(u - b)^+(x)\, dx$$

$$+ b \int_\Omega w(x) \cdot \nabla(u - b)^+(x)\, dx = \int_\Omega (u(x) - b)^+w(x) \cdot \nabla(u - b)^+(x)\, dx$$

$$+ b \int_\Omega w(x) \cdot \nabla(u - b)^+(x)\, dx.$$

Question 3 of part I gives $\int_\Omega (u(x) - b)^+w(x) \cdot \nabla(u - b)^+(x)\, dx = 0$. On the other hand, since $\operatorname{div} w = 0$ in $\mathcal{D}^\star(\Omega)$, we have $\int_\Omega w \cdot \nabla\varphi\, dx = 0$ for all $\varphi \in \mathcal{D}(\Omega)$. By density of $\mathcal{D}(\Omega)$ in $H_0^1(\Omega)$ we also have (we use here only the fact that $w \in L^2(\Omega)^N$) $\int_\Omega w \cdot \nabla\varphi\, dx = 0$ for all $\varphi \in H_0^1(\Omega)$ and therefore, in particular, for $\varphi = (u - b)^+$. Hence

$$\int_\Omega u(x)w(x) \cdot \nabla(u - b)^+(x)\, dx = 0.$$

Returning to (2.85), we finally obtain $a \left\|(u - b)^+\right\|_{H_0^1(\Omega)} \le 0$ and therefore $(u - b)^+ = 0$ a.e., that is, $u \le b$ a.e. in Ω.

Remark: assuming $g = 0$ and denoting by u the solution to (2.55), the previous proof shows that $u \le 0$ a.e. in Ω if $f \le 0$ a.e. in Ω. If now we assume $f \ge 0$, we note that $(-u)$ is the solution to (2.55) with $(-f)$ instead of f. Hence $(-u) \le 0$ a.e., that is, $u \ge 0$ a.e.

Part III

1. The fact that $u_n \geq 0$ a.e. in Ω is a direct consequence of the remark at the end of the proof of question 4 of part II.
2. The function u_n satisfies

$$u_n \in H_0^1(\Omega),$$

$$\frac{1}{n} \int_\Omega \nabla u_n(x) \cdot \nabla v(x)\, dx - \int_\Omega u_n(x) D_1 v(x)\, dx = \int_\Omega f(x) v(x)\, dx,$$

$$\forall v \in H_0^1(\Omega).$$

Set $\bar{u}_n = u_n + \beta \psi$ (so that $\nabla \bar{u}_n = \nabla u_n + \beta(1,0)^t$). The integration by parts formulas in $H^1(\Omega)$ (Theorem 1.33) give that, for a function $v \in H_0^1(\Omega)$, we have

$$\int_\Omega D_1 v\, dx = 0, \quad \int_\Omega \psi D_1 v\, dx = -\int_\Omega v \partial_1 \psi\, dx = -\int_\Omega v\, dx.$$

Hence the function $\bar{u}_n$ is a solution of

$$\bar{u}_n \in H^1(\Omega), \ \gamma(u_n) = \beta x_1 \ (in\ L^2(\partial\Omega)),$$

$$\frac{1}{n} \int_\Omega \nabla \bar{u}_n(x) \cdot \nabla v(x)\, dx - \int_\Omega \bar{u}_n(x) D_1 v(x)\, dx = \int_\Omega (f(x) + \beta) v(x)\, dx,$$

$$\forall v \in H_0^1(\Omega).$$

Choose $\beta = -\|f\|_{L^\infty(\Omega)}$, so that $f + \beta \leq 0$ a.e. in Ω and $\beta x_1 \leq 0$ on $\partial\Omega$. Applying question 4 of part II gives $\bar{u}_n \leq 0$ a.e. in Ω and therefore $u_n \leq \|f\|_{L^\infty(\Omega)}$ a.e. in Ω. By the previous question, this gives $0 \leq u_n \leq \|f\|_{L^\infty(\Omega)}$ a.e. in Ω. We can therefore choose $C_1 = \|f\|_{L^\infty(\Omega)}$.
3. Question 3 of part II gives

$$\frac{1}{n} \|u_n\|_{H_0^1(\Omega)}^2 \leq \int_\Omega f u_n\, dx.$$

Since (with C_1 given in the previous question)

$$\int_\Omega f u_n\, dx \leq \lambda_N(\Omega) \|u_n\|_\infty \|f\|_\infty \leq \lambda_N(\Omega) C_1 \|f\|_\infty,$$

we therefore have

$$\|u_n\|_{H_0^1(\Omega)}^2 \leq n \lambda_N(\Omega) C_1 \|f\|_\infty.$$

We can therefore take $C_2 = \sqrt{\lambda_N(\Omega) C_1 \|f\|_\infty}$.
4. Let $n \in \mathbb{N}^\star$. The function u_n is the weak solution of $-\Delta u_n = f - D_1 u_n$. Since Ω is convex and $f - D_1 u_n \in L^2(\Omega)$ (because $u_n \in H_0^1(\Omega)$), Remark 2.20 gives that $u \in H^2(\Omega)$.

5. Since $u_n \in C^1(\overline{\Omega})$ and $u_n = 0$ on $\partial\Omega$, we have, for all $x_2 \in]0, 1[$,

$$\partial_1 u_n(0, x_2) = \lim_{x_1 \to 0^+} \frac{u_n(x_1, x_2)}{x_1}.$$

Question 1 of part III gives that $u_n(x_1, x_2) \geq 0$ for all $(x_1, x_2) \in \Omega$ (since u_n is continuous, the fact that $u_n \geq 0$ a.e. in Ω implies that $u_n \geq 0$ everywhere in Ω). Hence $\partial_1 u_n(0, x_2) \geq 0$ for all $x_2 \in]0, 1[$. The three other requested properties are shown in a similar way.

6. Question 2 of part III gives that the sequence $(u_n)_{n \in \mathbb{N}}$ is bounded in $L^\infty(\Omega)$. Hence there exists a subsequence of the sequence $(u_n)_{n \in \mathbb{N}}$, still denoted $(u_n)_{n \in \mathbb{N}}$, and there exists a $u \in L^\infty(\Omega)$ such that $u_n \to u$ $\star$-weakly in $L^\infty(\Omega)$ when $n \to +\infty$. Taking $\varphi = \mathbb{1}_{u<0}$, we notice that $\int_\Omega u_n \varphi \, \mathrm{d}x \geq 0$ (because $u_n \geq 0$ a.e.) and therefore $\int_\Omega u \varphi \, \mathrm{d}x \geq 0$, that is

$$\int_{u<0} u(x) \, \mathrm{d}x \geq 0.$$

This indeed gives $u \geq 0$ a.e.

7. Let $\varphi \in \mathcal{D}(\Omega)$; for all $n \in \mathbb{N}^\star$ we have

$$\frac{1}{n} \int_\Omega \nabla u_n(x) \cdot \nabla \varphi(x) \, \mathrm{d}x - \int_\Omega u_n(x) \partial_1 \varphi(x) \, \mathrm{d}x = \int_\Omega f(x) \varphi(x) \, \mathrm{d}x. \qquad (2.86)$$

Since $\|u_n\|_{H_0^1(\Omega)} \leq C_2 \sqrt{n}$, we have

$$\left| \frac{1}{n} \int_\Omega \nabla u_n(x) \cdot \nabla \varphi(x) \, \mathrm{d}x \right| \leq \frac{1}{n} \|u_n\|_{H_0^1(\Omega)} \|\varphi\|_{H_0^1(\Omega)} \leq \frac{C_2}{\sqrt{n}} \|\varphi\|_{H_0^1(\Omega)}.$$

Hence $\lim_{n \to +\infty} \frac{1}{n} \int_\Omega \nabla u_n(x) \cdot \nabla \varphi(x) \, \mathrm{d}x = 0$. On the other hand, $u_n \to u$ $\star$-weakly in $L^\infty(\Omega)$, so that

$$\lim_{n \to +\infty} \int_\Omega u_n(x) \partial_1 \varphi(x) \, \mathrm{d}x = \int_\Omega u(x) \partial_1 \varphi(x) \, \mathrm{d}x.$$

Passing to the limit as $n \to +\infty$ in (2.86), we therefore obtain

$$-\int_\Omega u(x) \partial_1 \varphi(x) \, \mathrm{d}x = \int_\Omega f(x) \varphi(x) \, \mathrm{d}x,$$

which indeed gives $D_1 u = f$ in $\mathcal{D}^\star(\Omega)$.

8. Since $u_n \in H^2(\Omega)$, the derivative by transposition $-\Delta u_n$ belongs to $L^2(\Omega)$ and (2.86) gives

$$-\frac{1}{n} \Delta u_n + D_1 u_n = f \text{ a.e.}$$

Multiplying this equation by φ (we use here only the fact that $\varphi \in L^2(\Omega)$) we therefore have

$$-\frac{1}{n}\int_\Omega \Delta u_n \varphi \, \mathrm{d}x + \int_\Omega \varphi D_1 u_n \, \mathrm{d}x = \int_\Omega f\varphi \, \mathrm{d}x. \qquad (2.87)$$

Since the functions $D_1 u_n$ and $D_2 u_n$ belong to $H^1(\Omega)$, we may integrate by parts, owing to Theorem 1.33. Since $\varphi \in C^1(\overline{\Omega})$, we obtain

$$-\frac{1}{n}\int_\Omega (D_1 D_1 u_n)\varphi \, \mathrm{d}x = \frac{1}{n}\int_\Omega D_1 u_n \partial_1 \varphi \, \mathrm{d}x$$
$$+\frac{1}{n}\int_0^1 \gamma(D_1 u_n)(0,x_2)\varphi(0,x_2)\, \mathrm{d}x_2 - \frac{1}{n}\int_0^1 \gamma(D_1 u_n)(1,x_2)\varphi(1,x_2)\, \mathrm{d}x_2.$$

and

$$-\frac{1}{n}\int_\Omega (D_2 D_2 u_n)\varphi \, \mathrm{d}x = \frac{1}{n}\int_\Omega D_2 u_n \partial_2 \varphi \, \mathrm{d}x$$
$$+\frac{1}{n}\int_0^1 \gamma(D_2 u_n)(x_1,0)\varphi(x_1,0)\, \mathrm{d}x_1 - \frac{1}{n}\int_0^1 \gamma(D_2 u_n)(x_1,1)\varphi(x_1,1)\, \mathrm{d}x_1.$$

An integration by parts also gives (since $u_n \in H_0^1(\Omega)$)

$$\int_\Omega \varphi D_1 u_n \, \mathrm{d}x = -\int_\Omega u_n \partial_1 \varphi \, \mathrm{d}x.$$

Using these three integrations by parts in (2.87), we obtain the required equality.
9. Since $\varphi \geq 0$ on $\partial\Omega$, question 5 and the equality of question 8 give, for all $n \in \mathbf{N}^\star$,

$$\frac{1}{n}\int_\Omega \nabla u_n(x)\nabla \varphi(x)\, \mathrm{d}x - \int_\Omega u_n(x)\partial_1\varphi(x)\, \mathrm{d}x \leq \int_\Omega f(x)\varphi(x)\, \mathrm{d}x.$$

Passing to the limit as $n \to +\infty$, as in question 7, we indeed obtain (2.57).
10. Question 7 gives $D_1 u = f$ in $\mathcal{D}^\star(\Omega)$. Since u is of class C^1, $D_1 u$ is represented by the classical derivative of u. Then, since $\partial_1 u$ and f are continuous on Ω, we get that $\partial_1 u = f$ everywhere in Ω. Since $\partial_1 u$ and f are continuous on $\overline{\Omega}$, we even have $\partial_1 u = f$ everywhere in $\overline{\Omega}$. We now take $\varphi \in C^1(\overline{\Omega})$ such that $\varphi \geq 0$ on $\partial\Omega$, $\varphi(x) = 0$ if $x = (x_1, x_2) \in \partial\Omega$, $x_1 \neq 0$.. An integration by parts in (2.57) then gives

$$\int_0^1 u(0,x_2)\varphi(0,x_2)\, \mathrm{d}x_2 \leq 0.$$

In this inequality, the function $\varphi(0, \cdot)$ can be equal (for example) to any function belonging to $\mathcal{D}(]0,1[)$ and taking its values in $\mathbf{R}_+$. Since $u(0, \cdot)$ is a continuous function on $]0,1[$, we therefore deduce from this inequality that $u(0,x_2) \leq 0$ for all $x_2 \in]0,1[$.

Question 6 gives $u \geq 0$ a.e. on Ω. Since u is continuous on $\overline{\Omega}$, we therefore have $u \geq 0$ everywhere on $\overline{\Omega}$. So we finally obtain $u(0, x_2) = 0$ for all $x_2 \in]0, 1[$ (and even $[0, 1]$).

The function u is entirely determined by f. We have

$$u(x_1, x_2) = \int_0^{x_1} f(t, x_2) \, dt.$$

11. Set $w = (\alpha, \beta)$, with $\alpha, \beta \neq 0$. With the same method as the one developed above for the case $w = (-1, 0)$, the results of question 10 applied in this new setting show that u is the solution of the following problem:

$$-w \cdot \nabla u = f \text{ in } \Omega,$$

$u(0, x_2) = 0$ for all $x_2 \in]0, 1[$ if $\alpha < 0$ and $u(1, x_2) = 0$ for all $x_2 \in]0, 1[$ if $\alpha > 0$, $u(x_1, 0) = 0$ for all $x_1 \in]0, 1[$ if $\beta < 0$ and $u(x_1, 1) = 0$ for all $x_1 \in]0, 1[$ if $\beta > 0$.

Problem 2.24 (Non-homogeneous Dirichlet condition)

1. Let $G \in H^1(\Omega)$ such that $\gamma(G) = g$. The function u is a solution to (2.17) if and only if $u - G$ is a solution of

$$w = u - G \in H_0^1(\Omega),$$
$$\int_\Omega \left(\sum_{i=1}^N \sum_{j=1}^N a_{i,j}(x) D_j w(x) D_i v(x) \right) dx = S(v), \forall v \in H_0^1(\Omega),$$

with $S(v) = \displaystyle\int_\Omega f(x) v(x) \, dx - \int_\Omega \left(\sum_{i=1}^N \sum_{j=1}^N a_{i,j}(x) D_j G(x) D_i v(x) \right) dx.$

Since $(a_{i,j})_{i,j=1,\ldots,N} \subset L^\infty(\Omega)$ and $D_j G \in L^2(\Omega)$ for all j, we have $S \in H^{-1}(\Omega)$. The existence and uniqueness of u is then given by Theorem 2.9.

2. The proof is identical to the previous one, replacing $\int_\Omega f(x) v(x) \, dx$ with $T(v)$.

3. The Sobolev embedding theorem (Theorem 1.41) provides the existence of $C \geq 0$, depending only on Ω and q, $q < +\infty$, such that, for all $u \in H_0^1(\Omega)$,

$$\|v\|_{L^q(\Omega)} \leq C \|v\|_{H_0^1(\Omega)}.$$

Taking $q = \frac{p}{p-1}$ ($q = 1$ if $p = +\infty$), we obtain thanks to Hölder's inequality that for all $v \in H_0^1(\Omega)$,

$$\int_\Omega f(x) v(x) \, dx \leq \|f\|_{L^p(\Omega)} \|v\|_{L^q(\Omega)} \leq C \|f\|_{L^p(\Omega)} \|v\|_{H_0^1(\Omega)}.$$

The mapping $f \mapsto \int_{\Omega} f(x)v(x)\,\mathrm{d}x$ therefore belongs to $H^{-1}(\Omega)$ and therefore, by question (2), there exists a unique solution u to (2.17).

4. The Sobolev embedding theorem (Theorem 1.41) provides the existence of C, depending only on Ω such that, for all $u \in H_0^1(\Omega)$,

$$\|v\|_{L^{2^\star}(\Omega)} \le C \|v\|_{H_0^1(\Omega)}, \text{ with } 2^\star = \frac{2N}{N-2}.$$

Taking $q = \frac{p}{p-1}$, that is $q = \frac{2N}{N-2}$, we obtain, by Hölder's inequality,

$$\int_{\Omega} f(x)v(x)\,\mathrm{d}x \le \|f\|_{L^p(\Omega)} \|v\|_{L^q(\Omega)} \le C \|f\|_{L^p(\Omega)} \|v\|_{H_0^1(\Omega)},$$

$$\text{for all } v \in H_0^1(\Omega).$$

This shows that the mapping $f \mapsto \int_{\Omega} f(x)v(x)\,\mathrm{d}x$ belongs to $H^{-1}(\Omega)$ and therefore, by question (2), there exists a unique solution u to (2.17).

Problem 2.25 (The space $H^{\frac{1}{2}}(\partial\Omega)$)

1. We first show the existence of $\bar{u} \in H^1(\Omega)$ such that $\gamma(\bar{u}) = u$ and $\|u\|_{H^{\frac{1}{2}}(\partial\Omega)} = \|\bar{u}\|_{H^1(\Omega)}$. Let $(v_n)_{n\in\mathbb{N}}$ be a sequence in $H^1(\Omega)$ such that $\gamma(v_n) = u$ (for all n) and $\|v_n\|_{H^1(\Omega)} \to \|u\|_{H^{\frac{1}{2}}(\partial\Omega)}$ as $n \to +\infty$. The sequence $(v_n)_{n\in\mathbb{N}}$ is therefore bounded in $H^1(\Omega)$ and there exists a $\bar{u} \in H^1(\Omega)$ and a subsequence still denoted $(v_n)_{n\in\mathbb{N}}$ such that $v_n \to \bar{u}$ weakly in $H^1(\Omega)$ and (see Remark 1.20)

$$\|\bar{u}\|_{H^1(\Omega)} \le \liminf_{n\to+\infty} \|v_n\|_{H^1(\Omega)} = \|u\|_{H^{\frac{1}{2}}(\partial\Omega)}.$$

Since $\gamma(v_n) = u$ in $L^2(\partial\Omega)$ (for all n), we also have $\gamma(\bar{u}) = u$ (in $L^2(\partial\Omega)$). This can be proved thanks to the compactness of the operator γ from $H^1(\Omega)$ to $L^2(\partial\Omega)$ (Remark 2.33) but also using only the continuity of the operator γ from $H^1(\Omega)$ to $L^2(\partial\Omega)$ and Mazur's lemma[25][26] ; indeed, by this lemma, there exists a sequence $(w_n)_{n\in\mathbb{N}}$ in $H^1(\Omega)$, where for each $n \in \mathbb{N}$, w_n is a convex combination of the set $\{v_m, m \ge n\}$, and such that $w_n \to \bar{u}$ in $H^1(\Omega)$; we therefore have $\gamma(w_n) = u$ in $L^2(\partial\Omega)$. Since $\gamma(\bar{u}) = u$ and since

$$\|\bar{u}\|_{H^1(\Omega)} \le \liminf_{n\to+\infty} \|v_n\|_{H^1(\Omega)} = \|u\|_{H^{\frac{1}{2}}(\partial\Omega)}$$

$$= \inf\{\|w\|_{H^1(\Omega)} : \gamma(w) = u\} \le \|\bar{u}\|_{H^1(\Omega)}$$

---

[25] **Mazur's Lemma**: Let X be a normed vector space and $(x_n)_{n\in\mathbb{N}}$ a sequence weakly converging to $x \in X$, then there exists a sequence $(y_n)_{n\in\mathbb{N}}$ such that for all $n \in \mathbb{N}$, the term y_n is of the form $y_n = \sum_{k=n}^{P_n} \lambda_k x_k$, with $p_n \ge n$, $\lambda_k \ge 0$ for all $k = n, \dots, p_n$ and $\sum_{k=n}^{P_n} \lambda_k = 1$, and such that $\|y_n - x\| = 0$ as $n \to +\infty$.

[26] Stanislaw Mieczyslaw Mazur (1905–1981), Polish mathematician and politician.

we have $\|\bar{u}\|_{H^1(\Omega)} = \|u\|_{H^{\frac{1}{2}}(\partial\Omega)}$ and $v_n \to \bar{u}$ in $H^1(\Omega)$ as $n \to +\infty$. For all $\varphi \in H^1_0(\Omega)$ and $t > 0$, $\gamma(\bar{u} + t\varphi) = u$, we therefore have

$$\|\bar{u}\|^2_{H^1(\Omega)} = \|u\|^2_{H^{\frac{1}{2}}(\partial\Omega)} \leq \|\bar{u} + t\varphi\|^2_{H^1(\Omega)},$$

that is to say,

$$\int_\Omega (\nabla\bar{u}(x) \cdot \nabla\bar{u}(x) + \bar{u}(x)^2)\,dx$$

$$\leq \int_\Omega (\nabla(\bar{u} + t\varphi)(x) \cdot \nabla(\bar{u} + t\varphi)(x) + (\bar{u} + t\varphi)(x)^2)\,dx.$$

Expanding the right-hand side and letting t tend to 0, we get that

$$\int_\Omega (\nabla\bar{u}(x) \cdot \nabla\varphi(x) + \bar{u}(x)\varphi(x))\,dx \geq 0,$$

and, therefore (changing φ to $-\varphi$),

$$\int_\Omega (\nabla\bar{u}(x) \cdot \nabla\varphi(x) + \bar{u}(x)\varphi(x))\,dx = 0.$$

The function $\bar{u}$ is therefore a solution to (2.58)–(2.59). The uniqueness of the solution to (2.58)–(2.59) is an immediate consequence of the uniqueness for the homogeneous problem (that is, problem (2.58)–(2.59) with $u = 0$).

2. The mapping $u \mapsto \|u\|_{H^{\frac{1}{2}}(\partial\Omega)}$ is a norm that is induced by an inner product. Indeed, for $u, v \in H^{\frac{1}{2}}(\partial\Omega)$, let $\bar{u}, \bar{v}$ be the associated elements of $H^1(\Omega)$ (given by the previous question) and define the inner product of u with v by $(u \mid v)_{H^{\frac{1}{2}}(\partial\Omega)} = (\bar{u} \mid \bar{v})_{H^1(\Omega)}$.

It remains to show that $H^{\frac{1}{2}}(\partial\Omega)$ is complete. Let $(u_n)_{n\in\mathbb{N}}$ be a Cauchy sequence in $H^{\frac{1}{2}}(\partial\Omega)$ and let $(\bar{u}_n)_{n\in\mathbb{N}}$ be the associated sequence in $H^1(\Omega)$; this latter sequence is a Cauchy sequence in $H^1(\Omega)$, so it converges (in $H^1(\Omega)$) to some limit w. Since the trace operator is continuous from $H^1(\Omega)$ to $L^2(\partial\Omega)$, we have $u_n = \gamma(\bar{u}_n) \to \gamma(w)$ in $L^2(\partial\Omega)$, which proves that $\gamma(w) \in H^{\frac{1}{2}}(\partial\Omega)$ because $w \in H^1(\Omega)$. Finally, we note that $\|u_n - \gamma(w)\|_{H^{\frac{1}{2}}(\partial\Omega)} \leq \|\bar{u}_n - w\|_{H^1(\Omega)} \to 0$ as $n \to +\infty$. We have thus shown that $u_n \to \gamma(w)$ in $H^{\frac{1}{2}}(\partial\Omega)$. The space $H^{\frac{1}{2}}(\partial\Omega)$ is therefore complete.

3. If $u_n \to u$ in $H^{\frac{1}{2}}(\partial\Omega)$, we have (with the previous notation) $\bar{u}_n \to \bar{u}$ in $H^1(\Omega)$ and therefore, by continuity of the trace operator from $H^1(\Omega)$ to $L^2(\partial\Omega)$, $u_n \to u$ in $L^2(\partial\Omega)$.

Problem 2.26 (Normal trace of an element of H_{div})

1. The mapping $v \mapsto \|u\|_{H_{\mathrm{div}}(\Omega)}$ is indeed a norm that is induced by an inner product; this inner product is the mapping of $H_{\mathrm{div}}(\Omega)^2$ to $\mathbb{R}$ defined by, for $u = (u_1, u_2)$ and $v = (v_1, v_2)$,

$$(u \mid v)_{H_{\mathrm{div}}(\Omega)} = (u_1 \mid v_1)_{L^2(\Omega)} + (u_2 \mid v_2)_{L^2(\Omega)} + (\mathrm{div}\, u \mid \mathrm{div}\, v)_{L^2(\Omega)}.$$

It remains to show that $H_{\mathrm{div}}(\Omega)$ is complete. Let $(u_n)_{n \in \mathbb{N}}$ be a Cauchy sequence in $H_{\mathrm{div}}(\Omega)$. The sequences $(u_n)_{n \in \mathbb{N}}$ and $(\mathrm{div}\, u_n)_{n \in \mathbb{N}}$ are Cauchy sequences in $(L^2(\Omega))^2$ and $L^2(\Omega)$. They converge in $L^2(\Omega)^2$ and $L^2(\Omega)$ towards u and ξ. For all $\varphi \in \mathcal{D}(\Omega)$,

$$\int_\Omega \mathrm{div}\, u_n(x)\varphi(x)\, \mathrm{d}x = \langle (\mathrm{div}\, u_n), \varphi \rangle_{\mathcal{D}^\star(\Omega), \mathcal{D}(\Omega)}$$

$$= -\langle (u_n)_1, \partial_1 \varphi \rangle_{\mathcal{D}^\star(\Omega), \mathcal{D}(\Omega)} - \langle (u_n)_2, \partial_2 \varphi \rangle_{\mathcal{D}^\star(\Omega), \mathcal{D}(\Omega)}$$

$$= -\int_\Omega (u_n)_1(x)\partial_1 \varphi(x)\, \mathrm{d}x - \int_\Omega (u_n)_2(x)\partial_2 \varphi(x)\, \mathrm{d}x = -\int_\Omega u_n(x) \cdot \nabla \varphi(x)\, \mathrm{d}x.$$

Passing to the limit as $n \to +\infty$ we obtain $\int_\Omega \xi(x)\varphi(x)\, \mathrm{d}x = -\int_\Omega u(x) \cdot \nabla \varphi(x)\, \mathrm{d}x$. Since $-\int_\Omega u(x) \cdot \nabla \varphi(x)\, \mathrm{d}x = \langle \mathrm{div}\, u, \varphi \rangle_{\mathcal{D}^\star(\Omega), \mathcal{D}(\Omega)}$, this proves that $\mathrm{div}\, u = \xi$ and therefore $u \in H_{\mathrm{div}}(\Omega)$. Finally, we have indeed $u_n \to u$ in $H_{\mathrm{div}}(\Omega)$ and therefore $H_{\mathrm{div}}(\Omega)$ is complete.

2. a. As seen in the previous question (with u_n instead of v), for all $\varphi \in \mathcal{D}(\Omega)$,

$$\int_\Omega \mathrm{div}\, v(x)\varphi(x)\, \mathrm{d}x = \langle \mathrm{div}\, v, \varphi \rangle_{\mathcal{D}^\star(\Omega), \mathcal{D}(\Omega)} = -\int_\Omega v(x) \cdot \nabla \varphi(x)\, \mathrm{d}x.$$

Then, since $\mathcal{D}(\Omega)$ is dense in $H_0^1(\Omega)$, this equality holds for all $\varphi \in H_0^1(\Omega)$.

   b. Observe that $(u_1 - u_2) \in \ker(\gamma)$ and therefore $(u_1 - u_2) \in H_0^1(\Omega)$ (because $\ker(\gamma) = H_0^1(\Omega)$, see Theorem 1.32). Then apply the previous question with $\varphi = u_1 - u_2$.

3. Thanks to question 2 the right-hand term of (2.61) does not depend on the choice of $\overline{u}$. The mapping T is therefore well defined on $H^{\frac{1}{2}}(\partial\Omega)$. It is clearly a linear mapping. To show that it is continuous (and therefore that $T(v) \in H^{-\frac{1}{2}}(\partial\Omega)$) it is enough to notice that, for all $\overline{u} \in H^1(\Omega)$ such that $\gamma(\overline{u}) = u$,

$$\left| \int_\Omega \nabla \overline{u} \cdot v\, \mathrm{d}x + \int_\Omega \overline{u}\, \mathrm{div}\, v\, \mathrm{d}x \right| \leq \| \, |\nabla \overline{u}| \, \|_{L^2(\Omega)} \, \| \, |v| \, \|_{L^2(\Omega)}$$

$$+ \| \, |\overline{u}| \, \|_{L^2(\Omega)} \|\mathrm{div}\, v\|_{L^2(\Omega)}$$

$$\leq 2 \|\overline{u}\|_{H^1(\Omega)} \|v\|_{H_{\mathrm{div}}(\Omega)}.$$

Taking the lower bound of the set of functions $\overline{u} \in H^1(\Omega)$ satisfying $\gamma(\overline{u}) = u$, we obtain $T(v) \in H^{-\frac{1}{2}}(\partial\Omega)$ and $\|T(v)\|_{H^{-\frac{1}{2}}(\partial\Omega)} \leq 2 \|v\|_{H_{\mathrm{div}}(\Omega)}.$

4. The mapping T is clearly linear from $H_{\mathrm{div}}(\Omega)$ to $H^{-\frac{1}{2}}(\partial\Omega)$. It is continuous because the previous question gives $\|T(v)\|_{H^{-\frac{1}{2}}(\partial\Omega)} \leq 2\,\|v\|_{H_{\mathrm{div}}(\Omega)}$.

5. Applying the integration by parts theorem for elements of $H^1(\Omega)$ (Theorem 1.33) gives, for $u \in H^{\frac{1}{2}}(\partial\Omega)$ and $\overline{u} \in H^1(\Omega)$ such that $\gamma(\overline{u}) = u$,

$$\langle T(v), u\rangle_{H^{-\frac{1}{2}}(\partial\Omega), H^{\frac{1}{2}}(\partial\Omega)} = \int_\Omega \nabla\overline{u} \cdot v \, dx + \int_\Omega \overline{u}\, \mathrm{div}\, v \, dx$$

$$= \int_{\partial\Omega} \gamma(\overline{u})\gamma(v) \cdot n \, d\lambda = \int_{\partial\Omega} u\gamma(v) \cdot n \, d\lambda.$$

Problem 2.27 (Normal trace on a part of the boundary)

1. Since $u \in C^\infty(\Omega)$, $\mathrm{div}\, v = -\partial_1\partial_2 u + \partial_2\partial_1 u = 0$. Since $u \in H^1(\Omega)$, we also have $v_1, v_2 \in L^2(\Omega)$, and therefore $v \in H_{\mathrm{div}}(\Omega)$.
2. The condition $\left(a + \frac{1}{n}\right)\sqrt{2} < 1$ ensures that $x_1^2 + x_2^2 < 1$ for $(x_1, x_2) \in \Omega$ and therefore that $v^{(n)}$ is well defined and belongs to $C^\infty(\overline{\Omega})$; and that $\mathrm{div}\, v^{(n)} = 0$. Let us show that $v^{(n)} \to v$ in $H_{\mathrm{div}}(\Omega)$ as $n \to +\infty$. Let $n_0 > 0$ such that $\left(a + \frac{1}{n_0}\right)\sqrt{2} < 1$ and define, for $i = 1, 2$, g_i by $g_i = v_i$ in $]0, a + \frac{1}{n_0}[^2$ and $g_i = 0$ outside of $]0, a + \frac{1}{n_0}[^2$. Since $g_i \in L^2(\mathbb{R}^2)$, the theorem of continuity in mean in $L^2(\Omega)$ (see e.g. [26, Exercice 6.5]) gives that $\lim_{n\to+\infty} \left\|g_i(\cdot + \frac{1}{n}, \cdot) - g_i\right\|_{L^2(\mathbb{R}^2)} = 0$. But, for $n \geq n_0$,

$$\left\|v_i^{(n)} - v_i\right\|_{L^2(\Omega)} = \left\|g_i\left(\cdot + \frac{1}{n}, \cdot\right) - g_i\right\|_{L^2(\Omega)} \leq \left\|g_i\left(\cdot + \frac{1}{n}, \cdot\right) - g_i\right\|_{L^2(\mathbb{R}^2)}$$

and therefore $v^{(n)} \to v$ in $L^2(\Omega)^2$ as $n \to +\infty$, so that finally, $v^{(n)} \to v$ in $H_{\mathrm{div}}(\Omega)$ as $n \to +\infty$.

Hence the condition (2.65) gives that $S_{v^{(n)}}(\mathbb{1}_{\partial\Omega}) \to S_v(\mathbb{1}_{\partial\Omega})$ as $n \to +\infty$.
3. First observe that

$$\int_0^a \frac{\left(-\ln(x_1 + \frac{1}{n})\right)^{\beta-1}}{x_1 + \frac{1}{n}} \, dx_1 \geq \int_{\frac{1}{n}}^a \frac{\left(-\ln(x_1)\right)^{\beta-1}}{x_1} \, dx_1;$$

then apply the monotone convergence theorem to the function g_n defined by

$$g_n(x_1) = \frac{\left(-\ln(x_1)\right)^{\beta-1}}{x_1} \text{ if } \frac{1}{n} < x_1 < a \text{ and } g_n(x_1) = 0 \text{ if } 0 < \frac{1}{n}.$$

Therefore, since $\beta > 0$, $\psi_n \to +\infty$ as $n \to +\infty$.
4. The equality $S_{v^{(n)}}(\mathbb{1}_{\partial\Omega}) = \psi_n$ follows from the calculation of $\partial_1 u$. Passing to the limit as $n \to +\infty$ in this equality yields $S_v(\mathbb{1}_{\partial\Omega}) = +\infty$, which is impossible: indeed, since $v \in H_{\mathrm{div}}(\Omega)$, the mapping S_v is defined from $H^{\frac{1}{2}}(I)$ to $\mathbb{R}$ and therefore $S_v(\mathbb{1}_{\partial\Omega}) \in \mathbb{R}$ because $\mathbb{1}_{\partial\Omega} \in H^{\frac{1}{2}}(I)$.

5. The fact that $\langle T(v), \mathbb{1}_{\partial\Omega} \rangle_{H^{-\frac{1}{2}}(\partial\Omega), H^{\frac{1}{2}}(\partial\Omega)} = 0$ is due to the fact that $\mathbb{1}_{\partial\Omega}$ is the trace of a constant function and that $\operatorname{div} v = 0$. The fact that

$$\lim_{n \to +\infty} \int_{\partial\Omega} \mathbb{1}_{\partial\Omega}\, v^{(n)} \cdot n \, d\lambda = 0$$

is then a consequence of the continuity of T and of question 5 of Problem 2.26.

Problem 2.28 (A slight generalisation of the Liouville theorem)

1. If the theorem is true in the case $u \geq 0$ a.e., it is also true if there exists a $c \in \mathbb{R}$ such that $u \geq c$ a.e. Indeed, by considering the function $u - c$ we are reduced to the case $u \geq 0$.

   Let $(\rho_n)_{n \in \mathbb{N}^\star}$ be a sequence of regularising kernels, that is to say,

   $$\rho \in \mathcal{D}(\mathbb{R}^d), \int_{\mathbb{R}^d} \rho \, dx = 1, \ \rho \geq 0, \ \rho(x) = 0 \text{ if } |x| \geq 1,$$
   $$\text{and, for } n \in \mathbb{N}^\star, x \in \mathbb{R}^d, \ \rho_n(x) = n^d \rho(nx).$$

   For $p \in \mathbb{N}^\star$, let B_p be the ball (for the Euclidean norm) of $\mathbb{R}^d$ with centre 0 and radius p and $\mathbb{1}_{B_p}$ the characteristic function of B_p. For $n \in \mathbb{N}^\star$, observe that the function u_n by $u_n = u \star \rho_n$ is well defined. Note that on the ball B_p, $u_n = (u \mathbb{1}_{B_{p+1}}) \star \rho_n$ for all $n \in \mathbb{N}^\star$. The theorem of continuity in mean in $L^1(\mathbb{R}^N)$ [26, Exercice 5.21] gives that $u \mathbb{1}_{B_{p+1}} \star \rho_n \to u \mathbb{1}_{B_{p+1}}$ in $L^1(\mathbb{R}^N)$ as $n \to +\infty$ (see [26], theorem 5.21). Hence $u_n \to u$ in $L^1(B_p)$. Since p is arbitrary, this is equivalent to saying that $u_n \to u$ in $L^1_{\text{loc}}(\mathbb{R}^d)$. The theorems of continuity and differentiability under the integral sign give that $u_n \in C^\infty(\mathbb{R}^d)$ (see for example, [26], problem 7.22).
   Now note that for $x \in \mathbb{R}^d$ and all $n \in \mathbb{N}^\star$, since $\Delta u = 0$ in $\mathcal{D}^\star(\mathbb{R}^d)$ and $\rho_n(x - \cdot) \in \mathcal{D}(\mathbb{R}^d)$, denoting by dy the element of integration in $\mathbb{R}^d$ (for the Lebesgue measure),

   $$0 = \langle \Delta u, \rho_n(x - \cdot)\rangle_{\mathcal{D}^\star(\mathbb{R}^d), \mathcal{D}(\mathbb{R}^d)} = \int_{\mathbb{R}^d} u(y) \Delta \rho_n(x - y) \, dy = \Delta u_n(x).$$

   Finally, it is clear that $u_n \geq 0$. Hence, $u_n \in C^\infty(\mathbb{R}^d)$, $\Delta u_n = 0$ in $\mathbb{R}^n$ and $u_n \geq 0$. If the theorem is shown for such a function u_n, there exists then a $C_n \in \mathbb{R}$ such that $u_n = C_n$. But, as $u_n \to u$ in $L^1_{\text{loc}}(\mathbb{R}^d)$, the sequence $(C_n)_{n \in \mathbb{N}}$ converges in $\mathbb{R}$ to some limit C (for example, $C = \int_{B_1} u(x) dx$) and we obtain $u = C$ a.e.

2. Here we use Green's[27] formula. For a function $u \in C^2(\overline{B_r})$, since $\Delta u = 0$, it yields that

   $$0 = \int_{B_r} \Delta u(x) \, dx = \int_{C_r} \nabla u(x) \cdot n(x) \, d\gamma(x).$$

---

[27] George Green (1793–1841), an autodidact British physicist and mathematician.

Let $r > 0$. The change of variables $x = ry$ gives (noting that the Jacobian of the change of variables is constant)

$$h(r) = \frac{1}{r^{d-1}} \int_{C_r} u(x)\, d\gamma(x) = \int_{C_1} u(ry)\, d\gamma(y).$$

By differentiating the term on the right under the integral sign and making the change of variables $ry = x$ (note that $n(x) = x/r = y$),

$$h'(r) = \int_{C_1} \nabla u(ry) \cdot y\, d\gamma(y) = \frac{1}{r^{d-1}} \int_{C_r} \nabla u(x) \cdot n(x)\, d\gamma(y).$$

The function h is therefore constant over $\mathbb{R}_+$. We denote this constant by H in the following.

3. Thanks to the change of variables $x \mapsto (r, y)$ with $r = |x|$ and $y \in C_1$, we obtain (see for example [26], problem 7.28)

$$\int_{B_r} u(x)\, dx = \int_0^r \rho^{d-1} \Big(\int_{C_1} u(\rho y)\, d\gamma(y) \Big)\, d\rho = \int_0^r \Big(\int_{C_\rho} u(z)\, d\gamma(z) \Big)\, d\rho$$

$$= \int_0^r \rho^{d-1} H\, d\rho = \frac{r^d}{d-1}.$$

It is now sufficient to use the continuity of u at 0,

$$\Big| \frac{1}{|B_r|} \int_{B_r} u(x)\, dx - u(0) \Big| \leq \sup_{x \in B_r} |u(x) - u(0)| \to 0 \text{ when } r \to 0.$$

Hence, for all $r > 0$,

$$\frac{1}{|B_r|} \int_{B_r} u(x)\, dx = u(0).$$

Of course, 0 plays no particular role. We can translate the function u and we then obtain, for all $a \in \mathbb{R}^d$ and all $r > 0$,

$$\frac{1}{|B_r|} \int_{B_{a,r}} u(x)\, dx = u(a).$$

4. The proposed double inequality comes from $B_{r-\alpha} \subset B_{a,r} \subset B_{r+\alpha}$ and $u \geq 0$. Thus

$$\frac{|B_{r-\alpha}|}{|B_r|} \frac{1}{|B_{r-\alpha}|} \int_{B_{r-\alpha}} u(x)\, dx \leq \frac{1}{|B_r|} \int_{B_{a,r}} u(x)\, dx$$

$$\leq \frac{|B_{r+\alpha}|}{|B_r|} \frac{1}{|B_{r+\alpha}|} \int_{B_{r+\alpha}} u(x)\, dx.$$

Letting $r \to \infty$, we obtain (note that α is fixed) $u(a) = u(0)$. Hence u is constant.

Chapter 3
Quasi-Linear Elliptic Problems

In the previous chapter, the elliptic problems considered were linear. However, the modelling of many phenomena gives rise to elliptic problems that contain non-linearities and that deserve the attention of mathematicians. Semi-linear PDEs are the closest to linear PDEs, in the sense that higher order derivatives appear as linear terms, with coefficients that are functions of the space variables; non-linearities can appear in the terms of the lower order derivatives and of the right-hand side. For example, a semi-linear PDE is $-\Delta u = f(x, u)$, where x designates the space variable. In a quasi-linear PDE, the higher order derivatives also only appear as linear terms, but with coefficients that can be functions of the unknowns and their lower order derivatives. The Leray–Lions problem, which is written in the form $-\mathrm{div}(a(x, u, \nabla u)) = f(u)$ is an example of a quasi-linear PDE, and this is the type of equation that we explore in this chapter.

We examine three categories of methods aimed at establishing the existence of solutions for quasi-linear elliptic problems: *compactness* methods, *monotonicity* methods and *minimisation* methods. In addition, we discuss a technique for proving the uniqueness of solutions.

3.1 Compactness Methods

Compactness methods are based on the topological degree theory of Leray–Schauder and its corollary, the Schauder fixed point theorem. The are so-called because of the compactness assumption of the mapping considered for these results. In this section we present the degree theory and Schauder's fixed point theorem; we then give an example using Schauder's fixed point theorem for the existence of a solution to a quasi-linear diffusion problem with a bounded right-hand side, and an example using the Leray–Schauder topological degree for a quasi-linear convection-diffusion problem, with an unbounded right-hand side.

T. Gallouët, R. Herbin, *Weak Solutions of Partial Differential Equations*,
Mathématiques et Applications 90, https://doi.org/10.1007/978-3-031-98982-7_3

3.1.1 Topological Degree and Schauder's Theorem

Let Ω be an open bounded subset of $\mathbb{R}^N$, $N \geq 1$, or an open bounded subset of a Banach space E, and let $g \in C(\bar{\Omega}, \mathbb{R}^N)$ (or $g \in C(\bar{\Omega}, E)$) and $y \in \mathbb{R}^N$ (or $y \in E$). The first objective of this section is to show that there exists an $x \in \bar{\Omega}$ such that $g(x) = y$.

We begin by providing the existence (and uniqueness) of a mapping, called the topological degree, in finite dimension introduced in 1933 by Brouwer[1] then in infinite dimension by Leray and Schauder[2]. This mapping is a possible way to obtain an existence theorem of a solution to a quasi-linear problem.

Theorem 3.1 (Brouwer's topological degree). *Let $N \geq 1$. Let $\mathfrak{A}$ be the set of triplets (g, Ω, y) where Ω is an open bounded subset of $\mathbb{R}^N$, $g \in C(\bar{\Omega}, \mathbb{R}^N)$ and $y \in \mathbb{R}^N$ such that $y \notin \{g(x), x \in \partial\Omega\}$. There exists a mapping d from $\mathfrak{A}$ to $\mathbb{Z}$, called "topological degree", satisfying the following three properties:*

(d1) *(Normalisation) $d(\mathrm{Id}, \Omega, y) = 1$ if $y \in \Omega$.*

(d2) *(Degree of a union) If $\Omega_1 \cup \Omega_2 \subset \Omega$, $\Omega_1 \cap \Omega_2 = \emptyset$ and $y \notin \{g(x), x \in \bar{\Omega} \setminus (\Omega_1 \cup \Omega_2)\}$, then*

$$d(g, \Omega, y) = d(g, \Omega_1, y) + d(g, \Omega_2, y).$$

(d3) *(Invariance by homotopy) If $h \in C([0, 1] \times \bar{\Omega}, \mathbb{R}^N)$, $y \in C([0, 1], \mathbb{R}^N)$ and if, for all $t \in [0, 1]$, $y(t) \notin \{h(t, x), x \in \partial\Omega\}$, then*

$$d(h(t, \cdot), \Omega, y(t)) = d(h(0, \cdot), \Omega, y(0)) \text{ for all } t \in [0, 1].$$

The properties of the topological degree (d1)–(d3) given in Theorem 3.1 and more particularly the additivity property (d2) lead to the following properties as well as the existence result of Corollary 3.3.

Remark 3.2 (Important properties). 1. If $\Omega = \emptyset$, taking $\Omega_1 = \Omega_2 = \emptyset$ in (d2), we obtain that $d(g, \emptyset, y) = 0$.

2. Let A be an invertible $N \times N$ matrix, Ω an open bounded subset of $\mathbb{R}^N$ and $y \in \mathbb{R}^N$ such that $A^{-1}y \in \Omega$; for $x \in \mathbb{R}^N$, we set $g(x) = Ax$. We then have $(g, \Omega, y) \in \mathcal{A}$ and the degree $d(g, \Omega, y)$ is equal to the sign of the determinant of A, so $d(g, \Omega, y) \neq 0$.

Corollary 3.3 (Existence by Brouwer). *With the notation of Theorem 3.1, let $(g, \Omega, y) \in \mathfrak{A}$ such that $d(g, \Omega, y) \neq 0$, then there exists an $x \in \Omega$ such that $g(x) = y$.*

---

[1] Luitzen Egbertus Jan Brouwer (1881–1966), Dutch mathematician and philosopher, who worked in topology, set theory, measure theory and complex analysis.

[2] Juliusz Pawel Schauder (1899–1943), Polish mathematician of Jewish origin, known for his work in functional analysis, partial differential equations and mathematical physics. Murdered by the Gestapo, after denunciation by a German colleague.

Proof Let us reason by contradiction and suppose that there does not exist an $x \in \Omega$ such that $g(x) = y$. Taking $\Omega_1 = \Omega_2 = \emptyset$, since by hypothesis $y \notin \{g(x), x \in \partial\Omega\}$, we have $y \notin \{g(x), x \in \bar{\Omega} \setminus (\Omega_1 \cup \Omega_2)\}$ and therefore by (d2) and by item 1 of Remark 3.2,

$$d(g, \Omega, y) = d(g, \Omega_1, y) + d(g, \Omega_2, y) = 0.$$

$\blacksquare$

Corollary 3.3 provides a method, called the *topological degree method*, or *degree argument* for finding solutions to non-linear problems, which is as follows. Let $N \geq 1$, Ω be an open bounded subset of $\mathbb{R}^N$, $g \in C(\bar{\Omega}, \mathbb{R}^N)$ and $y \in \mathbb{R}^N$. The aim is to show that there exists an $x \in \Omega$ such that $g(x) = y$. To this end, we construct a mapping h from $[0, 1] \times \bar{\Omega}$ to $\mathbb{R}^N$ such that

1. $h(1, \cdot) = g$,
2. $h(0, \cdot) = \tilde{g}$, where $\tilde{g}$ is an invertible linear mapping such that $y \in \{\tilde{g}(x), x \in \Omega\}$.
3. $h(t, x) \neq y$ for all $t \in [0, 1]$ and all $x \in \partial\Omega$.

We then obtain that $d(g, \Omega, y) = d(\tilde{g}, \Omega, y) \neq 0$ and therefore that there exists an $x \in \Omega$ such that $g(x) = y$.

Remark 3.4 (The case $N = 1$). In the case $N = 1$, the degree argument that we just described is not really interesting, as it brings nothing more than the intermediate value theorem.

For a more detailed description of the degree, one can refer to the work [32]. A consequence of this topological degree method is Brouwer's fixed point theorem, see [14] for the original article in German, and [17, 52] for more modern versions.

Theorem 3.5 (Brouwer's Fixed Point). *Let $N \geq 1$, $R > 0$ and $f \in C(B_R, B_R)$ with $B_R = \{x \in \mathbb{R}^N, \|x\| \leq R\}$ (where $\mathbb{R}^N$ is equipped with a norm denoted $\|\cdot\|$). Then f has a fixed point, that is, there exists an $x \in B_R$ such that $f(x) = x$.*

Proof If there exists an $x \in \partial B_R$ (that is, $\|x\| = R$) such that $f(x) = x$, there is nothing more to prove. Now assume that $f(x) \neq x$ for all $x \in \partial B_R$, and set $\Omega = \{x \in \mathbb{R}^N, \|x\| < R\}$ (which gives $B_R = \bar{\Omega}$) and, for $t \in [0, 1]$ and $x \in B_R$, $h(t, x) = x - t f(x)$. Observe that $h(t, x) \neq 0$ for all $x \in \partial\Omega = \{x \in \mathbb{R}^N, \|x\| = R\}$. Hence $d(h(1, \cdot), \Omega, 0) = d(h(0, \cdot), \Omega, 0) = d(\mathrm{Id}, \Omega, 0) = 1$ and therefore there exists an $x \in \Omega$ such that $f(x) = x$. $\blacksquare$

Of course, the same results (Theorem 3.1, Corollary 3.3 and Theorem 3.5) are true if $\mathbb{R}^N$ is replaced by a Banach space E of finite dimension (we reduce to $\mathbb{R}^N$, $N = \dim E$ using a basis of E). Brouwer's theorem (Theorem 3.1), and Corollary 3.3, giving an existence result, also generalise to infinite dimension: in the definition of $\mathfrak{A}$ g is then required to be a compact perturbation of the identity, that is $g = \mathrm{Id} - f$ with f compact in the sense of Definition 3.6[3] This generalisation is due to Leray and Schauder [37]. A consequence of this generalisation is Schauder's theorem (Theorem

---

[3] In finite dimension, a continuous mapping is always compact, so the hypothesis "continuous" is identical to the hypothesis "compact perturbation of the identity".

3.11) which generalises Brouwer's theorem to infinite dimension by requiring that f is compact. Recall that in an infinite-dimensional space a continuous mapping is not necessarily compact, whatever the space considered. A counter example to Schauder's theorem if f is only continuous is given in Problem 3.2.

We have seen in previous chapters compact linear mappings. Definition 3.6 generalises this notion of compact mapping to non-linear mappings.

Definition 3.6 (Compact mapping). Let E be a (real) Banach space, B a subset of E and f a mapping from B to E. We say that f is compact if f satisfies the following two properties:

1. f is continuous,
2. $\{f(x), x \in C\}$ is relatively compact (in E) for every bounded subset C of B.

Note that in the original article by Leray–Schauder [37], the term "compact" is not used, the expression "completely continuous" is used instead.

In the previous definition, observe that if f is linear (and $B = E$) the second condition implies the first. But this is false for non-linear mappings.

Definition 3.7. Let E be a (real) Banach space. Define $\mathcal{A}$ as the set of triplets $(\mathrm{Id} - f, \Omega, y)$ where Ω is an open bounded subset of E, f is a compact mapping from $\bar{\Omega}$ to E, in the sense of Definition 3.6, and $y \in E$ is such that $y \notin \{x - f(x), x \in \partial\Omega\}$.

Theorem 3.8 (Leray–Schauder topological degree). *Let E be a (real) Banach space and $\mathcal{A}$ be given by Definition 3.7; there exists a mapping d from $\mathcal{A}$ to $\mathbb{Z}$, called the "topological degree", satisfying the following three properties:*

(d1) (Normalisation) $d(\mathrm{Id}, \Omega, y) = 1$ *if* $y \in \Omega$.
(d2) (Degree of a union) $d(\mathrm{Id} - f, \Omega, y) = d(\mathrm{Id} - f, \Omega_1, y) + d(\mathrm{Id} - f, \Omega_2, y)$ *if* $\Omega_1 \cup \Omega_2 \subset \Omega$, $\Omega_1 \cap \Omega_2 = \emptyset$ *and* $y \notin \{x - f(x), x \in \bar{\Omega} \setminus (\Omega_1 \cup \Omega_2)\}$.
(d3) (Invariance by homotopy) *If h is a compact mapping from $[0, 1] \times \bar{\Omega}$ to E (in the sense of the Definition 3.6), $y \in C([0, 1], E)$ and $y(t) \notin \{x - h(t, x), x \in \partial\Omega\}$ for all $t \in [0, 1]$, we then have $d(\mathrm{Id} - h(t, \cdot), \Omega, y(t)) = d(\mathrm{Id} - h(0, \cdot), \Omega, y(0))$ for all $t \in [0, 1]$.*

As in the case of a finite-dimensional space (see Corollary 3.3), the properties of the topological degree (given in Theorem 3.8) yield that if $d(\mathrm{Id} - f, \Omega, y) \neq 0$ then there exists an $x \in \Omega$ such that $x - f(x) = y$. To give the analogue of the second property of Remark 3.2, we need a second theorem due to Leray and Schauder.

Theorem 3.9 (Compact linear mapping, Leray–Schauder). *Let E be a (real) Banach space, L a compact linear mapping from E to E and let Ω an open bounded subset containing 0. Assume that*

$$x \in E, \; Lx = x \Rightarrow x \notin \partial\Omega. \tag{3.1}$$

Then $(\mathrm{Id} - L, \Omega, 0) \in \mathcal{A}$, where $\mathcal{A}$ is given by Definition 3.7, and $d(\mathrm{Id} - L, \Omega, 0) \neq 0$.

Remark 3.10 (An equivalent assumption to (3.1)). Note that, since L is linear, the hypothesis (3.1) is equivalent to saying $(x \in E,\ Lx = x) \Rightarrow x = 0$ (which is equivalent to saying that 1 is not an eigenvalue of L). Indeed, suppose that $Lx = x$ and that $x \neq 0$; let us show that this contradicts the hypothesis (3.1). Since L is linear, $L(tx) = tx$ for all $t \in \mathbb{R}$. Choose $t = \sup\{\alpha, sx \in \Omega\ \forall s \in [0, \alpha]\}$. Since Ω is an open bounded subset containing 0, there exist $\varepsilon > 0$ and $R > 0$ such that $B(0, \varepsilon) \subset \Omega \subset B(0, R)$; we thus have

$$\frac{\varepsilon}{\|x\|} \leq t \leq \frac{R}{\|x\|},$$

which gives $t \in \mathbb{R}^{\star}$. The definition of t gives $sx \in \Omega$ if $s \in [0, t[$ and therefore $t \in \overline{\Omega}$. Then since Ω is open, $t \notin \Omega$, we thus have $tx \in \partial\Omega$, hence the contradiction. The converse is immediate because $0 \notin \partial\Omega$.

We can now, as in the finite-dimensional case, provide a method to find solutions to non-linear problems. Let E be a Banach space, Ω an open bounded subset of E containing 0, f a mapping from $\overline{\Omega}$ to E. The aim is to show that there exists an $x \in \Omega$ such that $x - f(x) = 0$ (if necessary, by changing f, we can always reduce to this form). To this end, we construct a compact mapping h from $[0, 1] \times \overline{\Omega}$ to E such that

1. $h(1, \cdot) = f$,
2. $h(0, \cdot) = L$ with L linear from E to E,
3. $x - h(t, x) \neq 0$ for all $t \in [0, 1]$ and all $x \in \partial\Omega$.

We then obtain $d(\mathrm{Id} - f, \Omega, 0) = d(\mathrm{Id} - L, \Omega, 0) \neq 0$ and therefore there exists an $x \in \Omega$ such that $x - f(x) = 0$. Of course, to be able to construct such a function h, f must be a compact mapping and L must be a compact linear mapping.

As in the finite-dimensional case, a first consequence of the existence of the topological degree is the obtention of a fixed point theorem.

Theorem 3.11 (Schauder's fixed point). *Let E be a Banach space, $R > 0$, $B_R = \{x \in E,\ \|x\| \leq R\}$ and f a compact mapping of B_R to B_R (that is to say, f is continuous and $\{f(x), x \in B_R\}$ is relatively compact in E). Then f admits a fixed point, that is to say, there exists an $x \in B_R$ such that $f(x) = x$.*

Proof The proof is very close to that of Theorem 3.5. If there exists an $x \in \partial B_R$ (that is to say, such that $\|x\| = R$) such that $f(x) = x$, there is nothing more to prove. Assuming then that $f(x) \neq x$ for all $x \in \partial B_R$, set $\Omega = \{x \in E,\ \|x\| < R\}$ (which gives $B_R = \overline{\Omega}$) and, for $t \in [0, 1]$ and $x \in B_R$, $h(t, x) = tf(x)$. Observe that $x - h(t, x) \neq 0$ for all $x \in \partial\Omega = \{x \in E,\ \|x\| = R\}$. The compactness of h is deduced from that of f. Hence $d(\mathrm{Id} - h(1, \cdot), \Omega, 0) = d(\mathrm{Id} - h(0, \cdot), \Omega, 0) = d(\mathrm{Id}, \Omega, 0) = 1$ and therefore there exists an $x \in \Omega$ such that $f(x) = x$. $\blacksquare$

Schauder's theorem (Theorem 3.11) is false if we replace the assumption of compactness of f by the simple assumption of continuity. However, the main difficulty in using Schauder's theorem (or, more generally, in using the topological degree) is often to show the continuity of f (or, in the use of the topological degree, the continuity of the mapping h of Theorem 3.8).

3.1.2 Existence Results by Schauder's Theorem

The aim is now to use Schauder's fixed point theorem to prove the existence of a solution to a quasi-linear elliptic problem. Let us start by stating the assumptions, and for that, let us first give the definition of a Carathéodory[4] function.

Definition 3.12 (Carathéodory function). Let $N, p, q \in \mathbb{N}^\star$ and Ω be an open subset of $\mathbb{R}^N$. Let a be a function from $\Omega \times \mathbb{R}^p$ to $\mathbb{R}^q$. We say that a is a Carathéodory function if $a(\cdot, s)$ is Borel-measurable for all $s \in \mathbb{R}^p$ and $a(x, \cdot)$ is continuous for almost all $x \in \Omega$.

Remark 3.13. Let $N, p, q \in \mathbb{N}^\star$, Ω be an open subset of $\mathbb{R}^N$ and a a Carathéodory function from $\Omega \times \mathbb{R}^p$ to $\mathbb{R}^q$. The function a is then Borel-measurable from $\Omega \times \mathbb{R}^p$ to $\mathbb{R}^q$ (which could be false if a was only Borel-measurable with respect to each of its arguments). If v is a Borel-measurable function from Ω to $\mathbb{R}^p$, the function $x \mapsto a(x, v(x))$ is then Borel-measurable from Ω to $\mathbb{R}^q$. This property is used several times in the following (without recalling it) when v belongs to $L^r(\Omega)$ (for an $r \in [1, +\infty]$) by choosing a representative Borel-measurable of v (the function $x \mapsto a(x, v(x))$ does not depend on the representative chosen for v, modulo the equivalence relation "= a.e.").

We now assume the following conditions:

$$N \geq 1, \ \Omega \text{ is an open bounded subset of } \mathbb{R}^N, \tag{3.2a}$$

$$a : \Omega \times \mathbb{R} \to \mathbb{R} \text{ is a Carathéodory function}, \tag{3.2b}$$

$$\text{there exist } \alpha > 0 \text{ and } \beta > 0 \text{ such that } \alpha \leq a(\cdot, s) \leq \beta \text{ a.e. and for all } s \in \mathbb{R}, \tag{3.2c}$$

$$f : \Omega \times \mathbb{R} \to \mathbb{R} \text{ is a Carathéodory function and } f \in L^\infty(\Omega \times \mathbb{R}). \tag{3.2d}$$

Under the assumptions 3.2, we seek to prove the existence of a solution u to the following problem:

$$\begin{cases} u \in H_0^1(\Omega), \\ \displaystyle\int_\Omega a(x, u(x)) \nabla u(x) \cdot \nabla v(x)\, \mathrm{d}x = \int_\Omega f(x, u(x)) v(x)\, \mathrm{d}x, \ \forall v \in H_0^1(\Omega). \end{cases} \tag{3.3}$$

Theorem 3.14 (Existence, bounded right-hand side). *Under the assumptions* (3.2), *there exists a solution u to* (3.3).

Proof For $\bar{u} \in L^2(\Omega)$, the chapter on linear elliptic equations gives us the existence and uniqueness of a solution u to

$$\begin{cases} u \in H_0^1(\Omega), \\ \displaystyle\int_\Omega a(x, \bar{u}(x)) \nabla u(x) \cdot \nabla v(x)\, \mathrm{d}x = \int_\Omega f(x, \bar{u}(x)) v(x)\, \mathrm{d}x, \ \forall v \in H_0^1(\Omega). \end{cases} \tag{3.4}$$

---

[4] Constantin Carathéodory (1873–1950), German mathematician of Greek origin, whose research focused on the calculus of variations, and partial differential equations.

More precisely, to prove the existence and uniqueness of a solution u to (3.4), we apply Theorem 2.6. With this aim, we put problem (3.4) in the form of problem (2.6) by setting $a_{i,j} = 0$ if $i \neq j$, $a_{i,i} = a(\cdot, \bar{u})$ and $f = f(\cdot, \bar{u})$ (in this last equality, the function f on the left-hand side is that of problem (2.6) and the function f on the right-hand side is that of problem (3.4)). Theorem 2.6 indeed provides the existence and uniqueness of u as a solution to (3.4).

Set $T(\bar{u}) = u$; the mapping T is therefore a mapping from E to E with $E = L^2(\Omega)$. A fixed point of T is a solution to (3.3). To prove the existence of such a fixed point, we use Theorem 3.11.

First of all, using the assumption (3.2c), Poincaré's inequality and the L^∞ bound of f, we show that the image of T is in a bounded subset of $H_0^1(\Omega)$ and therefore (by Rellich's theorem, Theorem 1.36) in a compact subset of $L^2(\Omega)$. Taking R large enough, the mapping T therefore sends $B_R = \{v \in L^2(\Omega), \|v\|_2 \leq R\}$ to B_R and $\{T(\bar{u}), \bar{u} \in B_R\}$ is relatively compact in $L^2(\Omega)$. To use Theorem 3.11, it remains to show the continuity of T.

Let $(\bar{u}_n)_{n \in \mathbb{N}}$ be a sequence in E such that $\bar{u}_n \to \bar{u}$ in E, as $n \to +\infty$. Set $u_n = T(\bar{u}_n)$; we can assume that up to a subsequence, $\bar{u}_n \to \bar{u}$ a.e. and that there exists a $w \in H_0^1(\Omega)$ such that $u_n \to w$ weakly in $H_0^1(\Omega)$ (and therefore also $u_n \to w$ in $L^2(\Omega)$). We show that w is a solution to (3.4). Indeed, let $v \in H_0^1(\Omega)$. We have

$$\int_\Omega a(x, \bar{u}_n(x)) \nabla u_n(x) \cdot \nabla v(x) \, dx = \int_\Omega f(x, \bar{u}_n(x)) v(x) \, dx, \quad \text{for all } v \in H_0^1(\Omega).$$

We pass to the limit as $n \to +\infty$; using the dominated convergence theorem and passing to the limit in the product of a weak convergence and a convergence in L^2 (see Lemma 3.26 for a more general result containing this one), we obtain

$$\int_\Omega a(x, \bar{u}(x)) \nabla w(x) \cdot \nabla v(x) \, dx = \int_\Omega f(x, \bar{u}(x)) v(x) \, dx, \quad \text{for all } v \in H_0^1(\Omega).$$

This proves that $w = T(\bar{u})$. We have therefore proved that up to a subsequence, $T(\bar{u}_n) \to T(\bar{u})$ in $L^2(\Omega)$. A classical proof by contradiction shows that the whole sequence $(T(\bar{u}_n))_{n \in \mathbb{N}}$ converges; see e.g. the solution of question 4 of Problem 1.14. Hence T is continuous, and Theorem 3.11 allows us to conclude to the existence of a fixed point of T. $\blacksquare$

3.1.3 Existence Results by Topological Degree

We take up the problem of Section 3.1.2 by removing the assumption that f is bounded, which allowed a simple application of Schauder's theorem, and by adding a convection term. Consider the following diffusion-convection-reaction equation

$$\begin{cases} u \in H_0^1(\Omega), \\[2mm] \displaystyle\int a(x, u(x))\nabla u(x) \cdot \nabla v(x) + \int_\Omega G(x)\varphi(u(x)) \cdot \nabla v(x)\, dx \\[4mm] \qquad = \displaystyle\int_\Omega f(x, u(x))v(x)\, dx, \quad \forall v \in H_0^1(\Omega), \end{cases} \tag{3.5}$$

which is the weak formulation of the following problem:

$$\begin{cases} -\mathrm{div}(a(\cdot, u)\nabla u) - \mathrm{div}(G\varphi(u)) = f(\cdot, u), \ \text{in } \Omega, \\[2mm] u = 0 \ \text{on } \partial\Omega. \end{cases} \tag{3.6}$$

Note that this equation is nonlinear for three reasons: the diffusion, convection and reaction terms are nonlinear. The first term on the left-hand side is called the "diffusion term", the second term on the left-hand side is called the "convection term" and the right-hand side is called the "reaction term". We also recall that $\varphi(u)$ is a slightly incorrect notation for the mapping $x \mapsto \varphi(u(x))$ (the correct notation would rather be $\varphi \circ u$). Similarly, $a(\cdot, u)$ is the mapping $x \mapsto a(x, u(x))$ and $f(\cdot, u)$ is the mapping $x \mapsto f(x, u(x))$.

Assume the following hypotheses:

$$\Omega \text{ is an open bounded subset of } \mathbb{R}^N, \ N \geq 1, \tag{3.7a}$$

$$a \text{ is a Carathéodory function (see Definition 3.12)}, \tag{3.7b}$$

$$\exists \alpha, \beta > 0 \ \ \alpha \leq a(x, s) \leq \beta \ \ \forall s \in \mathbb{R} \qquad \text{a.e. } x \in \Omega, \tag{3.7c}$$

$$G \in C^1(\bar{\Omega}, \mathbb{R}^N), \ \mathrm{div}\, G = 0, \tag{3.7d}$$

$$\varphi \in C(\mathbb{R}, \mathbb{R}) \text{ and there exists } C_1 \geq 0 \ : \ |\varphi(s)| \leq C_1|s| \qquad \forall s \in \mathbb{R}, \tag{3.7e}$$

$$f \text{ is a Carathéodory function, and } \exists C_2 \geq 0 \text{ and } d \in L^2(\Omega) \tag{3.7f}$$

$$|f(x, s)| \leq d(x) + C_2|s| \ \ \forall s \in \mathbb{R} \qquad \text{a.e. } x \in \Omega, \tag{3.7g}$$

$$\lim_{s \to \pm\infty} \frac{f(x, s)}{s} = 0 \text{ for all } x \in \Omega. \tag{3.7h}$$

Remark 3.15 (Fredholm Alternative). In the case where $a \equiv 1$, $\varphi = 0$ and f is of the form $f(x, s) = d(x) + \lambda s$ where λ is an eigenvalue of the Laplacian on Ω with Dirichlet condition (that is, there exists a $w \in H_0^1(\Omega)$, $w \neq 0$, such that $-\Delta w = \lambda w$ in $\mathcal{D}^\star(\Omega)$) and d is an element of $L^2(\Omega)$, problem (3.6) becomes $-\Delta u = \lambda u + d$, with Dirichlet condition. This problem only admits a solution if d is orthogonal to the eigenspace associated with λ (and in this case there is no uniqueness). This property is known as the *Fredholm*[5] *Alternative*, see Problem 2.3. It is to ensure existence for all d in $L^2(\Omega)$ that we add the sub-linearity assumption on f (assumption (3.7h)).

Remark 3.16 (Coercivity). When $\operatorname{div} G \neq 0$, the problem can, under certain conditions, be treated in a similar way to that given in the proof of Theorem 3.17. This is the case, for example, if $\varphi(u) = u$ and if $\operatorname{div} G \leq \lambda_1$ a.e., where λ_1 is the first eigenvalue of $u \mapsto -\operatorname{div}(\alpha \nabla u)$ with Dirichlet condition (this eigenvalue is positive). Under the assumptions "G is of class C^1 and $\varphi(u) = u$" (but without the condition on $\operatorname{div} G$), the problem becomes more difficult (see Problem 3.6), even in the linear case, that is to say, the case where a and f do not depend on u. The main difficulty is due to the lack of coercivity of the operator $u \mapsto -\operatorname{div}(\alpha \nabla u) - \operatorname{div}(Gu)$.

Theorem 3.17 (Existence, unbounded right-hand side). *Under the assumptions (3.7), there exists a solution to (3.5).*

Proof The proof given here relies on a topological degree argument. This method requires *a priori* estimates, that is to say, estimates on u, without knowing its existence. So let us assume u is a solution to (3.5). We can (and will) show that there exists an $R > 0$ such that $\|u\|_{L^2} \leq R$. We establish the estimates from the non-linear problem, and not from the linearised problem, which would be the case by Theorem 3.11 of Schauder's fixed point, see Remark 3.18; this presents serious advantages. For example in the non-linear convection term, we can write (formally)

$$\int_\Omega G\varphi(u) \cdot \nabla u \quad \mathrm{d}x = \int_\Omega G \cdot \nabla \phi(u)\, \mathrm{d}x$$

$$= -\int_\Omega \operatorname{div} G\, \phi(u)\, \mathrm{d}x$$

$$= 0 \qquad \text{because } \operatorname{div} G = 0,$$

where ϕ is the primitive of φ vanishing at 0 (here again $\phi(u)$ should rather be written $\phi \circ u$). Note that the estimate $\|u\|_{L^2(\Omega)} \leq R$ amounts to showing that all solutions are in the ball B_R (closed ball centred at 0 and of radius R), which is a *uniform* estimate on all solutions.

We rewrite the problem in the form:

$$\begin{cases} u \in H_0^1(\Omega), \\ \displaystyle\int_\Omega a(\cdot, u)\nabla u \cdot \nabla v\, \mathrm{d}x = \langle F(u), v\rangle_{H^{-1}(\Omega), H_0^1(\Omega)}, \, \forall v \in H_0^1(\Omega), \end{cases}$$

where $F(u)$ is, for $u \in L^2(\Omega)$, the element of $H^{-1}(\Omega)$ defined by

$$\langle F(u), v\rangle_{H^{-1}(\Omega), H_0^1(\Omega)} = -\int_\Omega G\varphi(u) \cdot \nabla v\, \mathrm{d}x + \int_\Omega f(\cdot, u)v\, \mathrm{d}x.$$

---

[5] Erik Ivar Fredholm (1866–1927), Swedish mathematician known for his work on integral equations and spectral theory.

Since $G \in L^\infty(\Omega)^N$, $|\varphi(s)| \le C_1|s|$ and $|f(\cdot, s)| \le d + C_2|s|$, the mapping F which associates u with $F(u)$ is continuous from $L^2(\Omega)$ to $H^{-1}(\Omega)$.

For $S \in H^{-1}(\Omega)$, the linear problem

$$\begin{cases} w \in H_0^1(\Omega), \\ \displaystyle\int_\Omega a(\cdot, u)\nabla w \cdot \nabla v \, dx = \langle S, v\rangle_{H^{-1}(\Omega), H_0^1(\Omega)}, \end{cases} \tag{3.8}$$

admits a unique solution $w \in H_0^1(\Omega)$. Let B_u be the operator which associates S in $H^{-1}(\Omega)$ with the solution w to (3.8). The operator B_u is a continuous linear from $H^{-1}(\Omega)$ to $H_0^1(\Omega)$ and $H_0^1(\Omega)$ is compactly embedded in $L^2(\Omega)$. Hence the operator B_u is compact from $H^{-1}(\Omega)$ to $L^2(\Omega)$. The problem (3.5) is equivalent to solving the fixed point problem $u = B_u(F(u))$. We therefore show below, by a topological degree argument, that the following problem admits a solution

$$\begin{cases} u \in L^2(\Omega), \\ u = B_u(F(u)). \end{cases}$$

For $t \in [0, 1]$, we set $h(t, u) = B_u(t\, F(u)) \in L^2(\Omega)$. The mapping h is thus defined from $[0, 1] \times L^2(\Omega)$ to $L^2(\Omega)$. For $R > 0$, we set $B_R = \{u \in L^2(\Omega)$ such that $\|u\|_{L^2(\Omega)} < R\}$. Let us show that

1. there exists an $R > 0$ such that, for all $t \in [0, 1]$ and all $u \in L^2(\Omega)$, if $u - h(t, u) = 0$ then $\|u\|_{L^2(\Omega)} < R$. (This a priori estimate is the most difficult point to show.)
2. h is continuous from $[0, 1] \times \bar{B}_R$ to $\bar{B}_R$.
3. The set $\{h(t, u), t \in [0, 1], u \in \bar{B}_R\}$ is a relatively compact subset of $L^2(\Omega)$.

If we assume that items 1 and 3 are proven, we have no solution to the equation $u - h(t, u) = 0$ on the edge of the ball B_R, and we can therefore define the degree $d(\mathrm{Id} - h(t, .), B_R, 0)$. This degree does not depend on t, so we have:

$$\begin{aligned} d(\mathrm{Id} - h(t, \cdot), B_R, 0) &= d(\mathrm{Id} - h(0, \cdot), B_R, 0) \\ &= d(\mathrm{Id}, B_R, 0) = 1. \end{aligned}$$

Thus, there exists a $u \in B_R$ such that $u - h(1, u) = 0$, that is to say,

$$u = B_u(F(u)).$$

So u is a solution to (3.5) (and Theorem 3.17 is proven).

It remains to prove points 1–3. Let us start by proving item 3 (for all $R > 0$). Let $R > 0$ and assume that $\|u\|_{L^2} \le R$. Then:

$$F(u) \in H^{-1}(\Omega), \text{ and } \langle F(u), v\rangle_{H^{-1}(\Omega), H_0^1(\Omega)} = -\int_\Omega G\varphi(u) \cdot \nabla v \, dx + \int_\Omega f(\cdot, u)v \, dx.$$

Let us estimate $\langle F(u), v \rangle_{H^{-1}(\Omega), H^1_0(\Omega)}$:

$$\langle F(u), v \rangle_{H^{-1}(\Omega), H^1_0(\Omega)} \leq \| \, |G| \, \|_\infty \|\varphi(u)\|_{L^2(\Omega)} \|v\|_{H^1_0(\Omega)} + \|f(\cdot, u)\|_{L^2(\Omega)} \|v\|_{L^2(\Omega)}$$

$$\leq \| \, |G| \, \|_\infty \, C_1 \|u\|_{L^2(\Omega)} \|v\|_{H^1_0(\Omega)} + \|d\|_{L^2(\Omega)} \|v\|_{L^2(\Omega)} + C_2 \|u\|_{L^2(\Omega)} \|v\|_{L^2(\Omega)}$$

$$\leq (\| \, |G| \, \|_\infty \, C_1 \, R + C_\Omega \|d\|_{L^2(\Omega)} + C_2 \, C_\Omega \, R \,) \|v\|_{H^1_0(\Omega)},$$

where C_Ω depends only on Ω (and is given by the Poincaré inequality). So

$$t\|F(u)\|_{H^{-1}} \leq \| \, |G| \, \|_\infty \, C_1 R + C_\Omega \|d\|_{L^2(\Omega)} + C_2 \, C_\Omega \, R, \quad \forall t \in [0, 1].$$

Let us set $h(t, u) = B_u(tF(u)) = w$ and show that there exists an $\bar{R}$ depending only on $R, G, C_\Omega, C_1, C_2, \alpha$ such that

$$\|h(t, u)\|_{H^1_0(\Omega)} \leq \bar{R}.$$

By definition, w is a solution of

$$\begin{cases} \displaystyle\int_\Omega a(\cdot, u)\nabla w \cdot \nabla v = \langle t \, F(u), v \rangle_{H^{-1}(\Omega), H^1_0(\Omega)} & \forall v \in H^1_0(\Omega), \\ w \in H^1_0(\Omega). \end{cases} \tag{3.9}$$

Taking $v = w$ in (3.9), we obtain:

$$\alpha \|w\|^2_{H^1_0(\Omega)} \leq \|tF(u)\|_{H^{-1}(\Omega)} \|w\|_{H^1_0(\Omega)} \leq \widetilde{R} \|w\|_{H^1_0(\Omega)},$$

with $\widetilde{R} = \| \, |G| \, \|_\infty \, C_1 R + C_\Omega \|d\|_{L^2(\Omega)} + C_2 \, C_\Omega \, R$. Hence $\|h(t, u)\|_{H^1_0(\Omega)} = \|w\|_{H^1_0(\Omega)} \leq \frac{\widetilde{R}}{\alpha} = \bar{R}$.

It follows from Rellich's theorem (Theorem 1.36) that the set $\{h(t, u), t \in [0, 1], u \in \bar{B}_R\}$ is relatively compact in $L^2(\Omega)$, which clearly shows item 3.

Let us now prove item 2. Let $(t_n)_{n \in \mathbb{N}} \subset [0, 1]$ be a sequence such that $t_n \to t$ when $n \to +\infty$ and let $(u_n)_{n \in \mathbb{N}} \subset L^2(\Omega)$ be a sequence such that $u_n \to u$ in $L^2(\Omega)$. We want to show that $h(t_n, u_n) \to h(t, u)$ in $L^2(\Omega)$. Let $w_n = h(t_n, u_n)$ and $w = h(t, u)$. To show that $w_n \to w$ in $L^2(\Omega)$, we seek to pass to the limit in the following equation:

$$\begin{cases} \displaystyle\int_\Omega a(\cdot, u_n)\nabla w_n \cdot \nabla v \, dx = -t_n \int_\Omega G\varphi(u_n) \cdot \nabla v \, dx + t_n \int_\Omega f(\cdot, u_n)v \, dx \\ w_n \in H^1_0(\Omega). \end{cases} \tag{3.10}$$

We already know that $(w_n)_{n \in \mathbb{N}}$ is bounded in $H^1_0(\Omega)$, because the sequence $(u_n)_{n \in \mathbb{N}}$ is bounded in $L^2(\Omega)$ (this is what we showed in the previous step: if $\|u_n\|_{L^2(\Omega)} \leq R$ then $\|w_n\|_{H^1_0(\Omega)} \leq \bar{R}$). Since the sequence $(w_n)_{n \in \mathbb{N}}$ is bounded in $H^1_0(\Omega)$, up to a subsequence again denoted $(w_n)_{n\in\mathbb{N}}$,

$$w_n \to \bar{w} \text{ in } H_0^1 \text{ weak and } w_n \to \bar{w} \text{ in } L^2(\Omega),$$

$$u_n \to u \text{ a.e. and } \exists\, H \in L^2(\Omega) \; ; \; |u_n| \le H \text{ a.e.}$$

Let $v \in H_0^1(\Omega)$. Since $a(\cdot, u_n) \to a(\cdot, u)$ a.e., we have $a(\cdot, u_n)\nabla v \to a(\cdot, v)\nabla v$ a.e., and $|a(\cdot, u_n)\nabla v| \le \beta|\nabla v|$. Hence $a(\cdot, u_n)\nabla v \to a(\cdot, u)\nabla v$ in $L^2(\Omega)^N$. But $\nabla w_n \to \nabla\bar{w}$ in $(L^2(\Omega))^N$ weak. Therefore

$$\int_\Omega a(\cdot, u_n)\nabla w_n \cdot \nabla v \, dx \to \int_\Omega a(\cdot, u)\nabla\bar{w} \cdot \nabla v \, dx \text{ when } n \to +\infty.$$

Observe that $\varphi(u_n) \to \varphi(u)$ a.e. and that $|\varphi(u_n)| \le C_1|u_n| \le C_1 H$; therefore by the Lebesgue dominated convergence theorem, $\varphi(u_n) \to \varphi(u)$ in $L^2(\Omega)$ and $\int_\Omega G\varphi(u_n) \cdot \nabla v \, dx \to \int_\Omega G\varphi(u) \cdot \nabla v \, dx$ when $n \to +\infty$. Finally, for the last term, since $f(\cdot, u_n) \to f(\cdot, u)$ a.e. and $|f(\cdot, u_n)| \le |d| + C_2 H$ a.e., we therefore have by the dominated convergence theorem $f(\cdot, u_n) \to f(\cdot, u)$ in $L^2(\Omega)$. Hence $\int_\Omega f(\cdot, u_n)v \, dx \to \int_\Omega f(\cdot, u)v \, dx$ when $n \to +\infty$. By passing to the limit in (3.10), we obtain

$$\int_\Omega a(\cdot, u)\nabla\bar{w} \cdot \nabla v \, dx = -t \int_\Omega G\varphi(u) \cdot \nabla v \, dx + t \int_\Omega f(\cdot, u)v \, dx,$$

and therefore $\bar{w} = h(t, u) = w$. By reasoning by contradiction, we then show that (without extracting a subsequence) $w_n \to w$ weakly in H_0^1 and $w_n \to w$ in $L^2(\Omega)$, where $w_n = h(t_n, u_n)$ and $w = h(t, u)$; the mapping h is therefore continuous. We have therefore shown item 2.

It remains now to prove item 1. We want to show that there exists an $R > 0$ such that for all $t \in [0, 1]$ and all $u \in L^2(\Omega)$, if $u = h(t, u)$, then $\|u\|_{L^2(\Omega)} < R$.

Let $t \in [0, 1]$, and $u = h(t, u) = t\, B_u(F(u))$, that is to say,

$$\begin{cases} \displaystyle\int_\Omega a(\cdot, u)\nabla u \cdot \nabla v \, dx = -t \int_\Omega G\varphi(u) \cdot \nabla v \, dx + t \int_\Omega f(\cdot, u)v \, dx, \quad \forall v \in H_0^1(\Omega). \\[2mm] u \in H_0^1(\Omega). \end{cases}$$

$$(3.11)$$

For $s \in \mathbb{R}$, we set $\Phi(s) = \int_0^s \varphi(\xi) \, d\xi$ (Φ is therefore a primitive of φ). Since $u \in H_0^1(\Omega)$, it is not difficult to show that $\Phi(u) \in W_0^{1,1}(\Omega)$ and that

$$\int_\Omega G\varphi(u) \cdot \nabla u \, dx = \int_\Omega G \cdot \nabla\Phi(u) \, dx.$$

(This is left as an exercise, it is enough to approximate u, in $H_0^1(\Omega)$, by a sequence of functions belonging to $\mathcal{D}(\Omega)$.) Since $\mathrm{div}\, G = 0$, we then have

$$\int_{\Omega} G\varphi(u) \cdot \nabla u \, dx = \int_{\Omega} G \cdot \nabla \Phi(u) \, dx = -\int_{\Omega} (\operatorname{div} G) \, \Phi(u) \, dx = 0. \qquad (3.12)$$

We now choose $v = u$ in (3.11). By the hypotheses (3.7), we therefore have:

$$\alpha \|u\|^2_{H^1_0(\Omega)} \leq \int_{\Omega} |f(\cdot, u)u| \, dx.$$

Let us deduce from this last inequality that there exists $R > 0$ such that $\|u\|_{L^2(\Omega)} < R$. This is where we use the hypothesis (3.7h), i.e. $\lim_{s \to \pm\infty} \dfrac{f(x, s)}{s} = 0$. Let us reason by contradiction. Suppose that such an R does not exist. Then there exists a sequence $(u_n)_{n \in \mathbb{N}^\star}$ of elements of $H^1_0(\Omega)$ such that

$$\|u_n\|_{L^2(\Omega)} \geq n \quad \text{and} \quad \alpha \|u_n\|^2_{H^1_0} \leq \int_{\Omega} |f(\cdot, u_n)u_n| \, dx.$$

Let us show that this is impossible. Let us set $v_n = \dfrac{u_n}{\|u_n\|_{L^2(\Omega)}}$. Hence $\|v_n\|_{L^2(\Omega)} = 1$ and

$$\alpha \|v_n\|^2_{H^1_0(\Omega)} \leq \int_{\Omega} \left| \frac{f(\cdot, u_n)}{\|u_n\|_{L^2(\Omega)}} v_n \right| \, dx.$$

But $|f(s)| \leq |d| + C_2|s|$. We therefore have

$$\begin{aligned}
\alpha \|v_n\|^2_{H^1_0(\Omega)} &\leq \int_{\Omega} \frac{|d| + C_2|u_n|}{\|u_n\|_{L^2}} |v_n| \, dx \\
&\leq \int_{\Omega} \frac{|d||v_n|}{\|u_n\|_{L^2}} \, dx + C_2 \int_{\Omega} |v_n|^2 \, dx \\
&\leq \|d\|_{L^2(\Omega)} + C_2.
\end{aligned}$$

The sequence $(v_n)_{n \in \mathbb{N}^\star}$ is thus bounded in $H^1_0(\Omega)$, and therefore, up to a subsequence, $v_n \to v$ in $L^2(\Omega)$ as $n \to +\infty$. Hence, by passing to the limit as $n \to +\infty$, $\|v\|_{L^2} = 1$ (which gives $v \neq 0$). We also have, still up to a subsequence:

$$\begin{aligned}
v_n &\to v \quad \text{a.e.,} \\
|v_n| &\leq H \quad \text{with} \quad H \in L^2(\Omega).
\end{aligned}$$

Finally, using Poincaré's inequality, there exists a C_Ω, depending only on Ω, such that

$$\frac{\alpha}{C_\Omega} = \frac{\alpha}{C_\Omega} \|v_n\|^2_{L^2} \leq \alpha \|v_n\|^2_{H^1_0} \leq \int_{\Omega} \frac{|f(\cdot, u_n)|}{\|u_n\|_{L^2(\Omega)}} |v_n| \, dx.$$

Set

$$X_n = \int_{\Omega} \frac{|f(\cdot, u_n)||v_n|}{\|u_n\|_{L^2(\Omega)}} \, dx.$$

We now show that $X_n \to 0$ as $n \to +\infty$, which is impossible since $X_n \geq \dfrac{\alpha}{C_\Omega} > 0$.

Let us show that $\dfrac{f(\cdot, u_n)|v_n|}{\|u_n\|_{L^2(\Omega)}} \to 0$ a.e. with domination (in $L^1(\Omega)$); we then have by the dominated convergence theorem that $X_n \to 0$ as $n \to +\infty$. We first show the domination. We have

$$\frac{|f(\cdot, u_n)|}{\|u_n\|_{L^2(\Omega)}} \leq \frac{|d| + C_2|u_n|}{\|u_n\|_{L^2}} \leq |d| + C_2|v_n| \leq |d| + C_2 H \quad \text{a.e.,}$$

and so $|\dfrac{f(\cdot, u_n)}{\|u_n\|_{L^2(\Omega)}} v_n| \leq (|d| + C_2 H)H$ a.e. with $(|d| + C_2 H)H \in L^1(\Omega)$.

We now show the a.e. convergence. We have $v_n \to v$ a.e. and therefore there exists an $A \in \mathcal{B}(\mathbb{R}^N)$ such that $\lambda_N(A^c) = 0$ and $v_n(x) \to v(x)$ for all $x \in A$. The conditions on f also allow us to assume that $f(x, \cdot)$ is continuous for all $x \in A$ and $f(x, s) \leq d(x) + s$ for all $x \in A$ and all $s \in \mathbb{R}$.

Let $x \in A$,

1st case: if $v(x) > 0$; $v_n(x) \to v(x)$, but $\lim\limits_{n \to +\infty} \|u_n\|_{L^2(\Omega)} = +\infty$, therefore $u_n(x) = v_n(x)\|u_n\|_{L^2(\Omega)} \to +\infty$. By hypothesis (3.7h), $\lim_{s \to +\infty} \dfrac{f(x, s)}{s} = 0$, and therefore

$$\frac{f(x, u_n(x))}{\|u_n\|_{L^2(\Omega)}} v_n(x) = \frac{f(x, u_n(x))u_n(x)}{u_n(x)\|u_n\|_{L^2(\Omega)}} v_n(x)$$

$$= \frac{f(x, u_n(x))}{u_n(x)} (v_n(x))^2 \to 0 \text{ when } n \to \infty.$$

2nd case: if $v(x) < 0$, we have similarly $\lim\limits_{n \to +\infty} \dfrac{f(x, u_n(x))}{\|u_n\|_{L^2(\Omega)}} v_n(x) = 0$, because $\lim\limits_{s \to -\infty} \dfrac{f(x, s)}{s} = 0$.

3rd case: if $v(x) = 0$, we have

$$\left| \frac{f(x, u_n(x))}{\|u_n\|_{L^2(\Omega)}} v_n(x) \right| \leq \frac{|d(x)| + C_2|u_n(x)|}{\|u_n\|_{L^2(\Omega)}} |v_n(x)|$$

$$\leq (|d(x)| + C_2|v_n(x)|)|v_n(x)|$$

$$\to 0 \quad \text{because} \quad v(x) = 0.$$

In summary, we have $\dfrac{f(\cdot, u_n)}{\|u_n\|_{L^2(\Omega)}} v_n \to 0$ a.e. when $n \to +\infty$. We have thus shown that $\lim_{n \to +\infty} X_n = 0$, in contradiction with $X_n \geq \frac{\alpha}{C_\Omega}$ for all $n \in \mathbb{N}^\star$.

We have therefore shown that there exists an $R > 0$ such that $\big(u = h(t, u)\big) \Rightarrow \big(\|u\|_{L^2(\Omega)} < R\big)$, which concludes the proof of 1. Thus the existence of a solution to (3.5) has been established, and hence Theorem 3.17 is proved. ∎

Remark 3.18 (Proof of the existence theorem (Theorem 3.17) by Schauder). The proof of the existence theorem (Theorem 3.17) by Schauder's theorem is significantly more complicated than in the case when f is bounded, presented in Theorem 3.14. Consider the following linear problem:

$$\int_\Omega a(\bar{u})\nabla u \cdot \nabla v \, dx + \int_\Omega G\varphi(\bar{u}) \cdot \nabla v \, dx = \int_\Omega f(\bar{u})v \, dx \qquad \text{for all } v \in H_0^1(\Omega). \quad (3.13)$$

In equation (3.13), we have used the shorthand notation $a(\bar{u})$ and $f(\bar{u})$ to represent the functions $x \mapsto a(x, \bar{u}(x))$ and $x \mapsto f(x, \bar{u}(x))$, respectively. This abbreviated notation will be frequently used throughout the text.

Let T be the operator defined from $L^2(\Omega)$ to $L^2(\Omega)$ by $T(\bar{u}) = u$, where u is the solution to (3.13). Note that T is continuous and even compact. However, it is difficult to show that T sends a ball of $L^2(\Omega)$ into itself. With this aim, we need to obtain an estimate on u with respect to $\bar{u}$, and this is not easy. Let us take $v = u$ in (3.13), as we did in Section 3.1.2; we obtain, thanks to the assumptions (3.7),

$$\alpha\|u\|_{H_0^1(\Omega)}^2 \leq \| |G| \|_\infty C_1 \|\bar{u}\|_{L^2(\Omega)} \|u\|_{H_0^1(\Omega)}$$
$$+ \|d\|_{L^2(\Omega)} \|u\|_{L^2(\Omega)} + C_2\|\bar{u}\|_{L^2(\Omega)} \|u\|_{L^2(\Omega)}.$$

Let us show that the last term alone prevents us from easily obtaining estimates. Suppose $G = 0$ and $d = 0$. We then have, thanks to Poincaré's inequality:

$$\frac{\alpha}{C_\Omega^2}\|u\|_{L^2(\Omega)}^2 \leq \alpha\|u\|_{H_0^1(\Omega)}^2 \leq C_2\|\bar{u}\|_{L^2(\Omega)} \|u\|_{L^2(\Omega)},$$

that is to say, $\|u\|_{L^2(\Omega)} \leq \dfrac{C_\Omega^2}{\alpha}C_2\|\bar{u}\|_{L^2} \leq \dfrac{C_\Omega^2}{\alpha}C_2 R$ if $\bar{u}$ is in the ball of centre 0 and radius R of $L^2(\Omega)$, i.e. if $\|\bar{u}\|_{L^2(\Omega)} \leq R$. We cannot conclude that $\|u\|_{L^2(\Omega)} < R$ (unless $C_\Omega^2 C_2 < \alpha$, in which case the only solution to (3.5) can be $u = 0$. Indeed, if u is a solution to (3.5), the previous reasoning with $\bar{u} = u$ gives $\|u\|_{L^2(\Omega)} \leq C_\Omega^2 \frac{C_2}{\alpha}\|u\|_{L^2}$ and therefore $u = 0$ because $C_\Omega^2 \frac{C_2}{\alpha} < 1$). The method therefore cannot be applied directly but, with a bit of additional work, it can be used, thanks to the assumption of (3.7b). One solution is, for example, to consider a sequence of problems with truncated right-hand side and to pass to the limit.

It is now natural to wonder if, under the assumptions (3.7) (the assumptions of Theorem 3.17), we can show the uniqueness of the solution. In the case where f does not depend on u and where the equation is linear (i.e. when a does not depend on u and φ is linear), it is enough to take the difference of two solutions as a test function in the two weak formulations associated with these two solutions and to take the differences of the two equations obtained. We thus show that the two solutions are equal (a.e.). In the general case, the situation is more complicated.

Remark 3.19 (Lipschitz-continuity of f does not yield uniqueness). Assuming that f is Lipschitz-continuous (and independent of x) is not sufficient to obtain the uniqueness of the solution to (3.5) (under the assumptions (3.7) which give existence). It is enough to convince oneself of this by considering the following example, where f is Lipschitz-continuous and for which there is no uniqueness. Let us take $a = 1$, $\varphi = 0$ and $f(u) = \lambda u$, where λ is an eigenvalue of $(-\Delta)$ with Dirichlet condition. The problem is then

$$
\begin{cases}
\displaystyle\int_\Omega \nabla u \cdot \nabla v \, dx = \lambda \int_\Omega uv \, dx \quad \forall v \in H_0^1(\Omega), \\
u \in H_0^1(\Omega).
\end{cases}
$$

This problem admits two solutions: $u_1 = 0$, and an eigenfunction $u_2 \neq 0$ associated with λ. Obviously $f(u) = \lambda u$ does not satisfy the condition (3.7h) but we can slightly modify f so that this condition is met, while keeping u_1 and u_2 as solutions as soon as $u_2 \in L^\infty(\Omega)$ (the fact that $u_2 \in L^\infty(\Omega)$ is still true if $N < 6$; this can be proved, for example, from Problem 2.22). Indeed, by taking $\tilde{f}$ which is equal to f on $]-\gamma, \gamma[$ where $\gamma = \|u_2\|_{L^\infty}$ and which is connected to 0 afterwards, in such a way that $\tilde{f} \in C_c(\mathbb{R}, \mathbb{R})$ and $\tilde{f}$ is Lipschitz-continuous, we have the same solutions for $\tilde{f}$, that is to say, for the problem:

$$
\begin{cases}
\displaystyle\int_\Omega \nabla u \cdot \nabla v \, dx = \int_\Omega \tilde{f}(u)v \, dx \quad \forall v \in H_0^1(\Omega), \\
u \in H_0^1(\Omega)
\end{cases}
$$

we thus have 2 solutions with $\lim_{s \to \pm\infty} \frac{\tilde{f}(s)}{s} = 0$ (the hypotheses (3.7) are indeed satisfied for $\tilde{f}$). The fact that f is Lipschitz-continuous does not therefore give uniqueness.

Let us prove uniqueness in the case where f does not depend on s, i.e. $f(x, s) = d(x)$, under the existence assumptions (3.7) and assuming in addition that a and φ are Lipschitz continuous, that is to say:

$$
\exists C_3 > 0 \; ; \; \forall s, s_2 \in \mathbb{R}, \begin{cases} |a(x, s_1) - a(x, s_2)| \leq C_3 |s_1 - s_2| \text{ a.e. } x \in \mathbb{R}, \\ |\varphi(s_1) - \varphi(s_2)| \leq C_3 |s_1 - s_2|. \end{cases} \tag{3.14}
$$

Theorem 3.20 (Existence and uniqueness). *Under the hypotheses* (3.7) *and* (3.14), *there exists a unique solution to* (3.5).

Proof The technique used here to show uniqueness first appeared in an article by Artola in 1986 [5]. Let u_1 and u_2 be two solutions of (3.5). We have:

$$
\int_\Omega a(u_1)\nabla u_1 \cdot \nabla v \, dx + \int_\Omega G\varphi(u_1) \cdot \nabla v \, dx = \int_\Omega fv \, dx, \tag{3.15}
$$

$$
\int_\Omega a(u_2)\nabla u_2 \cdot \nabla v \, dx + \int_\Omega G\varphi(u_2) \cdot \nabla v \, dx = \int_\Omega fv \, dx. \tag{3.16}
$$

The idea is to take $v = T_\varepsilon(u_1 - u_2)$ in (3.15) and (3.16), where $\varepsilon > 0$ and T_ε is the truncation at level ε, that is to say

$$
T_\varepsilon(s) = \begin{cases} -\varepsilon & \text{if } s < -\varepsilon, \\ s & \text{if } -\varepsilon \le s \le \varepsilon, \\ \varepsilon & \text{if } s > \varepsilon. \end{cases}
$$

If u_1 and $u_2 \in H_0^1(\Omega)$, then $T_\varepsilon(u_1 - u_2) \in H_0^1(\Omega)$ and $\nabla T_\varepsilon(u_1 - u_2) = \nabla(u_1 - u_2)\mathbb{1}_{\{0<|u_1-u_2|<\varepsilon\}}$ (this is a simple generalisation of Lemma 2.26). Taking $v = T_\varepsilon(u_1 - u_2)$, and by taking the difference of (3.15) and (3.16), we obtain:

$$
\int_\Omega (a(u_1)\nabla u_1 \cdot \nabla(T_\varepsilon(u_1 - u_2)) - a(u_2)\nabla u_2 \cdot \nabla(T_\varepsilon(u_1 - u_2)))\, dx
$$
$$
= \int_\Omega G(\varphi(u_2) - \varphi(u_1)) \cdot \nabla(T_\varepsilon(u_1 - u_2))\, dx,
$$

which gives, by setting $A_\varepsilon = \{0 < |u_1 - u_2| < \varepsilon\}$,

$$
\int_{A_\varepsilon} a(u_1)\nabla(u_1 - u_2) \cdot \nabla(u_1 - u_2)\, dx = \int_{A_\varepsilon} (a(u_2) - a(u_1))\nabla u_2 \cdot \nabla(u_1 - u_2)\, dx
$$
$$
+ \int_{A_\varepsilon} G(\varphi(u_2) - \varphi(u_1)) \cdot \nabla(u_1 - u_2)\, dx.
$$

By hypothesis, $a(u_1) \ge \alpha$ a.e., and therefore:

$$
\alpha \int_{A_\varepsilon} |\nabla(u_1 - u_2)|^2\, dx \le \int_{A_\varepsilon} C_3 |u_2 - u_1|\, |\nabla u_2|\, |\nabla(u_2 - u_1)|\, dx
$$
$$
+ \int_{A_\varepsilon} C_3 |u_1 - u_2|\, |G|\, |\nabla(u_1 - u_2)|\, dx.
$$

We have $|u_1 - u_2| \le \varepsilon$ a.e. in A_ε. By applying the Cauchy–Schwarz inequality in the last two integrals, we obtain:

$$
\alpha \int_{A_\varepsilon} |\nabla(u_1 - u_2)|^2\, dx \le C_3\, \varepsilon \left(\int_{A_\varepsilon} |\nabla u_2|^2\, dx\right)^{\frac{1}{2}} \left(\int_{A_\varepsilon} |\nabla(u_2 - u_1)|^2\, dx\right)^{\frac{1}{2}}
$$
$$
+ C_3\, \varepsilon \left(\int_{A_\varepsilon} |G|^2\, dx\right)^{\frac{1}{2}} \left(\int_{A_\varepsilon} |\nabla(u_1 - u_2)|^2\, dx\right)^{\frac{1}{2}}.
$$

Therefore, we have

$$
\alpha \left(\int_{A_\varepsilon} (\nabla(u_1 - u_2))^2\, dx\right)^{\frac{1}{2}} \le C_3\, \varepsilon\, a_\varepsilon
$$

with

$$a_\varepsilon = \left(\int_{A_\varepsilon} |G|^2 \, dx \right)^{\frac{1}{2}} + \left(\int_{A_\varepsilon} |\nabla u_2|^2 \, dx \right)^{\frac{1}{2}}$$

or equivalently

$$\alpha \left(\int_\Omega |\nabla (T_\varepsilon(u_1 - u_2))|^2 \, dx \right)^{\frac{1}{2}} = \alpha \| \, |\nabla T_\varepsilon(u_1 - u_2)| \, \|_{L^2(\Omega)} \le C_3 \, \varepsilon \, a_\varepsilon,$$

and therefore, denoting by λ_N the Lebesgue measure on $\mathbb{R}^N$,

$$\frac{\alpha}{\lambda_N(\Omega)^{\frac{1}{2}}} \| \, |\nabla T_\varepsilon(u_1 - u_2))| \, \|_{L^1(\Omega)} \le \alpha \| \, |\nabla T_\varepsilon(u_1 - u_2)| \, \|_{L^2(\Omega)} \le C_3 \, \varepsilon \, a_\varepsilon.$$

Since $H_0^1(\Omega) \subset W_0^{1,1}(\Omega)$, we have $T_\varepsilon(u_1 - u_2) \in H_0^1(\Omega) \subset W_0^{1,1}(\Omega)$ and the Sobolev embedding gives

$$\|T_\varepsilon(u_1 - u_2)\|_{L^{1^\star}} \le \| \, |\nabla T_\varepsilon(u_1 - u_2)| \, \|_{L^1(\Omega)}, \text{ with } 1^\star = \frac{N}{N-1}$$

and therefore

$$\frac{\alpha}{\lambda_N(\Omega)^{\frac{1}{2}}} \|T_\varepsilon(u_1 - u_2)\|_{L^{1^\star}} \le C_3 \, \varepsilon \, a_\varepsilon.$$

If $N = 1$, we have $\frac{N}{N-1} = +\infty$ and we conclude that $u_1 = u_2$ a.e. The case $N \ge 2$ requires a slight additional development. We note that

$$\|T_\varepsilon(u_1 - u_2)\|_{L^{1^\star}} = \left(\int_\Omega |T_\varepsilon(u_1 - u_2)|^{1^\star} \, dx \right)^{\frac{1}{1^\star}}$$

$$\ge \left(\int_{B_\varepsilon} \varepsilon^{1^\star} \, dx \right)^{\frac{1}{1^\star}} \ge \varepsilon \, (\lambda_N(B_\varepsilon))^{\frac{N-1}{N}},$$

where $B_\varepsilon = \{x \; |u_1(x) - u_2(x)| \ge \varepsilon\}$. Hence

$$\varepsilon(\lambda_N(B_\varepsilon))^{\frac{N-1}{N}} \le \frac{(\lambda_N(\Omega))^{\frac{1}{2}}}{\alpha} C_3 \, \varepsilon \, a_\varepsilon$$

so that $(\lambda_N(B_\varepsilon))^{\frac{N-1}{N}} \le C_4 \, a_\varepsilon$. Let us take, for $n \in \mathbb{N}^\star$, $\varepsilon = \frac{1}{n}$. We have $A_{\frac{1}{n+1}} \subset A_{\frac{1}{n}}$ and $\cap_{n \in \mathbb{N}} A_{\frac{1}{n}} = \emptyset$, therefore $\lambda_N(A_{\frac{1}{n}}) \to 0$ when $n \to +\infty$ (by non-increasing continuity of a measure). Recall that

$$a_{\frac{1}{n}} = \left(\int_{A_{\frac{1}{n}}} |G|^2 \, dx \right)^{\frac{1}{2}} + \left(\int_{A_{\frac{1}{n}}} |\nabla u_2|^2 \, dx \right)^{\frac{1}{2}}.$$

Since $G, |\nabla u_2| \in L^2(\Omega)$, we get that $\lim_{n \to +\infty} a_{\frac{1}{n}} = 0$.

We also have $B_{\frac{1}{n+1}} \supset B_{\frac{1}{n}}$ and $\cup_{n\in\mathbb{N}} B_{\frac{1}{n}} = \{|u_1 - u_2| > 0\}$. Therefore $\lim_{n\to+\infty} \lambda_N(B_{\frac{1}{n}}) = \lambda_N\{|u_1 - u_2| > 0\}$ (by non-decreasing continuity of the measure). Since

$$\left(\lambda_N(B_{\frac{1}{n}})\right)^{\frac{N-1}{N}} \le C_4 \, a_{\frac{1}{n}},$$

passing to the limit as n tends to $+\infty$, we obtain $\lambda_N\{|u_1 - u_2| > 0\} \le 0$, and therefore $u_1 = u_2$ a.e., which concludes the proof. $\blacksquare$

3.2 Monotonicity Methods

If in the quasi-linear diffusion problem (3.3), we now consider a right-hand side f that depends on ∇u. We can still prove the existence of a solution with Schauder's theorem: this is the subject of Problem 3.5. The question is more difficult in the case where the mapping a depends on ∇u. More precisely, we place ourselves under the following assumptions:

$$\begin{cases} \Omega \text{ bounded open subset of } \mathbb{R}^N, \\ a \in C(\mathbb{R}^N, \mathbb{R}), \\ \exists \alpha, \beta \in \mathbb{R}_+^* \,; \alpha \le a(\xi) \le \beta, \forall \xi \in \mathbb{R}^N, \\ f \in L^2(\Omega). \end{cases}$$

We seek to prove the existence of a solution to the following problem:

$$\begin{cases} u \in H_0^1(\Omega), \\ \displaystyle\int_\Omega a(\nabla u)\nabla u \cdot \nabla v \, dx = \int_\Omega fv \, dx, \forall v \in H_0^1(\Omega). \end{cases}$$

Can we apply Schauder's theorem? To apply it, it must be used in $H_0^1(\Omega)$ so that $a(\nabla u)$ makes sense. Let $\tilde{u} \in H_0^1(\Omega)$, by the Lax–Milgram theorem, there exists a unique solution $u \in H_0^1(\Omega)$ of:

$$\begin{cases} u \in H_0^1(\Omega), \\ \displaystyle\int_\Omega a(\nabla \tilde{u})\nabla u \cdot \nabla v \, dx = \int_\Omega fv \, dx, \; \forall v \in H_0^1(\Omega). \end{cases} \tag{3.17}$$

Let T be the operator from $H_0^1(\Omega)$ to $H_0^1(\Omega)$ defined by $T(\tilde{u}) = u$, where u is the solution to (3.17). The mapping T is indeed a mapping from $H_0^1(\Omega)$ to $H_0^1(\Omega)$ and

(1) there exists an $R > 0$ s.t. $\|u\|_{H_0^1(\Omega)} \le R$ for all $\tilde{u} \in H_0^1(\Omega)$,

(2) the mapping T is continuous from $H_0^1(\Omega)$ to $H_0^1(\Omega)$. Indeed, if $\tilde{u}_n \to \tilde{u}$ in $H_0^1(\Omega)$, we have $\nabla \tilde{u}_n \to \nabla \tilde{u}$ in $L^2(\Omega)^N$ and it is not very difficult to show that $T(\tilde{u}_n) \to T(\tilde{u})$ in $H_0^1(\Omega)$.

But the mapping T is (in general) *not* compact (from $H_0^1(\Omega)$ to $H_0^1(\Omega)$). If we were in finite dimension, points (1) and (2) would be enough to show the existence of a solution. The idea is then to consider approximate problems in finite dimension and to pass to the limit using the monotonicity of the operator. In the case presented above, this consists of adding the hypothesis $(a(\xi)\xi - a(\eta)\eta) \cdot (\xi - \eta) \geq 0$, which here gives the monotonicity of the operator $u \mapsto \mathrm{div}(a(\nabla u)\nabla u)$ from $H_0^1(\Omega)$ to $H^{-1}(\Omega)$, see Remark 3.24.

3.2.1 Leray–Lions Operators

We consider here a somewhat simplified case of the Leray–Lions[6] operators, whose study is exposed in a famous article by the authors in 1965 [36]. We place ourselves under the following assumptions:

$$\Omega \text{ bounded open subset of } \mathbb{R}^N, \ N \geq 1, \ 1 < p < +\infty, \tag{3.18a}$$

$$a \ : \ \mathbb{R}^N \to \mathbb{R}^N \text{ continuous}, \tag{3.18b}$$

$$(\text{coercivity}) \ \exists \alpha > 0 \ | \ a(\xi) \cdot \xi \geq \alpha |\xi|^p, \ \forall \xi \in \mathbb{R}^N, \tag{3.18c}$$

$$(\text{growth}) \ \exists C \in \mathbb{R} \ |a(\xi)| \leq C(1 + |\xi|^{p-1}), \ \forall \xi \in \mathbb{R}^N, \tag{3.18d}$$

$$(\text{monotonicity}) \ (a(\xi) - a(\eta)) \cdot (\xi - \eta) \geq 0 \ \ \ \forall(\xi,\eta) \in (\mathbb{R}^N)^2, \tag{3.18e}$$

$$\sigma \in L^\infty(\Omega) \ \exists \sigma_0 > 0 \ \sigma \geq \sigma_0 \ \ \ \text{a.e.}, \tag{3.18f}$$

$$f \in L^{\frac{p}{p-1}}(\Omega). \tag{3.18g}$$

These assumptions allow us in particular to deal with the "LES" (Large Eddy Simulation) model used in fluid mechanics. We are then interested in the following problem:

$$\begin{cases} -\mathrm{div}(\sigma(x)a(\nabla u(x))) = f(x) \text{ in } \Omega, \\ \\ u = 0 \quad \text{on } \partial\Omega. \end{cases} \tag{3.19}$$

Example 3.21 (*p*-Laplacian and Smagorinsky operator). For $\sigma \equiv 1$ and $a(\xi) = |\xi|^{p-2}\xi$ $(1 < p < +\infty)$, the equation reads $-\mathrm{div}(|\nabla u|^{p-2}\nabla u) = f$. The operator $u \mapsto -\mathrm{div}(|\nabla u|^{p-2}\nabla u)$ is called the *p*-Laplacian. For $p = 2$, we find the classical Laplacian. The case $p = 3$ gives the Smagorinsky operator[7] which appears in the LES model.

---

[6] Jacques-Louis Lions (1928–2001), French mathematician, member of the Academy of Sciences, specialist in the theory of partial differential equations and control theory.

[7] Joseph Smagorinsky (1924–2005), American physicist specialising in turbulence models, dynamic meteorology and climatology.

Remark 3.22 (Leray–Lions Operator, general case). In the general framework of so-called Leray–Lions operators, the term $a(\nabla u)$ of problem (3.19) considered here is replaced by a term of the form $a(x, u, \nabla u)$, and the function a is then a function of three variables.

Let us look for an adequate weak form of (3.19). Note that if $w \in L^p(\Omega)^N$, then the (3.18d) growth hypothesis on a gives

$$|a(w)| \leq C(1 + |w|^{p-1})$$

$$\leq C + C|w|^{p-1} \in L^{\frac{p}{p-1}}(\Omega)$$

because $C \in L^\infty$ and $|w|^{p-1} \in L^{\frac{p}{p-1}}(\Omega) = L^{p'}(\Omega)$ with $p' = \frac{p}{p-1}$ (i.e. $\frac{1}{p} + \frac{1}{p'} = 1$). So if $u \in W_0^{1,p}(\Omega)$, we have $\nabla u \in (L^p(\Omega))^N$ and $a(\nabla u) \in (L^{p'}(\Omega))^N$. Let us then take $v \in W_0^{1,p}(\Omega)$. We have $\nabla v \in L^p(\Omega)^N$. So we have

$$a(\nabla u) \cdot \nabla v = \sum_{i=1}^{N} a_i(\nabla u) D_i v \in L^1(\Omega).$$

It is therefore natural to look for $u \in W_0^{1,p}(\Omega)$ and to take the test functions in $W_0^{1,p}(\Omega)$.

We also recall that if $f \in L^{p'}(\Omega)$, the mapping $v \mapsto \int_\Omega f(x)v(x)\,dx$ is a continuous linear function from $W_0^{1,p}(\Omega)$ to $\mathbb{R}$. It is therefore an element of the (topological) dual space of $W_0^{1,p}(\Omega)$ (this dual is denoted $W^{-1,p'}(\Omega)$). By abuse of language, we still denote this element of $W^{-1,p'}(\Omega)$ by f, that is, for $f \in L^{p'}(\Omega)$, we have

$$\langle f, v \rangle_{W^{-1,p'}(\Omega), W_0^{1,p}(\Omega)} = \int_\Omega f(x)v(x)\,dx \text{ for all } v \in W_0^{1,p}(\Omega).$$

The weak form of (3.19) that we consider is therefore:

$$\begin{cases} u \in W_0^{1,p}(\Omega), \\ \int_\Omega \sigma a(\nabla u) \cdot \nabla v \, dx = <f, v>_{W^{-1,p'}(\Omega), W_0^{1,p}(\Omega)}, \forall v \in W_0^{1,p}(\Omega). \end{cases} \tag{3.20}$$

Remark 3.23 (More general right-hand side). The natural framework of problem (3.20) is therefore not limited to the case $f \in L^{p'}(\Omega)$. We can replace in (3.20) f by any element of $W^{-1,p'}(\Omega)$. We can take for example $f = -\text{div}F$ with $F \in L^{p'}(\Omega)^N$, where $\text{div}F$ is defined by

$$\langle \text{div}F, v \rangle_{W^{-1,p'}(\Omega), W_0^{1,p}(\Omega)} = -\int_\Omega F(x) \cdot \nabla v(x)\,dx.$$

We can then replace $\langle f, v \rangle_{W^{-1,p'}(\Omega), W_0^{1,p}(\Omega)}$ by $\int_\Omega F(x) \cdot \nabla v(x) \, dx$ in (3.20). Under this assumption, the existence and uniqueness theorem given below (Theorem 3.25) remains true.

Remark 3.24 (Monotone operator). The assumptions made on a and σ (assumptions (3.18b)-(3.18f)) give the monotonicity of the operator $u \mapsto A(u) = -\mathrm{div}(\sigma a(\nabla u))$, that is to say, for all $u, v \in W_0^{1,p}(\Omega)$,

$$\langle A(u) - A(v), u - v \rangle_{W^{-1,p'}(\Omega), W_0^{1,p}(\Omega)} \geq 0.$$

This positivity comes from the fact that $\int_\Omega \sigma(a(\nabla u) - a(\nabla v))(\nabla u - \nabla v) \, dx \geq 0$. For the passage to the limit from finite dimension to infinite dimension, it will be essential to prove the equality (3.25).

Theorem 3.25 (Existence and uniqueness). *Under the assumptions* (3.18), *there exists a solution* $u \in W_0^{1,p}(\Omega)$ *to* (3.20). *If in addition a is strictly monotone, that is to say, if*

$$(a(\xi) - a(\eta)) \cdot (\xi - \eta) > 0 \text{ for all } (\xi, \eta) \in (\mathbb{R}^N)^2, \ \xi \neq \eta \qquad (3.21)$$

then the solution is unique.

For the proof of this theorem, we need some classical integration lemmas that we state here.

Lemma 3.26 (Product between convergence and weak convergence). *Let* $1 < p < +\infty$. *Set* $p' = \frac{p}{p-1}$. *We suppose that* $f_n \to f$ *in* $L^p(\Omega)$ *and* $g_n \to g$ *weakly in* $L^{p'}(\Omega)$. *Then*

$$\int_\Omega f_n \, g_n \, dx \to \int_\Omega f \, g \, dx \text{ when } n \to +\infty.$$

See e.g. [26, Proposition 6.81] for its proof.

However, recall that if $f_n \to f$ weakly in L^p and $g_n \to g$ weakly in $L^{p'}$, we do not generally have convergence of $\int_\Omega f_n \, g_n \, dx$ towards $\int_\Omega f g \, dx$.

Lemma 3.27. *If* $a \in C(\mathbb{R}^N, \mathbb{R}^N)$ *satisfies the growth assumption* (3.18d) *and if* $u_n \to u$ *in* $W_0^{1,p}(\Omega)$, *then* $a(\nabla u_n) \to a(\nabla u)$ *in* $L^{p'}(\Omega)^N$.

Lemma 3.27 is proved with Lebesgue's dominated convergence theorem and its partial converse: see for example [26, theorems 6.9 and 6.11].

Since the spaces $W_0^{1,p}(\Omega), L^p(\Omega)$ and $L^{p'}(\Omega)$ are reflexive for $1 < p < +\infty$, by virtue of Theorem 1.22, for any bounded sequence in one of these spaces, we can extract a weakly convergent subsequence in this space. Finally, recall that the spaces $W_0^{1,p}(\Omega), L^p(\Omega)$ and $L^{p'}(\Omega)$ are separable, i.e. they contain a countable dense subset, which will allow us to approximate the problem by finite-dimensional problems.

We also need the following result on coercive operators for the proof of Theorem 3.25:

Lemma 3.28 (Coercive operator in $\mathbb{R}^N$). *Let $T : \mathbb{R}^N \to \mathbb{R}^N$ be a continuous mapping and assume that T is coercive, that is to say,*

$$\frac{T(v) \cdot v}{|v|} \to +\infty \ \textit{when} \ |v| \to +\infty.$$

Let $b \in \mathbb{R}^N$. Then, there exists a $v \in \mathbb{R}^N$ such that $T(v) = b$. The operator T is therefore surjective.

Proof We use Brouwer's topological degree because we are in finite dimension. Set $h(t, v) = tT(v) + (1 - t)v$. For $t = 0$, we have $h(0, v) = v$ (thus $h(0, v) = I$, where I is the operator $v \mapsto v$). For $t = 1$ we have $h(1, v) = T(v)$. To apply the degree, we first note that the mapping $h : [0, 1] \times \mathbb{R}^N \to \mathbb{R}^N$ is continuous (because T is continuous).

We then want to show that there exists an $R > 0$ such that

$$t \in [0, 1], v \in \mathbb{R}^N \text{ and } h(t, v) = b \Rightarrow |v| < R. \tag{3.22}$$

Assume that we have proved (3.22). If necessary, by increasing R, we can also assume that $|b| < R$. Set $B_R = \{x \in \mathbb{R}^N \text{ such that } |x| < R\}$. By homotopy invariance of the degree, we therefore have that $d(h(t, \cdot), B_R, b)$ does not depend on t, and therefore:

$$d(T, B_R, b) = d(\mathrm{Id}, B_R, b).$$

Since $b \in B_R$, we have $d(\mathrm{Id}, B_R, b) = 1$ and therefore $d(T, B_R, b) = 1$. Thus there exists a $v \in B_R$ such that $T(v) = b$.

It remains to prove that there exists an $R > 0$ satisfying (3.22).

Let $t \in [0, 1]$ and $v \in \mathbb{R}^N$ such that $h(t, v) = b$, that is to say, $tT(v) + (1 - t)v = b$. Hence $tT(v) \cdot v + (1 - t)v \cdot v = b \cdot v \leq |b||v|$ and therefore, if $v \neq 0$,

$$t\frac{T(v) \cdot v}{|v|} + (1 - t)|v| = t\frac{T(v) \cdot v}{|v|} + (1 - t)\frac{v \cdot v}{|v|} \leq |b|.$$

Since $\frac{T(w) \cdot w}{|w|} \to +\infty$ when $|w| \to +\infty$, there exists an $R > 0$ such that

$$|w| \geq R \Rightarrow \min\left(\frac{T(w) \cdot w}{|w|}, |w|\right) > |b|.$$

Hence $|v| < R$, which concludes the proof.

∎

This lemma generalises to any finite-dimensional space:

Lemma 3.29 (Coercive operator in finite dimension). *Let E be a finite-dimensional space, and let $T : E \to E'$ be a continuous mapping (note that $\dim E' = \dim E < +\infty$). Assume that T is coercive, that is to say:*

$$\frac{\langle T(v), v \rangle_{E',E}}{\|v\|_E} \to +\infty \text{ when } \|v\|_E \to +\infty.$$

Then, for all $b \in E'$ there exists a $v \in E$ such that $T(v) = b$.

Proof We seek to reduce to $\mathbb{R}^N$. Let $N = \dim E$; we choose a basis of E, denoted $(e_1, \ldots, e_N)$, and introduce $(e_1^\star, \ldots, e_N^\star)$ the dual basis of E' (that is to say, such that $\langle e_i^\star, e_j \rangle_{E',E} = \delta_{ij}$). Define a mapping I from $\mathbb{R}^N$ to E and a mapping J from $\mathbb{R}^N$ to E' by

$$I(\alpha) = \sum_{i=1}^N \alpha_i e_i \text{ and } J(\beta) = \sum_{i=1}^N \beta_i e_i^\star \text{ for } \alpha, \beta \in \mathbb{R}^N.$$

The operator I is a linear bijection from $\mathbb{R}^N$ to E and the operator J is a linear bijection from $\mathbb{R}^N$ to E'. Let $\tilde{T} = J^{-1} \circ T \circ I$. The operator $\tilde{T}$ is continuous from $\mathbb{R}^N$ to $\mathbb{R}^N$.

Let $\alpha \in \mathbb{R}^N$. Set $v = I(\alpha)$ and $\beta = \tilde{T}(\alpha)$ (so $\beta = J^{-1}(T(v))$). We then have

$$\tilde{T}(\alpha) \cdot \alpha = \beta \cdot \alpha = \sum_{i=1}^N \beta_i \alpha_i = \langle \sum_{j=1}^N \beta_j e_j^\star, \sum_{i=1}^N \alpha_i e_i \rangle_{E',E}$$

$$= \langle J(\beta), I(\alpha) \rangle_{E',E} = J(\beta)(\sum_{i=1}^N \alpha_i e_i) = \langle T(v), v \rangle_{E',E}.$$

Taking as norm on $\mathbb{R}^N$, $\|\alpha\| = \|I(\alpha)\|_E$, the assumption of coercivity on T then gives

$$\lim_{\|\alpha\| \to +\infty} \frac{\tilde{T}(\alpha) \cdot \alpha}{\|\alpha\|} = +\infty.$$

By equivalence of norms in finite dimension, we also have

$$\lim_{|\alpha| \to +\infty} \frac{\tilde{T}(\alpha) \cdot \alpha}{|\alpha|} = +\infty.$$

We can therefore apply Lemma 3.28 to $\tilde{T}$.

Let $b \in E'$; we set $\beta = J^{-1}(b)$. Lemma 3.28 provides the existence of $\alpha \in \mathbb{R}^N$ such that $\tilde{T}(\alpha) = \beta$. Set $v = I(\alpha)$. We then have $T(v) = T \circ I(\alpha) = J \circ \tilde{T}(\alpha) = J(\beta) = b$. We have thus proved the existence of v in E such that $T(v) = b$. $\blacksquare$

Proof of Theorem 3.25

Step I Existence of the solution to a problem in finite dimension

The space $W_0^{1,p}(\Omega)$ is separable and therefore there exists a countable family $(f_n)_{n\in\mathbb{N}^\star}$ that is dense in $W_0^{1,p}(\Omega)$. Let $E_n = \text{Vect}\{f_1 \ldots f_n\}$ be the vector space generated by the first n functions of this family. Then $\dim E_n \leq n$ and $E_n \subset E_{n+1}$ for all $n \in \mathbb{N}^\star$ and we have $\overline{\cup_{n\in\mathbb{N}}E_n} = W_0^{1,p}$. Hence for all $v \in W_0^{1,p}(\Omega)$ there exists a sequence $(v_n)_{n\in\mathbb{N}^\star}$ such that $v_n \in E_n$ for all $n \in \mathbb{N}^\star$ and $v_n \to v$ in $W_0^{1,p}(\Omega)$ when $n \to +\infty$.

In this first step, we fix $n \in \mathbb{N}^\star$ and seek u_n solution of the following problem, posed in finite dimension:

$$\begin{cases} u_n \in E_n, \\ \displaystyle\int_\Omega \sigma a(\nabla u_n) \cdot \nabla v \, dx = \langle f, v\rangle_{W^{-1,p'},W_0^{1,p}}, & \forall v \in E_n. \end{cases} \tag{3.23}$$

The mapping $b_n : v \mapsto \langle f, v\rangle_{W^{-1,p'},W_0^{1,p}}$ is a linear mapping from E_n to $\mathbb{R}$; since $\dim E_n < +\infty$, it is therefore also continuous. Therefore, $b_n \in E_n'$ and

$$\langle b_n, v\rangle_{E_n',E_n} = \langle f, v\rangle_{W^{-1,p'},W_0^{1,p}}.$$

Let $u \in E_n$. We denote by $T_n(u)$ the mapping from E_n to $\mathbb{R}$ which associates to $v \in E_n$, $\int_\Omega \sigma a(\nabla u) \cdot \nabla v \, dx$. This mapping is linear, therefore it also belongs to E_n' and

$$\langle T_n(u), v\rangle_{E_n',E_n} = \int_\Omega \sigma a(\nabla u) \cdot \nabla v \, dx.$$

Thus, we have defined a function T_n from E_n to E_n'. Owing to Lemma 3.29, if T_n is continuous and coercive, then it is surjective, and therefore there exists a $u_n \in E_n$ satisfying $T_n(u_n) = b_n$, that is to say, u_n is a solution to the problem (3.23).

I-1 Continuity of T_n

Recall that n is fixed. To simplify the notation, we forget here the index n, that is to say, we set $E = E_n$ and $T = T_n$. We equip E with the norm defined by $\|\cdot\|_E = \|\cdot\|_{W_0^{1,p}}$. Let $u, \bar{u} \in E$. We have:

$$\|T(u) - T(\bar{u})\|_{E'} = \max_{v\in E,\, \|v\|_E=1} \langle T(u) - T(\bar{u}), v\rangle_{E',E}$$

$$= \max_{v\in E,\, \|v\|_{W_0^{1,p}}=1} \int_\Omega \sigma(a(\nabla u) - a(\nabla \bar{u})) \cdot \nabla v \, dx$$

$$\leq \max_{v\in W_0^{1,p},\, \|v\|_{W_0^{1,p}}=1} \int_\Omega \sigma(a(\nabla u) - a(\nabla \bar{u})) \cdot \nabla v \, dx.$$

Set $\beta = \|\sigma\|_{L^\infty(\Omega)}$, we then obtain

$$\|T(u) - T(\bar{u})\|_{E'} \leq \max_{v \in W_0^{1,p},\, \|v\|_{W_0^{1,p}}=1} \beta \| |a(\nabla u) - a(\nabla \bar{u})| \|_{L^{p'}(\Omega)} \| |\nabla v| \|_{L^p(\Omega)}$$

$$\leq \beta \| |a(\nabla u) - a(\nabla \bar{u})| \|_{L^{p'}(\Omega)}.$$

Therefore, if $(u_n)_{n \in \mathbb{N}}$ is a sequence in E such that $u_n \to \bar{u}$ in E, we have

$$\|T(u_n) - T(\bar{u})\|_{E'} \leq \beta \| |a(\nabla u_n) - a(\nabla \bar{u})| \|_{L^{p'}(\Omega)}.$$

By Lemma 3.27, we have $a(\nabla u_n) \to a(\nabla \bar{u})$ in $(L^{p'}(\Omega))^N$. We have thus shown that $T(u_n) \to T(\bar{u})$ in E', and therefore that T is continuous (from E to E').

I-2 Coercivity of T_n

Recall that we have set $E = E_n$ and $T = T_n$. We want to show that

$$\frac{\langle T(u), u\rangle_{E',E}}{\|u\|_E} \to +\infty \quad \text{when } \|u\|_E \to +\infty.$$

By definition, and thanks to the hypotheses (3.18),

$$\langle T(u), u\rangle_{E',E} = \int_\Omega \sigma \, a(\nabla u) \cdot \nabla u \ \mathrm{d}x \geq \sigma_0 \, \alpha \int_\Omega |\nabla u|^p \ \mathrm{d}x.$$

Therefore, we have

$$\langle T(u), u\rangle_{E',E} \geq C\|u\|_{W_0^{1,p}}^p = C\|u\|_E^p \quad \text{with } C = \sigma_0 \, \alpha.$$

Finally,

$$\frac{\langle T(u), u\rangle_{E',E}}{\|u\|_E} \geq C\|u\|_E^{p-1} \to +\infty \quad \text{when } \|u\|_E \to +\infty,$$

because we have assumed $p > 1$ (the case $p = 1$ is more difficult and requires additional tools, but it is interesting in geometry for the problem of minimal surfaces for example).

We have thus shown that T is coercive; we can therefore apply Lemma 3.29. It gives the existence of a solution to the finite-dimensional problem (3.23) (this technique also allows us, for example, to prove the existence of a solution for problem (3.20) with an approximation by $P1$ finite elements, see for example [2] for a presentation of this type of method).

Step II Existence of the solution to the problem (3.20)

We have shown the existence of a solution to problem (3.23). We now attempt (and succeed!) a passage to the limit in this problem when $n \to +\infty$, to show the existence of a solution to problem (3.20). To this end,

(1) we obtain an *estimate* on u_n, which yields some *compactness*,

(2) we perform a *passage to the limit* on problem (3.23) in order to show the existence of a solution u of problem (3.20) as the limit of the solutions u_n of problem (3.23); for this,

(3) we need to use a trick to show that the *limit of the non-linear term* is indeed the term we want.

II-1 Estimate on u_n

Taking $v = u_n$ in (3.23) yields

$$\sigma_0\,\alpha \int_\Omega |\nabla u_n|^p \, dx \le \|f\|_{W^{-1,p'}} \|u_n\|_{W_0^{1,p}}$$

by coercivity of a. Hence: $\sigma_0\,\alpha\|u_n\|_{W_0^{1,p}}^p \le \|f\|_{W^{-1,p'}}\|u_n\|_{W_0^{1,p}}$, from which

$$\|u_n\|_{W_0^{1,p}}^{p-1} \le \frac{1}{\sigma_0\alpha}\|f\|_{W^{-1,p'}}.$$

II-2 Passage to the limit

The sequence $(u_n)_{n\in\mathbb{N}}$ is therefore bounded in $W_0^{1,p}(\Omega)$, which is reflexive. Hence there exists a subsequence still denoted $(u_n)_{n\in\mathbb{N}}$ such that $u_n \to u$ weakly in $W_0^{1,p}(\Omega)$. By hypothesis, $|a(\nabla u_n)| \le C(1 + |\nabla u_n|^{p-1})$, therefore the sequence $(a(\nabla u_n))_{n\in\mathbb{N}}$ is bounded in $(L^{p'}(\Omega))^N$, which is reflexive. Hence there exists a $\zeta \in (L^{p'}(\Omega))^N$ such that, up to a subsequence,

$$a(\nabla u_n) \to \zeta \quad \text{weakly in} \quad (L^{p'}(\Omega))^N.$$

Let $v \in W_0^{1,p}(\Omega)$. We know that $\overline{\bigcup_{n\in\mathbb{N}} E_n} = W_0^{1,p}$, hence there exists $(v_n)_{n\in\mathbb{N}}$ s.t. $v_n \in E_n$ for all $n \in \mathbb{N}^\star$ and

$$v_n \to v \text{ in } W_0^{1,p}(\Omega),$$

$$\nabla v_n \to \nabla v \text{ in } (L^p(\Omega))^N.$$

Then using (3.23) with $v = v_n$ we obtain:

$$\int_\Omega \sigma a(\nabla u_n) \cdot \nabla v_n \, dx = \langle f, v_n\rangle_{W^{-1,p'},W_0^{1,p}}.$$

But $\langle f, v_n\rangle_{W^{-1,p'},W_0^{1,p}} \to \langle f, v\rangle_{W^{-1,p'},W_0^{1,p}}$, because v_n converges weakly towards v in $W_0^{1,p}(\Omega)$. Moreover $a(\nabla u_n) \to \zeta$ weakly in $(L^{p'}(\Omega))^N$ and $\nabla v_n \to \nabla v$ in $(L^p(\Omega))^N$. Therefore by Lemma 3.26, we have

$$\int_\Omega \sigma\zeta \cdot \nabla v \, dx = \langle f, v\rangle_{W^{-1,p'},W_0^{1,p}}, \quad \forall v \in W_0^{1,p}(\Omega). \tag{3.24}$$

We have thus proved the existence of $u \in W_0^{1,p}(\Omega)$ such that u is the weak limit in $W_0^{1,p}(\Omega)$ of the sequence $(u_n)_{n\in\mathbb{N}}$ and such that the weak limit in $(L^{p'}(\Omega))^N$ of the sequence $(a(\nabla u_n))_{n\in\mathbb{N}}$, denoted ζ, satisfies (3.24). If ζ were equal to $a(\nabla u)$, we would be finished, and this would be easy to establish if a were linear. Indeed, in this case, there exists a square matrix of size N, denoted A, such that $a(\xi) = A\xi$ for all $\xi \in \mathbb{R}^N$ (and therefore $p = p' = 2$). Since $u_n \to u$ in $\mathcal{D}^\star(\Omega)$ we also have $D_i u_n \to D_i u$ in $\mathcal{D}^\star(\Omega)$ (for all i). Since the sequence $(D_i u_n)_{n \in \mathbb{N}}$ is bounded in $L^2(\Omega)$, $D_i u_n \to D_i u$ weakly in $L^2(\Omega)$ (for all i) and therefore $A\nabla u_n \to A\nabla u$ weakly in $L^2(\Omega)^N$, which proves that $\zeta = a(\nabla u)$. Unfortunately, the situation is more complicated when a is non-linear.

II-3 Limit of the non-linear term

Finally, it remains to prove that

$$\int_\Omega \sigma \zeta \cdot \nabla v \, dx = \int_\Omega \sigma \, a(\nabla u) \cdot \nabla v \, dx \quad \text{for all } v \in W_0^{1,p}(\Omega). \tag{3.25}$$

To this end, we begin by studying the limit of $\displaystyle\int_\Omega \sigma \, a(\nabla u_n) \cdot \nabla u_n \, dx$. The trick is to use the equation! Indeed $\displaystyle\int_\Omega \sigma \, a(\nabla u_n) \cdot \nabla u_n \, dx = \langle f, u_n \rangle_{W^{-1,p'}, W_0^{1,p}} \to \langle f, u \rangle_{W^{-1,p'}, W_0^{1,p}}$ because $u_n \to u$ in $W_0^{1,p}(\Omega)$ weakly. But we know (from the previous step) that u satisfies (3.24), and therefore: $\langle f, u \rangle_{W^{-1,p'}, W_0^{1,p}} = \displaystyle\int_\Omega \sigma \zeta \cdot \nabla u \, dx$. Therefore

$$\lim_{n \to +\infty} \int_\Omega \sigma \, a(\nabla u_n) \cdot \nabla u_n \, dx = \langle f, u \rangle_{W^{-1,p'}, W_0^{1,p}}$$

$$= \int_\Omega \sigma \zeta \cdot \nabla u \, dx.$$

For the rest of the proof of (3.25), depending on the hypotheses used, we distinguish between the methods of Minty and of Leray–Lions (which are however very similar):

(a) Minty's[8] trick, which only uses the monotonicity of a, that is $(a(\xi) - a(\eta)) \cdot (\xi - \eta) \geq 0$. In this case only $u_n \to u$ weakly in $W_0^{1,p}(\Omega)$.

(b) The method of Leray–Lions, which uses the strict monotonicity of a, that is $(a(\xi) - a(\eta)) \cdot (\xi - \eta) > 0$ for all $\xi, \eta, \ \xi \neq \eta$. We then also have $u_n \to u$ in $W_0^{1,p}(\Omega)$.

II-3-a First method: Minty's Trick

Let $v \in W_0^{1,p}(\Omega)$; there exists $(v_n)_{n \in \mathbb{N}}$ such that $v_n \in E_n$ for all $n \in \mathbb{N}$ and $v_n \to v$ in $W_0^{1,p}(\Omega)$ as $n \to +\infty$. Thanks to the monotonicity assumption on a (and $\sigma > 0$),

$$0 \leq \int_\Omega \sigma(a(\nabla u_n) - a(\nabla v_n)) \cdot (\nabla u_n - \nabla v_n) \, dx$$

---

[8] George James Minty Jr. (1929–1986), American mathematician, specialised in analysis and discrete mathematics.

$$= \int_\Omega \sigma\, a(\nabla u_n) \cdot \nabla u_n \, dx - \int_\Omega \sigma\, a(\nabla u_n) \cdot \nabla v_n \, dx - \int_\Omega \sigma a(\nabla v_n) \cdot \nabla u_n \, dx$$

$$+ \int_\Omega \sigma\, a(\nabla v_n) \cdot \nabla v_n \, dx$$

$$= T_{1,n} - T_{2,n} - T_{3,n} + T_{4,n}.$$

We saw in (c)(i) that $T_{1,n} = \displaystyle\int_\Omega \sigma\, a(\nabla u_n) \cdot \nabla u_n \, dx \to \int_\Omega \sigma \zeta \cdot \nabla u \, dx$ when $n \to +\infty$.
We have

$$\lim_{n \to +\infty} T_{2,n} = \lim_{n \to +\infty} \int_\Omega \sigma\, a(\nabla u_n) \cdot \nabla v_n \, dx = \int_\Omega \sigma \, \zeta \cdot \nabla v \, dx$$

by product of a convergence in $(L^p)^N$ and a weak convergence in $(L^{p'}(\Omega)^N)$
(Lemma 3.26). Similarly,

$$\lim_{n \to +\infty} T_{3,n} = \lim_{n \to +\infty} \int_\Omega \sigma a(\nabla v_n) \cdot \nabla u_n \, dx = \int_\Omega \sigma\, a(\nabla v) \cdot \nabla u \, dx$$

by product of a convergence in $(L^{p'}(\Omega)^N)$ and a weak convergence in $(L^p(\Omega))^N$.
Finally, we also have

$$\lim_{n \to +\infty} T_{4,n} = \lim_{n \to +\infty} \int_\Omega \sigma\, a(\nabla v_n) \cdot \nabla v_n \, dx = \int_\Omega \sigma\, a(\nabla v) \cdot \nabla v \, dx$$

and this last term is the simplest because we have the product of a convergence in
$(L^{p'}(\Omega)^N)$ and a convergence in $(L^p(\Omega))^N$.

The limit in the inequality therefore gives:

$$\int_\Omega \sigma(\zeta - a(\nabla v)) \cdot (\nabla u - \nabla v) \, dx \geq 0 \text{ for all } v \in W_0^{1,p}(\Omega).$$

We now cleverly choose the test function v by setting $v = u + \dfrac{1}{n} w$, with $w \in W_0^{1,p}(\Omega)$
and $n \in \mathbb{N}^\star$. We thus obtain:

$$-\frac{1}{n} \int_\Omega \sigma \left(\zeta - a(\nabla u + \frac{1}{n}\nabla w) \right) \cdot \nabla w \, dx \geq 0$$

and therefore

$$\int_\Omega \sigma \left(\zeta - a(\nabla u + \frac{1}{n}\nabla w) \right) \cdot \nabla w \, dx \leq 0.$$

But $u + \dfrac{1}{n}\, w \to u$ in $W_0^{1,p}(\Omega)$, so by Lemma 3.27,

$$a\left(\nabla u + \frac{1}{n}\nabla w \right) \to a(\nabla u) \text{ in } (L^{p'}(\Omega))^N.$$

Passing to the limit as $n \to +\infty$, we then obtain

$$\int_\Omega \sigma(\zeta - a(\nabla u)) \cdot \nabla w \ dx \leq 0 \quad \forall w \in W_0^{1,p}(\Omega).$$

By linearity (we can change w to $-w$), we therefore have: $\int_\Omega \sigma(\zeta - a(\nabla u)) \cdot \nabla w \ dx = 0$, $\forall w \in W_0^{1,p}(\Omega)$. Hence

$$\int_\Omega \sigma \, \zeta \cdot \nabla w \ dx = \int_\Omega \sigma \, a(\nabla u) \cdot \nabla w \ dx \quad \forall w \in W_0^{1,p}(\Omega).$$

We have therefore indeed proved that u is a solution to (3.20).

Note that we showed this result by approximation, i.e., by first proving the existence of a solution to an approximate problem posed in finite dimensions, then by passing to the limit. This is also possible using an approximate problem obtained with numerical schemes, for example with a numerical scheme using $P1$ finite elements.

II-3-b Second method: Leray–Lions Trick

We now assume the strict monotonicity of a, that is to say:

$$(a(\xi) - a(\eta)) \cdot (\xi - \eta) > 0 \ \text{ if } \ \xi \neq \eta.$$

Let us show that u is a solution to (3.20) (we already know this, thanks to the first method) but also that $a(\nabla u) = \zeta$ and in particular that $u_n \to u$ in $W_0^{1,p}(\Omega)$.

Since $\overline{\cup_{n \in \mathbb{N}} E_n} = W_0^{1,p}$, there exists a sequence $(v_n)_{n \in \mathbb{N}}$ such that $v_n \in E_n$ for all $n \in \mathbb{N}$ and $v_n \to u$ in $W_0^{1,p}(\Omega)$. By the assumption of monotonicity, we have:

$$\int_\Omega \sigma(a(\nabla u_n) - a(\nabla v_n)) \cdot (\nabla u_n - \nabla v_n) \geq 0.$$

With the same reasoning as for Minty (but now $v_n \to u$ instead of $v_n \to v$), we have:

$$\lim_{n \to +\infty} \int_\Omega \sigma(a(\nabla u_n) - a(\nabla v_n)) \cdot (\nabla u_n - \nabla v_n) \ dx = \int_\Omega \sigma(\zeta - a(\nabla u)) \cdot \underbrace{(\nabla u - \nabla u)}_{= 0} \ dx,$$

therefore if we set $F_n(x) = \sigma(a(\nabla u_n) - a(\nabla v_n)) \cdot (\nabla u_n - \nabla v_n)$, we have $F_n \geq 0$ a.e. and $\int_\Omega F_n \ dx \to 0$ when $n \to +\infty$. Therefore $F_n \to 0$ in L^1, and therefore, up to a subsequence, $F_n \to 0$ a.e. We can also assume that, up to a subsequence, $\nabla v_n \to \nabla u$ a.e.

Let us now show that, up to a subsequence, $\nabla u_n \to \nabla u$ a.e. Let $x \in \Omega$ such that $\sigma(x) \geq \sigma_0$, $F_n(x) \to 0$ (therefore we have $a(\nabla u_n(x)) - a(\nabla v_n(x)) \cdot (\nabla u_n(x) - \nabla v_n(x)) \to 0$) and $\nabla v_n(x) \to \nabla u(x)$ as $n \to +\infty$. We also assume that for this value of x, the assumptions on a (growth, coercivity and monotonicity) are satisfied and that $\nabla v_n(x)$ converges (in $\mathbb{R}$). These assumptions on x only remove a set of measure zero points.

1. The sequence $(\nabla u_n(x))_{n\in\mathbb{N}}$ is bounded in $\mathbb{R}^N$. Indeed, thanks to the assumptions (3.18c) (3.18d) of coercivity and growth on a, we have

$$\frac{1}{\sigma_0}F_n(x) \geq (a(\nabla u_n(x)) - a(\nabla v_n(x)) \cdot (\nabla u_n(x) - \nabla v_n(x))$$

$$\geq \alpha|\nabla u_n(x)|^p - C(1 + |\nabla u_n(x)|^{p-1})|\nabla v_n(x)|$$

$$- C(1 + |\nabla v_n(x)|^{p-1})|\nabla u_n(x)| + \alpha|\nabla v_n(x)|^p.$$

Hence the sequence $(F_n(x))_{n\in\mathbb{N}}$ is unbounded if the sequence $(\nabla u_n(x))_{n\in\mathbb{N}}$ is unbounded. However, $F_n(x) \to 0$ when $n \to +\infty$; the sequence $(\nabla u_n(x))_{n\in\mathbb{N}}$ is therefore bounded.

2. Let $\xi \in \mathbb{R}^N$ be the limit of a subsequence of the sequence $(\nabla u_n(x))_{n\in\mathbb{N}}$, also denoted as $(\nabla u_n(x))_{n\in\mathbb{N}}$. Since x is fixed, this subsequence satisfies $\lim_{n\to+\infty} a(\nabla u_n(x)) = a(\xi)$. Since $\lim_{n\to+\infty} F_n(x) = 0$, it follows that

$$(a(\xi) - a(\nabla u(x)) \cdot (\xi - \nabla u(x)) = 0.$$

However, the first term of this equality is positive if $\xi \neq \nabla u(x)$, and therefore $\xi = \nabla u(x)$. Thus (without extracting a subsequence for this step) $\nabla u_n(x) \to \nabla u(x)$ when $n \to +\infty$.

We have therefore shown that $\nabla u_n \to \nabla u$ a.e. (up to a subsequence, since we extract a subsequence to have $F_n \to 0$ a.e. and $\nabla v_n \to \nabla u$ a.e.). It follows that $a(\nabla u_n) \to a(\nabla u)$ a.e. However, we already know that $a(\nabla u_n) \to \zeta$ in $(L^{p'})^N(\Omega)$ weakly. We then deduce that $\zeta = a(\nabla u)$ thanks to the following integration lemma (see [26, Problem 6.18] for the proof):

Lemma 3.30 (Convergence by dominated norm). *Let Ω be an open bounded subset and suppose that $f_n \to f$ a.e. and that $(f_n)_{n\in\mathbb{N}}$ is bounded in L^q, $q > 1$. Then $f_n \to f$ in $L^r(\Omega)$ for all r such that $1 \leq r < q$.*

Owing to Lemma 3.30, the sequence $(a(\nabla u_n))_{n\in\mathbb{N}}$ converges in $L^r(\Omega)^N$, for $1 \leq r < p'$, towards $a(\nabla u)$. Since this same sequence converges weakly in $(L^{p'})^N(\Omega)$ towards ζ, we can conclude (for example by using the uniqueness of the weak limit in $L^r(\Omega)^N$) that the limits are equal, that is

$$\zeta = a(\nabla u) \text{ a.e.}$$

It remains to show that $u_n \to u$ in $W_0^{1,p}(\Omega)$. We already know that $u_n \to u$ weakly in $W_0^{1,p}(\Omega)$ and therefore in $L^p(\Omega)$. Since, up to a subsequence, $\nabla u_n \to \nabla u$ a.e., Lemma 3.30 gives us $\nabla u_n \to \nabla u$ in $L^r(\Omega)^N$ for all $1 \leq r < p$ and therefore $u_n \to u$ in $W_0^{1,r}(\Omega)$ for all $1 \leq r < p$. By reasoning by contradiction we even have $u_n \to u$ in $W_0^{1,r}(\Omega)$ for all $1 \leq r < p$ without extracting a subsequence. But this does not give the convergence in $W_0^{1,p}(\Omega)$. To prove this convergence in $W_0^{1,p}(\Omega)$, we reuse step (c)(i) common to Minty and Leray–Lions. This step has given

$$\lim_{n \to +\infty} \int_\Omega \sigma a(\nabla u_n) \cdot \nabla u_n \, dx = \int_\Omega \sigma \zeta \cdot \nabla u \, dx.$$

But we now know that $\zeta = a(\nabla u)$ a.e., so we have

$$\lim_{n \to +\infty} \int_\Omega \sigma a(\nabla u_n) \cdot \nabla u_n \, dx = \int_\Omega \sigma a(\nabla u) \cdot \nabla u \, dx.$$

We now recall a classical integration lemma (see [26, problem 4.25]), valid in any measure space, but given here only in the case that interests us.

Lemma 3.31 (Convergence L^1 for a sequence of non-negative functions). *Let Ω be an open bounded subset of $\mathbb{R}^N$. Let $f \in L^1(\Omega)$ and assume that the sequence $(f_n)_{n \in \mathbb{N}}$ of functions from $L^1(\Omega)$ satisfies:*

1. *$f_n \geq 0$ a.e., for all $n \in \mathbb{N}$,*
2. *$f_n \to f$ a.e. as $n \to +\infty$,*
3. *$\lim_{n \to +\infty} \int_\Omega f_n(x) \, dx = \int_\Omega f(x) \, dx$.*

Then $f_n \to f$ in $L^1(\Omega)$.

We already know that $\nabla u_n \to \nabla u$ a.e.

We then apply Lemma 3.31 to the sequence $(f_n)_{n \in \mathbb{N}}$ defined by $f_n = \sigma a(\nabla u_n) \cdot \nabla u_n$. Lemma 3.31 gives the $L^1(\Omega)$ convergence of this sequence and therefore the equi-integrability[9] of the sequence $(f_n)_{n \in \mathbb{N}}$. With the coercivity assumption on a and the assumption on σ, we obtain the equi-integrability of the sequence $(|\nabla u_n|^p)_{n \in \mathbb{N}}$. It remains to apply Vitali's[10] theorem[11] to conclude that $\nabla u_n \to \nabla u$ in $L^p(\Omega)^N$ and therefore that $u_n \to u$ in $W_0^{1,p}(\Omega)$. As usual, a contradiction argument shows that this convergence in $W_0^{1,p}(\Omega)$ occurs without extraction of subsequence as soon as u_n converges weakly to u in $W_0^{1,p}(\Omega)$ (and to have this convergence, we had to extract a subsequence). In step III below, we prove the uniqueness of the solution to (3.20) if a is strictly monotone. This allows us to conclude that the whole sequence $(u_n)_{n \in \mathbb{N}}$ converges in $W_0^{1,p}(\Omega)$ towards the unique solution to (3.20).

---

[9] Let $(X, \mathcal{T}, m)$ be a measure space; we say that a sequence $(f_n)_{n \in \mathbb{N}} \subset L^1(X, \mathcal{T}, m)$ is equi-integrable if, for all $\varepsilon > 0$, there exists a $\delta > 0$ such that

$$A \in \mathcal{T}, n \in \mathbb{N}, m(A) \leq \delta \Rightarrow \int_A |f_n| \, dm \leq \varepsilon.$$

[10] Giuseppe Vitali (1875–1932), Italian mathematician, specialist in measure theory.

[11] **Vitali's Theorem:** Let $(X, \mathcal{T}, m)$ be a measure space; let L^1 denote the space $L^1_{\mathbb{R}}(X, T, m)$. Let $(f_n)_{n \in \mathbb{N}}$ be a sequence in L^1 such that $f_n \to f$ a.e., f taking its values in $\mathbb{R}$. Then, $f \in L^1$ and $f_n \to f$ in L^1 if and only if the following two conditions are met:

1. the sequence $(f_n)_{n \in \mathbb{N}}$ is equi-integrable.
2. For all $\varepsilon > 0$, there exists a $C \in \mathcal{T}$ such that $m(C) < +\infty$ and $\int_{C^c} |f_n| \, dm \leq \varepsilon$ for all $n \in \mathbb{N}$.

(See for example [26, Theorem 4.51] for the proof.)

We have thus completed the proof of the existence of a solution to problem (3.20). It remains to show the uniqueness of the solution.

Step III: Uniqueness

Assume that a is strictly monotone. Let u_1 and u_2 be two solutions,

$$\int_\Omega \sigma a(\nabla u_i) \cdot \nabla v \, dx = \langle f, v \rangle_{W^{-1,p'}, W_0^{1,p}} \quad i = 1, 2 \quad \forall v \in W_0^{1,p}(\Omega).$$

We take the difference of the two equations, and we take $v = u_1 - u_2$. We get:

$$\int_\Omega \sigma(a(\nabla u_1) - a(\nabla u_2)) \cdot \nabla(u_1 - u_2) \, dx = 0.$$

But $Y = \sigma(a(\nabla u_1) - a(\nabla u_2)) \cdot \nabla(u_1 - u_2) \geq 0$ and $Y > 0$ if $\nabla u_1 \neq \nabla u_2$; so we have $\nabla u_1 = \nabla u_2$ a.e. and therefore $u_1 = u_2$ because u_1 and $u_2 \in W_0^{1,p}(\Omega)$. ∎

Remark 3.32. In the case of Leray–Lions operators, when a depends on x and ∇u but also on u, we can still show uniqueness results if $p \leq 2$.

3.3 Method by Minimisation of a Functional

We saw in Chapter 2 that it is sometimes possible to obtain a solution to a linear elliptic problem by looking for a point where a functional reaches its minimum. Such a method is also possible for non-linear elliptic equations. This type of method is very well described in the book [34]. Two problems are proposed to illustrate this method, namely Problems 3.8 and 3.9. Problem 3.9 uses Lagrange's theorem of existence[12] multipliers, which we state (and prove) here in a simple case (sufficient for Problem 3.9).

Let us start by recalling (without proof) the implicit function theorem, which we will need later.

Theorem 3.33 (Implicit functions). *Let E be a normed vector space and F, G two Banach spaces. Let $f \in C(E \times F, G)$ and $(a, b) \in E \times F$ such that $f(a, b) = 0$ and assume that*

- *for all $x \in E$, the mapping $f_x : y \mapsto f(x, y)$ belongs to $C^1(F, G)$. Let $d_2 f(x, y)$ be the differential of f_x at point y,*
- *the mapping $(x, y) \mapsto d_2 f(x, y)$ belongs to $C(E \times F, \mathcal{L}(F, G))$,*
- *the mapping $d_2 f(a, b)$ is a continuous bijection from F to G (and therefore has a continuous inverse).*

---

[12] Joseph Louis de Lagrange (in Italian Giuseppe Luigi Lagrangia), 1736–1813, was an Italian mathematician, mechanic and astronomer, naturalised French.

Then, there exist $\varepsilon > 0$ and $\eta > 0$ such that for all $x \in B(a,\eta)$ (with $B(a,\eta)$ the open ball of centre 0 and radius η), there exists one and only one $y \in B(b,\varepsilon)$ (open ball of centre 0 and radius ε) such that $f(x,y) = 0$. Let $\phi(x)$ be this value of y. The function ϕ is of class C^1.

Theorem 3.34 (Lagrange Multiplier). *Let E be a Banach space, $f \in C(E,\mathbb{R})$, $g \in C^1(E,\mathbb{R})$ and $A = \{v \in E; \, g(v) = 0\}$. Let $u \in A$ such that $f(u) \le f(v)$ for all $v \in A$ and assume that f is differentiable at point u and that $\mathrm{Im}(dg(u)) = \mathbb{R}$. Then, denoting by $df(u)$ and $dg(u)$ the differentials of f and g at point u, there exists a $\lambda \in \mathbb{R}$ such that $df(u) = \lambda\, dg(u)$.*

Proof Note that $dg(u)$ and $df(u)$ are elements of E'. Since $\mathrm{Im}(dg(u)) = \mathbb{R}$, there exists a $v \in E$ such that $dg(u)(v) = 1$. Define the function G from $E \times \mathbb{R}$ to $\mathbb{R}$ by $G(w,t) = g(u + w + tv)$. Since $g \in C^1(E,\mathbb{R})$, $G \in C^1(E \times \mathbb{R}, \mathbb{R})$. Let $d_1 G$ (resp. $d_2 G$) be the differential of G with respect to the first variable (resp. second variable); thus, $d_1 G(w,t) \in \mathcal{L}(E,\mathbb{R}) = E'$ and $d_2 G(w,t) \in \mathcal{L}(\mathbb{R},\mathbb{R})$, which we identify with $\mathbb{R}$. Since $d_2 G(0,0) = dg(u)(v) \ne 0$, the implicit function theorem gives the existence of $\varepsilon > 0$ and $\eta > 0$ such that for all $w \in B(0,\eta)$ (open ball centred at 0 with radius η), there exists one and only one $t \in]-\varepsilon, \varepsilon[$ such that $G(w,t) = 0$. We denote by $\phi(w)$ this value of t. The implicit function theorem also gives that ϕ is of class C^1 (note that, for example, $d\phi(0) \in E'$). Since $g(u + w + \phi(w)v) = 0$ for all $w \in B(0,\eta)$, a first order development of g (at point u) and of ϕ (at point 0) gives,[13] since $dg(u)(v) = 1$,

$$dg(u)(w + d\phi(0)(w)v) = 0 \text{ and therefore } dg(u)(w) + d\phi(0)(w) = 0, \qquad (3.26)$$

from which we get

$$d\phi(0)(w) = -dg(u)(w) \text{ for all } w \in E.$$

For all $w \in B(0,\eta)$, $f(u) \le f(u + w + \phi(w)v)$ (because $u + w + \phi(w)v \in A$) and therefore with a first order development of f and ϕ (recall that f is differentiable at point u) $df(u)(w + d\phi(0)(w)v) = 0$, and then

$$0 = df(u)(w) - dg(u)(w)df(u)(v) \text{ for all } w \in E.$$

This gives $df(u) = \lambda\, dg(u)$ with $\lambda = df(u)(v)$. ∎

Remark 3.35 (Generalisation of Theorem 3.34). In Theorem 3.34, the function g has values in $\mathbb{R}$. We can generalise the theorem by taking a function g with values in $\mathbb{R}^p$ for $p > 1$. We then need to replace $\mathrm{Im}(dg(u)) = \mathbb{R}$ with $\mathrm{Im}(dg(u)) = \mathbb{R}^p$ and the conclusion becomes *there exist $\lambda_1, \ldots, \lambda_p \in \mathbb{R}$ such that $df(u) = \sum_{i=1}^{p} \lambda_i dg_i(u)$ where $g_1, \ldots, g_p$ are the p components of g.* The proof of this result is the subject of Problem 3.12.

---

[13] to obtain (3.26) one can, for example, use the first order developments with sw instead of w, divide by s ($s > 0$) and let s tend to 0.

3.4 Problem Set

Problem 3.1 (Compact mapping and compact perturbation of identity ($\star$)). *Solution on page 212.* Let E be a Banach space, and assume that $g : E \to E$ is compact and that $g = \mathrm{Id} - f$ with f compact. Show that E is of finite dimension.

Problem 3.2 (Counterexample to the fixed point theorem ($\star$)). *Solution on page 213.* Consider the Hilbert space ℓ^2 (see Problem 1.9). For $x = (x_n)_{n \in \mathbb{N}} \in \ell^2$, we define $T(x)$ by

$$T(x) = (1 - \|x\|_2, x_0, x_1, \ldots, x_n, \ldots).$$

Recall that $\|x\|_2 = \left(\sum_{n \in \mathbb{N}} x_n^2 \right)^{\frac{1}{2}}$. Let B be the closed unit ball of ℓ^2.

1. Show that T is a continuous mapping that sends B to B.
2. Show that T does not have a fixed point in B.
   What is the assumption violated in Brouwer's fixed point theorem (Theorem 3.5)? And in Schauder's fixed point theorem (Theorem 3.11)?

Problem 3.3 (Degree of an affine mapping ($\star$)). *Solution on page 213.* Let E be a (real) Banach space. For $R > 0$, we set $B_R = \{v \in E \mid \|v\|_E < R\}$.

1. Let f be a constant mapping from E to E. There exists therefore $a \in E$ such that $f(v) = a$ for all $v \in E$. Let $R > 0$ such that $\|a\|_E \neq R$. Show that $d(\mathrm{Id} - f, B_R, 0)$ is well defined and that $d(\mathrm{Id} - f, B_R, a) = 1$ if $R > \|a\|_E$ and $d(\mathrm{Id} - f, B_R, a) = 0$ if $R < \|a\|_E$.
2. Let L be a compact linear mapping from E to E and assume that 1 is not an eigenvalue of L. Let $a \in E$. Define f from E to E by setting $f(v) = Lv + a$ for all $v \in E$.

   a. Show that the equation $u - f(u) = 0$ has at most one solution.
   b. Show that the equation $u - f(u) = 0$ has a unique solution. Let b be this solution. Show that

   $$d(\mathrm{Id} - f, B_R, 0) \neq 0 \text{ if } R > \|b\|_E \text{ and}$$
   $$d(\mathrm{Id} - f, B_R, 0) = 0 \text{ if } R < \|b\|_E.$$

Problem 3.4 (Existence by Schauder ($\star\star\star$)). *Solution on page 214.* Let Ω be an open bounded subset of $\mathbb{R}^N$ ($N \geq 1$), $g \in L^2(\Omega)$, a a continuous function from $\mathbb{R}$ to $\mathbb{R}$ and h a continuous function from $\mathbb{R} \times \mathbb{R}^N$ to $\mathbb{R}$. Assume that

- $0 < \alpha = \inf_{s \in \mathbb{R}} a(s) \leq \sup_{s \in \mathbb{R}} a(s) = \beta < +\infty$,
- there exist $\delta \in [0, 1[$ and $C_1 \in \mathbb{R}$ such that $|h(s, \xi)| \leq C_1(1 + |s|^\delta + |\xi|^\delta)$ for all $(s, \xi) \in \mathbb{R} \times \mathbb{R}^N$.

1. *Existence and uniqueness for the linearised problem* – Let $\bar{u} \in H_0^1(\Omega)$. Show that $h(\bar{u}, \nabla\bar{u}) \in L^2(\Omega)$ and that there exists a unique solution u of

$$u \in H_0^1(\Omega),$$

$$\int_\Omega a(\bar{u}(x))\nabla u(x) \cdot \nabla v(x)\, dx + \int_\Omega h(\bar{u}(x), \nabla\bar{u}(x))v(x)\, dx \qquad (3.27)$$

$$= \int_\Omega g(x)v(x)\, dx, \ \forall v \in H_0^1(\Omega).$$

In the following, let T denote the mapping which associates u to $\bar{u}$, the unique solution to (3.27). The mapping T is therefore from $H_0^1(\Omega)$ to $H_0^1(\Omega)$.

2. *Estimate on the solution u of the linear problem* – Let $\bar{u} \in H_0^1(\Omega)$ and $u = T(\bar{u})$. Show that there exists a C_2 depending only on Ω, α, g and C_1 such that

$$\|u\|_{H_0^1(\Omega)} \leq C_2(1 + \|\bar{u}\|_{H_0^1(\Omega)}^\delta).$$

[Take $v = u$ in (3.27).] Deduce that there exists an $R \in \mathbb{R}_+^\star$ such that

$$\|\bar{u}\|_{H_0^1(\Omega)} \leq R \Rightarrow \|u\|_{H_0^1(\Omega)} \leq R.$$

3. *Continuity of the operator T* – Let $(\bar{u}_n)_{n\in\mathbb{N}}$ be a sequence in $H_0^1(\Omega)$ and $\bar{u} \in H_0^1(\Omega)$ and assume that $\bar{u}_n \to \bar{u}$ in $H_0^1(\Omega)$ as $n \to +\infty$. Set $f_n = h(\bar{u}_n, \nabla\bar{u}_n)$, $f = h(\bar{u}, \nabla\bar{u})$, $u_n = T(\bar{u}_n)$ and $u = T(\bar{u})$.

   a. Show that $f_n \to f$ in $L^2(\Omega)$.
   b. Show that $u_n \to u$ weakly in $H_0^1(\Omega)$.
   c. Show that $\displaystyle\int_\Omega a(\bar{u}_n(x))\nabla u_n(x) \cdot \nabla u_n(x)\, dx \to \int_\Omega a(\bar{u}(x))\nabla u(x) \cdot \nabla u(x)\, dx$.
   Deduce that $u_n \to u$ in $H_0^1(\Omega)$. [One may be inspired by Problem 2.19.]

4. *Compactness of the operator T* – Let $(\bar{u}_n)_{n\in\mathbb{N}}$ be a bounded sequence in $H_0^1(\Omega)$. Set $f_n = h(\bar{u}_n, \nabla\bar{u}_n)$ and $u_n = T(\bar{u}_n)$. Show that there exists a subsequence of the sequence $(\bar{u}_n)_{n\in\mathbb{N}}$, still denoted $(\bar{u}_n)_{n\in\mathbb{N}}$, and that there exist $f \in L^2(\Omega)$ and $b \in L^\infty(\Omega)$ such that

   $$f_n \to f \text{ weakly in } L^2(\Omega),$$
   $$a(\bar{u}_n) \to b \text{ a.e.}$$

   Show that the sequence $(u_n)_{n\in\mathbb{N}}$ is convergent in $H_0^1(\Omega)$. [One may reason as in Problem 2.19.]

5. *Existence for the non-linear problem* – Show that there exists a $u \in H_0^1(\Omega)$ such that $u = T(u)$ (and therefore u is a solution to (3.27) with $\bar{u} = u$).

Problem 3.5 (Existence by Schauder, generalisation of Problem 3.4 ($\star\star\star$)). *Solution on page 218.* Let Ω be an open bounded subset of $\mathbb{R}^N$ ($N \geq 1$), a a function from $\Omega \times \mathbb{R}$ to $\mathbb{R}$ and f a function from $\Omega \times \mathbb{R} \times \mathbb{R}^N$ to $\mathbb{R}$ satisfying:

- a is measurable with respect to $x \in \Omega$, for all $s \in \mathbb{R}$, and continuous with respect to $s \in \mathbb{R}$, a.e. in $x \in \Omega$.
- There exist $\alpha, \beta \in \mathbb{R}_+^\star$ such that $\alpha \leq a(x, s) \leq \beta$ for all $s \in \mathbb{R}$ and a.e. in $x \in \Omega$.
- f is measurable with respect to $x \in \Omega$, for all $s, p \in \mathbb{R} \times \mathbb{R}^N$, and continuous with respect to $(s, p) \in \mathbb{R} \times \mathbb{R}^N$, a.e. in $x \in \Omega$.
- There exist $d \in L^2(\Omega)$, $C \in \mathbb{R}$ and $\delta \in [0, 1[$ such that $|f(\cdot, s, p)| \leq C(d + |s|^\delta + |p|^\delta)$ a.s., for all $s, p \in \mathbb{R} \times \mathbb{R}^N$.

Show that there exists a unique solution u to the following problem:

$$\begin{cases} u \in H_0^1(\Omega) \ : \ \forall v \in H_0^1(\Omega), \\ \displaystyle\int_\Omega a(x, u(x)) \nabla u(x) \cdot \nabla v(x)\, dx = \int_\Omega f(x, u(x), \nabla u(x)) v(x)\, dx. \end{cases} \tag{3.28}$$

[One could construct a mapping from $H_0^1(\Omega)$ to $H_0^1(\Omega)$ and use Schauder's theorem.]

Problem 3.6 (Convection-diffusion, Dirichlet, existence ($\star\star\star\star$)). *Solution on page 222.* Let Ω be an open bounded subset of $\mathbb{R}^N$, $N = 2$ or 3, $p > N$, $W \in L^P(\Omega)^N$, φ a Lipschitz-continuous function from $\mathbb{R}$ to $\mathbb{R}$ such that $\varphi(0) = 0$ and $f \in L^2(\Omega)$. We are interested here in the following problem

$$\begin{cases} -\Delta u + \operatorname{div}(W\varphi(u)) = f \text{ in } \Omega, \\ u = 0 \text{ on } \partial\Omega. \end{cases} \tag{3.29}$$

The aim of this problem is to show the existence of a weak solution to problem (3.29). The uniqueness (and positivity if $f \geq 0$ a.e.) of the weak solution is shown in Problem 3.7.

1. Let $u \in H_0^1(\Omega)$. Show that $W\varphi(u) \in L^2(\Omega)^N$. [Use Sobolev's embedding theorem, Theorem 1.41.]

This first question allows us to define the weak formulation of problem (3.29). It consists in looking for a solution u to the following problem.

$$\begin{cases} u \in H_0^1(\Omega), \ : \ \forall v \in H_0^1(\Omega), \\ \displaystyle\int_\Omega \nabla u(x) \cdot \nabla v(x)\, dx - \int_\Omega \varphi(u(x)) W(x) \cdot \nabla v(x)\, dx = \int_\Omega f(x) v(x)\, dx. \end{cases} \tag{3.30}$$

To show the existence of a solution to (3.30) we use a topological degree argument, constructing a mapping h from $[0, 1] \times L^q(\Omega)$ to $L^q(\Omega)$ with $q = \frac{2p}{p-2}$ (so that $\frac{1}{p} + \frac{1}{q} = \frac{1}{2}$). We therefore set for the following questions $q = \frac{2p}{p-2}$. If $N = 3$, we set $2^\star = 6$ and if $N = 2$, we choose for $2^\star$ a number strictly greater than q (so that, for $N = 2$ or 3, $H_0^1(\Omega)$ is continuously embedded into $L^{2^\star}(\Omega)$ and compactly into $L^q(\Omega)$).

2. (Construction of the operators B and h) Let $\tilde{u} \in L^q$. Show that there exists a unique solution u to

$$u \in H_0^1(\Omega), \ : \ \forall v \in H_0^1(\Omega),$$

$$\int_\Omega \nabla u(x) \cdot \nabla v(x) \, dx - \int_\Omega \varphi(\tilde{u}(x)) W(x) \cdot \nabla v(x) \, dx = \int_\Omega f(x) v(x) \, dx. \quad (3.31)$$

Let B be the operator which associates to $\tilde{u}$ in $L^q(\Omega)$ the solution u to (3.31). Then, for $t \in [0, 1]$ and $\tilde{u} \in L^q(\Omega)$, we set $h(t, \tilde{u}) = B(t\tilde{u})$.

3. Show that h is continuous and compact from $[0, 1] \times L^q(\Omega)$ to $L^q(\Omega)$.
4. (*a priori* estimates) Let $u \in L^q(\Omega)$ s.t. $u = h(t, u)$; then

$$u \in H_0^1(\Omega), \ : \ \forall v \in H_0^1(\Omega),$$

$$\int_\Omega \nabla u(x) \cdot \nabla v(x) \, dx - \int_\Omega \varphi(t\, u(x)) W(x) \cdot \nabla v(x) \, dx = \int_\Omega f(x) v(x) \, dx. \quad (3.32)$$

For $s \in \mathbb{R}$, we set $\psi(s) = \int_0^s \frac{1}{(1+|\xi|)^2} \, d\xi$.

a. Show that $\psi(u) \in H_0^1(\Omega)$. Taking $v = \psi(u)$ in (3.32), show that there exists a C_ℓ depending only on Ω, W, φ and f such that

$$\|\ln(1 + |u|)\|_{H_0^1(\Omega)} \le C_\ell.$$

b. For $v \in L^{2^\star}(\Omega)$, show that for all $A \ge 0$ we have

$$\int_\Omega |v(x)|^q \, dx \le \left(\int_\Omega |v(x)|^{2^\star} \, dx \right)^{\frac{q}{2^\star}} \lambda_N(\{|v| \ge A\})^{1 - \frac{q}{2^\star}} + A^q \lambda_N(\Omega).$$

Recall that λ_N is the Lebesgue measure on the Borel sets of $\mathbb{R}^N$.
c. Using a. and b., show that there exists a $C > 0$, depending only on Ω, W, φ and f such that $\|u\|_{H_0^1(\Omega)} < C$.

5. (Topological degree) Show the existence of a solution to (3.30).
6. In this question, we remove the assumption $\varphi(0) = 0$ and we consider an element T from $H^{-1}(\Omega)$. Show that there exists a solution u to (3.30) with $\int_\Omega f(x) v(x) \, dx$ replaced by $\langle T, v \rangle_{H^{-1}(\Omega), H_0^1(\Omega)}$.

Problem 3.7 (Convection-diffusion, Dirichlet, uniqueness ($\star\star\star$)). *Solution on page 226.* We continue here with the same assumptions as in Problem 3.6, that is, we consider an open bounded subset Ω of $\mathbb{R}^N$, $N = 2$ or 3, $p > N$, $W \in L^p(\Omega)^N$, a Lipschitz-continuous function φ from $\mathbb{R}$ to $\mathbb{R}$ such that $\varphi(0) = 0$ and $f \in L^2(\Omega)$. Problem 3.6 showed that there exists a weak solution u of (3.29), that is, a solution u of

$$\begin{cases} u \in H_0^1(\Omega), \ : \ \forall v \in H_0^1(\Omega), \\ \int_\Omega \nabla u(x) \cdot \nabla v(x) \, dx - \int_\Omega \varphi(u(x)) W(x) \cdot \nabla v(x) \, dx = \int_\Omega f(x) v(x) \, dx. \end{cases} \quad (3.33)$$

The aim of this problem is to prove the uniqueness of the solution to (3.33) and to show that $u \geq 0$ a.e. if $f \geq 0$ a.e.

1. Demonstrate the uniqueness of the solution to (3.33).
2. In this question (and only in this question) we remove the assumption $\varphi(0) = 0$ and we consider an element T from $H^{-1}(\Omega)$. Show that the problem (3.30) with $\int_\Omega f(x)v(x)\,dx$ replaced by $\langle T, v \rangle_{H^{-1}(\Omega), H_0^1(\Omega)}$ has a unique solution (the existence was shown in Problem 3.6).
3. Assume $f \leq 0$ a.e. Let u be the solution to (3.33). Show that $u \leq 0$ a.e. [*For $n \in \mathbb{N}^\star$, take $v = S_n(u)$ in (3.33) with $S_n \in C(\mathbb{R}, \mathbb{R})$ defined by $S_n(s) = \max(0, \min(s, \frac{1}{n}))$ and let n tend to $+\infty$.*]

Problem 3.8 (Existence by minimisation ($\star\star\star$)). *Solution on page 229.* Let Ω be an open bounded subset of $\mathbb{R}^N$ ($N \geq 1$), a a function from Ω to $\mathbb{R}$ and f a function from $\Omega \times \mathbb{R}$ to $\mathbb{R}$ satisfying:

- $a \in L^\infty(\Omega)$.
- There exists an $\alpha \in \mathbb{R}_+^\star$ such that $\alpha \leq a$ a.e.
- f is measurable with respect to $x \in \Omega$, for all $s \in \mathbb{R}$, and continuous with respect to $s \in \mathbb{R}$, a.e. in $x \in \Omega$.
- There exist $d \in L^2(\Omega)$, $C \in \mathbb{R}$ and $\delta \in [0, 1[$ such that for almost all $x \in \Omega$, $|f(\cdot, s)| \leq C|s|^\delta + d$ for all $s \in \mathbb{R}$.

Define $F(x, s) = \int_0^s f(x, t)\,dt$. For almost all $x \in \Omega$, the function $s \mapsto F(x, s)$ is therefore a C^1 class function from $\mathbb{R}$ to $\mathbb{R}$.

1. Let $u \in H_0^1(\Omega)$; show that $F(\cdot, u) \in L^1(\Omega)$. [*Recall that $F(\cdot, u)$ denotes the function $x \mapsto F(x, u(x))$ which is defined a.e. Saying that it belongs to $L^1(\Omega)$ is to be understood in the sense that there exists $v \in \mathcal{L}^1(\Omega)$ such that $v = F(\cdot, u)$ a.e.*]

For $u \in H_0^1(\Omega)$, we define $E(u) = \frac{1}{2} \int_\Omega a(x)\nabla u(x) \cdot \nabla u(x)\,dx - \int_\Omega F(x, u(x))\,dx$.

2. Show that $E(u) \to +\infty$ when $\|u\|_{H^1(\Omega)} \to +\infty$.
3. Let $\eta = \inf\{E(u), u \in H_0^1(\Omega)\}$. Show that $\eta \in \mathbb{R}$ and that there exists a $u \in H_0^1(\Omega)$ such that $E(u) = \eta$.
4. Show that there exists a solution u of the following problem:

$$\begin{cases} u \in H_0^1(\Omega), \ : \ \forall v \in H_0^1(\Omega), \\ \displaystyle\int_\Omega a(x)\nabla u(x) \cdot \nabla v(x)\,dx = \int_\Omega f(x, u(x))v(x)\,dx. \end{cases} \tag{3.34}$$

Problem 3.9 (Minimisation with constraint ($\star\star\star\star$)). *Solution on page 232.* Let Ω be an open bounded subset of $\mathbb{R}^N$, $N \geq 2$, and $p \in\,]1, \frac{N+2}{N-2}[$ (with $\frac{N+2}{N-2} = +\infty$ if $N = 2$); we search for a non-zero solution to the following problem:

$$\begin{cases} -\Delta u = |u|^{p-1}u \text{ in } \Omega, \\ u = 0 \text{ on } \partial\Omega. \end{cases} \tag{3.35}$$

1. For $v \in H_0^1(\Omega)$, we set $E(v) = \frac{1}{2} \int_\Omega |\nabla v(x)|^2 \, dx$ and $F(v) = \int_\Omega |v|^{p+1} \, dx$. We also set $A = \{v \in H_0^1(\Omega), F(v) = 1\}$. Show that there exists a $u \in A$ such that $u \geq 0$ a.e. and $E(u) \leq E(v)$ for all $v \in A$.
2. Show that there exists a non-zero weak solution u, $u \geq 0$ a.e., to (3.35).

Problem 3.10 (Weak convergence and non-linearity ($\star\star\star$)). *Solution on page 235.* Preliminary remark: Let $\varphi \in C(\mathbb{R}, \mathbb{R})$; when a sequence $(u_n)_{n\in\mathbb{N}}$ tends weakly towards u in an L^p space and the sequence $\varphi(u_n)$ tends weakly towards f in an L^q space, it is generally false that $f = \varphi(u)$ a.e. We add the assumption that $\int u_n \varphi(u_n) \, dx$ converges towards $\int uf \, dx$ (with $1/p + 1/q = 1$). If φ is non-decreasing, Minty's trick then allows us to show that $f = \varphi(u)$ a.e. If φ is increasing, we even obtain the convergence of u_n towards u (this is the "Leray–Lions trick"). This problem details these ideas in the $p = q = 2$ framework, and with a finite measure.

Let $(X, \mathcal{T}, m)$ be a finite measure space (that is, $m(X) < +\infty$) and denote by L^2 the space $L_\mathbb{R}^2(X, \mathcal{T}, m)$. Let $(u_n)_{n\in\mathbb{N}}$ and $(v_n)_{n\in\mathbb{N}}$ be two bounded sequences of L^2 and $u, v \in L^2$ and assume that the sequences $(u_n)_{n\in\mathbb{N}}$ and $(v_n)_{n\in\mathbb{N}}$ converge weakly in L^2 towards u and v respectively.

$$\lim_{n\to+\infty} \int u_n w \, dm = \int uw \, dm \quad \text{and} \quad \lim_{n\to+\infty} \int v_n w \, dm = \int vw \, dm \text{ for all } w \in L^2.$$

1. Assume, in this question only, that $v_n = u_n$ a.e., for all $n \in \mathbb{N}$ (and therefore $u = v$ a.e.). Show that $u_n \to u$ in L^2 as $n \to +\infty$ if and only if $\int u_n^2 \, dm \to \int u^2 \, dm$ as $n \to +\infty$.

Assume for the rest of the problem that $\int u_n v_n \, dm \to \int uv \, dm$ as $n \to +\infty$ and that there exists a function φ from $\mathbb{R}$ to $\mathbb{R}$ such that

- φ is continuous, and there exists a $C \in \mathbb{R}$ such that $|\varphi(s)| \leq C + C|s|$ for all $s \in \mathbb{R}$.
- $v_n = \varphi(u_n)$ a.e., for all $n \in \mathbb{N}$.

2. Let $w \in L^2$, show that, as $n \to +\infty$,

$$\int (\varphi(u_n) - \varphi(w))(u_n - w) \, dm \to \int (v - \varphi(w))(u - w) \, dm. \tag{3.36}$$

3. Assume that φ is non-decreasing.

   a. Let $\bar{w} \in L^2$ and $t \in \mathbb{R}$. Show that

$$\int (v - \varphi(u + t\bar{w}))t\bar{w} \, dm \leq 0.$$

   [Use (3.36).] Deduce that $\int (v - \varphi(u))\bar{w} \, dm = 0$.
   b. Show that $v = \varphi(u)$ a.e.

4. Assume that φ is increasing. For $n \in \mathbb{N}$, we set $G_n = (\varphi(u_n) - \varphi(u))(u_n - u)$.

   a. Show that $G_n \to 0$ in L^1 as $n \to +\infty$ (use (3.36)).
   b. Show that there exists a subsequence of the sequence $(G_n)_{n \in \mathbb{N}}$, denoted $(G_{\psi(n)})_{n \in \mathbb{N}}$ (with ψ increasing from $\mathbb{N}$ to $\mathbb{N}$) such that $G_{\psi(n)} \to 0$ a.e. Deduce that $u_{\psi(n)} \to u$ a.e. (use the increase of φ).
   c. Show that $u_n \to u$ in L^p for all $p \in [1, 2[$.

Problem 3.11 (Leray–Lions Operator ($\star\star\star\star$)). *Solution on page 236.* Let Ω be an open bounded subset of $\mathbb{R}^N$ ($N \geq 1$), a a function from $\Omega \times \mathbb{R} \times \mathbb{R}^N$ to $\mathbb{R}^N$ and $p \in]1, +\infty[$ satisfying (with $p' = \frac{p}{p-1}$):

- (Carathéodory Function) a is measurable with respect to $x \in \Omega$, for all $s, \xi \in \mathbb{R} \times \mathbb{R}^N$, and continuous with respect to $(s, \xi) \in \mathbb{R} \times \mathbb{R}^N$, a.e. in $x \in \Omega$.
- (Coercivity) There exists an $\alpha \in \mathbb{R}_+^\star$ such that $a(x, s, \xi) \cdot \xi \geq \alpha |\xi|^p$ for all $(s, \xi) \in \mathbb{R} \times \mathbb{R}^N$ and a.e. in $x \in \Omega$.
- (Growth) There exist $d \in L^{p'}(\Omega)$ and $C \in \mathbb{R}$ such that $|a(\cdot, s, \xi)| \leq C(d + |s|^{p-1} + |\xi|^{p-1})$ a.e., for all $s, \xi \in \mathbb{R} \times \mathbb{R}^N$.
- (Strict Monotonicity) $(a(x, s, \xi) - a(x, s, \eta) \cdot (\xi - \eta) > 0$ for all $(s, \xi, \eta) \in \mathbb{R} \times \mathbb{R}^N \times \mathbb{R}^N$, $\xi \neq \eta$, and a.e. in $x \in \Omega$.

Let $f \in W^{-1,p'}(\Omega)$.

1. Show that there exists a solution u to the following problem:

$$\begin{cases} u \in W_0^{1,p}(\Omega) \; : \; \forall v \in W_0^{1,p}(\Omega), \\ \displaystyle\int_\Omega a(x, u(x), \nabla u(x)) \cdot \nabla v(x) \, dx = \langle f, v \rangle_{W^{-1,p'}(\Omega), W_0^{1,p}(\Omega)}. \end{cases} \tag{3.37}$$

   [Reproduce the proof of Theorem 3.25]
2. Show that $u_n \to u$ in $W_0^{1,p}(\Omega)$.

Problem 3.12 (Lagrange Multipliers ($\star\star$)). *Solution on page 240.* Let E be a Banach space, $f \in C(E, \mathbb{R})$, $g \in C^1(E, \mathbb{R}^p)$, $p > 1$ and $A = \{v \in E; g(v) = 0\}$. Let $u \in A$ such that $f(u) \leq f(v)$ for all $v \in A$. We denote by $g_1, \ldots, g_p$ the components of g (and therefore $g_i \in C^1(E, \mathbb{R})$ for $i = 1, \ldots, p$).

Assume that f is differentiable at point u and that $\mathrm{Im}(dg(u)) = \mathbb{R}^p$. Then, denoting by $df(u)$ and $dg_i(u)$ the differentials of f and g_i (for $i = 1, \ldots, p$) at point u, show that there exist $\lambda_1, \ldots, \lambda_p \in \mathbb{R}$ such that $df(u) = \sum_{i=1}^p \lambda_i \, dg_i(u)$.

   [One may adapt to the case $p > 1$ the proof of Theorem 3.34.]

Problem 3.13 (Pohozaev's identity in $\mathbb{R}^N$ ($\star\star\star$)). *Solution on page 241.* Let $g \in C(\mathbb{R}, \mathbb{R})$ and G the primitive of g that vanishes at 0. Let $u \in L^\infty_{\mathrm{loc}}(\mathbb{R}^N)$ ($N \geq 1$) and assume that $G(u) \in L^1(\mathbb{R}^N)$, $|\nabla u| \in L^2(\mathbb{R}^N)$ and

$$-\Delta u = g(u) \text{ in } \mathcal{D}^\star(\mathbb{R}^N).$$

The aim of the problem is to prove Pohozaev's identity, which reads

$$\frac{N-2}{2} \int_{\mathbb{R}^n} |\nabla u(x)|^2 \, dx = N \int_{\mathbb{R}^n} G(u(x)) \, dx. \tag{3.38}$$

1. Show that $u \in H^2_{\mathrm{loc}}(\mathbb{R}^N)$, i.e. that $\varphi u \in H^2(\mathbb{R}^n)$ for all $\varphi \in \mathcal{D}(\mathbb{R}^N)$ (and therefore $u \in H^2(B_R)$ for all $R > 0$, where $B_R = \{x \in \mathbb{R}^N, |x| < R\}$).
2. Show that there exists a sequence $(u_n)_{n \in \mathbb{N}^\star}$ of $C^\infty(\mathbb{R}^N, \mathbb{R})$ such that (recall that $B_R = \{x \in \mathbb{R}^N, |x| < R\}$)

   a. For all $R > 0$ the sequence $(u_n)_{n \in \mathbb{N}^\star}$ is bounded in $L^\infty(B_R)$.
   b. $u_n \to u$ a.e. as $n \to +\infty$.
   c. For all $R > 0$ and all $i, j \in \{1, \ldots, N\}$, $\partial_i u_n \to D_i u$ in $L^2(B_R)$ and $\partial^2_{i,j} u_n \to D^2_{i,j} u_n$ in $L^2(B_R)$ as $n \to +\infty$.

The following question uses the sequence $(u_n)_{n \in \mathbb{N}^\star}$ from question 2.

3. Let $\psi \in C^\infty_c(\mathbb{R}, \mathbb{R})$ such that $\psi(r) = 0$ if $r < 1$. Choose $R > 0$ such that $\psi(r) = 0$ if $r > R$ and set $h_n(x) = \psi(|x|) \sum_{j=1}^N x_j \partial_j u_n(x)$ (so that $h \in \mathcal{D}(\mathbb{R}^N)$).

   a. Show that

   $$\lim_{n \to +\infty} \langle g(u), h_n \rangle_{\mathcal{D}^\star(\mathbb{R}^N), \mathcal{D}(\mathbb{R}^N)} = \lim_{n \to +\infty} \int_{\mathbb{R}^N} g(u(x)) h_n(x) \, dx$$
   $$= -N \int_{\mathbb{R}^N} G(u(x)) \psi(|x|) \, dx - \int_{\mathbb{R}^N} G(u(x)) \psi'(|x|) |x| \, dx. \tag{3.39}$$

   b. Show that

   $$\lim_{n \to +\infty} \langle -\Delta u, h_n \rangle_{\mathcal{D}^\star(\mathbb{R}^N), \mathcal{D}(\mathbb{R}^N)} = \frac{2-N}{2} \int_{\mathbb{R}^N} |\nabla u(x)|^2 \psi(|x|) \, dx +$$
   $$\int_{\mathbb{R}^N} \psi'(|x|) \frac{x_i}{|x|} \sum_{i=1}^N \sum_{j=1}^N D_i u(x) x_j D_j u(x) \, dx - \int_{\mathbb{R}^N} \psi'(|x|) \frac{|x|}{2} |\nabla u(x)|^2 \, dx. \tag{3.40}$$

4. Prove the identity (3.38).

3.5 Problem Set Solutions

Problem 3.1 (Compact mapping and compact perturbation of the identity)

It is enough to notice that $\mathrm{Id} = \mathrm{Id} - f + f$ is then the sum of two compact mappings, therefore compact. Hence the unit ball of E is compact, which characterises a finite-dimensional space.

Problem 3.2 (Counterexample to Brouwer)

1. Let $x = (x_n)_{n\in\mathbb{N}} \in \ell^2$ and $y = (y_n)_{n\in\mathbb{N}} \in \ell^2$.

$$\|T(y) - T(x)\|_2^2 = \left(\|y\|_2 - \|x\|_2\right)^2 + \|x - y\|_2^2 \le 2\,\|x - y\|_2^2.$$

   This proves that T is continuous from ℓ^2 to ℓ^2.
   Then, $\|T(x)\|_2^2 = (1 - \|x\|_2)^2 + \|x\|_2^2 = 1 - 2\,\|x\|_2 + 2\,\|x\|_2^2 \le 1$ if $\|x\|_2 \le 1$.
   Hence $T(B) \subset B$.
2. If $T(x) = x$, then all the x_n are equal, so x is the null sequence and therefore $T(x) \ne 0$. Therefore, there is no solution to the equation $T(x) = x$. Theorem 3.5 does not apply because ℓ^2 is an infinite-dimensional space. Theorem 3.11 does not apply because T is not a compact mapping.

Problem 3.3 (Degree of an affine mapping)

1. It is clear that f is continuous and compact and that $u - f(u) = 0$ if and only if $u = a$. If $\|a\|_E \ne R$, $d(\mathrm{Id} - f, B_R, a)$ is therefore well defined.
   If $R < \|a\|_E$, the equation $u - f(u) = 0$ has no solution in B_R and therefore $d(\mathrm{Id} - f, B_R, 0) = 0$.
   If $R > \|a\|_E$, we set $h(t, v) = tf(v)$. The function h is continuous and compact from $[0, 1] \times E$ to E and the equation $u = tf(u)$ has no solution on ∂B_R for $t \in [0, 1]$ (because the unique solution of $u = tf(u)$ is ta). Hence $d(\mathrm{Id} - f, B_R, 0) = d(\mathrm{Id} - h(1, \cdot), B_R, 0) = d(\mathrm{Id} - h(0, \cdot), B_R, 0) = d(\mathrm{Id}, B_R, 0) = 1$.
2. a. Let $u_1, u_2 \in E$ such that $u_1 - f(u_1) = 0$ and $u_2 - f(u_2) = 0$. Setting $u = u_1 - u_2$ we therefore have $u - Lu = 0$. Since 1 is not an eigenvalue of L, we therefore have $u = 0$, which indeed proves that the equation $u - f(u) = 0$ has at most one solution.
   b. Set $h(t, u) = Lu + ta$. The function h is continuous and compact from $[0, 1] \times E$ to E. Let $R > 0$. The equation $u = h(0, u)$ has no solution on ∂B_R (because 1 is not an eigenvalue of L).

   On the one hand, if the equation $u = h(t, u)$ has no solution for $t \in]0, 1]$ on ∂B_R, we have, owing to Theorem 3.9,

$$d(\mathrm{Id} - f, B_R, 0) = d(\mathrm{Id} - h(1, 0), B_R, 0) = d(\mathrm{Id} - L, B_R, 0) \ne 0.$$

   Therefore, there exists a $u \in B_R$ such that $u - f(u) = 0$.

   On the other hand, if the equation $u = h(t, u)$ has a solution on ∂B_R for a t in $]0, 1]$, let c be this solution and observe that $(c/t) - f(c/t) = 0$.

   In all cases, we have therefore shown that there exists a $u \in E$ such that $u - f(u) = 0$.

   Finally, we note that $d(\mathrm{Id} - f, B_R, 0) = d(\mathrm{Id} - L, B_R, 0) \ne 0$ if $R > \|b\|_E$ and $d(\mathrm{Id} - f, B_R, 0) = 0$ if $R < \|b\|_E$.

Problem 3.4 (Existence by Schauder)

1. We note that $|h(\bar{u}, \nabla\bar{u})| \le C_1(1 + |\bar{u}|^\delta + |\nabla\bar{u}|^\delta) \in L^{\frac{2}{\delta}}(\Omega) \subset L^2(\Omega)$ because $\delta < 1$. For $u, v \in H_0^1(\Omega)$, we set

$$A(u, v) = \int_\Omega a(\bar{u})\nabla u \cdot \nabla v \, dx.$$

   Since $0 < \alpha \le a(\bar{u}) \le \beta$, the form A is continuous and coercive bilinear on $(H_0^1(\Omega))^2$. By the Lax–Milgram theorem (Theorem 2.3), there is therefore indeed a unique solution to (3.27).

2. Taking $v = u$ in (3.27), we obtain

$$\alpha\|u\|_{H_0^1(\Omega)}^2 \le \|g\|_{L^2(\Omega)}\|u\|_{L^2(\Omega)} + C_1|\Omega|^{\frac{1}{2}}\|u\|_{L^2(\Omega)} + \Big(C_1\||\bar{u}|^\delta\|_{L^2(\Omega)}$$

$$+ C_1\||\nabla\bar{u}|^\delta\|_{L^2(\Omega)}\Big)\|u\|_{L^2(\Omega)}.$$

   With C_Ω given by the Poincaré inequality, we get

$$\frac{\alpha}{C_\Omega}\|u\|_{H_0^1(\Omega)} \le \|g\|_{L^2(\Omega)} + C_1\left(|\Omega|^{\frac{1}{2}} + \||\bar{u}|^\delta\|_{L^2(\Omega)} + \||\nabla\bar{u}|^\delta\|_{L^2(\Omega)}\right). \quad (3.41)$$

   Owing to Hölder's inequality (or Jensen's[14] inequality[15] for the (concave) function from $\mathbb{R}_+$ to $\mathbb{R}_+$ defined by $s \mapsto s^\delta$), we obtain

$$\||\bar{u}|^\delta\|_{L^2(\Omega)}^2 = \int_\Omega |\bar{u}|^{2\delta} \, dx \le \Big(\int_\Omega \bar{u}^2 \, dx\Big)^\delta |\Omega|^{1-\delta}, \text{ and} \quad (3.42)$$

$$\||\nabla\bar{u}|^\delta\|_{L^2(\Omega)}^2 = \int_\Omega |\nabla\bar{u}|^{2\delta} \, dx \le \Big(\int_\Omega |\nabla\bar{u}|^2 \, dx\Big)^\delta |\Omega|^{1-\delta}. \quad (3.43)$$

   (If $\delta \in [\frac{1}{2}, 1[$, these two inequalities correspond to the classical embedding of $L^2(\Omega)$ in $L^{2\delta}(\Omega)$.)

   These two upper bounds together with Poincaré's inequality yield the existence of $\bar{C}$ depending only on Ω and δ such that

$$\||\bar{u}|^\delta\|_{L^2(\Omega)} \le \bar{C}\|\bar{u}\|_{H_0^1(\Omega)}^\delta \text{ and } \||\nabla\bar{u}|^\delta\|_{L^2(\Omega)} \le \bar{C}\|\bar{u}\|_{H_0^1(\Omega)}^\delta.$$

   Returning to (3.41), we obtain the existence of C_2 depending only on α, g, Ω and C_1 such that

$$\|u\|_{H_0^1(\Omega)} \le C_2\left(1 + \|\bar{u}\|_{H_0^1(\Omega)}^\delta\right).$$

---

[14] Johan Jensen (1859–1925), Danish mathematician and engineer.

[15] See for instance [26, Problem 4.43].

To conclude, we note that there exists an $R \in \mathbb{R}_+^\star$ such that $R > C_2(1 + R^\delta)$. Indeed, we have

$$R - C_2 - C_2 R^\delta = R\left(1 - \frac{C_2}{R} - C_2 R^{\delta-1}\right).$$

It is therefore sufficient to take $R > 2C_2$ and $R > (2C_2)^{\frac{1}{1-\delta}}$. (This is where the assumption $\delta < 1$ is used.) We then have

$$\|\bar{u}\|_{H_0^1(\Omega)} \le R \Rightarrow \|u\|_{H_0^1(\Omega)} \le C_2(1 + R^\delta) \le R.$$

3. a. If $f_n \not\to f$ in $L^2(\Omega)$, there exist $\varepsilon > 0$ and a subsequence, still denoted $(f_n)_{n\in\mathbb{N}}$, such that

$$\|f_n - f\|_{L^2(\Omega)} \ge \varepsilon \text{ for all } n \in \mathbb{N}. \tag{3.44}$$

We can assume that, up to a subsequence, $\bar{u}_n \to \bar{u}$ a.e., with $|\bar{u}_n| \le G \in L^2(\Omega)$ and that $\nabla\bar{u}_n \to \nabla\bar{u}$ a.e., with $|\nabla\bar{u}_n| \le H \in L^2(\Omega)$. We then have $f_n \to f$ a.e. and $|f_n| \le C_1(1 + |G|^\delta + |H|^\delta)$ a.e. (and for all $n \in \mathbb{N}$). Since $0 \le \delta \le 1$, we have $|G|^\delta + |H|^\delta \in L^2(\Omega)$. The dominated convergence theorem (in $L^2(\Omega)$) then gives $f_n \to f$ in $L^2(\Omega)$, in contradiction with (3.44). Note that it is necessary to reason by contradiction to show that the entire sequence $(f_n)_{n\in\mathbb{N}}$ converges.

   b. Question 2 shows that the sequence $(u_n)_{n\in\mathbb{N}}$ is bounded in $H_0^1(\Omega)$. If $u_n \not\to u$ weakly in $H_0^1(\Omega)$, there exist $\varepsilon > 0$, $\psi \in H^{-1}(\Omega)$ and a subsequence, still denoted $(u_n)_{n\in\mathbb{N}}$, such that

$$|\langle \psi, u_n - u \rangle_{H^{-1}(\Omega), H_0^1(\Omega)}| \ge \varepsilon \text{ for all } n \in \mathbb{N}. \tag{3.45}$$

Then we can assume that, up to a subsequence, there exists a $w \in H_0^1(\Omega)$ such that

$$u_n \to w \quad \text{weakly in } H_0^1(\Omega), \tag{3.46}$$

$$\bar{u}_n \to \bar{u} \quad \text{a.e.} \tag{3.47}$$

Let $v \in H_0^1(\Omega)$; since $u_n = T(\bar{u}_n)$, we have

$$\int_\Omega a(\bar{u}_n)\nabla u_n \cdot \nabla v \, \mathrm{d}x + \int_\Omega f_n v \, \mathrm{d}x = \int_\Omega g v \, \mathrm{d}x.$$

Since $a(\bar{u}_n) \to a(\bar{u})$ a.e. and $|a(\bar{u}_n)| \le \beta$ a.e., we have, by the dominated convergence theorem, $a(\bar{u}_n)\nabla v \to a(\bar{u})\nabla v$ in $L^2(\Omega)^N$. Passing to the limit as $n \to +\infty$ in the previous equality, we therefore obtain

$$\int_\Omega a(\bar{u})\nabla w \cdot \nabla v \, \mathrm{d}x + \int_\Omega f v \, \mathrm{d}x = \int_\Omega g v \, \mathrm{d}x,$$

which proves that $w = u$, in contradiction with (3.45). We have thus shown that $u_n \rightarrow u$ weakly in $H^1_0(\Omega)$ as $n \rightarrow +\infty$.

c. For all $n \in \mathbb{N}$ we have

$$\int_\Omega a(\bar{u}_n)\nabla u_n \cdot \nabla u_n \, dx = \int_\Omega g u_n \, dx - \int_\Omega f_n u_n \, dx.$$

Since $u_n \rightarrow u$ in $L^2(\Omega)$ and $f_n \rightarrow f$ in $L^2(\Omega)$, we get

$$\lim_{n\rightarrow+\infty} \int_\Omega a(\bar{u}_n)\nabla u_n \cdot \nabla u_n \, dx = \int_\Omega g u \, dx - \int_\Omega f u \, dx,$$

and therefore, since $u = T(\bar{u})$,

$$\lim_{n\rightarrow+\infty} \int_\Omega a(\bar{u}_n)\nabla u_n \cdot \nabla u_n \, dx = \int_\Omega a(\bar{u})\nabla u \cdot \nabla u \, dx.$$

We now note that

$$\alpha \, \|u_n - u\|^2_{H^1_0(\Omega)} \leq \int_\Omega a(\bar{u}_n)(\nabla u_n - \nabla u) \cdot (\nabla u_n - \nabla u) \, dx.$$

To show that $u_n \rightarrow u$ in $H^1_0(\Omega)$, it is therefore sufficient to show that

$$\lim_{n\rightarrow+\infty} \int_\Omega a(\bar{u}_n)(\nabla u_n - \nabla u) \cdot (\nabla u_n - \nabla u) \, dx = 0. \tag{3.48}$$

To establish (3.48), first note that a classical proof by contradiction shows that

$$a(\bar{u}_n)\nabla u \rightarrow a(\bar{u})\nabla u \text{ in } L^2(\Omega)^N, \text{ when } n \rightarrow +\infty.$$

Since $\nabla u_n \rightarrow \nabla u$ weakly in $L^2(\Omega)^N$, we get that

$$\lim_{n\rightarrow+\infty} \int_\Omega a(\bar{u}_n)\nabla u \cdot \nabla u_n \, dx = \int_\Omega a(\bar{u})\nabla u \cdot \nabla u \, dx$$
$$\text{and } \lim_{n\rightarrow+\infty} \int_\Omega a(\bar{u}_n)\nabla u \cdot \nabla u \, dx = \int_\Omega a(\bar{u})\nabla u \cdot \nabla u \, dx.$$

This allows us to show (3.48) and therefore conclude that $u_n \rightarrow u$ in $H^1_0(\Omega)$, which proves the continuity of the operator T from $H^1_0(\Omega)$ to $H^1_0(\Omega)$.

4. The sequence $(f_n)_{n\in\mathbb{N}}$ is bounded in $L^2(\Omega)$ and the sequence $(\bar{u}_n)_{n\in\mathbb{N}}$ is bounded in $H^1_0(\Omega)$ and therefore relatively compact in $L^2(\Omega)$. We can therefore assume that, up to a subsequence, there exist $f \in L^2(\Omega)$ and $\zeta \in L^2(\Omega)$ such that

$$\begin{aligned} f_n &\rightarrow f \text{ weakly in } L^2(\Omega), \\ \bar{u}_n &\rightarrow \zeta \text{ a.e.} \end{aligned} \tag{3.49}$$

Setting $b = a(\zeta)$, we therefore have $b \in L^\infty(\Omega)$ and $a(\bar{u}_n) \to b$ a.e. (We can show that $\zeta \in H_0^1(\Omega)$ but it is false to say that $f = h(\zeta, \nabla\zeta)$ a.e.)

Since $\alpha \le b \le \beta$ a.e., there exists a unique solution u of

$$\begin{cases} u \in H_0^1(\Omega), \\ \displaystyle\int_\Omega b\nabla u \cdot \nabla v \, dx = \int_\Omega (g - f)v \, dx \text{ for all } v \in H_0^1(\Omega). \end{cases} \tag{3.50}$$

Let us show that the sequence $(u_n)_{n\in\mathbb{N}}$ converges in $H_0^1(\Omega)$ towards a solution u to (3.50) (we are working here with the extracted subsequence that satisfies (3.49)).

We already know that the sequence $(u_n)_{n\in\mathbb{N}}$ is bounded in $H_0^1(\Omega)$. By reasoning by contradiction, we show that $u_n \to u$ weakly in $H_0^1(\Omega)$. Indeed, since the $(u_n)_{n\in\mathbb{N}}$ is bounded, there exists a $w \in H_0^1(\Omega)$ such that, up to a subsequence, $u_n \to w$ weakly in $H_0^1(\Omega)$. We then have, for all $n \in \mathbb{N}$ and all $v \in H_0^1(\Omega)$,

$$\int_\Omega a(\bar{u}_n)\nabla u_n \cdot \nabla v \, dx = \int_\Omega (g - f_n)v \, dx.$$

By passing to the limit as $n \to +\infty$ in this equation, thanks to the convergences given in (3.49), we obtain

$$\int_\Omega b\nabla w \cdot \nabla v \, dx = \int_\Omega (g - f)v \, dx.$$

This proves that $w = u$ and $u_n \to u$ weakly in $H_0^1(\Omega)$ as $n \to +\infty$. Therefore $u_n \to u$ in $L^2(\Omega)$ as $n \to +\infty$.

It remains to show the convergence of u_n towards u in $H_0^1(\Omega)$. With this aim we first note that

$$\int_\Omega a(\bar{u}_n)\nabla u_n \cdot \nabla u_n \, dx = \int_\Omega (g - f_n)u_n \, dx \to \int_\Omega (g - f)u \, dx \text{ when } n \to +\infty.$$

Since u is a solution to (3.50) we therefore have

$$\lim_{n\to+\infty} \int_\Omega a(\bar{u}_n)\nabla u_n \cdot \nabla u_n \, dx = \int_\Omega b\nabla u \cdot \nabla u \, dx.$$

Using this convergence and (3.49) we then show that

$$\lim_{n\to+\infty} \int_\Omega a(\bar{u}_n)\nabla(u_n - u) \cdot \nabla(u_n - u) \, dx = 0.$$

Since $\alpha \|u_n - u\|_{H_0^1(\Omega)}^2 \le \int_\Omega a(\bar{u}_n)\nabla(u_n - u) \cdot \nabla(u_n - u) \, dx$ we conclude that $u_n \to u$ in $H_0^1(\Omega)$.

5. Here it is sufficient to apply Schauder's theorem. The operator T is continuous and compact from $H_0^1(\Omega)$ to $H_0^1(\Omega)$. There exists an $R > 0$ s.t. T sends the ball

of centre 0 and radius R (from $H_0^1(\Omega)$) into itself. Schauder's theorem then gives that there exists a u in this ball (and therefore in $H_0^1(\Omega)$) such that $u = T(u)$. The function u thus found is a solution to (3.27) with $\bar{u} = u$.

Problem 3.5 (Existence by Schauder, generalisation of Problem 3.4)

We follow the same approach as for Problem 3.4 (the modifications are minor). We construct an operator, denoted by T, from $H_0^1(\Omega)$ to $H_0^1(\Omega)$. We show that, if R is well chosen, T sends the ball of centre 0 and radius R into itself. Finally, we show the continuity and compactness of T and we then conclude by Schauder's theorem (Theorem 3.11).

Construction of T

Let $\bar{u} \in H_0^1(\Omega)$. By choosing a representative of the class of functions $\bar{u}$, the function $x \mapsto a(x, \bar{u}(x))$ is Borel-measurable and only changes on a set of null measure if we change the representative of $\bar{u}$. This is due to the fact that a is a Carathéodory function (which is the first assumption on a). With the second assumption on a, we therefore have $a(\cdot, \bar{u}) \in L^\infty(\Omega)$ and $\alpha \le a(\cdot, \bar{u}) \le \beta$ a.e. Similarly, by choosing representatives of $\bar{u}$ and $\nabla \bar{u}$, the function $x \mapsto f(x, \bar{u}(x), \nabla \bar{u})$ is Borel-measurable and only changes on a set of null measure if we change the representatives of $\bar{u}$ and $\nabla \bar{u}$. This is also due to the fact that f is a Carathéodory function (which is the first assumption on f). With the second assumption on f, we have

$$|f(\cdot, \bar{u}, \nabla \bar{u})| \le C(d + |\bar{u}|^\delta + |\nabla \bar{u}|^\delta) \text{ a.e.}$$

and therefore $f(\cdot, \bar{u}, \nabla \bar{u}) \in L^2(\Omega)$ because $d, \bar{u}, |\nabla \bar{u}| \in L^2(\Omega)$ and $\delta \le 1$. We can now apply Theorem 2.6, which gives the existence and uniqueness of a solution u to the problem

$$\begin{cases} u \in H_0^1(\Omega), \ : \ \forall v \in H_0^1(\Omega), \\ \displaystyle\int_\Omega a(x, \bar{u}(x)) \nabla u(x) \cdot \nabla v(x) \, \mathrm{d}x = \int_\Omega f(x, \bar{u}(x), \nabla \bar{u}(x)) v(x) \, \mathrm{d}x. \end{cases} \tag{3.51}$$

Set $u = T(\bar{u})$. We have thus constructed a mapping T from $H_0^1(\Omega)$ to $H_0^1(\Omega)$. To conclude, it is now sufficient to show that T admits a fixed point.

Estimates on T

Let $\bar{u} \in H_0^1(\Omega)$ and $u = T(\bar{u})$; taking $v = u$ in (3.51) yields

$$\alpha \|u\|_{H_0^1(\Omega)}^2 \le C(\|d\|_{L^2(\Omega)} + \||\bar{u}|^\delta\|_{L^2(\Omega)} + C_1 \||\nabla \bar{u}|^\delta\|_{L^2(\Omega)}) \|u\|_{L^2(\Omega)}.$$

With C_Ω given by the Poincaré inequality, we get

$$\frac{\alpha}{C_\Omega} \|u\|_{H_0^1(\Omega)} \leq C(\|d\|_{L^2(\Omega)} + \||\bar{u}|^\delta\|_{L^2(\Omega)} + \||\nabla\bar{u}|^\delta\|_{L^2(\Omega)}). \tag{3.52}$$

But as in Problem 3.4, we have (using Hölder's inequality)

$$\||\bar{u}|^\delta\|_{L^2(\Omega)}^2 \leq \left(\int_\Omega \bar{u}^2 \, dx\right)^\delta |\Omega|^{1-\delta} \quad \text{and} \quad \||\nabla\bar{u}|^\delta\|_{L^2(\Omega)}^2 \leq \left(\int_\Omega |\nabla\bar{u}|^2 \, dx\right)^\delta |\Omega|^{1-\delta}.$$

Thus, owing to the Poincaré inequality, there exists a $\bar{C} \in \mathbb{R}_+$ depending only on Ω and δ such that

$$\||\bar{u}|^\delta\|_{L^2(\Omega)} \leq \bar{C} \|\bar{u}\|_{H_0^1(\Omega)}^\delta \quad \text{and} \quad \||\nabla\bar{u}|^\delta\|_{L^2(\Omega)} \leq \bar{C} \|\bar{u}\|_{H_0^1(\Omega)}^\delta.$$

With (3.52), we obtain the existence of C_2 depending only on α, Ω and C such that

$$\|u\|_{H_0^1(\Omega)} \leq C_2 \left(1 + \|\bar{u}\|_{H_0^1(\Omega)}^\delta\right). \tag{3.53}$$

To conclude this step, we note that there exists an $R \in \mathbb{R}_+^\star$ such that $R > C_2(1 + R^\delta)$ (because $\delta < 1$) and therefore

$$\|\bar{u}\|_{H_0^1(\Omega)} \leq R \implies \|u\|_{H_0^1(\Omega)} \leq C_2(1 + R^\delta) \leq R,$$

which shows that T sends B_R to B_R where B_R is the ball (closed) centred at 0 and of radius R.

Continuity of T

Let $(\bar{u}_n)_{n\in\mathbb{N}}$ be a sequence in $H_0^1(\Omega)$ and $\bar{u} \in H_0^1(\Omega)$ and assume that $\bar{u}_n \to \bar{u}$ in $H_0^1(\Omega)$ as $n \to +\infty$. Set $u_n = T(\bar{u}_n)$ and $u = T(\bar{u})$. We want to show that $u_n \to u$ in $H_0^1(\Omega)$ as $n \to +\infty$. We also set $g_n = f(\cdot, \bar{u}_n, \nabla\bar{u}_n)$, $g = f(\cdot, \bar{u}, \nabla\bar{u})$, $A_n = a(\cdot, \bar{u}_n)$ and $A = a(\cdot, \bar{u})$.

We start by noting that $g_n \to g$ in $L^2(\Omega)$ as $n \to +\infty$. The proof here is almost identical to that of $f_n \to f$ in Problem 3.4; the only modification is in the domination of g_n, which here is $|g_n| \leq C(d + |G|^\delta + |H|^\delta)$ a.e. (instead of $|f_n| \leq C_1(1 + |G|^\delta + |H|^\delta)$ a.e. in Problem 3.4).

We now show that $u_n \to u$ weakly in $H_0^1(\Omega)$. We still follow the same reasoning as in Problem 3.4. Inequality (3.53) gives that the sequence $(u_n)_{n\in\mathbb{N}}$ is bounded in $H_0^1(\Omega)$. If $u_n \nrightarrow u$ weakly in $H_0^1(\Omega)$, there exist $\varepsilon > 0$, $\psi \in H^{-1}(\Omega)$ and a subsequence, still denoted $(u_n)_{n\in\mathbb{N}}$, such that

$$|\langle \psi, u_n - u\rangle_{H^{-1}(\Omega), H_0^1(\Omega)}| \geq \varepsilon \text{ for all } n \in \mathbb{N}. \tag{3.54}$$

Then there exists a $w \in H_0^1(\Omega)$ such that, up to a subsequence,

$$u_n \rightharpoonup w \quad \text{weakly in } H_0^1(\Omega),$$
$$\bar{u}_n \to \bar{u} \quad \text{a.e.}$$

Let $v \in H_0^1(\Omega)$; since $u_n = T(\bar{u}_n)$, we have

$$\int_\Omega A_n \nabla u_n \cdot \nabla v \, dx = \int_\Omega g_n v \, dx. \tag{3.55}$$

Since $A_n \to A$ a.e. (since a is a.e. continuous with respect to its second argument) and $|A_n| \le \beta$ a.e., we have, by the dominated convergence theorem, $A_n \nabla v \to A \nabla v$ in $L^2(\Omega)^N$. Passing to the limit as $n \to +\infty$ in the previous equality, we therefore obtain

$$\int_\Omega A \nabla w \cdot \nabla v \, dx = \int_\Omega g v \, dx,$$

which proves that $w = u$ (thanks to the uniqueness of the solution to (3.51)) and therefore $u_n \rightharpoonup u$ weakly in $H_0^1(\Omega)$, in contradiction with (3.54). We have thus shown that the whole sequence $(u_n)_{n \in \mathbb{N}}$ converges to u weakly in $H_0^1(\Omega)$ as $n \to +\infty$.

We now want to show that $u_n \to u$ in $H_0^1(\Omega)$ (and not just weakly). Taking $v = u_n$ in (3.55), we get

$$\int_\Omega A_n \nabla u_n \cdot \nabla u_n \, dx = \int_\Omega g_n u_n \, dx.$$

Since $u_n \to u$ in $L^2(\Omega)$ and $g_n \to g$ in $L^2(\Omega)$, we get

$$\lim_{n \to +\infty} \int_\Omega A_n \nabla u_n \cdot \nabla u_n \, dx = \int_\Omega g u \, dx,$$

and therefore, since $u = T(\bar{u})$, (3.51) gives

$$\lim_{n \to +\infty} \int_\Omega A_n \nabla u_n \cdot \nabla u_n \, dx = \int_\Omega A \nabla u \cdot \nabla u \, dx. \tag{3.56}$$

We now note that

$$\alpha \|u_n - u\|_{H_0^1(\Omega)}^2 \le \int_\Omega A_n (\nabla u_n - \nabla u) \cdot (\nabla u_n - \nabla u) \, dx. \tag{3.57}$$

As in Problem 3.4, we use the fact that

$$A_n \nabla u \to A \nabla u \text{ in } L^2(\Omega)^N, \text{ when } n \to +\infty$$

and that $\nabla u_n \rightharpoonup \nabla u$ weakly in $L^2(\Omega)^N$. This gives

$$\lim_{n \to +\infty} \int_\Omega A_n \nabla u \cdot \nabla u_n \, dx = \int_\Omega A \nabla u \cdot \nabla u \, dx$$

$$\text{and} \lim_{n \to +\infty} \int_\Omega A_n \nabla u \cdot \nabla u \, dx = \int_\Omega A \nabla u \cdot \nabla u \, dx,$$

and therefore, with (3.56),

$$\lim_{n \to +\infty} \int_\Omega A_n (\nabla u_n - \nabla u) \cdot (\nabla u_n - \nabla u) \, dx = 0.$$

We conclude, by (3.57), that $u_n \to u$ in $H_0^1(\Omega)$, which proves the continuity of the operator T from $H_0^1(\Omega)$ to $H_0^1(\Omega)$.

Compactness of T

We still follow the reasoning made for Problem 3.4. Let $(\bar{u}_n)_{n \in \mathbb{N}}$ be a bounded sequence in $H_0^1(\Omega)$. Set $g_n = f(\cdot, \bar{u}_n, \nabla \bar{u}_n)$ and $u_n = T(\bar{u}_n)$.

The sequence $(g_n)_{n \in \mathbb{N}}$ is bounded in $L^2(\Omega)$ and the sequence $(\bar{u}_n)_{n \in \mathbb{N}}$ is bounded in $H_0^1(\Omega)$ and therefore relatively compact in $L^2(\Omega)$. We can therefore assume that, up to a subsequence, there exist $g \in L^2(\Omega)$ and $\zeta \in L^2(\Omega)$ such that

$$\begin{aligned} g_n &\to g \text{ weakly in } L^2(\Omega), \\ \bar{u}_n &\to \zeta \text{ a.e.} \end{aligned} \tag{3.58}$$

Setting $b = a(\cdot, \zeta)$, we therefore have $b \in L^\infty(\Omega)$ and $a(\cdot, \bar{u}_n) \to b$ a.e.

Since $\alpha \le b \le \beta$ a.e., there exists a unique solution u of

$$\begin{cases} u \in H_0^1(\Omega), \\ \displaystyle\int_\Omega b \nabla u \cdot \nabla v \, dx = \int_\Omega g v \, dx \text{ for all } v \in H_0^1(\Omega). \end{cases} \tag{3.59}$$

Let us show that the sequence $(u_n)_{n \in \mathbb{N}}$ converges in $H_0^1(\Omega)$ towards a solution u to (3.59) (we are working here with the extracted sequence which satisfies (3.58)). We already know that the sequence $(u_n)_{n \in \mathbb{N}}$ is bounded in $H_0^1(\Omega)$. By reasoning by contradiction, we show that $u_n \to u$ weakly in $H_0^1(\Omega)$. Indeed, suppose there exists a $w \in H_0^1(\Omega)$ such that, up to a subsequence, $u_n \to w$ weakly in $H_0^1(\Omega)$. We then have, for all $n \in \mathbb{N}$ and all $v \in H_0^1(\Omega)$,

$$\int_\Omega a(x, \bar{u}_n) \nabla u_n \cdot \nabla v \, dx = \int_\Omega g_n v \, dx.$$

Passing to the limit as $n \to +\infty$ in this equation, thanks to the convergences given in (3.58), we obtain

$$\int_\Omega b \nabla w \cdot \nabla v \, dx = \int_\Omega g v \, dx.$$

This proves that $w = u$, and therefore $u_n \to u$ weakly in $H_0^1(\Omega)$ as $n \to +\infty$. We also have $u_n \to u$ in $L^2(\Omega)$.

It remains to show the convergence of u_n towards u in $H_0^1(\Omega)$. With this aim we first note that

$$\int_\Omega a(x, \bar{u}_n)\nabla u_n \cdot \nabla u_n \, dx = \int_\Omega g_n u_n \, dx \to \int_\Omega g u \, dx \text{ when } n \to +\infty.$$

Since u is a solution to (3.59) we therefore have

$$\lim_{n \to +\infty} \int_\Omega a(x, \bar{u}_n)\nabla u_n \cdot \nabla u_n \, dx = \int_\Omega b \nabla u \cdot \nabla u \, dx.$$

Using this convergence and (3.58) we then show that

$$\lim_{n \to +\infty} \int_\Omega a(x, \bar{u}_n)\nabla(u_n - u) \cdot \nabla(u_n - u) \, dx = 0.$$

Since $\alpha \|u_n - u\|_{H_0^1(\Omega)}^2 \le \int_\Omega a(x, \bar{u}_n)\nabla(u_n - u) \cdot \nabla(u_n - u) \, dx$ we conclude that $u_n \to u$ in $H_0^1(\Omega)$. (Note that convergence is only obtained for the extracted sequence which satisfies (3.58).) We have indeed proved the compactness of T.

Conclusion It is sufficient here to apply Schauder's theorem. The operator T is continuous and compact from $H_0^1(\Omega)$ to $H_0^1(\Omega)$. According to question 2 of Problem 3.4, there exists an $R > 0$ such that T sends the ball of centre 0 and radius R (from $H_0^1(\Omega)$) into itself. Schauder's theorem then allows us to say that there exists a u in this ball (and therefore in $H_0^1(\Omega)$) such that $u = T(u)$. The function u thus found is a solution to (3.28).

Problem 3.6 (Convection-diffusion, Dirichlet, existence)

1. Theorem 1.41 states that $u \in L^6(\Omega)$ if $N = 3$ and that $u \in L^r(\Omega)$ for all $r \in [1, +\infty[$ if $N = 2$. Since φ is Lipschitz-continuous and $\varphi(0) = 0$, there exists a C_1 such that $|\varphi(s)| \le C_1|s|$ for all $s \in \mathbb{R}$. Hence $\varphi(u) \in L^6(\Omega)$ if $N = 3$ and $\varphi(u) \in L^r(\Omega)$ for all $r \in [1, +\infty[$ if $N = 2$.

   For $N = 3$, we have $W \in L^3(\Omega)^3$ and $\varphi(u) \in L^6(\Omega)$, which gives $W\varphi(u) \in L^2(\Omega)^3$ because $\frac{1}{6} + \frac{1}{3} = \frac{1}{2}$.

   For $N = 2$, we have $W \in L^p(\Omega)^2$ and $\varphi(u) \in L^{\frac{2p}{p-2}}(\Omega)$, which gives $W\varphi(u) \in L^2(\Omega)^2$ because $\frac{1}{p} + \frac{p-2}{2p} = \frac{1}{2}$.

2. The mapping $v \mapsto \int_\Omega \varphi(\tilde{u}(x))W(x) \cdot \nabla v(x) \, dx + \int_\Omega f(x)v(x) \, dx$ is a continuous linear function from $H_0^1(\Omega)$ into $\mathbb{R}$. The existence and uniqueness of a solution u to (3.31) is therefore a consequence of Theorem 2.9.

3. We first prove the continuity of h. Let $(t_n, \tilde{u}_n)_{n \in \mathbb{N}}$ be a sequence in $[0, 1] \times L^q(\Omega)$ s.t. $t_n \to t$ and $\tilde{u}_n \to \tilde{u}$ in $L^q(\Omega)$ as $n \to +\infty$. Set $u_n = h(t_n, \tilde{u}_n)$ and $u = h(t, \tilde{u})$.

We want to show that $u_n \to u$ in $L^q(\Omega)$. We reason by contradiction: if $u_n \not\to u$ in $L^q(\Omega)$, there exist $\varepsilon > 0$ and a subsequence, still denoted by $(u_n)_{n \in \mathbb{N}}$, s.t.

$$\|u_n - u\|_{L^q(\Omega)} \geq \varepsilon \text{ for all } n \in \mathbb{N}. \tag{3.60}$$

After a possible extraction of a subsequence (which does not change (3.60)), we can also suppose that

$$\tilde{u}_n \to \tilde{u} \text{ a.e. and } |\tilde{u}_n| \leq H \text{ a.e. for all } n \in \mathbb{N},$$

with $H \in L^q(\Omega)$. Since $|\varphi(s)| \leq C_1|s|$, the dominated convergence theorem in $L^q(\Omega)$ yields that $\varphi(t_n \tilde{u}_n) \to \varphi(t\tilde{u})$ in $L^q(\Omega)$ and therefore $\varphi(t_n \tilde{u}_n)W \to \varphi(t\tilde{u})W$ in $L^2(\Omega)^N$ (noting that $|HW| \in L^2(\Omega)$, because $|W| \in L^p(\Omega)$ and $\frac{1}{p} + \frac{1}{q} = \frac{1}{2}$).

Since the sequence $(\varphi(t_n \tilde{u}_n)W)_{n \in \mathbb{N}}$ is bounded in $L^2(\Omega)^N$ and u_n is a solution to (3.31) with $t_n \tilde{u}_n$ instead of $\tilde{u}$, we show (by taking $v = u_n$ in 3.31) that the sequence $(u_n)_{n \in \mathbb{N}}$ is bounded in $H_0^1(\Omega)$. We can therefore suppose (still up to a subsequence) that there exists a $\bar{u}$ s.t. $u_n \to \bar{u}$ weakly in $H_0^1(\Omega)$. We also have (since $H_0^1(\Omega)$ is compactly embedded in $L^q(\Omega)$) $u_n \to \bar{u}$ in $L^q(\Omega)$. We then show that $\bar{u}$ is a solution to (3.31) with $t\tilde{u}$ instead of $\tilde{u}$ (and therefore that $\bar{u} = u$). It is enough for this to pass to the limit, for all $v \in H_0^1(\Omega)$, in the following equation

$$\int_\Omega \nabla u_n(x) \cdot \nabla v(x)\, dx - \int_\Omega \varphi(t_n \tilde{u}_n(x))W(x) \cdot \nabla v(x)\, dx = \int_\Omega f(x)v(x)\, dx.$$

This passage to the limit results from the fact that $\nabla u_n \to \nabla \bar{u}$ weakly in $L^2(\Omega)^N$ and $\varphi(t_n \tilde{u}_n)W \to \varphi(t\tilde{u})W$ in $L^2(\Omega)^N$.

We thus obtain that $\bar{u} = B(t\tilde{u}) = u$, in contradiction with (3.60) (because $u_n \to \bar{u}$ in $L^q(\Omega)$). We have thus shown that $u_n \to u$ in $L^q(\Omega)$. In fact, a similar argument by contradiction would even show that $u_n \to u$ weakly in $H_0^1(\Omega)$, but this is unnecessary for the continuation.

We now prove the compactness of h (which is a bit easier). Assume that t is arbitrary in $[0, 1]$ and that $\tilde{u}$ remains in a bounded subset of $L^q(\Omega)$. Set $u = h(t, \tilde{u})$. The function u is therefore a solution to (3.31) with $t\tilde{u}$ instead of $\tilde{u}$. Thanks to $|\varphi(s)| \leq C_1|s|$, the function $\varphi(\tilde{u})$ remains in a bounded subset of $L^q(\Omega)$ and therefore $\varphi(\tilde{u})W$ remains in a bounded subset of $L^2(\Omega)^N$. Now taking $v = u$ in (3.31) (with $t\tilde{u}$ instead of $\tilde{u}$), we get that u remains in a bounded subset of $H_0^1(\Omega)$. Since $H_0^1(\Omega)$ is compactly embedded in $L^q(\Omega)$, u remains in a compact set of $L^q(\Omega)$, which proves the compactness of h.

Note: a probably slightly faster way to show the continuity and compactness of h is to note that h is the composition of B, which is a continuous and compact operator from $L^q(\Omega)$ to $L^q(\Omega)$, with the mapping $(t, u) \mapsto tu$, which is continuous from $[0, 1] \times L^q(\Omega)$ to $L^q(\Omega)$.

4. a. The function ψ (from $\mathbb{R}$ to $\mathbb{R}$) is of class C^1 on $\mathbb{R}$ and is Lipschitz-continuous (because $|\psi'(s)| \le 1$ for all $s \in \mathbb{R}$). Lemma 2.25 then gives that $\psi(u) \in H_0^1(\Omega)$ and $\nabla\psi(u) = \frac{\nabla u}{(1+|u|)^2}$. We also note that $|\psi(s)| \le 1$ for all s.

Taking $v = \psi(u)$ in (3.32) and using $|\varphi(s)| \le C_1|s|$ and $|\psi(s)| \le 1$, we obtain

$$\int_\Omega \frac{|\nabla u(x)|^2}{(1+|u(x)|)^2}\, dx \le C_1 \int_\Omega \frac{|tu(x)|}{(1+|u(x)|)^2}|W(x)||\nabla u(x)|\, dx + \|f\|_{L^1(\Omega)}$$
$$\le C_1 \int_\Omega |W(x)|\frac{|\nabla u(x)|}{1+|u(x)|}\, dx + \|f\|_{L^1(\Omega)}\,.$$

Using $ab \le \frac{a^2}{2C_1} + 2C_1 b^2$ for $a, b \in \mathbb{R}$, we obtain

$$\frac{1}{2}\int_\Omega \frac{|\nabla u(x)|^2}{(1+|u(x)|)^2}\, dx \le 2C_1^2\, \|\|W\|\|_{L^2(\Omega)}^2 + \|f\|_{L^1(\Omega)}\,.$$

We now note that (still by Lemma 2.25) $\ln(1+|u|) \in H_0^1(\Omega)$ and the previous inequality gives

$$\|\ln(1+|u|)\|_{H_0^1(\Omega)}^2 \le 2(2C_1^2\, \|\|W\|\|_{L^2(\Omega)}^2 + \|f\|_{L^1(\Omega)}),$$

which gives the desired upper bound with $C_\ell^2 = 2(2C_1^2\, \|\|W\|\|_{L^2(\Omega)}^2 + \|f\|_{L^1(\Omega)})$.

b. We have

$$\int_\Omega |v(x)|^q\, dx = \int_{\{|v|\ge A\}} |v(x)|^q\, dx + \int_{\{|v|<A\}} |v(x)|^q\, dx$$
$$\le \int_{\{|v|\ge A\}} |v(x)|^q\, dx + A^q \lambda_N(\Omega).$$

Then, Hölder's inequality (with $\frac{2^\star}{q}$ and its conjugate) gives

$$\int_{\{|v|\ge A\}} |v(x)|^q\, dx = \int_\Omega |v(x)|^q \mathbb{1}_{\{|v|\ge A\}}\, dx$$
$$\le \left(\int_\Omega |v(x)|^{2^\star} dx\right)^{\frac{q}{2^\star}} \lambda_N(\{|v|\ge A\})^{1-\frac{q}{2^\star}},$$

which indeed gives the desired inequality.

c. Taking $v = u$ in (3.32), we obtain, by Hölder's inequality,

$$\|u\|_{H_0^1(\Omega)}^2 \le C_1 \|u\|_{L^q(\Omega)} \|\|W\|\|_{L^p(\Omega)} \|u\|_{H_0^1(\Omega)}$$
$$+ \|f\|_{L^2} C_\Omega \|u\|_{H_0^1(\Omega)}\,, \quad (3.61)$$

where C_Ω only depends on Ω and is given by Poincaré's inequality.

Owing to the inequality given in 4b (raised to the power $\frac{1}{q}$), we have

$$\|u\|_{L^q(\Omega)} \le 2\,\|u\|_{L^{2\star}}\,\lambda_N(\{|u| \ge A\})^{\frac{1}{q}-\frac{1}{2\star}} + 2A\lambda_N(\Omega)^{\frac{1}{q}}.$$

Since $H_0^1(\Omega)$ is continuously embedded in $L^{2^\star}(\Omega)$, there exists a $\bar{C}_\Omega$ depending only on Ω such that $\|u\|_{L^{2\star}(\Omega)} \le \bar{C}_\Omega\,\|u\|_{H_0^1(\Omega)}$. Hence

$$\forall A > 0, \ \|u\|_{L^q(\Omega)} \le 2\bar{C}_\Omega\,\|u\|_{H_0^1(\Omega)}\,\lambda_N(\{|u| \ge A\})^{\frac{1}{q}-\frac{1}{2\star}} + 2A\lambda_N(\Omega)^{\frac{1}{q}}.$$

Thanks to the result proven in question 4a and owing to Poincaré's inequality, we have

$$\forall A > 0, \ \ln(1+A)\lambda_N(\{|u| \ge A\})^{\frac{1}{2}} \le \|\ln(1+|u|)\|_{L^2(\Omega)}$$
$$\le C_\Omega\,\|\ln(1+|u|)\|_{H_0^1(\Omega)} \le C_\Omega C_\ell.$$

Since $\lim_{A\to+\infty}\ln(1+A) = +\infty$, there exists therefore an A depending (like C_ℓ, also note that p and q are given by W) only on Ω, W, φ and f such that

$$\lambda_N(\{|u| \ge A\})^{\frac{1}{q}-\frac{1}{2\star}} \le \frac{1}{4\bar{C}_\Omega C_1\,\||W|\|_{L^p(\Omega)}}.$$

With this choice of A, (3.61) gives

$$\|u\|_{H_0^1(\Omega)}^2 \le \frac{1}{2}\,\|u\|_{H_0^1(\Omega)}^2 + (2AC_1\,\||W|\|_{L^p(\Omega)}\,\lambda_N(\Omega)^{\frac{1}{q}}$$
$$+ \|f\|_{L^2(\Omega)}\,C_\Omega)\,\|u\|_{H_0^1(\Omega)}.$$

Thus

$$\|u\|_{H_0^1(\Omega)} \le 2(2AC_1\,\||W|\|_{L^p(\Omega)}\,\lambda_N(\Omega)^{\frac{1}{q}} + \|f\|_{L^2(\Omega)}\,C_\Omega),$$

which is an estimate on $\|u\|_{H_0^1(\Omega)}$ depending only on Ω, W, φ and f.

5. Since $H_0^1(\Omega)$ is continuously embedded in $L^q(\Omega)$, question 4c gives the existence of $R > 0$ (depending only on Ω, W, φ, f and q) such that

$$t \in [0,1], \ u \in L^q(\Omega), \ u = h(t,u) \Rightarrow \|u\|_{L^q(\Omega)} < R.$$

Question 3 gives the continuity and compactness of h from $[0,1] \times L^q(\Omega)$ to $L^q(\Omega)$. We can therefore apply the homotopy invariance of the topological degree on the ball (open) of $L^p(\Omega)$ centred at 0 and of radius R with target point 0. We obtain

$$d(\mathrm{Id} - h(1,\cdot), B_R, 0) = d(\mathrm{Id} - h(0,\cdot), B_R, 0).$$

The mapping $\tilde{u} \mapsto h(0, \tilde{u})$ is constant ($h(0, \tilde{u})$ is, for all $\tilde{u}$, the weak solution of $-\Delta u = f$ in Ω with $u = 0$ on $\partial\Omega$). The solution of $v = h(0, v)$ is unique and belongs to B_R. This is enough to say that $d(\mathrm{Id} - h(0, \cdot), B_R, 0) \neq 0$ (we can bring the constant to 0 by homotopy noting, for example, that $d(\mathrm{Id} - th(0, \cdot), B_R, 0)$ does not depend on $t \in [0, 1]$, see Problem 3.3).

6. We start by replacing $\int_\Omega f(x)v(x)\,dx$ by $\langle T, v \rangle_{H^{-1}(\Omega), H_0^1(\Omega)}$ in (3.30). The proof is then very similar to the previous one. The only points requiring a small modification are in questions 4a and 4c. In question 4a, we have bounded $\int_\Omega |f\psi(u)|\,dx$ by $\|f\|_{L^1(\Omega)}$. We now need to bound

$$\left| \langle T, \psi(v) \rangle_{H^{-1}(\Omega), H_0^1(\Omega)} \right|.$$

This bounding is done by noting that

$$\left| \langle T, \psi(v) \rangle_{H^{-1}(\Omega), H_0^1(\Omega)} \right| \leq \|T\|_{H^{-1}(\Omega)} \, \|\psi(u)\|_{H_0^1(\Omega)}$$

$$\leq \|T\|_{H^{-1}(\Omega)} \left\| \frac{|\nabla u|}{(1 + |u|)^2} \right\|_{L^2(\Omega)} \leq \|T\|_{H^{-1}(\Omega)} \left\| \frac{|\nabla u|}{1 + |u|} \right\|_{L^2(\Omega)}$$

$$\leq 4 \|T\|_{H^{-1}(\Omega)}^2 + \frac{1}{4} \left\| \frac{|\nabla u|}{1 + |u|} \right\|_{L^2(\Omega)}^2.$$

In question 4c, we have bounded $\int_\Omega |fu|\,dx$ by $C_\Omega \|f\|_{L^2(\Omega)} \|u\|_{H_0^1(\Omega)}$. We now need to bound

$$\left| \langle T, u \rangle_{H^{-1}(\Omega), H_0^1(\Omega)} \right|,$$

which is obtained by noting that

$$\left| \langle T, u \rangle_{H^{-1}(\Omega), H_0^1(\Omega)} \right| \leq \|T\|_{H^{-1}(\Omega)} \, \|u\|_{H_0^1(\Omega)} .$$

We now need to remove the assumption $\varphi(0) = 0$. It is enough to return to the previous case by replacing φ by $\varphi - \varphi(0)$ and adding the term $-\int_\Omega \varphi(0)W(x) \cdot \nabla v(x)\,dx$ to the right-hand side of (3.30). We do indeed return to the previous case because the mapping $v \mapsto \int_\Omega \varphi(0)W(x) \cdot \nabla v(x)\,dx$ is an element of $H^{-1}(\Omega)$ (since $W \in L^2(\Omega)^N$).

Problem 3.7 (Convection-diffusion, Dirichlet, uniqueness)

1. The proof of uniqueness presented for Theorem 3.20 did not fully use the assumptions on G (which were $G \in C^1(\bar{\Omega}, \mathbb{R}^N)$ and $\mathrm{div}\, G = 0$). It only used the fact that $G \in L^2(\Omega)^N$. Here we have $W \in L^p(\Omega)^N$. Since $p > N$, this indeed gives $W \in L^2(\Omega)^N$ and the proof given for Theorem 3.20 is therefore also valid here. We quickly recall it.

Let u_1 and u_2 be two solutions of (3.33); we have therefore, for all $v \in H_0^1(\Omega)$,

$$\int_\Omega \nabla u_1 \cdot \nabla v \, dx - \int_\Omega \varphi(u_1) W \cdot \nabla v \, dx = \int_\Omega f v \, dx, \qquad (3.62)$$

$$\int_\Omega \nabla u_2 \cdot \nabla v \, dx - \int_\Omega \varphi(u_2) W \cdot \nabla v \, dx = \int_\Omega f v \, dx. \qquad (3.63)$$

For $n \in \mathbb{N}^\star$ we define $T_n \in C(\mathbb{R}, \mathbb{R})$ by $T_n(s) = \max(-\frac{1}{n}, \min(s, \frac{1}{n}))$. Lemma 2.26 (or rather its generalisation, see Remark 2.27) gives $T_n(u_1 - u_2) \in H_0^1(\Omega)$ and $\nabla T_n(u_1 - u_2) = \nabla(u_1 - u_2)\mathbb{1}_{A_n}$ with $A_n = \{0 < |u_1 - u_2| < \frac{1}{n}\}$.

We take $v = T_n(u_1 - u_2)$ in (3.62) and (3.63). We obtain

$$\int_\Omega \nabla(u_1 - u_2) \cdot \nabla T_n(u_1 - u_2) \, dx = \int_\Omega (\varphi(u_1) - \varphi(u_2)) W \cdot \nabla(T_n(u_1 - u_2)) \, dx.$$

With C_1 such that $|\varphi(s_1) - \varphi(s_2)| \le C_1 |s_1 - s_2|$ for all $s_1, s_2 \in \mathbb{R}$, this gives

$$\int_{A_n} |\nabla(u_1 - u_2)|^2 \, dx \le \int_{A_n} C_1 |u_1 - u_2| \, |W| \, |\nabla(u_1 - u_2)| \, dx.$$

Since $|u_1 - u_2| \le \frac{1}{n}$ a.e. in A_n, applying the Cauchy–Schwarz inequality in the last integral, we therefore obtain:

$$\int_{A_n} |\nabla(u_1 - u_2)|^2 \, dx \le \frac{C_1}{n} \left(\int_{A_n} |W|^2 \, dx \right)^{\frac{1}{2}} \left(\int_{A_n} |\nabla(u_1 - u_2)|^2 \, dx \right)^{\frac{1}{2}}.$$

Hence

$$\| \, |\nabla T_n(u_1 - u_2)| \, \|_{L^2(\Omega)} = \left(\int_{A_n} |\nabla(u_1 - u_2)|^2 \, dx \right)^{\frac{1}{2}} \le \frac{C_1}{n} a_n,$$

$$\text{with } a_n = \left(\int_{A_n} |W|^2 \, dx \right)^{\frac{1}{2}}.$$

Owing to the Sobolev and Hölder inequalities, with $1^\star = \frac{N}{N-1}$ and denoting by m the Lebesgue measure on $\mathbb{R}^N$, we get

$$\|T_n(u_1 - u_2)\|_{L^{1^\star}} \le \| \, |\nabla T_n(u_1 - u_2)| \, \|_{L^1(\Omega)}$$

$$\le m(\Omega)^{\frac{1}{2}} \| \, |\nabla T_n(u_1 - u_2)| \, \|_{L^2(\Omega)} \le \frac{C_1 m(\Omega)^{\frac{1}{2}}}{n} a_n.$$

Setting $B_n = \{|u_1 - u_2| \ge \frac{1}{n}\}$ yields

$$\frac{1}{n} (m(B_n))^{\frac{N-1}{N}} \le \left(\int_{B_n} |T_n(u_1 - u_2)|^{1^\star} \, dx \right)^{\frac{1}{1^\star}} \le \|T_n(u_1 - u_2)\|_{L^{1^\star}}.$$

Hence

$$(m(B_n))^{\frac{N-1}{N}} \le C_1 m(\Omega)^{\frac{1}{2}} a_n. \tag{3.64}$$

For $n \in \mathbb{N}^\star$, we have $A_{n+1} \subset A_n$. Since $\cap_{n \in \mathbb{N}^\star} A_n = \emptyset$, the non-increasing continuity of m gives that $\lim_{n \to +\infty} m(A_n) = 0$. Since $W \in L^2(\Omega)^N$ we get that $\lim_{n \to +\infty} a_n = 0$ and therefore, thanks to (3.64), that $\lim_{n \to +\infty} m(B_n) = 0$.

We finally note that $B_{n+1} \supset B_n$, for all $n \in \mathbb{N}^\star$, and $\cup_{n \in \mathbb{N}} B_n = \{|u_1 - u_2| > 0\}$. Therefore $\lim_{n \to +\infty} m(B_n) = m\{|u_1 - u_2| > 0\}$ (by non-decreasing continuity of a measure). Hence $m\{|u_1 - u_2| > 0\} = 0$ and thus, $u_1 = u_2$ a.e.

2. The proof is identical to the previous one. It is sufficient to replace, in (3.62) and (3.63), $\int_\Omega f(x)v(x)\,\mathrm{d}x$ by $\langle T, v \rangle_{H^{-1}(\Omega), H_0^1(\Omega)}$.

3. The proof is similar to that of the first question. For $n \in \mathbb{N}^\star$, we take $v = S_n(u)$ in (3.33) with $S_n \in C(\mathbb{R}, \mathbb{R})$ defined by $S_n(s) = \max(0, \min(s, \frac{1}{n}))$. This is possible because $S_n(u) \in H_0^1(\Omega)$. We also know that $\nabla S_n(u) = \mathbb{1}_{E_n} \nabla u$ with $E_n = \{0 < u < \frac{1}{n}\}$ (see Remark 2.27). Since $S_n(u) \ge 0$ a.e. and $f \le 0$ a.e., we obtain

$$\int_\Omega \nabla u \cdot \nabla S_n(u)\,\mathrm{d}x - \int_\Omega \varphi(u)W \cdot \nabla S_n(u)\,\mathrm{d}x \le 0.$$

So we have

$$\int_\Omega |\nabla S_n(u)|^2\,\mathrm{d}x = \int_\Omega \nabla u \cdot \nabla S_n(u)\,\mathrm{d}x \le \int_\Omega \varphi(u)W \cdot \nabla S_n(u)\,\mathrm{d}x.$$

There exists a $C_1 > 0$ such that $|\varphi(s_1) - \varphi(s_2)| \le C_1|s_1 - s_2|$ for all $s_1, s_2 \in \mathbb{R}$, this gives, using the Cauchy–Schwarz inequality in the last integral,

$$\| |\nabla S_n(u)| \|_{L^2(\Omega)} \le \frac{C_1}{n} \left(\int_{E_n} |W(x)|^2\,\mathrm{d}x \right)^{\frac{1}{2}}.$$

Set $\gamma_n = \left(\int_{E_n} |W(x)|^2\,\mathrm{d}x \right)^{\frac{1}{2}}$. We now use the Sobolev and Hölder inequalities to obtain, with $1^\star = \frac{N}{N-1}$ and denoting by m the Lebesgue measure on $\mathbb{R}^N$,

$$\|S_n(u)\|_{L^{1^\star}} \le \| |\nabla S_n(u)| \|_{L^1(\Omega)}$$

$$\le m(\Omega)^{\frac{1}{2}} \| |\nabla S_n(u)| \|_{L^2(\Omega)} \le \frac{C_1 m(\Omega)^{\frac{1}{2}}}{n} \gamma_n.$$

Set $D_n = \{u \ge \frac{1}{n}\}$, so that

$$\frac{1}{n}(m(D_n))^{\frac{N-1}{N}} \le \left(\int_{D_n} |S_n(u)|^{1^\star}\,\mathrm{d}x \right)^{\frac{1}{1^\star}} \le \|S_n(u)\|_{L^{1^\star}}.$$

So we have

$$(m(D_n))^{\frac{N-1}{N}} \le C_1 m(\Omega)^{\frac{1}{2}} \gamma_n. \tag{3.65}$$

We conclude as in the first question. For $n \in \mathbb{N}^{\star}$, we have $E_{n+1} \subset E_n$. Since $\cap_{n \in \mathbb{N}^{\star}} E_n = \emptyset$, the non-increasing continuity of m gives that $\lim_{n \to +\infty} m(E_n) = 0$. Since $W \in L^2(\Omega)^N$ one has $\lim_{n \to +\infty} \gamma_n = 0$ and therefore, thanks to (3.65), $\lim_{n \to +\infty} m(D_n) = 0$.

We finally note that $D_{n+1} \supset D_n$, for all $n \in \mathbb{N}^{\star}$, and $\cup_{n \in \mathbb{N}} D_n = \{u > 0\}$. Therefore $\lim_{n \to +\infty} m(D_n) = m\{u > 0\}$ (by non-decreasing continuity of a measure). We thus obtain $m\{u > 0\} = 0$ and therefore $u \leq 0$ a.e.

Problem 3.8 (Existence by minimisation)

1. For the solution of this question and the following questions, A denotes a set of null measure, that is to say such that $\lambda_N(A) = 0$ (recall that λ_N designates the Lebesgue measure on $\mathbb{R}^N$), and such that the mapping $s \mapsto f(x, s)$ is, for all $x \in A^c = \Omega \setminus A$, continuous from $\mathbb{R}$ to $\mathbb{R}$ and satisfies $|f(x, s)| \leq C|s|^{\delta} + d$ (for all $s \in \mathbb{R}$). The definition of F gives, for all $x \in A^c$ and for all $s \in \mathbb{R}$,

$$|F(x, s)| \leq \frac{C}{\delta + 1}|s|^{\delta+1} + d|s| \leq \frac{C}{\delta + 1}(|s|^2 + 1) + d(|s|^2 + 1) \qquad (3.66)$$

because $\delta \leq 1$ and therefore $|F(x, u(x))| \leq C_1|u(x)|^2 + C_1$, with $C_1 = d + \frac{C}{\delta+1}$. Since $u \in L^2(\Omega)$ and Ω is bounded (and therefore of finite Lebesgue measure), $F(\cdot, u) \in L^1(\Omega)$.

N.B. The above solution is not quite correct from the measurability point of view (problems of measurability are often somewhat forgotten in many works, including this one). In fact if we choose a representative for u, that is to say, an element of the class u, we can indeed show (thanks to the hypotheses on f) that there exists a measurable function v from Ω to $\mathbb{R}$ (endowed with Borel σ-algebras) such that $v = F(\cdot, u)$ a.e. and the previous reasoning indeed gives $v \in \mathcal{L}^1(\Omega)$. For more precision, one can consult, for example, [26], problem 7.14.

2. From the first inequality given in (3.66) we get that $\lim_{s \to \pm\infty} \frac{F(x,s)}{s^2} = 0$ uniformly with respect to $x \in A^c$. Hence, for all $\varepsilon > 0$, there exist C_{ε}, depending only on ε, C and δ such that $|F(x, s)| \leq \varepsilon|s|^2 + C_{\varepsilon}$ for all $s \in \mathbb{R}$ and all $x \in A^c$.

Let $u \in H_0^1(\Omega)$. Using this upper bound of F, the lower bound of a and the Poincaré inequality (2.8), we obtain

$$E(u) \geq \frac{\alpha}{2} \int_{\Omega} |\nabla u(x)|^2 \, dx - \varepsilon \int_{\Omega} |u(x)|^2 \, dx - C_{\varepsilon} \lambda_N(\Omega)$$

$$\geq \left(\frac{\alpha}{2} - \varepsilon C_{\Omega}^2\right) \int_{\Omega} |\nabla u(x)|^2 \, dx - C_{\varepsilon} \lambda_N(\Omega).$$

Choosing $\varepsilon > 0$ such that $\left(\frac{\alpha}{2} - \varepsilon C_{\Omega}^2\right) > 0$ and noting that $\| |\nabla u| \|_{L^2(\Omega)}^2 \geq \frac{1}{C_{\Omega}^2 + 1} \|u\|_{H^1(\Omega)}^2$, it follows that $E(u) \to +\infty$ when $\|u\|_{H^1(\Omega)} \to +\infty$.

3. We choose a minimising sequence, that is, a sequence $(u_n)_{n\in\mathbb{N}}$ from $H_0^1(\Omega)$ such that $\lim_{n\to+\infty} E(u_n) = \eta$. Question 2 shows that the sequence $(u_n)_{n\in\mathbb{N}}$ is bounded, and thus converges weakly in $H_0^1(\Omega)$ (up to a subsequence); thanks to Rellich's Theorem 1.36 it converges in $L^2(\Omega)$. Finally, we can assume that, still up to a subsequence, it converges a.e. while remaining dominated in $L^2(\Omega)$ (see, for example, [26], theorem 6.11). Hence, $D_i u_n \to D_i u$ weakly in $L^2(\Omega)$ as $n \to +\infty$, for all i (because $u_n \to u$ weakly in $H_0^1(\Omega)$, Problem 1.22), $u_n \to u$ a.e. and, for all $n \in \mathbb{N}$ $|u_n| \le g$ a.e. with $g \in L^2(\Omega)$.

For the first term of $E(u_n)$, the mapping $v \mapsto \int_\Omega a(x)\nabla u(x) \cdot \nabla v(x)\, dx$ belongs to the dual space of $H_0^1(\Omega)$. We therefore have

$$\left\| \sqrt{a}|\nabla u| \right\|_{L^2(\Omega)}^2 = \int_\Omega a(x)\nabla u(x) \cdot \nabla u(x)\, dx = \lim_{n\to+\infty} \int_\Omega a(x)\nabla u(x) \cdot \nabla u_n(x)\, dx.$$

But, by the Cauchy–Schwarz inequality,

$$\int_\Omega a(x)\nabla u(x) \cdot \nabla u_n(x)\, dx \le \left\| \sqrt{a}|\nabla u| \right\|_{L^2(\Omega)} \left\| \sqrt{a}|\nabla u_n| \right\|_{L^2(\Omega)},$$

and therefore $\left\| \sqrt{a}|\nabla u| \right\|_{L^2(\Omega)}^2 \le \liminf_{n\to+\infty} \left\| \sqrt{a}|\nabla u| \right\|_{L^2(\Omega)} \left\| \sqrt{a}|\nabla u_n| \right\|_{L^2(\Omega)}$, which gives

$$\int_\Omega a(x)\nabla u(x) \cdot \nabla u(x)\, dx \le \liminf_{n\to+\infty} . \int_\Omega a(x)\nabla u_n(x) \cdot \nabla u_n(x)\, dx.$$

For the second term of $E(u_n)$, since $s \mapsto F(x, s)$ is continuous for $x \in A^c$, we obtain with the number C_1 from the first question

$$F(\cdot, u_n) \to F(\cdot, u) \text{ a.e., when } n \to +\infty,$$
$$|F(\cdot, u_n)| \le C_1|u_n|^2 + C_1 \le C_1|g|^2 + C_1 \text{ a.e., for all } n \in \mathbb{N}.$$

Owing to the dominated convergence theorem,

$$\int_\Omega F(x, u_n(x))\, dx \to \int_\Omega F(x, u(x))\, dx \text{ as } n \to +\infty.$$

This gives

$$\eta \le E(u)$$
$$= \frac{1}{2} \int_\Omega a(x)\nabla u(x) \cdot \nabla u(x)\, dx - \int_\Omega F(x, u(x))\, dx \le \liminf_{n\to+\infty} E(u_n) = \eta,$$

and therefore $\eta \in \mathbb{R}$ and $E(u) = \eta$.

4. Let us show that the function u found in the previous question is a solution to (3.34). Let $v \in H_0^1(\Omega)$. Since $E(u) \leq E(u + tv)$, we obtain, for $0 < t$,

$$0 \leq \frac{E(u + tv) - E(u)}{t}$$

$$= \int_\Omega a(x)\nabla u(x) \cdot \nabla v(x)\, dx + t \int_\Omega a(x)\nabla v(x) \cdot \nabla v(x)\, dx$$

$$- \int_\Omega \frac{F(x, u(x) + tv(x)) - F(x, u(x))}{t}\, dx. \quad (3.67)$$

Since $s \mapsto F(x, s)$ is of class C^1 for $x \in A^c$,

$$\frac{F(\cdot, u + tv) - F(\cdot, u)}{t} \to f(\cdot, u)v \quad \text{a.e., when } t \to 0,$$

$$\left|\frac{F(\cdot, u + tv) - F(\cdot, u)}{t}\right| \leq |v|(C|u| + C|v| + C + d) \quad \text{a.e., } \forall t \in]0, 1[.$$

For the above upper bound, we used the mean value theorem. It gives for almost all $x \in \Omega$ and all $t \in]0, 1[$,

$$|F(x, u(x) + tv(x)) - F(x, u(x))| \leq \max_{\theta \in [0,1]} |f(x, u(x) + t\theta v(x))tv(x)|$$

$$\leq t|v(x)|(C(|u(x)| + |v(x)|)^\delta + d)$$

$$\leq t|v(x)|(C(|u(x)| + |v(x)|) + C + d).$$

We can now let $t \to 0$, with $t \in]0, 1[$, in (3.67) to obtain, by the dominated convergence theorem,

$$0 \leq \int_\Omega a(x)\nabla u(x) \cdot \nabla v(x)\, dx - \int_\Omega f(x, u(x))v(x)\, dx.$$

Since this inequality is also true for $-v$, we finally obtain that u is a solution to (3.34).

N.B. The proof given for this question only uses the (directional) differentiability of E at the point u in all directions v of $H_0^1(\Omega)$. Another possible proof would be to show that E is differentiable at u. This would give $Df(u) = 0$, that is to say u is a solution to (3.34).

A function can be (directionally) differentiable at point u in all directions without being differentiable at point u. This is the case for example for the function $u \mapsto \int_\Omega |u(x)|\, dx$ in $L^1(\Omega)$ at point 0.

Problem 3.9 (Minimisation with constraint)

1. Theorem 1.41 on Sobolev embeddings gives $H_0^1(\Omega) \subset L^{2^\star}(\Omega)$ if $N > 2$ and $H_0^1(\Omega) \subset L^q(\Omega)$ for all $q < +\infty$ if $N = 2$. Hence $F(u) \in L^1(\Omega)$ if $u \in H_0^1(\Omega)$ (for $N > 2$, $1 + p = 2^\star$) and therefore $A \neq \emptyset$.

   Let $\eta = \inf\{E(v), v \in A\}$. Let $(u_n)_{n \in \mathbb{N}}$ be a minimising sequence, that is to say, a sequence of elements of A such that $\lim_{n \to +\infty} E(u_n) = \eta$. The sequence $(u_n)_{n \in \mathbb{N}}$ is bounded in $H_0^1(\Omega)$ and we can therefore assume that (up to a subsequence) it converges weakly in $H_0^1(\Omega)$ and, thanks to Rellich's theorem (Theorem 1.36), that it converges in $L^2(\Omega)$ to some limit u.

   As in Problem 3.8, the weak convergence of u_n towards u in $H_0^1(\Omega)$ gives

   $$E(u) \leq \liminf_{n \to +\infty} E(u_n).$$

   We now show that $u \in A$. By Theorem 1.41, the sequence $(u_n)_{n \in \mathbb{N}}$ is bounded in $L^{2^\star}(\Omega)$ if $N > 2$ and in $L^q(\Omega)$ for all $q < +\infty$ if $N = 2$. Since it converges in $L^2(\Omega)$, a simple application of Hölder's inequality[16] gives that it converges in $L^q(\Omega)$ for all $q < 2^\star$ and therefore, in particular in $L^{p+1}(\Omega)$. Hence $\lim_{n \to +\infty} F(u_n) = F(u) = 1$ and therefore $u \in A$ and $E(u) = \eta$ because $\eta \leq E(u) \leq \liminf_{n \in \mathbb{N}} E(u_n) = \eta$.

   Finally, to have $u \geq 0$ a.e., it is enough to replace u by $|u|$ because $F(|u|) = F(u)$ and $E(|u|) = E(u)$ (using Lemma 2.26 or rather its generalisation, Remark 2.27).
2. We show that the function u found in the previous question is, up to a multiplicative constant, a weak solution to (3.35). With this aim, we apply Theorem 3.34. As

$$E(u + v) = \frac{1}{2} \int_\Omega |\nabla u(x)|^2 \, dx + \int_\Omega \nabla u(x) \cdot \nabla v(x) \, dx + \frac{1}{2} \int_\Omega |\nabla v(x)|^2 \, dx,$$

the function E is differentiable at the point u and its differential at the point u, denoted by $dE(u)$, is the element of $H^{-1}\Omega$ defined by

$$\langle dE(u), v \rangle_{H^{-1}(\Omega), H_0^1(\Omega)} = \int_\Omega \nabla u(x) \cdot \nabla v(x) \, dx.$$

Let us now show that $F \in C^1(H_0^1(\Omega), \mathbb{R})$. Let us start by showing that, for all $s \in \mathbb{R}$ and all $h \in \mathbb{R}$,

$$|s + h|^{p+1} = |s|^{p+1} + (p + 1)s^p \operatorname{sgn}(s)h + R(s, h), \, with |R(s, h)|$$
$$\leq (p + 1)^2 (|s|^{p-1}h^2 + |h|^{p+1}), \quad (3.68)$$

where $\operatorname{sgn}(s) = 1$ if $s \geq 0$ and -1 otherwise. Indeed,

---

[16] If $1 \leq p < r < q < +\infty$ and $u \in L^p(X, \mathcal{T}, m) \cap L^q(X, \mathcal{T}, m)$, $\|u\|_{L^r} \leq \|u\|_{L^p}^\theta \|u\|_{L^q}^{1-\theta}$, with $\theta = \frac{p}{r} \frac{q-r}{q-p}$.

- for $s > 0$ and $h \geq -s$, a Taylor expansion gives that there exists a $\theta \in]0, 1[$ such that

$$
\begin{aligned}
|R(s, h)| &= \frac{p(p+1)}{2}(s + \theta h)^{p-1}h^2 \\
&\leq \frac{p(p+1)}{2}(|s|^{p-1}h^2 + |h|^{p+1});
\end{aligned}
$$

- for $s > 0$ and $h < -s$, we have

$$
\begin{aligned}
|R(s, h)| &= |s + h|^{p+1} - |s|^{p+1} - (p + 1)s^p \operatorname{sgn}(s)h| \\
&\leq (p + 2)|h|^{p+1};
\end{aligned}
$$

- for $s < 0$, by changing s to $-s$, we obtain the same bounds for $|R(s, h)|$;
- finally, for $s = 0$, $|R(s, h)| = |h|^{p+1}$.

The bound of $|R(s, h)|$ in (3.68) is thus proven. Let us now consider $v \in H_0^1(\Omega)$. For all $w \in H_0^1(\Omega)$, we use (3.68) with $s = v(x)$ and $h = w(x)$ (for almost all $x \in \Omega$). Since $|v|^p \in L^{1+\frac{1}{p}}(\Omega)$ and $w \in L^{p+1}(\Omega)$, Hölder's inequality gives $|v|^p w \in L^1(\Omega)$. We then obtain with the bound of $|R(s, h)|$ and (again) by Hölder's inequality,

$$
F(v + w) = F(v) + \int_\Omega (p + 1)v(x)^p \operatorname{sgn}(v(x))w(x)\, dx + R(w),
$$

with

$$
\begin{aligned}
|R(w)| &\leq (p + 1)^2 \int_\Omega (|v(x)|^{p-1}w(x)^2 + |w(x)|^{p+1})\, dx \\
&\leq (p + 1)^2 (\|v\|_{L^{p+1}(\Omega)}^{p-1} \|w\|_{L^{p+1}(\Omega)}^2 + \|w\|_{L^{p+1}(\Omega)}^{p+1}).
\end{aligned}
$$

Since $H_0^1(\Omega)$ is continuously embedded into $L^{p+1}(\Omega)$, it follows that $R(w) = \|w\|_{L^{p+1}(\Omega)} \epsilon(w)$ with $\epsilon(w) \to 0$ (in $H_0^1(\Omega)$) when $w \to 0$ in $H_0^1(\Omega)$. This proves that F is differentiable at the point v and that

$$
\langle dF(v), w \rangle_{H^{-1}(\Omega), H_0^1(\Omega)} = \int_\Omega \int_\Omega (p + 1)v(x)^p \operatorname{sgn}(v(x))w(x)\, dx.
$$

Note that owing to Hölder's inequality,

$$
\begin{aligned}
|\langle dF(v), w \rangle_{H^{-1}(\Omega), H_0^1(\Omega)}| &\leq (p + 1) \||v|^p\|_{L^{1+\frac{1}{p}}(\Omega)}^{\frac{p}{p+1}} \|w\|_{L^{p+1}(\Omega)} \\
&\leq (p + 1) \|v\|_{L^{p+1}(\Omega)}^p C_p \|w\|_{H_0^1(\Omega)},
\end{aligned}
$$

where C_p is a number such that, for all $w \in H_0^1(\Omega)$, $\|w\|_{L^{p+1}(\Omega)} \leq C_p \|w\|_{H_0^1(\Omega)}$.

234   3 Quasi-Linear Elliptic Problems

Hence

$$\|dF(v)\|_{H^{-1}(\Omega)} \le (p+1) \|v\|_{L^{p+1}(\Omega)}^{p}$$
$$\le (p+1)C_p^p \|v\|_{H_0^1(\Omega)}^{p}.$$

We now show the continuity of $v \mapsto dF(v)$ from $H_0^1(\Omega)$ to $H^{-1}(\Omega)$. Let $(v_n)_{n\in\mathbb{N}}$ be a sequence in $H_0^1(\Omega)$ such that $v_n \to v$ in $H_0^1(\Omega)$ as $n \to +\infty$. We also have $v_n \to v$ in $L^{p+1}(\Omega)$ as $n \to +\infty$. From the definition of $dF(v_n)$ and $dF(v)$ we get

$$\|dF(v_n) - dF(v)\|_{H^{-1}(\Omega)}$$
$$\le (p+1)C_p \|v_n(x)^p \operatorname{sgn}(v_n(x)) - v(x)^p \operatorname{sgn}(v(x))\|_{L^{1+\frac{1}{p}}(\Omega)}^{\frac{p}{p+1}}.$$

We can assume that there exists a $g \in L^{p+1}(\Omega)$ such that (up to a subsequence)

$$v_n \to v \text{ a.e. when } n \to +\infty,$$
$$|v_n| \le g \text{ a.e. and for all } n \in \mathbb{N},$$

Thus, owing to the dominated convergence theorem,

$$v_n(x)^p \operatorname{sgn}(v_n(x)) \to v(x)^p \operatorname{sgn}(v(x)) \text{ in } L^{1+\frac{1}{p}}(\Omega) \text{ when } n \to +\infty.$$

Since the limit does not depend on the chosen subsequence, a classical proof by contradiction shows that the whole sequence converges. We have thus shown that $dF(v_n) \to dF(v)$ in $H^{-1}(\Omega)$ and therefore that $F \in C^1(H_0^1(\Omega), \mathbb{R})$.

We can now apply Theorem 3.34. Since u is not the null function (because $F(u) = 1$), $\operatorname{Im}(dF(u) = \mathbb{R})$ and Theorem 3.34 gives the existence of $\lambda \in \mathbb{R}$ such that $dE(u) = \lambda df(u)$, that is, as $u \ge 0$ a.e., for all $v \in H_0^1(\Omega)$,

$$\int_\Omega \nabla u(x) \cdot \nabla v(x)\, dx = \lambda \int_\Omega \int_\Omega (p+1)u(x)^p v(x)\, dx. \tag{3.69}$$

Since $u \in H_0^1(\Omega)$ and $u \ne 0$ (in $H_0^1(\Omega)$), the left-hand term of (3.69) is positive for $v = u$. Hence $\lambda > 0$. We now choose $\theta > 0$ such that $\theta = (p+1)\lambda\theta^p$ (which is possible because $p > 1$), the function $\bar{u} = \theta u$ is a weak solution to (3.35), that is, it satisfies

$$\bar{u} \in H_0^1(\Omega),$$
$$\int_\Omega \nabla \bar{u}(x) \cdot \nabla v(x)\, dx = \int_\Omega \int_\Omega \bar{u}(x)^{p-1}\bar{u}(x)v(x)\, dx \text{ for all } v \in H_0^1(\Omega).$$

Problem 3.10 (Weak convergence and non-linearity)

1. We note that

$$\|u_n - u\|_2^2 = \int u_n^2 \, dm + \int u^2 \, dm - 2 \int u_n u \, dm. \qquad (3.70)$$

Since $u_n \to u$ weakly in L^2, we have $\lim_{n \to +\infty} \int u_n u \, dm = \int u^2 \, dm$. We then deduce from (3.70) that $u_n \to u$ in L^2 if and only if $\lim_{n \to +\infty} \int u_n^2 \, dm = \int u^2 \, dm$.

2. We start by noting that $\varphi(w) \in L^2$ (thanks to the hypotheses on φ and $m(X) < +\infty$). We then have

$$\int (\varphi(u_n) - \varphi(w))(u_n - w) \, dm = \int (v_n u_n - v_n w - \varphi(w)u_n + \varphi(w)w) \, dm.$$

The weak convergences of u_n and v_n to u and v give

$$\lim_{n \to +\infty} \int u_n \varphi(w) \, dm = \int u\varphi(w) \, dm \text{ and } \lim_{n \to +\infty} \int v_n w \, dm = \int vw \, dm.$$

Finally, we have, by hypothesis, $\lim_{n \to +\infty} \int u_n v_n \, dm = \int uv \, dm$. Hence

$$\lim_{n \to +\infty} (\varphi(u_n) - \varphi(w))(u_n - w) \, dm = \int (v - \varphi(w))(u - w) \, dm.$$

3. a. We use here (3.36) with $w = u + t\bar{w}$. We obtain, as $n \to +\infty$,

$$\int (\varphi(u_n) - \varphi(u + t\bar{w}))(u_n - u - t\bar{w}) \, dm \to - \int (v - \varphi(u + t\bar{w}))t\bar{w} \, dm.$$

Since φ is non-decreasing, we have $(\varphi(u_n) - \varphi(u + t\bar{w}))(u_n - u - t\bar{w}) \geq 0$ a.e. and therefore $\int (\varphi(u_n) - \varphi(u + t\bar{w}))(u_n - u - t\bar{w}) \, dm \geq 0$. Passing to the limit as $n \to +\infty$ yields $\int (v - \varphi(u + t\bar{w}))t\bar{w} \, dm \leq 0$.

Taking $t = \frac{1}{m}$ ($m \in \mathbb{N}^\star$), we therefore have $\int (v - \varphi(u + \frac{\bar{w}}{m}))\bar{w} \leq 0$. By applying the dominated convergence theorem (notice that $|(v - \varphi(u + \frac{\bar{w}}{m}))\bar{w}| \leq F$ a.e. with $F = (|v| + C + C|u| + C|\bar{w}|)|\bar{w}| \in L^1$), we obtain, when $m \to \infty$,

$$\int (v - \varphi(u))\bar{w} \, dm \leq 0.$$

Similarly, taking $t = -\frac{1}{m}$, we show $\int (v - \varphi(u))\bar{w} \, dm \geq 0$. Hence $\int (v - \varphi(u))\bar{w} \, dm = 0$.

   b. Setting $\bar{w} = \mathbb{1}_A - \mathbb{1}_{A^c}$, with $A = \{x \in X ; (v - \varphi(u))(x) \geq 0\}$, the previous question then gives $\int |v - \varphi(u)| \, dm = 0$ and therefore $v = \varphi(u)$ a.e.

4. a. Taking $w = u$ in (3.36), we obtain $\lim_{n \to +\infty} \int G_n \, dm = 0$. Since φ is non-decreasing, we have $G_n \geq 0$ a.e., for all $n \in \mathbb{N}$ and therefore $\|G_n\|_1 = \int G_n \, dm$. We therefore deduce that $G_n \to 0$ in L^1.

b. Since $G_n \to 0$ in L^1, there exists a subsequence of the sequence $(G_n)_{n\in\mathbb{N}}$, denoted by $(G_{\psi(n)})_{n\in\mathbb{N}}$ (with ψ increasing from $\mathbb{N}$ to $\mathbb{N}$) such that $G_{\psi(n)} \to 0$ a.e. (this is a partial converse of the dominated convergence theorem). Therefore, there exists an $A \in \mathcal{T}$ such that $m(A) = 0$ and $G_{\psi(n)}(x) \to 0$ as $n \to +\infty$ if $x \in A^c$.

Let $x \in A^c$. Set $a = u(s)$. For $s \in \mathbb{R}$, we set $f(s) = (\varphi(s) - \varphi(a))|s - a|$. Since φ is increasing continuous, the function f is also increasing continuous. It therefore admits an inverse function, denoted by g, which is continuous. Since $|f(u_{\psi(n)}(x))| = G_{\psi(n)}(x) \to 0$, we have $f(u_{\psi(n)}(x)) \to 0$ and therefore $u_{\psi(n)}(x) = g(f(u_{\psi(n)}(x))) \to g(0) = a$. Hence $\lim_{n\to+\infty} u_{\psi(n)}(x) = u(x)$ for all $x \in A^c$, which proves that $u_{\psi(n)} \to u$ a.e.

c. We show that $u_n \to u$ in L^p for all $p \in [1, 2[$ by reasoning by contradiction. We suppose that there exists a $p \in [1, 2[$ such that $(u_n)_{n\in\mathbb{N}}$ does not converge to u in L^p. There then exist $\varepsilon > 0$ and a subsequence of the sequence $(u_n)_{n\in\mathbb{N}}$ which remains outside the ball (of L^p) centred at u and of radius ε. By the reasoning of the previous question, from this subsequence, one can extract a subsequence, denoted by $(u_n)_{\psi(n)}$, which converges a.e. to u. Since the sequence $(u_n)_{n\in\mathbb{N}}$ is bounded in L^2, one can then show that this subsequence converges in L^p to u (this is a consequence of Vitali's theorem, see footnote 11, page 202). This is in contradiction with the fact that this subsequence remains outside the ball (of L^p) centred at u and of radius ε. We have thus shown that $u_n \to u$ in L^p for all $p \in [1, 2[$.

Problem 3.11 (Leray–Lions Operator)

1. The first assumption on a (Carathéodory function) is used to ensure that by choosing measurable representatives for u and ∇u the function $x \mapsto a(x, u(x), \nabla u(x))$ is measurable. We do not detail this point here (see, for example, [26], problem 7.14, for a proof). The third assumption on a (growth) then ensures that $a(\cdot, u, \nabla u) \in L^{p'}(\Omega)$ as soon as $u \in W^{1,p}(\Omega)$. The equation requested in problem (3.37) makes sense (since we integrate the product of a function from $L^{p'}(\Omega)$ with a function from $L^p(\Omega)$).

   As in the proof of Theorem 3.25, we choose a countable family $(f_n)_{n\in\mathbb{N}^\star}$ dense in $W_0^{1,p}(\Omega)$ and we set $E_n = \mathrm{Vect}\{f_1, \ldots, f_n\}$ the vector space generated by the first n functions of this family.

   We fix $n \in \mathbb{N}^\star$ and we look for a solution u_n of the following problem, posed in finite dimension:

$$\begin{cases} u_n \in E_n, \ : \ \forall v \in E_n, \\ \displaystyle\int_\Omega a(x, u_n(x), \nabla u_n(x)) \cdot \nabla v(x)\, dx = \langle f, v \rangle_{W^{-1,p'}(\Omega), W_0^{1,p}(\Omega)}. \end{cases} \tag{3.71}$$

The mapping $b_n : v \mapsto \langle f, v \rangle_{W^{-1,p'}, W_0^{1,p}}$ is a linear (and therefore) continuous mapping of E_n to $\mathbb{R}$. For $u \in E_n$, let $T_n(u)$ the mapping from E_n to $\mathbb{R}$ which associates $v \in E_n$ with $\int_\Omega a(x, u_n(x), \nabla u_n(x)) \cdot \nabla v(x)\, dx$. This mapping is linear, so it also belongs to E_n' and the problem (3.71) consists in looking for $u_n \in E_n$ such that $T_n(u_n) = b_n$. The existence of u_n is therefore a consequence of Lemma 3.29 if we show that T_n is continuous and coercive.

The continuity of T_n is shown as in the proof of Theorem 3.25. It is enough to verify that $v_m \to v$ in $W_0^{1,p}(\Omega)$ when $m \to +\infty$ implies $a(\cdot, v_m, \nabla v_m) \to a(\cdot, v, \nabla v)$ in $L^{p'}(\Omega)$ when $m \to +\infty$. Let us then assume that $v_m \to v$ in $W_0^{1,p}(\Omega)$ when $m \to +\infty$, that is $v_m \to v$ and $D_i v_m \to D_i v$ (for all i) in $L^p(\Omega)$ when $m \to +\infty$. By the dominated convergence partial converse ([26], theorem 6.11), we may thus assume that there exists a $g \in L^p(\Omega)$ such that, up to a subsequence,

$$u_m \to u \ \text{a.e., when } m \to +\infty,$$

$$D_i u_m \to D_i u \ \text{a.e., when } m \to +\infty, \ \text{for } i = 1, \ldots, N,$$

$$|u_m| \leq g \ \text{a.e. and for all } m \in \mathbb{N},$$

$$|\nabla u_m| \leq g \ \text{a.e. and for all } m \in \mathbb{N}.$$

It is now sufficient (thanks to the growth assumption on a) to apply the dominated convergence theorem to deduce that $a(\cdot, v_m, \nabla v_m) \to a(\cdot, v, \nabla v)$ in $L^{p'}(\Omega)$ when $m \to +\infty$. Finally, since the limit does not depend on the chosen subsequence, this convergence occurs without extraction of a subsequence. As in the proof of Theorem 3.25, this allows us to show that T_n is continuous (from E_n to E_n).

The coercivity assumption of a gives the coercivity of T_n, exactly as in the proof of Theorem 3.25. The existence of a solution u_n to (3.71) is therefore a consequence of Lemma (3.29). It is now necessary to take the limit as $n \to +\infty$ to obtain a solution to (3.37).

Taking $v = u_n$ in (3.71) we show, as in step II-1 (page 197) of the proof of Theorem 3.25, that the sequence $(u_n)_{n \in \mathbb{N}}$ is bounded in $W_0^{1,p}(\Omega)$. We can therefore assume that, up to a subsequence, $u_n \to u$ weakly in $W_0^{1,p}(\Omega)$. The growth assumption on a shows that the sequence $(|a(\cdot, u_n, \nabla u_n)|)_{n \in \mathbb{N}}$ is bounded in $(L^{p'}(\Omega))^N$. Hence there exists $\zeta \in (L^{p'}(\Omega))^N$ such that, after a new extraction of a subsequence,

$$a(\cdot, u_n, \nabla u_n) \to \zeta \quad \text{weakly in} \quad (L^{p'}(\Omega))^N.$$

The proof of Theorem 3.25 then gives (without modification)

$$\int_\Omega \zeta \cdot \nabla v\, dx = \langle f, v \rangle_{W^{-1,p'}, W_0^{1,p}} \quad \forall v \in W_0^{1,p}(\Omega).$$

It is now necessary to show, as in the proof of Theorem 3.25, thanks to the strict monotonicity of a, that $\zeta = a(\cdot, u, \nabla u)$.

We start by noting that the proof of the common step to Minty and Leray–Lions (the beginning of step II-3 in the proof of Theorem 3.25) gives here

$$\lim_{n \to +\infty} \int_\Omega a(x, u_n(x), \nabla u_n(x)) \cdot \nabla u_n(x) \ dx = \int_\Omega \zeta(x) \cdot \nabla u(x) \ dx. \qquad (3.72)$$

The rest of the proof is close to that given for Theorem 3.25. The new difficulty here is the presence of u_n in $a(\cdot, u_n, \nabla u_n)$. With this aim, we note that $u_n \to u$ weakly in $W_0^{1,p}(\Omega)$ and therefore, by Rellich's theorem (Theorem 1.36), $u_n \to u$ in $L^p(\Omega)$ and we can therefore assume that $u_n \to u$ a.e. up to a subsequence, and that there exists a $g \in L^p(\Omega)$ such that, for all $n \in \mathbb{N}$, $|u_n| \leq g$ a.e..

We now resume the reasoning of the proof of Theorem 3.25. Since $\overline{\cup_{n \in \mathbb{N}} E_n} = W_0^{1,p}(\Omega)$, there exists a sequence $(v_n)_{n \in \mathbb{N}}$ such that $v_n \in E_n$ for all $n \in \mathbb{N}$ and $v_n \to u$ in $W_0^{1,p}(\Omega)$. We can also assume that, up to a subsequence, there exists a $G \in L^p(\Omega)$ such that

$$v_n \to u \ \text{ a.e., when } n \to +\infty,$$
$$D_i v_n \to D_i u \ \text{ a.e., when } n \to +\infty, \text{ for } i = 1, \dots, N,$$
$$|v_n| \leq G \ \text{ a.e. and for all } n \in \mathbb{N},$$
$$|\nabla v_n| \leq G \ \text{ a.e. and for all } n \in \mathbb{N}.$$

This gives, in particular, by the dominated convergence theorem that $a(\cdot, v_n, \nabla v_n) \to a(\cdot, u, \nabla u)$ in $L^{p'}(\Omega)$ but also that $a(\cdot, u_n, \nabla v_n) \to a(\cdot, u, \nabla u)$ in $L^{p'}(\Omega)$ as $n \to +\infty$. Let us set

$$F_n(x) = (a(x, u_n(s), \nabla u_n(x)) - a(x, u_n(x), \nabla v_n(x))) \cdot (\nabla u_n - \nabla v_n) \ dx.$$

By the monotonicity assumption on a, we have $\int_\Omega F_n(x) \ dx \geq 0$. Expanding, we get

$$\int_\Omega F_n(x) \ dx = \int_\Omega a(x, u_n(s), \nabla u_n(x)) \cdot \nabla u_n(x) \ dx$$
$$- \int_\Omega a(x, u_n(s), \nabla u_n(x)) \cdot \nabla v_n(x) \ dx - \int_\Omega (a(x, u_n(s), \nabla v_n(x)) \cdot \nabla u_n(x) \ dx$$
$$+ \int_\Omega a(x, u_n(x), \nabla v_n(x)) \cdot \nabla v_n(x) \ dx \geq 0.$$

We pass to the limit as $n \to +\infty$ in the four terms of the right-hand side of this equality.

- Thanks to (3.72), as $n \to +\infty$, the first term tends towards $\int_\Omega \zeta(x) \cdot \nabla u(x) \ dx$.
- The second term also tends towards $\int_\Omega \zeta(x) \cdot \nabla u(x) \ dx$ (by Lemma 3.26).
- The third term tends towards $\int_\Omega a(x, u(x), \nabla u(x)) \cdot \nabla u(x) \ dx$ (by Lemma 3.26).
- The fourth term also tends towards $\int_\Omega a(x, u(x), \nabla u(x)) \cdot \nabla u(x) \ dx$.

Therefore, we have that $F_n \geq 0$ a.e. and $\int_\Omega F_n(x)\,dx \to 0$ when $n \to +\infty$. Hence $F_n \to 0$ in $L^1(\Omega)$ and we can assume that, up to a subsequence, $F_n \to 0$ a.e.. We also know that $\nabla v_n \to \nabla u$ a.e. as $n \to +\infty$. The goal is to show that $\nabla u_n \to \nabla u$ a.e..

Let $x \in \Omega$ such that $F_n(x) \to 0$ and $\nabla v_n(x) \to \nabla u(x)$ as $n \to +\infty$. We also assume that for this value of x, the hypotheses on a (growth, coercivity and monotonicity) are satisfied and that $\nabla v_n(x)$, $u_n(x)$ and $v_n(x)$ converge (in $\mathbb{R}$) towards $\nabla u(x)$, $u(x)$ and $u(x)$. These hypotheses on x only remove a set of measure zero points, that is, they are satisfied for $x \in A$ with A of measure zero. We first show that the sequence $(\nabla u_n(x))_{n \in \mathbb{N}}$ is bounded in $\mathbb{R}^N$. Indeed, thanks to the hypotheses (3.18c) (3.18d) of coercivity and growth on a, we have

$$F_n(x) \geq \alpha|\nabla u_n(x)|^p - C(d(x) + |u_n(x)|^{p-1} + \nabla u_n(x)|^{p-1})|\nabla v_n(x)|$$
$$- C(d(x) + |u_n|^{p-1} + |\nabla v_n(x)|^{p-1})|\nabla u_n(x)| + \alpha|\nabla v_n(x)|^p.$$

Hence if the sequence $(\nabla u_n(x))_{n \in \mathbb{N}}$ is unbounded, then the sequence $(F_n(x))_{n \in \mathbb{N}}$ is unbounded, which contradicts the fact that $F_n(x) \to 0$ when $n \to +\infty$. We conclude that the sequence $(\nabla u_n(x))_{n \in \mathbb{N}}$ is bounded. Now let $\xi \in \mathbb{R}^N$ be the limit of a subsequence of the sequence $(\nabla u_n(x))_{n \in \mathbb{N}}$, still denoted as $(\nabla u_n(x))_{n \in \mathbb{N}}$. Since x is fixed, this subsequence satisfies $\lim_{n \to +\infty} a(x, u_n(x), \nabla u_n(x)) = a(x, u(x), \xi)$. Since $\lim_{n \to +\infty} F_n(x) = 0$, we get

$$(a(x, u(x), \xi) - a(x, u(x), \nabla u(x)) \cdot (\xi - \nabla u(x)) = 0.$$

However, the first term of this equality is positive if $\xi \neq \nabla u(x)$, so that $\xi = \nabla u(x)$. Therefore, (without extracting a subsequence) $\nabla u_n(x) \to \nabla u(x)$ when $n \to +\infty$.

In summary, we thus showed that $\nabla u_n \to \nabla u$ a.e. (more precisely $\nabla u_n(x) \to \nabla u(x)$ as $n \to +\infty$ for all $x \in A$). Hence $a(\cdot, u, \nabla u_n) \to a(\cdot, u, \nabla u)$ a.e., so that $\zeta = a(\cdot, u, \nabla u)$ exactly as in the proof of Theorem 3.25. We have thus shown that u is a solution of Problem (3.37).

2. Here again we reason as in the proof of the Theorem 3.25. Since $\zeta = a(\cdot, u, \nabla u)$, (3.72) gives

$$\lim_{n \to +\infty} \int_\Omega a(x, u_n(x), \nabla u_n(x)) \cdot \nabla u_n(x)\,dx = \int_\Omega a(x, u(x), \nabla u(x)) \cdot \nabla u(x)\,dx.$$

We apply Lemma 3.31 to the sequence $(f_n)_{n \in \mathbb{N}}$ defined by $f_n = a(\cdot, u_n, \nabla u_n) \cdot \nabla u_n$. It gives the convergence in $L^1(\Omega)$ of this sequence and therefore the equi-integrability of the sequence $(f_n)_{n \in \mathbb{N}}$. With the coercivity assumption on a, we obtain the equi-integrability of the sequence $(|\nabla u_n|^p)_{n \in \mathbb{N}}$. It remains to apply Vitali's theorem to conclude that $\nabla u_n \to \nabla u$ in $L^p(\Omega)^N$ and therefore that $u_n \to u$ in $W_0^{1,p}(\Omega)$.

Problem 3.12 (Lagrange multipliers)

Note that $dg_i(u)$ (for $i = 1, \ldots, p$) and $df(u)$ are elements of E' and $dg(u) \in \mathcal{L}(E, R^p)$. Since $\mathrm{Im}(dg(u)) = \mathbb{R}^p$, for all $i = 1, \ldots, p$ there exists a $v_i \in E$ such that $dg(u)(v_i) = e_i$, where $\{e_1, \ldots, e_p\}$ is the canonical basis of $\mathbb{R}^p$.

Define the function G from $E \times \mathbb{R}^p$ to $\mathbb{R}$ by $G(w, t) = g(u + w + \sum_{i=1}^{p} t_i v_i)$, where $t_1, \ldots t_p$ are the components of t.

Since $g \in C^1(E, \mathbb{R})$, $G \in C^1(E \times \mathbb{R}^p, \mathbb{R})$. We denote by $d_1 G$ (resp. $d_2 G$), the differential of G with respect to the first (resp. second) variable; thus $d_1 G(w, t) \in \mathcal{L}(E, \mathbb{R}) = E'$ and $d_2 G(w, t) \in \mathcal{L}(\mathbb{R}^p, \mathbb{R}^p)$ (which we can identify with a matrix of p rows and p columns).

For $h = (h_1, \ldots, h_p)^t \in \mathbb{R}^p$, $d_2 G(0, 0)(h) = dg(u)(h \cdot v) = \sum_{i=1}^{p} h_i dg(u)(v_i) = Ah = h$. Since $d_2 G(0, 0)$ is invertible, the implicit function theorem gives the existence of $\varepsilon > 0$ and $\eta > 0$ such that for all $w \in B(0, \eta)$ (open ball of E centred at 0 and radius η), there exists a unique $t \in B(0, \varepsilon)$ (open ball of $\mathbb{R}^p$ centred at 0 and radius ε) such that $G(w, t) = 0$. We denote this value of t by $\phi(w)$ (and $(\phi(w)_i$ are therefore the components of $\phi(w)$). The implicit function theorem also gives that ϕ is of class C^1 (note that, for example, $d\phi(0) \in \mathcal{L}(E, \mathbb{R}^p)$).

Since $g(u + w + \sum_{i=1}^{p} \phi(w)_i v_i) = 0$ for all $w \in B(0, \eta)$, a first order development of g (at point u) and of ϕ (at point 0) gives, denoting by $d\phi(0)(w)_i$ the components of $d\phi(0)(w)$ (which is a vector of $\mathbb{R}^p$),

$$dg(u)\left(w + \sum_{i=1}^{p} d\phi(0)(w)_i dg(u)(v_i)\right) = 0,$$

which gives, since $dg(u)(v_i) = e_i$,

$$d\phi(0)(w) = -dg(u)(w) \text{ for all } w \in E.$$

For all $w \in B(0, \eta)$, $f(u) \le f(u + w + \sum_{i=1}^{p} \phi(w)_i v_i)$ (since $u + w + \sum_{i=1}^{p} \phi(w)_i v_i \in A$) and therefore with a first order development of f and ϕ (recall that f is differentiable at point u) $df(u)(w + \sum_{i=1}^{p} d\phi(0)(w)_i v_i)) = 0$, that is

$$0 = df(u)(w) + \sum_{i=1}^{p} d\phi(0)(w)_i df(u)(v_i).$$

This gives, for all $w \in E$,

$$df(u)(w) = -\sum_{i=1}^{p} df(u)(v_i) dg(u)(w)_i = -\sum_{i=1}^{p} \lambda_i dg_i(u)(w)$$

with $\lambda_i = df(u)(v_i)$ (since $dg(u)(w)_i = dg_i(u)(w)$).

Problem 3.13 (Pohozaev's identity in $\mathbb{R}^N$)

1. It is sufficient to apply the regularity theorem in $\mathbb{R}^N$ given in Remark 2.23. Indeed, let $\varphi \in \mathcal{D}(\mathbb{R}^N)$. Since $u \in L^\infty_{\text{loc}}(\mathbb{R}^N)$ and $|\nabla u| \in L^2(\mathbb{R}^N)$, $\varphi u \in H^1(\mathbb{R}^N)$. (Note that $D_i(\varphi u) = u\partial_i\varphi + \varphi D_i u \in L^2(\mathbb{R}^N)$, i.e. that $D_i u$, an element of $\mathcal{D}^\star(\mathbb{R}^N)$, is identified with $u\partial_i\varphi + \varphi D_i u$, an element of $L^2(\mathbb{R}^N)$, see Remark 1.7.) Then, since $g(u)\varphi \in L^2(\mathbb{R}^N)$ and $\varphi u \in H^1(\mathbb{R}^N)$,

$$
\Delta(\varphi u) = u\Delta\varphi + \varphi\Delta u + 2\sum_{i=1}^N (D_i u)\partial_i\varphi
$$

$$
= u\Delta\varphi + \varphi g(u) + 2\sum_{i=1}^N (D_i u)\partial_i\varphi \in L^2(\mathbb{R}^N).
$$

Hence $\varphi u \in H^1(\mathbb{R}^N)$ and $\Delta(\varphi u) \in L^2(\mathbb{R}^N)$; the regularity theorem in $\mathbb{R}^N$ (Remark 2.23) then gives $\varphi u \in H^2(\mathbb{R}^N)$.

We conclude this question by briefly recalling the calculation of $\Delta(\varphi u)$. Let $\psi \in \mathcal{D}(\mathbb{R}^N)$,

$$
\langle -\Delta(\varphi u), \psi \rangle_{\mathcal{D}^\star(\mathbb{R}^N), \mathcal{D}(\mathbb{R}^N)} = -\int_{\mathbb{R}^N} \varphi(x)u(x)\Delta\psi(x)\,\mathrm{d}x.
$$

But, $\varphi(x)\Delta\psi(x) = \Delta(\varphi\psi)(x) - 2\sum_{i=1}^N \partial_i\varphi(x)\partial_i\psi(x) - \psi(x)\Delta\varphi(x)$ and therefore

$$
\langle -\Delta(\varphi u), \psi \rangle_{\mathcal{D}^\star(\mathbb{R}^N), \mathcal{D}(\mathbb{R}^N)}
$$

$$
= -\int_{\mathbb{R}^N} \left(u(x)\Delta(\varphi\psi)(x)\,\mathrm{d}x + 2\sum_{i=1}^N u(x)\partial_i\varphi(x)\partial_i\psi(x) + u(x)\psi(x)\Delta\varphi(x)\right)\mathrm{d}x
$$

$$
= -\langle \Delta u, \varphi\psi \rangle_{\mathcal{D}^\star(\mathbb{R}^N), \mathcal{D}(\mathbb{R}^N)}
$$

$$
- 2\sum_{i=1}^N \langle D_i(u\partial_i\varphi), \psi \rangle_{\mathcal{D}^\star(\mathbb{R}^N), \mathcal{D}(\mathbb{R}^N)} + \langle u\Delta\varphi, \psi \rangle_{\mathcal{D}^\star(\mathbb{R}^N), \mathcal{D}(\mathbb{R}^N)}.
$$

Finally, since $D_i(u\partial_i\varphi) = (D_i u)\partial_i\varphi + u\partial^2_{i,i}\varphi$ (see Remark 1.7),

$$
\langle -\Delta(\varphi u), \psi \rangle_{\mathcal{D}^\star(\mathbb{R}^N), \mathcal{D}(\mathbb{R}^N)}
$$

$$
= -\langle \varphi\Delta u, \psi \rangle_{\mathcal{D}^\star(\mathbb{R}^N), \mathcal{D}(\mathbb{R}^N)}
$$

$$
- 2\sum_{i=1}^N \langle (D_i u)\partial_i\varphi, \psi \rangle_{\mathcal{D}^\star(\mathbb{R}^N), \mathcal{D}(\mathbb{R}^N)} - \langle u\Delta\varphi, \psi \rangle_{\mathcal{D}^\star(\mathbb{R}^N), \mathcal{D}(\mathbb{R}^N)}.
$$

2. It is sufficient to use the classical regularisation process by convolution. Let $\rho \in C_c^\infty(\mathbb{R}^N, \mathbb{R})$ such that $\rho(x) \geq 0$ for all x, $\rho(x) = 0$ if $|x| > 1$ and $\int_{\mathbb{R}^N} \rho(x)\, dx = 1$. For $n \in \mathbb{N}^\star$, we define ρ_n by $\rho_n(x) = n^N \rho(nx)$ and $u_n = u \star \rho_n$.
   Property (2a) is due to the fact that $u \in L^\infty_{\text{loc}}$. The fact that the restriction of u_n to B_R is equal (in B_R) to the convolution with ρ_n of the restriction of u to B_{R+1} allows us to show that $u_n \to u$ in $H^2(B_R)$ (recall for this that $\partial_i u_n = D_i u \star \rho_n$ and $\partial^2_{i,j} u_n = D^2_{i,j} u \star \rho_n$) and in particular gives (2c). Finally, the convergence (up to a subsequence) of u_n to u in $L^2(B_R)$ gives a pointwise convergence. Thanks to the diagonal process, this subsequence can be chosen independent of n (and we therefore obtain (2b)) but this independence is in fact unimportant for this problem.

3. a. Since $g(u) \in L^1_{\text{loc}}(\mathbb{R}^N)$, we have

$$\langle g(u), h_n \rangle_{\mathcal{D}^\star(\mathbb{R}^N), \mathcal{D}(\mathbb{R}^N)} = \int_{\mathbb{R}^N} g(u(x)) h_n(x)\, dx.$$

   Then, the dominated convergence theorem gives $g(u_n) \to g(u)$ in $L^2(B_R)$. On the other hand, the sequence $(\partial_i u_n)_{n \in \mathbb{N}^\star}$ is bounded in $L^2(B_R)$. Hence

$$\lim_{n \to +\infty} \int_{\mathbb{R}^N} (g(u_n(x)) - g(u(x))\psi(|x|) \sum_{i=1}^N x_i \partial_i u_n(x)\, dx = 0.$$

   Then,

$$\int_{\mathbb{R}^N} g(u_n(x)) \psi(|x|) \sum_{i=1}^N x_i \partial_i u_n(x)\, dx = \int_{\mathbb{R}^N} \psi(|x|) \sum_{i=1}^N x_i \partial_i G(u_n)(x)\, dx$$

$$= -N \int_{\mathbb{R}^N} G(u_n)(x)\psi(|x|)\, dx - \int_{\mathbb{R}^N} G(u_n)(x)\psi'(|x|)|x|\, dx.$$

   (Recall that $G(u_n)$ denotes the function $G \circ u_n$.)
   The dominated convergence theorem gives $G(u_n) \to G(u)$ in $L^2(B_R)$. Thus (3.39) holds.

   b. We have

$$\langle -\Delta u, h_n \rangle_{\mathcal{D}^\star(\mathbb{R}^N), \mathcal{D}(\mathbb{R}^N)} = \sum_{i=1}^N \int_{\mathbb{R}^N} D_i u(x) \cdot \partial_i h_n(x)\, dx.$$

   Since $\partial_i u_n \to D_i u$ in $L^2(B_R)$ and the first and second derivatives of the sequence $(u_n)_{n \in \mathbb{N}^\star}$ are bounded in $L^2(B_R)$,

$$\lim_{n \to +\infty} \sum_{i=1}^N \int_{\mathbb{R}^N} (D_i u(x) - \partial_i u_n(x)) \cdot \partial_i h_n(x)\, dx = 0.$$

Then,

$$
\sum_{i=1}^{N} \int_{\mathbb{R}^N} \partial_i u_n(x) \cdot \partial_i h_n(x) \, dx
$$

$$
= \int_{\mathbb{R}^N} \psi(|x|) \sum_{i=1}^{N} \sum_{j=1}^{N} \partial_i u_n(x) \partial_i (x_j \partial_j u_n)(x) \, dx
$$

$$
+ \int_{\mathbb{R}^N} \psi'(|x|) \frac{x_i}{|x|} \sum_{i=1}^{N} \sum_{j=1}^{N} \partial_i u_n(x) x_j \partial_j u_n(x) \, dx,
$$

and

$$
\int_{\mathbb{R}^N} \psi(|x|) \sum_{i=1}^{N} \sum_{j=1}^{N} \partial_i u_n(x) \partial_i (x_j \partial_j u_n)(x) \, dx
$$

$$
= \int_{\mathbb{R}^N} \psi(|x|) \sum_{i=1}^{N} \sum_{j=1}^{N} \delta_{i,j} \partial_i u_n(x) \partial_j u_n(x) \, dx
$$

$$
+ \int_{\mathbb{R}^N} \psi(|x|) \sum_{i=1}^{N} \sum_{j=1}^{N} x_j \partial_j \frac{(\partial_i u_n)^2}{2}(x) \, dx
$$

$$
= \int_{\mathbb{R}^N} \psi(|x|) |\nabla u_n(x)|^2 \, dx - \frac{N}{2} \int_{\mathbb{R}^N} \psi(|x|) |\nabla u_n(x)|^2 \, dx
$$

$$
- \int_{\mathbb{R}^N} \psi'(|x|) \frac{|x|}{2} |\nabla u_n(x)|^2 \, dx.
$$

Since $|\nabla u_n(x)|^2 \to |\nabla u(x)|^2$ in $L^1(B_R)$ and $\partial_i u_n \to D_i u$ in $L^2(B_R)$ as $n \to +\infty$, (3.40) follows.

4. The convergences (3.39) and (3.40) give, for all $\psi \in C_c^\infty(\mathbb{R}, \mathbb{R})$ such that $\psi(r) = 0$ for $r > 1$,

$$
\frac{N-2}{2} \int_{\mathbb{R}^N} |\nabla u(x)|^2 \psi(|x|) \, dx = N \int_{\mathbb{R}^N} G(u(x)) \psi(|x|) \, dx + R_\psi, \qquad (3.73)
$$

with

$$
R_\psi = \int_{\mathbb{R}^N} \psi'(|x|) \frac{x_i}{|x|} \sum_{i=1}^{N} \sum_{j=1}^{N} D_i u(x) x_j D_j u(x) \, dx
$$

$$
- \int_{\mathbb{R}^N} \psi'(|x|) \frac{|x|}{2} |\nabla u(x)|^2 \, dx + \int_{\mathbb{R}^N} G(u(x)) \psi'(|x|) |x| \, dx.
$$

For $n \in \mathbb{N}^\star$, we choose $\psi_n \in C_c^\infty(\mathbb{R}, \mathbb{R})$ such that $\psi_n(r) = 1$ if $r < n$, $\psi_n(r) = 0$ if $r > 3n$ and $\psi(r) \in [0, 1]$, $\psi'(r) \leq \frac{1}{n}$ if $n < |x| < 3n$. The fact that $G(u) \in L^1(\mathbb{R}^N)$ and $|\nabla u| \in L^2(\mathbb{R})$ gives $\lim_{n \to +\infty} R_{\psi_n} = 0$. The equality (3.38) then follows from (3.73) written with $\psi = \psi_n$ using the dominated convergence theorem.

Chapter 4
Parabolic Problems

Parabolic PDEs are used to describe a wide variety of phenomena, particularly time-dependent phenomena, including heat conduction, particle diffusion and the valuation of financial products. The basic example of a parabolic PDE is the one-dimensional heat equation, $\partial_t u = \alpha \partial_{xx}^2 u$, where $u(x, t)$ is the temperature at time t and at position x along a thin rod, and $\alpha \in \mathbb{R}_+^\star$ is the thermal diffusivity. For a body in 2 or 3 dimensions, the heat equation reads $u_t = \alpha \Delta u$, where Δ is the Laplace operator. The fact that $-\Delta$ is an elliptic operator (see Chapter 2) suggests a broader definition of a parabolic PDE: $u_t = -Au$, where A is a second order elliptic operator.

In the following section, we give a quick overview of classical solutions to parabolic equations, as well as solutions obtained by semi-groups, also known as *mild* solutions. The solutions that interest us in the context of this book are weak solutions, and their study requires quite fine tools of vector-valued integration that we introduce in Section 4.2. Section 4.3 is devoted to the study of weak solutions to the heat equation. We present two methods for proving the existence and uniqueness of a weak solution: the classical Faedo–Galerkin method, and a more recent method, which we call *generalised coercivity*. Quasi-linear parabolic problems are discussed in Section 4.4, through two examples, one concerning a quasi-linear diffusion problem and the other a quasi-linear diffusion-convection problem. Finally, in Section 4.5, a number of powerful tools to ensure time compactness are given. Some of these tools are particularly suited to the study of the convergence of numerical approximations of non-linear parabolic equations.

4.1 Classical and Semi-Group Solutions

In the last century, various methods have been developed to solve parabolic equations. Before we turn to weak solutions, we explore three other approaches in the following subsections: one based on the Fourier transform, one based on an expansion on a basis of eigenfunctions, and finally, one based on semi-group theory. These concepts are extensively detailed in the book [22].

© The Author(s), under exclusive license to Springer Nature Switzerland AG 2025
T. Gallouët, R. Herbin, *Weak Solutions of Partial Differential Equations,*
Mathématiques et Applications 90, https://doi.org/10.1007/978-3-031-98982-7_4

In the case of semi-group solutions, also known under the name of *mild* solutions, it is interesting to highlight their difference compared to the weak solutions studied later (Section 4.3 for the heat equation). Indeed, mild solution are always unique, while for some problems, the uniqueness of weak solutions is not guaranteed, see Section 4.1.4 for an example in the steady-state case.

4.1.1 Solutions of the Heat Equation in $\mathbb{R}^N$ by Fourier Transform

Let $N \geq 1$ and $u_0 \in C(\mathbb{R}^N, \mathbb{R})$. We are interested here in looking for classical solutions to the following problem

$$
\begin{aligned}
&\partial_t u(x, t) - \Delta u(x, t) = 0, \; x \in \mathbb{R}^N, \; t \in \mathbb{R}_+^\star, \\
&u(x, 0) = u_0(x), \; x \in \mathbb{R}^N,
\end{aligned}
\tag{4.1}
$$

where $\partial_t u$ denotes the partial derivative of u with respect to time t, and Δu denotes the Laplacian of u defined by (2.4).

Definition 4.1 (Classical solution of the heat equation in $\mathbb{R}^N$). We say that a function u from $\mathbb{R}^N \times \mathbb{R}_+$ to $\mathbb{R}$ is a classical solution to problem (4.1) if $u \in C^2(\mathbb{R}^N \times \mathbb{R}_+^\star, \mathbb{R}) \cap C(\mathbb{R}^N \times \mathbb{R}_+, \mathbb{R})$ and u satisfies (4.1) in the classical sense of derivation and the initial condition.

We start with a small formal calculation (the term *formal* often meaning in mathematics "not necessarily justified"). We assume that we can use for the initial condition and for the solution the Fourier transform (in space). We recall the following formula for the Fourier transform of an integrable function on $\mathbb{R}^N$:

$$
\hat{f}(\xi) = \frac{1}{(2\pi)^{\frac{N}{2}}} \int e^{-ix \cdot \xi} f(x) \, dx.
$$

If u is a solution to (4.1), if allowed, the Fourier transform applied to (4.1) yields

$$
\partial_t \hat{u}(\xi, t) + |\xi|^2 \hat{u}(\xi, t) = 0, \; \text{for } \xi \in \mathbb{R}^N \text{ and } t > 0, \text{ and } \hat{u}(\xi, 0) = \hat{u}_0(\xi), \text{ for } \xi \in \mathbb{R}^N,
$$

which gives

$$
\hat{u}(\xi, t) = e^{-|\xi|^2 t} \hat{u}_0(\xi), \; \text{for } \xi \in \mathbb{R}^N \text{ and } t \geq 0.
$$

Choosing $g(t) \in L^1(\mathbb{R}^N)$ such that $\widehat{g(t)}(\xi) = e^{-|\xi|^2 t}$, we therefore have $\hat{u}(\cdot, t) = \widehat{g(t)} \hat{u}_0$ for all $t \geq 0$ and so (using the fact that the Fourier transform transforms convolution into product),

$$
\hat{u}(\cdot, t) = (2\pi)^{-\frac{N}{2}} \widehat{g(t) \star u_0} \text{ for } t \geq 0,
$$

or alternatively

$$
u(\cdot, t) = (2\pi)^{-\frac{N}{2}} g(t) \star u_0 \text{ for } t \geq 0.
$$

It remains to calculate $g(t)$; since $\widehat{\widehat{g(t)}} \in L^1(\mathbb{R}^N)$ for $t > 0$, the Fourier inversion theorem gives $g(t) = \widehat{\widehat{g(t)}}(-\cdot)$, that is to say,

$$g(t)(x) = \frac{1}{(2\pi)^{\frac{N}{2}}} \int e^{ix\cdot\xi} e^{-|\xi|^2 t} \, d\xi \quad \text{for } x \in \mathbb{R}^N \text{ and } t > 0.$$

The change of variables $\xi = \frac{\eta}{\sqrt{2t}}$ then gives

$$g(t)(x) = \frac{1}{(2\pi)^{\frac{N}{2}}} \frac{1}{(2t)^{\frac{N}{2}}} \int e^{ix\cdot\frac{\eta}{\sqrt{2t}}} e^{-\frac{|\eta|^2}{2}} \, d\eta \quad \text{for } x \in \mathbb{R}^N \text{ and } t > 0.$$

Finally, we obtain

$$g(t)(x) = \frac{1}{(2t)^{\frac{N}{2}}} e^{-|\frac{x}{\sqrt{2t}}|^2 \frac{1}{2}} = \frac{1}{(2t)^{\frac{N}{2}}} e^{-\frac{|x|^2}{4t}} \quad \text{for } x \in \mathbb{R}^N \text{ and } t > 0,$$

which gives

$$u(x,t) = \frac{1}{(4\pi t)^{\frac{N}{2}}} \int e^{-\frac{|x-y|^2}{4t}} u_0(y) \, dy \quad \text{for } x \in \mathbb{R}^N \text{ and } t > 0. \tag{4.2}$$

We can then state the following existence result.

Theorem 4.2 (Existence of classical solutions for the heat problem in $\mathbb{R}^N$). *Let $u_0 \in C(\mathbb{R}^N, \mathbb{R})$; if, furthermore, u_0 satisfies one of the following sufficient conditions*

* $u_0 \in L^1(\mathbb{R}^N)$, $\qquad\qquad\qquad\qquad\qquad\qquad\qquad\qquad\qquad$ (4.3a)
* $\exists C \in \mathbb{R}_+, \exists p \in \mathbb{N} \; : \; \forall x \in \mathbb{R}^N, \; |u_0(x)| \le C(1 + |x|^p), \qquad$ (4.3b)

then there exists a classical solution to (4.1) *in the sense of Definition 4.1.*

Proof For $t > 0$ and $x, y \in \mathbb{R}^N$, we set

$$\psi(x, y, t) = \frac{1}{(4\pi t)^{\frac{N}{2}}} e^{-\frac{|x-y|^2}{4t}} u_0(y)$$

and we define the function $u : \mathbb{R}^N \times \mathbb{R}_+ \mapsto \mathbb{R}$ by the formula (4.2) for $t > 0$ and by u_0 for $t = 0$, that is,

$$u(x,t) = \begin{cases} \int \psi(x, y, t) \, dy & \text{for } x \in \mathbb{R}^N \text{ and } t > 0, \\ u_0(x) & \text{for } x \in \mathbb{R}^N \text{ and } t = 0. \end{cases}$$

Assume that u_0 satisfies (4.3a) or (4.3b). In a first step we prove that $u \in C^2(\mathbb{R}^N \times \mathbb{R}_+^\star, \mathbb{R})$ and satisfies the partial derivative equation given in problem (4.1). Then, in a second step we prove that $u \in C(\mathbb{R}^N \times \mathbb{R}_+, \mathbb{R})$ and u satisfies the initial condition. This finally gives that u is a classical solution to problem (4.1) in the sense of Definition 4.1.

Step 1 For all $t > 0$ and $x \in \mathbb{R}^N$ the function $y \mapsto \psi(x, y, t)$ is integrable over $\mathbb{R}^N$. The function u is therefore perfectly defined in $\mathbb{R}^N \times \mathbb{R}_+^\star$ by the formula given in the theorem. An easy consequence of the theorems of continuity and differentiability under the integral sign then gives that $u \in C^\infty(\mathbb{R}^N \times \mathbb{R}_+^\star, \mathbb{R})$ and, since $\partial_t \psi(x, y, t) - \partial_{xx}^2 \psi(x, y, t) = 0$, we obtain that u satisfies the partial derivative equation given in problem (4.1).

Step 2 Let $x_0 \in \mathbb{R}^N$, we want to show that $u(t, x) \to u_0(x_0)$ when $(x, t) \to (x_0, 0)$ with $t > 0$. Without loss of generality, we can assume that $x_0 = 0$ (the point 0 plays no particular role in the equation of problem (4.1) or in the conditions on u_0 because the condition (4.3b) only means that u_0 grows at most polynomially). Without loss of generality, we can also assume that $u_0(0) = 0$; indeed, if u_0 is a constant function, the formula for u gives that u is equal to this constant function.

It is therefore now a matter of showing that $u(t, x) \to 0$ when $(x, t) \to (0, 0)$ with $t > 0$. Let $\varepsilon > 0$. Since u_0 is continuous, there exists a $\delta > 0$ such that $|u_0(y)| \le \varepsilon$ if $|y| \le \delta$. Hence, for $t > 0$ and $x \in \mathbb{R}^N$, $|x| \le \frac{\delta}{2}$ (so that $|x - y| \ge \frac{|y|}{2}$ if $|y| \ge \delta$),

$$|u(x, t)| \le \varepsilon \frac{1}{(4\pi t)^{\frac{N}{2}}} \int e^{-\frac{|x-y|^2}{4t}} \, dy + \frac{1}{(4\pi t)^{\frac{N}{2}}} \int_{|y|>\delta} e^{-\frac{|x-y|^2}{4t}} |u_0(y)| \, dy$$

$$\le \varepsilon + \frac{1}{(4\pi t)^{\frac{N}{2}}} \int_{|y|>\delta} e^{-\frac{|y|^2}{16t}} |u_0(y)| \, dy.$$

We now distinguish two cases for u_0, according to assumptions (4.3a) or (4.3b).

– if u_0 satisfies (4.3a): $u_0 \in L^1(\mathbb{R}^N)$. In this case, for $t > 0$ and $x \in \mathbb{R}^N$, $|x| \le \frac{\delta}{2}$,

$$|u(x, t)| \le \varepsilon + \frac{1}{(4\pi t)^{\frac{N}{2}}} e^{-\frac{\delta^2}{16t}} \|u_0\|_{L^1(\mathbb{R}^N)} \, .$$

Since $\dfrac{1}{(4\pi t)^{\frac{N}{2}}} e^{-\frac{\delta^2}{16t}} \to 0$ when $t \to 0$ (with $t > 0$), there exists an $\eta > 0$ such that $|u(x, t) \le 2\varepsilon$ if $|x| \le \frac{\delta}{2}$ and $0 < t < \eta$. This indeed gives $u(t, x) \to 0$ when $(x, t) \to (0, 0)$ with $t > 0$.

– if u_0 satisfies (4.3b): there exists a $C \in \mathbb{R}_+$ and $p \in \mathbb{N}$ such that $|u_0(x)| \le C(1 + |x|^p)$ for all $x \in \mathbb{R}^N$ (which includes the case $u_0 \in L^\infty(\mathbb{R}^N)$). We then have (still with $t > 0$ and $x \in \mathbb{R}^N$, $|x| \le \frac{\delta}{2}$), with the change of variables $y = z\sqrt{t}$ and for $0 < t < 1$,

$$|u(x, t)| \le \varepsilon + \frac{C}{(4\pi t)^{\frac{N}{2}}} \int_{|y|>\delta} e^{-\frac{|y|^2}{16t}} (1 + |y|^p) \, dy$$

$$= \varepsilon + \frac{C}{(4\pi)^{\frac{N}{2}}} \int_{|z|>\frac{\delta}{\sqrt{t}}} e^{-\frac{|z|^2}{16}} (1 + |z\sqrt{t}|^p) \, dz$$

$$\leq \varepsilon + \frac{C}{(4\pi)^{\frac{N}{2}}} \int_{|z|>\frac{\delta}{\sqrt{t}}} e^{-\frac{|z|^2}{16}} (1 + |z|^P)\, dz.$$

Since the function $z \mapsto e^{-\frac{|z|^2}{16}}(1 + |z|^P)$ is integrable, there exists an $\eta > 0$ such that $|u(x,t)| \leq 2\varepsilon$ if $|x| \leq \frac{\delta}{2}$ and $0 < t < \eta$. This indeed gives $u(t,x) \to 0$ when $(x,t) \to (0,0)$ with $t > 0$. ∎

Under the assumptions of Theorem 4.2, there is never uniqueness of the classical solution to problem (4.1) in the sense of Definition 4.1. More precisely, we can construct a non-zero classical solution to (4.1) with $u_0 = 0$: an example of a function $u \in C^\infty(\mathbb{R}^N \times [0, +\infty[)$, $u \neq 0$, such that

$$\begin{cases} \partial_t u - \Delta u = 0 \text{ in } \mathbb{R}^N \times \mathbb{R}_+, \\ u(x,0) = 0 \quad \forall x \in \mathbb{R}^N, \end{cases}$$

is given in Smoller's book [50].

However, if we add a suitable growth assumption on the solution (see Evans' book [22], for example), we obtain a uniqueness result, which allows us to state the following existence and uniqueness theorem.

Theorem 4.3 (Existence and uniqueness for the heat equation). *Let $u_0 \in C(\mathbb{R}^N, \mathbb{R})$ satisfying (4.3b). Then there exists a unique function u satisfying:*

1. *u is a classical solution to problem (4.1) in the sense of Definition 4.1,*
2. *$\forall T > 0, \exists C_T \in \mathbb{R}_+, P_T \in \mathbb{N}$ such that $|u(x,t)| \leq C_T(1 + |x|^{P_T})$, $\forall x \in \mathbb{R}^N$, $\forall t \in [0,T]$.*

Proof *Existence:* Theorem 4.2 states that the function u defined by the formula (4.2) for $t > 0$ and by u_0 for $t = 0$, is a classical solution to problem (4.1) in the sense of Definition 4.1. It is therefore now necessary to show that this function u satisfies condition 2 of the theorem.

First note that the growth assumption on u_0 given in (4.3b) is equivalent to asking for the existence of $C \in \mathbb{R}_+$ and $q \in \mathbb{N}$ such that

$$|u_0(x)| \leq C + C \sum_{i=1}^{N} x_i^{2q} \text{ for all } x = (x_1, \ldots, x_N)^t \in \mathbb{R}^N.$$

It is therefore sufficient to show that if $u_0(x) = x_i^{2q}$ (for all $x \in \mathbb{R}^N$) with $q \in \mathbb{N}$ and $i \in \{1, \ldots, N\}$, then u satisfies condition 2. Without loss of generality, we can limit ourselves to the case $i = 1$. Let $q \in \mathbb{N}$. For $u_0(x) = x_1^{2q}$ (for all $x \in \mathbb{R}^N$) we have, for all $x \in \mathbb{R}^N$ and all $t > 0$,

$$u(x,t) = \frac{1}{(4\pi t)^{\frac{N}{2}}} \int e^{-\frac{|x-y|^2}{4t}} y_1^{2q}\, dy = \frac{1}{(4\pi t)^{\frac{1}{2}}} \int e^{-\frac{(x_1-y_1)^2}{4t}} y_1^{2q}\, dy_1.$$

With the change of variables $(x_1 - y_1) = 2\sqrt{t}z_1$, we obtain

$$
\begin{aligned}
u(x,t) &= \frac{1}{\pi^{\frac{1}{2}}} \int e^{-z_1^2} |x_1 - 2\sqrt{t}z_1|^{2q}\, dz_1 \\
&\leq \frac{2}{\pi^{\frac{1}{2}}} \left(\int e^{-z_1^2} x_1^{2q}\, dz_1 + \int e^{-z_1^2} |2\sqrt{t}z_1|^{2q}\, dz_1 \right) \\
&\leq \frac{2}{\pi^{\frac{1}{2}}} \left(x_1^{2q} \int e^{-z_1^2}\, dz_1 + (2\sqrt{t})^{2q} \int e^{-z_1^2} z_1^{2q}\, dz_1 \right).
\end{aligned}
$$

It follows that u satisfies condition 2.

Uniqueness: We can prove uniqueness using the maximum principle for parabolic equations. We do not detail this point here but we shall return to this principle when studying weak solutions on an open bounded subset of $\mathbb{R}^N$ (Proposition 4.34). Using this principle, theorem 7, section 2.3 (*heat equation*) from [22] gives the uniqueness of the classical solution to (4.1) under the condition (weaker than condition 2) that for all $T > 0$, there exist $a_T, C_T \in \mathbb{R}$ such that

$$
|u(x,t)| \leq C_T e^{a_T |x|^2} \quad \text{for all } x \in \mathbb{R}^N \text{ and all } 0 \leq t \leq T.
$$

This concludes the proof of Theorem 4.3. $\blacksquare$

4.1.2 Solutions by Expansion on a Basis of Eigenfunctions

Let Ω be an open bounded subset of $\mathbb{R}^N$, and let $u_0 : \Omega \to \mathbb{R}$ be given. We are now interested in the problem

$$
\begin{cases}
\partial_t u - \Delta u = 0 & \text{in} \quad \Omega \times \mathbb{R}_+^\star, \\
u(x,t) = 0 \text{ for } x \in \partial\Omega \text{ and } t > 0, \\
u(x,0) = u_0(x) \text{ for } x \in \Omega.
\end{cases}
\tag{4.4}
$$

We recall (see (2.12)) that the Laplace operator $\mathcal{A}$ is defined on its domain $D(\mathcal{A}) = \{u \in H_0^1(\Omega) \, ; \Delta u \in L^2(\Omega)\}$ by $\mathcal{A}u = -\Delta u$.

We have already seen in Section 2.2 that $u \in D(\mathcal{A})$ means that $u \in H_0^1(\Omega)$ and there exists an $f \in L^2(\Omega)$ such that $\int_\Omega \nabla u \cdot \nabla v\, dx = \int_\Omega f v\, dx$ for all $v \in \mathcal{D}(\Omega)$ (or, equivalently, for all $v \in H_0^1(\Omega)$).

We now assume that $u_0 \in L^2(\Omega)$ and seek a solution to (4.4) in the following sense:

$$
\begin{cases}
u \in C^1(]0, +\infty[, L^2(\Omega)) \cap C([0, +\infty[, L^2(\Omega)), \\
u(t) \in D(\mathcal{A}) \text{ and } u'(t) + \mathcal{A}u(t) = 0 \text{ a.e., for all } t > 0, \\
u(0) = u_0 \text{ a.e.,}
\end{cases}
\tag{4.5}
$$

where $u(t)$ denotes the function $x \mapsto u(x,t)$. We recall (see Chapter 2, and more specifically Theorem 2.16) that there exists a Hilbert basis of $L^2(\Omega)$ (that is to say, an orthonormal family $(e_n)_{n \in \mathbb{N}^*}$ that is dense in $L^2(\Omega)$), formed of eigenfunctions of the operator $\mathcal{A}$. For all $n \in \mathbb{N}^*$, the function e_n is a weak solution of

$$\begin{cases} -\Delta e_n = \lambda_n e_n \text{ in } \Omega, \\ e_n = 0 \text{ on } \partial\Omega, \end{cases}$$

with $\lambda_n > 0$ and $\lambda_n \uparrow +\infty$ when $n \to +\infty$. For all $n \in \mathbb{N}^*$, we therefore have $e_n \in D(\mathcal{A})$ and $\mathcal{A}e_n = \lambda_n e_n$.

Let $u_0 \in L^2(\Omega)$; denoting by $(u|v)_2$ the inner product of u and v in $L^2(\Omega)$, we have

$$u_0 = \sum_{n=1}^{+\infty} (u_0|e_n)_2 e_n,$$

this series being convergent in $L^2(\Omega)$. We also have $\displaystyle\sum_{n=1}^{+\infty} (u_0|e_n)_2^2 = \|u_0\|_2^2 < +\infty$.

Let us set, for $t \geq 0$,

$$u(t) = \sum_{n=1}^{+\infty} e^{-\lambda_n t} (u_0|e_n)_2 e_n,$$

which is a series that is convergent in $L^2(\Omega)$. So we have $u(t) \in L^2(\mathbb{R}^N)$ for all $t \geq 0$ and since the sequence $(\lambda_n)_{n \in \mathbb{N}^*}$ is bounded below by λ_1 with $\lambda_1 > 0$, we even have $u \in C([0, +\infty[, L^2(\Omega))$ and $u(0) = u_0$ a.e. On the other hand, since $\lambda_n \uparrow +\infty$ as $n \to +\infty$, it follows that the function u is differentiable from $]0, +\infty[$ to $L^2(\Omega)$ and that the derivative of u is obtained by differentiating the series term by term (this is a consequence of the fact that the series differentiated term by term is, on $]0, +\infty[$, locally uniformly convergent in $L^2(\Omega)$). So we have, for all $t > 0$,

$$u'(t) = \frac{du}{dt} = \sum_{n=1}^{+\infty} (-\lambda_n) e^{-\lambda_n t} (u_0|e_n)_2 e_n;$$

note that the series on the right-hand side is convergent in $L^2(\Omega)$.

Recall that (according to Section 2.2) $u(t) \in D(\mathcal{A})$ for all $t > 0$ because the series $\displaystyle\sum_{n=1}^{+\infty} \lambda_n e^{-\lambda_n t} (u_0|e_n)_2 e_n$ is convergent in $L^2(\Omega)$. Section 2.2 also gives, for all $t > 0$,

$$\mathcal{A}u(t) = \sum_{n=1}^{+\infty} \lambda_n e^{-\lambda_n t} (u_0|e_n)_2 e_n.$$

So we have $u \in C^1(]0, +\infty[, L^2(\Omega))$ and $u'(t) + \mathcal{A}(u(t)) = 0$ a.e. for all $t \in]0, +\infty[$. We have thus found a solution to problem (4.5).

Let us now show that this solution is unique. To this end, let us show that if u is a solution to the problem with initial data zero, that is

$$\begin{cases} u \in C^1(]0, +\infty[, L^2(\Omega)) \cap C([0, +\infty[, L^2(\Omega)), \\ u(t) \in D(\mathcal{A}) \text{ and } u'(t) + \mathcal{A}u(t) = 0 \text{ a.e., for all } t > 0, \\ u(0) = 0 \text{ a.e.,} \end{cases}$$

then $u(t) = 0$ a.e. for all $t \geq 0$. Let $t > 0$. Since $u'(t) + \mathcal{A}u(t) = 0$ a.e., by taking the inner product with $u(t) \in L^2(\Omega)$, we get $(u'(t)|u(t))_2 + (\mathcal{A}u(t)|u(t))_2 = 0$. Since $(\mathcal{A}u(t)|u(t))_2 = \int_\Omega |\nabla u(t)|^2 \, dx$, we therefore have

$$(u'(t)|u(t))_2 = -\int_\Omega |\nabla u(t)|^2 \, dx \leq 0.$$

Let ψ be the mapping defined by $t \mapsto \psi(t) = \|u(t)\|_{L^2(\Omega)}^2$. The mapping ψ is continuous on $\mathbb{R}_+$, differentiable on $\mathbb{R}_+^\star$ and $\psi'(t) = 2(u'(t)|u(t))_2$ for all $t > 0$. Hence $\psi'(t) \leq 0$ for all $t > 0$. The mapping ψ is therefore non-increasing on $\mathbb{R}_+$ and, since $\psi(0) = 0$, it follows that $\psi(t) \leq 0$ for all $t \geq 0$. Therefore $\psi(t) = 0$ for all $t \geq 0$. We finally have $u(t) = 0$ a.e. for all $t \geq 0$.

We have thus proved the existence and uniqueness of the solution to (4.5). We could call this solution "almost classical" because the time derivative is taken in the classical sense (since u is of class C^1 with values in $L^2(\Omega)$) and the initial condition is taken in the classical sense in $L^2(\Omega)$. However, the Laplacian of $u(t)$ is taken in the sense of weak derivatives. Additional work on the regularity of the functions e_n is required to show that the equation $\partial_t u - \Delta u$ is satisfied in the classical sense on $\mathbb{R}^N \times \mathbb{R}_+^\star$. This work is carried out for the case $N = 1$ in Problem 4.1.

Remark 4.4 (Generalisation)**.** We can also prove the existence and uniqueness (for $u_0 \in L^2(\Omega)$) of the solution to (4.5) by replacing, in the definition of $\mathcal{A}$, Δu by $\sum_{i,j=1}^N D_j(a_{ij}D_i u)$, with $a_{ij} \in L^\infty(\Omega)$, under a coercivity assumption, that is if there exists an $\alpha > 0$ such that $\alpha|\xi|^2 \leq \sum a_{ij}\xi_i \, \xi_j$ a.e. in Ω and for all $\xi \in \mathbb{R}^N$. We can also replace $u'(t) + \mathcal{A}u(t) = 0$ by $u'(t) + \mathcal{A}u(t) = f(t)$ if $f \in C([0, \infty[, L^2(\Omega))$.

4.1.3 Solutions by Semigroups

Let E be a real Banach space and $\mathcal{A} : D(\mathcal{A}) \subset E \to E$ a linear operator. The set $D(\mathcal{A})$ is therefore the v.s.s. of E on which $\mathcal{A}$ is defined.

Definition 4.5 (*m*-accretive operator)**.** We say that $\mathcal{A}$ is *m*-accretive if $\mathcal{A}$ satisfies:

(i) $D(\mathcal{A})$ is dense in E,
(ii) $\forall \lambda > 0$, $(\mathrm{Id} + \lambda\mathcal{A})$ is invertible, with a continuous inverse and

$$\left\|(I + \lambda\mathcal{A})^{-1}\right\|_{\mathcal{L}(E,E)} \leq 1.$$

Recall that $\mathcal{L}(E, E)$ designates the set of continuous linear operators from E to E, and that, for all $T \in \mathcal{L}(E, E)$,

$$\|T\|_{\mathcal{L}(E,E)} = \sup_{u \in E, u \neq 0} \frac{\|T(u)\|_E}{\|u\|_E}.$$

Remark 4.6 (Maximal monotone operator). Let E be a real Hilbert space and $\mathcal{A} : D(\mathcal{A}) \subset E \to E$ a linear operator. The operator $\mathcal{A}$ is m-accretive if and only if it satisfies:

$$\begin{cases} (\mathcal{A}u|u)_E \geq 0 \quad \forall u \in D(\mathcal{A}), \\ (\mathrm{Id} + \mathcal{A}) \text{ surjective.} \end{cases}$$

In this case, we say that $\mathcal{A}$ is maximal monotone.

Example: The Laplacian with homogeneous Dirichlet condition Let Ω be an open bounded subset of $\mathbb{R}^N$, $E = L^2(\Omega)$ and $\mathcal{A}$ be defined by $D(\mathcal{A}) = \{u \in H_0^1(\Omega); \ \Delta u \in L^2(\Omega)\}$ and $\mathcal{A}u = -\Delta u$ if $u \in D(\mathcal{A})$. The operator $\mathcal{A}$ is then m-accretive (or maximal monotone since we are in the case of a Hilbert space).

Remark 4.7 (Graph of an m-accretive operator). The graph of an m-accretive operator is closed.

We admit the following theorem due to Hille[1] and Yosida[2] (see for example [13] for the proof in the case of Hilbert spaces and [15] for applications):

Theorem 4.8 (Hille–Yosida). *Let E be a Banach space, $\mathcal{A} : D(\mathcal{A}) \subset E \to E$ an m-accretive linear operator and $u_0 \in D(\mathcal{A})$. Then there exists a unique $u : \mathbb{R}_+ \to E$ such that*

$$u \in C^1([0, +\infty[, E), \tag{4.6a}$$

$$u(t) \in D(\mathcal{A}) \, \forall t \geq 0, \tag{4.6b}$$

$$u'(t) + \mathcal{A}(u(t)) = 0 \, \forall t \geq 0, \tag{4.6c}$$

$$u(0) = u_0. \tag{4.6d}$$

The proof of this theorem can be carried out by a time discretisation: we consider the problem $\frac{u_{n+1} - u_n}{\delta t} + \mathcal{A}u_{n+1} = 0$, which can also be written as $u_{n+1} = (\mathrm{Id} + \delta t \mathcal{A})^{-1} u_n$, where $\delta t > 0$ is the step of the discretisation. We then perform a passage to the limit as $\delta t \to 0$.

Definition 4.9 (Semi-group). Let E be a Banach space, $\mathcal{A} : D(\mathcal{A}) \subset E \to E$ an m-accretive linear operator. For $u_0 \in D(\mathcal{A})$ and $t \geq 0$, we set $S(t)u_0 = u(t)$, where u is the unique function satisfying (4.6). Then the operator $S(t)$ is a continuous linear operator from $D(\mathcal{A}) \subset E$ to E that satisfies

---

[1] Carl Einar Hille (1894–1980), American mathematician of Swedish origin, specialist in analysis.

[2] Kōsaku Yosida (1909–1990), Japanese mathematician, specialist in functional analysis.

$$\begin{cases} S(t + s) = S(t) \circ S(s) \text{ for all } t, s \geq 0 \\ S(0) = \mathrm{Id} \\ \|S(t)u_0\|_E \leq \|u_0\|_E, \text{ for all } u_0 \in D(\mathcal{A}). \end{cases}$$

We say that $\{S(t), t \geq 0\}$ is a contraction semi-group.

For $t \geq 0$, since $\overline{D(\mathcal{A})} = E$ and $\|S(t)u_0\|_E \leq \|u_0\|_E$ for all $u_0 \in D(\mathcal{A})$, the operator $S(t)$ extends uniquely (by density) to all E to an operator $\overline{S(t)}$ and $\overline{S(t)} \in \mathcal{L}(E, E)$. The set $\{\overline{S(t)}, t \geq 0\}$ is then a contraction semi-group on E.

Definition 4.10 (Solution by semi-group or *mild* solution). Let E be a Banach space, $\mathcal{A} : D(\mathcal{A}) \subset E \to E$ an m-accretive linear operator, and $u_0 \in E$. The function $u(t) = \overline{S(t)}u_0$, uniquely defined, is called a semi-group solution, or *mild* solution, of the problem

$$\begin{cases} \partial_t u + \mathcal{A}u = 0, \\ u(0) = u_0. \end{cases}$$

In the case where the operator $\mathcal{A}$ is the Laplacian with homogeneous Dirichlet condition and $E = L^2(\Omega)$, one might wonder in what sense the semi-group solution satisfies (4.4). It can be shown that the mild solution is a solution in a weak sense that we define in Section 4.3. Moreover, this mild solution is, in this particular case, the unique weak solution. However, this situation is not completely general. The mild solution given in Definition 4.10 is obtained by density from the (unique) solution given by Theorem 4.8. The mild solution is therefore always unique (as soon as $u_0 \in E$, whatever the Banach space E and the m-accretive operator $\mathcal{A}$). For problems arising from partial differential equations, this mild solution is generally a weak solution to the problem we want to solve; however, the problem of the uniqueness of the weak solution is much more difficult. There can be non-uniqueness of the weak solution even if this latter is defined in a sense that seems reasonable. There is then no equivalence between the notion of mild solution and weak solution, because the mild solution is unique while there is no uniqueness of the weak solution.

To try to clarify this difficulty, we are interested, in the following subsection, in an elliptic problem for which we show that the solution obtained by density, in the manner of the mild solution introduced above, is unique; we then show that by taking a natural definition of weak solution, there is no uniqueness of weak solutions. This shows in particular that in the parabolic case, there is also no equivalence between mild solution and weak solution: we show as an example the case of stationary solutions in the following section.

4.1.4 Elliptic Problems with L^1 Data

Consider an open bounded subset Ω of $\mathbb{R}^N$, $N \geq 2$, and functions a_{ij}, for $i, j = 1, \ldots, N$, belonging to $L^\infty(\Omega)$ and satisfying the coercivity assumption:

$$\exists\, \alpha > 0 \qquad \sum_{i,j=1}^{N} \alpha_{ij}\, \xi_i\, \xi_j \geq \alpha |\xi|^2 \text{ a.e. in } \Omega, \quad \forall \xi \in \mathbb{R}^N. \qquad (4.7)$$

We denote by A the function with values in $\mathcal{M}_N(\mathbb{R})$, defined by the functions a_{ij}, $i, j = 1, \ldots, N$. By the Lax–Milgram theorem, for every $f \in L^2(\Omega)$ there exists a unique solution u of the following problem:

$$\begin{cases} u \in H_0^1(\Omega), \\ \displaystyle\int_\Omega A\nabla u \cdot \nabla v \, dx = \int f v \, dx, \quad \forall v \in H_0^1(\Omega). \end{cases} \qquad (4.8)$$

We now consider the same problem with a less regular data $f \in L^1(\Omega)$.

Step 1 - Estimate, regular case - For all $1 \leq q < \frac{N}{N-1}$, we show that there exists a $C_q \in \mathbb{R}$ (only depending on A and q) such that if u is the unique solution to (4.8) with $f \in L^2(\Omega)$ then $\|u\|_{W_0^{1,q}(\Omega)} \leq C_q \|f\|_{L^1(\Omega)}$. We then set $T_q(f) = u$.

Step 2 - Mild solution - By the previous estimate, for all $1 \leq q < \frac{N}{N-1}$, the mapping T_q from $L^2(\Omega)$ to $W_0^{1,q}(\Omega)$ is continuous from $L^2(\Omega)$, equipped with the $L^1(\Omega)$-norm, to $W_0^{1,q}(\Omega)$ equipped with its natural norm. By density of $L^2(\Omega)$ in $L^1(\Omega)$, we can therefore extend T_q in a unique way over all $L^1(\Omega)$. Let $\overline{T}_q$ denote this extension. Observe that $\overline{T}_{q_1} f = \overline{T}_{q_2} f$ for all q_1, q_2 such that $1 \leq q_1 < q_2 < \frac{N}{N-1}$. We can thus define a linear operator T from L^1 to $\bigcap_{1 \leq q < \frac{N}{N-1}} W_0^{1,q}(\Omega)$ by $T(f) = \overline{T}_q(f)$ for all $1 \leq q < \frac{N}{N-1}$.

Definition 4.11 (Mild solution for an elliptic problem with L^1 data). Under the previous assumptions on Ω and A, let $f \in L^1(\Omega)$. The function Tf defined above is the mild solution of

$$\begin{cases} -\mathrm{div}(A\nabla u) = f \text{ in } \Omega, \\ u = 0 \text{ on } \partial\Omega. \end{cases} \qquad (4.9)$$

Step 3 - Mild solution and weak solution - It is natural to wonder what the relationship is between the mild solution $u = Tf$ and a weak solution of the homogeneous Dirichlet problem. We can show [10] that u is a weak solution in the following natural sense.[3] To show this result, we use the density of $L^2(\Omega)$ in the set of measures on Ω in the sense of weak-$\star$ convergence in $C(\bar{\Omega})'$; indeed, the set of measures on Ω can be seen as a subset of the dual space of $C(\overline{\Omega})$). We also use the fact that the elements of $W_0^{1,r}(\Omega)$ are continuous functions for $r > N$. The interested reader may consult [10]:

$$\begin{cases} \displaystyle u \in \bigcap_{1 \leq q < \frac{N}{N-1}} W_0^{1,q}(\Omega), \\ \displaystyle\int_\Omega A\nabla u \cdot \nabla v \, dx = \int_\Omega f v \, dx \quad \forall v \in \bigcup_{r>N} W_0^{1,r}(\Omega). \end{cases} \qquad (4.10)$$

---

[3] A similar result of existence of a mild solution and therefore of existence of a weak solution is still true if f is a measure on Ω (in (4.10), we then replace $f v \, dx$ by $v \, df$).

Step 4 - Mild uniqueness, weak non-uniqueness - The mild solution is unique by construction. Let us now show that the weak solution is unique in the case $N = 2$, but that there are counterexamples if $N > 2$.

- *Uniqueness in the case $N = 2$* - A duality argument and a regularity result will allow us to show that if $N = 2$, the weak solution is unique. Let u be a solution of

$$\begin{cases} u \in \displaystyle\bigcap_{1 \leq q < 2} W_0^{1,q}(\Omega), \\ \displaystyle\int A\nabla u \cdot \nabla v \; dx = 0, \quad \forall v \in \bigcup_{r>2} W_0^{1,r}(\Omega). \end{cases} \tag{4.11}$$

We want to show that $u = 0$. We cannot take as a test function $v = u$ in (4.10) or (4.11), due to the lack of regularity of u. To circumvent this problem, we reason using the regularity of the solutions of an associated problem. First note that u is a solution to (4.11) and that (4.11) can be reformulated as follows:

$$\begin{cases} u \in \displaystyle\bigcap_{1 \leq q < 2} W_0^{1,q}(\Omega), \\ \displaystyle\int A^t \nabla v \cdot \nabla u \; dx = 0, \quad \forall v \in \bigcup_{r>2} W_0^{1,r}(\Omega). \end{cases} \tag{4.12}$$

By the Lax–Milgram theorem, if $g \in L^2(\Omega)$, there exists a unique solution v of the following problem:

$$\begin{cases} v \in H_0^1(\Omega), \\ \displaystyle\int A^t \nabla v \cdot \nabla w \, dx = \int gw \, dx, \quad \forall w \in H_0^1(\Omega). \end{cases} \tag{4.13}$$

We would like to be able to take $w = u$ in (4.13), because we would then have, thanks to (4.12), $\int_\Omega gu \, dx = 0$ for all $g \in L^2(\Omega)$, which would allow us to conclude that $u = 0$. This is possible if $v \in W_0^{1,r}(\Omega)$ for some $r > 2$ (in order to use (4.12)) and if $u \in H_0^1(\Omega)$ (in order to use (4.13)). But, in general, $u \notin H_0^1(\Omega)$. The trick is to consider a more regular second term g in (4.13) and to use the regularity result stated in the following theorem.

Theorem 4.12 (Meyers, [40]). *Let Ω be an open bounded subset of $\mathbb{R}^N$, $N \geq 2$, and let A be a function with values in $\mathcal{M}_N(\mathbb{R})$ defined by the functions a_{ij}, $i, j = 1, \ldots, N$, belonging to $L^\infty(\Omega)$ and satisfying the assumption (4.7). Then there exists a $p^\star > 2$, depending only on A and Ω, such that if $g \in L^\infty(\Omega)$ and v is a solution to (4.13) then $v \in W_0^{1,p^\star}(\Omega)$.*

Let $g \in L^\infty(\Omega)$ and v be a solution to (4.13). Thanks to Meyers' theorem, $v \in W_0^{1,p^\star}(\Omega)$. Let $(p^\star)'$ be the conjugate exponent of $p^\star$, so that $(p^\star)' = \frac{p^\star}{p^\star - 1} < 2$. By density of $\mathcal{D}(\Omega)$ in $W_0^{1,(p^\star)'}(\Omega)$, (4.13) yields that

$$\int A^t \nabla v \cdot \nabla w \, \mathrm{d}x = \int g w \, \mathrm{d}x, \quad \forall w \in W_0^{1,(p^\star)'}(\Omega). \tag{4.14}$$

However, since $(p^\star)' = \frac{p^\star}{p^\star - 1} < 2$, any solution u of (4.12) belongs to $W_0^{1,(p^\star)'}(\Omega)$. We can therefore take $w = u$ in (4.14). But since $v \in W_0^{1,p^\star}(\Omega)$ with $p^\star > 2$, we also have, by (4.12), $\int_\Omega A^t \nabla v \cdot \nabla u \, \mathrm{d}x = 0$. Hence $\int_\Omega g u \, \mathrm{d}x = 0$, for all $g \in L^\infty(\Omega)$. We obtain that $u = 0$ by successively taking $g = \mathbb{1}_{\{u>0\}}$ and then $g = \mathbb{1}_{\{u<0\}}$.

- *Non-uniqueness if $N > 2$* - For $N = 3$, Meyers' regularity theorem is still true, so the uniqueness proof is still valid if A is such that $p^\star > N = 3$, a condition that allows us to take v (solution of (4.13)) as a test function in (4.10). But there are matrix functions $A = (a_{ij})_{i,j=1,N}$ satisfying the assumptions of Theorem 4.12 for which $p^\star < 3$. Uniqueness is then no longer guaranteed and we can indeed construct cases for which the solution to (4.10) is not unique (see [45]).

Step 5 - Uniqueness of the entropy weak solution - What could we add to the weak solution definition (4.10) to ensure its uniqueness? For $f \in L^1(\Omega)$, it can be shown [8] that there exists one and only one solution to a new formulation, known as the entropic formulation, due to Ph. Bénilan[4], which reads:

$$\begin{cases} u \in W_0^{1,q}(\Omega), \text{ for all } 1 \le q < \frac{N}{N-1}, \quad T_k(u) \in H_0^1(\Omega), \ \forall k > 0, \\ \displaystyle\int_\Omega A \nabla u \cdot \nabla T_k(u - \varphi) \, \mathrm{d}x = \int f T_k(u - \varphi) \, \mathrm{d}x \quad \forall \varphi \in \mathcal{D}(\Omega), \ \forall k \ge 0, \end{cases} \tag{4.15}$$

where T_k is the truncation function, defined by: $T_k(s) = \max(\min(k, s), -k)$. The article [8] by Ph. Bénilan *et al.* proves that there exists a unique solution to the problem (4.15), which is therefore the mild solution to (4.9). Ph. Bénilan conjectured that adding the condition $T_k(u) \in H_0^1(\Omega)$ (for all $k > 0$) to the formulation (4.10) should be sufficient to ensure uniqueness. Indeed, in the articles by Serrin and Prignet [48, 45], the counterexample consists of constructing a non-zero solution to (4.10) with $f = 0$, but this solution does not satisfy $T_k(u) \in H_0^1(\Omega)$ for $k > 0$. Bénilan's conjecture therefore remains a conjecture to this day...

Another open question is to find a formulation similar to (4.15) that gives uniqueness when f is a measure on Ω.

Remark 4.13 (Limitation of the semi-group method). A significant difficulty of this semi-group method is that of generalising to the case where the operator $\mathcal{A}$ in Theorem 4.8 depends on t.

---

[4] Philippe Bénilan (1941–2001), French mathematician, specialist in PDEs, professor at the University of Besançon.

4.2 Integration of Functions with Vector Values

The weak formulation of linear parabolic equations, such as the heat equation, requires the integration of functions with values in an infinite-dimensional Banach space, such as the space $L^2(\Omega)$. In order to define spaces such as the spaces $L^2(]0,T[, L^2(\Omega))$, $L^2(]0,T[, H_0^1(\Omega))$ and $L^2(]0,T[, H^{-1}(\Omega))$, it is first necessary to define the integral in such a framework. Let us first recall how the integral over $\mathbb{R}$ is defined (see also [26, Chapter 4]. We start by defining the notion of measurability. Let $(X, \mathcal{T}, m)$ be a σ-finite measure space and let f be a function from X to $\mathbb{R}$. We can give two equivalent definitions of measurability of f (Proposition 4.15 below provides this equivalence):

1. A function f from X to $\mathbb{R}$ is measurable if $f^{-1}(B) \in \mathcal{T}$ for all $B \in \mathcal{B}(\mathbb{R})$.
2. A function f from X to $\mathbb{R}$ is measurable if f is a pointwise limit of step functions, i.e. if there exists a sequence $(f_n)_{n \in \mathbb{N}}$ of step functions such that $f_n(x) \to f(x)$ as $n \to +\infty$, for all $x \in X$.

Recall that a step function from X to $\mathbb{R}$ (or to E in the sequel) is a function taking a finite number of values such that, for all a, $\{x \in X; f(x) = a\} \in \mathcal{T}$.

Consider now a function f from X to a Banach space E; Definitions 1 and 2 with E instead of $\mathbb{R}$ then give two definitions of the measurability of f. These two definitions are equivalent if E is separable (see Proposition 4.15). But it is Definition 2 (pointwise limit of step functions) that is interesting for defining the integral. More precisely, since the integral does not see the difference between two functions on a set of measure zero, the really useful notion for integration is the notion of *m-measurability*, which we define as follows (see also [26, Chapter 3]):

Definition 4.14 (*m*-measurable function). Let $(X, \mathcal{T}, m)$ be a measure space. We say that a function f, defined from X with values in $\mathbb{R}$ or with values in a Banach E, is *m*-measurable if it is a.e. a limit of step functions.

Proposition 4.15 (Measurability and *m*-measurability). *Let $(X, \mathcal{T}, m)$ be a σ-finite measure space.*

1. *A function f from X with values in $\mathbb{R}$ is m-measurable if and only if there exists a measurable function g such that $f = g$ a.e.*
2. *Let E be a separable Banach space and f a function from X to E. Then,*

   a. *f is measurable in the sense "$f^{-1}(B) \in \mathcal{T}$ for all $B \in B(E)$" if and only if f is a pointwise limit of step functions.*
   b. *As in the real case, the function f from X to E is m-measurable if and only if there exists a measurable function g such that $f = g$ a.e.*

Proof The proof of points 1 and 2a for $E = \mathbb{R}$ can be found in the book [26, chapter 3], and that of points 2a for separable E and 2b in the monograph [19, chapter 1].   ■

Let $(X, \mathcal{T}, m)$ be a σ-finite measure set, E a Banach space and f an m-measurable function from X to E. We note that the function $x \mapsto \|f(x)\|_E$ is m-measurable from X to $\mathbb{R}_+$. The term $\int_X \|f(x)\|_E \, dm(x)$ is therefore perfectly defined in $\mathbb{R}_+ \cup \{+\infty\}$; we can therefore define the space $\mathcal{L}^1_E(X, \mathcal{T}, m)$ as follows.

Definition 4.16 (The space $\mathcal{L}^1_E(X, \mathcal{T}, m)$). Let $(X, \mathcal{T}, m)$ be a σ-finite measure space, E a Banach space and f an m-measurable function from X to E. The function f belongs to $\mathcal{L}^1_E(X, \mathcal{T}, m)$ if

$$\int_X \|f(x)\|_E \, dm(x) < +\infty.$$

We now want to define the integral of an element of $\mathcal{L}^1_E(X, \mathcal{T}, m)$ where $(X, \mathcal{T}, m)$ is a σ-finite measure space and E a Banach space. Let $f \in \mathcal{L}^1_E(X, \mathcal{T}, m)$. We cannot use the same technique[5] as for $E = \mathbb{R}$. However, since f is m-measurable, we know that there exists a sequence $(f_n)_{n \in \mathbb{N}}$ of step functions that converges to f a.e. Therefore, there exists a set $A \in \mathcal{T}$ such that $m(A^c) = 0$ and $f_n(x) \to f(x)$ for all $x \in A$. We then set

$$A_n = \{x \in A \ \|f_n(x)\|_E \leq 2\|f(x)\|_E\} \text{ and } g_n = f_n \, \mathbb{1}_{A_n}.$$

Note that the number 2 could be replaced here by any number greater than 1 and A could be replaced by another set satisfying the same properties without changing the definition of the integral given in Definition 4.17. The sequence $(g_n)_{n \in \mathbb{N}}$ satisfies:

g_n is a step function for all $n \in \mathbb{N}$,

$g_n \to f$ a.e.,

$\int_X \|g_n - f\|_E \, dm \to 0$ by the dominated convergence theorem.

The definition of the integral of a step function is immediate: we set $I_n = \int_X g_n \, dm$. We can then show that $(I_n)_{n \in \mathbb{N}}$ is a Cauchy sequence in E, which is therefore convergent. We can also show that the limit of this sequence depends only on f (and not on the choice of f_n and A). It is therefore natural to define the integral of f as the limit of this sequence as follows.

Definition 4.17 (Integral with values in E). Let $f : X \to E$ where $(X, \mathcal{T}, m)$ is a σ-finite measure space and E a Banach space. Let $f \in \mathcal{L}^1_E(X, \mathcal{T}, m)$ (see Definition 4.16), let $(f_n)_{n \in \mathbb{N}}$ be a sequence of step functions such that $f_n \to f$ a.e., and let $A \in \mathcal{T}$ such that $m(A^c) = 0$ and $f_n(x) \to f(x)$ for all $x \in A$. Set $A_n = \{x \in A \ \|f_n(x)\|_E \leq 2\|f(x)\|\}$ and $g_n = f_n \, \mathbb{1}_{A_n}$. Define the integral of f by:

$$\int_X f \, dm = \lim_{n \to +\infty} \int_X g_n \, dm \in E.$$

---

[5] Recall that the construction of the integral in $\mathbb{R}$ consists of decomposing a function f into f^+ and f^- and starting by integrating measurable non-negative functions, see [26].

As in the case $E = \mathbb{R}$, we then define $L^1_E(X, \mathcal{T}, m)$ as the set of equivalence classes of $\mathcal{L}^1_E(X, \mathcal{T}, m)$ for the equivalence relation $=$ a.e. We can then define the spaces $\mathcal{L}^p_E(X, \mathcal{T}, m)$ and $L^p_E(X, \mathcal{T}, m)$ for all $1 \leq p \leq +\infty$. An element of $L^p_E(X, \mathcal{T}, m)$ is therefore a set of elements of $\mathcal{L}^p_E(X, \mathcal{T}, m)$ (and two functions belonging to this set are equal a.e.). As in the case $E = \mathbb{R}$, if $F \in L^p_E(X, \mathcal{T}, m)$, we confuse F and f if $f \in F$ (so that we reason as if $F \in \mathcal{L}^p_E(X, \mathcal{T}, m)$).

Proposition 4.18 (The spaces $L^p_E(X, \mathcal{T}, m)$). *Let* $1 \leq p \leq +\infty$, $(X, \mathcal{T}, m)$ *a σ-finite measure space and E a Banach space; then*

1. $L^p_E(X, \mathcal{T}, m)$ *(with its natural norm) is complete: it is therefore a Banach space.*
2. *If $p < +\infty$ and E is separable, $L^p_E(X, \mathcal{T}, m)$ is separable.*
3. *If $p = 2$ and E is a Hilbert space, then $L^2_E(X, \mathcal{T}, m)$ is also a Hilbert space, whose inner product is defined by:*

$$(u|v)_{L^2_E(X, \mathcal{T}, m)} = \int_X (u(x)|v(x))_E \, \mathrm{d}m(x).$$

Consequently, from any bounded sequence in $L^2_E(X, \mathcal{T}, m)$ one can extract a weakly convergent subsequence in $L^2_E(X, \mathcal{T}, m)$.
4. *If $1 < p < +\infty$ and if E is a reflexive separable Banach space, the space $L^p_E(X, \mathcal{T}, m)$ is then a reflexive separable Banach space.*
5. *(Duality in $L^p_E(X, \mathcal{T}, m)$) Let $1 \leq p \leq +\infty$, $p' = \frac{p}{p-1}$ and $v \in L^{p'}_{E'}(X, \mathcal{T}, m)$. The mapping $T_v : u \mapsto \int \langle v, u \rangle_{E', E} \, \mathrm{d}m$ is well defined, linear and continuous from $L^p_E(X, \mathcal{T}, m)$ to $\mathbb{R}$ so that $T_v \in (L^p_E(X, \mathcal{T}, m))'$. Moreover, the mapping $T : v \mapsto T_v$ is an isometry of $L^{p'}_{E'}(X, \mathcal{T}, m)$ onto its image (which is therefore a subset of $L^p_E(X, \mathcal{T}, m)'$). If E' is separable and $p < +\infty$, the image of T is $(L^p_E(X, \mathcal{T}, m))'$ in its entirety.*
6. *(Dominated convergence) Let $1 \leq p < +\infty$ and $(u_n)_{n \in \mathbb{N}}$ a sequence in $L^p_E(X, \mathcal{T}, m)$. If*

   a. $u_n \to u$ *a.e.,*
   b. $\|u_n\|_E \leq G$ *a.e. for all $n \in \mathbb{N}$ with $G \in \mathcal{L}^p_{\mathbb{R}}(X, \mathcal{T}, m)$,*

   then $u_n \to u$ in $L^p_E(X, \mathcal{T}, m)$.

Proof (of Proposition 4.18) Properties 1–4 are not proved here, we refer to e.g. [19]. Property 5 is partially proved in Problem 4.2. Property 6 is easier; one simply needs to notice that

$$\|u_n - u\|_E \leq \|u_n\|_E + \|u\|_E \leq 2G.$$

Therefore, we have $\|u_n - u\|^p_E \leq 2^p |G|^p \in \mathcal{L}^1_{\mathbb{R}}(X, \mathcal{T}, m)$. By the dominated convergence theorem for functions with values in $\mathbb{R}$, $\int_X \|u_n - u\|^p_E \, \mathrm{d}m \to 0$ and therefore that $u_n \to u$ in $L^p_E(X, \mathcal{T}, m)$. $\blacksquare$

Remark 4.19 (Uniformly convex space). Let E be a Banach space; we say that E is uniformly convex if

$$\forall \eta > 0, \exists \varepsilon > 0 \text{ such that } (\|x\|_E = 1, \|y\|_E = 1, \|x - y\| \geq \eta) \Rightarrow \left\|\frac{x + y}{2}\right\| \leq 1 - \varepsilon.$$

The spaces $L^p(\Omega)$ with $1 < p < +\infty$ and Ω open subset of $\mathbb{R}^N$ are uniformly convex. An important consequence of the fact that E is a uniformly convex Banach space is that if $(x_n)_{n \in \mathbb{N}}$ is a sequence in E such that $x_n \to x$ weakly in E and $\|x_n\|_E \to \|x\|_E$ as $n \to +\infty$, then $x_n \to x$ in E as $n \to +\infty$ (see for example [25, Problem 4.3]). This property can potentially simplify certain proofs of Proposition 4.18 (but is not necessarily required). Note also that a uniformly convex Banach space is always reflexive, see for example [25, theorem 2.1].

Let us now introduce different notions of derivative for a vector-valued function in the context of parabolic equations, that is to say:

$$(X, \mathcal{T}, m) = (]0, T[, \mathcal{B}(]0, T[), \lambda),$$

where $T > 0$, $\mathcal{B}(]0, T[)$ denotes the σ-algebra of Borel sets of the interval $]0, T[$ and λ the Lebesgue measure on $\mathbb{R}$. Let us first properly define the derivative with respect to time of a function from $]0, T[$ to E which is not differentiable in the classical sense (that is to say in the sense of the existence of the limit, in E, of the usual differential quotient); we denote this derivative by $\partial_t u$. We now denote by $L_E^p(]0, T[)$ or $L^p(]0, T[, E)$ the space $L_E^p(]0, T[, \mathcal{B}(]0, T[), \lambda)$ and $L_{E, loc}^1(]0, T[)$ or $L_{loc}^1(]0, T[, E)$ the set of (classes of) functions with values in E locally integrable on $]0, T[$ (with the Lebesgue measure).

The following lemma is the counterpart for the integral with values in a Banach space of the fundamental lemma (Lemma 1.2) which allowed us to define the derivative by transposition of functions with values in $\mathbb{R}$. Its proof is similar.

Lemma 4.20. *Let E be a Banach space. Let $u \in L_{loc}^1(]0, T[, E)$ and $\mathcal{D}(]0, T[)$ be the set of functions from $]0, T[$ with values in $\mathbb{R}$, of class C^∞ and with compact support. Assume that*

$$\forall \varphi \in \mathcal{D}(]0, T[), \int_0^T u(t)\varphi(t) \, dt = 0.$$

Then $u = 0$ a.e.

We can thus define the derivative by transposition of a function u from $]0, T[$ with values in a Banach space E.

Definition 4.21 (Derivative by transposition). Let E be a Banach space, $1 \le p \le +\infty$ and $u \in L_E^p(]0, T[)$. Let $\mathcal{D}$ be the space $\mathcal{D}(]0, T[)$ (recall that $\mathcal{D}(]0, T[)$ is the space of C^∞ class functions with compact support from $]0, T[$ with values in $\mathbb{R}$), and $\mathcal{D}_E^\star$ the set of linear mappings from $\mathcal{D}$ to E. Define $\partial_t u \in \mathcal{D}_E^\star$ by:

$$\langle \partial_t u, \varphi \rangle_{\mathcal{D}_E^\star, \mathcal{D}} = -\int_0^T u(t)\varphi'(t)\, \mathrm{d}t \in E.$$

Remark 4.22 (Classical derivative and derivative by transposition). If $u \in C^1(]0, T[, E)$ we have, for all $\varphi \in \mathcal{D}(]0, T[)$,

$$-\int_0^T u(t)\varphi'(t)\, \mathrm{d}t = \int_0^T u'(t)\varphi(t)\, \mathrm{d}t = \langle \partial_t u, \varphi \rangle_{\mathcal{D}_E^\star, \mathcal{D}}.$$

We then confuse u' (classical derivative) with $\partial_t u$ (derivative by transposition), that is, the function u' which belongs to $C([0, T], E)$ with the linear mapping from $\mathcal{D}$ to E denoted by $\partial_t u$. Indeed, thanks to Lemma 4.20, the function u' is entirely determined by $\partial_t u$, which justifies the confusion between u' and $\partial_t u$.

Definition 4.23 (Weak derivative). Let E and F be two Banach spaces and $1 \le p, q \le +\infty$ and assume that there exists a vector space G such that $E \subset G$ and $F \subset G$. Let $u \in L_E^p(]0, T[)$ (so we have $\partial_t u \in \mathcal{D}_E^\star$); we say that $\partial_t u \in L_F^q(]0, T[)$ is the weak derivative of u if there exists a function $v \in L_F^q(]0, T[)$ such that

$$\langle \partial_t u, \varphi \rangle_{\mathcal{D}_E^\star, \mathcal{D}} = -\underbrace{\int_0^T u(t)\varphi'(t)\, \mathrm{d}t}_{\in E} = \underbrace{\int_0^T v(t)\varphi(t)\, \mathrm{d}t}_{\in F}.$$

Let us emphasize that this equality only makes sense if there exists a vector space G such that $E \subset G$ and $F \subset G$. In this case, we confuse $\partial_t u \in \mathcal{D}_E^\star$ and $v \in L_F^q(]0, T[)$. Here too, Lemma 4.20 is used for this identification because it tells us that if the function v exists then it is unique.

In summary, we therefore have three notions of the derivative of a function u from $]0, T[$ to values in E (Banach space).

- Classical derivative: $u' :]0, T[\to E$ (rarely exists...).
- Derivative by transposition: $\partial_t u \in \mathcal{D}_E^\star$ (exists as soon as $u \in L_E^p(]0, T[)$).
- Weak derivative: $\partial_t u \in L_F^q(]0, T[)$ where F is a Banach space such that there exists a vector space G such that $E \subset G$ and $F \subset G$.

Let us now give some examples of derivatives by transposition and weak derivatives. Let Ω be an open bounded subset of $\mathbb{R}^N$.

1. $E = H_0^1(\Omega)$, $F = H^{-1}(\Omega)$. We then have $E, F \subset G = \mathcal{D}^\star(\Omega)$.
2. $1 \le p < +\infty$, $q = \frac{p}{p-1}$, $E = W_0^{1,p}(\Omega)$, $F = W^{-1,q}(\Omega)$. We also have $E, F \subset G = \mathcal{D}^\star(\Omega)$.

3. $E = H^1(\Omega)$ $F = (H^1(\Omega))'$ and we assume that Ω has a Lipschitz boundary. For this example, we have $F \not\subset \mathcal{D}^\star(\Omega)$. Indeed, we take for example Φ defined, for $v \in H^1(\Omega)$, by

$$\Phi(v) = \int_{\partial\Omega} v \, d\gamma(x).$$

The mapping Φ is indeed a continuous linear mapping on $H^1(\Omega)$, therefore $\Phi \in H^1(\Omega)'$. But $\Phi = 0$ on $\mathcal{D}(\Omega)$, and therefore, since Φ is not the null mapping, F is not embedded in $\mathcal{D}^\star(\Omega)$. In this example, to embed E and F into the same space G, we start by identifying the topological dual $(L^2(\Omega))'$ of $L^2(\Omega)$ with $L^2(\Omega)$. Indeed, by the Riesz representation theorem, if $\Phi \in L^2(\Omega)'$, there exists a unique function $u \in L^2(\Omega)$ such that $\Phi(v) = (v|u)_2$ for all $v \in L^2(\Omega)$. We then confuse u and Φ. By this identification, we therefore have $L^2(\Omega)' = L^2(\Omega)$. We now note that, if E, H are two Banach spaces such that E is dense and compactly embedded in H, we then have $H' \subset E'$. We take here $E = H^1(\Omega)$ and $H = L^2(\Omega)$ and we therefore have, thanks to the identification between $L^2(\Omega)$ and $L^2(\Omega)'$,

$$H' = L^2(\Omega) \subset H^1(\Omega)'.$$

Finally, we therefore have $E \subset G$ and $F \subset G$ with $G = (H^1(\Omega))'$.

4. In the previous example, the space H is a Hilbert space, and by identifying H with H', we therefore have $E \subset E'$. But if the goal is only to have $E \subset E'$, it is not necessary to have $H' \subset E'$ and we can remove the assumption of density of E in H. More precisely, let E be a Banach space and H a Hilbert space such that E is continuously imbedded in H. We identify H with H'. Now let $u \in E$. Since $u \in H$, we therefore have an identification between u and Φ_u which is the mapping $v \mapsto (v|u)_H$ from H to $\mathbb{R}$. We can then also identify $u \in E$ with the restriction of Φ_u to E, which belongs to E'. This identification is legitimate because if u_1 and u_2 are two different elements of E, the restrictions of Φ_{u_1} and Φ_{u_2} to E are different mappings. We thus have $E \subset E'$. In this example, however, it should be noted that (in the absence of density of E in H) different elements of H' have the same restriction to E (and can therefore correspond to the same element of E). An interesting example of this situation is obtained by taking $H = L^2(\Omega)^N$ where Ω is an open bounded subset of $\mathbb{R}^N$, $N > 1$, and $E = \{u \in H : \operatorname{div} u = 0\}$.

Remark 4.24 (Comparison of $L^p(]0, T[, L^p(\Omega))$ and $L^p(\Omega \times]0, T[)$). Let $T > 0$, Ω an open subset of $\mathbb{R}^N$, $1 \leq p < +\infty$. Let $u \in L^p(]0, T[, L^p(\Omega))$. There then exists a $v \in L^p(\Omega \times]0, T[)$ such that $u(t) = v(\cdot, t)$ a.e. (in Ω) and for almost all $t \in]0, T[$. (Note that this equality is true whatever the representatives chosen for u and v.) Conversely, if $v \in L^p(\Omega \times]0, T[)$, there exists a $u \in L^p(]0, T[, L^p(\Omega))$ such that $u(t) = v(\cdot, t)$ a.e. (in Ω) and for almost all $t \in]0, T[$. These two assertions allow us to identify $L^p(\Omega \times]0, T[)$ and $L^p(]0, T[, L^p(\Omega))$. Note however that the second assertion is false for $p = +\infty$ (see [19], page 28, for an example). Keeping these notations, we can then compare $\partial_t u$ (which applies to an element of $\mathcal{D}(]0, T[)$) and $\partial_t v$ (which applies to an element of $\mathcal{D}(\Omega \times]0, T[)$). Indeed, for all $\varphi \in \mathcal{D}(]0, T[)$ and all $\psi \in \mathcal{D}(\Omega)$ we have

$$\int_{\Omega} \langle \partial_t u, \varphi \rangle_{\mathcal{D}_E^{\star}, \mathcal{D}}(x) \psi(x)\, dx = -\int_{\Omega} \Big(\int_0^T u(t)\varphi'(t)\, dt \Big)(x)\psi(x)\, dx$$

$$= -\int_{\Omega} \int_0^T v(x,t)\varphi'(t)\psi(x)\, dx\, dt$$

$$= \langle \partial_t v, \varphi\psi \rangle_{\mathcal{D}^{\star}(\Omega \times]0,T[), \mathcal{D}(\Omega \times]0,T[)}.$$

Proposition 4.25 (Commutation of the duality action and the integral). *Let E be a Banach space, $u \in L_E^1(]0,T[)$, $\psi \in E'$ and $\varphi \in \mathcal{D}(]0,T[)$. Then,*

$$\langle \psi, \int_0^T u(t)\varphi(t)\, dt \rangle_{E',E} = \int_0^T \langle \psi, u(t) \rangle_{E',E}\, \varphi(t)\, dt.$$

The proof of Proposition 4.25 is left as an exercise.

Lemma 4.26 (Time continuity). *Let E be a Banach space, $1 \le p \le +\infty$ and $u \in L_E^p(]0,T[)$ and assume that $\partial_t u \in L_E^p(]0,T[)$ (then $u \in W_E^{1,p}(]0,T[)$. Then, $u \in C([0,T],E)$, and even $u \in C^{0,1-\frac{1}{p}}([0,T],E)$. More precisely, there exists an $a \in E$ such that $u(t) = a + \int_0^t \partial_t u(s)\, ds$ for almost all $t \in]0,T[$; we can then identify u with the continuous function on $[0,T]$ defined by $t \mapsto a + \int_0^t \partial_t u(s)\, ds$. In particular, for all $0 \le t_1 < t_2 \le T$,*

$$u(t_2) - u(t_1) = \int_{t_1}^{t_2} \partial_t u(s)\, ds.$$

Proof The proof is similar to the case $E = \mathbb{R}$, see Problem 1.3. Set $\partial_t u = v \in L_E^p(]0,T[)$ and define w by $w(t) = \int_0^t v(s)\, ds$, so that $w \in C([0,T],E)$. Noting that $\partial_t w = v = \partial_t u$ and therefore $\partial_t(w - u) = 0$, we have

$$\int_0^T (w-u)\partial_t \varphi\, dt = 0 \text{ for all } \varphi \in \mathcal{D}(]0,T[). \qquad (4.16)$$

We now choose a function $\varphi_0 \in \mathcal{D}(]0,T[)$ s.t. $\int_0^T \varphi_0(s)\, ds = 1$. For $\psi \in \mathcal{D}(]0,T[)$ we define φ by

$$\varphi(t) = \int_0^t \psi(s)\, ds - \int_0^t \varphi_0(s)\, ds \int_0^T \psi(s)\, ds,$$

so that $\varphi \in \mathcal{D}(]0,T[)$ and we can take φ in (4.16). Thus, there exists an $a \in E$ such that $w - u = a$ a.e. and therefore $u \in C([0,T],E)$ (the proof is left as an exercise). Thanks to Hölder's inequality, we then show that $u \in C^{0,1-\frac{1}{p}}$. $\blacksquare$

We now prove a more difficult lemma also providing the continuity of u. Assume that E is a Banach space and F a Hilbert space such that $E \subset F$, with continuous embedding, and that E is dense in F. We therefore also have $F' \subset E'$. Since F is a Hilbert space, we can identify F with its dual by the Riesz representation theorem, i.e. we identify $v \in F$ with the mapping $T_v : u \mapsto (v \mid u)_F$ (which belongs to F'). The mapping $v \mapsto T_v$ is an isometry (bijective) from F to F'. With this identification, we therefore have $E \subset F = F' \subset E'$. Therefore, every element of E is then an element of E', and for $u, v \in E$, we write that

$$\langle v, u \rangle_{E',E} = (v \mid u)_F,$$

that is to say, in fact $\langle T_v, u \rangle_{E',E} = (v \mid u)_F$, where T_v is the element of F' identified with $v \in F$. With this identification of F with F', we now give a continuity result of u in F if $u \in L^2(]0, T[, E)$ and $\partial_t u \in L^2(]0, T[, E')$. Be careful, this last hypothesis only makes sense because of the identification of F with F': if we change the space F, then we change the meaning of $\partial_t u \in L^2(]0, T[, E')$, even though E' has not changed!

Lemma 4.27 (Sufficient condition for continuity). *Let E be a Banach space and F a Hilbert space such that $E \subset F$, with continuous embedding, and E dense in F. We identify F with F' (so that $E \subset F = F' \subset E'$). Let $u \in L^2_E(]0, T[)$, we assume that $\partial_t u \in L^2_{E'}(]0, T[)$. Then $u \in C([0, T], F)$ and, for all $t_1, t_2 \in [0, T]$, $t_1 > t_2$, we have*

$$\|u(t_1)\|_F^2 - \|u(t_2)\|_F^2 = 2 \int_{t_2}^{t_1} \langle \partial_t u, u \rangle_{E',E} \, dt. \tag{4.17}$$

Proof Let us show that $u \in C([0, T[, F)$. Let $\rho \in \mathcal{D}(] - 2, -1[)$ such that $\int_{\mathbb{R}} \rho \, dx = 1$ and $\rho \geq 0$. For $n \in \mathbb{N}^\star$, we set $\rho_n = n\rho(n\cdot)$ so that the support of ρ_n is included in $] - \dfrac{2}{n}, -\dfrac{1}{n}[$ and that $\int \rho_n \, dx = 1$: the sequence $(\rho_n)_{n \in \mathbb{N}}$ is a sequence of regularising kernels. Define u_n by convolution with ρ_n, that is, $u_n = \tilde{u} \star \rho_n$ with

$$\tilde{u} = u \text{ on }]0, T[$$
$$u = 0 \text{ on }]0, T[^c.$$

Hence $u_n \to u$ in $L^2_E(]0, T[)$, $u_n \in C_c^\infty(\mathbb{R}, E)$ and $u'_n = \tilde{u} \star \rho'_n$. Let $t \in \mathbb{R}$, we have

$$u'_n(t) = \int_0^T u(s)\rho'_n(t - s) \, ds.$$

Note that $\rho_n(t - s) = 0$ if $t - s \notin \left] \dfrac{-2}{n}, \dfrac{-1}{n} \right[$, that is, $s \notin \left] t + \dfrac{1}{n}, t + \dfrac{2}{n} \right[$. Let $\varepsilon > 0$ and $t \in [0, T - \varepsilon[$. For $n \geq n_0$ with n_0 such that $\dfrac{2}{n_0} < \varepsilon$ we have $t + \dfrac{2}{n} < T$ and therefore

$$\rho_n(t - \cdot) \in \mathcal{D}(]0, T[).$$

We then have

$$u'_n(t) = \langle \partial_t u, \rho_n(t - \cdot) \rangle_{\mathcal{D}_E^\star, \mathcal{D}}$$
$$= \int_0^T \partial_t u(s)\rho_n(t - s) \, ds \in E', \text{ because } \partial_t u \in L^2_{E'}(]0, T[).$$

Consequently, $u'_n = \widetilde{\partial_t u} \star \rho_n$ on $[0, T - \varepsilon[$, with $\widetilde{\partial_t u} = \begin{cases} \partial_t u \text{ on }]0, T[\\ 0 \text{ on }]0, T[^c. \end{cases}$

But $\widetilde{\partial_t u} \star \rho_n \to \widetilde{\partial_t u}$ in $L^2(\mathbb{R}, E')$, and therefore $u'_n = \widetilde{\partial_t u} \star \rho_n \to \partial_t u$ in $L^2_{E'}(]0, T - \varepsilon[)$.

In summary, $u_n \to u$ in $L^2_E(]0, T[)$ and $u'_n \to \partial_t u$ in $L^2_{E'}(]0, T - \varepsilon[)$ $\quad \forall \varepsilon > 0$.

Let us now show that $(u_n(t))_{n \in \mathbb{N}}$ is a Cauchy sequence in F for all $t \in [0, T[$, and even, for all $\varepsilon > 0$, uniformly if $t \in [0, T - \varepsilon]$. Let $\varepsilon > 0$ and $\varphi \in C^1([0, T], \mathbb{R})$ be defined by $\varphi(t) = \|u_n(t) - u_m(t)\|_F^2$. Hence, for $t \in]0, T[$,

$$\varphi'(t) = 2(\underbrace{(u'_n - u'_m)(t)}_{\in E} \mid \underbrace{(u_n - u_m)(t)}_{\in E})_F.$$

Recall that $u_n \in C_c^\infty(\mathbb{R}, E)$ and therefore $(u'_n - u'_m)(t) \in E$. Let $t_1, t_2 \in [0, T - \varepsilon]$. We have

$$\varphi(t_2) - \varphi(t_1) = \int_{t_1}^{t_2} 2(u'_n(s) - u'_n(s) \mid u_n(s) - u_m(s))_F \, ds$$

$$= 2 \int_{t_1}^{t_2} \langle \partial_t(u_n - u_m)(s), u_n(s) - u_m(s) \rangle_{E', E} \, ds.$$

Therefore,

$$\varphi(t_2) - \varphi(t_1) \leq 2 \left(\int_0^{T-\varepsilon} \|\partial_t(u_n - u_m)\|_{E'}^2 \, ds \right)^{\frac{1}{2}} \left(\int_0^T \|u_n - u_m\|_E^2 \, ds \right)^{\frac{1}{2}}$$

$$\leq 2 \|\partial_t(u_n - u_m)\|_{L^2_{E'}} \|u_n - u_m\|_{L^2_E}.$$

Let $\eta > 0$. Since $u_n \to u$ in $L^2_E(]0, T[)$ and since the sequence $(\partial_t u_n)_{n \in \mathbb{N}^\star}$ is bounded in $L^2_{E'}(]0, T - \varepsilon[)$, there exists an n_0 such that

$$\varphi(t_2) - \varphi(t_1) \leq \eta \text{ for } m, n \geq n_0.$$

Hence

$$\varphi(t_2) \leq \varphi(t_1) + \eta \quad \text{for } n, m \geq n_0.$$

Integrating this inequality for $t_1 \in]0, T - \varepsilon[$ yields

$$(T - \varepsilon)\varphi(t_2) \leq \int_0^{T-\varepsilon} \|u_n - u_m\|_F^2 \, dt_1 + T\eta$$

$$\leq \|u_n - u_m\|_{L^2_F}^2 + T\eta$$

Using once again that $u_n \to u$ in $L^2_E(]0, T[)$, there exists an n_1 such that

$$(T - \varepsilon)\varphi(t_2) \leq (T + 1)\eta \text{ for } n, m \geq n_1.$$

Therefore, we have $\varphi(t) \le \frac{(T+1)\eta}{T-\varepsilon}$ for $n, m \ge n_1$ and $t \in [0, T - \varepsilon]$. We have thus shown that, for all $t \in [0, T - \varepsilon]$,

$$n, m \ge n_1 \Rightarrow \|u_n(t) - u_m(t)\|_F^2 \le \frac{(T+1)\eta}{T-\varepsilon}.$$

This clearly shows that the sequence $(u_n(t))_{n \in \mathbb{N}}$ is a Cauchy sequence in F uniformly with respect to t, if $t \in [0, T - \varepsilon]$. Therefore, there exists a function w from $[0, T[$ to F such that $u_n(t) \to w(t)$ in F for all $t \in [0, T[$. Since this convergence is uniform on $[0, T - \varepsilon]$ for all $\varepsilon > 0$, the function w is continuous on $[0, T - \varepsilon]$ for all $\varepsilon > 0$. Therefore, we have $w \in C([0, T[, F)$. Since $u_n \to u$ in $L_E^2(]0, T[)$ (and therefore a.e. on $]0, T[$, up to a subsequence), we therefore have $u = w$ a.e., and therefore $u \in C([0, T[, F)$ (since we identify, as usual, the class of functions u with its continuous representative, which is precisely w).

In a similar way, by shifting the regularising kernels to the other side, we show that $u \in C(]0, T], F)$. Therefore, we have $u \in C([0, T], F)$.

Finally, we also have for $t_1, t_2 \in [0, T]$,

$$\|u_n(t_1)\|_F^2 - \|u_n(t_2)\|_F^2 = 2 \int_{t_2}^{t_1} (u_n'(s) \mid u_n(s))_F \, ds = 2 \int_{t_2}^{t_1} \langle \partial_t u_n(s), u_n(s) \rangle_{E',E} \, ds.$$

By passing to the limit in n, we obtain

$$\|u(t_1)\|_F^2 - \|u(t_2)\|_F^2 = 2 \int_{t_2}^{t_1} \langle \partial_t u(s), u(s) \rangle_{E',E} \, ds.$$

Remark 4.28. A classical example of Lemma 4.27 is to take $E = H_0^1(\Omega)$ (where Ω is an open subset of $\mathbb{R}^N$, $N \ge 1$), $F = L^2(\Omega)$, F' identified with F, and therefore $E' = H^{-1}(\Omega)$. The choice of E' for the space in which is $\partial_t u$ is crucial in the proof of Lemma 4.27, but generalisations of this lemma are possible. Problem 4.4 shows that $u \in L^2(]0, 1[, H^1(]0, +\infty[)$ and $\partial_t u \in L^2(]0, 1[, H^{-1}(]0, +\infty[)$ (the space L^2 being identified with itself) are sufficient assumptions to obtain $u \in C([0, T], L^2(\Omega))$. The Lemma 4.27 does not apply here directly because $H^{-1}(]0, +\infty[)$ is the dual space of $H_0^1(]0, +\infty[)$ and is therefore not the dual space of $H^1(]0, +\infty[)$. An extension argument allows us to return to Lemma 4.27 (Problem 4.4).

An interesting consequence of Lemma 4.27 is the formula for integration by parts in time, which is the subject of the following lemma.

Lemma 4.29 (Integration by parts in time). *Let E be a Banach space and F a Hilbert space such that $E \subset F$, with continuous embedding, and E dense in F. We identify F with F' (so that $E \subset F = F' \subset E'$). Let $u, v \in L_E^2(]0, T[)$. We assume that $\partial_t u, \partial_t v \in L_{E'}^2(]0, T[)$. Then $u, v \in C([0, T], F)$ and*

$$\int_0^T \langle \partial_t u, v \rangle_{E',E} \, dt + \int_0^T \langle \partial_t v, u \rangle_{E',E} \, dt = (u(T) \mid v(T))_F - (u(0) \mid v(0))_F.$$

Proof Lemma 4.27 gives $u, v \in C([0,T], F)$. Now applying formula (4.17) to the functions u, v and $u + v$ (with $t_1 = T$ and $t_2 = 0$), we get

$$\|(u+v)(T)\|_F^2 - \|(u+v)(0)\|_F^2 = 2\int_0^T \langle \partial_t(u+v), u+v\rangle_{E',E}\, dt,$$

$$\|u(T)\|_F^2 - \|u(0)\|_F^2 = 2\int_0^T \langle \partial_t u, u\rangle_{E',E}\, dt,$$

$$\|v(T))\|_F^2 - \|v(0)\|_F^2 = 2\int_0^T \langle \partial_t u, u\rangle_{E',E}\, dt.$$

By subtracting the first equation from the following two, this gives (noting that $\partial_t(u+v) = \partial_t u + \partial_t v$)

$$2(u(T)\,|\,v(T)))_F - 2(u(0)\,|\,v(0))_F = 2\int_0^T \left(\langle \partial_t u, v\rangle_{E',E} + \langle \partial_t v, u\rangle_{E',E}\right)\, dt,$$

which gives the desired formula. $\blacksquare$

4.3 Weak Solutions of the Heat Equation

In this section, we are interested in the heat equation posed on open bounded subset Ω of $\mathbb{R}^N$, and during a time interval $]0,T[$, $T \in \mathbb{R}_+^\star$. Let f be a function from $\Omega \times]0,T[$ to $\mathbb{R}$ and u_0 a function from Ω to $\mathbb{R}$. We are looking for u such that

$$\begin{cases} \partial_t u - \Delta u = f & \text{in } \Omega \times]0,T[, \\ u = 0 & \text{on } \partial\Omega \times]0,T[, \\ u(\cdot, 0) = u_0. \end{cases}$$

This linear problem admits a unique weak solution; this result is the subject of Theorem 4.30 below. We provide two proofs of Theorem 4.30, of very different spirit:

- by the classical Faedo–Galerkin method, which consists of approximating the initial PDE by a differential system,
- by a more direct method using a technique we call here generalised coercivity, which calls upon results of abstract functional analysis.

Throughout the following, if E is a Banach space and $T > 0$, we often denote $L^2(]0,T[, E)$ (instead of $L_E^2(]0,T[)$) the space $L_E^2(]0,T[, \mathcal{B}(]0,T[, \lambda)$.

Theorem 4.30 (Existence and uniqueness of the weak solution of the heat equation). *Let Ω be an open bounded subset of $\mathbb{R}^N$, $T > 0$ and $u_0 \in L^2(\Omega)$. We identify $L^2(\Omega)$ with its dual and we suppose that $f \in L^2(]0,T[, H^{-1}(\Omega))$. Then, there exists a unique function u such that*

$$\begin{cases} u \in L^2(]0,T[, H_0^1(\Omega)), \ \partial_t u \in L^2(]0,T[, H^{-1}(\Omega)), \\ \displaystyle\int_0^T \langle \partial_t u(s), v(s) \rangle_{H^{-1}, H_0^1} \, ds + \int_0^T \int_\Omega \nabla u(s) \cdot \nabla v(s) \, dx \, ds \\ \displaystyle\quad = \int_0^T \langle f(s), v(s) \rangle_{H^{-1}, H_0^1} \, ds, \ \forall v \in L^2(]0,T[, H_0^1(\Omega)), \\ u(0) = u_0 \ \ a.e. \end{cases} \tag{4.18}$$

(Recall that $u(s)$ (resp. $v(s)$) denotes the function $x \mapsto u(x,s)$ (resp. $v(x,s)$). We also have the following estimates on u and $\partial_t u$:

$$\|u\|_{L^2(]0,T[, H_0^1(\Omega))} \le \|u_0\|_2 + \|f\|_{L^2(]0,T[, H^{-1}(\Omega))} \,, \tag{4.19}$$

$$\|\partial_t u\|_{L^2(]0,T[, H^{-1}(\Omega))} \le \|u_0\|_2 + 2\,\|f\|_{L^2(]0,T[, H^{-1}(\Omega))} \,, \tag{4.20}$$

$$\|u(t)\|_2^2 \le \|u_0\|_2^2 + \|\partial_t u\|_{L^2(]0,T[, H^{-1}(\Omega))}^2 + \|u\|_{L^2(]0,T[, H_0^1(\Omega))}^2 \,,$$

$$\text{for all } t \in [0,T[. \tag{4.21}$$

Recall that we have identified $L^2(\Omega)$ with $L^2(\Omega))'$, so that $H_0^1(\Omega) \subset L^2(\Omega) = L^2(\Omega)' \subset H^{-1}(\Omega)$. Since we are looking for u such that $u \in L^2(]0,T[, H_0^1(\Omega))$ and $\partial_t u \in L^2(]0,T[, H^{-1}(\Omega))$, we necessarily have $u \in C([0,T], L^2(\Omega))$ (according to Lemma 4.27). The function u is therefore defined for all $t \in [0,T]$, and thus the condition $u(0) = u_0$ a.e. makes sense.

4.3.1 Proof of the Existence and Uniqueness Theorem by the Faedo–Galerkin Method

The idea of this method is to first solve an approximate problem which is written in the form of a differential system. With this aim, we can look for an approximate solution in space using for instance a finite element method; however in the case of the Laplace operator, the easiest way is to use a Hilbert basis formed by its eigenfunctions, that is to say, a Hilbert basis of $L^2(\Omega)$, denoted by $\{e_n, n \in \mathbb{N}^\star\}$, such that e_n is (for all n) a weak solution of

$$\begin{cases} -\Delta e_n = \lambda_n e_n & \text{in } \Omega, \\ e_n = 0 & \text{on } \partial\Omega, \end{cases}$$

with $\lambda_n \in \mathbb{R}$.

Step 1. Preliminary remarks. By Theorem 2.16, there exists a family $(e_n)_{n \in \mathbb{N}^\star}$ which is a Hilbert basis of $L^2(\Omega)$ and which satisfies

$$\begin{cases} e_n \in H_0^1(\Omega), \\ \displaystyle\int_\Omega \nabla e_n \cdot \nabla v \, dx = \lambda_n \int_\Omega e_n v \, dx, \text{ for all } v \in H_0^1(\Omega), \end{cases}$$

with $\lambda_n > 0$ for all $n \in \mathbb{N}^\star$ and $\lambda_n \uparrow +\infty$ as $n \to +\infty$.

Since $(e_n)_{n\in\mathbb{N}^\star}$ is a Hilbert basis of $L^2(\Omega)$, we have, for all $w \in L^2(\Omega)$, $w = \sum_{n\in\mathbb{N}^\star}(w|e_n)_2 e_n$ in the sense of the convergence in $L^2(\Omega)$, i.e. $\sum_{i=1}^{n}(w|e_i)e_i \to w$ in $L^2(\Omega)$ as $n \to +\infty$.

Let us now show that the family $(\lambda_n^{-\frac{1}{2}}e_n)_{n\in\mathbb{N}^\star}$ is a Hilbert basis of $H_0^1(\Omega)$. Recall that $H_0^1(\Omega)$ is a Hilbert space; the norm of u in $H_0^1(\Omega)$ is defined by $\|u\|_{H_0^1(\Omega)} = \| |\nabla u| \|_{L^2(\Omega)}$ and therefore the inner product of u and v in $H_0^1(\Omega)$ is given by $(u|v)_{H_0^1(\Omega)} = \int_\Omega \nabla u \cdot \nabla v \, dx$.

First note that for all $n, m \geq 1$,

$$\int_\Omega \nabla e_n \cdot \nabla e_m \, dx = \lambda_n \int_\Omega e_n e_m \, dx = \lambda_n \delta_{n,m}.$$

Hence $(e_n|e_m)_{H_0^1(\Omega)} = 0$ if $n \neq m$ and

$$\left\| \frac{e_n}{\sqrt{\lambda_n}} \right\|_{H_0^1(\Omega)}^2 = \int_\Omega \frac{\nabla e_n \cdot \nabla e_n}{\lambda_n} \, dx = 1.$$

Then, observe that the vector space generated by the family $(e_n)_{n\in\mathbb{N}^\star}$, denoted by $\mathrm{Vect}\{e_n, n \in \mathbb{N}^\star\}$, is dense in $H_0^1(\Omega)$. Indeed, let $v \in H_0^1(\Omega)$ such that $(v|e_n)_{H_0^1(\Omega)} = 0$ for all $n \in \mathbb{N}^\star$, then, for all $n \in \mathbb{N}^\star$,

$$0 = (v|e_n)_{H_0^1(\Omega)} = \int_\Omega \nabla e_n \cdot \nabla v \, dx = \lambda_n \int_\Omega e_n v \, dx.$$

Since $(e_n)_{n\in\mathbb{N}^\star}$ is a Hilbert basis of $L^2(\Omega)$ (and $\lambda_n \neq 0$ for all $n \in \mathbb{N}^\star$), we get that $v = 0$ a.e. This shows that the orthogonal complement of $\mathrm{Vect}\{e_n, n \in \mathbb{N}^\star\}$ in $H_0^1(\Omega)$ is reduced to $\{0\}$ and therefore that $\mathrm{Vect}\{e_n, n \in \mathbb{N}^\star\}$ is dense in $H_0^1(\Omega)$.

Finally, we thus obtain that the family $(\lambda_n^{-\frac{1}{2}}e_n)_{n\in\mathbb{N}^\star}$ is a Hilbert basis of $H_0^1(\Omega)$.

Step 2. Approximate solution. Let $n \in \mathbb{N}^\star$; we set $E_n = \mathrm{Vect}\{e_p, p = 1, \ldots n\}$, and we look for an approximate solution u_n in the form $u_n(t) = \sum_{i=1}^{n}\alpha_i(t)e_i$ with $\alpha_i \in C([0,T], \mathbb{R})$. Assuming that the α_i are differentiable for all t (which is not generally true), we therefore have

$$u_n'(t) = \sum_{i=1}^{n}\alpha_i'(t)e_i,$$

so that, for all $\varphi \in H_0^1(\Omega)$ and all $t \in]0, T[$, we have (taking into account that $H_0^1(\Omega)$ is embedded in $L^2(\Omega)$),

$$\langle u_n'(t), \varphi \rangle_{H^{-1}(\Omega), H_0^1(\Omega)} = \sum_{i=1}^{n}\alpha_i'(t)\int_\Omega e_i \varphi \, dx.$$

On the other hand, for all $t \in [0, T]$ we have

$$-\Delta u_n(t) = -\sum_{i=1}^{n} \alpha_i(t)\Delta e_i = \sum_{i=1}^{n} \lambda_i \alpha_i(t) e_i \text{ in } \mathcal{D}^\star(\Omega) \text{ and in } H^{-1}(\Omega),$$

that is, for all $\varphi \in H_0^1(\Omega)$,

$$\langle -\Delta u_n(t), \varphi \rangle_{H^{-1}(\Omega), H_0^1(\Omega)} = \int_\Omega \nabla u_n(t) \cdot \nabla \varphi \, dx = \sum_{i=1}^{n} \lambda_i \alpha_i(t) \int_\Omega e_i \varphi \, dx.$$

Finally, since $f \in L^2(]0, T[, H^{-1})$, we have for all $\varphi \in H_0^1(\Omega)$, $\langle f(\cdot), \varphi \rangle_{H^{-1}(\Omega), H_0^1(\Omega)}$ $\in L_{\mathbb{R}}^1(]0, T[)$. The quantity $\langle f(t), \varphi \rangle_{H^{-1}(\Omega), H_0^1(\Omega)}$ is therefore defined for almost all t and we finally obtain, for almost all t and for all $\varphi \in H_0^1(\Omega)$,

$$\langle u_n'(t) - \Delta u_n(t) - f(t), \varphi \rangle_{H^{-1}(\Omega), H_0^1(\Omega)}$$
$$= \sum_{i=1}^{n} (\alpha_i'(t) + \lambda_i \alpha_i(t)) \int_\Omega e_i \varphi \, dx - \langle f(t), \varphi \rangle_{H^{-1}(\Omega), H_0^1(\Omega)}.$$

To obtain u_n, a natural idea is to choose the functions α_i so that

$$\langle u_n'(t) - \Delta u_n(t) - f(t), \varphi \rangle_{H^{-1}(\Omega), H_0^1(\Omega)} = 0$$

for all $\varphi \in E_n$. Setting $f_i(t) = \langle f(t), e_i \rangle_{H^{-1}(\Omega), H_0^1(\Omega)}$, this is equivalent to asking for all $i \in \{1, \ldots, n\}$,

$$\alpha_i'(t) + \lambda_i \alpha_i(t) = f_i(t).$$

Taking into account the initial condition and setting $\alpha_i^{(0)} = (u_0|e_i)_2$, this therefore suggests taking

$$\alpha_i(t) = \alpha_i^{(0)} e^{-\lambda_i t} + \int_0^t e^{-\lambda_i(t-s)} f_i(s) \, ds. \tag{4.22}$$

The functions α_i thus defined belong to $C([0, T], \mathbb{R})$ and we therefore have $u_n \in C([0, T], E_n) \subset C([0, T], H_0^1(\Omega))$ with $u_n(t) = \sum_{i=1}^n \alpha_i(t) e_i$.

Step 3. Precision on the time derivative. Let $n \in \mathbb{N}^\star$ and u_n be the approximate solution given by the previous step. The functions α_i are not necessarily differentiable. Let us specify what the derivative by transposition is. This derivative is denoted by $\partial_t u_n$ or, more simply, $\partial_t u_n$. By definition of differentiation by transposition, $\partial_t u_n$ belongs to $\mathcal{D}_E^\star$ with $E = H_0^1(\Omega)$. Let $\varphi \in \mathcal{D}(]0, T[)$ we have

$$\langle \partial_t u_n, \varphi \rangle_{\mathcal{D}_E^\star, \mathcal{D}} = -\int_0^T u_n(t)\varphi'(t) \, dt \in E_n \subset H_0^1(\Omega).$$

Since $u_n = \sum_{i=1}^n \alpha_i e_i$, we therefore have

$$\langle \partial_t u_n, \varphi \rangle_{\mathcal{D}_E^\star, \mathcal{D}} = -\sum_{i=1}^n \int_0^T \alpha_i(t) e_i \varphi'(t) \, dt = -\sum_{i=1}^n \left(\int_0^T \alpha_i(t) \varphi'(t) \, dt \right) e_i.$$

We now use (4.22),

$$\int_0^T \alpha_i(t) \varphi' \, dt = T_i + S_i,$$

with

$$T_i = \int_0^T \alpha_i^{(0)} e^{-\lambda_i t} \varphi'(t) \, dt = \int_0^T \alpha_i^{(0)} \lambda_i e^{-\lambda_i t} \varphi(t) \, dt.$$

$$S_i = \int_0^T \left(\int_0^t e^{-\lambda_i(t-s)} f_i(s) \, ds \right) \varphi'(t) \, dt.$$

Owing to Fubini's theorem,

$$S_i = \int_0^T \left(\int_0^T \mathbb{1}_{[0,t]}(s) e^{-\lambda_i(t-s)} f_i(s) \, ds \right) \varphi'(t) \, dt$$

$$= \int_0^T \left(\int_0^T \mathbb{1}_{[s,T]}(t) e^{-\lambda_i(t-s)} \varphi'(t) \, dt \right) f_i(s) \, ds$$

$$= \int_0^T \left(\int_s^T e^{-\lambda_i(t-s)} \varphi'(t) \, dt \right) f_i(s) \, ds$$

$$= \int_0^T \left(\int_s^T \lambda_i e^{-\lambda_i(t-s)} \varphi(t) \, dt \right) f_i(s) \, ds - \int_0^T \varphi(s) f_i(s) \, ds$$

$$= \int_0^T \left(\int_0^t \lambda_i e^{-\lambda_i(t-s)} f_i(s) \, ds \right) \varphi(t) \, dt - \int_0^T f_i(t) \varphi(t) \, dt.$$

Hence $T_i + S_i = \int_0^T \lambda_i \alpha_i(t) \varphi(t) \, dt - \int_0^T f_i(t) \varphi(t) \, dt$, and therefore

$$\langle \partial_t u_n, \varphi \rangle_{\mathcal{D}^\star, \mathcal{D}} = -\sum_{i=1}^n \int_0^T \lambda_i \alpha_i(t) e_i \varphi(t) \, dt + \sum_{i=1}^n \int_0^T f_i(t) e_i \varphi(t) \, dt.$$

Since this equality holds for all $\varphi \in \mathcal{D}(]0, T[)$, we finally have

$$\partial_t u_n = -\sum_{i=1}^n \lambda_i \alpha_i e_i + \sum_{i=1}^n f_i e_i \in L^2(]0, T[, E_n),$$

which can also be written, with $f^{(n)} = \sum_{i=1}^n f_i e_i$,

$$\partial_t u_n = \Delta u_n + f^{(n)} \in L^2(]0, T[, E_n) \subset L^2(]0, T[, H_0^1(\Omega)) \subset L^2(]0, T[, H^{-1}(\Omega)).$$

Now let $v \in L^2(]0, T[, H_0^1(\Omega))$. Since $\partial_t u_n \in L^2(]0, T[, H^{-1}(\Omega))$, we have $\langle \partial_t u_n, v \rangle_{H^{-1}(\Omega), H_0^1(\Omega)} \in L^1(]0, T[)$ and

$$\int_0^T \langle \partial_t u_n(t), v(t) \rangle_{H^{-1}(\Omega), H_0^1(\Omega)} \, dt$$

$$= -\int_0^T \int_\Omega \nabla u_n \cdot \nabla v \, dx \, dt + \sum_{i=1}^n \int_0^T \int_\Omega f_i e_i v \, dx \, dt.$$

This gives, returning to the definition of f_i,

$$\int_0^T \langle \partial_t u_n(t), v(t) \rangle_{H^{-1}(\Omega), H_0^1(\Omega)} \, dt + \int_0^T \int_\Omega \nabla u_n \cdot \nabla v \, dx \, dt$$

$$= \sum_{i=1}^n \int_0^T \langle f(t), e_i \rangle_{H^{-1}(\Omega), H_0^1(\Omega)} \left(\int_\Omega e_i v \, dx \right) dt$$

$$= \int_0^T \langle f(t), \sum_{i=1}^n (v|e_i)_2 e_i \rangle_{H^{-1}(\Omega), H_0^1(\Omega)} \, dt.$$

Let P_n be the orthogonal projection operator in $L^2(\Omega)$ on the v.s.s. E_n. The operator P_n can therefore be seen as an operator from $L^2(\Omega)$ to $H_0^1(\Omega)$ (because $E_n \subset H_0^1(\Omega)$). We then denote by P_n^t the transposed operator, which is therefore an operator from $H^{-1}(\Omega)$ to $(L^2(\Omega))'$ which is itself identified with $L^2(\Omega)$ and is also a v.s.s. of $H^{-1}(\Omega)$. We then obtain (for all $v \in L^2(]0, T[, H_0^1(\Omega)))$

$$\int_0^T \langle \partial_t u_n, v \rangle_{H^{-1}, H_0^1} \, dt + \int_0^T \int_\Omega \nabla u_n \cdot \nabla v \, dx \, dt$$

$$= \int_0^T \langle f, P_n v \rangle_{H^{-1}, H_0^1} \, dt = \int_0^T \langle P_n^t f, v \rangle_{H^{-1}, H_0^1} \, dt. \quad (4.23)$$

We also have $u_n \in C([0, T], H_0^1(\Omega))$ and $u_n(0) = P_n u_0$.

Step 4. Estimates on the approximate solution. For $n \in \mathbb{N}^\star$, we have

$$u_n \in C([0, T], H_0^1(\Omega)) \subset L^2(]0, T[, H_0^1(\Omega)) \text{ and}$$

$$\partial_t u_n = \Delta u_n + f^{(n)} \in L^2(]0, T[, H^{-1}(\Omega)).$$

According to Section 4.2, we therefore have

$$\frac{1}{2} \|u_n(T)\|_2^2 - \frac{1}{2} \|u_0\|_2^2 = \int_0^T \langle \partial_t u_n, u_n \rangle_{H^{-1}, H_0^1} \, dt.$$

Taking $v = u_n$ in (4.23) yields

$$\frac{1}{2} \|u_n(T)\|_2^2 - \frac{1}{2} \|u_0\|_2^2 + \int_0^T \int_\Omega |\nabla u_n|^2 \, dx \, dt = \int_0^T \langle f, P_n u_n \rangle_{H^{-1}, H_0^1} \, dt,$$

and therefore

$$\|u_n\|_{L^2(]0,T[,H_0^1(\Omega))}^2 = \int_0^T \int_\Omega |\nabla u_n|^2 \, dx \, dt \leq \frac{1}{2} \|u_0\|_2^2 + \int_0^T \langle f, P_n u_n \rangle_{H^{-1},H_0^1} \, dt.$$

Since $P_n u_n = u_n$, we get

$$\|u_n\|_{L^2(]0,T[,H_0^1(\Omega))}^2 \leq \frac{1}{2} \|u_0\|_2^2 + \int_0^T \langle f, u_n \rangle_{H^{-1},H_0^1} \, dt$$

$$\leq \frac{1}{2} \|u_0\|_2^2 + \|f\|_{L^2(]0,T[,H^{-1}(\Omega))} \|u_n\|_{L^2(]0,T[,H_0^1(\Omega))}$$

$$\leq \frac{1}{2} \|u_0\|_2^2 + \frac{1}{2} \|f\|_{L^2(]0,T[,H^{-1}(\Omega))}^2 + \frac{1}{2} \|u_n\|_{L^2(]0,T[,H_0^1(\Omega))}^2 .$$

Hence

$$\|u_n\|_{L^2(]0,T[,H_0^1(\Omega))}^2 \leq \|u_0\|_2^2 + \|f\|_{L^2(]0,T[,H^{-1}(\Omega))}^2 ,$$

which also gives

$$\|u_n\|_{L^2(]0,T[,H_0^1(\Omega))} \leq \|u_0\|_2 + \|f\|_{L^2(]0,T[,H^{-1}(\Omega))} .$$

Since $\partial_t u_n = \Delta u_n + P_n^t f$ (equality (4.23)) and $\|P_n w\|_{H_0^1(\Omega)} \leq \|w\|_{H_0^1(\Omega)}$ for all $w \in H_0^1(\Omega)$, we also obtain a bound on $\partial_t u_n$:

$$\|\partial_t u_n\|_{L^2(]0,T[,H^{-1}(\Omega))} \leq \|u_n\|_{L^2(]0,T[,H_0^1(\Omega))} + \|f\|_{L^2(]0,T[,H^{-1}(\Omega))}$$

and therefore

$$\|\partial_t u_n\|_{L^2(]0,T[,H^{-1}(\Omega))} \leq \|u_0\|_2 + 2 \|f\|_{L^2(]0,T[,H^{-1}(\Omega))} .$$

The sequence $(u_n)_{n\in\mathbb{N}^\star}$ is thus bounded in $L^2(]0,T[,H_0^1(\Omega))$ and the sequence $(\partial_t u_n)_{n\in\mathbb{N}^\star}$ is bounded in $L^2(]0,T[,H^{-1}(\Omega))$.

Step 5. Passage to the limit. Thanks to the estimates obtained in the previous step, we can assume, up to a subsequence, that as $n \to +\infty$,

$$u_n \to u \text{ weakly in } L^2(]0,T[,H_0^1(\Omega)),$$

$$\partial_t u_n \to w \text{ weakly in } L^2(]0,T[,H^{-1}(\Omega))$$

and the estimates on u_n and $\partial_t u_n$ also give the following estimates on u and w:

$$\|u\|_{L^2(]0,T[,H_0^1(\Omega))} \leq \|u_0\|_2 + \|f\|_{L^2(]0,T[,H^{-1}(\Omega))} ,$$

$$\|w\|_{L^2(]0,T[,H^{-1}(\Omega))} \leq \|u_0\|_2 + 2 \|f\|_{L^2(]0,T[,H^{-1}(\Omega))} .$$

We first show that $w = \partial_t u$ (then we show that u is a solution of $\partial_t u = \Delta u + f$ in the sense requested by (4.18)).

By definition of $\partial_t u$, we have, for all $\varphi \in \mathcal{D}(]0,T[)$,

$$\int_0^T \partial_t u(t)\varphi(t)\,\mathrm{d}t = -\int_0^T u(t)\varphi'(t)\,\mathrm{d}t.$$

To prove that $\partial_t u = w$, it is therefore sufficient to show that we have, for all $\varphi \in \mathcal{D}(]0,T[)$,

$$\int_0^T w(t)\varphi(t)\,\mathrm{d}t = -\int_0^T u(t)\varphi'(t)\,\mathrm{d}t. \tag{4.24}$$

Recall that the left term of the equality (4.24) belongs to $H^{-1}(\Omega)$ while the right term belongs to $H_0^1(\Omega)$. This equality therefore uses the fact that $H_0^1(\Omega) \subset H^{-1}(\Omega)$, this inclusion being due to the fact that we have identified $L^2(\Omega)'$ with $L^2(\Omega)$.

Let $\varphi \in \mathcal{D}(]0,T[)$ and let us show that (4.24) is satisfied. For $\psi \in H_0^1(\Omega)$ given, we consider the mapping S from $L^2(]0,T[,H_0^1(\Omega))$ to $\mathbb{R}$ defined by

$$S(v) = \int_\Omega \left(-\int_0^T v(t)\varphi'(t)\,\mathrm{d}t\right)\psi(x)\,\mathrm{d}x \text{ for } v \in L^2(]0,T[,H_0^1(\Omega)).$$

The mapping S is a continuous linear mapping from $L^2(]0,T[,H_0^1(\Omega))$ to $\mathbb{R}$. Since $u_n \to u$ weakly in $L^2(]0,T[,H_0^1(\Omega))$, we therefore have $S(u_n) \to S(u)$ as $n \to +\infty$. However, for $v = u_n$ and for $v = u$, we have

$$S(v) = -\left(\int_0^T v(t)\varphi'(t)\,\mathrm{d}t\,|\,\psi\right)_2 = -\left\langle \int_0^T v(t)\varphi'(t)\,\mathrm{d}t, \psi\right\rangle_{H^{-1},H_0^1}.$$

Hence, as $n \to +\infty$,

$$-\left\langle \int_0^T u_n(t)\varphi'(t)\,\mathrm{d}t, \psi\right\rangle_{H^{-1},H_0^1} \to -\left\langle \int_0^T u(t)\varphi'(t)\,\mathrm{d}t, \psi\right\rangle_{H^{-1},H_0^1}.$$

Thanks to the fact that $-\int_0^T u_n(t)\varphi'(t)\,\mathrm{d}t = \int_0^T \partial_t u_n(t)\varphi(t)\,\mathrm{d}t$ (by definition of $\partial_t u_n$),

$$\left\langle \int_0^T \partial_t u_n(t)\varphi(t)\,\mathrm{d}t, \psi\right\rangle_{H^{-1},H_0^1} \to -\left\langle \int_0^T u(t)\varphi'(t)\,\mathrm{d}t, \psi\right\rangle_{H^{-1},H_0^1}.$$

Consider now the mapping $\bar{S}$ from $L^2(]0,T[,H^{-1}(\Omega))$ to $\mathbb{R}$ defined by

$$\bar{S}(v) = \left\langle \int_0^T v(t)\varphi(t)\,\mathrm{d}t, \psi\right\rangle_{H^{-1},H_0^1} \text{ for } v \in L^2(]0,T[,H^{-1}(\Omega)).$$

The mapping $\bar{S}$ is a continuous linear mapping from $L^2(]0,T[,H^{-1}(\Omega))$ to $\mathbb{R}$. Since $\partial_t u_n \to w$ weakly in $L^2(]0,T[,H^{-1}(\Omega))$, we have therefore $\bar{S}(\partial_t u_n) \to \bar{S}(w)$ as $n \to +\infty$, that is,

$$\langle \int_0^T \partial_t u_n(t)\varphi(t)\,dt, \psi \rangle_{H^{-1},H_0^1} \to \langle \int_0^T w(t)\varphi(t)\,dt, \psi \rangle_{H^{-1},H_0^1}.$$

Hence for all $\psi \in H_0^1(\Omega)$, we have

$$-\langle \int_0^T u(t)\varphi'(t)\,dt, \psi \rangle_{H^{-1},H_0^1} = \langle \int_0^T w(t)\varphi(t)\,dt, \psi \rangle_{H^{-1},H_0^1}.$$

We have therefore shown that $-\int_0^T u(t)\varphi'(t)\,dt = \int_0^T w(t)\varphi(t)\,dt$ for all $\varphi \in \mathcal{D}(]0,T[)$, that is, $\partial_t u = w$.

We now know that $u_n \to u$ weakly in $L^2(]0,T[, H_0^1(\Omega))$ and that $\partial_t u_n \to \partial_t u$ weakly in $L^2(]0,T[, H^{-1}(\Omega))$. To show that u is a solution of $\partial_t u = \Delta u + f$ in the sense required by (4.18), it is then sufficient to take the limit in (4.23). Let $v \in L^2(]0,T[, H_0^1(\Omega))$. We have for all $n \in \mathbb{N}^\star$, according to (4.23),

$$\int_0^T \langle \partial_t u_n, v \rangle_{H^{-1},H_0^1}\,dt + \int_0^T (u_n|v)_{H_0^1}\,dt = \int_0^T \langle f, P_n v \rangle_{H^{-1},H_0^1}\,dt.$$

We may pass to the limit in the two terms on the left of this equality as $n \to +\infty$ thanks to the convergence of u_n and $\partial_t u_n$. For the term on the right, we use the preliminary step. We note that $P_n v(t) \to v(t)$ in $H_0^1(\Omega)$ for almost all t and $\|P_n v(t)\|_{H_0^1(\Omega)} \le \|v(t)\|_{H_0^1(\Omega)}$ for almost all t. Thus we may pass to the limit in the term on the right-hand side, thanks to the dominated convergence theorem. We thus obtain

$$\int_0^T \langle \partial_t u, v \rangle_{H^{-1},H_0^1}\,dt + \int_0^T (u|v)_{H_0^1}\,dt = \int_0^T \langle f, v \rangle_{H^{-1},H_0^1}\,dt,$$

which is indeed the formulation (4.18).

Step 6. Initial condition. Since

$$u \in L^2(]0,T[, H_0^1(\Omega)) \quad \text{and} \quad \partial_t u \in L^2(]0,T[, H^{-1}(\Omega)),$$

we know that $u \in C([0,T], L^2(\Omega))$ (see Section 4.2). To complete the proof that u is a solution to (4.18), it only remains to show that $u(0) = u_0$ a.e. (that is, $u(0) = u_0$ in $L^2(\Omega)$).

We know that $u(t) \to u(0)$ in $L^2(\Omega)$ when $t \to 0$. We also know that $u_n(0) = \sum_{i=1}^n \alpha_i^{(0)} e_i \to u_0$ in $L^2(\Omega)$ as $n \to +\infty$. To deduce that $u(0) = u_0$, it is sufficient to show that the sequence $(u_n)_{n\in\mathbb{N}^\star}$ is relatively compact in $C([0,T], H^{-1}(\Omega))$. Indeed, if the sequence $(u_n)_{n\in\mathbb{N}^\star}$ is relatively compact in $C([0,T], H^{-1}(\Omega))$, there exists a $w \in C([0,T], H^{-1}(\Omega))$ and a subsequence, still denoted by $(u_n)_{n\in\mathbb{N}^\star}$, such that $u_n(t) \to w(t)$ in $H^{-1}(\Omega)$ uniformly with respect to $t \in [0,T]$ (and therefore also in $L^2(]0,T[, H^{-1}(\Omega)))$. In particular, we therefore have $w(0) = u_0$. But we already know that $u_n \to u$ weakly in $L^2(]0,T[, H_0^1(\Omega))$ and therefore also weakly in $L^2(]0,T[, H^{-1}(\Omega))$. By the uniqueness of the limit, we therefore have $u = w$ a.e.

on $]0, T[$ and therefore $u(t) = w(t)$ for all $t \in [0, T]$ because u and w are continuous on $[0, T]$. We thus finally obtain, $u(0) = w(0) = u_0$.

Let us then show that the sequence $(u_n)_{n \in \mathbb{N}^\star}$ is relatively compact in $C([0, T], H^{-1}(\Omega))$. By the Ascoli theorem (Theorem 1.35), it suffices to show that

1. For all $t \in [0, T]$, $(u_n(t))_{n \in \mathbb{N}^\star}$ is relatively compact in $H^{-1}(\Omega)$.
2. $\|u_n(t) - u_n(s)\|_{H^{-1}} \to 0$, when $s \to t$, uniformly with respect to $n \in \mathbb{N}^\star$ (and for all $t \in [0, T]$).

To prove the second point, we use the fact that $\partial_t u_n \in L^1([0, T], H^{-1}(\Omega))$. Owing to Lemma 4.26), for all $t_1, t_2 \in [0, T]$, $t_1 > t_2$, and all $n \in \mathbb{N}^\star$, we have

$$u_n(t_1) - u_n(t_2) = \int_{t_2}^{t_1} \partial_t u_n(s)\,\mathrm{d}s\,(\text{in } H^{-1}(\Omega))$$

and therefore

$$\|u_n(t_1) - u_n(t_2)\|_{H^{-1}} \leq \int_{t_2}^{t_1} \|\partial_t u_n(s)\|_{H^{-1}}\,\mathrm{d}s$$

$$\leq \Big(\int_0^T \|\partial_t u_n(s)\|_{H^{-1}}^2\,\mathrm{d}s \Big)^{\frac{1}{2}} \sqrt{t_1 - t_2} \leq \|\partial_t u_n\|_{L^2(]0,T[,H^{-1})} \sqrt{t_1 - t_2}.$$

Since the sequence $(\partial_t u_n)_{n \in \mathbb{N}^\star}$ is bounded in $L^2(]0, T[, H^{-1}(\Omega))$, we get that $\|u_n(t) - u_n(s)\|_{H^{-1}} \to 0$, when $s \to t$, uniformly with respect to $n \in \mathbb{N}^\star$ (and for all $t \in [0, T]$).

To prove the first item, we again use Section 4.2 (Lemma 4.27). Since $u_n \in L^2(]0, T[, H_0^1(\Omega))$ and $\partial_t u_n \in L^2(]0, T[, H^{-1}(\Omega))$, we have, for all $t, s \in [0, T]$,

$$\|u_n(t)\|_2^2 = \|u_n(s)\|_2^2 + 2 \int_s^t \langle \partial_t u_n(\xi), u_n(\xi) \rangle_{H^{-1}, H_0^1}\,\mathrm{d}\xi,$$

and therefore

$$\|u_n(t)\|_2^2 \leq \|u_n(s)\|_2^2 + 2 \int_s^t |\langle \partial_t u_n(\xi), u_n(\xi) \rangle_{H^{-1}, H_0^1}|\,\mathrm{d}\xi$$

$$\leq \|u_n(s)\|_2^2 + 2 \|\partial_t u_n\|_{L^2(]0,T[,H^{-1})} \|u_n\|_{L^2(]0,T[,H_0^1)} \cdot$$

Integrating this inequality with respect to s over $[0, T]$ yields

$$T \|u_n(t)\|_2^2 \leq \|u_n\|_{L^2(]0,T[,L^2(\Omega))}^2 + 2T \|\partial_t u_n\|_{L^2(]0,T[,H^{-1})} \|u_n\|_{L^2(]0,T[,H_0^1)} \cdot$$

This shows that the sequence $(u_n(t))_{n \in \mathbb{N}^\star}$ is bounded in $L^2(\Omega)$ for all $t \in [0, T]$ (and even uniformly with respect to t). Hence the sequence $(u_n(t))_{n \in \mathbb{N}^\star}$ is relatively compact in $H^{-1}(\Omega)$ for all $t \in [0, T]$. We can therefore apply the Ascoli theorem (Theorem 1.35) and conclude, as indicated at the beginning of this step, that $u(0) =$

u_0 a.e. This concludes the proof that u is a solution to (4.18) and therefore the proof of the "existence" part of Theorem 4.30.

In the proof, we obtained the following estimates

$$\|u\|_{L^2(]0,T[,H_0^1(\Omega))} \leq \|u_0\|_2 + \|f\|_{L^2(]0,T[,H^{-1}(\Omega))},$$

$$\|\partial_t u\|_{L^2(]0,T[,H^{-1}(\Omega))} \leq \|u_0\|_2 + 2\,\|f\|_{L^2(]0,T[,H^{-1}(\Omega))}.$$

Finally, since $u \in L^2(]0,T[,H_0^1(\Omega))$ and $\partial_t u \in L^2(]0,T[,H^{-1}(\Omega))$, Lemma 4.27 gives, for all t,

$$\|u(t)\|_2^2 = \|u_0\|_2^2 + 2 \int_0^t \langle \partial_t u(s), u(s) \rangle_{H^{-1},H_0^1} \, ds,$$

we get that

$$\|u(t)\|_2^2 \leq \|u_0\|_2^2 + \|\partial_t u\|_{L^2(]0,T[,H^{-1}(\Omega))}^2 + \|u\|_{L^2(]0,T[,H_0^1(\Omega))}^2,$$

for all $t \in [0,T[$.

Step 7. Uniqueness. We still have to prove the uniqueness part of the Theorem 4.30. Let u_1, u_2 be two solutions of (4.18); we set $u = u_1 - u_2$. Taking the difference of the equations satisfied by u_1 and u_2 and taking, for $t \in [0,T]$, $v = u\mathbb{1}_{]0,t[}$ as a test function, we obtain

$$\int_0^t \langle \partial_t u(s), u(s) \rangle_{H^{-1},H_0^1} \, ds + \int_0^t \int_\Omega \nabla u(s) \cdot \nabla u(s) \, dx \, ds = 0.$$

Since $u \in L^2(]0,T[,H_0^1(\Omega))$ and $\partial_t u \in L^2(]0,T[,H^{-1}(\Omega))$, we have, according to Section 4.2,

$$\frac{1}{2}(\|u(t)\|_2^2 - (\|u(0)\|_2^2)) = \int_0^t \langle \partial_t u(s), u(s) \rangle_{H^{-1},H_0^1} \, ds.$$

From this, for all $t \in [0,T]$,

$$(\|u(t)\|_2^2 - (\|u(0)\|_2^2)) + 2 \int_0^t \int_\Omega \nabla u(s) \cdot \nabla u(s) \, dx \, ds = 0.$$

Finally, since $u(0) = 0$, we indeed obtain, ultimately, $u(t) = 0$ a.e. in Ω, for all $t \in [0,T]$. This concludes the proof of the uniqueness part of Theorem 4.30.

4.3.2 Proof of the Existence and Uniqueness Theorem by Generalised Coercivity

Another proof of Theorem 4.30 consists in applying a general theorem which gives the bijectivity of a mapping from a Banach space into the dual space of a reflexive Banach space.

Theorem 4.31 (Bijective linear mapping from a Banach space into a dual space). *Let E be a Banach space, F a reflexive Banach space and B a continuous linear mapping from E to F'. If the condition*

$$\exists \beta > 0 \text{ such that } \forall u \in E, \|Bu\|_{F'} \geq \beta \|u\|_E , \tag{4.25}$$

is satisfied, then $\mathrm{Im}(B)$ is closed and B is injective. Moreover, if the condition

$$\left[\forall u \in E, \langle Bu, v \rangle_{F',F} = 0\right] \implies v = 0 \tag{4.26}$$

is satisfied, then B is bijective from E to F'.

Proof Assume that the condition (4.25) is satisfied. Let $(y_n)_{n \in \mathbb{N}} \subset \mathrm{Im}(B)$ be a sequence converging in F' towards $y \in F'$. Therefore, there exists a sequence $(u_n)_{n \in \mathbb{N}} \subset E$ such that $y_n = Bu_n$ for all $n \in \mathbb{N}$. The sequence $(y_n)_{n \in \mathbb{N}}$ is a Cauchy sequence in F', and therefore, due to condition (4.25), the sequence $(u_n)_{n \in \mathbb{N}}$ is a Cauchy sequence in E. It converges towards a certain $u \in E$; moreover, by continuity of B, we therefore have $y = Bu$, which shows that $\mathrm{Im}(B)$ is closed. Thanks to condition (4.25), it is also clear that if $Bu = Bv$ then $u = v$, which shows that B is injective.

Now suppose that condition (4.26) is satisfied, and let us show that B is bijective. Since we have already shown that B is injective and that $\mathrm{Im}(B)$ is closed, it suffices to show that $\mathrm{Im}(B)$ is dense in F'. To do this, we apply the characterisation of the density of a v.s.s. of a Banach space seen in Problem 1.10, question 2a with $G = \mathrm{Im}(B)$, which applies because we have assumed F to be reflexive. This characterisation is exactly condition (4.26), which concludes the proof. ∎

Remark 4.32 (The *inf-sup* condition). Condition (4.25) is equivalently written

$$\exists \beta > 0 \text{ such that } \inf_{\substack{u \in E \\ \|u\|_E = 1}} \sup_{\substack{v \in F \\ \|v\|_F = 1}} \langle Bu, v \rangle_{F',F} \geq \beta \|u\|_E ,$$

which explains the usual "inf-sup" name.

Let us now give a direct proof of Theorem 4.30 by applying Theorem 4.31, with

- $E = \{u \in L^2(]0,T[, H_0^1(\Omega)) : \partial_t u \in L^2(]0,T[, H^{-1}(\Omega))\}$,

$$\|u\|_E^2 = \|u\|_{L^2(]0,T[,H_0^1(\Omega))}^2 + \|u\|_{L^2(]0,T[,H^{-1}(\Omega))}^2$$

(note that, for the definition of the space E, $L^2(\Omega)$ is identified with its dual and that $E \subset C([0,T], L^2(\Omega))$ by Lemma 4.27),

- $F = L^2(]0,T[, H_0^1(\Omega)) \times L^2(\Omega)$ (with the natural associated norm),
- the linear mapping B from E to F' defined by

$$\forall u \in E, \forall (v,z) \in F,$$
$$\langle Bu, (v,z) \rangle_{F',F} = \langle \partial_t u - \Delta u, v \rangle_{L^2(]0,T[,H^{-1}(\Omega)),L^2(]0,T[,H_0^1(\Omega))}$$
$$+ \int_\Omega u_0(x) z(x)\, dx, \quad \text{with } u_0 = u(0).$$

If we show that B is bijective, we immediately have the existence and uniqueness of the solution to (4.18), which gives the proof of Theorem 4.30. Let us therefore show that B satisfies the conditions of Theorem 4.31. We start by showing that

$$\forall u \in E, \|Bu\|_{F'} \geq \|u\|_E, \tag{4.27}$$

inspired by the proof of [4, Theorem 3.6, second proof]. Recall that $-\Delta$ is a continuous linear operator bijective from $H_0^1(\Omega)$ to $H^{-1}(\Omega)$. Let R be its inverse, which is therefore a continuous linear mapping from $H^{-1}(\Omega)$ to $H_0^1(\Omega)$. For all $u, v \in H_0^1(\Omega)$,

$$\langle -\Delta u, v \rangle_{H^{-1}(\Omega), H_0^1(\Omega)} = (\nabla u \mid \nabla v)_{L^2(\Omega)^d} = \int_\Omega \nabla u(x) \cdot \nabla v(x)\, dx$$

and therefore for all $\varphi \in H^{-1}(\Omega)$ and $v \in H_0^1(\Omega)$ (since $R\varphi = u$ with $-\Delta u = \varphi$)

$$\int_\Omega \nabla(R\varphi) \cdot \nabla v\, dx = \langle \varphi, v \rangle_{H^{-1}(\Omega), H_0^1(\Omega)}.$$

We can therefore write

$$\langle Bu, (v,z) \rangle_{F',F} = \int_0^T \int_\Omega \nabla(R\partial_t u + u) \cdot \nabla v\, dx\, dt + \int_\Omega u(0) z\, dx.$$

Let us choose $(v,z) = (R\partial_t u + u, u(0))$; we then have

$$\langle Bu, (v,z) \rangle_{F',F} = \|\nabla(R\partial_t u + u)\|^2_{L^2(]0,T[,L^2(\Omega)^d)} + \|u(0)\|^2_{L^2(\Omega)}$$
$$= \|\nabla(R\partial_t u)\|^2_{L^2(]0,T[,L^2(\Omega)^d)} + \|\nabla u\|^2_{L^2(]0,T[,L^2(\Omega)^d)}$$
$$+ 2\int_0^T \int_\Omega \nabla(R\partial_t u) \cdot \nabla u\, dx\, dt + \|u(0)\|^2_{L^2(\Omega)}.$$

By choosing as norm in $H_0^1(\Omega)$ the norm in $L^2(\Omega)^d$ of the gradient of the function and for norm in $H^{-1}(\Omega)$ the dual norm, we obtain that $\|R\varphi\|_{H_0^1(\Omega)} = \|\varphi\|_{H^{-1}(\Omega)}$. Hence $\|\nabla(R\partial_t u)\|_{L^2(]0,T[,L^2(\Omega)^d)} = \|\partial_t u\|_{L^2(]0,T[,H^{-1}(\Omega))}$, and therefore

$$\langle Bu, (v, z)\rangle_{F',F} = \|\partial_t u\|^2_{L^2(]0,T[,H^{-1}(\Omega))} + \|u\|^2_{L^2(]0,T[,H_0^1(\Omega))} +$$

$$2 \int_0^T \langle \partial_t u, u \rangle_{H^{-1}(\Omega),H_0^1(\Omega)} + \|u(0)\|^2_{L^2(\Omega)} \, .$$

By Lemma (4.27), we therefore have

$$\langle Bu, (v, z)\rangle_{F',F} = \|\partial_t u\|^2_{L^2(]0,T[,H^{-1}(\Omega))} + \|u\|^2_{L^2(]0,T[,H_0^1(\Omega))}$$

$$+ \|u(T)\|^2_{L^2(\Omega)} - \|u(0)\|^2_{L^2(\Omega)} + \|u(0)\|^2_{L^2(\Omega)} \geq \|u\|^2_E \, .$$

Since $\|(v, z)\|^2_F = \langle Bu, (v, z)\rangle_{F',F}$ and $\|Bu\|_{F'} \|(v, z)\|_F \geq \langle Bu, (v, z)\rangle_{F',F}$, we therefore have $\langle Bu, (v, z)\rangle_{F',F} \leq \|Bu\|^2_{F'}$ and we have therefore shown that inequality (4.27) is satisfied. Thus, owing to Theorem 4.31, $\mathrm{Im}(B)$ is closed (because F is a reflexive Banach space) and B is injective. To show that B is bijective, it remains to show that $\mathrm{Im}(B)$ is dense. With this aim, it is sufficient, still by Theorem 4.31, that condition (4.26) is satisfied. Let $(v, z) \in F$ such that

$$\forall u \in E, \langle Bu, (v, z)\rangle_{F',F} = 0,$$

which can be written as

$$\forall u \in E, \langle \partial_t u - \Delta u, v \rangle_{L^2(H^{-1}),L^2(H_0^1)} + \int_\Omega u_0(x)z(x) \, \mathrm{d}x = 0. \qquad (4.28)$$

We want to show that if this condition is satisfied, then $v = 0$ and $z = 0$.

- We first choose $u \in E$ such that $u(x, t) = \varphi(t)w(x)$ with $\varphi \in \mathcal{D}(]0,T[)$ and $w \in H_0^1(\Omega)$. Substituting this into (4.28), we obtain

$$\forall \varphi \in \mathcal{D}(]0,T[), \forall w \in H_0^1(\Omega),$$

$$\int_0^T \left[\varphi'(t) \int_\Omega w(x)v(x,t) \, \mathrm{d}x + \varphi(t) \int_\Omega \nabla w(x) \cdot \nabla v(x,t) \, \mathrm{d}x \right] \mathrm{d}t = 0. \quad (4.29)$$

Since $v \in L^2(]0,T[, H_0^1(\Omega))$, its derivative (in time) by transposition $\partial_t v$ is defined by

$$\langle \partial_t v, \varphi \rangle_{\mathcal{D}^*,\mathcal{D}} = - \int_0^T v(\cdot,t)\varphi'(t) \, \mathrm{d}t \in H_0^1(\Omega), \forall \varphi \in \mathcal{D}(]0,T[).$$

Recall that this derivative by transposition is a weak derivative belonging to $L^2(]0,T[, H^{-1}(\Omega))$ if there exists a $\psi \in L^2(]0,T[, H^{-1}(\Omega))$ such that

$$\underbrace{- \int_0^T v(\cdot,t)\varphi'(t) \, \mathrm{d}t}_{\in H_0^1(\Omega)} = \underbrace{\int_0^T \psi(t)\varphi(t) \, \mathrm{d}t}_{\in H^{-1}(\Omega)} \, .$$

The function ψ is then still denoted by $\partial_t v$. Note that this equality makes sense because $L^2(\Omega)$ is identified with its dual, which allows us to identify $H_0^1(\Omega)$ with a subspace of $H^{-1}(\Omega)$. This last equality can also be written as

$$\forall w \in H_0^1(\Omega),\ \Big(-\int_0^T v(\cdot,t)\varphi'(t)\,dt\Big|w\Big)_{L^2} = \Big\langle \int_0^T \psi(t)\varphi(t)\,dt, w\Big\rangle_{H^{-1},H_0^1}.$$

Then, thanks to the commutativity property of the integral with the duality action (Proposition 4.25),

$$\forall w \in H_0^1(\Omega),$$

$$-\int_0^T \varphi'(t)(v(\cdot,t)|w)_{L^2}\,dt = \int_0^T \varphi(t)\langle\psi(t),w\rangle_{H^{-1},H_0^1}\,dt. \quad (4.30)$$

However, noting that $-\Delta v \in L^2(]0,T[,H^{-1}(\Omega))$ and that, for almost all $t \in]0,T[$,

$$\langle -\Delta v(\cdot,t), w\rangle_{H^{-1},H_0^1} = \int_\Omega \nabla v(\cdot,t)\cdot\nabla w\,dx,$$

we see that the equality (4.29) gives exactly (4.30) for $\psi = -\Delta v$. Hence $\partial_t v = -\Delta v$ and therefore $v \in E$ and

$$\partial_t v + \Delta v = 0 \text{ in } L^2(]0,T[,H^{-1}(\Omega)).$$

- Now let $u \in E$ be arbitrary. Since $v \in E$, we can use Lemma 4.29. It gives

$$\int_0^T \langle\partial_t u, v\rangle_{H^{-1},H_0^1}\,dt + \int_0^T \langle\partial_t v, u\rangle_{H^{-1},H_0^1}\,dt = (u(T)|v(T))_{L^2} - (u(0)|v(0))_{L^2}.$$

On the other hand, since $u, v \in L^2(]0,T[,H_0^1(\Omega))$,

$$\int_0^T \langle-\Delta u, v\rangle_{H^{-1},H_0^1}\,dt = \int_0^T \langle-\Delta v, u\rangle_{H^{-1},H_0^1}\,dt.$$

Substituting this into (4.28), we obtain

$$(u(T)|v(T))_{L^2} - (u(0)|v(0))_{L^2} - \int_0^T \langle\partial_t v, u\rangle_{H^{-1},H_0^1}\,dt$$

$$+ \int_0^T \langle-\Delta v, u\rangle_{H^{-1},H_0^1}\,dt + (u(0)|z)_{L^2} = 0.$$

And, since $\partial_t v + \Delta v = 0$, we obtain

$$(u(T)|v(T))_{L^2} - (u(0)|v(0))_{L^2} + (u(0)|z)_{L^2} = 0.$$

By choosing u in the form $u(t) = \frac{T-t}{T}w$ with w arbitrary in $H_0^1(\Omega)$, we obtain $(w|z - u(0))_{L^2} = 0$ and therefore $z = v(0)$. The previous equality then becomes $(u(T)|v(T))_{L^2} = 0$ and therefore $v(T) = 0$ by taking $u = v$. Finally, integrating over $[0, t]$ rather than $[0, T]$, we show that $v(t) = 0$ for all $t \in [0, T]$; we have thus indeed shown that $v = 0$ and $z = 0$, which concludes the proof.

4.3.3 Other Properties of the Solution to the Heat Equation

Let us now provide some additional properties of the weak solution to the heat equation.

Proposition 4.33 (Continuous dependence). *Let Ω be an open bounded subset of $\mathbb{R}^N$ and $T > 0$. For $u_0 \in L^2(\Omega)$ and $f \in L^2(]0, T[, H^{-1}(\Omega))$, let $T(u_0, f)$ denote the solution to problem (4.18). The operator T is a continuous linear mapping from $L^2(\Omega) \times L^2(]0, T[, H^{-1}(\Omega))$ to $L^2(]0, T[, H_0^1(\Omega))$ and to $C([0, T], L^2(\Omega))$.*

Proof It is sufficient to revisit the estimates seen in Theorem 4.30, for u a solution to (4.18):

$$\|u\|_{L^2(]0,T[,H_0^1)} \le \|f\|_{L^2(]0,T[,H^{-1})} + \|u_0\|_2 ,$$

$$\|u(t)\|_2^2 \le \|u_0\|_2^2 + \|\partial_t u\|_{L^2(]0,T[,H^{-1})}^2 + \|u\|_{L^2(]0,T[,H_0^1)}^2 \quad \text{for all } t \in [0,T],$$

$$\|\partial_t u\|_{L^2(]0,T[,H^{-1})} \le \|u_0\|_2 + 2\|f\|_{L^2(]0,T[,H^{-1})} .$$

Thus T is continuous in the announced spaces. $\blacksquare$

Proposition 4.34 (Non-negativity and maximum principle). *Let Ω be an open bounded subset of $\mathbb{R}^N$, $T > 0$ and $u_0 \in L^2(\Omega)$. Let u be the solution to (4.18) with $f = 0$.*

1. *Assume $u_0 \ge 0$ a.e. We then have $u(t) \ge 0$ a.e. and for all $t \in [0, T]$.*
2. *Assume that $u_0 \in L^\infty(\Omega)$. Let $A, B \in \mathbb{R}$ such that $A \le 0 \le B$ and $A \le u_0 \le B$ a.e. We then have $A \le u(t) \le B$ a.e. and for all $t \in [0, T]$.*

Proof To prove the first item, we rather show (which is equivalent) that $u_0 \le 0$ a.e. implies $u(t) \le 0$ a.e. for all $t \in [0, T]$. Assume then that $u_0 \le 0$ a.e. Owing to Lemma 4.35 with $\varphi(s) = s^+$, $u^+ \in L^2(]0, T[, H_0^1(\Omega))$. For $t \in [0, T]$, we can therefore take $v = u^+ \mathbb{1}_{]0,t[}$ in equation (4.18) and obtain

$$\int_0^t \langle \partial_t u(s), u^+(s) \rangle_{H^{-1}, H_0^1} \, ds + \int_0^t \int_\Omega \nabla u(s) \cdot \nabla u^+(s) \, dx \, ds = 0.$$

Using Lemma 4.35 again, we have

$$\frac{1}{2}\|u^+(t)\|_2^2 - \frac{1}{2}\|u^+(0)\|_2^2 + \int_0^t \|u^+(s)\|_{H_0^1}^2 \, ds = 0.$$

Since $u^+(0) = u_0^+ = 0$ a.e., we get that $\|u^+(t)\|_2 = 0$ and therefore $u(t) \leq 0$ a.e.

The proof of the second item is similar by using Lemma 4.35 with $\varphi(s) = (s-B)^+$ and $\varphi(s) = (s-A)^-$. $\blacksquare$

Lemma 4.35. *Let Ω be an open bounded subset of $\mathbb{R}^N$ and $T > 0$ and let φ be a Lipschitz-continuous function from $\mathbb{R}$ to $\mathbb{R}$ such that $\varphi(0) = 0$. Define Φ by*

$$\Phi(\xi) = \int_0^\xi \varphi(\alpha)\,d\alpha \ \text{for}\ \xi \in \mathbb{R}.$$

Let $u \in L^2(]0,T[,H_0^1(\Omega))$ such that $\partial_t u \in L^2(]0,T[,H^{-1}(\Omega))$. We then have $\varphi(u) = L^2(]0,T[,H_0^1(\Omega))$, $\Phi(u) \in C([0,T], L^1(\Omega))$ and, for all $t_1, t_2 \in [0,T]$,

$$\int_\Omega \Phi(u(t_2))\,dx - \int_\Omega \Phi(u(t_1))\,dx = \int_{t_1}^{t_2} \langle \partial_t u(s), \varphi(u(s)) \rangle_{H^{-1},H_0^1}\,ds.$$

We also have for almost all $t \in]0,T[$, $\varphi(u(t)) \in H_0^1(\Omega)$ and $\nabla \varphi(u(t)) = \varphi'(u(t))\nabla u$ a.e., that is, to be more precise, $\nabla \varphi(u)(x,t) = \varphi'(u(x,t))\nabla u(x,t)$ for almost all $x \in \Omega$. In this equality, we can take for $\varphi'(u(x,t))$ any value if φ is not differentiable at the point $u(x,t)$. In particular, this shows that, for all $a \in \mathbb{R}$, $\nabla u = 0$ a.e. on the set $\{u = a\}$.

Proof This lemma is the "parabolic" version of Lemmas 2.25 and 2.26 seen previously. The proof consists in first considering the case (as in Lemma 2.25) where φ of class C^1, and regularising u. Then we approximate φ by C^1 (at least when φ is differentiable except at a finite number of points, the general case being more difficult); this proof is not detailed here. $\blacksquare$

We now give the equivalence between the weak formulation (4.18) and the other weak formulation (4.31). This second formulation is, in particular, interesting when we are looking to prove the convergence of approximate solutions obtained by a discretisation in space and time. This equivalence is still true in the non-linear case, see for example the proof of existence by numerical convergence of the Stefan problem in Problem 4.9.

Proposition 4.36 (Equivalence between two weak formulations). *Let Ω be an open bounded subset of $\mathbb{R}^N$, $T > 0$, $u_0 \in L^2(\Omega)$ and $f \in L^2(]0,T[,H^{-1}(\Omega))$ (we identify, as usual $L^2(\Omega)$ with $L^2(\Omega)'$). Then u is a solution to (4.18) if and only if u satisfies:*

$$\begin{cases} u \in L^2(]0,T[,H_0^1(\Omega)), \\[2mm] -\displaystyle\int_0^T \int_\Omega u \partial_t \varphi\,dx\,dt - \int_\Omega u_0(x)\varphi(x,0)\,dx + \int_0^T \int_\Omega \nabla u \cdot \nabla \varphi\,dx\,dt \\[2mm] \qquad = \displaystyle\int_0^T \langle f(s), \varphi(\cdot,s) \rangle_{H^{-1},H_0^1}\,ds \ \text{for all}\ \varphi \in \mathcal{D}([0,T[\times\Omega). \end{cases} \tag{4.31}$$

Proof Let us first show that "u is a solution to (4.18) $\Rightarrow u$ is a solution to (4.31)". We therefore assume that u is a solution to (4.18). Let $\varphi \in \mathcal{D}(\Omega \times [0, T[)$, and set, for $n \in \mathbb{N}^{\star}$,

$$\varphi_n(x, t) = \sum_{i=0}^{n-1} \mathbb{1}_{]t_i, t_{i+1}[}(t) \varphi(x, t_i), \tag{4.32}$$

where $t_i = \frac{i}{n} T$. Since φ is a regular function, it is clear that $\varphi_n \in L^2(]0, T[, H_0^1(\Omega))$ and that $\varphi_n \to \varphi$ in $L^2(]0, T[, H_0^1(\Omega))$ as $n \to +\infty$. Since $\varphi_n \in L^2(]0, T[, H_0^1(\Omega))$ and u is a solution to (4.18), we have

$$\int_0^T \langle \partial_t u, \varphi_n \rangle_{H^{-1}, H_0^1} \, dt + \int_0^T \int_\Omega \nabla u \cdot \nabla \varphi_n \, dx \, dt = \int_0^T \langle f, \varphi_n \rangle_{H^{-1}, H_0^1} \, dt.$$

Set $T_n = \int_0^T \langle \partial_t u, \varphi_n \rangle_{H^{-1}, H_0^1} \, dt$. Since $\varphi_n \to \varphi$ in $L^2(]0, T[, H_0^1(\Omega))$ as $n \to +\infty$, the previous equality gives

$$\lim_{n \to +\infty} T_n = \int_0^T \langle \partial_t u, \varphi \rangle_{H^{-1}, H_0^1} \, dt$$

$$= -\int_0^T \int_\Omega \nabla u \cdot \nabla \varphi \, dx \, dt + \int_0^T \langle f, \varphi \rangle_{H^{-1}, H_0^1} \, dt. \tag{4.33}$$

We now calculate $\lim_{n \to +\infty} T_n$ using (4.32). We have

$$T_n = \int_0^T \left\langle \partial_t u(t), \sum_{i=0}^{n-1} \mathbb{1}_{]t_i, t_{i+1}[}(t) \varphi(\cdot, t_i) \right\rangle_{H^{-1}, H_0^1} dt$$

$$= \sum_{i=0}^{n-1} \int_{t_i}^{t_{i+1}} \langle \partial_t u(t), \varphi(\cdot, t_i) \rangle_{H^{-1}, H_0^1} \, dt$$

$$= \sum_{i=0}^{n-1} \left\langle \int_{t_i}^{t_{i+1}} \partial_t u(t) \, dt, \varphi(\cdot, t_i) \right\rangle_{H^{-1}, H_0^1}.$$

Since $u, \partial_t u \in L^2(]0, T[, H^{-1}(\Omega))$ (recall that $H_0^1(\Omega) \subset H^{-1}(\Omega)$ by the identification of $L^2(\Omega)$ with its dual), we have (according to Section 4.2) $u \in C([0, T], H^{-1}(\Omega))$ and

$$\int_{t_i}^{t_{i+1}} \partial_t u(t) \, dt = u(t_{i+1}) - u(t_i) \in H^{-1}(\Omega).$$

Hence

$$T_n = \sum_{i=0}^{n-1} \langle u(t_{i+1}) - u(t_i), \varphi(\cdot, t_i) \rangle_{H^{-1}, H_0^1} = \sum_{i=0}^{n-1} \int_\Omega (u(x, t_{i+1}) - u(x, t_i)) \varphi(x, t_i) \, dx.$$

The last equality comes from the way in which an element of $H_0^1(\Omega)$ is considered as an element of $H^{-1}(\Omega)$. A discrete integration by parts then gives (noting that $\varphi(\cdot, t_n) = 0$)

$$T_n = -\int_\Omega u(x,0)\varphi(x,0)\,dx + \sum_{i=1}^n \int_\Omega (\varphi(x,t_{i-1}) - \varphi(x,t_i))u(x,t_i)\,dx.$$

Then, since φ is a regular function,

$$T_n = -\int_\Omega u_0(x)\varphi(x,0)\,dx - \sum_{i=1}^n \int_\Omega \Big(\int_{t_{i-1}}^{t_i} \partial_t\varphi(x,t)\,dt\Big)u(x,t_i)\,dx$$

$$= -\int_\Omega u_0(x)\varphi(x,0)\,dx - \int_0^T \int_\Omega \partial_t\varphi(x,t)\Big(\sum_{i=1}^n \mathbb{1}_{]t_{i-1},t_i[}u(x,t_i)\Big)\,dx\,dt$$

$$= -\int_\Omega u_0(x)\varphi(x,0)\,dx - \int_0^T \int_\Omega \partial_t\varphi(x,t)(u(x,t) + R_n(x,t))\,dx\,dt,$$

with $R_n(x,t) = \sum_{i=1}^n \mathbb{1}_{]t_{i-1},t_i[}u(x,t_i) - u(x,t)$. For all $t \in [0,T]$, we have

$$\|R_n(\cdot,t)\|_{L^2(\Omega)} \le \max\{\|u(s_1) - u(s_2)\|_{L^2(\Omega)}, \; s_1, s_2 \in [0,T], \; |s_1 - s_2| \le \frac{T}{n}\}.$$

Since $u \in C([0,T], L^2(\Omega))$, it follows that $\lim_{n\to+\infty} \|R_n(\cdot,t)\|_{L^2(\Omega)} = 0$ uniformly with respect to $t \in [0,T]$ and therefore that

$$\lim_{n\to+\infty} \int_0^T \int_\Omega \partial_t\varphi(x,t)R_n(x,t)\,dx\,dt = 0.$$

In summary, we have

$$\lim_{n\to+\infty} T_n = -\int_\Omega u_0(x)\varphi(x,0)\,dx - \int_0^T \int_\Omega \partial_t\varphi(x,t)u(x,t)\,dx\,dt.$$

With (4.33) we therefore have, for all $\varphi \in \mathcal{D}([0,T[\times\Omega)$,

$$-\int_0^T \int_\Omega \partial_t\varphi(x,t)u(x,t)\,dx\,dt - \int_\Omega u_0(x)\varphi(x,0)\,dx + \int_0^T \int_\Omega \nabla u \cdot \nabla\varphi\,dx\,dt$$

$$= \int_0^T \langle f, \varphi\rangle_{H^{-1},H_0^1}\,dt.$$

This shows that u is indeed a solution to (4.31).

We now show that "u is a solution to (4.31) $\Rightarrow u$ is a solution to (4.18)". Assuming that u is a solution to (4.31), let us show that u is a solution to (4.18) in two steps: We first show that $\partial_t u \in L^2(]0,T[, H^{-1}(\Omega))$ and $\partial_t u = \Delta u + f$ (note that $\Delta u, f \in L^2(]0,T[, H^{-1}(\Omega))$ then that $u(0) = u_0$.

Step 1 We show here that $\partial_t u \in L^2(]0, T[, H^{-1}(\Omega))$ and $\partial_t u = \Delta u + f$. We use the definition of $\partial_t u$. We have $\partial_t u \in \mathcal{D}_E^\star$, with $E = H_0^1(\Omega)$, and for all $\phi \in C^\infty(]0, T[, \mathbb{R})$

$$\langle \partial_t u, \phi \rangle_{\mathcal{D}_E^\star, \mathcal{D}} = - \int_0^T u(t)\phi'(t)\, \mathrm{d}t \in H_0^1(\Omega).$$

To show that $\partial_t u \in L^2(]0, T[, H^{-1}(\Omega))$ and $\partial_t u = \Delta u + f$, we therefore need to show that, for all $\phi \in \mathcal{D}(]0, T[)$,

$$- \int_0^T u(t)\phi'(t)\, \mathrm{d}t = \int_0^T (\Delta u(t) + f(t))\phi(t)\, \mathrm{d}t.$$

Note that the left-hand side of this equality belongs to $H_0^1(\Omega)$ and therefore in $H^{-1}(\Omega)$ (thanks to the identification between $L^2(\Omega)$ and its dual) and that the right-hand side belongs to $H^{-1}(\Omega)$. To prove the equality of these two terms, it is sufficient to show that

$$\langle - \int_0^T u(t)\phi'(t)\, \mathrm{d}t, \psi \rangle_{H^{-1}, H_0^1} = \langle \int_0^T (\Delta u(t) + f(t))\phi(t)\, \mathrm{d}t, \psi \rangle_{H^{-1}, H_0^1}$$

$$\text{for all } \psi \in H_0^1(\Omega),$$

that is to say,

$$- \int_0^T \langle u(t)\phi'(t), \psi \rangle_{H^{-1}, H_0^1}\, \mathrm{d}t = \int_0^T \langle (\Delta u(t) + f(t))\phi(t), \psi \rangle_{H^{-1}, H_0^1}\, \mathrm{d}t$$

$$\text{for all } \psi \in H_0^1(\Omega).$$

By density of $\mathcal{D}(\Omega)$ in $H_0^1(\Omega)$, it is sufficient to consider $\psi \in \mathcal{D}(\Omega)$. Using the way in which $H_0^1(\Omega)$ is included in $H^{-1}(\Omega)$, we have

$$- \int_0^T \langle u(t)\phi'(t), \psi \rangle_{H^{-1}, H_0^1}\, \mathrm{d}t = - \int_0^T \int_\Omega u(x, t)\phi'(t)\psi(x)\, \mathrm{d}x\, \mathrm{d}t.$$

On the other hand, we have

$$\int_0^T \langle (\Delta u(t) + f(t))\phi(t), \psi \rangle_{H^{-1}, H_0^1}\, \mathrm{d}t$$

$$= \int_0^T \phi(t)\langle \Delta u(t) + f(t), \psi \rangle_{H^{-1}, H_0^1}\, \mathrm{d}t$$

$$= - \int_0^T \phi(t) \int_\Omega \nabla u(x, t) \cdot \nabla \psi(x)\, \mathrm{d}x\, \mathrm{d}t + \int_0^T \phi(t)\langle f, \psi \rangle_{H^{-1}, H_0^1}\, \mathrm{d}t.$$

In summary, to complete step 1, it is therefore sufficient to show that

$$
-\int_0^T \int_\Omega u(x,t)\phi'(t)\psi(x)\,\mathrm{d}x\,\mathrm{d}t
$$

$$
= \int_0^T \phi(t)\left(-\int_\Omega \nabla u(x,t)\cdot\nabla\psi(x)\,\mathrm{d}x\,\mathrm{d}t + \langle f,\psi\rangle_{H^{-1},H_0^1}\right)\mathrm{d}t
$$

$$
\text{for all } \phi \in \mathcal{D}(]0,T[) \text{ and all } \psi \in \mathcal{D}(\Omega). \quad (4.34)
$$

As already mentioned, this gives $\partial_t u \in L^2(]0,T[,H^{-1}(\Omega))$ and $\partial_t u = \Delta u + f$.

To show (4.34), we use (4.31). Let $\phi \in \mathcal{D}(]0,T[)$ and $\psi \in \mathcal{D}(\Omega)$. Since $\varphi \in \mathcal{D}(]0,T[\times\Omega)$, we may choose $\varphi(x,t) = \phi(t)\psi(x)$ in (4.31) and obtain

$$
-\int_0^T \int_\Omega u(x,t)\phi'(t)\psi(x)\,\mathrm{d}x\,\mathrm{d}t + \int_0^T \int_\Omega (\nabla u(x,t)\cdot\nabla\psi(x))\phi(t)\,\mathrm{d}x\,\mathrm{d}t
$$

$$
= \int_0^T \langle f(t),\phi(t)\psi\rangle_{H^{-1},H_0^1}\,\mathrm{d}t.
$$

This gives (4.34) and therefore completes step 1, that is, $\partial_t u \in L^2(]0,T[,H^{-1}(\Omega))$ and $\partial_t u = \Delta u + f$ (which is the equation requested in (4.18)).

Step 2 Since $u \in L^2(]0,T[,H_0^1(\Omega))$ and $\partial_t u \in L^2(]0,T[,H^{-1}(\Omega))$, we have, according to Section 4.2, $u \in C([0,T],L^2(\Omega))$. We show in this second step that $u(0) = u_0$. For $n \in \mathbb{N}^\star$, we choose a function r_n from $\mathcal{D}(]0,T[)$ decreasing and such that $|r_n'(t)| \le 2n$ for all t, $r_n(0) = 1$ and $r_n(t) = 0$ if $t \ge \frac{1}{n}$.

Let $\psi \in \mathcal{D}(\Omega)$ and $n \in \mathbb{N}^\star$ such that $\frac{1}{n} < T$. We take $\varphi(x,t) = r_n(t)\psi(x)$ in (4.31). We obtain

$$
-\int_0^{\frac{1}{n}} \int_\Omega u(x,t)r_n'(t)\psi(x)\,\mathrm{d}x\,\mathrm{d}t - \int_\Omega u_0(x)\psi(x)\,\mathrm{d}x
$$

$$
+ \int_0^{\frac{1}{n}} \int_\Omega \nabla u(x,t)\cdot\nabla\psi(x)r_n(t)\,\mathrm{d}x\,\mathrm{d}t = \int_0^{\frac{1}{n}} \langle f(t),\psi\rangle_{H^{-1},H_0^1}r_n(t)\,\mathrm{d}t,
$$

which can still be written as

$$
T_n = \int_\Omega u_0(x)\psi(x)\,\mathrm{d}x - R_n + S_n, \quad (4.35)
$$

with

$$
T_n = -\int_0^{\frac{1}{n}} \int_\Omega u(x,t)r_n'(t)\psi(x)\,\mathrm{d}x\,\mathrm{d}t, \; R_n = \int_0^{\frac{1}{n}} \int_\Omega \nabla u(x,t)\cdot\nabla\psi(x)r_n(t)\,\mathrm{d}x\,\mathrm{d}t
$$

$$
\text{and } S_n = \int_0^{\frac{1}{n}} \langle f(t),\psi\rangle_{H^{-1},H_0^1}r_n(t)\,\mathrm{d}t.
$$

We first show that R_n and S_n tend towards 0. Indeed, we have

$$
|R_n| \leq \int_0^{\frac{1}{n}} \int_\Omega |\nabla u(x,t)||\nabla \psi(x)| \, dx \, dt
$$

$$
\leq \left(\int_0^{\frac{1}{n}} \int_\Omega |\nabla u(x,t)|^2 \, dx \, dt \right)^{\frac{1}{2}} \left(\int_0^{\frac{1}{n}} \int_\Omega |\nabla \psi(x)|^2 \, dx \, dt \right)^{\frac{1}{2}}
$$

$$
\leq \|u\|_{L^2(]0,T[,H_0^1)} \frac{1}{\sqrt{n}} \|\psi\|_{H_0^1} .
$$

Hence $\lim_{n \to +\infty} R_n = 0$. We also have $\lim_{n \to +\infty} S_n = 0$ because

$$
|S_n| \leq \|f\|_{L^2(]0,T[,H^{-1})} \frac{1}{\sqrt{n}} \|\psi\|_{H_0^1} .
$$

We now note that

$$
T_n = -\int_0^{\frac{1}{n}} \int_\Omega u(x,0) r_n'(t)\psi(x) \, dx \, dt - \int_0^{\frac{1}{n}} \int_\Omega (u(x,t) - u(x,0)) r_n'(t)\psi(x) \, dx \, dt
$$

$$
= \int_\Omega u(x,0)\psi(x) \, dx - \int_0^{\frac{1}{n}} \int_\Omega (u(x,t) - u(x,0)) r_n'(t)\psi(x) \, dx \, dt.
$$

(We used here $r_n(0) = 1$ and $r_n(\frac{1}{n}) = 0$.) We bound the last term of this equality:

$$
\left| \int_0^{\frac{1}{n}} \int_\Omega (u(x,t) - u(x,0)) r_n'(t)\psi(x) \, dx \, dt \right|
$$

$$
\leq \frac{1}{n} 2n \|\psi\|_{L^2(\Omega)} \max_{t \in [0,\frac{1}{n}]} \|u(\cdot,t) - u(\cdot,0)\|_{L^2(\Omega)} .
$$

Since $u \in C([0,T], L^2(\Omega))$ this last term tends towards 0 as $n \to +\infty$ and we therefore have, finally,

$$
\lim_{n \to +\infty} T_n = \int_\Omega u(x,0)\psi(x) \, dx.
$$

With (4.35), since $\lim_{n \to +\infty} R_n = \lim_{n \to +\infty} S_n = 0$, we get

$$
\int_\Omega u(x,0)\psi(x) \, dx = \int_\Omega u_0(x)\psi(x) \, dx \quad \text{for all } \psi \in \mathcal{D}(\Omega).
$$

This allows us to conclude that $u(0) = u_0$ and ends the proof of Proposition 4.36. ∎

We now give a compactness theorem which is very useful for the resolution of non-linear problems like those in Section 4.4. This is the parabolic equivalent of the compactness theorems seen for elliptic problems. Let Ω be an open bounded subset of $\mathbb{R}^N$. For $f \in H^{-1}(\Omega)$, let $T(f)$ be the weak solution of the equation $-\Delta u = f$ with

$u \in H_0^1(\Omega)$. We have already shown that the operator T is compact from $H^{-1}(\Omega)$ to $L^2(\Omega)$.

Proposition 4.37 (Compactness). *Let Ω be an open bounded subset of $\mathbb{R}^N$ and $T > 0$ and let us identify $L^2(\Omega)'$ with $L^2(\Omega)$. For $f \in L^2(]0, T[, H^{-1}(\Omega))$ and $u_0 \in L^2(\Omega)$ let $T(f, u_0)$ be the solution to (4.18). The operator T is compact from $L^2(]0, T[, H^{-1}(\Omega)) \times L^2(\Omega)$ to $L^2(]0, T[, L^2(\Omega))$.*

Proof Proposition 4.33 already gives the continuity of T. It remains therefore to show that T maps the bounded subsets of $L^2(]0, T[, H^{-1}(\Omega)) \times L^2(\Omega)$ to relatively compact subsets of $L^2(]0, T[, L^2(\Omega))$. Let A be a bounded subset of $L^2(]0, T[, H^{-1}(\Omega)) \times L^2(\Omega)$. Set $B = \{T(f, u_0), (f, u_0) \in A\}$. The estimates seen in Theorem 4.30 show that B is a bounded part of $L^2(]0, T[, H_0^1(\Omega))$ and that $\{\partial_t u, u \in B\}$ is a bounded part of $L^2(]0, T[, H^{-1}(\Omega))$. The compactness lemma given below (Lemma 4.38), due to J.L. Lions, then gives the relative compactness of A in $L^2(]0, T[, L^2(\Omega))$. ∎

Lemma 4.38 (Space-time compactness, L^2 framework). *Let Ω be an open bounded subset of $\mathbb{R}^N$ and $(u_n)_{n \in \mathbb{N}}$ a sequence in $L^2(]0, T[, L^2(\Omega))$ and identify $L^2(\Omega)'$ with $L^2(\Omega)$. Assume that*

1. *the sequence $(u_n)_{n \in \mathbb{N}}$ is bounded in $L^2(]0, T[, H_0^1(\Omega))$,*
2. *the sequence $((\partial_t u_n))_{n \in \mathbb{N}}$ is bounded in $L^2(]0, T[, H^{-1}(\Omega))$.*

Then, the sequence $(u_n)_{n \in \mathbb{N}}$ is relatively compact in $L^2(]0, T[, L^2(\Omega))$.

Proof This lemma is proved later in a more general framework, see Theorem 4.42. ∎

Remark 4.39 (General elliptic operators). The results of this section, that is to say Theorem 4.30 and Propositions 4.33, 4.34, 4.36 and 4.37, are still true when the operator Δu is replaced by $\mathrm{div}(A \nabla u)$, if the matrix A has coefficients in $L^\infty(\Omega)$ and there exists an $\alpha > 0$ such that $A\xi \cdot \xi \geq \alpha |\xi|^2$ a.e. and for all $\xi \in \mathbb{R}^N$ (see Problem 4.5). It is also possible to consider the case where the coefficients of the matrix A also depend on t. One possibility is then to make a time discretisation and to replace on each time interval the matrix A by its mean value over the considered interval and solve the problem on each of these time intervals. It is then sufficient to obtain estimates on the approximate solution and to take the limit on the discretisation step.

4.4 Existence and Uniqueness for Quasi-Linear Parabolic Problems

As in the elliptic case, the existence of the solution of a nonlinear parabolic problem can be proven by Schauder's fixed point theorem or by a topological degree argument. We give two examples below.

4.4.1 First Example: Nonlinear Diffusion

Let us first consider the following example: Let Ω be an open bounded subset of $\mathbb{R}^N$ ($N \geq 1$) and $A : \mathbb{R} \to M_N(\mathbb{R})$ (where $M_N(\mathbb{R})$ designates the set of $N \times N$ matrices with real coefficients) such that

$$\forall s \in \mathbb{R}, A(s) = (a_{i,j}(s))_{i,j=1,\ldots,N} \text{ where } a_{i,j} \in L^\infty(\mathbb{R}) \cap C(\mathbb{R}, \mathbb{R}), \qquad (4.36)$$

$$\exists \alpha > 0 \ A(s)\xi \cdot \xi \geq \alpha |\xi|^2, \forall \xi \in \mathbb{R}^N, \forall s \in \mathbb{R}, \qquad (4.37)$$

$$f \in L^2(]0, T[, H^{-1}(\Omega)) \text{ and } u_0 \in L^2(\Omega). \qquad (4.38)$$

Then we can show by the Schauder theorem that there exists a solution u of (still with $L^2(\Omega)$ identified with $L^2(\Omega)'$):

$$\begin{cases} u \in L^2(]0, T[, H_0^1(\Omega)), \partial_t u \in L^2(]0, T[, H^{-1}(\Omega)) \\ \text{(and therefore u } \in C([0, T], L^2(\Omega))), \\ \displaystyle\int_0^T \langle \partial_t u, v \rangle_{H^{-1}, H_0^1} \, dt + \int_0^T \int_\Omega A(u)\nabla u \cdot \nabla v \, dx \, dt \\ \qquad = \displaystyle\int_0^T \langle f, v \rangle_{H^{-1}, H_0^1} \, dt, \quad \forall v \in L^2(]0, T[, H_0^1(\Omega)), \\ u(\cdot, 0) = u_0. \end{cases} \qquad (4.39)$$

To use the Schauder theorem, we use the resolution of linear problems: let $\bar{u} \in L^2(]0, T[, L^2(\Omega))$, and let T be the operator from $L^2(]0, T[, L^2(\Omega))$ to $L^2(]0, T[, L^2(\Omega))$ defined by $T(\bar{u}) = u$, where u is *the* solution to the problem (4.39) where we replaced $A(u)$ by $A(\bar{u})$, that is to say:

$$\begin{cases} u \in L^2(]0, T[, H_0^1(\Omega)), \partial_t u \in L^2(]0, T[, H^{-1}(\Omega)), \\ \displaystyle\int_0^T \langle \partial_t u, v \rangle_{H^{-1}, H_0^1} \, dt + \int_0^T \int_\Omega A(\bar{u})\nabla u \cdot \nabla v \, dx \, dt \\ \qquad = \displaystyle\int_0^T \langle f, v \rangle_{H^{-1}, H_0^1} \, dt, \quad \forall v \in L^2(]0, T[, H_0^1(\Omega)), \\ u(\cdot, 0) = u_0. \end{cases}$$

One then shows that there exists an $R > 0$ such that the image of T is included in the ball B_R of radius R and centre 0 in $L^2(]0, T[, L^2(\Omega)))$. The operator T is continuous and compact thanks to the space-time compactness lemma (Lemma 4.38). Applying the Schauder theorem concludes the proof, see Problem 4.6. Note that we can also show uniqueness if A is Lipschitz-continuous.

4.4.2 Second Example: Non-Linear Convection Diffusion

Let us now study a second example using the topological degree for the existence of the solution. Let Ω be an open bounded subset of $\mathbb{R}^N$ ($N \geq 2$, the case $N = 1$ is rather simpler). Consider the following convection-diffusion equation (with possibly non-linear convection):

$$\partial_t u + \operatorname{div}(b f(u)) - \Delta u = 0,$$

$$\text{with } b \in L^2(]0, T[, (L^2(\Omega)^N) \text{ and } u_0 \in L^2(\Omega).$$

Assume that f is a bounded Lipschitz-continuous function from $\mathbb{R}$ to $\mathbb{R}$ (and $f(u)$ denotes, as usual, the function $(x, t) \mapsto f(u(x, t))$). A weak formulation of the problem reads (where $L^2(\Omega)$ is identified with $L^2(\Omega)'$):

$$\begin{cases}
u \in L^2(]0, T[, H_0^1(\Omega)), \partial_t u \in L^2(]0, T[, H^{-1}(\Omega)) \\
\text{(and therefore } u \in C([0, T], L^2(\Omega))), \\
\displaystyle\int_0^T \langle \partial_t u, v \rangle_{H^{-1}, H_0^1} \, dt - \int_0^T \int_\Omega b f(u) \cdot \nabla v \, dx \, dt \\
\displaystyle + \int_0^T \int_\Omega \nabla u \cdot \nabla v \, dx \, dt = 0, \ \forall v \in L^2(]0, T[, H_0^1(\Omega)), \\
u(\cdot, 0) = u_0.
\end{cases} \tag{4.40}$$

Let us show the existence of a solution to problem (4.40) by a degree argument. With this aim, we show that problem (4.40) can be written in the form $u - h(1, u) = 0$, where the mapping h, defined from $[0, 1] \times E$ to E, with $E = L^2(]0, T[, L^2(\Omega))$, satisfies:

1. h is compact,
2. there exists an $R \in \mathbb{R}_+$ such that

$$u - h(s, u) = 0, \ s \in [0, 1], \ u \in E \Rightarrow u \notin \partial B_R,$$

3. the mapping defined from E to E by $u \mapsto u - h(0, u)$ is linear.

(This is sufficient to obtain the existence of a solution to problem (4.40).)

Let $s \in [0, 1]$ and $u \in E$, and let $h(s, u)$ be the solution to the following problem:

$$\begin{cases}
w \in L^2(]0, T[, H_0^1(\Omega)), \partial_t w \in L^2(]0, T[, H^{-1}(\Omega)), \\
\displaystyle\int_0^T \langle \partial_t w, v \rangle_{H^{-1}, H_0^1} \, dt - \int_0^T \int_\Omega s \, b f(u) \cdot \nabla v \, dx \, dt \\
\displaystyle + \int_0^T \int_\Omega \nabla w \cdot \nabla v \, dx \, dt = 0, \ \forall v \in L^2(]0, T[, H_0^1(\Omega)), \\
w(\cdot, 0) = s u_0.
\end{cases} \tag{4.41}$$

Note that $h(s, u) = w$ is well defined because $b f(u) \in L^2(]0, T[, (L^2(\Omega)^N)$. Furthermore, item 3 above is satisfied because $h(0, u) = 0$ and therefore $u - h(0, u) = u$. It remains to show items 1 and 2, that is, that h is continuous, that $\{h(s, u), s \in [0, 1], u \in B\}$ is a relatively compact subset of E, if B is a bounded subset of E, and that there exists an $R > 0$ such that $u - h(s, u) = 0$ has no solution with $s \in [0, 1]$ and $u \in \partial B_R$.

Let us first show that h is continuous. Let $(s_n)_{n\in\mathbb{N}}$ be a sequence in $[0,1]$ and $(u_n)_{n\in\mathbb{N}}$ a sequence in E such that $s_n \to s$ in $\mathbb{R}$ and $u_n \to u$ in E when $n \to +\infty$. Set $w_n = h(s_n, u_n)$, $w = h(s, u)$ and we want to show that $w_n \to w$ in E. The functions w_n satisfy:

$$\begin{cases} w_n \in L^2(]0,T[,H_0^1(\Omega)),\, \partial_t w_n \in L^2(]0,T[,H^{-1}(\Omega)), \\ \displaystyle\int_0^T \langle \partial_t w_n, v\rangle_{H^{-1},H_0^1}\, \mathrm{d}t - \int_0^T \int_\Omega s_n\, bf(u_n)\cdot\nabla v\, \mathrm{d}x\, \mathrm{d}t \\ \displaystyle + \int_0^T \int_\Omega \nabla w_n \cdot \nabla v\, \mathrm{d}x\, \mathrm{d}t = 0,\ \forall v \in L^2(]0,T[,H_0^1(\Omega)), \\ w_n(\cdot,0) = s_n u_0. \end{cases} \tag{4.42}$$

Taking $v = w_n$ in (4.42) gives that $(w_n)_{n\in\mathbb{N}}$ is bounded in $L^2(]0,T[,H_0^1(\Omega))$ and that the sequence $(\partial_t w_n))_{n\in\mathbb{N}}$ is bounded in $L^2(]0,T[,H^{-1}(\Omega))$. Thus, owing to Lemma 4.38, $w_n \to \tilde{w}$ in E up to a subsequence. But the bounds on w_n and $\partial_t w_n$ also give $w_n \to \tilde{w}$ weakly in $L^2(]0,T[,H_0^1(\Omega))$ and $\partial_t w_n \to \partial_t \tilde{w}$ weakly in $L^2(]0,T[,H^{-1}(\Omega))$. Finally, we can assume, still up to a subsequence, that $u_n \to u$ a.e. By passing to the limit in the equation satisfied by w_n and by taking the limit on the initial condition, we then show that $\tilde{w} = w$ (as in the proof of Theorem 4.30). This allows us to conclude thanks to the uniqueness part of Theorem 4.30, that, up to a subsequence, the whole sequence $(w_n)_{n\in\mathbb{N}}$ converges to w in E. We can thus conclude to the continuity of h.

We now set $A = \{h(s,u),\ u \in L^2(]0,T[,L^2(\Omega)),\ s \in [0,1]\}$. We note that A is bounded in $L^2(]0,T[,H_0^1(\Omega))$ and that $\{\partial_t w,\ w \in A\}$ is bounded in $L^2(]0,T[,H^{-1}(\Omega))\}$. Lemma 4.38 then gives the relative compactness of A in E. Finally, the existence of $R > 0$ such that $u - h(s,u) = 0$ does not have a solution with $s \in [0,1]$ and $u \in \partial B_R$ comes from the fact that A is bounded in $L^2(]0,T[,H_0^1(\Omega))$ and therefore also in E.

In conclusion, we can use the homotopy invariance of the degree. We obtain $d(\mathrm{Id} - h(1,\cdot), B_R, 0) = d(\mathrm{Id} - h(0,\cdot), B_R, 0) = d(\mathrm{Id}, B_R, 0) = 1$. Therefore, there exists a $u \in B_R$ such that $u - h(1,u) = 0$, that is to say, u is a solution to (4.40).

We now show the uniqueness of the solution to (4.40) in a slightly more general framework: we consider the equation

$$\partial_t u + \mathrm{div}(bf(u)) - \mathrm{div}(a(u)\nabla u) = 0,$$

with a Lipschitz-continuous function a from $\mathbb{R}$ to $\mathbb{R}$ satisfying $\alpha \le a(s) \le \beta$ for all $s \in \mathbb{R}$, with $\alpha, \beta > 0$. Note that we could also show the existence in this framework with a proof similar to the previous one, replacing ∇w by $(sa(u) + (1-s)\alpha)\nabla w$ in (4.41) and Δu by $\mathrm{div}(a(u)\nabla u)$.

Let u_1, u_2 be two solutions of (4.40) (with $\mathrm{div}(a(u)\nabla u)$ instead of Δu). Setting $u = u_1 - u_2$, let us show that $u = 0$ a.e.

Let $\varepsilon > 0$ and define the function T_ε from $\mathbb{R}$ to $\mathbb{R}$ by $T_\varepsilon(s) = \max\{-\varepsilon, \min\{s, \varepsilon\}\}$. Denoting by ϕ_ε the primitive of T_ε vanishing at 0 and taking $v = T_\varepsilon(u)$ in the weak formulations satisfied by u_1 and u_2 yields

$$\int_0^T \langle \partial_t u, T_\varepsilon(u) \rangle_{H^{-1}, H_0^1} \, dt - \int_0^T \int_\Omega b(f(u_1) - f(u_2)) \cdot \nabla T_\varepsilon(u) \, dx \, dt$$

$$+ \int_0^T \int_\Omega a(u_1) \nabla u \cdot \nabla T_\varepsilon(u) \, dx \, dt = \int_0^T \int_\Omega (a(u_2) - a(u_1)) \nabla u_2 \cdot \nabla T_\varepsilon(u) \, dx \, dt.$$

Since $\nabla T_\varepsilon(u) = \nabla u \mathbb{1}_{0 < |u| < \varepsilon}$ a.e., we get

$$\int_\Omega \phi_\varepsilon(u(x, T)) \, dx - \int_\Omega \phi_\varepsilon(u(x, 0)) \, dx + \alpha \int_0^T \int_\Omega \nabla u \cdot \nabla u \mathbb{1}_{0 < |u| < \varepsilon} dx \, dt$$

$$\leq \int_0^T \int_\Omega |b| |f(u_1) - f(u_2)| |\nabla u| \mathbb{1}_{0 < |u| < \varepsilon} dx \, dt$$

$$+ \int_0^T \int_\Omega |a(u_1) - a(u_2)| |\nabla u_2| |\nabla u| \mathbb{1}_{0 < |u| < \varepsilon} dx \, dt. \quad (4.43)$$

Since a and f are Lipschitz-continuous functions, there exists an L s.t.

$$|f(s_1) - f(s_2)| \leq L|s_1 - s_2| \text{ and } |a(s_1) - a(s_2)| \leq L|s_1 - s_2|, \text{ for all } s_1, s_2 \in \mathbb{R}.$$

We then use the fact that $u_0 = 0$ a.e. and $\phi_\varepsilon \geq 0$ to deduce from (4.43), with $A_\varepsilon = \{0 < |u| < \varepsilon\}$ and $y = (x, t)$,

$$\alpha \int_0^T \int_\Omega |\nabla T_\varepsilon(u)|^2 \, dx \, dt \leq L\varepsilon \left(\int_{A_\varepsilon} |b|^2 \, dy \right)^{\frac{1}{2}} \left(\int_0^T \int_\Omega |\nabla T_\varepsilon(u)|^2 \, dx \, dt \right)^{\frac{1}{2}}$$

$$+ L\varepsilon \left(\int_{A_\varepsilon} |\nabla u_2|^2 \, dy \right)^{\frac{1}{2}} \left(\int_0^T \int_\Omega |\nabla T_\varepsilon(u)|^2 \, dx \, dt \right)^{\frac{1}{2}}.$$

Hence $\alpha ||| \nabla T_\varepsilon(u) |||_{L^2(Q)} \leq a_\varepsilon \varepsilon$, with $Q =]0, T[\times \Omega$ and

$$a_\varepsilon = L \left(\int_{A_\varepsilon} |b|^2 \, dy \right)^{\frac{1}{2}} + L \left(\int_{A_\varepsilon} |\nabla u_2|^2 \, dy \right)^{\frac{1}{2}}.$$

Since $\cap_{\varepsilon > 0} A_\varepsilon = \emptyset$ the continuity of a measure for a non-increasing sequence of sets gives that the $(N + 1)$-dimensional Lebesgue measure of A_ε tends to 0 as $\varepsilon \to 0$ and since b and ∇u_2 belong[6] to $L^2(Q)^N$, we get

$$\lim_{\varepsilon \to 0} \int_{A_\varepsilon} |b|^2 \, dy = \lim_{\varepsilon \to 0} \int_{A_\varepsilon} |\nabla u_2|^2 \, dy = 0,$$

---

[6] Recall that $L^2(Q)$ can be identified with $L^2(]0, T[, L^2(\Omega))$, see Remark 4.24.

which gives $\lim_{\varepsilon \to 0} a_\varepsilon = 0$. We now have to use, for example, the embedding of $W_0^{1,1}(\Omega)$ to $L^{1^\star}(\Omega)$ (recall that $1^\star = \frac{N}{N-1}$) to obtain that, for $t \in]0, T[$,

$$\|T_\varepsilon(u(t))\|_{L^{1^\star}(\Omega)} \leq \||\nabla T_\varepsilon(u(t))\|\|_{L^1(\Omega)} . \tag{4.44}$$

Denoting by λ_N the Lebesgue measure on $\mathbb{R}^N$, observe that for $t \in]0, T[$

$$\varepsilon \lambda_N \{|u(t)| \geq \varepsilon\}^{\frac{1}{1^\star}} \leq \left(\int_\Omega |T_\varepsilon(u)|^{1^\star} \, dx \right)^{\frac{1}{1^\star}} .$$

Hence, with (4.44),

$$\varepsilon \lambda_N \{|u(t)| \geq \varepsilon\}^{\frac{1}{1^\star}} \leq \||\nabla T_\varepsilon(u(t))\|\|_{L^1(\Omega)} = \int_\Omega |\nabla T_\varepsilon(u(x,t))| \, dx;$$

integrating with respect to t, noting that $\frac{1}{1^\star} = \frac{N-1}{N}$ and using the Cauchy–Schwarz inequality,

$$\varepsilon \int_0^T \lambda_N \{|u(t)| \geq \varepsilon\}^{\frac{N-1}{N}} \, dt \leq \int_0^T \int_\Omega |\nabla T_\varepsilon(u(x,t))| \, dx \, dt$$

$$\leq \||\nabla T_\varepsilon(u)|\|_{L^2(Q)} (T\lambda_N(\Omega))^{\frac{1}{2}} \leq \frac{(T\lambda_N(\Omega))^{\frac{1}{2}}}{\alpha} a_\varepsilon \varepsilon.$$

Hence

$$\int_0^T \lambda_N \{|u(t)| \geq \varepsilon\}^{\frac{N-1}{N}} \, dt \leq \frac{(T\lambda_N(\Omega))^{\frac{1}{2}}}{\alpha} a_\varepsilon.$$

Owing to the dominated convergence theorem, passing to the limit as $\varepsilon \to 0$, we get (since $\lim_{\varepsilon \to 0} a_\varepsilon = 0$)

$$\int_0^T \lambda_N \{|u(t)| > 0\}^{\frac{N-1}{N}} \, dt \leq 0.$$

Hence $\lambda_N \{|u(t)| > 0\} = 0$ a.e. in $t \in]0, T[$ and therefore $u = 0$ a.e., which concludes this proof of uniqueness.

Remark 4.40 (Additional assumptions on b and u_0). Here are three additional properties (whose proof is left to the reader) to the existence and uniqueness result that we just gave in this section for the convection diffusion problem (4.40), under additional assumptions on b and u_0. This supplement is also useful for the study of hyperbolic equations, see Theorem 5.29.

1. Let us add the assumptions $\text{div} \, b = 0$ and $u_0 \in L^\infty(\Omega)$; then there exist $A \leq 0$ and $B \geq 0$ such that $A \leq u_0 \leq B$ a.e. Let u be the solution to (4.40), then we have $A \leq u \leq B$ a.e. To prove this result, take $v = (u - B)^+$ (to show $u \leq B$ a.e.) and $v = (A - u)^+$ (to show $u \geq A$ a.e.) in the weak formulation (4.40).

2. We still add the assumption div $b = 0$. In the study of the convection diffusion problem (4.40), we assumed f to be Lipschitz-continuous and bounded. In fact, if $u_0 \in L^\infty(\Omega)$, and therefore $A \leq u_0 \leq B$ a.e. with $A \leq 0 \leq B$, we can replace this assumption on f by "f is locally Lipschitz-continuous" (for instance: $f(s) = s^2$). The proof consists of replacing in (4.40) f by $\bar{f}$, where $\bar{f}$ is defined by $\bar{f}(s) = f(\max\{A, \min\{s, B\}\})$. The obtained solution, u, then satisfies $A \leq u \leq B$ a.e. (according to the previous item) and is therefore a solution to (4.40) (with f).

3. Finally, if $\operatorname{div} b \in L^\infty(\Omega)$, $\|\operatorname{div} b\|_\infty = C$, if f is Lipschitz-continuous with Lipschitz-continuous constant L (but f is not necessarily bounded) and $u_0 \in L^\infty(\Omega)$, $\|u_0\|_\infty = M$, we can also show an existence and uniqueness result for the solution to (4.40). The solution u then satisfies $\|u(t)\|_\infty \leq M \exp(CLt)$ for all $t \in [0, T]$.

4.5 Compactness in Time

In this section, we generalise the compactness lemma in L^2 (Lemma 4.38) used in the previous section as a consequence of Theorem 4.46, and we give its proof. This compactness result is useful for solving many parabolic problems.

Let us start with the abstract version of Lemma 4.38, obtained by replacing the space L^2 with any Hilbert space.

Lemma 4.41 (Compactness in time, Hilbert framework). *Assume that X and H are Hilbert spaces, that X is compactly embedded in H, and that X is dense in H; we identify H with H' (by the Riesz representation theorem). Assume that $(u_n)_{n\in\mathbb{N}}$ is a bounded sequence in $L^2(]0, T[, X)$ ($T > 0$ is given) and that $(\partial_t u_n)_{n\in\mathbb{N}}$ is a bounded sequence in $L^2(]0, T[, X')$. Then, the sequence $(u_n)_{n\in\mathbb{N}}$ is relatively compact in $L^2(]0, T[, H)$.*

Note that in Lemma 4.38 we have $X = H_0^1(\Omega)$ and $H = L^2(\Omega)$. This compactness result in the Hilbert framework (attributed to J.L. Lions) was then generalised by Aubin in the context of Banach spaces [6]:

Theorem 4.42 (Time compactness, L^p framework, $p > 1$). *Let X, B, Y be three Banach spaces and $1 < p < +\infty$. Assume that X is compactly embedded in B and that B is continuously embedded in Y. We now assume that $(u_n)_{n\in\mathbb{N}}$ is a bounded sequence in $L^p(]0, T[, X)$ ($T > 0$ is given) and that $(\partial_t u_n)_{n\in\mathbb{N}}$ is a bounded sequence in $L^p(]0, T[, Y)$. Then, the sequence $(u_n)_{n\in\mathbb{N}}$ is relatively compact in $L^p(]0, T[, B)$.*

This result was itself extended by J. Simon [49] to the case $p = 1$. It is this latter result that we state under the name of the Aubin–Simon theorem (see Theorem 4.46) and prove in this section (which therefore also proves the Lions lemmas (Lemmas 4.38, 4.41) and the Aubin theorem (Theorem 4.42)).

We finally provide some generalisations of the Aubin–Simon theorem, which prove useful in particular for the mathematical analysis of numerical schemes (see Problem 4.9).

We first state the fundamental theorem of compactness of Kolmogorov (in two slightly different forms), from which we then deduce Theorems 4.46 and 4.55.

Theorem 4.43 (Kolmogorov (1)). *Let B be a Banach space, $1 \le p < +\infty$, $T > 0$ and $A \subset L^p(]0,T[,B)$. The subset A is relatively compact in $L^p(]0,T[,B)$ if A satisfies the following conditions:*

1. *For every $f \in A$, there exists a $Pf \in L^p(\mathbb{R},B)$ such that $Pf = f$ a.e. in $]0,T[$ and $\|Pf\|_{L^p(\mathbb{R},B)} \le C$, where C only depends on A.*
2. *For every $\varphi \in \mathcal{D}(\mathbb{R})$, the family $\{\int_{\mathbb{R}}(Pf)\varphi\,\mathrm{d}t,\ f \in A\}$ is relatively compact in B.*
3. *$\|Pf(\cdot + h) - Pf\|_{L^p(\mathbb{R},B)} \to 0$, when $h \to 0$, uniformly with respect to $f \in A$.*

These three conditions are therefore sufficient conditions for relative compactness; in fact, these conditions are also necessary but this is not the interesting part of the theorem.

Proof (of Theorem 4.43) Let $(\rho_n)_{n\in\mathbb{N}^\star}$ be a sequence of (symmetric) regularising kernels, that is to say:

$$\rho_n(x) = n\rho(nx),\ \forall n \in \mathbb{N}^\star,\ \forall x \in \mathbb{R},$$
$$\text{with } \rho \in \mathcal{D}(\mathbb{R}),\ \int_{\mathbb{R}} \rho\,\mathrm{d}x = 1,\ \rho \ge 0,\ \rho(-x) = \rho(x) \text{ for all } x \in \mathbb{R}. \tag{4.45}$$

Set $K = [0,T]$ and, for $n \in \mathbb{N}^\star$, $A_n = \{(Pf \star \rho_n)_{|K},\ f \in A\}$. The proof is done in two steps. In step 1, we use the Ascoli theorem (Theorem 1.35) and hypothesis 2 of Theorem 4.43 to show that, for $n \in \mathbb{N}^\star$, the set A_n is relatively compact in the space of continuous functions $C(K,B)$ equipped with its usual topology. Hence A_n is also relatively compact in $L^p(]0,T[,B)$. In step 2, we show that hypotheses 1 and 3 of Theorem 4.43 imply that $Pf \star \rho_n \to Pf$ in $L^p(\mathbb{R},B)$, as $n \to +\infty$, uniformly with respect to $f \in A$. This allows us to conclude that the family A is relatively compact in $L^p(]0,T[,B)$.

Step 1. Let $n \in \mathbb{N}^\star$; to show that A_n is relatively compact in $C(K,B)$, we use the Ascoli theorem (Theorem 1.35); we must therefore prove that the following properties are satisfied:

(a) for all $t \in K$, the set $\{Pf \star \rho_n(t),\ f \in A\}$ is relatively compact in B,
(b) the sequence $\{Pf \star \rho_n,\ f \in A\}$ is equicontinuous from K to B (i.e. the continuity is uniform with respect to $f \in A$).

We first show that property (a) holds. Let $t \in K$;

$$Pf \star \rho_n(t) = \int_{\mathbb{R}} Pf(s)\rho_n(t - s)\,\mathrm{d}s = \int_{\mathbb{R}} Pf(s)\varphi(s)\,\mathrm{d}s.$$

Hypothesis 2 of Theorem 4.43 gives property (a), namely that $\{Pf \star \rho_n(t),\ f \in A\}$ is relatively compact in B.

Let us now show that property (b) is satisfied. Let $t_1, t_2 \in K$ and $q = \frac{p}{p-1}$; thanks to Hölder's inequality, we have

$$\|Pf \star \rho_n(t_2) - Pf \star \rho_n(t_1)\|_B \leq \int_{\mathbb{R}} \|Pf(s)\|_B \, |\rho_n(t_2 - s) - \rho_n(t_1 - s)| \, ds$$

$$\leq \|Pf\|_{L^p(\mathbb{R},B)} \, \|\rho_n(t_2 - \cdot) - \rho_n(t_1 - \cdot)\|_{L^q(\mathbb{R},\mathbb{R})} \, .$$

Since ρ_n is uniformly continuous with compact support, we get from this inequality (and from hypothesis 1 of Theorem 4.43) that the family $(A_n)_{n \in \mathbb{N}}$ is uniformly equicontinuous from K to B. This gives property (b), which allows us to conclude that the family $(A_n)_{n \in \mathbb{N}}$ is relatively compact in $C(K, B)$. This relative compactness is equivalent to saying that for all $\varepsilon > 0$, there exists a finite number of balls of radius ε (for the natural norm on $C(K, B)$) covering the set A_n. Since $\|\cdot\|_{L^p(]0,T[,B)} \leq T^{\frac{1}{p}} \|\cdot\|_{C(K,B)}$, we also obtain the relative compactness of A_n in $L^p(]0, T[, B)$.

Step 2.

Let $t \in \mathbb{R}$. Since $\int_{\mathbb{R}} \rho_n(s) ds = 1$ and introducing $\bar{s} = ns$, we have

$$Pf \star \rho_n(t) - Pf(t) = \int_{\mathbb{R}} (Pf(t - s) - Pf(t))\rho_n(s) \, ds$$

$$= \int_{-1}^{1} (Pf(t - \frac{\bar{s}}{n}) - Pf(t))\rho(\bar{s}) \, d\bar{s}.$$

Then, by Hölder's inequality, we have, with $q = \frac{p}{p-1}$,

$$\|Pf \star \rho_n(t) - Pf(t)\|_B^p \leq \left(\int_{-1}^{1} \left\| Pf(t - \frac{s}{n}) - Pf(t) \right\|_B^p \, ds \right) \|\rho\|_{L^q}^p \, .$$

Integrating with respect to $t \in \mathbb{R}$ and using the Fubini–Tonelli theorem, we obtain

$$\|Pf \star \rho_n - Pf\|_{L^p(]0,T[,B)}^p \leq 2 \sup\{\|Pf(\cdot + h) - Pf\|_{L^p(\mathbb{R},B)}^p, \ |h| \leq 1\} \|\rho\|_{L^q}^p \, .$$

Finally, the third hypothesis of Theorem 4.43 implies that

$$\|Pf \star \rho_n - Pf\|_{L^p(]0,T[,B)} \to 0 \text{ when } n \to +\infty,$$

uniformly with respect to $f \in A$.

Since the relative compactness of A_n in $L^p(]0, T[, B)$ for all $n \in \mathbb{N}$ (proved in step 1) gives the relative compactness of A in $L^p(]0, T[, B)$, this concludes the proof of Theorem 4.43. ∎

Let us now give an alternative form of the previous theorem which has the advantage of not requiring the extension of f outside $[0, T]$.

Theorem 4.44 (Kolmogorov (2)). *Let B be a Banach space, $1 \le p < +\infty$, $T > 0$ and $\mathcal{A} \subset L^p(]0, T[, B)$. The subset $\mathcal{A}$ is relatively compact in $L^p(]0, T[, B)$ if the following conditions are satisfied:*

1. *The subset $\mathcal{A}$ is bounded in $L^p(]0, T[, B)$.*
2. *For all $\varphi \in \mathcal{D}(\mathbb{R})$, the family $\{\int_0^T f\varphi \, dt, \ f \in \mathcal{A}\}$ is relatively compact in B.*
3. *There exists a non-decreasing function η from $]0, T[$ to $\mathbb{R}_+$ such that $\eta(h) \to 0$ as $h \to 0^+$ and, for all $h \in]0, T[$ and all $f \in \mathcal{A}$,*

$$\int_0^{T-h} \|f(t+h) - f(t)\|_B^p \, dt \le \eta(h).$$

Proof: The proof uses Theorem 4.43 with $Pf = 0$ on $[0, T]^c$. The first two items of the assumptions of Theorem 4.43 are clearly satisfied. The only difficulty is to prove the third item. We do it in two steps.

Step 1. We first show that $\lim_{\delta \to 0} \int_0^\delta \|f(t)\|_B^p \, dt \to 0$ when $\delta \to 0^+$, uniformly with respect to $f \in \mathcal{A}$.

Let $\delta, h \in]0, T[$ such that $\delta + h \le T$. For all $t \in]0, \delta[$ we have $\|f(t)\|_B \le \|f(t+h)\|_B + \|f(t+h) - f(t)\|_B$ and therefore

$$\|f(t)\|_B^p \le 2^p \|f(t+h)\|_B^p + 2^p \|f(t+h) - f(t)\|_B^p.$$

Integrating this inequality for $t \in]0, \delta[$, we get

$$\int_0^\delta \|f(t)\|_B^p \, dt \le 2^p \int_0^\delta \|f(t+h)\|_B^p \, dt + 2^p \int_0^\delta \|f(t+h) - f(t)\|_B^p \, dt. \quad (4.46)$$

Now let $h_0 \in]0, T[$ and $\delta \in]0, T - h_0[$. For all $h \in]0, h_0[$, the inequality (4.46) gives, since $\eta(h) \le \eta(h_0)$,

$$\int_0^\delta \|f(t)\|_B^p \, dt \le 2^p \int_0^\delta \|f(t+h)\|_B^p \, dt + 2^p \eta(h_0).$$

Integrating this inequality for $h \in]0, h_0[$, we get:

$$h_0 \int_0^\delta \|f(t)\|_B^p \, dt \le 2^p \int_0^{h_0} \left(\int_0^\delta \|f(t+h)\|_B^p \, dt \right) dh + 2^p h_0 \eta(h_0).$$

Then, thanks to the Fubini–Tonelli theorem,

$$\int_0^{h_0} \left(\int_0^\delta \|f(t+h)\|_B^p \, dt \right) dh = \int_0^\delta \left(\int_0^{h_0} \|f(t+h)\|_B^p \, dh \right) dt$$

$$\le \int_0^\delta \left(\int_0^T \|f(s)\|_B^p \, ds \right) dt \le \delta \|f\|_{L^p(]0,T[,B)}^p,$$

from which we get that

$$h_0 \int_0^\delta \|f(t)\|_B^p \, dt \le \delta 2^p \, \|f\|_{L^p(]0,T[,B)}^p + 2^p h_0 \eta(h_0). \tag{4.47}$$

Now let $\varepsilon > 0$; we first choose $h_0 \in]0,T[$ such that $2^p \eta(h_0) \le \varepsilon$. Then, we choose $\overline{\delta} = \min\{T - h_0, \varepsilon \frac{h_0}{2^p C}\}$, with $C = \sup_{f \in \mathcal{A}} \|f\|_{L^p(]0,T[,B)}^p$. We then obtain, for all $f \in \mathcal{A}$,

$$0 \le \delta \le \overline{\delta} \Rightarrow \int_0^\delta \|f(t)\|_B^p \, dt \le 2\varepsilon,$$

which implies that $\int_0^\delta \|f(t)\|_B^p \, dt \to 0$ when $\delta \to 0^+$, uniformly with respect to $f \in \mathcal{A}$. We have thus completed step 1.

Note that similar arguments give that $\int_{T-\delta}^T \|f(t)\|_B^p \, dt \to 0$ when $\delta \to 0^+$, uniformly with respect to $f \in \mathcal{A}$ (this is proved using g defined by $g(t) = f(T - t)$ instead of f).

Step 2. We now show that the third item of the assumptions of Theorem 4.43 is satisfied, which concludes the proof of Theorem 4.44.

Recall that $Pf(t) = 0$ if $t \in [0,T]^c$, and therefore, for all $h \in]0,T[$ and $f \in \mathcal{A}$,

$$\int_{\mathbb{R}} \|Pf(t+h) - Pf(t)\|_B^p \, dt$$

$$\le \int_{-h}^0 \|f(t+h)\|_B^p \, dt + \int_0^{T-h} \|f(t+h) - f(t)\|_B^p \, dt + \int_{T-h}^T \|f(t)\|_B^p \, dt$$

$$\le \int_0^h \|f(t)\|_B^p \, dt + \int_0^{T-h} \|f(t+h) - f(t)\|_B^p \, dt + \int_{T-h}^T \|f(t)\|_B^p \, dt.$$

Let $\varepsilon > 0$. There exists an $h_1 > 0$ such that $\eta(h_1) \le \varepsilon$. Thanks to step 1, there exists an $h_2 > 0$ such that for all $f \in \mathcal{A}$,

$$0 \le h \le h_2 \Rightarrow \int_0^h \|f(t)\|_B^p \, dt \le \varepsilon \text{ and } \int_{T-h}^T \|f(t)\|_B^p \, dt \le \varepsilon.$$

By defining $h_3 = \min\{h_1, h_2\}$, we have for all $f \in \mathcal{A}$,

$$0 \le h \le h_3 \Rightarrow \int_{\mathbb{R}} \|Pf(t+h) - Pf(t)\|_B^p \, dt \le 3\varepsilon,$$

which concludes step 2 and the proof of Theorem 4.44.   ∎

We then deduce from Theorem 4.44 a corollary which is useful for the proof of the Aubin–Simon theorem (Theorem 4.46).

Corollary 4.45 (Compactness in time). *Let B be a Banach space, $1 \le p < +\infty$, $T > 0$ and $\mathcal{A} \subset L^p(]0,T[, B)$. Let X be a Banach space included in B with compact embedding. Assume that $\mathcal{A}$ satisfies the following conditions:*

1. *$\mathcal{A}$ is bounded in $L^p(]0, T[, B)$.*
2. *$\mathcal{A}$ is bounded in $L^1(]0, T[, X)$.*

Note that these two conditions are satisfied if $\mathcal{A}$ is bounded in $L^p(]0, T[, X)$.

3. *There exists a non-decreasing function $\eta :]0, T[\to \mathbb{R}_+$ such that $\lim_{h \to 0^+} \eta(h) = 0$ and, for all $h \in]0, T[$ and all $f \in \mathcal{A}$,*

$$\int_0^{T-h} \|f(t + h) - f(t)\|_B^p \, dt \le \eta(h).$$

Then the set $\mathcal{A}$ is relatively compact in $L^p(]0, T[, B)$.

Proof To apply Theorem 4.44, we must verify that, for all $\varphi \in \mathcal{D}(\mathbb{R})$, the family $\{\int_0^T f\varphi \, dt, \ f \in \mathcal{A}\}$ is relatively compact in B.

Let $\varphi \in \mathcal{D}(\mathbb{R})$. If $f \in \mathcal{A}$, we have, denoting by $\|\varphi\|_u = \max_{t \in \mathbb{R}} |\varphi(t)|$,

$$\left\| \int_0^T f\varphi \, dt \right\|_X \le \|\varphi\|_u \|f\|_{L^1(]0, T[, X)} \cdot$$

Since $\mathcal{A}$ is bounded in $L^1(]0, T[, X)$, the family $\{\int_0^T f\varphi \, dt, \ f \in \mathcal{A}\}$ is bounded in X and therefore relatively compact in B. ∎

We now give a theorem, essentially due to J.P. Aubin (for $p > 1$) [6] and J. Simon (for $p = 1$) [49], which generalises the Lions lemma (Lemma 4.38) used in the previous section to prove the existence of a solution for parabolic equations.

Theorem 4.46 (Aubin–Simon). *Let $1 \le p < +\infty$, and let X, B, Y be three Banach spaces such that*

(i) *$X \subset B$ with compact embedding,*
(ii) *$X \subset Y$ and if $(f_n)_{n \in \mathbb{N}}$ is a sequence of elements of X such that the sequence $(\|f_n\|_X)_{n \in \mathbb{N}}$ is bounded in $\mathbb{R}$, $f_n \to f$ in B and $f_n \to 0$ in Y as $n \to +\infty$, then $f = 0$.*

Let $T > 0$ and $(u_n)_{n \in \mathbb{N}}$ be a sequence of elements of $L^p(]0, T[, X)$ such that

1. *$(u_n)_{n \in \mathbb{N}}$ is bounded in $L^p(]0, T[, X)$,*
2. *$(\partial_t u_n)_{n \in \mathbb{N}}$ is bounded in $L^1(]0, T[, Y)$.*

Then there exists a $u \in L^p(]0, T[, B)$ such that, up to a subsequence, $u_n \to u$ in $L^p(]0, T[, B)$.

Remark 4.47. Note that the assumption (ii) of Theorem 4.46 is weaker than the assumption of the original articles by Aubin and Simon, which is:

$$B \subset Y \text{ with continuous embedding.}$$

In particular, the assumption (ii) does not require that $B \subset Y$; this is interesting for example in applications to solutions of numerical schemes involving discrete spaces not necessarily included in the continuous space. Note that the assumption (i) can also be written as

If the sequence $(\|w_n\|_X)_{n\in\mathbb{N}}$ is bounded, then, up to a subsequence, there exists a $w \in B$ such that $w_n \to w$ in B.

This reformulation leads to a very useful generalisation for the proof of convergence of numerical approximations of the solution of parabolic equations: in this generalisation, X is replaced by a sequence $(X_n)_{n\in\mathbb{N}}$, see Theorem 4.55 below.

Proof (of Theorem 4.46) The proof uses Corollary 4.45 with $\mathcal{A} = \{u_n,\, n \in \mathbb{N}\}$. The set $\mathcal{A}$ is bounded in $L^p(]0, T[, B)$ (because X is continuously embedded in B) and therefore in $L^1(]0, T[, X)$, so that we only have to prove the third item of the assumptions of Corollary 4.45, that is

$$\int_0^{T-h} \|u_n(\cdot + h) - u_n\|_B^p \, dt \to 0 \text{ when } h \to 0,$$

$$\text{uniformly with respect to } n \in \mathbb{N}. \quad (4.48)$$

Let $0 < h < T$ and $\varepsilon > 0$. For $t \in]0, T - h[$, we have, using Lemma 4.48 given further on,

$$\|u_n(t + h) - u_n(t)\|_B \le \varepsilon \|u_n(t + h) - u_n(t)\|_X + C_\varepsilon \|u_n(t + h) - u_n(t)\|_Y$$
$$\le \varepsilon \|u_n(t + h)\|_X + \varepsilon \|u_n(t)\|_X + C_\varepsilon \|u_n(t + h) - u_n(t)\|_Y,$$

and therefore

$$\|u_n(t + h) - u_n(t)\|_B^p \le$$
$$(3\varepsilon)^p \|u_n(t + h)\|_X^p + (3\varepsilon)^p \|u_n(t)\|_X^p + (3C_\varepsilon)^p \|u_n(t + h) - u_n(t)\|_Y^p.$$

Integrating this inequality with respect to t (between 0 and $T - h$) leads to

$$\int_0^{T-h} \|u_n(t + h) - u_n(t)\|_B^p \, dt \le$$

$$2(3\varepsilon)^p \|u_n\|_{L^p(]0,T[,X)}^p + (3C_\varepsilon)^p \int_0^{T-h} \|u_n(t + h) - u_n(t)\|_Y^p \, dt. \quad (4.49)$$

We now use the second item of Remark 4.50, also given in Section 4.2, which tells us that $u_n \in C([0, T], Y)$ and $u_n(t_1) - u_n(t_2) = \int_{t_1}^{t_2} \partial_t u_n(s) \, ds$ for all $t_1, t_2 \in [0, T]$. This allows us to bound the second term of the right-hand side of (4.49):

$$\int_0^{T-h} \|u_n(t + h) - u_n(t)\|_Y^p \, dt \le \int_0^{T-h} \left(\int_t^{t+h} \|\partial_t u_n(s)\|_Y \, ds \right)^p dt$$

$$\le M^{p-1} \int_0^{T-h} \left(\int_t^{t+h} \|\partial_t u_n(s)\|_Y \, ds \right) dt$$

$$\le M^{p-1} \int_0^{T-h} \left(\int_0^T \mathbb{1}_{[t,t+h]}(s) \|\partial_t u_n(s)\|_Y \, ds \right) dt,$$

where M is an upper bound of the norm $L^1(]0, T[, Y)$ of $\partial_t u_n$, that is

$$\|\partial_t u_n\|_{L^1(]0,T[,Y)} \leq M \text{ for all } n.$$

Using the fact that $\mathbb{1}_{[t,t+h]}(s) = \mathbb{1}_{[s-h,s]}(t)$ and the Fubini–Tonelli theorem, we obtain

$$\int_0^{T-h} \|u_n(t+h) - u_n(t)\|_Y^p \, dt \leq h M^p. \tag{4.50}$$

Therefore, taking into account (4.50), (4.49) gives

$$\int_0^{T-h} \|u_n(t+h) - u_n(t)\|_B^p \, dt \leq 2(3\varepsilon)^p \|u_n\|_{L^p(]0,T[,X)}^p + (3C_\varepsilon)^p h M^p. \tag{4.51}$$

We can then conclude: Let $\eta > 0$, we choose $\varepsilon > 0$ to upper bound the first term of the right-hand side of (4.51) by η (independently of $n \in \mathbb{N}$). Since C_ε is given, there exists an $h_0 \in]0, T[$ such that the second term of the right-hand side of (4.51) is upper bounded by η (independently of $n \in \mathbb{N}$) if $0 < h < h_0$. Finally, we obtain

$$0 < h < h_0 \Rightarrow \int_0^{T-h} \|u_n(t+h) - u_n(t)\|_B^p \, dt \leq 2\eta,$$

which concludes the proof of Theorem 4.46. ∎

Lemma 4.48 (Essentially due to J.L. Lions [38]). *Let X, B, Y be three Banach spaces such that*

(i) *$X \subset B$ with compact embedding,*
(ii) *$X \subset Y$ and if $(u_n)_{n \in \mathbb{N}}$ is a sequence of elements of X such that the sequence $(\|u_n\|_X)_{n \in \mathbb{N}}$ is bounded, $u_n \to u$ in B and $u_n \to 0$ in Y as $n \to +\infty$, then $u = 0$.*

Then for every $\varepsilon > 0$, there exists a C_ε such that, for $w \in X$,

$$\|w\|_B \leq \varepsilon \|w\|_X + C_\varepsilon \|w\|_Y.$$

Remark 4.49 (On the assumptions of Lemma 4.48). In Theorem 4.46, assumption (ii) of Lemma 4.48 can be replaced by the stronger assumption

(ii)' *$B \subset Y$ with continuous embedding.*

Proof (of Lemma 4.48) We reason by contradiction: We suppose that there exist $\varepsilon > 0$ and a sequence $(u_n)_{n \in \mathbb{N}}$ such that $(u_n)_{n \in \mathbb{N}}$ such that $u_n \in X$ and $1 = \|u_n\|_B > \varepsilon \|u_n\|_X + n \|u_n\|_Y$, for all $n \in \mathbb{N}$. Then $(u_n)_{n \in \mathbb{N}}$ is bounded in X and therefore relatively compact in B. We can therefore assume (up to a subsequence) that $u_n \to u$ in B and $\|u\|_B = 1$. Moreover $u_n \to 0$ in Y (because $\|u_n\|_Y \leq \frac{1}{n}$). Hypothesis (ii) of Lemma 4.48 therefore gives that $u = 0$, which contradicts $\|u\|_B = 1$. ∎

Remark 4.50 (On the weak derivative of u). We recall here (items 1 and 2) some results essentially given in Section 4.2 (with a different proof for the second item). Let X, Y be two Banach spaces and Z a vector space such that X, $Y \subset Z$ (of course, a simple case is $Z = Y$). Let $p, q \in [1, \infty]$.

1. Assuming that $u \in L^p(]0, T[, X)$, the weak derivative of u, denoted by $\partial_t u$, is defined by its action on the test functions, that is to say its action on φ for all $\varphi \in \mathcal{D}(]0, T[)$ (note that φ takes its values in $\mathbb{R}$). More precisely, if $\varphi \in \mathcal{D}(]0, T[)$ (and φ' is the classical derivative of φ), the function $\varphi' u$ belongs to $L^p(]0, T[, X)$ and therefore to $L^1(]0, T[, X)$ and the action of $\partial_t u$ on φ is defined as

$$\langle \partial_t u, \varphi \rangle_{\mathcal{D}_X^\star, \mathcal{D}} = - \int_0^T \varphi'(t) u(t) \, \mathrm{d}t.$$

Note that $\langle \partial_t u, \varphi \rangle_{\mathcal{D}_X^\star, \mathcal{D}} \in X$.

Then $\partial_t u \in L^q(]0, T[, Y)$ means that there exists a $v \in L^q(]0, T[, Y)$ (and this v is unique) such that

$$- \int_0^T \varphi'(t) u(t) \, \mathrm{d}t = \int_0^T \varphi(t) v(t) \, \mathrm{d}t \text{ for all } \varphi \in \mathcal{D}(]0, T[). \tag{4.52}$$

Note that $\int_0^T \varphi'(t) u(t) \, \mathrm{d}t \in X \subset Z$ and $\int_0^T \varphi(t) v(t) \, \mathrm{d}t \in Y \subset Z$, so the equality in (4.52) is in the space Z. Finally, we identify the weak derivative of u (which is a linear form on $\mathcal{D}(]0, T[)$) with the function v (which belongs to $L^q(]0, T[, Y)$).
2. Let us show that if $u \in L^p(]0, T[, X)$ and $\partial_t u \in L^q(]0, T[, Y)$, then $u \in C([0, T], Y)$ and

$$u(t) = u(0) + \int_0^t \partial_t u(s) \, \mathrm{d}s, \text{ for all } t \in [0, T].$$

Indeed, let us set $v = \partial_t u$ and define $w \in C([0, T], Y)$ by

$$w(t) = \int_0^t v(s) \, \mathrm{d}s \text{ for all } t \in [0, T].$$

Extending u by 0 outside of $[0, T]$, let $u_n = u \star \rho_n$ for $n \in \mathbb{N}^\star$, with ρ_n defined by (4.45). Since $u \in L^p(\mathbb{R}, X)$, we have $u_n \to u$ in $L^p(\mathbb{R}, X)$ and therefore, up to a subsequence, $u_n(t) \to u(t)$ in X as $n \to +\infty$ for almost all $t \in]0, T[$.
Let $t_1, t_2 \in]0, T[$ such that $0 < t_1 < t_2 < T$ and $u_n(t_i) \to u(t_i)$ in X as $n \to +\infty$ for $i = 1, 2$. For $n \in \mathbb{N}^\star$ such that $\frac{1}{n} < t_1$ and $\frac{1}{n} < T - t_2$, we define φ_n by

$$\varphi_n(t) = -\mathbb{1}_{]t_1, t_2[} \star \rho_n(t) = - \int_{t_1}^{t_2} \rho_n(t - s) \, \mathrm{d}s.$$

We have $\varphi_n \in \mathcal{D}(]0, T[)$ and $\varphi'_n(t) = \rho_n(t - t_2) - \rho_n(t - t_1)$ for $t \in]0, T[$.

Since $u_n(t) = \int_{\mathbb{R}} u(s)\rho_n(t-s)\,\mathrm{d}s$ (and $\rho(-\cdot) = \rho$),

$$u_n(t_2) - u_n(t_1) = \int_0^T u(s)\varphi_n'(s)\,\mathrm{d}s = -\int_0^T v(s)\varphi_n(s)\,\mathrm{d}s.$$

Moreover, $\varphi_n \to -\mathbb{1}_{]t_1,t_2[}$ a.e. and since $|\varphi_n| \le 1$ a.e. (for all n), the dominated convergence theorem gives

$$-\int_0^T v(s)\varphi_n(s)\,\mathrm{d}s \to \int_{t_1}^{t_2} v(s)\,\mathrm{d}s \text{ in } Y.$$

We then obtain, by letting $n \to +\infty$,

$$u(t_2) - u(t_1) = \int_{t_1}^{t_2} v(s)\,\mathrm{d}s = w(t_2) - w(t_1).$$

This proves that there exists a $c \in \mathbb{R}$ such that $u = w + c$ a.e. on $]0, T[$ and then that $u \in C(]0, T[, Y)$ (since, as usual, we identify u with the continuous function $w + c$).

3. In Theorem 4.46, thanks to the previous item, we have $u_n \in C([0, T], Y)$ for all $n \in \mathbb{N}$. However, the limit u is not necessarily continuous. A simple counterexample is obtained with $p = 1$, $X = B = Y = \mathbb{R}$ and, for example, $T = 2$. We can then construct a bounded sequence $(u_n)_n$ in $W^{1,1}(]0, T[)$ whose limit is the characteristic function of $[1, 2]$.

Remark 4.51 (A simpler classical case). There is a classical case where Lemma 4.48 is simpler. Let B be a Hilbert space and X a Banach space $X \subset B$. Define on X the dual norm of $\|\cdot\|_X$ for the inner product of B, namely

$$\|u\|_Y = \sup\{(u|v)_B, \; ; v \in X, \|v\|_X \le 1\}.$$

The space X is therefore equipped with two norms, the norm $\|\cdot\|_X$ and the norm $\|\cdot\|_Y$. Then, for all $\varepsilon > 0$ and $u \in X$, we have

$$\|u\|_B \le \varepsilon \|u\|_X + \frac{1}{\varepsilon} \|u\|_Y .$$

The proof is simple, since

$$\|u\|_B = (u|u)_B^{\frac{1}{2}} \le (\|u\|_Y \|u\|_X)^{\frac{1}{2}} \le \varepsilon \|u\|_X + \frac{1}{\varepsilon} \|u\|_Y .$$

Note that the compactness of X in B is not necessary here (but, even in this simple case of Lemma 4.48, the compactness of X in B is necessary for Theorem 4.46).

We now give a generalisation of Corollary 4.45 and Theorem 4.46, using a sequence of subspaces of B instead of spaces X and Y, which allows us to use this result for the compactness of sequences obtained by numerical approximation. An example is given in Problem 4.9 (see [21, chapter 6] for more general examples).

Definition 4.52 (Compactly embedded sequence). Let B be a Banach space and $(X_n)_{n\in\mathbb{N}}$ a sequence of Banach spaces embedded in B. We say that the sequence $(X_n)_{n\in\mathbb{N}}$ is compactly embedded in B if every subsequence of a sequence $(u_n)_{n\in\mathbb{N}}$ such that

- $u_n \in X_n$ for all $n \in \mathbb{N}$,
- the sequence $(\|u_n\|_{X_n})_{n\in\mathbb{N}}$ is bounded,

is relatively compact in B.

A simple example of a sequence $(X_n)_{n\in\mathbb{N}}$ compactly embedded in B is given in Problem 4.7.

Proposition 4.53 (Compactness in time with a sequence of subspaces). *Let $1 \le p < +\infty$ and $T > 0$. Let $(X_n)_{n\in\mathbb{N}}$ be a sequence (of Banach spaces) compactly embedded in B and let $(f_n)_{n\in\mathbb{N}}$ be a sequence in $L^p(]0,T[,B)$ satisfying the following conditions*

1. *The sequence $(f_n)_{n\in\mathbb{N}}$ is bounded in $L^p(]0,T[,B)$.*
2. *The sequence $(\|f_n\|_{L^1(]0,T[,X_n)})_{n\in\mathbb{N}}$ is bounded.*
3. *There exists a non-decreasing function $\eta :]0,T[\to \mathbb{R}_+$ such that $\lim_{h\to 0^+}\eta(h) = 0$ and, for all $h \in]0,T[$ and $n \in \mathbb{N}$,*

$$\int_0^{T-h} \|f_n(t+h) - f_n(t)\|_B^p\, dt \le \eta(h),$$

then the sequence $(f_n)_{n\in\mathbb{N}}$ is relatively compact in $L^p(]0,T[,B)$.

Proof In Corollary 4.45, to apply Theorem 4.44, it is enough to prove that, for all $\varphi \in \mathcal{D}(\mathbb{R})$, the sequence $\{\int_0^T f_n\varphi\, dt, n \in \mathbb{N}\}$ is relatively compact in B.

Let $\varphi \in \mathcal{D}(\mathbb{R})$. For $n \in \mathbb{N}$, we have, with $\|\varphi\|_u = \max_{t\in\mathbb{R}}|\varphi(t)|$,

$$\left\|\int_0^T f_n\varphi\, dt\right\|_{X_n} \le \|\varphi\|_u \|f_n\|_{L^1(]0,T[,X_n)} \cdot$$

Since the sequence $(\|f_n\|_{L^1(]0,T[,X_n)})_{n\in\mathbb{N}}$ is bounded, the sequence $\{\left\|\int_0^T f_n\varphi\, dt\right\|_{X_n}, n \in \mathbb{N}\}$ is also bounded. Therefore, the sequence $\{\int_0^T f_n\varphi\, dt, n \in \mathbb{N}\}$ is relatively compact in B. $\blacksquare$

Here is a generalisation of Theorem 4.46 using a sequence of subspaces of B.

Definition 4.54 (Compact-continuous sequence). Let B be a Banach space, $(X_n)_{n\in\mathbb{N}}$ a sequence of Banach spaces embedded in B and $(Y_n)_{n\in\mathbb{N}}$ a sequence of Banach spaces. We say that the sequence $(X_n, Y_n)_{n\in\mathbb{N}}$ is compact-continuous in B if the following conditions are met

1. The sequence $(X_n)_{n\in\mathbb{N}}$ is compactly embedded in B (see Definition 4.52).
2. $X_n \subset Y_n$ (for all $n \in \mathbb{N}$) and if the sequence $(u_n)_{n\in\mathbb{N}}$ is such that $u_n \in X_n$ (for all $n \in \mathbb{N}$), $(\|u_n\|_{X_n})_{n\in\mathbb{N}}$ bounded and $\|u_n\|_{Y_n} \to 0$ as $n \to +\infty$, then any subsequence of $(u_n)_{n\in\mathbb{N}}$ converging in B converges (in B) to 0.

Theorem 4.55 (Aubin–Simon with a sequence of subspaces). *Let $1 \leq p < +\infty$. Let B be a Banach space, $(X_n)_{n \in \mathbb{N}}$ a sequence of Banach spaces embedded in B and $(Y_n)_{n \in \mathbb{N}}$ a sequence of Banach spaces and assume that the sequence $(X_n, Y_n)_{n \in \mathbb{N}}$ is compact-continuous in B (see Definition 4.54). Let $T > 0$ and $(f_n)_{n \in \mathbb{N}}$ a sequence of elements of $L^p(]0, T[, B)$ satisfying the following conditions*

1. *the sequence $(f_n)_{n \in \mathbb{N}}$ is bounded in $L^p(]0, T[, B)$,*
2. *$f_n \in L^p(]0, T[, X_n)$ (for all $n \in \mathbb{N}$) and the sequence $(\|f_n\|_{L^p(]0,T[,X_n)})_{n \in \mathbb{N}}$ is bounded,*
3. *$\partial_t f_n \in L^1(]0, T[, Y_n)$ (for all $n \in \mathbb{N}$) and the sequence $(\|\partial_t f_n\|_{L^1(]0,T[,Y_n)})_{n \in \mathbb{N}}$ is bounded,*

then there exists an $f \in L^p(]0, T[, B)$ such that, up to a subsequence, $f_n \to f$ in $L^p(]0, T[, B)$.

Proof The proof uses Proposition 4.53. It is sufficient to prove the third hypothesis on $(f_n)_{n \in \mathbb{N}}$ of Proposition 4.53, that is,

$$\int_0^{T-h} \|f_n(\cdot + h) - f_n\|_B^p \, dt \to 0 \text{ as } h \to 0, \text{ uniformly with respect to } n \in \mathbb{N}.$$

For all $n \in \mathbb{N}$, since $f_n \in L^p(]0, T[, B)$, we have $\int_0^{T-h} \|f_n(\cdot + h) - f_n\|_B^p \, dt \to 0$ as $h \to 0$. The only difficulty is to prove the uniformity of this convergence with respect to $n \in \mathbb{N}$. Thus, it is sufficient to prove that for all $\eta > 0$ there exist $n_0 \in \mathbb{N}$ and $0 < h_0 < T$ such that

$$n \geq n_0, \; 0 < h \leq h_0 \Rightarrow \int_0^{T-h} \|f_n(\cdot + h) - f_n\|_B^p \, dt \leq \eta. \tag{4.53}$$

Let $\varepsilon > 0$; Lemma 4.56 gives the existence of $n_0 \in \mathbb{N}$ and of $C_\varepsilon \in \mathbb{R}$ such that

$$n \geq n_0, \; u \in X_n \Rightarrow \|u\|_B \leq \varepsilon \|u\|_{X_n} + C_\varepsilon \|u\|_{Y_n}.$$

Hence, for $n \geq n_0$, $0 < h < T$ and $t \in]0, T - h[$,

$$
\begin{aligned}
\|f_n(t+h) &- f_n(t)\|_B \\
&\leq \varepsilon \|f_n(t+h) - f_n(t)\|_{X_n} + C_\varepsilon \|f_n(t+h) - f_n(t)\|_{Y_n} \\
&\leq \varepsilon \|f_n(t+h)\|_{X_n} + \varepsilon \|f_n(t)\|_{X_n} + C_\varepsilon \|f_n(t+h) - f_n(t)\|_{Y_n},
\end{aligned}
$$

then

$$
\begin{aligned}
\|f_n(t+h) &- f_n(t)\|_B^p \\
&\leq (3\varepsilon)^p \|f_n(t+h)\|_{X_n}^p + (3\varepsilon)^p \|f_n(t)\|_{X_n}^p + (3C_\varepsilon)^p \|f_n(t+h) - f_n(t)\|_{Y_n}^p.
\end{aligned}
$$

The integration of this inequality with respect to t leads to

$$\int_0^{T-h} \|f_n(t+h) - f_n(t)\|_B^P \, dt$$

$$\leq 2(3\varepsilon)^P \|f_n\|_{L^P(]0,T[,X_n)}^P + (3C_\varepsilon)^P \int_0^{T-h} \|f_n(t+h) - f_n(t)\|_{Y_n}^P \, dt. \quad (4.54)$$

Let us recall (see Remark 4.50) that $f_n \in C([0,T], Y_n)$ and $f_n(t_1) - f_n(t_2) = \int_{t_1}^{t_2} \partial_t f_n(s) \, ds$ for all $t_1, t_2 \in [0,T]$. We can therefore bound the second term of the right-hand side of (4.54):

$$\int_0^{T-h} \|f_n(t+h) - f_n(t)\|_{Y_n}^P \, dt$$

$$\leq \int_0^{T-h} \left(\int_t^{t+h} \|\partial_t f_n(s)\|_{Y_n} \, ds \right)^P \, dt$$

$$\leq \int_0^{T-h} M^{P-1} \left(\int_t^{t+h} \|\partial_t f_n(s)\|_{Y_n} \, ds \right) \, dt$$

$$\leq M^{P-1} \int_0^{T-h} \left(\int_0^T \mathbb{1}_{[t,t+h]}(s) \|\partial_t f_n(s)\|_{Y_n} \, ds \right) \, dt,$$

where M is a bound of the norm $L^1(]0,T[,Y_n)$ of $\partial_t f_n$. Using $\mathbb{1}_{[t,t+h]}(s) = \mathbb{1}_{[s-h,s]}(t)$ and the Fubini–Tonelli theorem, we obtain

$$\int_0^{T-h} \|f_n(t+h) - f_n(t)\|_{Y_n}^P \, dt \leq hM^P. \quad (4.55)$$

Thanks to this last inequality, (4.54) gives

$$\int_0^{T-h} \|f_n(t+h) - f_n(t)\|_B^P \, dt \leq 2(3\varepsilon)^P \|f_n\|_{L^P(]0,T[,X_n)}^P + (3C_\varepsilon)^P hM^P. \quad (4.56)$$

We can now conclude. Let $\eta > 0$; we choose $\varepsilon > 0$ in order to bound by η the first term of the right-hand side of (4.56) (independently of $n \in \mathbb{N}$). This choice of ε imposes n_0 and C_ε. For this given C_ε, there exists an $h_0 \in]0,T[$ such that the second term of the right-hand side of (4.56) is bounded by η (independently of $n \in \mathbb{N}$) if $0 < h < h_0$. Finally, we obtain

$$n \geq n_0, \ 0 < h < h_0 \Rightarrow \int_0^{T-h} \|f_n(t+h) - f_n(t)\|_B^P \, dt \leq 2\eta.$$

This concludes the proof of Theorem 4.55. $\blacksquare$

Lemma 4.56 (Lemma 4.48 for a sequence of subspaces). *Let B be a Banach space, $(X_n)_{n\in\mathbb{N}}$ a sequence of Banach spaces embedded in B and $(Y_n)_{n\in\mathbb{N}}$ a sequence of Banach spaces. Suppose that the sequence $(X_n, Y_n)_{n\in\mathbb{N}}$ is compact-continuous in B (see Definition 4.54); then, for all $\varepsilon > 0$, there exist $n_0 \in \mathbb{N}$ and C_ε such that, for $n \geq n_0$ and $w \in X_n$, we have*

$$\|w\|_B \leq \varepsilon \|w\|_{X_n} + C_\varepsilon \|w\|_{Y_n}.$$

Proof The proof can be done by contradiction. Suppose there exists an $\varepsilon > 0$ such that for all n_0 and all C there exist $m \geq n_0$ and $w \in X_m$ such that

$$\|w\|_B > \varepsilon \|w\|_{X_m} + C \|w\|_{Y_m}.$$

By homogeneity, we can assume $\|w\|_B = 1$. Then, we take, for all $n \in \mathbb{N}$, $C = n$ and $m = \varphi(n)$ with $\varphi(n) > \varphi(n-1)$ for $n > 1$ (and, for instance, $\varphi(0) = 0$).

This gives an increasing function φ from $\mathbb{N}$ to $\mathbb{N}$ and, for all $n \in \mathbb{N}$, $u_{\varphi(n)} \in X_{\varphi(n)}$ and

$$1 = \left\|u_{\varphi(n)}\right\|_B > \varepsilon \left\|u_{\varphi(n)}\right\|_{X_{\varphi(n)}} + n \left\|u_{\varphi(n)}\right\|_{Y_{\varphi(n)}}.$$

In order to define u_n for all n, we set $u_n = 0$ for $n \notin \mathrm{Im}(\varphi)$. The sequence $(u_n)_{n\in\mathbb{N}}$ is such that $u_n \in X_n$ (for all $n \in \mathbb{N}$), the sequence $(\|u_n\|_{X_n})_{n\in\mathbb{N}}$ is bounded (by $1/\varepsilon$) and $\|u_n\|_{Y_n} \to 0$ as $n \to +\infty$; the second assumption of Definition 4.54 gives that any subsequence (of the sequence $(u_n)_{n\in\mathbb{N}}$) converging in B converges (in B) to 0. But, since $(\|u_n\|_{X_n})_{n\in\mathbb{N}}$ is bounded, the first assumption of Definition 4.54 gives that the subsequence $(u_{\varphi(n)})_{n\in\mathbb{N}}$ has a subsequence converging in B. The limit of this subsequence must be 0 and its norm in B must be equal to 1, which is impossible. This concludes the proof of Lemma 4.56. ∎

It is also possible to replace in Theorem 4.55 the time derivative by a discrete derivative. This is interesting for proving the convergence of the approximate solution of a parabolic problem using a numerical scheme, as proposed in Problem 4.9 for the Stefan problem. This is the purpose of Theorem 4.57

Theorem 4.57 (Aubin–Simon for a sequence of subspaces and a discrete time derivative). *Let $1 \leq p < +\infty$, let B be a Banach space, $(X_n)_{n\in\mathbb{N}}$ a sequence of Banach spaces embedded in B and $(Y_n)_{n\in\mathbb{N}}$ a sequence of Banach spaces and assume that the sequence $(X_n, Y_n)_{n\in\mathbb{N}}$ is compact-continuous in B (see Definition 4.54). Let $T > 0$ and $(u_n)_{n\in\mathbb{N}}$ be a sequence in $L^p(]0, T[, B)$ satisfying the following conditions:*

1. *For all $n \in \mathbb{N}$, there exist $N \in \mathbb{N}^\star$ and $k_1, \ldots, k_N$ in $\mathbb{R}_+^\star$ such that $\sum_{i=1}^N k_i = T$ and $u_n(t) = v_i$ for $t \in (t_{i-1}, t_i)$, $i \in \{1, \ldots, N\}$, $t_0 = 0$, $t_i = t_{i-1} + k_i$, $v_i \in X_n$. (Of course, the values N, k_i and v_i depend on n). The discrete time derivative $\eth_t u_n$ is defined a.e. by*

$$\eth_t u_n(t) = \frac{v_i - v_{i-1}}{k_i} \text{ for } t \in]t_{i-1}, t_i[.$$

2. *The sequence $(u_n)_{n\in\mathbb{N}}$ is bounded in $L^p(]0,T[,B)$.*
3. *The sequence $(\|u_n\|_{L^p(]0,T[,X_n)})_{n\in\mathbb{N}}$ is bounded.*
4. *The sequence $(\|\delta_t u_n\|_{L^1(]0,T[,Y_n)})_{n\in\mathbb{N}}$ is bounded.*

Then, there exists a $u \in L^p(]0,T[,B)$ such that, up to a subsequence, $u_n \to u$ in $L^p(]0,T[,B)$.

Proof The beginning of the proof closely follows that of Theorem 4.55. As for the latter, we use Proposition 4.53 and it only remains to prove the third hypothesis on $(u_n)_{n\in\mathbb{N}}$ of this proposition, that is

$$\int_0^{T-h} \|u_n(\cdot + h) - u_n\|_B^p \, dt \to 0 \text{ as } h \to 0, \text{ uniformly with respect to } n \in \mathbb{N}.$$

Here too, since $u_n \in L^p(]0,T[,B)$, we have $\int_0^{T-h} \|u_n(\cdot + h) - u_n\|_B^p \, dt \to 0$ as $h \to 0$ for all $n \in \mathbb{N}$. The only difficulty is to prove the uniformity of this convergence with respect to $n \in \mathbb{N}$. Thus, it suffices to prove that for all $\eta > 0$ there exist $n_0 \in \mathbb{N}$ and $0 < h_0 < T$ such that

$$n \geq n_0, \, 0 < h \leq h_0 \Rightarrow \int_0^{T-h} \|u_n(\cdot + h) - u_n\|_B^p \, dt \leq \eta. \tag{4.57}$$

Let $\varepsilon > 0$; Lemma 4.56 gives the existence of $n_0 \in \mathbb{N}$ and $C_\varepsilon \in \mathbb{R}$ such that

$$n \geq n_0, \, u \in X_n \Rightarrow \|u\|_B \leq \varepsilon \|u\|_{X_n} + C_\varepsilon \|u\|_{Y_n}.$$

For $n \geq n_0$, $0 < h < T$ and $t \in]0, T - h[$, we have

$$\|u_n(t + h) - u_n(t)\|_B \leq \varepsilon \|u_n(t + h) - u_n(t)\|_{X_n} + C_\varepsilon \|u_n(t + h) - u_n(t)\|_{Y_n}$$
$$\leq \varepsilon \|u_n(t + h)\|_{X_n} + \varepsilon \|u_n(t)\|_{X_n} + C_\varepsilon \|u_n(t + h) - u_n(t)\|_{Y_n},$$

and so

$$\|u_n(t + h) - u_n(t)\|_B^p \leq$$
$$(3\varepsilon)^p \|u_n(t + h)\|_{X_n}^p + (3\varepsilon)^p \|u_n(t)\|_{X_n}^p + (3C_\varepsilon)^p \|u_n(t + h) - u_n(t)\|_{Y_n}^p.$$

Integrating this inequality with respect to t leads to

$$\int_0^{T-h} \|u_n(t + h) - u_n(t)\|_B^p \, dt \leq$$
$$2(3\varepsilon)^p \|u_n\|_{L^p(]0,T[,X_n)}^p + (3C_\varepsilon)^p \int_0^{T-h} \|u_n(t + h) - u_n(t)\|_{Y_n}^p \, dt. \tag{4.58}$$

The proof now differs from that of Theorem 4.55, because we are dealing with the discrete derivative of u_n instead of the derivative of u_n. Note that for almost all $t \in]0, T - h[$ (n and h are fixed)

$$u_n(t+h) - u_n(t) = \sum_{\substack{i \ \ t_i \in]t,t+h[}} (v_{i+1} - v_i)$$

$$= \sum_{i=1}^{N-1} (v_{i+1} - v_i)\mathbb{1}_{]t,t+h[}(t_i)$$

$$= \sum_{i=1}^{N-1} \frac{v_{i+1} - v_i}{k_{i+1}} k_{i+1}\mathbb{1}_{]t,t+h[}(t_i),$$

where $\mathbb{1}_{]t,t+h[}(t_i) = 1$ if $t_i \in]t,t+h[$ and 0 if $t_i \notin]t,t+h[$.

Let M be a bound of the norm $L^1(]0,T[,Y_n)$ of $\eth_t u_n$: $M \geq \sum_{i=1}^{N-1} \left\| \frac{v_{i+1}-v_i}{k_{i+1}} \right\|_{Y_n} k_{i+1}$ for all n. We then obtain

$$\|u_n(t+h) - u_n(t)\|_{Y_n}^p \leq \Big(\sum_{i=1}^{N-1} \left\| \frac{v_{i+1}-v_i}{k_{i+1}} \right\|_{Y_n} k_{i+1}\mathbb{1}_{]t,t+h[}(t_i)\Big)^p$$

$$\leq M^{p-1}\Big(\sum_{i=1}^{N-1} \left\| \frac{v_{i+1}-v_i}{k_{i+1}} \right\|_{Y_n} k_{i+1}\mathbb{1}_{]t,t+h[}(t_i)\Big)$$

Integrating this inequality with respect to $t \in (0,T-h)$ gives, since $\mathbb{1}_{]t,t+h[}(t_i) = \mathbb{1}_{]t_i-h,t_i[}(t)$,

$$\int_0^{T-h} \|u_n(t+h) - u_n(t)\|_{Y_n}^p \, dt \leq hM^p.$$

Using this inequality in (4.58), we get

$$\int_0^{T-h} \|u_n(t+h) - u_n(t)\|_B^p \, dt \leq 2(3\varepsilon)^p \|u_n\|_{L^p(]0,T[,X_n)}^p + (3C_\varepsilon)^p M^p h. \quad (4.59)$$

We can then conclude as in Theorem 4.55: let $\eta > 0$, we choose $\varepsilon > 0$ so that the first term of the right-hand side of (4.59) is bounded by η (independently of $n \in \mathbb{N}$). This choice of ε imposes n_0 and C_ε. For a given C_ε, there exists an $h_0 \in]0,T[$ such that the second term of the right-hand side of (4.59) is bounded by η (independently of $n \in \mathbb{N}$) if $0 < h < h_0$. Finally, we obtain

$$n \geq n_0, \ 0 < h < h_0 \Rightarrow \int_0^{T-h} \|u_n(t+h) - u_n(t)\|_B^p \, dt \leq 2\eta.$$

This concludes the proof of Theorem 4.57. ∎

Under the assumptions of Theorem 4.55 or Theorem 4.57, another interesting question is to prove an additional regularity for u, namely that $u \in L^p(]0,T[,X)$ where X is a space linked to the spaces X_n (and included in B). The following definition introduces the notion of a B-limit-embedded sequence, which specifies the link between the space X and the spaces X_n. The regularity result for such a sequence is stated in Theorem 4.59.

Definition 4.58 (*B*-limit-embedded sequence). Let B be a Banach space, $(X_n)_{n\in\mathbb{N}}$ a sequence of Banach spaces embedded in B and X a Banach space embedded in B. The sequence $(X_n)_{n\in\mathbb{N}}$ is said to be *B*-limit-embedded in X if there exists a $C \in \mathbb{R}$ such that if u is the limit in B of a subsequence of a sequence $(u_n)_{n\in\mathbb{N}}$ satisfying $u_n \in X_n$ and $\|u_n\|_{X_n} \leq 1$, then $u \in X$ and $\|u\|_X \leq C$.

Theorem 4.59 (Regularity of the limit). *Let* $1 \leq p < +\infty$, $T > 0$, B *be a Banach space,* X *be a Banach space embedded in* B, *and* $(X_n)_{n\in\mathbb{N}}$ *be a sequence of Banach spaces embedded in* B *and* B-*limit-embedded in* X *in the sense of Definition 4.58. For* $n \in \mathbb{N}$, *let* $u_n \in L^p(]0,T[, X_n)$ *such that the sequence* $(\|u_n\|_{L^p(]0,T[,X_n)})_{n\in\mathbb{N}}$ *is bounded and* $u_n \to u$ *in* $L^p(]0,T[, B)$ *as* $n \to +\infty$. *Then* $u \in L^p(]0,T[, X)$.

Proof Since $u_n \to u$ in $L^p(]0,T[, B)$ as $n \to +\infty$, we can assume, up to a subsequence, that $u_n \to u$ in B a.e. Since the sequence $(X_n)_{n\in\mathbb{N}}$ is B-limit-embedded in X, we have obtained, with C given by Definition 4.58,

$$\|u\|_X \leq C \liminf_{n\to+\infty} \|u_n\|_{X_n} \quad \text{a.e.}$$

By Fatou's lemma, we have

$$\int_0^T \|u(t)\|_X^p \, dt \leq C^p \int_0^T \liminf_{n\to+\infty} \|u_n(t)\|_{X_n}^p \, dt \leq C^p \liminf_{n\to+\infty} \int_0^T \|u_n(t)\|_{X_n}^p \, dt.$$

Since the sequence $(\|u_n\|_{L^p(]0,T[,X_n)})_{n\in\mathbb{N}}$ is bounded, it follows that u belongs to $L^p(]0,T[, X)$. $\blacksquare$

4.6 Problem Set

Problem 4.1 (Classical solution in dimension 1 ($\star\star$)). *Solution on page 322.* Let $u_0 \in L^2(]0,1[)$; we are interested here in the following problem:

$$\begin{cases} \partial_t u(x,t) - \partial_{xx}^2 u\,(x,t) = 0, \ x \in]0,1[, \ t \in]0,+\infty[, \\ u(0,t) = u(1,t) = 0, \ t \in]0,+\infty[, \\ u(x,0) = u_0(x), \ x \in [0,1]. \end{cases} \tag{4.60}$$

The notation $\partial_t u$ denotes the derivative of u with respect to t and $\partial_{xx}^2 u$ denotes the second derivative of u with respect to x. We say that u is a classical solution to (4.60) if u satisfies the following three conditions

(c1) u is of class C^2 on $]0,1[\times]0,+\infty[$ and satisfies in the classical sense $\partial_t u - \partial_{xx}^2 u = 0$ at every point of $]0,1[\times]0,+\infty[$,

(c2) for all $t > 0$, the functions u, $\partial_x u$ and $\partial_{xx}^2 u$ are continuous on $[0,1] \times [t,+\infty[$ and $u(0,t) = u(1,t) = 0$,

(c3) $u(\cdot,t) \to u_0$ in $L^2(]0,1[)$ as $t \to 0 \ (t > 0)$.

1. Show that the problem (4.60) admits at most one classical solution. [Suggestion: use the method developed in Section 4.1.]
2. Show that the problem (4.60) admits a classical solution. [Suggestion: use the method developed in Section 4.1 by specifying a suitable Hilbertian basis of $L^2(]0,1[)$, see Problem 2.3.]

Problem 4.2 (Dual of L_E^p (★★★)). *Solution on page 323.* Let $(X, \mathcal{T}, m)$ be a measure space, E a Banach space and $1 < p < +\infty$; we set $p' = \frac{p}{p-1}$.

1. Let $v \in L_{E'}^{p'}(X, \mathcal{T}, m)$ and $u \in L_E^p(X, \mathcal{T}, m)$.

   a. Show that the mapping $x \mapsto \langle v(x), u(x) \rangle_{E',E}$ is m-measurable from X to $\mathbb{R}$.
   b. Show that $\langle v, u \rangle_{E',E} \in L_{\mathbb{R}}^1(X, \mathcal{T}, m)$ and

$$\int |\langle v, u \rangle_{E',E}| \, dm \leq \|v\|_{L_{E'}^{p'}} \|u\|_{L_E^p}.$$

2. Let $v \in L_{E'}^{p'}(X, \mathcal{T}, m)$.

   a. Show that the mapping $T_v : u \mapsto \int \langle v, u \rangle_{E',E} \, dm$ is well defined, linear and continuous from $L_E^p(X, \mathcal{T}, m)$ to $\mathbb{R}$ (and hence $T_v \in L_E^p(X, \mathcal{T}, m)'$).
   b. Show that $\|T_v\|_{L_E^p(X,\mathcal{T},m)'} \leq \|v\|_{L_{E'}^{p'}(X,\mathcal{T},m)}$.
   c. (More difficult question) Show that $\|T_v\|_{L_E^p(X,\mathcal{T},m)'} = \|v\|_{L_{E'}^{p'}(X,\mathcal{T},m)}$.

Problem 4.3 (Weak derivative for a union of domains (★)). *Solution on page 326.* Assume that Ω_1 and Ω_2 are two disjoint open subsets of $\mathbb{R}^N$ ($N \geq 1$) and denote by Ω the interior of the closure of $\Omega_1 \cup \Omega_2$ (the open subset Ω can therefore be different from $\Omega_1 \cup \Omega_2$).

Let $T > 0$ and $f \in L^2(]0, T[, L^2(\Omega))$; for $i = 1, 2$, let f_i be the function obtained by restricting f to Ω_i, so $f_i \in L^2(]0, T[, L^2(\Omega_i))$. We identify (as usual) $L^2(\Omega)$ with its dual and $L^2(\Omega_i)$ with its dual (for $i = 1, 2$).

Assume that $\partial_t f_i \in L^2(]0, T[, H^1(\Omega_i)')$ (for $i = 1, 2$). Show that

$$\partial_t f \in L^2(]0, T[, H^1(\Omega)').$$

Problem 4.4 (On L^2 valued continuity (★★)). *Solution on page 327.* Set $T > 0$ and $\Omega =]0, +\infty[$; identify, as usual, $L^2(\Omega)$ with its dual and assume that $u \in L^2(]0, T[, H^1(\Omega))$ and $\partial_t u \in L^2(]0, T[, H^{-1}(\Omega))$. The notation $u(\cdot, t)$ designates the function $x \mapsto u(x, t)$. Note that $\int_0^T u(\cdot, t)\varphi'(t) \, dt \in H^1(\Omega)$; recall that $\partial_t u \in L^2(]0, T[, H^{-1}(\Omega))$ means that there exists a function v (still denoted by $\partial_t u$ and called the "weak derivative of u") belonging to $L^2(]0, T[, H^{-1}(\Omega))$ such that

$$\langle \partial_t u, \varphi \rangle_{\mathcal{D}_{H^1(\Omega)}^\star, \mathcal{D}} = -\int_0^T u(\cdot, t)\varphi'(t) \, dt = \int_0^T v(\cdot, t)\varphi(t) \, dt.$$

Note that $\int_0^T v(\cdot, t)\varphi(t) \, dt \in H^{-1}(\Omega)$, and that the equality makes sense because $H^1(\Omega) \subset L^2(\Omega) = L^2(\Omega)' \subset H^{-1}(\Omega)$. The purpose is now to show that $u \in$

$C([0,T], L^2(\Omega))$. Lemma 4.27 does not apply directly because $H^{-1}(\Omega)$ is not the dual space of $H^1(\Omega)$ (denoted by $H^1(\Omega)'$). Also note that the mapping which associates to $T \in H^1(\Omega)'$ its restriction to $H_0^1(\Omega)$, and which is therefore an element of $H^{-1}(\Omega)$, is not injective. The method proposed here consists in applying Lemma 4.27 (with $E = H^1(\mathbb{R})$ and $F = L^2(\mathbb{R})$) using a suitable extension of u. Let $\alpha > 0$, $\beta > 0$ and $\gamma > 0$ satisfying $\alpha - \beta = 1$ and $(\alpha + 1) - \frac{\beta}{\gamma} = 0$ (a possible example is $\alpha = 2, \beta = 1, \gamma = \frac{1}{3}$). Define $\bar{u}$ from $\mathbb{R} \times]0, T[$ by

$$\bar{u}(x,t) = \begin{cases} u(x,t) \text{ if } x \geq 0, \\ \alpha u(-x,t) - \beta u(-\gamma x, t) \text{ if } x < 0. \end{cases}$$

1. Show that $\bar{u} \in L^2(]0,T[, H^1(\mathbb{R}))$.
2. The aim here is to derive a crucial equality for the computation of $\partial_t \bar{u}$, that is to say $\int_0^T \bar{u}(\cdot, t)\varphi'(t)\, dt$. Let $\varphi \in \mathcal{D}(]0,T[)$ and $\psi \in \mathcal{D}(\mathbb{R})$. For $x > 0$, let

$$\bar{\psi}(x) = \psi(x) + \alpha\psi(-x) - \frac{\beta}{\gamma}\psi\left(-\frac{x}{\gamma}\right).$$

a. Show that

$$\left\langle \int_0^T \bar{u}(\cdot, t)\varphi'(t)\, dt, \psi \right\rangle_{H^{-1}(\mathbb{R}), H^1(\mathbb{R})} =$$

$$\int_0^T \left(\int_0^\infty u(x,t)\bar{\psi}(x)\, dx \right)\varphi'(t)\, dt.$$

[Use Proposition 4.25 then some changes of variables.]
b. Show that $\bar{\psi} \in H_0^1(\Omega)$ and that

$$\left\langle \int_0^T \bar{u}(\cdot, t)\varphi'(t)\, dt, \psi \right\rangle_{H^{-1}(\mathbb{R}), H^1(\mathbb{R})} =$$

$$\int_0^T \langle v(\cdot, t), \bar{\psi} \rangle_{H^{-1}(\Omega), H_0^1(\Omega)}\varphi(t)\, dt,$$

where $v = \partial_t u$ (and therefore $v \in L^2(]0,T[, H^{-1}(\Omega)))$.

3. Show that $\partial_t \bar{u} \in L^2(]0,T[, H^{-1}(\mathbb{R}))$. Deduce that $\bar{u} \in C([0,T], L^2(\mathbb{R}))$ and that $u \in C([0,T], L^2(\Omega))$.

N.B. By a local maps argument (see for example [11, Annexe C1]), the result proved here remains true if Ω is an open bounded subset of $\mathbb{R}^N$, $N \geq 1$, with a strongly Lipschitz boundary (Remark 1.24) [31].

Problem 4.5 (Non-homogeneous and non-isotropic diffusion ($\star\star\star\star$)). *Solution on page 329.* We use the hypotheses of Theorem 4.30. Let Ω be an open bounded subset of $\mathbb{R}^N$, $T > 0$ and $u_0 \in L^2(\Omega)$, identify $L^2(\Omega)$ with its dual space and assume that $f \in L^2(]0,T[, H^{-1}(\Omega))$.

Let A be a mapping from Ω to $\mathcal{M}_N(\mathbb{R})$ (the set of real square matrices with N rows and N columns). The coefficients of A are assumed to belong to $L^\infty(\Omega)$ and are such that there exists an $\alpha > 0$ such that

$$A\xi \cdot \xi \geq \alpha|\xi|^2 \text{ a.e., for all } \xi \in \mathbb{R}^N.$$

Following the proof of Theorem 4.30, show that there exists a unique function u such that

$$\begin{cases} u \in L^2(]0,T[,H_0^1(\Omega)), \partial_t u \in L^2(]0,T[,H^{-1}(\Omega)), \\ \displaystyle\int_0^T \langle \partial_t u(s), v(s)\rangle_{H^{-1},H_0^1}\, ds + \int_0^T \Big(\int_\Omega A\nabla u(s)\cdot\nabla v(s)\, dx\Big)\, ds = \\ \qquad\displaystyle\int_0^T \langle f(s), v(s)\rangle_{H^{-1},H_0^1}\, ds \ \text{ for all } v \in L^2(]0,T[,H_0^1(\Omega)), \\ u(0) = u_0 \text{ a.e.} \end{cases} \qquad (4.61)$$

Problem 4.6 (Existence by Schauder's theorem ($\star\star\star\star$)). *Solution on page 336.* Let Ω be an open bounded subset of $\mathbb{R}^N$ and $A : \mathbb{R} \to \mathcal{M}_N(\mathbb{R})$ (where $\mathcal{M}_N(\mathbb{R})$ denotes the $N \times N$ matrices with real coefficients) such that

$$\forall s \in \mathbb{R}, A(s) = (a_{i,j}(s))_{i,j=1,\dots,N} \text{ where } a_{i,j} \in L^\infty(\mathbb{R}) \cap C(\mathbb{R},\mathbb{R}), \qquad (4.62)$$

$$\exists \alpha > 0 \ A(s)\xi \cdot \xi \geq \alpha|\xi|^2, \forall \xi \in \mathbb{R}^N, \forall s \in \mathbb{R}, \qquad (4.63)$$

$$f \in L^2(]0,T[,H^{-1}(\Omega)) \text{ and } u_0 \in L^2(\Omega). \qquad (4.64)$$

As usual $L^2(\Omega)$ is identified with $L^2(\Omega)'$. The objective here is to prove the existence of a solution to problem (4.39).

Let $\bar{u} \in L^2(]0,T[,L^2(\Omega))$, and let $T : L^2(]0,T[,L^2(\Omega)) \to L^2(]0,T[,L^2(\Omega))$ be defined by $T(\bar{u}) = u$ where u is *the* solution (given by Problem 4.5) of the problem

$$\begin{cases} u \in L^2(]0,T[,H_0^1(\Omega)), \partial_t u \in L^2(]0,T[,H^{-1}(\Omega)), \\ \displaystyle\int_0^T \langle \partial_t u, v\rangle_{H^{-1},H_0^1}\, dt + \int_0^T \int_\Omega A(\bar{u})\nabla u \cdot \nabla v\, dx\, dt \\ \qquad = \displaystyle\int_0^T \langle f, v\rangle_{H^{-1},H_0^1}\, dt, \ \forall v \in L^2(]0,T[,H_0^1(\Omega)), \\ u(\cdot,0) = u_0. \end{cases}$$

1. Show that T is continuous from $L^2(]0,T[,L^2(\Omega))$ to $L^2(]0,T[,L^2(\Omega))$.
2. Show that T is compact from $L^2(]0,T[,L^2(\Omega))$ to $L^2(]0,T[,L^2(\Omega))$.
3. Show that there exists an $R > 0$ such that $\|T(\bar{u})\|_{L^2(]0,T[,L^2(\Omega))} \leq R$ for all $u \in L^2(]0,T[,L^2(\Omega))$.
4. Show that there exists a solution u to (4.39).
5. We now also assume that $a_{i,j}$ is, for all i,j, a Lipschitz-continuous function. Show that (4.39) has a unique solution.

Problem 4.7 (Example of a compactly embedded sequence ($\star$)). *Solution on page 338.* For $m \in \mathbb{N}^\star$, let $h = \frac{1}{m}$, $x_i = ih$ for $i \in \{0, \ldots, m\}$,

$$X_m = \{u \in L^2(]0, 1[), \ u(x) = u_i \in \mathbb{R} \text{ for all } x \in]x_{i-1}, x_i[, \ i \in \{1, \ldots, m\}\}$$

and for $u \in X_m$,

$$\|u\|_{X_m}^2 = \sum_{i=1}^{m-1} \frac{1}{h} |u_{i+1} - u_i|^2 + \|u\|_{L^2(]0,1[)}^2 .$$

Using Theorem 4.44, show that the sequence $(X_m)_{m\in\mathbb{N}^\star}$ is compactly embedded in $L^2(]0, 1[)$ in the sense from Definition 4.52.

Problem 4.8 (Kolmogorov's theorem, with $B = \mathbb{R}$ ($\star\star$)). *Solution on page 339.* The aim of this problem is to give a proof of Theorem 4.44 in the (simpler) case $B = \mathbb{R}$ and $p = 1$. Let $T > 0$; in the following the space $L^1(]0, T[, \mathbb{R})$ is denoted by L^1. Let $(u_n)_{n\in\mathbb{N}}$ be a bounded sequence in L^1 (so we have $\sup_{n\in\mathbb{N}} \|u_n\|_1 < +\infty$) and assume that for all $h \in]0, T[$ and all $n \in \mathbb{N}$,

$$\int_0^{T-h} |u_n(t + h) - u_n(t)| dt \leq \eta(h),$$

where η is a non-decreasing function from $]0, T[$ to $\mathbb{R}_+$ such that $\lim_{h\to 0^+} \eta(h) = 0$. The objective is to prove that the sequence $(u_n)_{n\in\mathbb{N}}$ is relatively compact in L^1.

1. Let $\delta, h \in]0, T[$ such that $\delta + h \leq T$. Show that

$$\int_0^\delta |u_n(t)| dt \leq \int_0^\delta |u_n(t + h)| dt + \int_0^\delta |u_n(t + h) - u_n(t)| dt. \qquad (4.65)$$

2. Let $h_0 \in]0, T[$ and $\delta \in]0, T - h_0[$, show that

$$h_0 \int_0^\delta |u_n(t)| dt \leq \delta \|u_n\|_1 + h_0 \eta(h_0). \qquad (4.66)$$

3. Show that $\int_0^\delta |u_n(t)| dt \to 0$ as $\delta \to 0^+$, uniformly with respect to n.
4. Show that the sequence $(u_n)_{n\in\mathbb{N}}$ is relatively compact in L^1. [Apply Kolmogorov's theorem, Theorem 4.43, using the extension of u_n by 0.]

Problem 4.9 (Existence for the Stefan problem by a numerical scheme ($\star\star\star$)). *Solution on page 341.* The idea here is to prove the existence of a weak solution to a non-linear parabolic problem by passing to the limit on a sequence of approximate solutions given by a numerical scheme. We consider the following problem.

$$\partial_t u(x, t) - \partial_{xx}^2 \varphi(u)(x, t) = v(x, t), \ x \in]0, 1[, \ t \in]0, T[, \qquad (4.67a)$$
$$\partial_x \varphi(u)(0, t) = \partial_x \varphi(u)(1, t) = 0, \ t \in]0, T[, \qquad (4.67b)$$
$$u(x, 0) = u_0(x), \ x \in]0, 1[, \qquad (4.67c)$$

where ∂_t (resp. ∂_x, ∂^2_{xx}) denotes the first order derivative with respect to t (resp. first order and second order with respect to x), and where φ, v, T, u_0 are given and are such that

1. $T > 0$, $v \in L^\infty(]0, 1[\times]0, T[)$,
2. φ is non-decreasing, Lipschitz-continuous from $\mathbb{R}$ to $\mathbb{R}$,
3. u_0 is Lipschitz-continuous from $[0, 1]$ to $\mathbb{R}$ (and therefore, in particular, $u_0 \in L^\infty(]0, 1[)$).

An important example is given by $\varphi(s) = \alpha_1 s$ if $s \leq 0$, $\varphi(s) = 0$ if $0 \leq s \leq L$ and $\varphi(s) = \alpha_2(s - L)$ if $s \geq L$, with α_1, α_2 and L given in $\mathbb{R}^\star_+$. Note that for this example, $\varphi' = 0$ on $]0, L[$.

The sets $]0, 1[$ and $]0, 1[\times]0, T[$ are equipped with their Borel σ-algebra and the Lebesgue measure on this σ-algebra. A weak solution to (4.67) is a function u satisfying

$$u \in L^\infty(]0, 1[\times]0, T[), \tag{4.68a}$$

$$\int_0^T \int_0^1 \left[u(x, t)\partial_t\psi(x, t) + \varphi(u(x, t))\partial^2_{xx}\psi(x, t) + v(x, t)\psi(x, t) \right] dx\, dt$$

$$+ \int_0^1 u_0(x)\psi(x, 0)\, dx = 0, \quad \forall \psi \in C^\infty_T(\mathbb{R}^2), \tag{4.68b}$$

where

$$C^\infty_T(\mathbb{R}^2) = \{\psi \in C^\infty(\mathbb{R}^2, \mathbb{R}) \quad \partial_x\psi(0, t) = \partial_x\psi(1, t) = 0 \text{ for all } t \in [0, T]$$
$$\text{and } \psi(x, T) = 0 \text{ for all } x \in [0, 1]\}. \tag{4.69}$$

1. (This question is independent of the following ones.) Assume, in this question only, that φ is of class C^2, v is continuous on $[0, 1]\times[0, T]$ and $u_0 \in C^2([0, 1], \mathbb{R})$. Let $w \in C^2(\mathbb{R}^2, \mathbb{R})$ and u the restriction of w to $]0, 1[\times]0, T[$; show that u is a solution to (4.68) if and only if u satisfies (4.67) in the classical sense (that is to say, for all $(x, t) \in [0, 1] \times [0, T]$).
2. (Passage to the limit in a non-linearity) Let $(u_n)_{n\in\mathbb{N}}$ be a bounded sequence in $L^\infty(]0, 1[\times]0, T[)$. Let $u \in L^\infty(]0, 1[\times]0, T[)$ and $f \in L^1(]0, 1[\times]0, T[)$. Assume that as $n \to +\infty$,

   a. $u_n \to u$ $\star$-weakly in $L^\infty(]0, 1[\times]0, T[)$,
   b. $\varphi(u_n) \to f$ in $L^1(]0, 1[\times]0, T[)$.

   (Recall that $\varphi(u_n)$ is the usual though incorrect notation for $\varphi \circ u_n$.) Show that $\int_0^T \int_0^1 \varphi(u_n)u_n\, dx\, dt \to \int_0^T \int_0^1 fu\, dx\, dt$ as $n \to +\infty$. Deduce that $\varphi(u) = f$ a.e. on $]0, 1[\times]0, T[$. [Hint: Use Minty's trick, described in Section 3.2.1 page 198 or in Problem 3.10]

   Let us now look for an approximate solution to (4.67). Let $N, M \in \mathbb{N}^\star$, and set $h = \dfrac{1}{N}$ and $k = \dfrac{T}{M}$. An approximate solution of (4.67) is constructed starting from the family $\{u_i^n, \; i = 1, \ldots, N, \; n = 0, \ldots, M\}$ satisfying the following equations

$$u_i^0 = \frac{1}{h} \int_{(i-1)h}^{ih} u_0(x) \, dx, \text{ for all } i = 1, \ldots, N, \tag{4.70a}$$

$$\frac{u_i^{n+1} - u_i^n}{k} - \frac{\varphi(u_{i-1}^{n+1}) - 2\varphi(u_i^{n+1}) + \varphi(u_{i+1}^{n+1})}{h^2} = v_i^n, \tag{4.70b}$$

$$\text{for all } i = 1, \ldots, N, \text{ for all } n = 0, \ldots, M - 1,$$

$$u_0^{n+1} = u_1^{n+1}, \quad u_{N+1}^{n+1} = u_N^{n+1} \text{ for all } n = 0, \ldots, M - 1, \tag{4.70c}$$

with $v_i^n = \dfrac{1}{kh} \displaystyle\int_{nk}^{(n+1)k} \int_{(i-1)h}^{ih} v(x, t) \, dx \, dt$, for all $i = 1, \ldots, N$, for all $n = 0, \ldots, M$.

(Note that this family is obtained by a classical finite difference discretisation in space and by the implicit Euler scheme in time.)

3. (Existence and uniqueness of the approximate solution.) Let $n \in \{0, \ldots, M - 1\}$ and assume the family $\{u_i^n, \; i = 1, \ldots, N\}$ is known. Let us now prove the existence and uniqueness of the family $\{u_i^{n+1}, \; i = 1, \ldots, N\}$ satisfying (4.70) with this value of n.

  a. Let $a > 0$. For $s \in \mathbb{R}$, set $g_a(s) = s + a\varphi(s)$. Show that g_a is an increasing bijective mapping from $\mathbb{R}$ to $\mathbb{R}$.

  b. Let $\overline{w} = (\overline{w}_i)_{i=1,\ldots,N} \in \mathbb{R}^N$ and set $\overline{w}_0 = \overline{w}_1$ and $\overline{w}_{N+1} = \overline{w}_N$. Show that there exists one and only one pair $(u, w) \in \mathbb{R}^N \times \mathbb{R}^N$, $u = (u_i)_{i=1,\ldots,N}$, $w = (w_i)_{i=1,\ldots,N}$ such that:

$$\varphi(u_i) = w_i, \; \forall i \in \{1, \ldots, N\}, \tag{4.71}$$

$$u_i + \frac{2k}{h^2} w_i = \frac{k}{h^2} (\overline{w}_{i-1} + \overline{w}_{i+1}) + u_i^n + kv_i^n, \; \forall i = 1, \ldots, N. \tag{4.72}$$

  c. The space $\mathbb{R}^N$ is equipped with the usual norm $\|\cdot\|_\infty$. Show that the mapping F is strictly contracting. [Suggestion: use the monotonicity of φ and note that, if $a = \varphi(\alpha)$ and $b = \varphi(\beta)$ then $|\alpha - \beta| \geq \frac{1}{L_\varphi}|a - b|$, where L_φ only depends on φ.]

  d. Let $\{u_i^{n+1}, \; i = 0, \ldots, N + 1\}$ be a solution to (4.70). Set $w = (w_i)_{i=1,\ldots,N}$, with $w_i = \varphi(u_i^{n+1})$ for $i \in \{1, \ldots, N\}$; show that $w = F(w)$.

  e. Let $w = (w_i)_{i=1,\ldots,N}$ such that $w = F(w)$. Show that there exists a solution $\{u_i^{n+1}, \; i = 0, \ldots, N + 1\}$ to (4.70) with $w_i = \varphi(u_i^{n+1})$ for all $i \in \{1, \ldots, N\}$.

  f. Show that there exists a unique solution $\{u_i^{n+1}, \; i = 0, \ldots, N + 1\}$ to (4.70).

The next three questions provide estimates on the approximate solution whose existence and uniqueness we have just proved.

4. ($L^\infty(]0,1[\times]0,T[)$ estimate on the approximate solution)
   Set $A = \|u_0\|_{L^\infty(]0,1[)}$ and $B = \|v\|_{L^\infty(]0,1[\times]0,T[)}$. Show, by induction on n, that
   $u_i^n \in [-A - nkB, A + nkB]$ for all $i = 1, \ldots, N$ and all $n = 0, \ldots, M$. [One could,
   for example, consider (4.70b) with i such that $u_i^{n+1} = \min\{u_j^{n+1}, j = 1, \ldots, N\}$.]

5. ($L^2(]0,T[, H_d^1)$ estimate of $\varphi(u)$, where H_d^1 is a discrete equivalent of the space
   H^1) Show that there exists a C_{T,φ,v,u_0} (depending only on T, φ, v and u_0) such
   that, for all $n = 0, \ldots, M - 1$,

$$\sum_{n=0}^{M-1} \sum_{i=1}^{N-1} (\varphi(u_{i+1}^{n+1}) - \varphi(u_i^{n+1}))^2 \le C_{T,\varphi,v,u_0} \frac{h}{k}. \tag{4.73}$$

   [Hint: multiply (4.70b) by u_i^{n+1} and sum over $i = 1, \ldots, N$ and $n = 0, \ldots, M$.]

6. ($L^2(]0,T[, L^2)$ estimate of a discrete equivalent of $\partial_t \varphi(u)$ and $L^\infty(]0,T[, H_d^1)$
   estimate of $\varphi(u)$) It is for these estimates that we use the fact that u_0 is Lipschitz-
   continuous (the previous estimates only used $u_0 \in L^\infty(]0,1[)$).

   a. Preliminary question. Let $\{w_i^n, i = 0, \ldots, N + 1, n = 0, \ldots, M\}$ be a family of
      real numbers such that $w_0^n = w_1^n$ and $w_N^n = w_{N+1}^n$ for all $n = 0, \ldots, M$. Show
      that

$$\sum_{n=0}^{M-1} \sum_{i=1}^{N} (2w_i^{n+1} - w_{i-1}^{n+1} - w_{i+1}^{n+1})(w_i^{n+1} - w_i^n)$$

$$\ge \frac{1}{2}\left[\sum_{i=1}^{N-1} (w_{i+1}^M - w_i^M)^2 - \sum_{i=1}^{N-1} (w_{i+1}^0 - w_i^0)^2 \right]. \tag{4.74}$$

      [Hint: notice that this formula is the discrete counterpart (with $\ge$ instead of
      $=$), of the equality

$$-\int_0^T \int_0^1 \partial_{xx} w(x,t)\, \partial_t w(x,t)\, dx\, dt =$$

$$\frac{1}{2}\left(\int_0^1 (\partial_x w(x,T))^2\, dx - \int_0^1 \partial_x(w(x,0))^2\, dx \right)$$

      for regular functions w satisfying the homogeneous Neumann conditions at 0
      and 1, and adapt the proof of this equality.]

   b. ($L^2(]0,T[, L^2)$ estimate of $\partial_t \varphi(u)$) Show that there exists a C_2 (depending
      only on T, φ, v and u_0) such that

$$\sum_{n=0}^{M-1} h \sum_{i=1}^{N} (\varphi(u_i^{n+1}) - \varphi(u_i^n))^2 \le C_2 k. \tag{4.75}$$

      [Hint: multiply (4.70b) by $\varphi(u_i^{n+1}) - \varphi(u_i^n)$, sum over i and n and use the
      preliminary question with $w_i^n = \varphi(u_i^n)$.]

c. $(L^\infty(]0, T[, H_d^1)$ estimate of $\varphi(u))$ Show that there exists a C_3 (depending only on T, φ, v and u_0) such that, for all $n = 1, \ldots, M$

$$\sum_{i=1}^{N-1} (\varphi(u_{i+1}^n) - \varphi(u_i^n))^2 \le C_3 h. \tag{4.76}$$

This estimate is interesting *per se* but is not used in the rest of the problem.

For given $M \in \mathbb{N}^\star$ and $N \in \mathbb{N}^\star$ (and thus h and k are given by $h = \frac{1}{N}$ and $M = \frac{k}{M}$), we define, a.e. on $[0, 1] \times [0, T]$, for a solution $\{u_i^n, i \in \{1, \ldots, N\}$ $n \in \{0, \ldots, M\}\}$ to (4.70), a function u by

$$u(x, t) = u^{(n+1)}(x), \text{ if } t \in]nk, (n + 1)k[$$

and
$$u^{(n)}(x) = u_i^n, \text{ if } x \in](i - 1)h, ih[, \ i = 1, \ldots, N, \ n = 0, \ldots, M.$$

The objective is now to pass to the limit in the discretisation parameters to obtain a solution to (4.68). Let $(h_n, k_n)_{n \in \mathbb{N}}$ be a sequence of discretisation parameters (and thus $N_n = \frac{1}{h_n}$, $M_n = \frac{T}{k_n} \in \mathbb{N}^\star$) such that $\lim_{n \to +\infty} h_n = \lim_{n \to +\infty} k_n = 0$. Let u_n be the solution (4.70) obtained with these parameters.

7. Show that the sequence $(u_n)_{n \in \mathbb{N}}$ admits a subsequence that converges $\star$-weakly in $L^\infty(]0, 1[\times]0, T[)$ and that the sequence $(\varphi(u_n))_{n \in \mathbb{N}}$ admits a subsequence that converges in $L^2(]0, 1[\times]0, T[)$. If necessary extracting a subsequence, show that as a consequence,

   a. $u_n \to u$ $\star$-weakly in $L^\infty(]0, 1[\times]0, T[)$,
   b. $\varphi(u_n) \to \varphi(u)$ in $L^p(]0, 1[\times]0, T[)$, for all $p \in [1, \infty[$.

8. Show that the function u found in the previous question is a solution to (4.68).

Remark 4.60. The solution to (4.68) can be proved to be unique. Another method to prove the existence of the solution to the problem (4.68) using the approximate solution given by a numerical scheme is described in [23]. It only uses $u_0 \in L^\infty(]0, T[)$ instead of Lipschitz-continuous u_0.

Problem 4.10 (Existence for the Stefan problem by regularisation ($\star\star$)). *Solution on page 352.* Let Ω be an open bounded subset of $\mathbb{R}^N$ ($N \ge 1$), $0 < T < +\infty$, $f \in L^2(]0, T[, (L^2(\Omega))$, $u_0 \in L^2(\Omega)$ and φ a Lipschitz-continuous function from $\mathbb{R}$ to $\mathbb{R}$, assumed to be non-decreasing (but not necessarily increasing, for instance the function φ can be constant over an interval of $\mathbb{R}$ with positive measure). This problem studies an alternative to the classical method of proving the existence of the solution of Alt and Lukhaus [3]; it consists of a regularisation technique, of a (weak) solution to the equation $\partial_t u - \Delta(\varphi(u)) = f$ (in $\Omega \times]0, T[$) with homogeneous Dirichlet boundary conditions and u_0 as the initial condition. More precisely, the function u is a solution in the following sense:

$$
\begin{cases}
u \in L^\infty(]0,T[, L^2(\Omega)), \, \partial_t u \in L^2(]0,T[, H^{-1}(\Omega)), \\[4pt]
u \in C([0,T], H^{-1}(\Omega)), \, \varphi(u) \in L^2(]0,T[, H_0^1(\Omega)), \\[4pt]
\displaystyle \int_0^T \langle \partial_t u(s), v(s)\rangle_{H^{-1}, H_0^1} \, ds + \int_0^T \int_\Omega \nabla\varphi(u(x,s)) \cdot \nabla v(x,s) \, dx \, ds \\[10pt]
\qquad\qquad = \displaystyle \int_0^T \int_\Omega f(x,s) v(x,s) \, dx \, ds, \quad \forall v \in L^2(]0,T[, H_0^1(\Omega)),
\end{cases}
\tag{4.77a}
$$

$$
u(\cdot, 0) = u_0,
\tag{4.77b}
$$

The space $L^2(\Omega)$ is, as usual, identified with $L^2(\Omega)'$. In (4.77), $u(s)$ (resp. $v(s)$) denotes the function $x \mapsto u(x,s)$ (resp. $v(x,s)$). Since $L^2(\Omega)$ is identified with $L^2(\Omega)'$, we can write $H_0^1(\Omega) \subset L^2(\Omega) = L^2(\Omega)' \subset H^{-1}(\Omega)$. The function $\partial_t u$ is the weak derivative of u (Definition 4.23). The fact that $u \in L^2(]0,T[, H^{-1}(\Omega))$ and $\partial_t u \in L^2(]0,T[, H^{-1}(\Omega))$ gives $u \in C([0,T], H^{-1}(\Omega))$ (Lemma 4.26). The function u is defined for all $t \in [0,T]$, which gives meaning to the initial condition $u(0) = u_0$.

To solve this problem, we introduce for $n > 0$ the function φ_n defined by $\varphi_n(s) = \varphi(s) + \dfrac{s}{n}$. We show the existence of a solution to the problem with φ_n instead of φ. Then, we take the limit as $n \to +\infty$.

1. Show that the result of Problem 4.6 gives the existence of a solution u_n of (4.77) with φ_n instead of φ, with the additional regularity $u_n \in L^2(]0,T[, H_0^1(\Omega))$ (and thus $u_n \in C([0,T]), L^2(\Omega))$. [Recall that Lemma 4.35 gives $\nabla\varphi(v) = \varphi'(v)\nabla v$ a.e. if $v \in L^2(]0,T[, H_0^1(\Omega))$.]

The next two questions are devoted to the study of the sequence $(u_n)_{n\in\mathbb{N}^\star}$.

2. Show that the sequence $(\varphi(u_n))_{n\in\mathbb{N}^\star}$ is bounded in $L^2(]0,T[, H_0^1(\Omega))$, that the sequence $(\partial_t u_n)_{n\in\mathbb{N}^\star}$ is bounded in $L^2(]0,T[, H^{-1}(\Omega))$ and that the sequence $(u_n)_{n\in\mathbb{N}^\star}$ is bounded in $C(]0,T[, L^2(\Omega))$. [Lemma 4.35 is still useful here.]
3. Show that there exist u and ζ such that, up to a subsequence, $\varphi(u_n) \to \zeta$ weakly in $L^2(]0,T[, H_0^1(\Omega))$ and $u_n \to u$ in $L^2(]0,T[, H^{-1}(\Omega))$ and weakly in $L^2(]0,T[, L^2(\Omega))$. [For the convergence of u_n, use Theorem 4.46.]

In the following questions, we show that this function u is the sought solution.

4. Show that $\lim_{n\to+\infty} \int_0^T \int_\Omega \varphi(u_n)u_n \, dx \, dt = \int_0^T \int_\Omega \zeta u \, dx \, dt$ as $n \to +\infty$.
5. Use Minty's trick (see for example in Chapter 3 the proof of (3.25)), to show that $\varphi(u) = \zeta$ a.e.
6. Show that $u \in C([0,T], H^{-1})$ and deduce that $u(0) = u_0$. [Then reason as in step 6 of Theorem 4.30.]
7. Show that u is a solution of (4.77).

4.7 Problem Set Solutions

Problem 4.1 (Classical solution in dimension 1)

1. If u and $\bar{u}$ are two classical solutions of (4.60), the function $(u - \bar{u})$ is then a classical solution to (4.60) with $u_0 = 0$ a.e. (on $]0, 1[$). To show that $u = \bar{u}$, it is therefore sufficient to show that the null function is the unique solution to (4.60) when $u_0 = 0$ a.e. Assuming then that u is a classical solution to (4.60) with $u_0 = 0$ a.e., let us show that $u(x, t) = 0$ for all $(x, t) \in [0, 1] \times]0, +\infty[$.

   Let $t > 0$, owing to the hypotheses (c1) and (c2),

   $$\int_0^1 \partial_t u(x, t) u(x, t)\, \mathrm{d}x = \frac{1}{2} \int_0^1 \partial_t (u^2)(x, t)\, \mathrm{d}x = \frac{1}{2} \frac{\mathrm{d}}{\mathrm{d}t} \Big(\int_0^1 u^2(x, t)\, \mathrm{d}x \Big)$$

   and

   $$\int_0^1 \partial_{xx}^2 u\,(x, t) u(x, t)\, \mathrm{d}x = - \int_0^1 \partial_x u(x, t)^2\, \mathrm{d}x.$$

   Since $\partial_t u\, u - \partial_{xx}^2 u\, u = 0$ on $]0, 1[\times]0, +\infty[$, we get that

   $$\frac{1}{2} \frac{\mathrm{d}}{\mathrm{d}t} \Big(\int_0^1 u^2(x, t)\, \mathrm{d}x \Big) + \int_0^1 \partial_x u(x, t)^2\, \mathrm{d}x = 0.$$

   Let $0 < \varepsilon < T < +\infty$. Integrating the previous equation between ε and T yields

   $$\frac{1}{2} \int_0^1 u^2(x, T)\, \mathrm{d}x + \int_0^T \int_0^1 \partial_x u^2(x, t)\, \mathrm{d}x\, \mathrm{d}t = \frac{1}{2} \int_0^1 u(x, \varepsilon)\, \mathrm{d}x,$$

   which gives

   $$\int_0^1 u^2(x, T)\, \mathrm{d}x \leq \int_0^1 u^2(x, \varepsilon)\, \mathrm{d}x.$$

   As $\varepsilon \to 0$, the right-hand side of this inequality tends towards 0 (by (c3)), so we have $\int_0^1 u^2(x, T)\, \mathrm{d}x = 0$, which gives $u(\cdot, T) = 0$ a.e. and therefore $u(x, T) = 0$ for all $x \in [0, 1]$ (since $u(\cdot, T)$ is assumed to be continuous on $[0, 1]$). Since $T > 0$ is arbitrary, we have indeed shown that $u(x, t) = 0$ for all $(x, t) \in [0, 1] \times]0, +\infty[$.

2. We use a result from Problem 2.3. For $n \in \mathbb{N}^\star$, we set $c_n = 2 \int_0^1 u_0(t) \sin(n\pi t)\, \mathrm{d}t$. Problem 2.3. gives that

   $$\Big\| u_0 - \sum_{p=1}^{n} c_p \sin(p\pi \cdot) \Big\|_2 \to 0, \quad \text{as } n \to \infty.$$

(Recall that $\|\cdot\|_2$ designates $\|\cdot\|_{L^2(]0,1[)}$.) Since $|c_n| \leq 2 \|u_0\|_2$ for all $n \in \mathbb{N}^\star$, the sequence $(c_n)_{n \in \mathbb{N}^\star}$ is bounded. * Hence the series

$$\sum_{n=1}^{+\infty} e^{-n^2\pi^2 t} c_n \sin n\pi x$$

is convergent (in $\mathbb{R}$) for all $x \in \mathbb{R}$ and $t > 0$. Moreover, for all $\varepsilon > 0$, the series

$$\sum_{n=1}^{+\infty} n e^{-n^2\pi^2\varepsilon} |c_n| \quad \text{and} \quad \sum_{n=1}^{+\infty} n^2 e^{-n^2\pi^2\varepsilon} |c_n|$$

are convergent. This shows that the function

$$x, t \mapsto \sum_{n=1}^{+\infty} e^{-n^2\pi^2 t} c_n \sin n\pi x$$

is, for all $\varepsilon > 0$, of class C^2 on $\mathbb{R} \times]\varepsilon, +\infty[$ (it is even of class C^∞) and that the series can be differentiated term by term, once in t and twice in x. Hence we may set, for $x \in]0, 1[$ and $t > 0$,

$$u(x, t) = \sum_{n=1}^{+\infty} e^{-n^2\pi^2 t} c_n \sin n\pi x.$$

The function u thus defined indeed satisfies conditions (c1) and (c2).

It remains to show that u satisfies (c3). Let $\varepsilon > 0$; for all $t > 0$ and $n_0 > 0$,

$$\|u(\cdot, t) - u_0\|_2 \le \left\| \sum_{n=1}^{n_0} c_n (1 - e^{-n^2 t^2}) \sin(n\pi\cdot) \right\|_2 + 2 \left\| u_0 - \sum_{n=1}^{n_0} c_n \sin(n\pi\cdot) \right\|_2.$$

We first choose n_0 so that the second term on the right-hand side of this inequality is less than ε. Then, since n_0 is now fixed, we note that there exists a $t_0 > 0$ such that the first term on the right-hand side of this inequality is less than ε as soon as $t \in]0, t_0]$. Hence u indeed satisfies (c3).

Problem 4.2 (Dual of L_E^p)

1. a. We choose for u and v representatives, so that $v \in \mathcal{L}_{E'}^{p'}(X, \mathcal{T}, m)$ and $u \in \mathcal{L}_E^p(X, \mathcal{T}, m)$. The m-measurability of the mapping

$$\langle v, u \rangle_{E',E} : x \mapsto \langle u(x), v(x) \rangle_{E',E}$$

is quite simple to prove (and does not depend on the representatives chosen for u and v). Indeed, it is enough to notice that there exist two sequences of step functions $(v_n)_{n \in \mathbb{N}}$ and $(u_n)_{n \in \mathbb{N}}$ (v_n is a function from X to E' and u_n from X to E) such that $v_n \to v$ a.e. and $u_n \to u$ a.e. as $n \to +\infty$. For all $n \in \mathbb{N}$, the function $\langle v_n, u_n \rangle_{E',E}$ is then a step function from X to $\mathbb{R}$. When $n \to +\infty$, it converges a.e. to the function $\langle v, u \rangle_{E',E}$, which is therefore m-measurable.

b. We then notice that $|\langle v(x), u(x) \rangle_{E',E}| \leq \|v(x)\|_{E'} \|u(x)\|_E$ for all $x \in X$. Integrating with respect to the measure m (we use here the monotonicity of the integral for m-measurable functions with values in $\mathbb{R}$ and Hölder's inequality), we obtain that $\langle v, u \rangle_{E',E} \in \mathcal{L}^1_{\mathbb{R}}(X, \mathcal{T}, m)$ (and therefore $\langle v, u \rangle_{E',E} \in L^1_{\mathbb{R}}(X, \mathcal{T}, m)$ with the usual confusion between $\mathcal{L}^1$ and L^1) and

$$\int |\langle v, u \rangle_{E',E}| \, dm \leq \int \|v\|_{E'} \|u\|_E \, dm \leq \|v\|_{L^{p'}_{E'}} \|u\|_{L^p_E}. \tag{4.78}$$

2. a. The previous question indeed shows that $\langle v, u \rangle_{E',E} \in L^1_E(X, \mathcal{T}, m)$ for all $u \in L^p_E(X, \mathcal{T}, m)$. The mapping $u \mapsto \int \langle v, u \rangle_{E',E} \, dm$ is therefore well defined from $L^p_E(X, \mathcal{T}, m)$ to $\mathbb{R}$. It is trivially linear. Its continuity is a consequence of the inequality (4.78).

b. By definition of the norm in the dual space of a Banach space, we have

$$\|T_v\|_{(L^p_E)'} = \sup\{|T_v(u)|, \, u \in L^p_E \text{ s.t. } \|u\|_{L^p_E} = 1\}.$$

The previous question shows that $|T_v(u)| \leq \|v\|_{L^{p'}_{E'}}$ if $\|u\|_{L^p_E} = 1$, so we have

$$\|T_v\|_{L^p_E(X, \mathcal{T}, m)'} \leq \|v\|_{L^{p'}_{E'}(X, \mathcal{T}, m)}.$$

c. Set $q = p'$. We reason here in two steps. We start by showing the requested equality if v is a step function. Then, we deal with the general case.

Step 1 Assume, in this step, that v is a non-zero step function. So there exists $n \in \mathbb{N}^\star$, a family $A_1, \ldots, A_n$ of elements from the σ-algebra $\mathcal{T}$ and a family $b_1, \ldots, b_n$ of elements from E', non-zero, such that

$$v = \sum_{i=1}^{n} b_i \mathbb{1}_{A_i}.$$

Let $\varepsilon > 0$; from the definition of the norm in E', we get that, for all $i \in \{1, \ldots, n\}$, there exists an $\bar{a}_i \in E$ such that

$$\|\bar{a}_i\|_E = 1 \text{ and } \langle b_i, \bar{a}_i \rangle_{E',E} \geq (1 - \varepsilon) \|b_i\|_{E'}.$$

Set $a_i = \|b_i\|_{E'}^{q-1} \bar{a}_i$ for all $i \in \{1, \ldots, n\}$ and

$$u = \sum_{i=1}^{n} a_i \mathbb{1}_{A_i}.$$

The function u is a step function (from X to E) and we have

$$\int \langle v, u \rangle_{E',E} \, dm = \sum_{i=1}^{n} m(A_i) \langle b_i, a_i \rangle_{E',E}$$

$$\geq (1 - \varepsilon) m(A_i) \, \|b_i\|_{E'}^{q} = (1 - \varepsilon) \, \|v\|_{L_{E'}^{q}}^{q} .$$

On the other hand, we have (since $p(q - 1) = q$)

$$\|u\|_{L_E^p}^{p} = \sum_{i=1}^{n} m(A_i) \, \|b_i\|_{E'}^{p(q-1)} = \sum_{i=1}^{n} m(A_i) \, \|b_i\|_{E'}^{q} = \|v\|_{L_{E'}^{q}}^{q} .$$

Hence

$$\frac{1}{\|u\|_{L_E^p}} \int \langle v, u \rangle_{E',E} \, dm \geq (1 - \varepsilon) \frac{\|v\|_{L_{E'}^{q}}^{q}}{\|v\|_{L_{E'}^{q}}^{\frac{q}{p}}} = (1 - \varepsilon) \, \|v\|_{L_{E'}^{q}} .$$

Setting $\bar{u} = \dfrac{u}{\|u\|_{L_E^p}}$, we therefore have $\|\bar{u}\|_{L_E^p} = 1$ and $\int \langle v, \bar{u} \rangle_{E',E} \, dm \geq$ $(1 - \varepsilon) \, \|v\|_{L_{E'}^{q}}$. This proves that

$$\|T_v\|_{(L_E^p)'} \geq (1 - \varepsilon) \, \|v\|_{L_{E'}^{q}} .$$

Since $\varepsilon > 0$ is arbitrary, owing to the result of question 2(b), we therefore obtain that $\|T_v\|_{(L_E^p)'} = \|v\|_{L_{E'}^{q}}$.

Step 2 We now deal with the general case. Let $v \in L_{E'}^{q}$ be non-zero (if $v = 0$ a.e. the requested equality is trivial). There exists a sequence $(v_n)_{n \in \mathbb{N}}$ of step functions (from X to E') such that $v_n \to v$ in $L_{E'}^{q}$ (this sequence can be constructed as in the definition of L^1, that is, by constructing v_n such that $\|v_n(x)\|_{E'} \leq 2 \, \|v(x)\|_{E'}$ for almost all $x \in E'$ and for all $n \in \mathbb{N}$). Hence, for any $\varepsilon > 0$, there exists a step function w such that $\|w - v\|_{L_{E'}^{p}} \leq \varepsilon$. By step 1, there exists a $u \in L_E^P$ such that $\|u\|_{L_E^p} = 1$ and

$$\int \langle w, u \rangle_{E',E} \, dm \geq \|w\|_{L_{E'}^{q}} - \varepsilon.$$

Owing to the result of question 1,

$$\int \langle v, u \rangle_{E',E} \, dm = \int \langle w, u \rangle_{E',E} \, dm + \int \langle v - w, u \rangle_{E',E} \, dm$$

$$\geq \|w\|_{L_{E'}^{q}} - \varepsilon - \|v - w\|_{L_{E'}^{q}} .$$

Since $\|v - w\|_{L_{E'}^q} \le \varepsilon$ and $\|w\|_{L_{E'}^q} \ge \|v\|_{L_{E'}^q} - \varepsilon$, we therefore have

$$\int \langle v, u \rangle_{E',E} \, \mathrm{d}m \ge \|v\|_{L_{E'}^q} - 3\varepsilon.$$

Since $\varepsilon > 0$ is arbitrary, $\|T_v\|_{(L_E^p)'} \ge \|v\|_{L_{E'}^q}$. Finally, by question 2(b), we indeed have $\|T_v\|_{(L_E^p)'} = \|v\|_{L_{E'}^q}$.

Problem 4.3 (Weak derivative for a union of domains)

Let us show that there exists a $u \in L^2(]0, T[, H^1(\Omega)')$ such that, for all $\varphi \in \mathcal{D}(]0, T[)$,

$$-\int_0^T f(t)\varphi'(t) \, \mathrm{d}t = \int_0^T u(t)\varphi(t) \, \mathrm{d}t.$$

The left-hand side of this equality belongs to $L^2(\Omega)$ and the term on the right belongs to $H^1(\Omega)'$. Since $L^2(\Omega)$ is identified with its dual and $H^1(\Omega)$ is dense in $L^2(\Omega)$, we have $L^2(\Omega) \subset H^1(\Omega)'$ and showing this equality therefore consists in showing that for all $\psi \in H^1(\Omega)$ we have

$$\int_\Omega \left(-\int_0^T f(t)\varphi'(t) \, \mathrm{d}t\right)(x)\psi(x) \, \mathrm{d}x = \langle \int_0^T u(t)\varphi(t) \, \mathrm{d}t, \psi \rangle_{H^1(\Omega)', H^1(\Omega)}. \quad (4.79)$$

We thus seek $u \in L^2(]0, T[, H^1(\Omega)')$ satisfying (4.79) for all $\varphi \in \mathcal{D}(]0, T[)$ and all $\psi \in H^1(\Omega)$.

For $i = 1, 2$, we know that $\partial_t f_i \in L^2(]0, T[, H^1(\Omega_i)')$, hence there exists $u_i \in L^2(]0, T[, H^1(\Omega_i)')$ (which we can confuse with $\partial_t f_i$) such that, for all $\varphi \in \mathcal{D}(]0, T[)$ and all $\psi \in H^1(\Omega_i)$ on a

$$\int_{\Omega_i} \left(-\int_0^T f_i(t)\varphi'(t) \, \mathrm{d}t\right)(x)\psi(x) \, \mathrm{d}x = \langle \int_0^T u_i(t)\varphi(t) \, \mathrm{d}t, \psi \rangle_{H^1(\Omega_i)', H^1(\Omega_i)}. \quad (4.80)$$

Let $\varphi \in \mathcal{D}(]0, T[)$ and $\psi \in H^1(\Omega)$. Let ψ_i be the restriction of ψ to Ω_i. Hence $\psi_i \in H^1(\Omega_i)$ and $\|\psi\|_{H^1(\Omega)}^2 \le \sum_{i=1}^2 \|\psi_i\|_{H^1(\Omega_i)}^2$. Using (4.80) (and Proposition 4.25) we have

$$\int_\Omega \left(-\int_0^T f(t)\varphi'(t) \, \mathrm{d}t\right)(x)\psi(x) \, \mathrm{d}x = \sum_{i=1}^2 \int_{\Omega_i} \left(-\int_0^T f_i(t)\varphi'(t) \, \mathrm{d}t\right)(x)\psi_i(x) \, \mathrm{d}x$$

$$= \sum_{i=1}^2 \langle \int_0^T u_i(t)\varphi(t) \, \mathrm{d}t, \psi_i \rangle_{H^1(\Omega_i)', H^1(\Omega_i)}$$

$$= \int_0^T \sum_{i=1}^2 \langle u_i(t)\varphi(t), \psi_i \rangle_{H^1(\Omega_i)', H^1(\Omega_i)} \, \mathrm{d}t$$

$$= \int_0^T \sum_{i=1}^2 \langle u_i(t), \psi_i \rangle_{H^1(\Omega_i)', H^1(\Omega_i)} \varphi(t) \, dt$$

$$= \int_0^T \langle u(t), \psi \rangle_{H^1(\Omega_i)', H^1(\Omega_i)} \varphi(t) \, dt$$

$$= \langle \int_0^T u(t)\varphi(t) \, dt, \psi \rangle_{H^1(\Omega)', H^1(\Omega)},$$

where $u(t)$ is defined (for almost every t) by

$$\langle u(t), \psi \rangle_{H^1(\Omega)', H^1(\Omega)} = \sum_{i=1}^2 \langle u(t), \psi \rangle_{H^1(\Omega_i)', H^1(\Omega_i)}.$$

Since $u_i \in L^2(]0, T[, H^1(\Omega_i)')$ and $\|\psi\|_{H^1(\Omega)}^2 \leq \sum_{i=1}^2 \|\psi_i\|_{H^1(\Omega_i)}^2$, we indeed have $u \in L^2(]0, T[, H^1(\Omega)')$ and this concludes the proof.

Problem 4.4 (On L^2 valued continuity)

1. The condition $\alpha - \beta = 1$ gives that $\bar{u}(\cdot, t)$ is continuous at the point $x = 0$ (for almost every t). This continuity at 0 then allows us to show that $\bar{u} \in L^2(]0, T[, H^1(\mathbb{R}))$.
2. a. Since $\bar{u} \in L^1(]0, T[, H^1(\mathbb{R}))$, $\int_0^T \bar{u}(\cdot, t)\varphi'(t) \, dt \in H^1(\mathbb{R})$. Thanks to the identification of $L^2(\mathbb{R})$ with its dual, the following continuous embeddings hold:
$$H^1(\mathbb{R}) \subset L^2(\mathbb{R}) = L^2(\mathbb{R})' \subset H^1(\mathbb{R})' = H^{-1}(\mathbb{R}).$$

We now express $\langle \int_0^T \bar{u}(\cdot, t)\varphi'(t) \, dt, \psi \rangle_{H^{-1}(\mathbb{R}), H^1(\mathbb{R})}$ using Proposition 4.25, the fact that $\bar{u}(\cdot, t) \in H^{-1}(\mathbb{R})$ (by the identification of $L^2(\mathbb{R})$ with its dual) and a change of variables,

$$\langle \int_0^T \bar{u}(\cdot, t)\varphi'(t) \, dt, \psi \rangle_{H^{-1}(\mathbb{R}), H^1(\mathbb{R})}$$

$$= \int_0^T \langle \bar{u}(\cdot, t)\varphi'(t), \psi \rangle_{H^{-1}(\mathbb{R}), H^1(\mathbb{R})} \, dt$$

$$= \int_0^T \langle \bar{u}(\cdot, t), \psi \rangle_{H^{-1}(\mathbb{R}), H^1(\mathbb{R})} \varphi'(t) \, dt$$

$$= \int_0^T \left(\int_{\mathbb{R}} \bar{u}(x, t)\psi(x) \, dx \right) \varphi'(t) \, dt$$

$$= \int_0^T \left(\int_0^\infty u(x, t)\psi(x) \, dx + \int_{-\infty}^0 \alpha u(-x, t)\psi(x) \, dx \right.$$

$$\left. - \int_{-\infty}^0 \beta u(-\gamma x, t)\psi(x) \, dx \right) \varphi'(t) \, dt$$

$$
= \int_0^T \Big(\int_0^\infty u(x,t)\psi(x)\,\mathrm{d}x + \int_0^\infty \alpha u(x,t)\psi(-x)\,\mathrm{d}x
$$

$$
- \int_0^\infty \frac{\beta}{\gamma} u(x,t)\psi(-\frac{x}{\gamma})\,\mathrm{d}x \Big)\varphi'(t)\,\mathrm{d}t
$$

$$
= \int_0^T \Big(\int_0^\infty u(x,t)\bar\psi(x)\,\mathrm{d}x \Big)\varphi'(t)\,\mathrm{d}t.
$$

b. The function $\bar\psi$ belongs to $\mathcal{D}(\bar\Omega)$ (that is, it is the restriction to Ω of an element of $\mathcal{D}(\mathbb{R})$) and its trace at $x = 0$ is null thanks to the fact that $(\alpha + 1) - \frac{\beta}{\gamma} = 0$. The function $\bar\psi$ therefore belongs to $H_0^1(\Omega)$ and there exists a $C \in \mathbb{R}_+$, depending only on α, β and γ, such that $\|\bar\psi\|_{H_0^1(\Omega)} \le C \|\psi\|_{H^1(\mathbb{R})}$. In the equality obtained in the previous question, we use the fact that $u(\cdot,t) \in H^{-1}(\Omega)$ (by the identification of $L^2(\Omega)$ with its dual), again Proposition 4.25 and the fact that $\partial_t u = v \in L^2(]0,T[, H^{-1}(\Omega))$. We obtain

$$
\langle \int_0^T \bar u(\cdot,t)\varphi'(t)\,\mathrm{d}t, \psi \rangle_{H^{-1}(\mathbb{R}),H^1(\mathbb{R})} = \int_0^T \Big(\int_0^\infty u(x,t)\bar\psi(x)\,\mathrm{d}x \Big)\varphi'(t)\,\mathrm{d}t
$$

$$
= \int_0^T \langle u(\cdot,t), \bar\psi \rangle_{H^{-1}(\Omega),H_0^1(\Omega)}\varphi'(t)\,\mathrm{d}t = \int_0^T \langle u(\cdot,t)\varphi'(t), \bar\psi \rangle_{H^{-1}(\Omega),H_0^1(\Omega)}\,\mathrm{d}t
$$

$$
= \langle \int_0^T u(\cdot,t)\varphi'(t)\,\mathrm{d}t, \bar\psi \rangle_{H^{-1}(\Omega),H_0^1(\Omega)} = \langle - \int_0^T v(\cdot,t)\varphi(t)\,\mathrm{d}t, \bar\psi \rangle_{H^{-1}(\Omega),H_0^1(\Omega)}
$$

$$
= - \int_0^T \langle v(\cdot,t), \bar\psi \rangle_{H^{-1}(\Omega),H_0^1(\Omega)}\varphi(t)\,\mathrm{d}t.
$$

3. For $\varphi \in \mathcal{D}(]0,T[)$ and $\psi \in \mathcal{D}(\mathbb{R})$, let $\psi \otimes \varphi$ denote the function $(x,t) \mapsto \psi(x)\varphi(t)$, and let S be the mapping from $\mathcal{D}(\mathbb{R}) \times \mathcal{D}(]0,T[)$ to $\mathbb{R}$ defined by

$$
S(\psi \otimes \varphi) = \int_0^T \langle v(\cdot,t), \bar\psi \rangle_{H^{-1}(\Omega),H_0^1(\Omega)}\varphi(t)\,\mathrm{d}t,
$$

where $\bar\psi$ is defined as in question 2. By linearity, S extends to the vector space G generated by the functions $\psi \otimes \varphi$. Hence, if $\xi \in G$, $\xi(x,t) = \sum_{i=1}^n a_i\varphi_i(t)\psi_i(x)$, with $n \in \mathbb{N}$, $\varphi \in \mathcal{D}(]0,T[)$ and $\psi_i \in \mathcal{D}(\mathbb{R})$ (for all i), and

$$
S(\xi) = \int_0^T \langle v(\cdot,t), \sum_{i=1}^n a_i\varphi_i(t)\bar\psi_i \rangle_{H^{-1}(\Omega),H_0^1(\Omega)}\,\mathrm{d}t.
$$

We have, for all t, with C given in question 2b,

$$
\Big\| \sum_{i=1}^n a_i\varphi_i(t)\bar\psi_i \Big\|_{H_0^1(\Omega)} \le C \Big\| \sum_{i=1}^n a_i\varphi_i(t)\psi_i \Big\|_{H^1(\mathbb{R})} = \|\xi(\cdot,t)\|_{H^1(\mathbb{R})}.
$$

Hence

$$|S(\xi)| \le \int_0^T \|v(\cdot,t)\|_{H^{-1}(\Omega)} \left\| \sum_{i=1}^n a_i \varphi_i(t)\bar\psi_i \right\|_{H_0^1(\Omega)} \, \mathrm{d}t$$

$$\le C \int_0^T \|v(\cdot,t)\|_{H^{-1}(\Omega)} \, \|\xi(\cdot,t)\|_{H^1(\mathbb{R})} \, \mathrm{d}t$$

$$\le C \, \|v\|_{L^2(]0,T[,H^{-1}(\Omega))} \, \|\xi\|_{L^2(]0,T[,H^1(\mathbb{R}))} \, .$$

The mapping S therefore extends to a continuous linear mapping from $L^2(]0,T[,$ $H^1(\mathbb{R}))$ to $\mathbb{R}$ (this extension is even unique because G is dense in $L^2(]0,T[,$ $H^1(\mathbb{R}))$). This shows that for any $\xi \in G$, there exists a $w \in L^2(]0,T[,(H^1(\mathbb{R})')$ (which is the dual space of $L^2(]0,T[,H^1(\mathbb{R})))$ such that

$$S(\xi) = \int_0^T \langle w(\cdot,t), \xi(\cdot,t)\rangle_{H^{-1}(\mathbb{R}),H^1(\mathbb{R})} \, \mathrm{d}t.$$

In particular this gives for all $\varphi \in \mathcal{D}(]0,T[)$ and $\psi \in \mathcal{D}(\mathbb{R})$,

$$\left\langle \int_0^T \bar u(\cdot,t)\varphi'(t)\,\mathrm{d}t, \psi \right\rangle_{H^{-1}(\mathbb{R}),H^1(\mathbb{R})} = -S(\psi \otimes \varphi)$$

$$= - \int_0^T \langle w(\cdot,t), \varphi(t)\psi\rangle_{H^{-1}(\mathbb{R}),H^1(\mathbb{R})} \, \mathrm{d}t$$

$$= \left\langle - \int_0^T w(t)\varphi(t)\,\mathrm{d}t, \psi \right\rangle_{H^{-1}(\mathbb{R}),H^1(\mathbb{R})} \, .$$

Hence $\int_0^T \bar u(\cdot,t)\varphi'(t)\,\mathrm{d}t = - \int_0^T w(t)\varphi(t)\,\mathrm{d}t$ for all $\varphi \in \mathcal{D}(]0,T[)$, that is to say

$$\partial_t \bar u = w \in L^2(]0,T[,(H^1(\mathbb{R})').$$

Lemma 4.27 then gives $\bar u \in C([0,T],L^2(\mathbb{R}))$ and therefore

$$u \in C([0,T],L^2(\Omega)).$$

Problem 4.5 (Non-homogeneous and non-isotropic diffusion)

If A is symmetric, the simplest approach is to use a Hilbert basis of $L^2(\Omega)$ formed by eigenfunctions of the operator $u \mapsto -\mathrm{div}(A\nabla u)$ (with Dirichlet condition, that is to say, $u = 0$ on $\partial\Omega$). We saw in Section 2.2 that such a basis existed. Consider then a family $\{e_n, n \in \mathbb{N}^\star\}$ such that e_n is (for all n) a weak solution of

$$\begin{cases} -\mathrm{div}\, A\nabla e_n = \lambda_n e_n & \text{in } \Omega, \\ e_n = 0 & \text{on } \partial\Omega, \end{cases}$$

with $\lambda_n \in \mathbb{R}$. The proof of the existence (and uniqueness) of a solution to (4.61) is then very similar to that given for the case where A is the identity matrix.

The case where A is non-symmetric is more difficult because we do not necessarily have a Hilbert basis of $L^2(\Omega)$ formed by eigenfunctions of the operator $u \mapsto -\operatorname{div}(A\nabla u)$. It is then necessary to slightly modify the proof. We then consider the Hilbert basis $\{e_n, n \in \mathbb{N}^\star\}$ associated to the Laplacian (with Dirichlet condition). For $n \in \mathbb{N}^\star$, the function e_n is therefore a weak (non-zero) solution of

$$-\Delta e_n = \lambda_n e_n \text{ in } \Omega,$$
$$e_n = 0 \text{ on } \partial\Omega.$$

Recall that $\|e_n\|_{L^2(\Omega)} = 1, \lambda_n > 0$ and $\lim_{n \to +\infty} \lambda_n = +\infty$. We also have $\|e_n\|_{H_0^1(\Omega)} = \sqrt{\lambda_n}$ and the family $\{\frac{e_n}{\sqrt{\lambda_n}}, n \in \mathbb{N}^\star\}$ is a Hilbert basis of $H_0^1(\Omega)$. This was step 1 of the proof of Theorem 4.30).

Remark 4.61 (Orthonormal family). Given the way $H_0^1(\Omega)$ is embedded in $H^{-1}(\Omega)$ (by identifying $L^2(\Omega)'$ with $L^2(\Omega)$) and the definition of the inner product in $H^{-1}(\Omega)$ (from the inner product in $H_0^1(\Omega)$), we also note that

$$(e_n|e_m)_{H^{-1}} = 0 \text{ if } n \neq m,$$
$$(e_n|e_m)_{H^{-1}} = \frac{1}{\lambda_n} \text{ if } n = m. \tag{4.81}$$

The family $\{\sqrt{\lambda_n}e_n, n \in \mathbb{N}^\star\}$ is therefore an orthonormal family of $H^{-1}(\Omega)$ (it is even a Hilbert basis of $H^{-1}(\Omega)$).

To show (4.81), we first recall the definition of the inner product in $H^{-1}(\Omega)$. If $u \in H_0^1(\Omega)$, we define T_u in $H^{-1}(\Omega)$ by

$$\langle T_u, \varphi \rangle_{H^{-1}, H_0^1} = (u|\varphi)_{H_0^1} = \int_\Omega \nabla u \cdot \nabla \varphi \, dx.$$

The mapping T is bijective from $H_0^1(\Omega)$ to $H^{-1}(\Omega)$. For $u, v \in H_0^1(\Omega)$, we then define the inner product in $H^{-1}(\Omega)$ by

$$(T_u|T_v)_{H^{-1}(\Omega)} = (u|v)_{H_0^1(\Omega)} = \int_\Omega \nabla u \cdot \nabla v \, dx.$$

Let $n \in \mathbb{N}^\star$. Given the way $H_0^1(\Omega)$ is embedded into $H^{-1}(\Omega)$ (by identifying $L^2(\Omega)'$ with $L^2(\Omega)$), we have, for all $\varphi \in H_0^1(\Omega)$,

$$\lambda_n \langle e_n, \varphi \rangle_{H^{-1}, H_0^1} = \lambda_n \int_\Omega e_n \varphi \, dx = \int_\Omega \nabla e_n \cdot \nabla \varphi \, dx,$$

which shows that $T_{e_n} = \lambda_n e_n$. Hence, for $n, m \in \mathbb{N}^\star$,

$$\lambda_n \lambda_m (e_n | e_m)_{H^{-1}(\Omega)} = (T_{e_n} | T_{e_m})_{H^{-1}(\Omega)} = (e_n | e_m)_{H_0^1(\Omega)}$$

$$= \int_\Omega \nabla e_n \cdot \nabla e_m \, \mathrm{d}x = \lambda_n \delta_{n,m}.$$

We thus deduce (4.81).

Let us now construct an approximate solution (step 2 of the proof of Theorem 4.30). For $n \in \mathbb{N}^\star$, we set $E_n = \mathrm{Vect}\{e_p, p = 1, \ldots n\}$, and we seek an approximate solution u_n in the form $u_n(t) = \sum_{i=1}^n \alpha_i(t) e_i$ with $\alpha_i \in C([0, T], \mathbb{R})$. The formal calculation made in the proof of Theorem 4.30 gives here (assuming that the α_i are differentiable for all t), for all $\varphi \in H_0^1(\Omega)$ and almost all $t \in]0, T[$,

$$\langle u_n'(t) - \mathrm{div}(A\nabla u_n(t)) - f(t), \varphi \rangle_{H^{-1}(\Omega), H_0^1(\Omega)}$$

$$= \sum_{i=1}^n \left(\alpha_i'(t) \int_\Omega e_i \varphi \, \mathrm{d}x + \alpha_i(t) \int_\Omega A\nabla e_i \cdot \nabla \varphi \, \mathrm{d}x \right) - \langle f(t), \varphi \rangle_{H^{-1}(\Omega), H_0^1(\Omega)}.$$

We then wish to choose the functions α_i so that

$$\varphi \in E_n, \quad \langle u_n'(t) - \mathrm{div}(A\nabla u_n)(t) - f(t), \varphi \rangle_{H^{-1}(\Omega), H_0^1(\Omega)} = 0.$$

Set $f_i(t) = \langle f(t), e_i \rangle_{H^{-1}(\Omega), H_0^1(\Omega)}$ and let $F(t)$ be the vector whose components are $f_i(t)$, $i = 1, \ldots, n$. Let M be the $n \times n$ matrix with real coefficients $M_{i,j} = \int_\Omega A\nabla e_j \cdot \nabla e_i \, \mathrm{d}x$. Denoting by $\alpha(t)$ the vector whose components are $\alpha_i(t)$, $i = 1, \ldots, n$, we therefore wish that

$$\alpha'(t) + M\alpha(t) = F(t).$$

Taking into account the initial condition, let $\alpha_i^{(0)} = (u_0 | e_i)_2$ and let $\alpha^{(0)}$ be the vector whose components are $\alpha_i^{(0)}$, $i = 1, \ldots, n$; the above equality suggests taking

$$\alpha(t) = \mathrm{e}^{-Mt} \alpha^{(0)} + \int_0^t \mathrm{e}^{-M(t-s)} F(s) \, \mathrm{d}s. \tag{4.82}$$

The functions α_i thus defined belong to $C([0, T], \mathbb{R})$, so that $u_n \in C([0, T], E_n) \subset C([0, T], H_0^1(\Omega))$ with $u_n(t) = \sum_{i=1}^n \alpha_i(t) e_i$.

We have reasoned here at a fixed n; the matrix M and the functions F and α therefore depend on n.

We now adapt step 3 of Theorem 4.30. Let $n \in \mathbb{N}^\star$ and u_n be the approximate solution given by the previous step. By definition, the derivative by transposition $\partial_t u_n$ belongs to $\mathcal{D}_E^\star$ with $E = H_0^1(\Omega)$. Let $\varphi \in \mathcal{D}(]0, T[)$ then

$$\langle \partial_t u_n, \varphi \rangle_{\mathcal{D}_E^\star, \mathcal{D}} = -\int_0^T u_n(t) \varphi'(t) \, \mathrm{d}t \in E_n \subset H_0^1(\Omega).$$

Since $u_n = \sum_{i=1}^{n} \alpha_i e_i$, this implies that

$$\langle \partial_t u_n, \varphi \rangle_{\mathcal{D}_E^\star, \mathcal{D}} = -\sum_{i=1}^{n} \int_0^T \alpha_i(t) e_i \varphi'(t)\, dt = -\sum_{i=1}^{n} \Big(\int_0^T \alpha_i(t) \varphi'(t)\, dt \Big) e_i.$$

We now use (4.82),

$$\int_0^T \alpha_i(t) \varphi'\, dt = T_i + S_i,$$

with

$$T_i = \int_0^T (e^{-Mt} \alpha^{(0)})_i \varphi'(t)\, dt = \int_0^T (M e^{-Mt} \alpha^{(0)})_i \varphi(t)\, dt,$$

$$S_i = \int_0^T \Big(\int_0^t e^{-M(t-s)} F(s)\, ds \Big)_i \varphi'(t)\, dt.$$

To transform S_i we use Fubini's theorem and we get

$$S_i = \int_0^T \Big(\int_0^t (M e^{-M(t-s)} F(s))_i\, ds \Big) \varphi(t)\, dt - \int_0^T f_i(t) \varphi(t)\, dt.$$

Hence $T_i + S_i = \int_0^T (M\alpha(t))_i \varphi(t)\, dt - \int_0^T f_i(t)\varphi(t)\, dt$, and therefore

$$\langle \partial_t u_n, \varphi \rangle_{\mathcal{D}^\star, \mathcal{D}} = -\sum_{i=1}^{n} \int_0^T (M\alpha(t))_i e_i \varphi(t)\, dt + \sum_{i=1}^{n} \int_0^T f_i(t) e_i \varphi(t)\, dt.$$

Since φ is arbitrary in $\mathcal{D}(]0,T[)$, we therefore have (a.e. in $]0,T[$)

$$\partial_t u_n = -\sum_{i=1}^{n} (M\alpha)_i e_i + \sum_{i=1}^{n} f_i e_i \in L^2(]0,T[, E_n).$$

The first term on the right-hand side of this equality is even continuous from $[0,T]$ to E_n. By definition of M, this equality gives (a.e. in $]0,T[$)

$$\partial_t u_n = -\sum_{i=1}^{n} \Big(\int_\Omega A\nabla u_n \cdot \nabla e_i\, dx \Big) e_i + \sum_{i=1}^{n} f_i e_i. \tag{4.83}$$

The situation is slightly different from that of Theorem 4.30 but we still have

$$\partial_t u_n \in L^2(]0,T[, E_n) \subset L^2(]0,T[, H_0^1(\Omega)) \subset L^2(]0,T[, H^{-1}(\Omega)).$$

Given the way $H_0^1(\Omega)$ is embedded in $H^{-1}(\Omega)$ (by identifying $L^2(\Omega)'$ with $L^2(\Omega)$) we have for all $v \in L^2(]0,T[, H_0^1(\Omega))$,

$$\int_0^T \langle \partial_t u_n(t), v(t) \rangle_{H^{-1}(\Omega), H_0^1(\Omega)} \, dt = - \int_0^T \int_\Omega \sum_{i=1}^n (M\alpha(t))_i e_i v \, dx \, dt$$

$$+ \sum_{i=1}^n \int_0^T \int_\Omega f_i e_i v \, dx \, dt.$$

This is particularly interesting when $v \in L^2(]0, T[, E_n)$ because owing to the definition of M we get

$$\int_0^T \langle \partial_t u_n(t), v(t) \rangle_{H^{-1}(\Omega), H_0^1(\Omega)} \, dt = - \int_0^T \int_\Omega A \nabla u_n \cdot \nabla v \, dx \, dt + \sum_{i=1}^n \int_0^T \int_\Omega f_i e_i v \, dx \, dt.$$

This gives, by definition of f_i, still for $v \in L^2(]0, T[, E_n)$,

$$\int_0^T \langle \partial_t u_n(t), v(t) \rangle_{H^{-1}(\Omega), H_0^1(\Omega)} \, dt + \int_0^T \int_\Omega A \nabla u_n \cdot \nabla v \, dx \, dt$$

$$= \sum_{i=1}^n \int_0^T \langle f(t), e_i \rangle_{H^{-1}(\Omega), H_0^1(\Omega)} \left(\int_\Omega e_i v \, dx \right) dt$$

$$= \int_0^T \langle f(t), v \rangle_{H^{-1}(\Omega), H_0^1(\Omega)} \, dt. \quad (4.84)$$

We also recall that $u_n \in C([0, T], H_0^1(\Omega))$ and $u_n(0) = P_n u_0$, where P_n is the orthogonal projection operator in $L^2(\Omega)$ on the subspace E_n.

We are now looking for estimates on the approximate solution (step 4). For $n \in \mathbb{N}^\star$, we have

$$u_n \in C([0, T], H_0^1(\Omega)) \subset L^2(]0, T[, H_0^1(\Omega))$$

and

$$\partial_t u_n \in L^2(]0, T[, E_n) \subset L^2(]0, T[, H^{-1}(\Omega)).$$

According to Section 4.2, we therefore have

$$\frac{1}{2} \|u_n(T)\|_2^2 - \frac{1}{2} \|u_0\|_2^2 = \int_0^T \langle \partial_t u_n, u_n \rangle_{H^{-1}, H_0^1} \, dt.$$

Taking $v = u_n$ in (4.84) yields

$$\frac{1}{2} \|u_n(T)\|_2^2 - \frac{1}{2} \|u_0\|_2^2 + \int_0^T \int_\Omega A \nabla u_n \cdot \nabla u_n \, dx \, dt = \int_0^T \langle f, u_n \rangle_{H^{-1}, H_0^1} \, dt,$$

and therefore

$$\alpha \|u_n\|_{L^2(]0, T[, H_0^1(\Omega))}^2 \leq \int_0^T \int_\Omega A \nabla u_n \cdot \nabla u_n \, dx \, dt \leq \frac{1}{2} \|u_0\|_2^2 + \int_0^T \langle f, u_n \rangle_{H^{-1}, H_0^1} \, dt.$$

Thus

$$\alpha \|u_n\|^2_{L^2(]0,T[,H^1_0(\Omega))} \leq \frac{1}{2} \|u_0\|^2_2 + \int_0^T \langle f, u_n\rangle_{H^{-1},H^1_0}\, dt$$

$$\leq \frac{1}{2} \|u_0\|^2_2 + \|f\|_{L^2(]0,T[,H^{-1}(\Omega))} \|u_n\|_{L^2(]0,T[,H^1_0(\Omega))}$$

$$\leq \frac{1}{2} \|u_0\|^2_2 + \frac{1}{2\alpha} \|f\|^2_{L^2(]0,T[,H^{-1}(\Omega))} + \frac{\alpha}{2} \|u_n\|^2_{L^2(]0,T[,H^1_0(\Omega))}\, .$$

Hence

$$\alpha \|u_n\|^2_{L^2(]0,T[,H^1_0(\Omega))} \leq \|u_0\|^2_2 + \frac{1}{\alpha} \|f\|^2_{L^2(]0,T[,H^{-1}(\Omega))}\, ,$$

which gives a bound on u_n in $L^2(]0,T[, H^1_0(\Omega))$.

To obtain the bound on $\partial_t u_n$, we use (4.83). For almost every t we have

$$\partial_t u_n(t) = -\sum_{i=1}^n \Big(\int_\Omega A\nabla u_n(t) \cdot \nabla e_i \, dx\Big) e_i + \sum_{i=1}^n f_i(t) e_i.$$

Let $v \in H^1_0(\Omega)$, we have $v = \sum_{i=1}^\infty (v|e_i)_{L^2} e_i$ and this series is convergent in $L^2(\Omega)$ and in $H^1_0(\Omega)$. Setting $P_n(v) = \sum_{i=1}^n (v|e_i)_{L^2} e_i$ we have $\|P_n v\|_{H^1_0(\Omega)} \leq \|v\|_{H^1_0(\Omega)}$. Since

$$\langle \partial_t u_n(t), P_n v\rangle_{H^{-1},H^1_0} = \langle \partial_t u_n(t), v\rangle_{H^{-1},H^1_0} = \int_\Omega \partial_t u_n(t) v \, dx$$

$$= -\sum_{i=1}^n \Big(\int_\Omega A\nabla u_n(t) \cdot \nabla e_i \, dx\Big)(v|e_i)_{L^2} + \sum_{i=1}^n f_i(t)(v|e_i)_{L^2}, \qquad (4.85)$$

we have,

$$\|\partial_t u_n(t)\|_{H^{-1}(\Omega)} = \sup\{\langle \partial_t u_n(t), v\rangle_{H^{-1},H^1_0},\ v \in E_n,\ \|v\|_{H^1_0(\Omega)} \leq 1\}.$$

However, for $v \in E_n$, we obtain with (4.85)

$$\langle \partial_t u_n(t), v\rangle_{H^{-1},H^1_0} = \int_\Omega A\nabla u_n(t) \cdot \nabla v \, dx + \langle f(t), v\rangle_{H^{-1},H^1_0}.$$

Using the fact that the coefficients of A belong to $L^\infty(\Omega)$, we get the existence of β, depending only on A, such that for almost all t,

$$\|\partial_t u_n(t)\|_{H^{-1}(\Omega)} \leq \beta \|u_n(t)\|_{H^1_0(\Omega)} + \|f(t)\|_{H^{-1}(\Omega)}\, .$$

The sequence $(u_n)_{n\in\mathbb{N}^\star}$ is bounded in $L^2(]0,T[, H^1_0(\Omega))$, therefore the sequence $(\partial_t u_n)_{n\in\mathbb{N}^\star}$ is bounded in $L^2(]0,T[, H^{-1}(\Omega))$.

We now pass to the limit as $n \to +\infty$ (step 5). Thanks to the estimates obtained in the previous step, we can assume that, up to a subsequence,

$$u_n \rightarrow u \text{ weakly in } L^2(]0,T[,H_0^1(\Omega)) \atop \partial_t u_n \rightarrow w \text{ weakly in } L^2(]0,T[,H^{-1}(\Omega))\Bigg\} \text{ as } n \rightarrow +\infty.$$

Similarly as in the proof of Theorem 4.30, we get $w = \partial_t u$.

Let $v \in L^2(]0,T[,H_0^1(\Omega))$, let $n \in \mathbb{N}^\star$ and $v_n(t) = P_n(v(t))$, so that, by the dominated convergence theorem, $v_n \rightarrow v$ in $L^2(]0,T[,H_0^1(\Omega))$. We then use (4.84) with v_n instead of v. We obtain

$$\int_0^T \langle \partial_t u_n, v_n \rangle_{H^{-1},H_0^1} \, dt + \int_0^T \int_\Omega A\nabla u_n \cdot \nabla v_n \, dx \, dx \, dt = \int_0^T \langle f, v_n \rangle_{H^{-1},H_0^1} \, dt.$$

Thanks to the convergence of u_n, v_n and $\partial_t u_n$, we may pass to the limit as $n \rightarrow +\infty$ in this equality and obtain

$$\int_0^T \langle \partial_t u, v \rangle_{H^{-1},H_0^1} \, dt + \int_0^T A\nabla u \cdot \nabla v \, dx \, dt = \int_0^T \langle f, v \rangle_{H^{-1},H_0^1} \, dt,$$

which is the desired result.

Since $u \in L^2(]0,T[,H_0^1(\Omega))$ and $\partial_t u \in L^2(]0,T[,H^{-1}(\Omega))$, we know that $u \in C([0,T],L^2(\Omega))$. To complete the proof that u is a solution to (4.61), it remains only to show that $u(0) = u_0$ a.e. (that is, $u(0) = u_0$ in $L^2(\Omega)$); the proof is identical to that given in the proof of Theorem 4.30. This completes the proof of the existence of a solution to (4.61).

We now prove the uniqueness of the solution to (4.61). The proof is very similar to that given for Theorem 4.30. Let u_1, u_2 be two solutions of (4.61). Set $u = u_1 - u_2$. Taking the difference of the equations satisfied by u_1 and u_2 and taking, for $t \in [0,T]$, $v = u\mathbb{1}_{]0,t[}$ as a test function, we obtain

$$\int_0^t \langle \partial_t u(s), u(s) \rangle_{H^{-1},H_0^1} \, ds + \int_0^t \int_\Omega A\nabla u(s) \cdot \nabla u(s) \, dx \, ds = 0.$$

Since $u \in L^2(]0,T[,H_0^1(\Omega))$ and $\partial_t u \in L^2(]0,T[,H^{-1}(\Omega))$, we have, according to Section 4.2,

$$\frac{1}{2}(\|u(t)\|_2^2 - (\|u(0)\|_2^2) = \int_0^t \langle \partial_t u(s), u(s) \rangle_{H^{-1},H_0^1} \, ds.$$

Thus, for all $t \in [0,T]$,

$$(\|u(t)\|_2^2 - (\|u(0)\|_2^2) + 2\int_0^t \int_\Omega A\nabla u(s) \cdot \nabla u(s) \, dx \, dt = 0.$$

Finally, since $u(0) = 0$ and $A\nabla u \cdot \nabla u \geq 0$ a.e., we obtain $u(t) = 0$ a.e. in Ω, for all $t \in [0,T]$. We have thus completed the proof of the uniqueness of the solution to (4.61).

Problem 4.6 (Existence by Schauder's theorem)

1. Let $(\bar{u}_n)_{n\in\mathbb{N}}$ be a sequence in $L^2(]0,T[,L^2(\Omega))$ and $\bar{u} \in L^2(]0,T[,L^2(\Omega))$ such that $\bar{u}_n \to \bar{u}$ in $L^2(]0,T[,L^2(\Omega))$ as $n \to +\infty$. Set $u_n = T(\bar{u}_n)$ and $u = T(\bar{u})$. To show that $u_n \to u$ in $L^2(]0,T[,L^2(\Omega))$ as $n \to +\infty$, we reason by contradiction. If $u_n \nrightarrow u$ in $L^2(]0,T[,L^2(\Omega))$, there exists $\varepsilon > 0$ and a subsequence, still denoted by $(u_n)_{n\in\mathbb{N}}$, such that

$$\|u_n - u\|_{L^2(]0,T[,L^2(\Omega))} \geq \varepsilon \text{ for all } n \in \mathbb{N}. \tag{4.86}$$

For $n \in \mathbb{N}$, u_n is the solution to (4.6) with $\bar{u}_n$ instead of $\bar{u}$. Taking $v = u_n$ as a test function yields, thanks to (4.64), that the sequence $(u_n)_{n\in\mathbb{N}}$ is bounded in $L^2(]0,T[,H_0^1(\Omega))$. Then, since

$$\partial_t u_n = \mathrm{div}(A(\bar{u}_n)\nabla u_n) + f \text{ in } L^2(]0,T[,H^{-1}(\Omega)),$$

we get that the sequence $(\partial_t u_n)_{n\in\mathbb{N}}$ is bounded in $L^2(]0,T[,H^{-1}(\Omega))$. We can therefore assume that there exist $w \in L^2(]0,T[,H_0^1(\Omega))$ and $\zeta \in L^2(]0,T[,H^{-1}(\Omega))$ such that, up to a subsequence,

$$u_n \to w \text{ weakly in } L^2(]0,T[,H_0^1(\Omega)) \text{ as } n \to +\infty,$$
$$\partial_t u_n \to \zeta \text{ weakly in } L^2(]0,T[,H^{-1}(\Omega)) \text{ as } n \to +\infty.$$

By Theorem 4.46, we also have $u_n \to u$ in $L^2(]0,T[,L^2(\Omega))$. As shown in the proof of Theorem 4.30 we necessarily have $\zeta = \partial_t w$.

We can also assume that, up to a subsequence, $\bar{u}_n \to \bar{u}$ a.e. as $n \to +\infty$, so that $A(\bar{u}_n) \to A(\bar{u})$ a.e. (thanks to the continuity of $a_{i,j}$).

Let $v \in L^2(]0,T[,H_0^1(\Omega))$, we can then pass to the limit as $n \to +\infty$, in (4.6) written with $\bar{u}_n$ and u_n (instead of $\bar{u}$ and u). We find that w is *the* solution to (4.6), that is, $w = u = T(\bar{u})$. Hence $u_n \to u$ in $L^2(]0,T[,L^2(\Omega))$, in contradiction with (4.86).

We have thus shown the continuity of T.

2. The first arguments of the previous question show that $\mathrm{Im}(T)$ is bounded in $L^2(]0,T[,H_0^1(\Omega))$ and that $\{\partial_t u, u \in \mathrm{Im}(T)\}$ is a bounded subset of $L^2(]0,T[,H^{-1}(\Omega))$. The compactness lemma (Lemma 4.38) then gives that $\mathrm{Im}(T)$ is relatively compact in $L^2(]0,T[,L^2(\Omega))$, which gives the compactness of T.

3. Here again, this follows from the first question. Indeed, we know that $\mathrm{Im}(T)$ is bounded in $L^2(]0,T[,H_0^1(\Omega))$ and therefore also in $L^2(]0,T[,L^2(\Omega))$.

4. Let B_R be the ball of $L^2(]0,T[,L^2(\Omega))$ centred at 0 with radius R given in the previous question. The operator T is continuous and compact from B_R to B_R. The Schauder theorem then gives the existence of $u \in B_R$ (and therefore $u \in L^2(]0,T[,L^2(\Omega))$) such that $u = T(u)$, which is therefore a solution to (4.39).

5. The proof here is very close to that made to show the uniqueness of the solution to (4.40).

   Let u_1, u_2 be two solutions of (4.39); let us set $u = u_1 - u_2$ and show that $u = 0$ a.e.

   For $\varepsilon > 0$ we define the function T_ε from $\mathbb{R}$ to $\mathbb{R}$ by

   $$T_\varepsilon(s) = \max\{-\varepsilon, \min\{s, \varepsilon\}\}.$$

   We also denote by ϕ_ε the primitive of T_ε that vanishes at 0. Taking $v = T_\varepsilon(u)$ in the weak formulations satisfied by u_1 and u_2, we obtain

   $$\int_0^T \langle \partial_t u, T_\varepsilon(u) \rangle_{H^{-1}, H_0^1} \, dt + \int_0^T \int_\Omega A(u_1) \nabla u \cdot \nabla T_\varepsilon(u) \, dx \, dt$$
   $$= \int_0^T \int_\Omega (A(u_2) - A(u_1)) \nabla u_2 \cdot \nabla T_\varepsilon(u) \, dx \, dt.$$

   Since $\nabla T_\varepsilon(u) = \nabla u \mathbb{1}_{0 < |u| < \varepsilon}$ a.e., we get that

   $$\int_\Omega \phi_\varepsilon(u(x, T)) \, dx - \int_\Omega \phi_\varepsilon(u(x, 0)) \, dx + \alpha \int_0^T \int_\Omega \nabla u \cdot \nabla u \mathbb{1}_{0 < |u| < \varepsilon} \, dx \, dt$$
   $$\leq \int_0^T \int_\Omega |(A(u_1) - A(u_2)) \nabla u_2| |\nabla u| \mathbb{1}_{0 < |u| < \varepsilon} \, dx \, dt. \quad (4.87)$$

   Since the functions $a_{i,j}$ are Lipschitz-continuous, there exists an L such that, for all $i, j \in \{1, \dots, N\}$ and $s_1, s_2 \in \mathbb{R}$,

   $$|a_{i,j}(s_1) - a_{i,j}(s_2)| \leq L |s_1 - s_2|.$$

   We then use the fact that $u_0 = 0$ a.e. and $\phi_\varepsilon \geq 0$ to deduce from (4.87), with $A_\varepsilon = \{0 < |u| < \varepsilon\}$ and $y = (x, t)$,

   $$\alpha \int_0^T \int_\Omega |\nabla T_\varepsilon(u)|^2 \, dx \, dt$$
   $$\leq N^2 L \varepsilon \left(\int_{A_\varepsilon} |\nabla u_2|^2 \, dy \right)^{\frac{1}{2}} \left(\int_0^T \int_\Omega |\nabla T_\varepsilon(u)|^2 \, dx \, dt \right)^{\frac{1}{2}}.$$

   Hence $\alpha \, ||| \nabla T_\varepsilon(u) |||_{L^2(Q)} \leq a_\varepsilon \varepsilon$, with $Q =]0, T[\times \Omega$ and

   $$a_\varepsilon = N^2 L \left(\int_{A_\varepsilon} |\nabla u_2|^2 \, dy \right)^{\frac{1}{2}}.$$

   Since $\cap_{\varepsilon > 0} A_\varepsilon = \emptyset$ the non-increasing continuity of a measure gives that the $(N + 1$-dimensional) Lebesgue measure of A_ε tends to 0 as $\varepsilon \to 0$. Since $\nabla u_2 \in L^2(Q)^N$ (note that $L^2(Q)$ can be identified with $L^2(]0, T[, L^2(\Omega)))$, we therefore have

$$\lim_{\varepsilon \to 0} \int_{A_\varepsilon} |\nabla u_2|^2 \, dy = 0,$$

which gives $\lim_{\varepsilon \to 0} a_\varepsilon = 0$. We now need to use, for example, the embedding of $W_0^{1,1}(\Omega)$ in $L^{1^\star}(\Omega)$; it gives that for $t \in]0, T[$,

$$\|T_\varepsilon(u(t))\|_{L^{1^\star}(\Omega)} \leq \|\,|\nabla T_\varepsilon(u(t))|\,\|_{L^1(\Omega)}. \tag{4.88}$$

We now note that for $t \in]0, T[$ (recall that λ_N designates the Lebesgue measure on $\mathbb{R}^N$),

$$\varepsilon \lambda_N \{|u(t)| \geq \varepsilon\}^{\frac{1}{1^\star}} \leq \Big(\int_\Omega |T_\varepsilon(u)|^{1^\star} \, dx\Big)^{\frac{1}{1^\star}}.$$

Hence, with (4.88),

$$\varepsilon \lambda_N \{|u(t)| \geq \varepsilon\}^{\frac{1}{1^\star}} \leq \|\,|\nabla T_\varepsilon(u(t))|\,\|_{L^1(\Omega)} = \int_\Omega |\nabla T_\varepsilon(u(x,t))| \, dx,$$

and, integrating with respect to t, knowing that $\frac{1}{1^\star} = \frac{N-1}{N}$ and using the Cauchy–Schwarz inequality,

$$\varepsilon \int_0^T \lambda_N \{|u(t)| \geq \varepsilon\}^{\frac{N-1}{N}} \, dt \leq \int_0^T \int_\Omega |\nabla T_\varepsilon(u(x,t))| \, dx \, dt$$

$$\leq \|\,|\nabla T_\varepsilon(u)|\,\|_{L^2(Q)} \, (T\lambda_N(\Omega))^{\frac{1}{2}} \leq \frac{(T\lambda_N(\Omega))^{\frac{1}{2}}}{\alpha} a_\varepsilon \varepsilon.$$

Therefore, we have

$$\int_0^T \lambda_N \{|u(t)| \geq \varepsilon\}^{\frac{N-1}{N}} \, dt \leq \frac{(T\lambda_N(\Omega))^{\frac{1}{2}}}{\alpha} a_\varepsilon.$$

Letting $\varepsilon \to 0$ and noting that $\lim_{\varepsilon \to 0} a_\varepsilon = 0$, the dominated convergence theorem yields

$$\int_0^T \lambda_N \{|u(t)| > 0\}^{\frac{N-1}{N}} \, dt \leq 0,$$

which gives $\lambda_N \{|u(t)| > 0\} = 0$ a.e. in $t \in]0, T[$; hence $u = 0$ a.e., which concludes this proof of uniqueness.

Problem 4.7 (Example of compactly embedded sequence)

Let $(u_m)_{m \in \mathbb{N}^\star}$ be a sequence such that

- $u_m \in X_m$ for all $m \in \mathbb{N}^\star$,
- the sequence $(\|u_m\|_{X_m})_{m \in \mathbb{N}^\star}$ is bounded,

We prove that the family $\mathcal{A} = \{u_m, \, m \in \mathbb{N}^\star\}$ satisfies the three hypotheses of Theorem 4.44 with $T = 1$, $p = 2$ and $B = \mathbb{R}$.

The first hypothesis of Theorem 4.44 is satisfied because $\|u_m\|_{X_m} \geq \|u_m\|_{L^2(]0,1[)}$. The second hypothesis of Theorem 4.44 is satisfied because $B = \mathbb{R}$ and the sequence $(u_m)_{m \in \mathbb{N}^\star}$ is bounded in $L^2(]0,1[)$ and therefore also in $L^1(]0,1[)$.

It remains to show that the third hypothesis is also satisfied.

Let $0 < \eta < 1$; we define the function χ_i by $\chi_i(x) = 1$ if $x_i \in]x, x+\eta[$ and 0 otherwise, so that, for all $u \in X_m$ and all $x \in]0, 1-\eta[$ such that $x \neq x_i$ and $x+\eta \neq x_i$ for all i,

$$u(x + \eta) - u(x) = \sum_{i=1}^{m-1} (u_{i+1} - u_i)\chi_i(x),$$

and therefore, since $\int_0^{1-\eta} \chi_i(x)^2 \, dx \leq \eta$ (noticing that $x < x_i < x+\eta$ if and only if $x_i - \eta < x < x_i$), the Cauchy–Schwarz inequality implies that

$$\int_0^{1-\eta} |u(x + \eta) - u(x)|^2 \, dx \leq m\eta \sum_{i=1}^{m-1} |u_{i+1} - u_i|^2 \leq \eta \|u\|_{X_m}^2.$$

Therefore,

$$\forall m \in \mathbb{N}^\star, \int_0^{1-\eta} |u_m(x + \eta) - u_m(x)|^2 \, dx \leq \eta \|u_m\|_{X_m}^2,$$

which proves that the third hypothesis of Theorem 4.44 is satisfied and therefore the sequence $(u_m)_{m \in \mathbb{N}^\star}$ admits a convergent subsequence in $L^2(]0,1[)$.

But, a subsequence of the sequence $(u_m)_{m \in \mathbb{N}^\star}$ also satisfies the three hypotheses of Theorem 4.44 (with $T = 1$, $p = 2$ and $B = \mathbb{R}$) and thus admits a convergent subsequence in $L^2(]0,1[)$.

This prove that the sequence $(X_m)_{m \in \mathbb{N}^\star}$ is compactly embedded in $L^2(]0,1[)$.

Problem 4.8 (Kolmogorov's theorem, with $B = \mathbb{R}$)

1. For all $t \in]0, \delta[$ we have

$$|u_n(t)| \leq |u_n(t + h)| + |u_n(t + h) - u_n(t)|.$$

Integrating this inequality between 0 and δ, we indeed obtain (4.65).

2. As usual, we choose for u_n one of the representatives, so that

$$u_n \in \mathcal{L}_{\mathbb{R}}^1(]0, T[, \mathcal{B}(]0, T[), \lambda), \forall n \in \mathbb{N}.$$

The inequality (4.65) is true for all $h \in]0, h_0[$. Integrating (4.65) between 0 and h_0 and noting that $\int_0^\delta |u_n(t + h) - u_n(t)| \, dt \leq \eta(h) \leq \eta(h_0)$ (because $h \leq h_0$ and $\delta \leq T - h_0 \leq T - h$), we obtain

$$h_0 \int_0^\delta |u_n(t)|\,\mathrm{d}t$$

$$\leq \int_0^{h_0} \left(\int_0^\delta |u_n(t+h)|\,\mathrm{d}t \right) \mathrm{d}h + \int_0^{h_0} \left(\int_0^\delta |u_n(t+h) - u_n(t)|\,\mathrm{d}t \right) \mathrm{d}h$$

$$\leq \int_0^{h_0} \left(\int_0^\delta |u_n(t+h)|\,\mathrm{d}t \right) \mathrm{d}h + h_0 \eta(h_0).$$

The Lebesgue measure is σ-finite and the mapping $(t,h) \mapsto u_n(t+h)$ is Borel-measurable from $]0,\delta[\times]0,h_0[$ to $\mathbb{R}$ (because it is the composition of $(t,h) \mapsto t+h$, which is continuous and therefore Borel-measurable, and of $s \mapsto u_n(s)$, which is Borel-measurable). We can therefore apply the Fubini–Tonelli theorem to obtain that

$$\int_0^{h_0} \left(\int_0^\delta |u_n(t+h)|\,\mathrm{d}t \right) \mathrm{d}h = \int_0^\delta \left(\int_0^{h_0} |u_n(t+h)|\,\mathrm{d}h \right) \mathrm{d}t$$

$$\leq \int_0^\delta \left(\int_0^T |u_n(s)|\,\mathrm{d}s \right)$$

$$\leq \delta \, \|u_n\|_1 \,.$$

Thus (4.66) holds.

3. Let $\varepsilon > 0$ and let $h_0 \in]0,T[$ such that $\eta(h_0) \leq \varepsilon$. Then, with $C = \sup_{n\in\mathbb{N}} \|u_n\|_1$, we set $\overline{\delta} = \min\{T - h_0, \varepsilon \frac{h_0}{C}\}$. We then obtain, for all $n \in \mathbb{N}$,

$$0 \leq \delta \leq \overline{\delta} \Rightarrow \int_0^\delta |u_n(t)|\,\mathrm{d}t \leq 2\varepsilon.$$

Hence $\int_0^\delta |u_n(t)|\,\mathrm{d}t \to 0$ as $\delta \to 0^+$, uniformly with respect to n.

A similar proof also shows that $\int_{T-\delta}^T |u_n(t)|\,\mathrm{d}t \to 0$ as $\delta \to 0^+$, uniformly with respect to n (introducing $v_n(t) = u_n(T-t)$).

Let us extend u_n by 0 outside $]0,T[$ (and still denote by u_n the extended function). To apply Theorem 4.44, let us show that

$$\int_{\mathbb{R}} |u_n(t+h) - u_n(t)|\,\mathrm{d}t \to 0 \text{ as } h \to 0^+, \text{ uniformly with respect to } n \in \mathbb{N}.$$

With this aim, we note that for $h > 0$ and $n \in \mathbb{N}$,

$$\int_{\mathbb{R}} |u_n(t+h) - u_n(t)|\,\mathrm{d}t$$

$$\leq \int_{-h}^0 |u_n(t+h)|\,\mathrm{d}t + \int_0^{T-h} |u_n(t+h) - u_n(t)|\,\mathrm{d}t + \int_{T-h}^T |u_n(t)|\,\mathrm{d}t$$

$$= \int_0^h |u_n(t)|\, dt + \int_0^{T-h} |u_n(t+h) - u_n(t)|\, dt + \int_{T-h}^T |u_n(t)|\, dt.$$

Let $\varepsilon > 0$; there exists an $h_1 > 0$ such that $\eta(h_1) \le \varepsilon$. Then, by the previous question, there exists an $h_2 > 0$ such that (for all $n \in \mathbb{N}$)

$$0 \le h \le h_2 \implies \int_0^h |u_n(t)|\, dt \le \varepsilon \ \text{ and } \ \int_{T-h}^T |u_n(t)|\, dt \le \varepsilon.$$

With $h_3 = \min\{h_1, h_2\}$, we therefore have (for all $n \in \mathbb{N}$)

$$0 \le h \le h_3 \implies \int_{\mathbb{R}} |u_n(t+h) - u_n(t)|\, dt \le 3\varepsilon,$$

which concludes the proof.

Problem 4.9 (Existence for the Stefan problem, by numerical scheme)

1. Since $w \in C^2(\mathbb{R}^2, \mathbb{R})$, we indeed have $u \in L^\infty(]0, 1[\times]0, T[)$, that is to say, (4.68a). Let $D =]0, 1[\times]0, T[$. Suppose that u satisfies (4.67) and let us show that u satisfies (4.68b). Let $\psi \in C_T^\infty(\mathbb{R}^2)$. Multiplying (4.67a) by ψ and integrating over D yields:

$$\int_D \partial_t u(x, t)\psi(x, t)\, dx\, dt - \int_D \partial_{xx}^2(\varphi(u))(x, t)\psi(x, t)\, dx\, dt$$
$$= \int_D v(x, t)\psi(x, t)\, dx\, dt. \quad (4.89)$$

An integration by parts yields that

$$\int_D \partial_t u(x, t)\psi(x, t)\, dx\, dt$$
$$= \int_0^1 u(x, T)\psi(x, T)\, dx - \int_0^1 u(x, 0)\psi(x, 0)\, dx - \int_D u(x, t)\partial_t \psi(x, t)\, dx\, dt.$$

Since $\psi \in C_T^\infty(\mathbb{R}^2)$ we therefore have $\psi(x, T) = 0$ for all $x \in [0, 1]$ and since u satisfies (4.67c), $u(x, 0) = u_0(x)$. Hence

$$\int_D \partial_t u(x, t)\psi(x, t)\, dx\, dt$$
$$= -\int_0^1 u_0(x)\psi(x, 0)\, dx - \int_D \partial_t \psi(x, t)u(x, t)\, dx\, dt. \quad (4.90)$$

Integrating by parts the second term of (4.89) yields

$$\int_D \partial^2_{xx}(\varphi(u))(x,t)\psi(x,t)\,dx\,dt$$
$$= \int_0^T \left[\partial_x(\varphi(u))(1,t)\psi(1,t) - \partial_x(\varphi(u))(0,t)\psi(0,t)\right]dt$$
$$- \int_D \partial_x(\varphi(u))(x,t)\partial_x\psi(x,t)\,dx\,dt, \quad (4.91)$$

and since u satisfies (4.67b), we have

$$\partial_x(\varphi(u))(0,t) = \partial_x(\varphi(u))(1,t) = 0 \quad t \in\,]0,T[.$$

Taking into account these relations and integrating by parts once more yields that

$$\int_D \partial^2_{xx}(\varphi(u))(x,t)\psi(x,t)\,dx\,dt = - \int_D \varphi(u)(x,t)\partial^2_{xx}\psi(x,t)\,dx\,dt. \quad (4.92)$$

Replacing in (4.90) and (4.92) in (4.89), we obtain (4.67a).

Conversely, assume that u satisfies (4.68b) and let $\psi \in C^\infty_c(D)$. Integrating (4.68b) by parts and taking into account that ψ and all its derivatives are null at the edge of D gives that

$$\int_D \left[-\partial_t u(x,t) + \partial^2_{xx}(\varphi(u))(x,t) - v(x,t)\right]\psi(x,t)\,dx\,dt = 0, \forall\psi \in C^\infty_c(D).$$

Since u is of class C^2 on D, this implies that equation (4.67a) is satisfied by u. We then take $\psi \in C^\infty_T(\mathbb{R}^2)$, and we integrate (4.68b) by parts; taking into account (4.69), we obtain:

$$-\int_0^1 u(x,0)\psi(x,0)\,dx - \int_D \partial_t u(x,t)\psi(x,t)\,dx\,dt$$
$$- \int_0^T \partial_x(\varphi(u))(1,t)\psi(1,t)\,dt + \int_0^T \partial_x(\varphi(u))(0,t)\psi(0,t)\,dt$$
$$+ \int_D [\partial^2_{xx}\varphi(u)(x,t) - v(x,t)]\psi(x,t)\,dx\,dt$$
$$+ \int_0^1 u_0(x)\psi(x,0)\,dx = 0, \quad (4.93)$$

which yields, since u satisfies (4.67a),

$$\int_0^1 (u_0(x) - u(x,0))\psi(x,0)\,dx - \int_0^T \partial_x(\varphi(u))(1,t)\psi(1,t)\,dt$$
$$+ \int_0^T \partial_x(\varphi(u))(0,t)\psi(0,t)\,dt = 0.$$

Since ψ can be arbitrarily taken as the trace on D of a function belonging to $C_c^\infty(]0,1[\times] - \infty, T[)$, it follows that u satisfies the initial condition (4.67c). Then, ψ can be arbitrarily taken as the trace on D of a function belonging to $C_c^\infty(] - \infty, 1[\times]0, T[)$ or of a function belonging to $C_c^\infty(]0, +\infty[\times]0, T[)$ satisfying (4.69). It follows that u satisfies the boundary conditions (4.67b), which concludes the question.

2. Note that

$$|\int_0^T \int_0^1 \varphi(u_n)u_n \, dx \, dt - \int_0^T \int_0^1 fu \, dx \, dt| \le$$

$$|\int_0^T \int_0^1 (\varphi(u_n) - f)u_n \, dx \, dt| + |\int_0^T \int_0^1 f(u_n - u) \, dx \, dt|$$

$$\le \|u_n\|_{L^\infty(]0,1[\times]0,T[)} \, \|\varphi(u_n) - f\|_{L^1(]0,1[\times]0,T[)}$$

$$+ |\int_0^T \int_0^1 f(u_n - u) \, dx \, dt|.$$

The first term on the right-hand side of this inequality tends to 0 because the sequence $(u_n)_{n\in\mathbb{N}}$ is bounded in $L^\infty(]0, 1[\times]0, T[)$ and $\varphi(u_n) \to f$ in $L^1(]0, 1[\times]0, T[)$. The second term on the right-hand side of this inequality tends to 0 because $u_n \to u$ $\star$-weakly in $L^\infty(]0, 1[\times]0, T[)$ and $f \in L^1(]0, 1[\times]0, T[)$. We have thus shown

$$\int_0^T \int_0^1 \varphi(u_n)u_n \, dx \, dt \to \int_0^T \int_0^1 fu \, dx \, dt \text{ as } n \to +\infty. \tag{4.94}$$

Let $v \in L^\infty(]0, 1[\times]0, T[)$; since φ is non-decreasing, for all $n \in \mathbb{N}$,

$$\int_0^T \int_0^1 (\varphi(u_n) - \varphi(v))(u_n - v) \, dx \, dt \ge 0.$$

Passing to the limit as $n \to +\infty$ yields, thanks to (4.94)

$$\int_0^T \int_0^1 (f - \varphi(v))(u - v) \, dx \, dt \ge 0. \tag{4.95}$$

We now choose $v = u - s\psi$ (so that $u - v = s\psi$) with $s > 0$ and $\psi \in C_c^\infty(D)$. By passing to the limit in (4.95) as $s \to 0$ and owing to the dominated convergence theorem, we get that

$$\int_0^T \int_0^1 (f - \varphi(u))\psi \, dx \, dt \ge 0.$$

Since ψ is arbitrary in $C_c^\infty(D)$ this indeed gives $\varphi(u) = f$ a.e. in D.

3. (Existence and uniqueness of the approximate solution.)

   a. The function $s \mapsto s$ is increasing, and by hypothesis on φ, the function $s \mapsto a\varphi(s)$ is non-decreasing; hence g_a is increasing as the sum of an increasing function and a non-decreasing function. On the other hand, since φ is non-decreasing, we have $\varphi(s) \leq \varphi(0)$, $\forall s \leq 0$, and therefore $\lim_{s \to -\infty} g_a(s) = -\infty$. Similarly, $\varphi(s) \geq \varphi(0)$, $\forall s \geq 0$, and therefore $\lim_{s \to +\infty} g_a(s) = +\infty$. The function g_a is continuous and therefore takes all the values of the interval $]-\infty, +\infty[$; since it is increasing, it is bijective.

   b. Let us set $a = \frac{2k}{h^2}$; the equation (4.72) then reads:

   $$g_a(u_i) = \frac{k}{h^2}(\overline{w}_{i-1} + \overline{w}_{i+1}) + u_i^n + kv_i^n, \text{ for all } i = 1, \ldots, N.$$

   By the previous question, there exists a unique $u_i \in \mathbb{R}$ that satisfies this equation; setting $\varphi(u_i) = w_i$ uniquely determines the solution to (4.71)–(4.72). We can therefore define a function F from $\mathbb{R}^N$ to $\mathbb{R}^N$ by $\overline{w} \mapsto F(\overline{w}) = w$ where w is, with u, the solution to (4.71)–(4.72).

   c. Let $\overline{w}^1$ and $\overline{w}^2 \in \mathbb{R}^N$ and let $w^1 = F(\overline{w}^1)$ and $w^2 = F(\overline{w}^2)$. By definition of F, we have:

   $$u_i^1 - u_i^2 + \frac{2k}{h^2}(w_i^1 - w_i^2) = \frac{k}{h^2}\left((\overline{w}_{i-1}^1 + \overline{w}_{i+1}^1) - (\overline{w}_{i-1}^2 + \overline{w}_{i+1}^2)\right),$$

   $$\forall i = 1, \ldots, N. \quad (4.96)$$

   Since φ is monotone, the sign of $w_i^1 - w_i^2 = \varphi(u_i^1) - \varphi(u_i^2)$ is the same as that of $u_i^1 - u_i^2$, and therefore

   $$|u_i^1 - u_i^2 + \frac{2k}{h^2}(w_i^1 - w_i^2)| = |u_i^1 - u_i^2| + \frac{2k}{h^2}|w_i^1 - w_i^2|. \quad (4.97)$$

   Moreover, since φ is Lipschitz-continuous, denoting by L_φ a Lipschitz constant of φ (with $L_\varphi > 0$), we have

   $$|w_i^1 - w_i^2| = |\varphi(u_i^1) - \varphi(u_i^2)| \leq L_\varphi |u_i^1 - u_i^2|,$$

   from which:

   $$|u_i^1 - u_i^2| \geq \frac{1}{L_\varphi}|w_i^1 - w_i^2|. \quad (4.98)$$

   Owing to (4.96)-(4.98), we get that

   $$\frac{1}{L_\varphi}|w_i^1 - w_i^2| + \frac{2k}{h^2}|w_i^1 - w_i^2| \leq \frac{k}{h^2}\left(|\overline{w}_{i-1}^1 - \overline{w}_{i-1}^2| + |\overline{w}_{i+1}^1 - \overline{w}_{i+1}^2|\right),$$

   $$\forall i = 1, \ldots, N.$$

Hence

$$|w_i^1 - w_i^2| \leq \frac{1}{1 + \frac{h^2}{2kL_\varphi}} \max_{i=1,\ldots,N} |\overline{w}_i^1 - \overline{w}_i^2| \quad \forall i = 1,\ldots,N.$$

from which we get that $\|w^1 - w^2\|_\infty \leq C\|\overline{w}^1 - \overline{w}^2\|_\infty$ with $C = \frac{1}{1+\frac{h^2}{2kL_\varphi}} < 1$.

The mapping F is therefore strictly contracting.

(Note that we used $\overline{w}_0 = \overline{w}_1$ and $\overline{w}_{N+1} = \overline{w}_N$ to obtain the upper bound of $|w_i^1 - w_i^2|$ for $i = 1$ and $i = N$.)

d. Observe that

$$u_i^{n+1} + \frac{2k}{h^2}\varphi(u_i^{n+1}) = \frac{k}{h^2}(\varphi(u_{i-1}^{n+1}) + \varphi(u_{i+1}^{n+1})) + u_i^n + kv_i^n,$$

$$\text{for all } i = 1,\ldots,N,$$

and so, setting $w_0 = \varphi(u_0^{n+1})$ and $w_{N+1} = \varphi(u_{N+1}^{n+1})$ leads to $w_0 = w_1, w_{N+1} = w_N$ (by (4.70c)) and

$$u_i^{n+1} + \frac{2k}{h^2}w_i = \frac{k}{h^2}(w_{i-1} + w_{i+1}) + u_i^n + kv_i^n, \text{ for all } i = 1,\ldots,N,$$

which clearly shows that $F(w) = w$ and also that, for all $i \in \{1,\ldots,N\}$, u_i^{n+1} is the unique solution of the following equation with unknown z:

$$z + \frac{2k}{h^2}\varphi(z) = \frac{k}{h^2}(w_{i-1} + w_{i+1}) + u_i^n + kv_i^n.$$

e. Since $w = F(w)$, the definition of F gives $w_0 = w_1, w_{N+1} = w_0$,

$$\varphi(u_i) = w_i, \text{ for all } i \in \{1,\ldots,N\},$$

and

$$u_i + \frac{2k}{h^2}w_i = \frac{k}{h^2}(w_{i-1} + w_{i+1}) + u_i^n + kv_i^n, \text{ for all } i = 1,\ldots,N.$$

Taking $u_i^{n+1} = u_i$ for all $i \in \{1,\ldots,N\}$ and $u_0^{n+1} = u_1^{n+1}$, $u_{N+1}^{n+1} = u_N^{n+1}$, the family $\{u_i^{n+1}, i = 0,\ldots,N+1\}$ is a solution to (4.70) with $w_i = \varphi(u_i^{n+1})$ for all $i \in \{1,\ldots,N\}$.

f. The existence of a solution $\{u_i^{n+1}, i = 0,\ldots,N+1\}$ to (4.70) is given by question 3e (because F has a fixed point).

To show uniqueness, we note that if $\{u_i^{n+1}, i = 0,\ldots,N+1\}$ is a solution to (4.70) question 3d shows that w is the unique fixed point of F with $w_i = \varphi(u_i^{n+1})$ for all $i = 1,\ldots,N$. Hence by (4.70)

$$u_i^{n+1} = \frac{k}{h^2}(w_{i-1} - 2w_i + w_{i+1}) + u_i^n + kv_i^n, \quad \text{for all } i = 1, \ldots, N, \qquad (4.99)$$

and the uniqueness of $\{u_i^{n+1}, i = 0, \ldots, N+1\}$ follows.

4. ($L^\infty(]0, 1[\times]0, T[)$ estimate on the approximate solution) The relation to be proved by induction is clearly satisfied at rank $n = 0$, by definition of A. Suppose it is true up to rank n, and let us prove it at rank $n + 1$. The relation (4.70b) still reads:

$$u_i^{n+1} = u_i^n + \frac{k}{h^2}(\varphi(u_{i-1}^{n+1}) - \varphi(u_i^{n+1})) + \frac{k}{h^2}(\varphi(u_{i+1}^{n+1}) - \varphi(u_i^{n+1})) + kv_i^n$$

for $i = 1, \ldots, N$ and $n = 0, \ldots, M - 1$. Suppose that i is such that $u_i^{n+1} = \min_{j=1,\ldots,N} u_j^{n+1}$. Since φ is non-decreasing, in this case:

$$\varphi(u_{i-1}^{n+1}) - \varphi(u_i^{n+1}) \geq 0 \quad \text{and} \quad \varphi(u_{i+1}^{n+1}) - \varphi(u_i^{n+1}) \geq 0,$$

and we get that

$$\min_{j=1,\ldots,N} u_j^{n+1} \geq u_i^n - kB$$

from which, by induction hypothesis

$$\min_{j=1,\ldots,N} u_j^{n+1} \geq -A - nkB - kB.$$

Similarly, considering now i such that $u_i^{n+1} = \max_{j=1,\ldots,N} u_j^{n+1}$ leads to:

$$\max_{j=1,\ldots,N} u_j^{n+1} \leq u_i^n + kB \leq A + nkB + kB.$$

Hence:

$$-A - (n+1)kB \leq u_i^{n+1} \leq A + (n+1)kB \quad \forall i = 1, \ldots, N, \quad \forall n = 0, \ldots, M - 1.$$

Hence, for all $n = 0, \ldots, M$, $\|u^n\|_{L^\infty(]0,1[)} \leq c_{u_0,v,T}$ with $c_{u_0,v,T} = A + BT$.

5. ($L^2(]0, T[, H_d^1)$ estimate of $\varphi(u)$, where H_d^1 is a discrete equivalent of the space H^1) Since φ is non-decreasing and Lipschitz-continuous,

$$\sum_{i=1}^{N-1}(\varphi(u_{i+1}^{n+1}) - \varphi(u_i^{n+1}))^2 \leq L_\varphi X^n$$

$$\text{with } X^n = \sum_{i=1}^{N-1}(\varphi(u_{i+1}^{n+1}) - \varphi(u_i^{n+1}))(u_{i+1}^{n+1} - u_i^{n+1}),$$

where L_φ denotes the Lipschitz constant of φ. We now show that there exists a C_{T,v,u_0} (depending only on T, v and u_0) such that $X^n \le C_{T,v,u_0}\frac{h}{k}$ for all $n = 0, \ldots, M - 1$. Multiplying (4.70b) by u_i^{n+1} and summing over $i = 1, \ldots, N$ yields

$$X_1^n + X_2^n = X_3^n$$

with

$$X_1^n = \sum_{i=1}^{N} \frac{u_i^{n+1} - u_i^n}{k} u_i^{n+1} = \frac{1}{k}\Big[\sum_{i=1}^{N} \frac{1}{2}(u_i^{n+1} - u_i^n)^2 + \frac{1}{2}(u_i^{n+1})^2 - \frac{1}{2}(u_i^n)^2\Big]$$

$$\ge \frac{1}{2k}\sum_{i=1}^{N}\big[(u_i^{n+1})^2 - (u_i^n)^2\big] := \widetilde{X}_1^n, \tag{4.100}$$

$$X_2^n = -\frac{1}{h^2}\sum_{i=1}^{N}(\varphi(u_{i-1}^{n+1}) - 2\varphi(u_i^{n+1}) + \varphi(u_{i+1}^{n+1}))u_i^{n+1}$$

$$= \frac{1}{h^2}\sum_{i=1}^{N-1}(\varphi(u_{i+1}^{n+1})\varphi(u_i^{n+1}))(u_{i+1}^{n+1} - u_i^{n+1}) = \frac{1}{h^2}X^n \tag{4.101}$$

and $X_3^n = \sum_{i=1}^{N} v_i^n u_i^{n+1}$. From (4.100), (4.101) and the previous question, we get that

$$\widetilde{X}_1^n + \frac{1}{h^2}X^n \le \sum_{i=1}^{N} v_i^n u_i^{n+1} \le \frac{1}{h}B(A + (n+1)kB).$$

Summing over $n = 0, \ldots, M - 1$ and noting that $\sum_{n=0}^{M-1} \widetilde{X}_1^n = \frac{1}{2k}\sum_{i=1}^{N}(u_i^M)^2 - \sum_{i=1}^{N}\frac{1}{2k}(u_i^0)^2$, we obtain that

$$\sum_{n=0}^{M-1} X^n \le \frac{h}{k}BT(A + TB) + \frac{h}{2k}A^2,$$

from which we get (4.73) with $C_{T,\varphi,v,u_0} = L_\varphi(BT(A + TB) + A^2)$.

6. ($L^2(]0, T[, L^2)$ estimate of a discrete equivalent of $\partial_t \varphi(u)$ and $L^\infty(]0, T[, H_d^1)$ estimate of $\varphi(u)$)

   a. Integrating by parts and owing to the Fubini–Tonelli theorem, a regular solution w of (4.67) satisfies

$$-\int_0^T \int_0^1 \partial_{xx}^2 w(x, t)\, \partial_t w(x, t)\, \mathrm{d}x\, \mathrm{d}t$$

$$= \int_0^T \int_0^1 \partial_x w(x, t)\, \partial_{xt}^2 w(x, t)\, \mathrm{d}x\, \mathrm{d}t$$

$$= \frac{1}{2} \int_0^1 \int_0^T \partial_t \left((\partial_x w(x,t))^2 \right) dx\, dt$$

$$= \frac{1}{2} \int_0^1 \left[(\partial_x w(x,T))^2) - (\partial_x w(x,0))^2) \right] dx\, dt.$$

Let us now write the discrete counterpart of this equality. Setting

$$Y = \sum_{n=1}^{M-1} \sum_{i=0}^{N} (2w_i^{n+1} - w_{i-1}^{n+1} - w_{i+1}^{n+1})(w_i^{n+1} - w_i^n),$$

since $w_0^n = w_1^n$ and $w_N^n = w_{N+1}^n$, a discrete integration by parts yields

$$Y = \sum_{n=0}^{M-1} \sum_{i=1}^{N} \left[(w_i^{n+1} - w_{i-1}^{n+1}) + (w_i^{n+1} - w_{i+1}^{n+1}) \right] (w_i^{n+1} - w_i^n)$$

$$= \sum_{n=0}^{M-1} \sum_{i=1}^{N-1} (w_i^{n+1} - w_{i+1}^{n+1}) \left[(w_i^{n+1} - w_i^n) - (w_{i+1}^{n+1} - w_{i+1}^n) \right]$$

$$= \sum_{n=0}^{M-1} \sum_{i=1}^{N-1} (w_i^{n+1} - w_{i+1}^{n+1}) \left[(w_i^{n+1} - w_{i+1}^{n+1}) - (w_i^n - w_{i+1}^n) \right]$$

Using the inequality $a(a-b) = \frac{1}{2}(a-b)^2 + \frac{1}{2}a^2 - \frac{1}{2}b^2 \geq \frac{1}{2}a^2 - \frac{1}{2}b^2$, we then obtain

$$Y \geq \frac{1}{2} \sum_{n=0}^{M-1} \sum_{i=1}^{N-1} \left[(w_i^{n+1} - w_{i+1}^{n+1})^2 - (w_i^n - w_{i+1}^n)^2 \right],$$

which indeed gives the inequality (4.74).

b. Multiply (4.70b) by $\varphi(u_i^{n+1}) - \varphi(u_i^n)$ and sum over i and n to obtain:

$$X + Y = Z, \tag{4.102}$$

with

$$X = \sum_{n=0}^{M-1} \sum_{i=1}^{N} \frac{u_i^{n+1} - u_i^n}{k} (\varphi(u_i^{n+1}) - \varphi(u_i^n))$$

$$Y = - \sum_{n=0}^{M-1} \sum_{i=1}^{N} \frac{\varphi(u_{i-1}^{n+1}) - 2\varphi(u_i^{n+1}) + \varphi(u_{i+1}^{n+1})}{h^2} (\varphi(u_i^{n+1}) - \varphi(u_i^n))$$

$$Z = \sum_{n=0}^{M-1} \sum_{i=1}^{N} v_i^n (\varphi(u_i^{n+1}) - \varphi(u_i^n)).$$

Denoting by L_φ a Lipschitz constant of φ ($L_\varphi > 0$), we obtain the following lower bound:

$$X \geq \frac{1}{L_\varphi k} \sum_{n=0}^{M-1} \sum_{i=1}^{N} (\varphi(u_i^{n+1}) - \varphi(u_i^n))^2. \tag{4.103}$$

Using the inequality (4.74) with $w_i^n = \varphi(u_i^n)$ for Y, we obtain:

$$Y \geq \frac{1}{2h^2} \Big[\sum_{i=1}^{N} (\varphi(u_{i+1}^M) - \varphi(u_i^M))^2 - \sum_{i=1}^{N} (\varphi(u_{i+1}^0) - \varphi(u_i^0))^2 \Big]. \tag{4.104}$$

Finally, Young's inequality yields the following upper bound on Z:

$$Z \leq \frac{1}{2L_\varphi k} \sum_{n=0}^{M-1} \sum_{i=1}^{N} (\varphi(u_i^{n+1}) - \varphi(u_i^n))^2 + \frac{1}{2} L_\varphi k \sum_{n=0}^{M-1} \sum_{i=1}^{N} (v_i^n)^2$$

$$\leq \frac{1}{2L_\varphi k} \sum_{n=0}^{M-1} \sum_{i=1}^{N} (\varphi(u_i^{n+1}) - \varphi(u_i^n))^2 + \frac{1}{2} \frac{L_\varphi}{h} B^2. \tag{4.105}$$

Using (4.102), (4.103), (4.104) and (4.105), we obtain:

$$\frac{1}{2L_\varphi k} \sum_{n=0}^{M-1} \sum_{i=1}^{N} (\varphi(u_i^{n+1}) - \varphi(u_i^n))^2$$

$$\leq \frac{1}{2h^2} \sum_{i=1}^{N} (\varphi(u_{i+1}^0) - \varphi(u_i^0))^2 + \frac{1}{2} \frac{L_\varphi}{h} B^2. \tag{4.106}$$

From this, we get that

$$\sum_{n=0}^{M-1} h \sum_{i=1}^{N} (\varphi(u_i^{n+1}) - \varphi(u_i^n))^2$$

$$\leq 2L_\varphi k \Big(\frac{1}{2h} \sum_{i=1}^{N} (\varphi(u_{i+1}^0) - \varphi(u_i^0))^2 + \frac{1}{2} L_\varphi B^2 \Big).$$

However, $|\varphi(u_{i+1}^0) - \varphi(u_i^0))| \leq L_\varphi c_0 h$, where c_0 depends only on u_0 (since u_0 is Lipschitz-continuous), thus we have

$$\sum_{n=0}^{M-1} h \sum_{i=1}^{N} (\varphi(u_i^{n+1}) - \varphi(u_i^n))^2 \leq 2L_\varphi k \Big(\frac{1}{2h^2} L_\varphi^2 c_0^2 h^2 + \frac{1}{2} L_\varphi B^2 \Big) = C_2 k,$$

which gives (4.75).

c. In the inequality (4.106), we omitted the non-negative term on the right of (4.104). If we take into account this term, (4.106) becomes

$$\frac{1}{2h^2} \sum_{i=1}^{N} (\varphi(u_{i+1}^M) - \varphi(u_i^M))^2 + \frac{1}{2L_\varphi k} \sum_{n=0}^{M-1} \sum_{i=1}^{N} (\varphi(u_i^{n+1}) - \varphi(u_i^n))^2$$

$$\leq \frac{1}{2h^2} \sum_{i=1}^{N} (\varphi(u_{i+1}^0) - \varphi(u_i^0))^2 + \frac{1}{2} \frac{L_\varphi}{h} B^2. \quad (4.107)$$

The multiplication of (4.107) by $2hL_\varphi k$ then yields

$$\frac{L_\varphi k}{h} \sum_{i=1}^{N} (\varphi(u_{i+1}^M) - \varphi(u_i^M))^2 \leq C_2 k,$$

which gives (4.76) for $n = M$. The same estimate for $n < M$ is obtained by summing from 0 to $n - 1$ instead of from 0 to $M - 1$.

7. Question 4 shows that

$$\forall n \in \mathbb{N}, \ \|u_n\|_{L^\infty(]0,1[\times]0,T[)} \leq c_{u_0,v,T}.$$

The sequence $(u_n)_{n \in \mathbb{N}}$ is therefore bounded in $L^\infty(]0, 1[\times]0, T[)$ and it admits a subsequence that $\star$-weakly converges in $L^\infty(]0, 1[\times]0, T[)$. Since φ is continuous, the sequence $(\varphi(u_n)_{n \in \mathbb{N}}$ is also bounded in $L^\infty(]0, 1[\times]0, T[)$ and therefore bounded in $L^2(]0, 1[\times]0, T[)$ (which we identify with $L^2(]0, T[, L^2]0, 1[)$).
We now apply Theorem 4.57 to the sequence $(\varphi(u_n))_{n \in \mathbb{N}}$ with for X_n the space X_m with $m = N_n$ defined in Problem 4.7, $Y_n = B = L^2(]0, 1[)$ and $p = 2$.
First note that the sequence of spaces $(X_n, Y_n)_{n \in \mathbb{N}}$ is compact-continuous in B. Indeed, the fact that the sequence $(X_n)_{n \in \mathbb{N}}$ is compactly embedded in B is shown in Problem 4.7. This gives property 1 of Definition 4.54. Observe that property 2 of Definition 4.54 is immediate because $Y_n = B$. Then, the function $\varphi(u_n)$ indeed has the form required by Theorem 4.57. It has already been noted that hypothesis 2 of Theorem 4.57 is satisfied. Hypothesis 3 of Theorem 4.57 is given by the estimate (4.73). Finally, hypothesis 4 of Theorem 4.57 is given by the estimate (4.75). Theorem 4.57 therefore gives that the sequence $(\varphi(u_n))_{n \in \mathbb{N}}$ admits a convergent subsequence in $L^2(]0, T[, L^2(]0, 1[)$ (identified with $L^2(]0, 1[\times]0, T[))$. From this sequence $(u_n)_{n \in \mathbb{N}}$ we can therefore extract a subsequence, still denoted by $(u_n)_{n \in \mathbb{N}}$, such that

- $u_n \to u$ $\star$-weakly in $L^\infty(]0, 1[\times]0, T[)$,
- $\varphi(u_n) \to f$ in $L^2(]0, 1[\times]0, T[)$.

Since the sequence $(\varphi(u_n))_{n \in \mathbb{N}}$ is bounded in L^∞, we also have $\varphi(u_n) \to f$ in $L^p(]0, 1[\times]0, T[)$, for all $p \in [1, \infty[$.
Finally, question 2 shows that $f = \varphi(u)$ a.e.

8. For given h, k, let u be the solution to (4.70) obtained with these parameters. Let $\psi \in C_T^\infty(\mathbb{R}^2)$. Set $\psi_i^{n+1} = \psi(x_i, (n+1)k)$. Multiplying (4.70b) by ψ_i^{n+1} and summing over i and n, we get

$$T_1 + T_2 = T_3,$$

$$T_1 = hk \sum_{n=0}^{M-1} \sum_{i=1}^{N} \frac{u_i^{n+1} - u_i^n}{k} \psi_i^{n+1},$$

$$T_2 = -hk \sum_{n=0}^{M-1} \sum_{i=1}^{N} \frac{\varphi(u_{i-1}^{n+1}) - 2\varphi(u_i^{n+1}) + \varphi(u_{i+1}^{n+1})}{h^2} \psi_i^{n+1},$$

$$T_3 = hk \sum_{n=0}^{M-1} \sum_{i=1}^{N} v_i^n \psi_i^{n+1}.$$

A discrete integration by parts in time gives

$$T_1 = hk \sum_{n=1}^{M-1} \sum_{i=1}^{N} u_i^n \frac{\psi_i^n - \psi_i^{n+1}}{k} + \sum_{i=1}^{N} hu_i^M \psi_i^M - \sum_{i=1}^{N} hu_i^0 \psi_i^1.$$

Hence there exists a C_1 depending only on $c_{u_0,v,T}$ (bound on u), u_0 and ψ such that

$$T_1 = -\int_0^T \int_0^1 u(x,t)\psi(x,t)\,dx\,dt + R_{1,1} + R_{1,2}$$
$$-\int_0^1 u(x,t)\psi(x,0)\,dx + R_{1,3},$$

with $|R_{1,1}| \leq C_1(k+h)$, $|R_{1,2}| \leq C_1 k$ and $|R_{1,3}| \leq C_1(k+h)$. A double integration by parts in time gives, by setting $\psi_0^n = \psi(-h, nk)$ and $\psi_{N+1}^n = \psi(1+h, nk)$,

$$T_2 = -hk \sum_{n=0}^{M-1} \sum_{i=1}^{N} \frac{\psi_{i-1}^{n+1} - 2\psi_i^{n+1} + \psi_{i+1}^{n+1}}{h^2} \varphi_i^{n+1}$$
$$-kh \sum_{n=0}^{M-1} \frac{\psi_0^{n+1} - \psi_1^{n+1}}{h^2} \varphi_1^{n+1}$$
$$-kh \sum_{n=0}^{M-1} \frac{\psi_N^{n+1} - \psi_{N+1}^{n+1}}{h^2} \varphi_N^{n+1}.$$

It follows that there exists a C_2 depending only on $c_{u_0,v,T}$, φ and ψ such that

$$T_1 = -\int_0^T \int_0^1 \Delta\psi(x,t)\varphi(u)(x,t)\,dx\,dt + R_{2,1} + R_{2,2} + R_{2,3},$$

with $|R_{2,1}| \leq C_2(k+h)$, $|R_{2,2}| \leq C_2 h$ and $|R_{2,2}| \leq C_2 h$.

The upper bound of $R_{2,1}$ is due to the C^3 character of ψ. For the upper bound of $R_{2,2}$ and $R_{2,3}$ it is enough to note that the Neumann condition on ψ gives an upper bound in Ch^2 (C depending only on ψ) of $|\psi_N^{n+1} - \psi_{N+1}^{n+1}|$ and $|\psi_0^{n+1} - \psi_1^{n+1}|$. Finally,

$$T_3 = \int_0^T \int_0^1 v(x,t)\psi(x,t)\,\mathrm{d}x\,\mathrm{d}t + R_3$$

and there exists a C_3 depending only on v and ψ such that $R_3 \le C_3(h+k)$.

It is now enough to write $T_1 + T_2 = T_3$ with the function u_n constructed with the parameters h_n and k_n. When $n \to +\infty$, since $\lim_{n\to+\infty} h_n = \lim_{n\to+\infty} k_n = 0$ and that $u_n \to u$ and $\varphi(u_n) \to \varphi(u)$ in $L^1(]0,1[\times]0,T[)$, we obtain (4.68b) (and we already knew that u satisfies (4.68a)).

Problem 4.10 (Existence for the Stefan problem, by regularisation)

1. The result of Problem 4.6 gives the existence of a solution u_n of

$$\begin{cases} u_n \in L^2(]0,T[,H_0^1(\Omega)), \ \partial_t u_n \in L^2(]0,T[,H^{-1}(\Omega)), \\[2mm] \displaystyle\int_0^T \langle \partial_t u_n(s), v(s)\rangle_{H^{-1},H_0^1}\,\mathrm{d}s \\[2mm] \qquad \displaystyle + \int_0^T \int_\Omega (\varphi'(u_n(x,s)) + \frac{1}{n})\nabla u_n(x,s)\cdot \nabla v(x,s)\,\mathrm{d}x\,\mathrm{d}s \\[2mm] \displaystyle = \int_0^T (\int_\Omega f(s,x)v(s,x)\,\mathrm{d}x)\,\mathrm{d}s, \qquad \forall v \in L^2(]0,T[,H_0^1(\Omega)), \\[2mm] u_n(0) = u_0 \ \text{ a.e.} \end{cases} \qquad (4.108)$$

Thanks to Lemma 4.35, we indeed obtain that u_n is solution of (4.77) with φ_n instead of φ, with the additional regularity $u_n \in L^2(]0,T[,H_0^1(\Omega))$ (and thus $u_n \in C([0,T]),L^2(\Omega))$.

2. Let Φ be a primitive of φ. Since $u_n \in L^2(]0,T[,H_0^1(\Omega))$ and $\partial_t u_n \in L^2(]0,T[,H^{-1}(\Omega))$, Lemma 4.35 gives that $\Phi(u_n) \in C([0,T],L^1(\Omega))$ and, for $0 \le t_1 < t_2 \le T$,

$$\int_\Omega \Phi(u_n(t_2))\,\mathrm{d}x - \int_\Omega \Phi(u_n(t_1))\,\mathrm{d}x = \int_{t_1}^{t_2} \langle \partial_t u_n(s), \varphi(u_n(s))\rangle_{H^{-1},H_0^1}\,\mathrm{d}s.$$

Then taking $v = \varphi(u_n)$ in (4.108), we obtain a control of $\Phi(u_n)$ in $C([0,T],L^1(\Omega))$ and a control of $\varphi(u_n)$ in $L^2(]0,T[,H_0^1(\Omega))$, in the sense that there exists a $C \in \mathbb{R}_+$, depending only on f and u_0 such that for all $t \in [0,T]$,

$$\|\Phi(u_n(t))\|_{L^1(\Omega)} \le C, \qquad (4.109)$$

$$\|\varphi(u_n)\|_{L^2(]0,T[,H_0^1(\Omega))} \le C \ \text{ and } \ \|u_n\|_{L^2(]0,T[,H_0^1(\Omega))} \le C\sqrt{n}. \qquad (4.110)$$

By virtue of (4.110) and (4.108), we then obtain an estimate on $\partial_t u_n$ in $L^2(]0,T[,H^{-1}(\Omega))$ which reads

$$\|\partial_t u_n\|_{L^2(]0,T[,H^{-1}(\Omega))} \leq C. \tag{4.111}$$

We now obtain an estimate on u_n in $C([0,T], L^2(\Omega))$ thanks to the fact that $f \in L^2(]0,T[, L^2(\Omega))$. Let us first recall that $u_n \in C([0,T], L^2(\Omega))$; by taking $v = u_n$ in (4.108) we obtain (with Lemma 4.27), that for all $t \in [0,T]$,

$$\frac{1}{2}\|u_n(t)\|^2_{L^2(\Omega)} - \frac{1}{2}\|u_0\|^2_{L^2(\Omega)} \leq \int_0^t \int_\Omega f(x,s)u_n(x,s)\,\mathrm{d}x\,\mathrm{d}s$$

$$\leq \int_0^t \|f(s)\|_{L^2(\Omega)}\|u_n(s)\|_{L^2(\Omega)}\,\mathrm{d}s$$

$$\leq \int_0^T \|f(s)\|^2_{L^2(\Omega)}\,\mathrm{d}s + \int_0^t \|u_n(s)\|^2_{L^2(\Omega)}\,\mathrm{d}s,$$

which gives an estimate on $\|u_n(t)\|^2_{L^2(\Omega)}$ by the classical Grönwall's[7] lemma[8]. We can then still assume that for all $t \in [0,T]$,

$$\|u_n(t)\|_{L^2(\Omega)} \leq C. \tag{4.112}$$

3. The estimate (4.110) allows us to assume that, up to a subsequence, which we still denote by $(u_n)_{n\in\mathbb{N}}$, we have $\varphi(u_n) \rightarrow \zeta$ weakly in $L^2(]0,T[, H_0^1(\Omega))$. Moreover, since $L^2(\Omega)$ is compactly embedded in $H^{-1}(\Omega)$, we can also assume that, up to a subsequence, still denoted by $(u_n)_{n\in\mathbb{N}}$, we have $u_n \rightarrow u$ in $L^2(]0,T[, H^{-1}(\Omega))$, thanks to (4.111)-(4.112)) and by applying Theorem 4.46, with $B = Y = H^{-1}(\Omega)$ and $X = L^2(\Omega)$.
   By virtue of (4.112), we also have $u_n \rightarrow u$ weakly in $L^2(]0,T[, L^2(\Omega))$ and $u \in L^\infty(]0,T[, L^2(\Omega))$.
4. Question 3 shows that $\varphi(u_n) \rightarrow \zeta$ weakly in $L^2(]0,T[, H_0^1(\Omega))$ and that $u_n \rightarrow u$ in $L^2(]0,T[, H^{-1}(\Omega))$ as $n \rightarrow +\infty$.
   Since $u \in L^2(]0,T[, L^2(\Omega))$, we also have

$$\int_0^T \int_\Omega \varphi(u_n(x,y))u_n(x,t)\,\mathrm{d}x\,\mathrm{d}t = \int_0^T \langle u_n(t), \varphi(u_n(t))\rangle_{H^{-1},H_0^1}\,\mathrm{d}t$$

$$\rightarrow \int_0^T \langle u(t), \zeta(t)\rangle_{H^{-1},H_0^1}\,\mathrm{d}t \text{ as } n \rightarrow +\infty.$$

---

[7] Thomas Hakon Grönwall (1877–1932), Swedish mathematician.

[8] Grönwall's Lemma: Let $\alpha > 0$, $\psi \in C([0,\alpha],\mathbb{R})$ and $a, b \geq 0$; we assume that, for all $t \in [0,\alpha]$, $0 \leq \psi(t) \leq a + b\int_0^t \psi(s)\,\mathrm{d}s$. Then, for all $t \in [0,\alpha]$, $\psi(t) \leq ae^{bt}$.

Hence

$$\int_0^T \int_\Omega \varphi(u_n(x,y))u_n(x,t)\,dx\,dt$$

$$\rightarrow \int_0^T \int_\Omega \zeta(x,t)u(x,t)\,dx\,dt \text{ as } n \rightarrow +\infty.$$

5. Since φ is non-decreasing, we have for all $v \in L^2(]0,T[, L^2(\Omega))$,

$$0 \le \int_0^T \int_\Omega \varphi(u_n(x,t)) - \varphi(v(x,t))(u_n(x,t) - v(x,t))\,dx\,dt$$

$$\rightarrow \int_0^T \int_\Omega (\zeta(x,t) - \varphi(v(x,t)))(u(x,t) - v(x,t))\,dx\,dt \text{ as } n \rightarrow +\infty,$$

and therefore

$$\int_0^T \int_\Omega (\zeta(x,t) - \varphi(v(x,t)))(u(x,t) - v(x,t))\,dx\,dt \ge 0. \qquad (4.113)$$

Let $w \in C_c^\infty(\Omega \times]0,T[)$ and $t > 0$. Taking $v = u + \tau w$ in (4.113) yields

$$\int_0^T \int_\Omega (\zeta(x,t) - \varphi(u(x,t) + \tau w(x,t))w(x,t)\,dx\,dt \le 0,$$

and by the dominated convergence theorem (which applies because φ is Lipschitz-continuous, by letting τ tend to 0, we obtain that

$$\int_0^T \int_\Omega (\zeta(x,t) - \varphi(u(x,t))w(x,t)\,dx\,dt \le 0.$$

Since w is arbitrary, this implies that $\varphi(u) = \zeta$ a.e.

6. Since $u_n \in L^2(]0,T[, L^2(\Omega)) \subset L^2(]0,T[, H^{-1}(\Omega))$ and $\partial_t u_n \in L^2(]0,T[, H^{-1}(\Omega))$, the function u_n belongs to $C([0,T], H^{-1}(\Omega))$, and by Lemma 4.26, also to $C^{0,\frac{1}{2}}([0,T], H^{-1}(\Omega))$. We now show that the sequence $(u_n)_{n \in \mathbb{N}^\star}$ is relatively compact in $C([0,T], H^{-1}(\Omega))$ by the Ascoli theorem (Theorem 1.35); to this end, let us prove that:

a. For all $t \in [0,T]$, the sequence $(u_n(t))_{n \in \mathbb{N}^\star}$ is relatively compact in $H^{-1}(\Omega)$.
b. $\|u_n(t) - u_n(s)\|_{H^{-1}} \rightarrow 0$ as $s \rightarrow t$, uniformly with respect to $n \in \mathbb{N}^\star$ (and for all $t \in [0,T]$).

Point 6b is a consequence of the fact that $\partial_t u_n \in L^1([0,T], H^{-1}(\Omega))$, thanks to Lemma 4.26, which states that for all $t_1, t_2 \in [0,T]$, $t_1 > t_2$ and all $n \in \mathbb{N}^\star$,

$$u_n(t_1) - u_n(t_2) = \int_{t_2}^{t_1} \partial_t u_n(s)\,ds,$$

and therefore

$$\|u_n(t_1) - u_n(t_2)\|_{H^{-1}} \leq \int_{t_2}^{t_1} \|\partial_t u_n(s)\|_{H^{-1}} \, ds$$

$$\leq \left(\int_0^T \|\partial_t u_n(s)\|_{H^{-1}}^2 \, ds \right)^{\frac{1}{2}} \sqrt{t_1 - t_2}$$

$$\leq \|\partial_t u_n\|_{L^2(]0,T[,H^{-1})} \sqrt{t_1 - t_2}.$$

The sequence $(\partial_t u_n)_{n \in \mathbb{N}^\star}$ is bounded in $L^2(]0, T[, H^{-1}(\Omega))$. We therefore deduce from this inequality that $\|u_n(t) - u_n(s)\|_{H^{-1}} \to 0$, as $s \to t$, uniformly with respect to $n \in \mathbb{N}^\star$ (and for all $t \in [0, T]$).

To prove point 6a, we use the estimate (4.112), which implies that the sequence $(u_n(t))_{n \in \mathbb{N}^\star}$ is bounded in $L^2(\Omega)$ for all $t \in [0, T]$ and is therefore relatively compact in $H^{-1}(\Omega)$ for all $t \in [0, T]$. We can therefore apply Ascoli's theorem (Theorem 1.35) and obtain, as announced, that $u(0) = u_0$.

7. By the estimate (4.111) and the linearity of the operator ∂_t, it follows that $\partial_t u_n \to \partial_t u$ weakly in $L^2(]0, T[, H^{-1}(\Omega))$. The estimate (4.110) also gives that

$$\left| \frac{1}{n} \int_0^T \int_\Omega \nabla u_n(x, s) \cdot \nabla v(x, s) \, dx \, ds \right| \leq \frac{C}{\sqrt{n}} \|v\|_{L^2(]0,T[,H_0^1(\Omega))},$$

and therefore

$$\lim_{n \to \infty} \frac{1}{n} \int_0^T \int_\Omega \nabla u_n(x, s) \cdot \nabla v(x, s) \, dx \, ds = 0.$$

It is now possible to pass to the limit in (4.108); we obtain that for all $v \in L^2(]0, T[, H_0^1(\Omega))$,

$$\int_0^T \langle \partial_t u(s), v(s) \rangle_{H^{-1}, H_0^1} \, ds + \int_0^T \int_\Omega \nabla \zeta(x, s) \cdot \nabla v(x, s) \, dx \, ds$$

$$= \int_0^T \left(\int_\Omega f(s, x) v(s, x) \, dx \right) ds.$$

Question 5 gives that $\zeta = \varphi(u)$ a.e. and question 6 gives that $u(0) = u_0$. Therefore, u is a solution to (4.77).

Chapter 5
Hyperbolic Problems

Hyperbolic PDEs occur for example in fluid mechanics (aeronautics, two-phase flows, modelling of dam breakage and avalanches). They are often obtained by neglecting the diffusion phenomena (because these are small at the considered scale) in the conservation equations of mechanics (for example, conservation of momentum or energy). The basic example of a linear hyperbolic PDE is the transport equation which, in one space dimension, reads $\partial_t u + a\partial_x u = 0$. Another example is the one-dimensional wave equation $\partial_{tt}^2 u - c^2 \partial_{xx}^2 u = 0$, which can be rewritten as a system of two first order PDEs.

In fact, the solutions of hyperbolic equations are said to be of the "wave" type, in the sense that a perturbation in the initial data of a hyperbolic PDE do not affect all points in space at the same time. Relative to a fixed temporal coordinate, perturbations have a finite propagation speed. This property qualitatively distinguishes hyperbolic PDEs from parabolic PDEs, for which a perturbation in the initial data affects all points in the domain at the same instant.

In this chapter, we first focus on linear and non-linear scalar hyperbolic equations, for which the mathematical theory is well advanced. In particular, we prove the existence and uniqueness of the entropy weak solution (Theorem 5.29, due to Kruzhkov[1]). The one-dimensional case is first studied in Section 5.1, while the multi-dimensional case is addressed in Section 5.2. Section 5.3 gives some elements for the study of non-linear hyperbolic systems, whose mathematical theory is far from complete. In this chapter, we study in detail the Riemann problem, which is one of the main tools of the classical numerical methods dedicated to these problems.

---

[1] Stanislav Nikolaevich Kruzhkov (1936–1997), Russian mathematician, specialist in the analysis of non-linear PDEs.

5.1 Scalar Equations: the One-Dimensional Case

We begin by studying scalar equations in the one-dimensional space case. The solutions that we seek are thus functions of $x \in \mathbb{R}$ and $t \in \mathbb{R}_+$. Let us first take the simple example of the transport (or advection) equation.

$$\partial_t u - c\partial_x u = 0, \; t \in \mathbb{R}_+^{\star}, \; x \in \mathbb{R},$$

where the real number c is the transport velocity, with initial condition:

$$u(x, 0) = u_0(x).$$

In the case where the initial condition u_0 is sufficiently regular, observe that the function:

$$u(x, t) = u_0(x + ct),$$

is a classical solution to this problem. If u_0 is non-regular (for example discontinuous), we will see that there is still a way to show that the function thus defined is a solution in a sense that we qualify as a *weak solution.*

If we now consider a non-linear equation, i.e. $\partial_t u + \partial_x(f(u)) = 0$, $t \in \mathbb{R}_+$, $x \in \mathbb{R}$, with for example $f(u) = u^2$ (the so-called Burgers[2] equation), and the same initial condition, we can still define weak solutions, but their calculation requires more work, as we shall see later.

Let $f \in C^1(\mathbb{R}, \mathbb{R})$ and $u_0 \in C^1(\mathbb{R})$ and consider the following Cauchy problem, consisting of a non-linear hyperbolic equation and an initial condition:

$$\partial_t u + \partial_x(f(u)) = 0, \quad (x, t) \in \mathbb{R} \times \mathbb{R}_+^{\star}, \tag{5.1a}$$

$$u(x, 0) = u_0(x), \quad x \in \mathbb{R}. \tag{5.1b}$$

Remark 5.1 (Conservation law and regularity of the flux function). The equation (5.1a) is called a non-linear hyperbolic conservation law. A conservation law describes the temporal evolution of a quantity conserved over time. The function f is the flux of the conservation law. For simplicity, we consider here that f is continuously differentiable. However, most of the results presented in this section are still valid if f is only Lipschitz-continuous, or even locally Lipschitz-continuous, but this last assumption generates more technical challenges.

5.1.1 Classical Solutions and Characteristic Curves

Let us start by giving the definition of a classical solution to the problem (5.1) even if, as we will see later, this problem does not generally have a classical solution.

---

[2] Johannes Martinus Burgers (1895–1981), Dutch physicist.

Notation. Let Q be a subset of $\mathbb{R}^m$ $(m \geq 1)$ and $k \in \mathbb{N}$. Recall that $u \in C^k(\bar{Q})$ if u is the restriction to Q of a function of class C^k on $\mathbb{R}^m$ (see Definition 1.28); recall that this is equivalent to the usual definition of continuity if $k = 0$ (see Problem 1.17). Let us also introduce the notation $u \in C_c^k(Q)$, which means that $u \in C^k(\bar{Q})$ and that there exists a compact set $K \subset Q$, such that $u = 0$ on K^c. This notation is used for example for $Q = \mathbb{R} \times \mathbb{R}_+$ or $Q = \mathbb{R} \times [0, T[$.

Definition 5.2 (Classical solution). Assume that $u_0 \in C^1(\mathbb{R})$ and $f \in C^1(\mathbb{R}, \mathbb{R})$. Then u is said to be a classical solution to (5.1) if $u \in C^1(\mathbb{R} \times \mathbb{R}_+, \mathbb{R})$ and if u satisfies the equations (5.1a) and (5.1b).

Before giving a non-existence result of a classical solution (Proposition 5.6), let us recall the classical result of existence and local uniqueness of solutions for a non-linear differential equation (Cauchy–Lipschitz theorem).

Theorem 5.3 (Cauchy–Lipschitz). *Let a be a locally Lipschitz-continuous function from $\mathbb{R} \times \mathbb{R}_+$ to $\mathbb{R}$ and $x_0 \in \mathbb{R}$. We consider the following Cauchy problem.*

$$\begin{cases} x'(t) = a(x(t), t), & t > 0, \\ x(0) = x_0. \end{cases} \tag{5.2}$$

For all $T > 0$, the problem (5.2) admits at most one (classical) solution defined on $[0, T[$. There exist $T_{\max} > 0$ (possibly equal to $+\infty$) and a function x continuous on $[0, T_{\max}[$, of class C^1 on $]0, T_{\max}[$, which is a (classical) solution to (5.2). Moreover, if $T_{\max} < +\infty$ then $|x(t)| \to +\infty$ as $t \to T_{\max}$.

Let u be a classical solution to (5.1). We then define

$$a \in C(\mathbb{R} \times \mathbb{R}_+, \mathbb{R})$$
$$(x, t) \mapsto a(x, t) = f'(u(x, t)). \tag{5.3}$$

It is clear that the function u is then a classical solution of the following problem:

$$\begin{cases} \partial_t u(x, t) + a(x, t) \partial_x u(x, t) = 0, & (x, t) \in \mathbb{R} \times \mathbb{R}_+^\star, \\ u(x, 0) = u_0(x). \end{cases} \tag{5.4}$$

The notion of characteristic curves for the equation (5.4) gives a link between a linear hyperbolic equation and an ordinary differential equation.

Definition 5.4 (Characteristic curve). Assume that the function a defined by (5.3) is locally Lipschitz-continuous from $\mathbb{R} \times \mathbb{R}_+$ to $\mathbb{R}$ and let $x_0 \in \mathbb{R}$. The characteristic curve of the problem (5.4) originating from $x_0 \in \mathbb{R}$ is defined as the curve obtained by solving the Cauchy problem (5.2) with a defined by (5.3).

Proposition 5.5 (Classical solutions and characteristic curves). *Let $f \in C^2(\mathbb{R}, \mathbb{R})$, $u_0 \in C^1(\mathbb{R})$ and u be a classical solution to (5.1). Then, for all $x_0 \in \mathbb{R}$ and all $t \in \mathbb{R}_+$, we have $u(x_0 + f'(u_0(x_0))t, t) = u_0(x_0)$. In other words, for all $x_0 \in \mathbb{R}$, the function u is constant on the line $t \mapsto x(t) = x_0 + f'(u_0(x_0))t$. (This line is the characteristic curve of the problem (5.4) originating from $x_0 \in \mathbb{R}$ with $a(x, t) = f'(u(x, t))$.)*

Proof Set $a(x,t) = f'(u(x,t))$. Since $f \in C^2(\mathbb{R}, \mathbb{R})$ and $u \in C^1(\mathbb{R} \times \mathbb{R}_+, \mathbb{R})$, the function a is locally Lipschitz-continuous from $\mathbb{R} \times \mathbb{R}_+$ to $\mathbb{R}$. The Cauchy–Lipschitz theorem therefore applies to the problem (5.4). Let $x_0 \in \mathbb{R}$. The problem (5.4) then admits a maximum solution $x(t)$ defined on $[0, T_{\max}[$, and $|x(t)|$ tends to infinity as t tends towards $T_{\max}$ if $T_{\max} < +\infty$. The three steps of the proof are as follows:

1. *Every classical solution u is constant on the characteristics.* Indeed, let φ be defined by $\varphi(t) = u(x(t), t)$; differentiating φ with respect to t yields

$$\varphi'(t) = \partial_t u(x(t), t) + \partial_x u(x(t), t) x'(t).$$

   Since the function x satisfies (5.2) with a defined by (5.3), this leads to:

$$\varphi'(t) = \partial_t u(x(t), t) + f'(u(x(t), t)) \partial_x u(x(t), t)$$
$$= \partial_t u(x(t), t) + \partial_x (f(u))(x(t), t) = 0.$$

   The function φ is therefore constant, and we have:

$$u(x(t), t) = \varphi(t) = \varphi(0) = u(x(0), 0) = u(x_0, 0) = u_0(x_0), \forall t \in [0, T_{\max}[.$$

   Therefore u (solution to (5.1)) is constant on the characteristic curve originating from x_0.
2. *The characteristic curves are straight lines*; indeed, $u(x(t), t) = u_0(x_0)$, for all $t \in [0, T_{\max}[$, and therefore $x'(t) = f'(u_0(x_0))$. Integrating, we find that the solution to (5.2) satisfies:

$$x(t) = f'(u_0(x_0))t + x_0. \tag{5.5}$$

3. $T_{\max} = +\infty$ *and therefore* $u(x,t) = u_0(x_0)$ *for all* $t \in [0, +\infty[$. Indeed, since x satisfies (5.5), if $T_{\max} < +\infty$ then $\lim_{t \to T_{\max}} |x(t)| = +\infty$. Hence $T_{\max} = +\infty$.

$\blacksquare$

Thanks to the concept of characteristic curve, we now show that classical solutions do not always exist, even if we start from a regular data u_0. Indeed, we show below that two characteristic curves can intersect. Since we have just shown that a classical solution is constant along a characteristic, we get the non-existence of a classical solution.

Proposition 5.6 (Non-existence of a classical solution). *Let* $f \in C^2(\mathbb{R}, \mathbb{R})$. *If* f' *is not constant, then there exists a* $u_0 \in C_c^\infty(\mathbb{R})$ *such that (5.1) does not admit a classical solution.*

Proof Since f' is not constant, there exist v_0 and $v_1 \in \mathbb{R}$ such that $f'(v_0) > f'(v_1)$, and we can construct $u_0 \in \mathcal{D}(\mathbb{R})$ such that $u_0(x_0) = v_0$ and $u_0(x_1) = v_1$, where x_0 and x_1 are given and such that $x_0 < x_1$, see figure 5.1.

Suppose that u is a classical solution with this initial data. Then, by Proposition 5.5:

$$u(x_0 + f'(u_0(x_0))t, t) = u_0(x_0) = v_0 \text{ and } u(x_1 + f'(u_0(x_1))t, t) = u_0(x_1) = v_1.$$

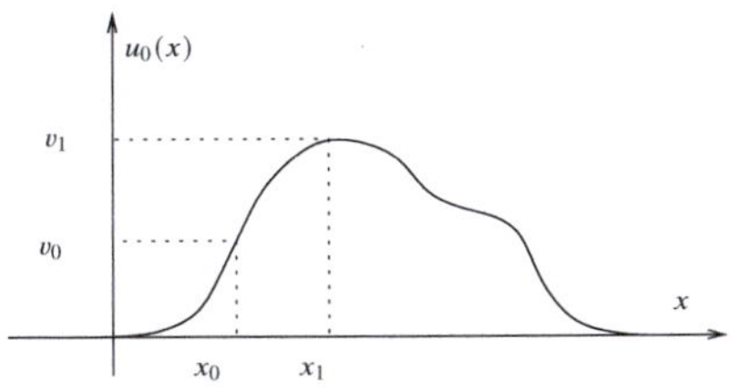
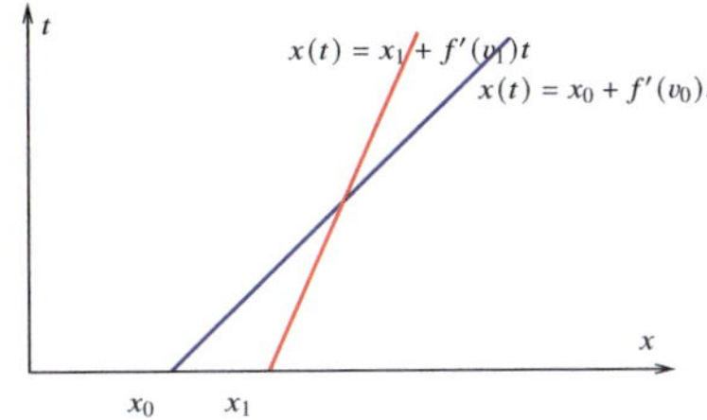

Fig. 5.1 Characteristic lines, non-linear case

Let T such that $x_0 + f'(v_0)T = x_1 + f'(v_1)T = \bar{x}$, that is to say,

$$T = \frac{x_1 - x_0}{f'(v_0) - f'(v_1)}.$$

We then have:

$$u(\bar{x}, T) = u_0(x_0) = v_0 = u_0(x_1) = v_1,$$

which is impossible (because $v_0 \neq v_1$). We conclude that (5.1) does not admit a classical solution for this initial data. ∎

5.1.2 Weak Solutions

Therefore, there does not always exist a classical solution to the problem (5.1). Let us then weaken the sense of solutions and define so-called weak solutions. We give this definition in a slightly more general framework, assuming that f is locally Lipschitz-continuous (instead of being of class C^1). Let $\mathrm{Lip}_{\mathrm{loc}}(\mathbb{R}, \mathbb{R})$ be the set of locally Lipschitz-continuous functions from $\mathbb{R}$ to $\mathbb{R}$. Note that if $f \in \mathrm{Lip}_{\mathrm{loc}}(\mathbb{R}, \mathbb{R})$, the function f is then differentiable a.e., its derivative is locally bounded and we have $f(d) - f(c) = \int_c^d f'(t)\, dt$ for all $c, d \in \mathbb{R}$.

Definition 5.7 (Weak solution). Let $u_0 \in L^\infty(\mathbb{R})$ and $f \in \mathrm{Lip}_{\mathrm{loc}}(\mathbb{R}, \mathbb{R})$. A weak solution to (5.1) is a function $u \in L^\infty(\mathbb{R} \times \mathbb{R}_+)$ such that

$$\int\!\!\int_{\mathbb{R} \times \mathbb{R}_+} [u(x,t)\partial_t \varphi(x,t) + f(u(x,t))\partial_x \varphi(x,t)]\, dx\, dt + \int_{\mathbb{R}} u_0(x)\varphi(x,0)\, dx = 0,$$

$$\forall \varphi \in C_c^1(\mathbb{R} \times \mathbb{R}_+, \mathbb{R}). \quad (5.6)$$

Note that the initial condition is taken into account in these conditions; indeed, the fact that $\varphi \in C_c^1(\mathbb{R} \times \mathbb{R}_+, \mathbb{R})$ implies that $\varphi(x,0)$ can take any value, since φ is compactly supported on $\mathbb{R}_+ = [0, +\infty[$. Let us now give the links between classical and weak solutions.

Proposition 5.8 (Classical solution and weak solution). *Let $f \in C^1(\mathbb{R}, \mathbb{R})$ and $u_0 \in L^\infty(\mathbb{R})$.*

1. *If u is a classical solution to (5.1) (and therefore $u_0 \in C^1(\mathbb{R}, \mathbb{R})$) then u is a weak solution to (5.1).*
2. *If $u \in C^1(\mathbb{R} \times \mathbb{R}_+)$ is a weak solution to (5.1) then $u_0 \in C^1(\mathbb{R}, \mathbb{R})$ (in the sense that the class of functions u_0 admits a representative of class C^1 and is then identified with this representative) and u is a classical solution to (5.1).*

Proof

1. Suppose that u is a classical solution to (5.1), that is to say of:

$$\begin{cases} \partial_t u + \partial_x(f(u)) = 0, & (x,t) \in \mathbb{R} \times \mathbb{R}_+^\star, \\ u(x,0) = u_0(x), & x \in \mathbb{R}. \end{cases}$$

Let $\varphi \in C_c^1(\mathbb{R} \times \mathbb{R}_+, \mathbb{R})$. Multiply (5.1) by φ and integrate over $\mathbb{R} \times \mathbb{R}_+$. We obtain:

$$\int_{\mathbb{R}} \int_{\mathbb{R}_+} \partial_t u(x,t)\varphi(x,t)\, dt\, dx + \int_{\mathbb{R}} \int_{\mathbb{R}_+} \partial_x(f(u))(x,t)\varphi(x,t)\, dt\, dx = 0.$$

Owing to Fubini's theorem, an integration by parts then gives (since the support of φ is a compact of $\mathbb{R} \times \mathbb{R}_+$):

$$-\int_{\mathbb{R}} u(x,0)\varphi(x,0)\, dx - \int_{\mathbb{R}} \int_{\mathbb{R}_+} u(x,t)\partial_t \varphi(x,t)\, dt\, dx$$
$$-\int_{\mathbb{R}_+} \int_{\mathbb{R}} f(u)(x,t)\partial_x \varphi(x,t)\, dx\, dt = 0,$$

and the relation (5.6) is thus obtained thanks to the initial condition $u(x,0) = u_0(x)$.

2. Let u be a weak solution to (5.1), which also satisfies $u \in C^1(\mathbb{R} \times [0, +\infty[)$. The regularity of u allows an integration by parts in (5.6).
   Let us start by taking φ with compact support in $\mathbb{R} \times]0, +\infty[$. Hence $\varphi(x,0) = 0$, and an integration by parts in (5.6) gives:

$$-\int_{\mathbb{R}} \int_{\mathbb{R}_+} \partial_t u(x,t)\varphi(x,t)\, dt\, dx - \int_{\mathbb{R}_+} \int_{\mathbb{R}} \partial_x(f(u))(x,t)\varphi(x,t)\, dx\, dt = 0.$$

Hence:

$$\int_{\mathbb{R}} \int_{\mathbb{R}_+} \left(\partial_t u(x,t) + \partial_x(f(u))(x,t)\right) \varphi(x,t)\, dt\, dx = 0, \forall \varphi \in C_c^1(\mathbb{R} \times]0, +\infty[).$$

Since $\partial_t u + \partial_x(f(u))$ is continuous on $\mathbb{R} \times \mathbb{R}_+^\star$, we get that $\partial_t u + \partial_x(f(u)) = 0$. Indeed, by Lemma 1.2, if $h \in L_{\text{loc}}^1(\mathbb{R} \times]0, +\infty[)$ and $\int_{\mathbb{R}} \int_{\mathbb{R}_+} h(x,t))\varphi(x,t)\, dt\, dx = 0$ for any function φ belonging to $C_c^1(\mathbb{R} \times]0, +\infty[)$, then $h = 0$ a.e.; if moreover h is continuous on $\mathbb{R} \times]0, +\infty[$, then $h = 0$ everywhere on $\mathbb{R} \times]0, +\infty[$.

We then take $\varphi \in C_c^1(\mathbb{R} \times \mathbb{R}_+)$. In this case, an integration by parts in (5.6) gives

$$\int_{\mathbb{R}} u(x,0)\varphi(x,0)\,\mathrm{d}x - \int_{\mathbb{R}} \int_{\mathbb{R}_+} (\partial_t u(x,t) + \partial_x(f(u))(x,t))\,\varphi(x,t)\,\mathrm{d}t\,\mathrm{d}x$$

$$- \int_{\mathbb{R}} u_0(x)\varphi(x,0)\,\mathrm{d}x = 0.$$

But we have just shown that $\partial_t u + \partial_x(f(u)) = 0$. Hence

$$\int_{\mathbb{R}} (u_0(x) - u(x,0))\varphi(x,0)\,\mathrm{d}x = 0, \forall \varphi \in C_c^1(\mathbb{R} \times \mathbb{R}_+).$$

This gives $u_0 = u(\cdot,0)$ a.e. Since u is continuous, we therefore have u_0 continuous (in the sense where we identify u_0 and $u(\cdot,0)$) and u is a classical solution to (5.1).

∎

If the function u is piecewise continuous and satisfies the equation (5.1a) piecewise, then we can give a necessary and sufficient condition for it to be a weak solution (but it is not necessarily a solution in the classical sense). This is the object of the following proposition.

Proposition 5.9 (Rankine–Hugoniot condition). *Let $\sigma \in \mathbb{R}$, $D_1 = \{(x,t) \in \mathbb{R} \times \mathbb{R}_+^\star \ x < \sigma t\}$ and $D_2 = \{(x,t) \in \mathbb{R} \times \mathbb{R}_+^\star \ x > \sigma t\}$.*

1. *Assume that $u \in C(\mathbb{R} \times \mathbb{R}_+^\star)$, that $u_{|D_i} \in C^1(\bar{D}_i, \mathbb{R})$, $i = 1, 2$ (in the sense of the Definition 1.28), that the equation (5.1a) is satisfied for all $(x,t) \in D_i$ $(i = 1, 2)$ and that the initial condition (5.1b) is satisfied a.e. Then u is a weak solution to (5.1).*
2. *More generally, we assume that $u_{|D_i} \in C^1(\bar{D}_i, \mathbb{R})$ $(i = 1, 2)$, that the equation (5.1a) is satisfied for all $(x,t) \in D_i$ $(i = 1, 2)$ and that the initial condition (5.1b) is satisfied a.e. For $t \in \mathbb{R}_+$, we set*

$$u_+(\sigma t, t) = \lim_{x \downarrow \sigma t} u(x,t) \ and \ u_-(\sigma t, t) = \lim_{x \uparrow \sigma t} u(x,t),$$

$$[u](\sigma t, t) = u_+(\sigma t, t) - u_-(\sigma t, t),$$

$$[f(u)](\sigma t, t) = f(u_+(\sigma t, t)) - f(u_-(\sigma t, t)).$$

Then u is a weak solution to (5.1) if and only if

$$\sigma[u](\sigma t, t) = [f(u)](\sigma t, t) \ for \ all \ t \in \mathbb{R}_+. \tag{5.7}$$

This is called the Rankine[3]–Hugoniot[4] condition.

---

[3] William John Macquorn Rankine (1820–1872), Scottish engineer who also contributed to physics and mathematics.

[4] Pierre-Henri Hugoniot (1851–1887), French inventor, mathematician and physicist specialising in fluid mechanics, particularly shocks.

Proof We directly prove point 2, which contains point 1. Assume that $u_{|D_i} \in C^1(\bar{D}_i, \mathbb{R})$ $(i = 1, 2)$, that the first equation of (5.1) is satisfied for all $(x, t) \in D_i$ $(i = 1, 2)$ and that the initial condition (of (5.1)) is satisfied a.e. on $\mathbb{R}$. We show that u is a weak solution to (5.1) if and only if (5.7) is satisfied. With this aim, we set

$$X = \int_{\mathbb{R}} \int_{\mathbb{R}_+} u(x, t) \partial_t \varphi(x, t) \, dt \, dx + \int_{\mathbb{R}_+} \int_{\mathbb{R}} f(u)(x, t) \partial_x \varphi(x, t) \, dx \, dt.$$

So we have $X = X_1 + X_2$, with

$$X_1 = \int_{\mathbb{R}} \int_{\mathbb{R}_+} u(x, t) \partial_t \varphi(x, t) \, dt \, dx \text{ and}$$

$$X_2 = \int_{\mathbb{R}_+} \int_{\mathbb{R}} (f(u))(x, t) \partial_x \varphi(x, t) \, dx \, dt.$$

Let us calculate X_1; since u is of class C^1 only on each of the domains D_i, it is not possible to integrate by parts on the whole domain $\mathbb{R} \times \mathbb{R}_+$. Let us then decompose the integral over D_1 and D_2; suppose for example that $\sigma < 0$, see figure 5.2. (The case $\sigma > 0$ is treated in a similar way and the case $\sigma = 0$ is rather simpler.) We then have $D_1 = \{(x, t) \ x \in \mathbb{R}_- \text{ and } 0 < t < \frac{x}{\sigma}\}$ and $D_2 = \mathbb{R}_+ \times \mathbb{R}_+ \cup \{(x, t) \ x \in \mathbb{R}_- \text{ and } \frac{x}{\sigma} < t < +\infty\}$. Hence:

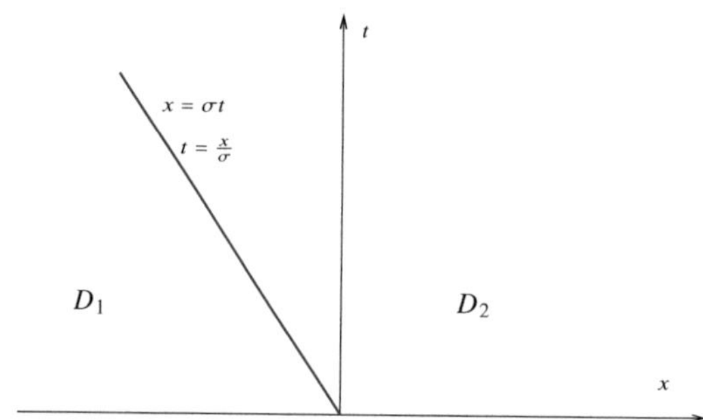

Fig. 5.2 The domains D_1 and D_2

$$X_1 = \int_{\mathbb{R}_-} \int_0^{\frac{x}{\sigma}} u(x, t) \partial_t \varphi(x, t) \, dt \, dx + \int_{\mathbb{R}_-} \int_{\frac{x}{\sigma}}^{+\infty} u(x, t) \partial_t \varphi(x, t) \, dt \, dx$$

$$+ \int_{\mathbb{R}_+} \int_{\mathbb{R}_+} u(x, t) \partial_t \varphi(x, t) \, dt \, dx.$$

Since u is of class C^1 on each of the domains, an integration by parts yields:

$$X_1 = \int_{\mathbb{R}_-} u_-(x, \frac{x}{\sigma})\varphi(x, \frac{x}{\sigma})\, dx - \int_{\mathbb{R}_-} u(x, 0)\varphi(x, 0)\, dx$$

$$- \int_{\mathbb{R}_-} \int_0^{\frac{x}{\sigma}} \partial_t u(x, t)\varphi(x, t)\, dt\, dx - \int_{\mathbb{R}_-} u_+(x, \frac{x}{\sigma})\varphi(x, \frac{x}{\sigma})\, dx$$

$$- \int_{\mathbb{R}_-} \int_{\frac{x}{\sigma}}^{+\infty} \partial_t u(x, t)\varphi(x, t)\, dt\, dx - \int_{\mathbb{R}_+} u(x, 0)\varphi(x, 0)\, dx$$

$$- \int_{\mathbb{R}_+} \int_{\mathbb{R}_+} \partial_t u(x, t)\varphi(x, t)\, dt\, dx. \quad (5.8)$$

Therefore,

$$X_1 = - \int_{\mathbb{R}} u(x, 0)\varphi(x, 0)\, dx - \int_{D_1} \partial_t u(x, t)\varphi(x, t)\, d(x, t)$$

$$- \int_{D_2} \partial_t u(x, t)\varphi(x, t)\, d(x, t) - \int_{\mathbb{R}_-} [u](x, \frac{x}{\sigma})\varphi(x, \frac{x}{\sigma})\, dx.$$

In the last integral, the change of variables $t = \frac{x}{\sigma}$ leads to

$$X_1 = - \int_{\mathbb{R}} u(x, 0)\varphi(x, 0)\, dx - \int_{D_1} \partial_t u(x, t)\varphi(x, t)\, d(x, t)$$

$$- \int_{D_2} \partial_t u(x, t)\varphi(x, t)\, d(x, t) + \sigma \int_{\mathbb{R}_+} [u](\sigma t, t)\varphi(\sigma t, t)\, dt.$$

Decomposing X_2 in the same way over $D_1 \cup D_2$ and noting now that $D_1 = \{(x, t) \in \mathbb{R} \times \mathbb{R}_+^\star \ x < \sigma t\}$ and $D_2 = \{(x, t) \in \mathbb{R} \times \mathbb{R}_+^\star \ x > \sigma t\}$ yields:

$$X_2 = \int_{\mathbb{R}_+} \int_{-\infty}^{\sigma t} f(u)(x, t)\partial_x \varphi(x, t)\, dx\, dt + \int_{\mathbb{R}_+} \int_{\sigma t}^{+\infty} f(u)(x, t)\partial_x \varphi(x, t)\, dx\, dt.$$

The function u is of class C^1 on each of the domains, so we can still integrate by parts. Since φ has compact support on $\mathbb{R} \times \mathbb{R}_+$, we obtain:

$$X_2 = - \int_{D_1} \partial_x f(u)(x, t)\varphi(x, t)\, d(x, t) - \int_{D_2} \partial_x f(u)(x, t)\varphi(x, t)\, d(x, t)$$

$$- \int_{\mathbb{R}_+} [f(u)](\sigma t, t)\varphi(\sigma t, t)\, dt.$$

Since $\partial_t u + \partial_x(f(u)) = 0$ on D_1 and D_2, we therefore have:

$$X = X_1 + X_2 = - \int_{\mathbb{R}} u(x, 0)\varphi(x, 0)\, dx + \int_{\mathbb{R}_+} (\sigma[u](\sigma t, t) - [f(u)](\sigma t, t))\varphi(\sigma t, t)\, dt.$$

Hence u is a weak solution to (5.1) if and only if (5.7) is satisfied. $\blacksquare$

Note that there are often multiple weak solutions. The concept of entropy weak solution allows us to obtain uniqueness. Let us first give an example of non-uniqueness of the weak solution. With this aim, we consider the Cauchy problem associated with the so-called Burgers equation, which reads

$$\partial_t u + \partial_x(u^2) = 0, \tag{5.9a}$$

$$u_0(x) = u_0(x) \tag{5.9b}$$

for this, we consider a particular initial data, in the form

$$u_0(x) = \begin{cases} u_g \text{ if } x < 0, \\ u_d \text{ if } x > 0, \end{cases}$$

with $u_g, u_d \in \mathbb{R}$. These initial data define a particular Cauchy problem, which we call the Riemann[5] problem.

We now consider the simple example obtained with $u_g = -1$ and $u_d = 1$. The problem considered is therefore the following, with $f(u) = u^2$, $u_g = -1$, $u_d = 1$:

$$\begin{cases} \partial_t u + \partial_x(f(u)) = 0, \\ u_0(x) = \begin{cases} u_g \text{ if } x < 0, \\ u_d \text{ if } x > 0. \end{cases} \end{cases} \tag{5.10}$$

We first look for a weak solution of the form:

$$u(x,t) = \begin{cases} u_g \text{ if } x < \sigma t, \\ u_d \text{ if } x > \sigma t. \end{cases} \tag{5.11}$$

This potential solution is discontinuous across the line with equation $x = \sigma t$ in the (x,t) plane. We replace $u(x,t)$ with these values in (5.6). According to Proposition 5.9 we know that u is a weak solution if the following condition (Rankine–Hugoniot condition) is satisfied:

$$\sigma(u_d - u_g) = (f(u_d) - f(u_g)),$$

which, with the particular initial condition chosen here, gives $2\sigma = 1^2 - (-1)^2 = 0$. But we can find other weak solutions. If u is a regular solution, we know that on the characteristic curves, which have the equation $x(t) = x_0 + f'(u_0(x_0))t$, the function u is constant. Since $f'(u) = 2u$, the characteristic curves are therefore straight lines with a slope of -2 if $x_0 < 0$, and with a slope of 2 if $x_0 > 0$. Let us construct these characteristics in figure 5.3: In the middle zone, which we have indicated by a question mark, we are looking for u in the form $u(x,t) = \varphi\left(\frac{x}{t}\right)$ and such that u is continuous on $\mathbb{R} \times \mathbb{R}_+^\star$. The following function u is suitable:

---

[5] Georg Friedrich Bernhard Riemann (1826–1866), German mathematician who made many important contributions, particularly in topology, analysis, and differential geometry.

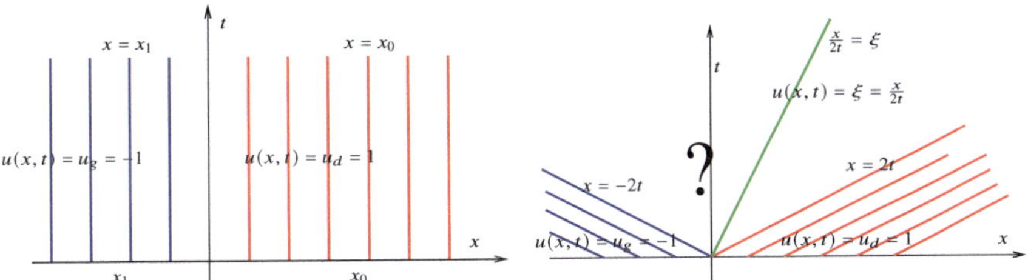

Fig. 5.3 Riemann problem for the Burgers equation

$$u(x,t) = \begin{cases} -1 & \text{if } x < -2t, \\[2mm] \dfrac{x}{2t} & \text{if } -2t < x < 2t, \\[2mm] 1 & \text{if } x > 2t. \end{cases} \tag{5.12}$$

5.1.3 Entropy Solution

We have just seen that weak solutions may be non-unique. How to choose the "right" weak solution, between (5.11) and (5.12)? Since hyperbolic problems are often obtained by neglecting diffusion terms in parabolic equations, a technique to choose the solution is to look for the limit of the associated diffusion problem, which reads:

$$\partial_t u + \partial_x (f(u)) - \varepsilon \partial_{xx} u = 0, \tag{5.13}$$

when the diffusion term becomes negligible, that is, as ε tends to 0. Let u_ε be the solution to (5.13) with the initial condition $u_\varepsilon(\cdot, 0) = u_0(\cdot)$ (assume for now the existence and uniqueness of u_ε). It can be shown that u_ε tends towards u (in a suitable sense) as ε tends to 0, where u is the "entropy weak solution" of (5.1), defined as follows.

Remark 5.10. The fact that the entropy weak solution to (5.1) is the limit (in a suitable sense) of the solution to (5.13) also allows us to show interesting properties about this solution. This is, for example, a way to prove the maximum principle, Proposition 5.23.

Definition 5.11 (Entropy weak solution). Let $u_0 \in L^\infty(\mathbb{R})$ and $f \in \text{Lip}_{\text{loc}}(\mathbb{R}, \mathbb{R})$. Let $u \in L^\infty(\mathbb{R} \times \mathbb{R}_+)$. We say that u is an entropy weak solution to (5.1) if for every function $\eta \in C^1(\mathbb{R}, \mathbb{R})$, called the "entropy", and for Φ defined by $\Phi(s) = \int_0^s f'(\tau)\eta'(\tau)\, d\tau$ (for $s \in \mathbb{R}$), called the "entropy flux", we have:

$$\int_{\mathbb{R}_+} \int_{\mathbb{R}} (\eta(u)\partial_t\varphi + \Phi(u)\partial_x\varphi)(x,t)\,dx\,dt + \int_{\mathbb{R}} \eta(u_0(x))\varphi(x,0)\,dx \geq 0,$$

$$\forall \varphi \in C_c^1(\mathbb{R} \times \mathbb{R}_+, \mathbb{R}_+). \quad (5.14)$$

Since the function η is convex from $\mathbb{R}$ to $\mathbb{R}$, it is locally Lipschitz-continuous and hence Φ is well defined. It is also interesting to note that in Definition 5.11 we can limit ourselves to functions η of class C^2 (it is enough to regularise η with a family of regularising kernels to convince oneself). If f and η are functions of class C^1, the function Φ is simply a function of class C^1 such that $\Phi' = \eta' f'$. Finally, of course, if u is an entropy weak solution then u is a weak solution (Proposition 5.14).

Remark 5.12 (Initial condition). Note that in Definition 5.11, we once again take $\varphi \in C_c^1(\mathbb{R} \times \mathbb{R}_+, \mathbb{R}_+)$ in order to properly take into account the initial condition, a formulation introduced in [24]; this is not always done in this way in older works on the subject, where the initial condition was ensured by the additional condition, $u(t) \rightarrow u_0$ in L_{loc}^1 as $t \rightarrow 0$ (see for example [16, Definition 5.11]). If the initial condition is taken into account only in the definition of weak solution (and is not repeated in the entropy condition), the choice of the functional space in which the solution is sought becomes crucial in order not to lose the uniqueness of the entropy weak solution. An example is given in Problem 5.8. It can be noted that if u is an entropy weak solution in the sense of Definition 5.11, then $u \in C([0, +\infty[, L_{\text{loc}}^1(\mathbb{R}))$ and $u(t) \rightarrow u_0$ in L_{loc}^1 as $t \rightarrow 0$.

Later we prove Theorem 5.29 in the multidimensional framework (but with the spatial variable in a bounded domain rather than in the whole space). This theorem asserts that if $u_0 \in L^\infty(\mathbb{R})$ and $f \in \text{Lip}_{\text{loc}}(\mathbb{R}, \mathbb{R})$ then there exists a unique entropy weak solution to (5.1) in the sense of Definition 5.11. Let us now see the links between classical solution, weak solution and entropy weak solution.

Proposition 5.13 (Classical solution and entropy weak solution). *Let $f \in C^1(\mathbb{R}, \mathbb{R})$ and $u_0 \in L^\infty(\mathbb{R}) \cap C^1(\mathbb{R}, \mathbb{R})$. If u is a classical solution to (5.1), then u is an entropy weak solution.*

Proof Let u be a classical solution to (5.1). Let $\eta \in C^1(\mathbb{R})$ (the convexity of η is unnecessary here) and Φ such that $\Phi' = f'\eta'$ (Φ is the flux function associated with η). Multiply (5.1) by $\eta'(u)$:

$$\eta'(u)\partial_t u + f'(u)\partial_x u\eta'(u) = 0,$$

or, since $\Phi' = f'\eta'$,

$$\partial_t(\eta(u)) + \Phi'(u)\partial_x u = 0.$$

We therefore finally have:

$$\partial_t(\eta(u)) + \partial_x(\Phi(u)) = 0. \quad (5.15)$$

Moreover, since $u(x,0) = u_0(x)$, we also have: $\eta(u(x,0)) = \eta(u_0(x))$. Let $\varphi \in C_c^1(\mathbb{R} \times \mathbb{R}_+, \mathbb{R})$. Multiplying (5.15) by φ, integrating over $\mathbb{R} \times \mathbb{R}_+$ and integrating by parts yields (5.14) (with equality). In the case of a classical solution, the entropy inequality is an equality. $\blacksquare$

An entropy weak solution is a weak solution:

Proposition 5.14. *Let* $f \in \mathrm{Lip}_{\mathrm{loc}}(\mathbb{R}, \mathbb{R})$ *and* $u_0 \in L^\infty(\mathbb{R})$. *If* u *is an entropy weak solution to* (5.1)*, then* u *is a weak solution to* (5.1)*.*

Proof It is enough to take $\eta(u) = u$ and $\eta(u) = -u$ in (5.14) to convince oneself of the result. $\blacksquare$

Owing to Proposition 5.13 and Kruzhkov's theorem (Theorem 5.29), if there are several weak solutions to problem (5.1) and one of them is regular, then the latter is necessarily the entropy weak solution. The following characterisation, which we admit, is often used in practice:

Proposition 5.15 (Kruzhkov's Entropies). *Let* $u_0 \in L^\infty(\mathbb{R})$ *and* $f \in \mathrm{Lip}_{\mathrm{loc}}(\mathbb{R}, \mathbb{R})$. *Let* $u \in L^\infty(\mathbb{R} \times \mathbb{R}_+)$. *The function* u *is an entropy weak solution to* (5.1) *(in the sense of Definition 5.11) if and only if for all* $k \in \mathbb{R}$ (5.14) *is satisfied with* η *defined by* $\eta(s) = |s - k|$, *and* Φ, *the associated entropy flux, defined by:*

$$\Phi(u) = f(\max(u, k)) - f(\min(u, k)).$$

Note that the function η, *called "Kruzhkov's entropy", is not of class* C^1.

We now examine the particular case of solutions having a discontinuity line, as in Proposition 5.9.

Proposition 5.16 (Discontinuity and entropy). *Let* $f \in C^1(\mathbb{R}, \mathbb{R})$ *and* $u_0 \in L^\infty(\mathbb{R})$. *Let* $\sigma \in \mathbb{R}$, $D_1 = \{(x,t) \in \mathbb{R} \times \mathbb{R}_+^\star \; x < \sigma t\}$ *and* $D_2 = \{(x,t) \in \mathbb{R} \times \mathbb{R}_+^\star \; x > \sigma t\}$. *Assume that* $u_{|D_i} \in C^1(\bar{D}_i, \mathbb{R})$ $(i = 1, 2)$, *that the first equation of* (5.1) *is satisfied for all* $(x,t) \in D_i$ $(i = 1, 2)$ *and that the initial condition (of* (5.1)*) is satisfied a.e. For* $t \in \mathbb{R}_+$, *we set*

$$u_+(\sigma t, t) = \lim_{x \downarrow \sigma t} u(x,t) \text{ and } u_-(\sigma t, t) = \lim_{x \uparrow \sigma t} u(x,t),$$

$$[u](\sigma t, t) = u_+(\sigma t, t) - u_-(\sigma t, t),$$

$$[f(u)](\sigma t, t) = f(u_+(\sigma t, t)) - f(u_-(\sigma t, t)).$$

Then u *is an entropy weak solution to* (5.1) *if and only if*

- *the Rankine–Hugoniot condition* (5.7) *is satisfied,*
- *for any convex function* $\eta \in C^1(\mathbb{R})$ *and* $\Phi \in C^1$ *such that* $\Phi' = f'\eta'$,

$$\sigma[\eta(u)](\sigma t, t) \geq [\Phi(u)](\sigma t, t) \text{ for all } t \in \mathbb{R}_+. \tag{5.16}$$

Proof Proposition 5.9 shows that u is a weak solution if and only if the Rankine–Hugoniot condition (5.7) is satisfied. By revisiting the proof of Proposition 5.9, it is shown that u is an entropy weak solution if and only if (5.7) and (5.16) are satisfied. This is the subject of Problem 5.8. ∎

In the case where the function f is strictly convex, Proposition 5.16 can be improved: see Proposition 5.18 below, whose proof is based on the following small technical lemma.

Lemma 5.17 (A result for convex functions). *Let f and η be two convex functions from $\mathbb{R}$ to $\mathbb{R}$. Let $a, b \in \mathbb{R}$, $a < b$, and $\sigma = \frac{f(b)-f(a)}{b-a}$. Let Φ be defined by $\Phi(s) = \int_0^s \eta'(t) f'(t)\, dt$ for $s \in \mathbb{R}$ (so that $\Phi' = \eta' f'$ a.e. on $\mathbb{R}$). Then,*

1. $\sigma(\eta(b) - \eta(a)) \le (\Phi(b) - \Phi(a))$,
2. *if η is strictly convex and f is convex and not affine between a and b, then*
$$\sigma(\eta(b) - \eta(a)) < (\Phi(b) - \Phi(a)).$$

Proof Let us first recall that if φ is a convex function from $\mathbb{R}$ to $\mathbb{R}$, then it is a locally Lipschitz-continuous function. It is therefore differentiable a.e., its derivative is locally bounded and $\varphi(\beta) - \varphi(\alpha) = \int_\alpha^\beta \varphi'(t)\, dt$ for all $(\alpha, \beta) \in \mathbb{R}^2$. For all $\gamma \in \mathbb{R}$,

$$(\Phi(b) - \Phi(a)) - \sigma(\eta(b) - \eta(a)) = \int_a^b \eta'(t)(f'(t) - \sigma)\, dt$$

$$= \int_a^b (\eta'(t) - \gamma)(f'(t) - \sigma)\, dt. \quad (5.17)$$

Since f is convex, the function f' is non-decreasing. Since σ is the mean value of f' on $]a, b[$, there exists a $c \in]a, b[$ such that

$$f'(t) \le \sigma \text{ for almost all } t \in]a, c[, \qquad f'(t) \ge \sigma \text{ for almost all } t \in]c, b[.$$

Now let $\gamma = \sup\{\eta'(s), s \le c\}$ in (5.17) so that $\eta'(s) \le \gamma$ if $s \le c$ and $\eta'(s) \ge \gamma$ if $s > c$; of course, if η' is continuous, we have $\gamma = \eta'(c)$. Since $(\eta'(t) - \gamma)(f'(t) - \sigma) \ge 0$ for almost all $t \in]a, b[$, we obtain

$$(\Phi(b) - \Phi(a)) - \sigma(\eta(b) - \eta(a)) = \int_a^b (\eta'(t) - \gamma)(f'(t) - \sigma)\, dt \ge 0,$$

which gives the first point of the lemma.

For the second point, we note that $\sigma(\eta(b) - \eta(a)) = (\Phi(b) - \Phi(a))$ gives $(\eta'(t) - \gamma)(f'(t) - \sigma) = 0$ a.e. on $]a, b[$. Since η is strictly convex, we have $(\eta' - \gamma) \ne 0$ a.e. on $]a, b[$. We then have $f' = \sigma$ a.e. on $]a, b[$ and this implies that f is affine on $]a, b[$, which contradicts the hypothesis. ∎

Proposition 5.18 (Entropy weak solution, strictly convex case). *Under the assumptions of Proposition 5.16, let u be a weak solution to (5.1), and assume that f is strictly convex. The following three conditions are then equivalent:*

1. u is an entropy weak solution,
2. $u_-(\sigma t, t) \geq u_+(\sigma t, t)$ for all $t \in \mathbb{R}^+$,
3. there exists an $\eta \in C^1(\mathbb{R}, \mathbb{R})$, strictly convex, such that (5.16) is satisfied (with Φ such that $\Phi' = f'\eta'$).

Proof Let us first prove the equivalence between the first two points.

If u is an entropy weak solution, we have (5.16) for all t and for any C^1 convex function η. Taking for η a strictly convex function, Lemma 5.17 necessarily gives, thanks to the fact that f is also strictly convex, $u_-(\sigma t, t) \geq u_+(\sigma t, t)$ for $t \in \mathbb{R}^+$.

Conversely, if u satisfies $u_-(\sigma t, t) \geq u_+(\sigma t, t)$ for $t \in \mathbb{R}^+$, then Lemma 5.17 gives (5.16) for all t and any C^1 convex function η (and this is also true if f is only a convex function). This concludes the equivalence between points 1 and 2.

To conclude the proof of Proposition 5.18, we note that the first point obviously implies the third. Conversely, if u satisfies the third point, Lemma 5.17 necessarily gives, thanks to the fact that f is also strictly convex, that $u_-(\sigma t, t) \geq u_+(\sigma t, t)$ for $t \in \mathbb{R}^+$, and u is therefore an entropy weak solution.   ∎

Remark 5.19 (Counterexample if f is not strictly convex). The equivalence between the first two points of Proposition 5.18 is false if we replace the assumption "f strictly convex" by "f convex". Of course, this is obvious if u_0 takes its values in an interval where f is an affine function, but this is also the case for more general u_0. We start by giving an example that appears in some articles concerning the modelling of road traffic. Then, we slightly adapt this example to be exactly within the assumptions of Proposition 5.18. Let $\alpha > 0$, $\beta < 0$ and $a = -\frac{\beta}{\alpha-\beta}$ (note that $a \in]0, 1[$ and $a\alpha = \beta(a - 1)$). Define f by $f(s) = \alpha s$ for $s \in [0, a]$, $f(s) = \beta(s - 1)$ $s \in]a, 1[$. Let $u_g \in]a, 1[$ and $u_d \in]0, a[$ and $u_0 = u_g$ in $\mathbb{R}_-$, $u_0 = u_d$ in $\mathbb{R}_+$. In this case, we leave it to the reader to prove that the entropy weak solution to (5.1) is the function u defined by

$$u(x, t) = \begin{cases} u_g & \text{if } x < \beta t, \\ a & \text{if } \beta t < x < \alpha t, \\ u_d & \text{if } x > \alpha t. \end{cases}$$

Since $u_g > a$ (and also $a > u_d$) and since f is concave, this solution seems to contradict Proposition 5.18. In this example, the function f is Lipschitz-continuous and the solution has two lines of discontinuity. By slightly modifying this example, we obtain a function $f \in C^1(\mathbb{R})$ and a single discontinuity. We take $a = \frac{1}{2}$ and $f(s) = \alpha s$ for $s \in [0, a]$, $f(s) = \beta s - \gamma s^2 + \delta$ with $\alpha = \gamma = \frac{4}{3}$, $\beta = \frac{8}{3}$, $\delta = -\frac{1}{3}$. The function f is of class C^1, strictly concave on $[a, 1]$, affine on $[0, a]$. As before, we take $u_g \in]a, 1[$ and $u_d \in]0, a[$ and $u_0 = u_g$ in $\mathbb{R}_-$, $u_0 = u_d$ in $\mathbb{R}_+$. The entropy weak solution to (5.1) is then the function u defined by (since $f'(a) = \alpha$)

$$u(x, t) = \begin{cases} u_g & \text{if } x < f'(u_g)t, \\ \xi & \text{if } f'(u_g)t < x < f'(a)t \text{ and } f'(\xi) = \frac{x}{t}, \ \xi \in]a, u_g[, \\ u_d & \text{if } x > f'(a)t = \alpha t. \end{cases}$$

Here too, since $a > u_d$ and f is concave, this solution seems to contradict Proposition 5.18; in fact it does not, since the assumptions of the proposition are not respected.

In the case where f is strictly convex or concave, we can verify that a weak solution is entropic thanks to the Lax condition [33]: this condition states that a weak discontinuous solution is entropic if the characteristics from either side of a discontinuity curve meet the line of discontinuity. It is stated in the following theorem.

Theorem 5.20 (Lax Condition, Scalar Equation). *Under the assumptions and notations of Proposition 5.16, we further assume that $u_g \neq u_d$ and that f is strictly convex or strictly concave; a weak solution u (i.e., satisfying (5.7)) is entropic if and only if it satisfies the Lax condition, which reads*

$$f'(u_g) > \sigma > f'(u_d). \tag{5.18}$$

If the Lax condition is satisfied, we say that the line of discontinuity is a shock (or an entropic discontinuity).

The proof of the fact that a weak solution satisfying the Lax condition is an entropy weak solution is the subject of Problem 5.3. Be careful, it is fundamental to assume that the solution is weak. If this is not the case, the Lax condition does not imply that the solution is entropic, see question 3 of the aforementioned problem. Another entropy condition is the *Oleinik*[6] *condition*, which has the advantage of not requiring the assumption of strict concavity or convexity of the function f (see [42] for the proof).

Theorem 5.21 (Oleinik Condition). *Under the assumptions and notations of Proposition 5.16, assume $u_g \neq u_d$ and define the interval $I(u_g, u_d)$ by*

$$I(u_g, u_d) = \{\theta u_g + (1 - \theta)u_d, \theta \in]0, 1[.\}$$

Then u is an entropy weak solution to (5.1) if and only if it satisfies the Oleinik condition, which reads

$$\forall u \in I(u_g, u_d), \frac{f(u_d) - f(u)}{u_d - u} \leq \sigma = \frac{f(u_d) - f(u_g)}{u_d - u_g}. \tag{5.19}$$

Note that if $f \in C^1(\mathbb{R}, \mathbb{R})$ and f is strictly convex or concave, then the Oleinik condition implies the Lax condition.

Propositions 5.9, 5.16 and 5.18 can be generalised to the case of discontinuous curves.

Proposition 5.22 (Rankine–Hugoniot, curve case). *Let $f \in C^1(\mathbb{R}, \mathbb{R})$ and $u_0 \in L^\infty(\mathbb{R})$ and assume that there exists a finite number of open subsets with Lipschitz boundary, D_i, $i = 1, \ldots, N$, such that*

1. $\mathbb{R} \times \mathbb{R}_+ = \cup_{i=1}^{N} \bar{D}_i.$

---

[6] Olga Oleinik (1925–2001), Russian mathematician who made significant contributions to the theoretical study of PDEs.

2. *For $i \neq j$, $\bar{D}_i \cap \bar{D}_j = \{(\sigma_{i,j}(t), t), t \in I_{i,j}\}$, where $I_{i,j}$ is an interval of $\mathbb{R}_+$ and $\sigma_{i,j}$ a Lipschitz-continuous function of $I_{i,j}$ in $\mathbb{R}$.*
3. *For all i, $u_{|D_i}$ belongs to $C^1(\bar{D}_i, \mathbb{R})$, is a (classical) solution of the first equation of (5.1) and satisfies (a.e.) the initial condition (from (5.1)) when $\bar{D}_i$ meets the axis $t = 0$.*

For $i, j \in \{1, \ldots, N\}$ and $t \in I_{i,j}$, we set

$$u_+(\sigma_{i,j}(t), t) = \lim_{x \downarrow \sigma_{i,j}(t)} u(x, t) \text{ and } u_-(\sigma_{i,j}(t), t) = \lim_{x \uparrow \sigma_{i,j}(t)} u(x, t),$$

$$[u](\sigma_{i,j}(t), t) = u_+(\sigma_{i,j}(t), t) - u_-(\sigma_{i,j}(t), t),$$

$$[f(u)](\sigma_{i,j}(t), t) = f(u_+(\sigma_{i,j}(t), t)) - f(u_-(\sigma_{i,j}(t)t)).$$

Then u is an entropy weak solution to (5.1) if and only if

$$\sigma'_{i,j}(t)[u](\sigma_{i,j}(t), t) = [f(u)](\sigma_{i,j}(t), t) \text{ for almost all } t \in I_{i,j}, \tag{5.20}$$

and, for any convex function $\eta \in C^1(\mathbb{R})$ and $\Phi \in C^1$ such that $\Phi' = f'\eta'$,

$$\sigma'_{i,j}(t)[\eta(u)](\sigma_{i,j}(t)t, t) \geq [\Phi(u)](\sigma_{i,j}(t), t) \text{ for almost all } t \in I_{i,j}. \tag{5.21}$$

Proof The proof is very similar to those of Propositions 5.9, 5.16 and 5.18. We only explain below why (5.20) is satisfied if u is a weak solution (5.1).

Let $i \neq j$ and $D_{i,j}$ be the interior of $\bar{D}_i \cup \bar{D}_j$. Let $\varphi \in \mathcal{D}(D_{i,j})$. Denoting by u_i the extension by continuity of u on $\bar{D}_i$, an integration by parts (space-time) on the domain D_i gives, denoting by $d(x, t)$ integration with respect to the Lebesgue space-time measure,

$$\int_{D_i} (\partial_t u(x, t) + \partial_x f(u)(x, t))\varphi(x, t) \, d(x, t) =$$

$$- \int_{D_i} (u(x, t)\partial_t \varphi(x, t) + f(u(x, t))\partial_x \varphi(x)) \, d(x, t) +$$

$$\int_{\bar{D}_i \cap \bar{D}_j} \begin{bmatrix} f(u_i(x, t)) \\ u_i(x, t) \end{bmatrix} \cdot n_{i,j} \, d\gamma(x, t),$$

where γ denotes the 1-dimensional Lebesgue measure on $\bar{D}_i \cap \bar{D}_j$ and $n_{i,j}$ is the normal vector to $\bar{D}_i \cap \bar{D}_j$ outward D_i. Of course, the last term is (possibly) non-zero only if $\bar{D}_i \cap \bar{D}_j = \{(\sigma_{i,j}(t), t), t \in I_{i,j}\}$ where $I_{i,j}$ is an interval of positive length of $\mathbb{R}_+$. An analogous formula exists for j. Since u is a classical solution on D_i and D_j,

$$\int_{D_i \cup D_j} (\partial_t u(x, t) + \partial_x f(u)(x, t))\varphi(x, t) \, d(x, t) = 0.$$

But, since u is a weak solution to (5.1), we also have

$$\int_{D_i \cup D_j} (\partial_t u(x,t) + \partial_x f(u)(x,t))\varphi(x,t)\,\mathrm{d}(x,t) = 0.$$

This shows that

$$\int_{\bar{D}_i \cap \bar{D}_j} \begin{bmatrix} f(u_i(x,t)) \\ u_i(x,t) \end{bmatrix} \cdot n_{i,j}\,\mathrm{d}\gamma(x,t) + \int_{\bar{D}_i \cap \bar{D}_j} \begin{bmatrix} f(u_j(x,t)) \\ u_j(x,t) \end{bmatrix} \cdot n_{j,i}\,\mathrm{d}\gamma(x,t) = 0.$$

This equality is true for all $\varphi \in \mathcal{D}(D_{i,j})$. Since $n_{i,j}(\sigma_{i,j}(t),t)$ is collinear to the vector $\begin{bmatrix} -1 \\ \sigma'_{i,j}(t) \end{bmatrix}$ (and that $n_{i,j} = -n_{j,i}$), we obtain the condition (5.20). $\blacksquare$

Note that the solutions of a non-linear hyperbolic equation respect the bounds of the initial solution.

Proposition 5.23 (Maximum principle). *Let $u_0 \in L^\infty(\mathbb{R})$ and let A and $B \in \mathbb{R}$ such that $A \le u_0 \le B$ a.e. Let $f \in \mathrm{Lip}_{\mathrm{loc}}(\mathbb{R},\mathbb{R})$, then the entropy weak solution $u \in L^\infty(\mathbb{R} \times \mathbb{R}_+)$ of (5.1) satisfies: $A \le u(x) \le B$ a.e. in $\mathbb{R} \times \mathbb{R}_+$.*

This property is proved by taking the limit either on the solutions of the associated viscous equation, equation (5.13) (Remark 5.10), or on the solutions approximated by a numerical scheme. In this latter case, the scheme should be designed so as to respect the bounds, and in any case, it is desirable that the numerical solutions respect the natural bounds associated to the problem.

Remark 5.24 (Bounded domain). Now, what if the spatial domain is different from $\mathbb{R}$, for example if the problem (5.1) is posed for $x \in I$ where I is an interval of $\mathbb{R}$? If f' does not change sign, we can give a good definition of entropy weak solution and prove existence and uniqueness of the entropy weak solution. In the case where f' changes sign (and this case is very interesting for several problems), the problem is much more difficult. The first result on the question is that of Bardos–Leroux–Nedelec [7]. In his thesis [43], F. Otto gives a very nice formulation for the boundary conditions, which is in particular very practical for proving the convergence of numerical schemes; this formulation is addressed in Section 5.1.4 in the one-dimensional case and in Section 5.2.2 in the multidimensional case.

We conclude this section by introducing the notions of "contact discontinuity", "shock wave" and "rarefaction wave".

If f is linear (or affine, which amounts to the same thing as we can assume $f(0) = 0$) and if $u_0 \in L^\infty(\mathbb{R})$, the weak solution to (5.1) is unique (see Problem 5.5), it is therefore the entropy weak solution. We can also show in this case that the entropy inequalities (5.14) are equalities. If the solution u has a discontinuity curve (necessarily a half-line in fact), we then speak of a "contact discontinuity".

If f is strictly convex (or concave) and the entropy weak solution u of (5.1) has a discontinuity curve, we speak of a shock or a shock wave. We can show in this case that the entropy inequalities (5.14) are strict for some η and φ. Still in the case where f is strictly convex or concave, if u_0 has a discontinuity at a point but this discontinuity does not propagate in the entropy weak solution, we speak of a "rarefaction wave".

5.1.4 Boundary Conditions

We now give a result of existence and uniqueness for a hyperbolic equation with boundary conditions using the formulation due to Otto [44]. Consider the problem:

$$\partial_t u + \partial_x (f(u)) = 0, \quad (x, t) \in]0, 1[\times \mathbb{R}_+, \tag{5.22a}$$

$$u(x, 0) = u_0(x), \quad x \in]0, 1[, \tag{5.22b}$$

$$u(0, t) = \bar{u}(t), \; u(1, t) = \bar{\bar{u}}(t), \quad t \in \mathbb{R}_+. \tag{5.22c}$$

As we shall see, the boundary conditions are only partially taken into account in the entropy weak formulation of the problem.

Definition 5.25 (Entropy weak solution with boundary conditions). Let $u_0 \in L^\infty([0, 1])$, $\bar{u}, \bar{\bar{u}} \in L^\infty(]0, +\infty[)$ and $f \in \mathrm{Lip}_{\mathrm{loc}}(\mathbb{R}, \mathbb{R})$. Let A, B such that $A \le u_0$, $\bar{u}, \bar{\bar{u}} \le B$ a.e. We say that $u \in L^\infty(]0, 1[\times \mathbb{R}_+)$ is an entropy weak solution to (5.22) if there exists an $M \ge 0$ such that

$$\int_0^{+\infty} \int_0^1 \eta(u(x, t)) \partial_t \varphi(x, t) \, dx \, dt + \int_0^1 \eta(u_0(x)) \varphi(x, 0) \, dx$$

$$+ \int_0^{+\infty} \int_0^1 \Phi(u(x, t)) \partial_x \varphi(x, t) \, dx \, dt + M \int_{\mathbb{R}_+} \varphi(0, t) \eta(\bar{u}(t)) \, dt \tag{5.23}$$

$$+ M \int_0^{+\infty} \varphi(1, t) \eta(\bar{\bar{u}}(t)) \, dt \ge 0,$$

for all $\varphi \in C_c^1([0, 1] \times \mathbb{R}_+, \mathbb{R}_+)$ and for any non-negative convex function η such that there exists an $s_0 \in [A, B]$ with $\eta(s_0) = 0$, and Φ such that $\Phi(s) = \int_{s_0}^s \eta'(t) f'(t) \, dt$. Recall that a convex function from $\mathbb{R}$ to $\mathbb{R}$ is locally Lipschitz-continuous and therefore differentiable a.e. by Rademacher's[7] theorem[8].

Theorem 5.26 (Existence and uniqueness, with boundary conditions). *Let* $u_0 \in L^\infty([0, 1])$, $\bar{u}, \bar{\bar{u}} \in L^\infty(]0, +\infty[)$ *and* $f \in \mathrm{Lip}_{\mathrm{loc}}(\mathbb{R}, \mathbb{R})$. *Let* A, B *such that* $A \le u_0 \le B$ *a.e. and* $A \le \bar{u}, \bar{\bar{u}} \le B$ *a.e. Then there exists a unique function* $u \in L^\infty(\mathbb{R} \times \mathbb{R}_+)$ *that is an entropy weak solution to (5.22). Moreover,*

$$A \le u(x, t) \le B \; a.e. \; (x, t) \in]0, 1[\times \mathbb{R}_+.$$

---

[7] Hans Rademacher (1892–1969), American mathematician, known for his work in analysis and number theory.

[8] Rademacher's theorem: If U is an open subset of $\mathbb{R}^n$ and $f : U \to \mathbb{R}^m$ is Lipschitz continuous, then f is differentiable almost everywhere in U.

The proof of existence in Theorem 5.26 is obtained with $M \geq \max_{s \in [A,B]} |f'(s)|$. A fairly simple way to prove this existence is to take the limit on monotone flux numerical schemes, see the proof of Theorem 5.37 in the multidimensional case. Uniqueness is true for any value of M but is obviously only interesting if a solution exists; not that the solution, which exists if $M \geq \max_{s \in [A,B]} |f'(s)|$, does not depend on M. If the solution is of class C^1, this solution is an entropy weak solution on $]0, 1[$, that is, it is a solution to (5.23) for all $\varphi \in C_c^1(]0, 1[\times \mathbb{R}_+, \mathbb{R}_+)$ and for any convex function η, and satisfies the boundary conditions in the sense given by [7]. This is still true if the solution is also of class BV (also said to be of bounded variation) in space for all t, so that the solution has, for all t, a limit at $x = 0$ and $x = 1$). The space $BV([0, 1])$ is defined as follows, with $\Omega =]0, 1[$.

Definition 5.27 (Function of bounded variation, BV space). Let Ω be an open subset of $\mathbb{R}^N$, $N \geq 1$; a function $v : \Omega \to \mathbb{R}$ is of bounded variation, i.e. $v \in BV(\overline{\Omega})$, if $v \in L^1(\Omega)$ and if $|v|_{BV(\overline{\Omega})} < +\infty$, with

$$|v|_{BV(\overline{\Omega})} = \sup\left\{ \int_{\Omega} v \operatorname{div} \varphi \, dx, \, \varphi \in C_c^{\infty}(\mathbb{R}^N, \mathbb{R}^N), \|\varphi\|_{L^{\infty}}(\Omega) \leq 1 \right\}. \tag{5.24}$$

Of course, if $u \in L^{\infty}(]0, 1[\times \mathbb{R}_+)$ is a solution to (5.23) as in Definition 5.25, the function u is also a solution to (5.23) with the Kruzhkov entropies, that is, η defined (for $k \in [A, B]$) by $\eta(s) = |s - k|$ (and the associated entropy flux Φ defined by $\Phi(s) = f(\max(s, k)) - f(\min(s, k))$). However, the fact that $u \in L^{\infty}(\mathbb{R} \times \mathbb{R}_+)$ satisfies (5.23) for all Kruzhkov entropies is not sufficient to ensure uniqueness (whereas it was sufficient in the case of the problem posed on all $\mathbb{R}$, without boundary conditions). An example of non-uniqueness is given in Remark 5.28. However, we obtain uniqueness if $u \in L^{\infty}(\mathbb{R} \times \mathbb{R}_+)$ satisfies (5.23) for all the "Kruzhkov semi-entropies", that is, the functions η defined (for $k \in [A, B]$) by $\eta(s) = (s - k)^+$ (and therefore $\Phi(s) = f(s) - f(k)$ if $s \geq k$ and 0 otherwise) and $\eta(s) = (s - k)^-$ (and therefore $\Phi(s) = f(k) - f(s)$ if $s \leq k$ and 0 otherwise).

Remark 5.28 (Counterexample to uniqueness with Kruzhkov's entropies). In this remark, we give an example of non-uniqueness if we limit ourselves in Definition 5.25 to Kruzhkov's entropies, that is, to functions η defined (for $k \in [A, B]$) by $\eta(s) = |s - k|$ (and therefore Φ, the associated entropy flux, defined by $\Phi(s) = f(\max(s, k)) - f(\min(s, k))$).

For this example, $f(s) = s^2$, $u_0 = 1$ a.e. in $]0, 1[$, $\underline{u} = -1$ a.e. on $\mathbb{R}_+$ and $\overline{u} = 1$ a.e. on $\mathbb{R}_+$. It is therefore possible to take $A = -1$, $B = 1$ and $M = 2$. With these values, Theorem 5.26 gives us an entropy weak solution u to (5.22) which reads

$$u(x, t) = \frac{x}{2t} \text{ if } x < 2t, u(x, t) = 1, \text{ if } x > 2t,$$

and corresponds to a rarefaction wave. This solution therefore remains a solution to (5.23) when limited to Kruzhkov entropies. We now consider the constant function $u = 1$ a.e. in $]0, 1[\times \mathbb{R}_+$. This constant function is also a solution to (5.23) if we limit ourselves in this definition to Kruzhkov entropies. This solution actually consists of propagating a non-entropic discontinuity at the point $x = 0$. Thus, there are two functions that satisfy (5.23) when limiting this definition to the Kruzhkov entropies.

5.2 Scalar Equations: the Multidimensional Case

Let $\Omega \subset \mathbb{R}^N$, $N = 2, 3$, $T > 0$, $b \in C^1(\bar{\Omega} \times [0, T])^N$ and $f \in C^1(\mathbb{R}, \mathbb{R})$ (but we could also consider the case $f \in \mathrm{Lip}_{\mathrm{loc}}(\mathbb{R}, \mathbb{R})$). Consider the following problem:

$$\begin{cases} \partial_t u + \mathrm{div}(b f(u)) = 0 \text{ in } \Omega \times]0, T[, \\ u(x, 0) = u_0(x) \text{ in } \Omega. \end{cases} \tag{5.25}$$

More precisely, we prove below, with suitable assumptions on the data, the existence and uniqueness of entropy weak solutions to this problem, first in the case without boundary conditions (Section 5.2.1), then in the case of a problem with boundary conditions (Section 5.2.2).

5.2.1 Case without Boundary Condition

Assume here that $b = 0$ on $\partial\Omega \times [0, T]$, so that boundary conditions on $\partial\Omega$ are superfluous.

Theorem 5.29 (Kruzhkov, case of a bounded domain, no boundary condition). *Let Ω be a bounded open subset of $\mathbb{R}^N$ ($N > 1$) with a Lipschitz boundary. Let $T > 0$ and $b \in C^1(\bar{\Omega} \times [0, T])^N$ be a function such that $b = 0$ on $\partial\Omega \times [0, T]$ and $\mathrm{div}, b = 0$ in $\Omega \times [0, T]$. Let $u_0 \in L^\infty(\Omega)$ and $f \in C^1(\mathbb{R}, \mathbb{R})$. There exists a unique entropy solution of (5.25), that is, a solution of*

$$u \in L^\infty(\Omega \times]0, T[),$$

$$\int_0^T \int_\Omega (\eta(u)\partial_t\varphi + \Phi(u)\, b \cdot \nabla\varphi)\, \mathrm{d}x\, \mathrm{d}t + \int_\Omega \eta(u_0(x))\varphi(x, 0)\, \mathrm{d}x \geq 0,$$

$$\forall \varphi \in C_c^\infty(\Omega \times [0, T[, \mathbb{R}_+), \tag{5.26}$$

for any convex function $\eta \in C^2(\mathbb{R}, \mathbb{R})$ and Φ such that $\Phi' = \eta' f'$.

Let $A \leq 0$ and $B \geq 0$ such that $A \leq u_0 \leq B$ almost everywhere in Ω, then $A \leq u \leq B$ almost everywhere in $\Omega \times]0, T[$.

The proof of this theorem is constructive: we construct a sequence $(u^{(n)})_{n \in \mathbb{N}}$ converging to a limit u that satisfies (5.26). Here, we choose $u^{(n)}$ to be the solution of a parabolic problem "close" to problem (5.25), in the sense that a diffusion term is added to the latter, which tends to 0 as $n \to +\infty$. Another possible approach is to construct a sequence $(u^{(n)})n \in \mathbb{N}$ via a numerical scheme, see [23, Section 29] and the proof of Theorem 5.37 below.

Under the assumptions of Theorem 5.29, let $n \in \mathbb{N}^\star$ and $u^{(n)}$ be the solution of

$$u^{(n)} \in L^2(]0, T[, H_0^1(\Omega)),$$

$$-\int_0^T \int_\Omega u^{(n)} \partial_t \varphi \, dx \, dt - \int_0^T \int_\Omega b f(u^{(n)}) \cdot \nabla\varphi \, dx \, dt + \frac{1}{n} \int_0^T \int_\Omega \nabla u^{(n)} \cdot \nabla\varphi \, dx \, dt$$

$$- \int_\Omega u_0(x)\varphi(x, 0) \, dx = 0, \forall \varphi \in C_c^\infty(\Omega \times [0, +\infty[). \quad (5.27)$$

The existence and uniqueness of a solution $u^{(n)}$ to (5.27) stems from the study of the convection-diffusion equation in Chapter 4. (The fact that $\frac{1}{n}$ appears instead of 1 does not pose any difficulty in this study.)

It was also proven in Chapter 4 that this formulation is equivalent to the following problem:

$$u^{(n)} \in L^2(]0, T[, H_0^1(\Omega)), \partial_t u^{(n)} \in L^2(]0, T[, H^{-1}(\Omega)), u^{(n)}(0) = u_0,$$

$$\int_0^T \langle \partial_t u^{(n)}, v \rangle_{H^{-1}, H_0^1} \, dt - \int_0^T \int_\Omega b f(u^{(n)}) \cdot \nabla v \, dx \, dt + \frac{1}{n} \int_0^T \int_\Omega \nabla u^{(n)} \cdot \nabla v \, dx \, dt = 0,$$

$$\forall v \in L^2(]0, T[, H_0^1(\Omega)). \quad (5.28)$$

This equivalence is crucial in the sequel. In order to pass to the limit in the sequence $(u^{(n)})_{n \in \mathbb{N}}$, let us first establish some estimates on this sequence.

Lemma 5.30 (Estimates on the approximate solutions). *Under the assumptions of Theorem 5.29:*

- *the sequence $(u^{(n)})_{n \in \mathbb{N}}$ of solutions to (5.27) is bounded in $L^\infty(\Omega \times]0, T[)$;*
- *the sequence $(\frac{1}{\sqrt{n}} \nabla u^{(n)})_{n \in \mathbb{N}}$ is bounded in $L^2(\Omega \times]0, T[)^d$;*

- *if, moreover, $u_0 \in BV(\overline{\Omega})$, then the sequence $(u^{(n)})_{n \in \mathbb{N}}$ is bounded in $BV(\overline{\Omega} \times [0, T])$.*

Proof Under the assumptions of Theorem 5.29, we have $A \leq u_0 \leq B$ almost everywhere, and the results of Chapter 4 (see Remark 4.40) show that for all $t \in [0, T]$, $A \leq u^{(n)} \leq B$ almost everywhere. The sequence $(u^{(n)})_{n \in \mathbb{N}}$ is thus bounded in $L^\infty(\Omega \times]0, T[)$.

Now, taking $v = u^{(n)}$ in (5.28), we obtain, using $\int_\Omega b f(u^{(n)}) \cdot \nabla u^{(n)} \, dx = 0$ almost everywhere (thanks to div, $b = 0$, see (3.12)):

$$\frac{1}{2}\left(\|u^{(n)}(T)\|_{L^2(\Omega)}^2 - \|u_0\|_{L^2(\Omega)}^2 \right) + \frac{1}{n} \int_0^T \int_\Omega |\nabla u^{(n)}|^2 \, dx \, dt = 0.$$

From this, we deduce

$$\frac{1}{n}\int_0^T\int_\Omega |\nabla u^{(n)}|^2 \, dx\, dt \le \frac{1}{2}\|u_0\|^2_{L^2(\Omega)} < +\infty \tag{5.29}$$

which provides an $L^2(\Omega\times\,]0,T[)^d$ estimate on $(\frac{1}{\sqrt{n}}\nabla u^{(n)})_{n\in\mathbb{N}}$. Note that this estimate does not yield compactness, but it is useful for passing to the limit (as $n\to +\infty$).

Finally, if $u_0 \in BV(\overline{\Omega})$, by differentiating the first equation of (5.25) with respect to x_i and multiplying by $\mathrm{sgn}(\partial_i u)$, we can show that the sequence $(u^{(n)})_{n\in\mathbb{N}}$ is bounded in $BV(\overline{\Omega}\times[0,T])$ (see Definition 5.27). $\blacksquare$

The following theorem, due to Helly[9], allows us to obtain the compactness of a bounded sequence in BV, which will be useful in one of the proofs of Theorem 5.29.

Theorem 5.31 (Helly). *Let $d \ge 1$ and $(u_n)_{n\in\mathbb{N}}$ be a bounded sequence in $L^1(Q)$ and bounded in $BV(Q)$, where Q is a compact set of $\mathbb{R}^d$, then $(u_n)_{n\in\mathbb{N}}$ is relatively compact in $L^1(Q)$.*

Let us now give the main ideas of the proof of Theorem 5.29.

Proof (of Theorem 5.29).
Existence. We assume here, for the sake of simplicity, that $u_0 \in BV(\overline{\Omega})$ and refer to Remark 5.32 for the main lines of the proof of the existence of a solution without this assumption.

Thanks to the estimate on $u^{(n)}$ in $L^\infty(\Omega\times]0,T[)$ given in Lemma 5.30, we can assume that up to a subsequence, $u^{(n)} \to u$ $\star$-weakly in $L^\infty(\Omega\times]0,T[)$.

If $f(u) = u$, then we leave it to the reader to show that u is a weak solution to (5.25) that is, a solution to 5.26 with only $\eta(s) = s$ but with all φ in $C_c^\infty(\Omega\times[0,T[,\mathbb{R})$ and with $=$ instead of $\ge$. Then we can show (but it is a bit more difficult) that u is a solution to (5.26) and this ends the "existence" part of Theorem 5.29.

If the function f' is not constant, the situation is much more difficult, even to show only that u is a weak solution to (5.25), because the convergence of $u^{(n)}$ to u is only weak and so we do not know if $f(u^{(n)})$ tends to $f(u)$ (and, more generally, we do not know if $\eta(u^{(n)})$ tends to $\eta(u)$ and $\Phi(u^{(n)})$ tends to $\Phi(u)$).

Thanks to the assumption $u_0 \in BV(\overline{\Omega})$, owing to Lemma 5.30, the sequence $(u^{(n)})_{n\in\mathbb{N}}$ is bounded in $BV(\overline{\Omega}\times[0,T])$, and we may thus invoke Helly's theorem with $Q = \overline{\Omega}\times[0,T]$ (and $d = N+1$). Since $u^{(n)} \to u$ in $L^1(Q)$, up to a subsequence, $u^{(n)} \to u$ in $L^p(Q)$ for any $p < +\infty$ and, again up to a subsequence, $u^{(n)} \to u$ a.e. We therefore apply Helly's theorem with $Q = \overline{\Omega}\times[0,T]$ (and $d = N+1$).

Let us then show that u is a solution to (5.26), which gives the existence of a solution to (5.26) if $u_0 \in BV(\overline{\Omega})$.

---

[9] Eduard Helly (1884–1943), Austrian mathematician, prisoner in Siberia during and after World War I. He was unable to obtain an academic position due to his Jewish background and exiled to the USA after the Anschluss in 1938. He made numerous contributions to functional analysis.

1. Let us show that u is a weak solution. Let $\varphi \in C_c^\infty(\Omega \times [0, +\infty[)$. We have

$$\int_0^T \int_\Omega (u^{(n)} \partial_t \varphi + bf(u^{(n)}) \cdot \nabla \varphi) \, dx \, dt + \int_\Omega u_0(x)\varphi(x,0) \, dx$$
$$- \frac{1}{n} \int_0^T \int_\Omega \nabla u^{(n)} \cdot \nabla \varphi \, dx \, dt = 0.$$

We first notice that the last term of the left-hand side tends towards 0, thanks to the estimate (5.29) and the Cauchy–Schwarz inequality, indeed

$$\left| \frac{1}{n} \int_0^T \int_\Omega \nabla u^{(n)} \cdot \nabla \varphi \, dx \, dt \right| \leq \frac{1}{\sqrt{n}} \| \frac{1}{\sqrt{n}} |\nabla u^{(n)}| \|_{L^2(\Omega \times]0,T[)} \| |\nabla \varphi| \|_{L^2(\Omega \times]0,T[)}$$

and $\| \frac{1}{\sqrt{n}} |\nabla u^{(n)}| \|_{L^2(\Omega \times]0,T[)} \leq \frac{1}{\sqrt{2}} \|u_0\|_{L^2(\Omega)}$ by (5.29) and therefore $\frac{1}{n} \int \int \nabla u^{(n)} \cdot \nabla \varphi \to 0$ as $n \to 0$. The other terms converge by the dominated convergence theorem, and therefore by passing to the limit, we obtain

$$\int_0^T \int_\Omega (u\partial_t \varphi + bf(u) \cdot \nabla \varphi) \, dx \, dt + \int_\Omega u_0(x)\varphi(x,0) \, dx = 0. \tag{5.30}$$

2. Let us show that u is an entropy weak solution. Since $u^{(n)}$ is a weak solution of

$$u_t^{(n)} + \operatorname{div}(bf(u^{(n)})) - \frac{1}{n}\Delta u^{(n)} = 0, \tag{5.31}$$

we can show (we admit it) that $u^{(n)} \in C^2(\Omega \times]0,T[)$ (this is what we call the regularising effect for a parabolic equation). The function $u^{(n)}$ is therefore a classical solution to (5.31). We can then multiply this equation by $\eta'(u^{(n)})$ with $\eta \in C^2(\mathbb{R}, \mathbb{R})$ convex. Since $\operatorname{div}b = 0$, we obtain

$$\partial_t(\eta(u^{(n)})) + bf'(u^{(n)})\eta'(u^{(n)}) \cdot \nabla u_n - \frac{1}{n}\eta'(u^{(n)})\Delta u^{(n)} = 0 \text{ on } \Omega \times]0,T[.$$

Thus:

$$\partial_t(\eta(u^{(n)})) + b \cdot \nabla(\Phi(u^{(n)})) - \frac{1}{n}\operatorname{div}(\eta'(u^{(n)})\nabla u^{(n)}) + \frac{1}{n}\eta''(u^{(n)})|\nabla u^{(n)}|^2 = 0.$$

But $\frac{1}{n}\eta''(u^{(n)})|\nabla u^{(n)}|^2 \geq 0$, we therefore have

$$\partial_t(\eta(u^{(n)})) + b \cdot \nabla(\Phi(u^{(n)})) - \frac{1}{n}\operatorname{div}(\eta'(u^{(n)})\nabla u^{(n)})) \leq 0.$$

Multiplying this equation by φ, with $\varphi \in C_c^\infty(\Omega \times [0,T[, \mathbb{R}_+)$, we obtain, still on $\Omega \times]0,T[$,

$$\varphi\partial_t(\eta(u^{(n)})) + \varphi b \cdot \nabla(\Phi(u^{(n)})) - \frac{1}{n}\varphi\operatorname{div}(\eta'(u^{(n)})\nabla u^{(n)})) \leq 0.$$

We integrate over $[\varepsilon, T[\times\Omega$ with $\varepsilon > 0$ and, after integration by parts, we obtain:

$$
-\int_\varepsilon^T\!\!\int_\Omega (\eta(u^{(n)}))\partial_t\varphi - \int_\Omega \eta(u^{(n)}(\varepsilon))\varphi(x,\epsilon)\,dx
$$
$$
-\int_\varepsilon^T\!\!\int_\Omega \Big(b\Phi(u^{(n)})\cdot\nabla\varphi + \frac{1}{n}\eta'(u^{(n)})\nabla u^{(n)}\cdot\nabla\varphi\Big)\,dx\,dt \leq 0.
$$

But we have, as $\varepsilon \to 0$, $u^{(n)}(\varepsilon) \to u^{(n)}(0) = u_0$ in $L^2(\Omega)$ and $\eta(u^{(n)}(\varepsilon)) \to \eta(u_0)$ in $L^2(\Omega)$ and therefore

$$
-\int_0^T\!\!\int_\Omega (\eta(u^{(n)}))\partial_t\varphi - \int_\Omega \eta(u_0)\varphi(x,0)\,dx
$$
$$
-\int_0^T\!\!\int_\Omega \Big(b\Phi(u^{(n)})\cdot\nabla\varphi + \frac{1}{n}\eta'(u^{(n)})\nabla u^{(n)}\cdot\nabla\varphi\Big)\,dx\,dt \leq 0.
$$

When $n \to \infty$, we have $\eta(u^{(n)}) \to \eta(u)$ and $\Phi(u^{(n)}) \to \Phi(u)$ in $L^2(\Omega\times]0,T[)$ and, with $C_{\eta,A,B} = \max\{|\eta'(s)|, A \leq s \leq B\}$,

$$
\Big|\frac{1}{n}\int_0^T\!\!\int_\Omega \eta'(u^{(n)})\nabla u^{(n)}\cdot\nabla\varphi\,dx\,dt\Big|
$$
$$
\leq C_{\eta,A,B}\frac{1}{\sqrt{n}}\Big\|\frac{1}{\sqrt{n}}|\nabla u^{(n)}|\Big\|_{L^2(\Omega\times]0,T[)}\||\nabla\varphi|\|_{L^2(\Omega\times]0,T[)} \to 0
$$

as $n \to +\infty$. We thus finally obtain,

$$
\int_0^T\!\!\int_\Omega (\eta(u))\partial_t\varphi + b\cdot\Phi(u^{(n)})\nabla\varphi\,dx\,dt + \int_\Omega \eta(u_0)\varphi(x,0)\,dx \geq 0
$$

for all $\varphi \in C_c^\infty(\Omega \times [0,T[, \mathbb{R}_+)$, which concludes the proof of existence.

Uniqueness Let u be a solution to (5.26); let us first show that we can take $\varphi \in C_c^\infty(\mathbb{R}^N \times [0,T[, \mathbb{R}_+)$ in (5.26). This is where the assumption $b = 0$ on the boundary of Ω is useful.

We admit here that we can construct a sequence $(\varphi_n)_{n\in\mathbb{N}}$ belonging to $\mathcal{D}(\Omega)$ such that $\varphi_n = 1$ on $K_n = \{x \in \Omega \; d(x,\partial\Omega) \geq \frac{1}{n}\}$, $0 \leq \varphi_n \leq 1$ and $|\nabla\varphi_n| \leq C_\Omega n$, where C_Ω only depends on Ω (the Lipschitz regularity of Ω is important here). Let $\varphi \in C_c^\infty(\mathbb{R}^N \times [0,T[, \mathbb{R}_+)$, we then take $\varphi(x,t)\varphi_n(x)$ as a test function in (5.26), we obtain

$$
\int_0^T\!\!\int_\Omega (\varphi_n\eta(u)\partial_t\varphi + \varphi_n\Phi(u)b\cdot\nabla\varphi)\,dx\,dt + \int_\Omega \eta(u_0(x))\varphi_n(x)\varphi(x,0)\,dx
$$
$$
+ \int_0^T\!\!\int_\Omega b\Phi(u)\varphi\cdot\nabla\varphi_n\,dx\,dt \geq 0.
$$

The first terms converge by the dominated convergence theorem. Let R_n denote the last term. The convergence of R_n is not too difficult, thanks to the fact that we have assumed b to be zero on the boundary.

$$R_n = \int_0^T \int_\Omega \Phi(u)\varphi \cdot \nabla \varphi_n \, \mathrm{d}x \, \mathrm{d}t = \int_0^T \int_{C_n} b\Phi(u)\varphi \cdot \nabla \varphi_n \, \mathrm{d}x \, \mathrm{d}t,$$

where $C_n = \Omega \setminus K_n$. Hence

$$|R_n| \le T\|b\|_{L^\infty(C_n)} C_{u,\Phi} \|\varphi\|_\infty C_\Omega n\lambda_N(C_n),$$

where $C_{u,\Phi} = \max\{|\Phi(s)|, s \in [-\gamma, \gamma]\}$, with $\gamma = \|u\|_{L^\infty(\Omega\times]0,T[)}$.

Since $b = 0$ on $\partial\Omega \times [0, T]$ (and b continuous), we have $\|b\|_{L^\infty(C_n)} \to 0$ as $n \to +\infty$. Finally, the sequence $(n\lambda_N(C_n))_{n\in\mathbb{N}}$ is bounded, and we therefore have $\lim_{n\to+\infty} R_n = 0$, so

$$\int_0^T \int_\Omega \eta(u)\partial_t\varphi + b\Phi(u) \cdot \nabla\varphi \, \mathrm{d}x \, \mathrm{d}t + \int_\Omega \eta(u_0)\varphi(x, 0) \, \mathrm{d}x \ge 0,$$

$$\forall \varphi \in C_c^\infty(\mathbb{R}^N \times [0, T[, \mathbb{R}_+),$$

for all $\eta \in C^2(\mathbb{R}, \mathbb{R})$, η convex.

By a regularisation process, we show that the regularity assumption on η, (i.e. η of class C^2) can be replaced by the weaker assumption "locally Lipschitz-continuous η", which has the advantage of being able to use Kruzhkov's entropies.

We can now prove the uniqueness of the solution to (5.26). Let u and v be two solutions to (5.26). Let us use (5.26) by taking for η a Kruzhkov entropy and functions $\varphi \in C_c^\infty(\mathbb{R}^N \times [0, T[, \mathbb{R}_+)$ (we have just shown that this is possible). We are revisiting here an idea from Kruzhkov, known as variable splitting. It first consists of choosing, in (5.26), $k = v(y, s)$ and taking $\varphi(x, t) = \psi(t)\rho_n(x - y)\bar\rho_n(t - s)$ with $\psi \in C_c^\infty([0, T[, \mathbb{R}_+)$, $\rho_n(x) = n^N\rho(nx)$ and $\bar\rho_n(t) = n\bar\rho(nt)$, where ρ and $\bar\rho$ are regularising kernels, and integrating with respect to y and s. The function ρ is non-negative, of class C^∞ on $\mathbb{R}^N$, and its support lies in the ball of radius 1 and its integral over $\mathbb{R}^N$ is 1. Similarly, the function $\bar\rho$ is non-negative, it is of class C^∞ on $\mathbb{R}$, and its support lies in the ball of radius 1 and its integral over $\mathbb{R}$ is 1. Furthermore, we choose $\bar\rho$ so that its support is in $\mathbb{R}_-$. With this choice of test function (and n large enough for the test function to be admissible in (5.26)) written with elements $\varphi \in C_c^\infty(\mathbb{R}^N \times [0, T[, \mathbb{R})$, we obtain:

$$A_{1,n} + A_{2,n} + A_{3,n} + A_{4,n} \ge 0, \tag{5.32}$$

with

$$A_{1,n} = \int_0^T \int_\Omega \int_0^T \int_\Omega |u(x, t) - v(y, s)|\psi'(t)\rho_n(x - y)\bar\rho_n(t - s) \, \mathrm{d}x \, \mathrm{d}t \, \mathrm{d}y \, \mathrm{d}s,$$

$$A_{2,n} = \int_0^T \int_\Omega \int_0^T \int_\Omega |u(x,t) - v(y,s)|\psi(t)\rho_n(x-y)\bar{\rho}_n'(t-s)\,dx\,dt\,dy\,ds,$$

$$A_{3,n} = \int_0^T \int_\Omega \int_0^T \int_\Omega (f(u(x,t)) - f(v(y,s)))$$
$$(\mathrm{sgn}(u(x,t) - v(y,s)))\psi(t)b \cdot \nabla\rho_n(x-y)\bar{\rho}_n(t-s)\,dx\,dt\,dy\,ds,$$

$$A_{4,n} = \int_0^T \int_\Omega \int_\Omega |u_0(x) - v(y,s)|\psi(0)\rho_n(x-y)\bar{\rho}_n(-s)\,dx\,dy\,ds.$$

We now take the limit as $n \to +\infty$ in (5.32). It is not difficult to show that

$$\lim_{n\to+\infty} A_{1,n} = \int_0^T \int_\Omega |u(x,t) - v(x,t)||\psi'(t)|\,dx\,dt.$$

Let us then show that $A_{2,n} + A_{3,n} \le 0$. To do so, consider the entropic formulation
for v, written with y and s as variables, and choose the Kruzhkov entropy associated
with $k = u(x,t)$ and $\varphi(y,s) = \psi(t)\rho_n(x-y)\bar{\rho}_n(t-s)$. Finally, integrating with
respect to $x \in \Omega$ and $t \in \mathbb{R}_+$ yields

$$-\int_0^T \int_\Omega \int_0^T \int_\Omega \Big[|v(y,s) - u(x,t)|\psi(t)\rho_n(x-y)\bar{\rho}_n'(t-s)$$
$$- (f(v(y,s)) - f(u(x,t)))(\mathrm{sgn}(v(y,s) - u(x,t))\psi(t)b \cdot \nabla\rho_n(x-y)\bar{\rho}_n(t-s)\Big]$$
$$dy\,ds\,dx\,dt \ge 0,$$

which gives $A_{2,n} + A_{3,n} \le 0$. Note that the term associated with the initial condition
is zero because $\bar{\rho}_n(t) = 0$ if $t \ge 0$.

It is now sufficient to show that $\lim_{n\to+\infty} A_{4,n} = 0$ to conclude by passing to the
limit in (5.32) that

$$\int_0^T \int_\Omega |u(x,t) - v(x,t)||\psi'(t)|\,dx\,dt \ge 0. \tag{5.33}$$

To show that $\lim_{n\to+\infty} A_{4,n} = 0$, let us return to the entropic formulation for v,
written with y and s as variables, and choose the Kruzhkov entropy associated with
$k = u_0(x)$ and $\varphi(y,s) = \psi(0)\rho_n(x-y) \int_s^\infty \bar{\rho}_n(-\tau)\,d\tau$, with n large enough for this
test function φ to be admissible. Finally, integrating with respect to $x \in \Omega$ yields

$$-A_{4,n} - \int_0^T \int_\Omega \int_\Omega (f(v(y,s)) - f(u_0(x)))(\mathrm{sgn}(v(y,s) - u_0(x))\psi(0)b \cdot \nabla\rho_n(x-y)$$
$$\times \int_s^\infty \bar{\rho}_n(-\tau)\,d\tau\,dy\,ds\,dx + \int_\Omega \int_\Omega |u_0(y) - u_0(x)|\psi(0)\rho_n(x-y)\,dx\,dy \ge 0.$$

Hence

$$0 \le A_{4,n} \le A_{5,n} + A_{6,n},$$

with

$$A_{5,n} = -\int_0^T \int_\Omega \int_\Omega (f(v(y,s)) - f(u_0(x))(\mathrm{sgn}(v(y,s) - u_0(x)) \, \psi(0)$$

$$\times b \cdot \nabla \rho_n(x-y) \int_s^\infty \bar{\rho}_n(-\tau) \, d\tau \; dy \, dx \, ds,$$

$$A_{6,n} = \int_\Omega \int_\Omega |u_0(y) - u_0(x)| \psi(0) \rho_n(x-y) \, dx \, dy.$$

It is not difficult to show that $\lim_{n \to +\infty} A_{5,n} = \lim_{n \to +\infty} A_{6,n} = 0$. It follows that $\lim_{n \to +\infty} A_{4,n} = 0$ and, finally, we obtain (5.33).

We can now conclude. Let $0 < \varepsilon < T$, we choose $\psi \in C_c^\infty([0,T[,\mathbb{R}_+)$ such that $\psi' < 0$ on $]0, T-\varepsilon]$. The inequality (5.33) then gives $u = v$ a.e. on $\Omega \times]0, T-\varepsilon[$. Since ε is arbitrarily small, we conclude that $u = v$ a.e. on $\Omega \times]0, T[$, which completes the proof of uniqueness. ∎

Remark 5.32 (Existence in the case where $u_0 \notin BV$). If u_0 belongs only to $L^\infty(\Omega)$, one initial method, employed by Kruzhkov, consists of approximating u_0 with a sequence of elements from $L^\infty(\Omega) \cap BV(\overline{\Omega})$ and proving that the sequence of associated entropy solutions converges (in a suitable sense, after extracting a subsequence) to an entropy solution associated with u_0.

The major drawback of this method is that it does not appear adaptable for proving the convergence of numerical schemes. Indeed, even if the initial condition is assumed to be in $BV(\overline{\Omega})$, the approximated solution obtained through a numerical scheme is not bounded in $BV(\overline{\Omega} \times [0,T])$ independently of the discretization parameters (except in the case of Cartesian grids).

For this reason, one may prefer an alternative method that does not rely on the BV estimate and was historically developed for the convergence of numerical schemes [24]. This method is used later in the proof of Theorem 5.37. The idea is as follows: With $u_0 \in L^\infty(\Omega)$, we no longer aim to directly obtain the compactness of the sequence $u^{(n)}$ in $L^1(\Omega \times]0, T[)$. Instead, using the estimate from Lemma 5.30 for $u^{(n)} \in L^\infty(\Omega)$, we show the convergence, up to a subsequence and in a suitable sense, of the sequence $(u^{(n)})_{n \in \mathbb{N}}$ to a limit $\tilde{u}$ that depends on an additional variable belonging to the interval $[0,1]$. This results in an unconventional compactness theorem providing a type of convergence we call "nonlinear weak-$\star$ convergence".

Next, we prove that $\tilde{u}$ is a solution of the problem in a more general sense than (5.26), which we refer to as a "process solution". This proof is very similar to that of the existence result in Theorem 5.29. We then prove the uniqueness of the process solution and that this process solution is indeed an entropy solution (i.e., a solution of (5.26)). The proof of uniqueness for the process solution is very close to that of Theorem 5.29, see also [24].

Note that the uniqueness proof from Theorem 5.29 remains valid even in the case where u_0 is only in $L^\infty(\Omega)$.

A byproduct of this proof is the convergence of $u^{(n)}$ to u in all spaces $L^p(\Omega\times]0,T[)$, $p < +\infty$, including the case where f is linear (or is linear over intervals of $\mathbb{R}$). The essential idea has thus been to replace Helly's compactness theorem with a weaker compactness theorem combined with a uniqueness result for the process solution of (5.25) (see, for example, [23, Chapter 5]).

Remark 5.33 (Case where Ω is unbounded). In the "uniqueness" part of the proof of Theorem 5.29, it would have been possible to take a function ψ also depending on x. We would then have obtained

$$\int_0^T \int_\Omega |u - v|\psi_t \, dx \, dt + \int_0^T \int_\Omega b(f(u) - f(v)\mathrm{sgn}(u - v) \cdot \nabla\psi \, dx \, dt \geq 0,$$

$$\forall \psi \in C_c^\infty(\mathbb{R}^N \times [0, T[, \mathbb{R}_+). \quad (5.34)$$

This is interesting for showing uniqueness in the case where the open subset Ω is unbounded (for example, $\Omega = \mathbb{R}^N$) by taking advantage of the "finite speed propagation" property for hyperbolic problems. More precisely, we take in (5.34)

$$\psi(x, t) = r(t)\varphi_a(|x| + \omega t) \text{ with } \omega = L_f\|b\|_\infty,$$

where L_f is an upper bound of $|f'|$ on the interval $[-\gamma, \gamma]$, with

$$\gamma = \max\{\|u\|_\infty, \|v\|_\infty\}, \ r(t) = \frac{1}{T}(T - t)^+,$$

$$\varphi_a \in C_c^\infty([0, \infty[, \mathbb{R}_+),$$

where $\varphi_a = 1$ on $[0, a]$ with $a > 0$ given and φ_a non-increasing.

Note that a simple regularisation argument allows us to take such a function ψ in (5.34). We then obtain

$$-\frac{1}{T}\int_0^T \int_{\mathbb{R}^N} |u - v|\varphi_a(|x| + \omega t) \, dx \, dt + \int_0^T \int_{\mathbb{R}^N} |u - v|r(t)\varphi_a'(|x| + \omega t)\omega \, dx \, dt$$

$$+ \int_0^T \int_{\mathbb{R}^N} b(f(u) - f(v))\mathrm{sgn}(u - v)r(t)\varphi_a'(|x| + \omega t)\frac{x}{|x|} \, dx \, dt \geq 0.$$

But

$$\int_0^T \int_{\mathbb{R}^N} b(f(u) - f(v)\mathrm{sgn}(u - v)r(t)\varphi_a'(|x| + \omega t)\frac{x}{|x|}$$

$$\leq -\int_0^T \int_{\mathbb{R}^N} \|b\|L_f|u - v|r(t)\varphi_a'(|x| + \omega t) \, dx \, dt$$

$$\leq -\int_0^T \int_{\mathbb{R}^N} |u - v|r(t)\varphi_a'(|x| + \omega t)\omega \, dx \, dt,$$

because $\omega = L_f \|b\|$. Hence

$$-\frac{1}{T} \int_0^T \int_{\mathbb{R}^N} |u - v| \varphi_a(|x| + \omega t) \, dx \, dt \geq 0.$$

Hence, introducing $B_{a,t} = \{x \text{ such that } |x| + \omega t \leq a\}$, we get that

$$\int_0^T \left(\int_{B_{a,t}} |u - v| \, dx \right) dt = 0.$$

We now let a tend to $+\infty$ and obtain, by monotone convergence, $\int_0^T \int_\Omega |u-v| \, dx \, dt = 0$ and therefore $u = v$ a.e. on $\Omega \times]0, T[$.

Remark 5.34 (assumptions on b). 1. We used the C^1 regularity of b to obtain the BV estimate on the approximate solutions. Even if we do not use the BV estimate, we still use the C^1 regularity of b for uniqueness. In fact, the proofs of the BV estimate and uniqueness remain correct as long as b is locally Lipschitz-continuous.

2. We assumed $\operatorname{div} b = 0$. We could replace this assumption with $\operatorname{div} b \in L^\infty$ provided that f is Lipschitz-continuous. We also assumed that $b = 0$ on $\partial\Omega$ (to avoid dealing with the difficult case of boundary conditions) but we could replace this condition with $b \cdot n = 0$ without much additional difficulty. The problem of boundary conditions arises when $b \cdot n \neq 0$. This is the object of the next section.

5.2.2 Case of Boundary Conditions

This section is dedicated to a generalisation of Theorem 5.26 in the multidimensional scalar case. Consider the problem (5.25), where we no longer assume that $b = 0$ on the boundary.

Definition 5.35 (Entropy weak solution, with boundary conditions, [44]). Let Ω be an open bounded subset of $\mathbb{R}^N$ ($N \geq 1$) with a Lipschitz boundary. Let $T > 0$, $f \in C^1(\mathbb{R}, \mathbb{R})$ (or $f : \mathbb{R} \to \mathbb{R}$ Lipschitz-continuous) and $b \in C^1(\bar{\Omega} \times [0, T])^N$. Let $u_0 \in L^\infty(\Omega)$ and $\bar{u} \in L^\infty(\partial\Omega \times]0, T[)$. Let $A, B \in \mathbb{R}$ such that $A \leq u_0 \leq B$ a.e. on Ω and $A \leq \bar{u} \leq B$ a.e. on $\partial\Omega \times]0, T[$.

A function $u : \Omega \times]0, T[\to \mathbb{R}$ is an entropy weak solution to (5.25) satisfying (weakly) the boundary condition $\bar{u}$ if

$$u \in L^\infty(\Omega \times]0, T[) \text{ and } \forall \kappa \in [A, B], \ \forall \varphi \in C_c^1(\bar{\Omega} \times [0, T), \mathbb{R}_+),$$

$$\int_0^T \int_{\Omega_T} [(u - \kappa)^\pm \partial_t \varphi + \operatorname{sign}_\pm(u - \kappa)(f(u) - f(\kappa)) b \cdot \operatorname{grad}\varphi] \, dx \, dt \tag{5.35}$$

$$+ M \int_0^T \int_{\partial\Omega} (\bar{u}(t) - \kappa)^\pm \varphi(x, t) \, d\gamma(x) \, dt + \int_\Omega (u_0 - \kappa)^\pm \varphi(x, 0) \, dx \geq 0,$$

where $d\gamma(x)$ represents integration with respect to the $(N-1)$-dimensional Lebesgue measure on the boundary of Ω introduced in Section 1.5, and M is such that $\|b\|_\infty |f(s_1) - f(s_2)| \le M|s_1 - s_2|$ for all $s_1, s_2 \in [A, B]$, where $\|b\|_\infty = \sup_{(x,t)\in\Omega\times[0,T]} |b(x,t)|$, with $|\cdot|$ denoting here the Euclidean norm in $\mathbb{R}^N$).

Remark 5.36.

1. If u satisfies the family of inequalities (5.35)), we can prove that u is a solution of a weak form of (5.25) and that it satisfies certain entropy inequalities in $\Omega\times]0,T[$, namely

$$|u - \kappa|_t + \mathrm{div}(b(f(\max(u,\kappa)) - f(\min(u,\kappa)))) \le 0 \text{ for all } \kappa \in \mathbb{R},$$

   but also on the boundary $\partial\Omega$ and at time $t = 0$. The entropy weak solution u satisfies the initial condition $(u(\cdot,0) = u_0)$ and partially satisfies the boundary conditions. For example, if $f' > 0$ and if u and Ω are sufficiently regular, then $u(x,t) = \overline{u}(x,t)$ if $x \in \partial\Omega$, $t \in]0,T[$ and $b(x,t) \cdot n(x,t) < 0$, where n is the outward normal unit vector to $\partial\Omega$.
2. Let $\overline{M} \ge 1$. It is interesting to note that u is a solution to (5.35) if and only if u is a solution to (5.35) where the term $\int_\Omega (u_0 - \kappa)^\pm \varphi(x,0)\,dx$ is replaced by $\overline{M} \int_\Omega (u_0 - \kappa)^\pm \varphi(x,0)\,dx$.

Theorem 5.37 (Existence and uniqueness, Otto, 1996). *Under the assumptions of Definition 5.35, if* $\mathrm{div}\, b = 0$ *in* $\Omega\times[0,T]$; *then there exists a unique entropy solution* $u \in L^\infty(\mathbb{R}\times\mathbb{R}_+)$ *satisfying (5.35). Furthermore,*

$$A \le u(x,t) \le B \text{ a.e. } (x,t) \in]0,1[\times\mathbb{R}_+.$$

Proof We provide here a sketch of the proof in the case where Ω is a bounded polygonal (or polyhedral) open subset of $\mathbb{R}^N$; this proof is based on the convergence of numerical approximations [53].

Step 1: Approximate solutions. Considering a fairly general mesh of Ω (with triangles, for example in the two-dimensional case), denoted by $\mathcal{T}$, and a time step k, an approximate solution $u_{\mathcal{T},k}$ of the problem (5.25) can be defined using two-point numerical fluxes (on the edges of the cells) constructed with a "numerical flux" function $g : \mathbb{R}^2 \to \mathbb{R}$ such that

- g is non-decreasing with respect to its first argument and non-increasing with respect to its second argument,
- $g(s,s) = f(s)$, for all $s \in [A, B]$,
- g is locally Lipschitz-continuous (or locally "Lip-diag", see [27, Definition 3.1]).

Let h be the mesh size, defined as the upper bound of the diameters of the elements of the mesh. Under a so-called Courant–Friedrichs–Lewy (CFL) condition, of the type $k \le (1 - \zeta)\frac{h}{L}$ with $\zeta > 0$, it can be shown that

$$A \le u_{\mathcal{T},k} \le B \text{ a.e. on } \Omega\times]0,T[.$$

Unfortunately, it does not seem easy to directly obtain a compactness result on the family of approximate solutions (although this compactness result is true, as we will see later).

Step 2: Weak compactness. Using only this L^∞ bound on $u_{\mathcal{T},k}$, we can assume (up to a subsequence) that $u_{\mathcal{T},k} \to u$, as the mesh step tends to 0 (with the CFL condition), in a "weak non-linear sense" (similar to convergence to a Young measure, see [23, Section 30] for example), that is, $u \in L^\infty(\Omega \times]0, T[\times (0, 1))$ and

$$\int_0^T \int_\Omega \Phi(u_{\mathcal{T},k}(x,t))\varphi(x,t)\,\mathrm{d}x\,\mathrm{d}t \to \int_0^1 \int_0^T \int_\Omega \Phi(u(x,t,\alpha))\varphi(x,t)\,\mathrm{d}x\,\mathrm{d}t\,\mathrm{d}\alpha,$$
$$\forall \varphi \in L^1(\Omega \times]0, T[), \; \forall \Phi \in C(\mathbb{R}, \mathbb{R}).$$

Step 3: passage to the limit. Using the monotonicity of numerical fluxes, the approximate solutions satisfy certain discrete entropy inequalities. By passing to the limit on these inequalities, we obtain that u (defined in step 2) satisfies certain inequalities very similar to (5.35), namely:

$$u \in L^\infty(\Omega \times]0, T[\times (0, 1)),$$
$$\int_0^1 \int_0^T \int_\Omega [(u - \kappa)^\pm \partial_t \varphi + \mathrm{sign}_\pm (u - \kappa)(f(u) - f(\kappa))b \cdot \mathrm{grad}\varphi]\,\mathrm{d}x\,\mathrm{d}t\,\mathrm{d}\alpha$$
$$+ M \int_0^T \int_{\partial\Omega} (\bar{u}(t) - \kappa)^\pm \varphi(x,t)\,\mathrm{d}\gamma(x)\,\mathrm{d}t + \int_\Omega (u_0 - \kappa)^\pm \varphi(x,0)\,\mathrm{d}x \geq 0, \tag{5.36}$$
$$\forall \kappa \in [A, B], \; \forall \varphi \in C_c^1(\overline{\Omega} \times [0, T), \mathbb{R}_+).$$

We choose here M not only larger than the Lipschitz constant of $\|b\|_\infty f$ on $[A, B]$, but also larger than the Lipschitz constant (on $[A, B]^2$) of the numerical fluxes associated with the edges of the cells. This choice of M is possible because the unique solution of (5.35) does not depend on M provided that M is greater than the Lipschitz constant of $\|b\|_\infty f$ on $[A, B]$ and because the numerical flux function can be chosen with a Lipschitz constant bounded by the Lipschitz constant of $\|b\|_\infty f$ (the Godunov flux, for example). This method leads to an existence result with M only larger than the Lipschitz constant of $\|b\|_\infty f$ on $s \in [A, B]$, by passing to the limit on the approximate solutions given with these numerical fluxes.

Step 4: Uniqueness of the solution to (5.36). In this step, the *variable dedoubling* method of Krushkov is used to prove the uniqueness of the solution of (5.36). Indeed, if u and w are two solutions of (5.36), the variable dedoubling method leads to:

$$\int_0^1 \int_0^1 \int_0^T \int_\Omega |u(x,t,\alpha) - w(x,t,\beta)|\partial_t \varphi\,\mathrm{d}x\,\mathrm{d}t\,\mathrm{d}\alpha\,\mathrm{d}\beta +$$
$$\int_0^1 \int_0^1 \int_0^T \int_\Omega (f(\max(u, w)) - f(\min(u, w)))b \cdot \mathrm{grad}\varphi\,\mathrm{d}x\,\mathrm{d}t\,\mathrm{d}\alpha\,\mathrm{d}\beta \geq 0, \tag{5.37}$$
$$\forall \varphi \in C_c^1(\overline{\Omega} \times [0, T), \mathbb{R}_+).$$

Taking $\varphi(x, t) = (T - t)^+$ in (5.37) (which is indeed possible), we find that u does not depend on α, w does not depend on β and $u = w$ a.e. on $\Omega \times]0, T[$. Consequently, u is also the unique solution of (5.35).

Step 5: Conclusion. Step 4 provides, in particular, the uniqueness of the solution of (5.35). It also shows that the nonlinear weak-$\star$ limit of the sequences of approximate solutions is a solution of (5.35) and, therefore, the existence of the solution of (5.35). Moreover, since the nonlinear weak-$\star$ limit of the sequences of approximate solutions does not depend on α, it is quite easy to deduce that this limit is "strong" in $L^p(\Omega \times]0, T[)$ for all $p \in [1, \infty)$ (see [23], for example); thanks to the uniqueness of the limit, the convergence takes place for the whole subsequence.

5.3 Hyperbolic Systems

The theory of hyperbolic systems is much less developed than that of scalar equations; we focus here on the aspects of this theory that we believe are most used in applications. As in the case of scalar equations, the solutions of nonlinear hyperbolic conservation law systems can exhibit discontinuities that propagate in the form of shock waves. The mathematical theory is particularly difficult, especially because the uniqueness criterion for weak solutions remains, for general systems, an open question. We refer to the works [47] and [28] for further details on both the theory and the applications.

5.3.1 Definitions

In this section, we are interested in hyperbolic systems in the one-dimensional case, in the sense that the so-called "space" variable, generally denoted by x, belongs to $\mathbb{R}$; as usual, the so-called "time" variable, denoted by t, belongs to $\mathbb{R}_+$. We refer to [9] for the study of multidimensional systems.

Let $p \in \mathbb{N}$ be the number of (scalar) partial differential equations in the system under study ($p = 1$ in the case of the scalar equations considered in the previous sections) and let D be the domain of admissible values, defined as the subset of $\mathbb{R}^p$ in which the unknown vector of this system of p equations takes its values. Let $F \in C^1(D, \mathbb{R}^p)$ and $U_0 \in (L^\infty(\mathbb{R}))^p$, with values in D, which we will sometimes denote $L^\infty(\mathbb{R}; D)$; we are looking for a vector function $U : \mathbb{R} \times \mathbb{R}_+ \to D$ solution, in a sense to be defined, of the system

$$\partial_t U + \partial_x(F(U)) = 0, \ x \in \mathbb{R}, t \in \mathbb{R}_+, \tag{5.38a}$$

$$U(x, 0) = U_0(x), \ x \in \mathbb{R}. \tag{5.38b}$$

For example, in the case of Euler's equations for an isentropic compressible flow, we have $p = 2$, $U = \begin{bmatrix} \rho \\ \rho u \end{bmatrix}$ where ρ is the density ($\rho > 0$) and u the speed (ρu is therefore the momentum) and therefore the domain of admissible values is $D = \mathbb{R}_+ \times \mathbb{R} \subset \mathbb{R}^2$. The function F is given by $F(U) = \begin{bmatrix} \rho u \\ \rho u^2 + p \end{bmatrix}$, with $p = \rho^\gamma$, where $\gamma > 1$ is a given real number.

The system (5.38) is not always well posed, and its hyperbolic nature depends on the eigenvalues of the Jacobian matrix of F, which is specified in the following definition.

Definition 5.38 (Hyperbolic and strictly hyperbolic system). Let $p > 1$, D a domain of $\mathbb{R}^p$ and $F \in C^1(D, \mathbb{R}^p)$. The system (5.38a) is said to be:

- *hyperbolic* (more precisely hyperbolic in D) if, for all $U \in D$, the Jacobian matrix $J_F(U) \in M_p(\mathbb{R})$ of the mapping F at point U is diagonalisable in $\mathbb{R}$ (recall that the coefficients of the Jacobian matrix of F are $(J_F(U))_{i,j} = \partial_j F_i(U)$);
- *strictly hyperbolic* if, for all $U \in D$, the Jacobian matrix $J_F(U)$ has p distinct real eigenvalues.

The simplest example of function F is: $F(U) = AU$ where $A \in M_p(\mathbb{R})$ and $D = \mathbb{R}^p$, which yields a system of linear equations. According to Definition 5.38, the system (5.38a) is then hyperbolic if A is diagonalisable in $\mathbb{R}$, that is, if there exists a basis $(\varphi_1, \ldots, \varphi_p)$ of $\mathbb{R}^p$ and a family $(\lambda_1, \ldots, \lambda_p) \in \mathbb{R}^p$ such that $A\varphi_i = \lambda_i \varphi_i$ for all $i = 1, \ldots, p$. In this case, we can decompose the initial condition U_0 on the basis $(\varphi_1, \ldots, \varphi_p)$; denoting by $\alpha_1, \ldots, \alpha_p$ the components of U_0 in this basis, we therefore have:

$$U_0(x) = \sum_{i=1}^{p} \alpha_i(x)\varphi_i$$

and the unique weak solution of the system (5.38) is then written

$$U(x,t) = \sum_{i=1}^{p} \alpha_i(x - \lambda_i t)\varphi_i,$$

see on this subject Problem 5.4.

5.3.2 Weak Solutions, Entropy Solutions

We place ourselves in this section in the context of Definition 5.38. As in the scalar case (which is in fact a particular example), problem (5.38) generally does not admit a classical solution (that is to say a regular function that satisfies the initial condition (5.38b) and the system (5.38a)). We therefore define the notion of weak solution, which, as in the scalar case (see Definition 5.7), involves carrying the derivatives on the test functions.

Definition 5.39 (Weak solution of a hyperbolic system). Let $p \geq 1$, $U_0 \in (L^\infty(\mathbb{R}))^p$ be a vector function from $\mathbb{R}$ with values in the domain $D \subset \mathbb{R}^p$ and $F \in C^1(D, \mathbb{R}^p)$. We say that U is a weak solution of the system (5.38) if $U \in L^\infty(\mathbb{R} \times \mathbb{R}_+; D)$ and satisfies

$$\int_{\mathbb{R}_+} \int_{\mathbb{R}} U(x,t)\partial_t \varphi(x,t) \, dx \, dt + \int_{\mathbb{R}_+} \int_{\mathbb{R}} F(U(x,t))\partial_x \varphi(x,t) \, dx \, dt$$
$$+ \int_{\mathbb{R}} U_0(x)\varphi(x,0) \, dx = 0, \quad \forall \varphi \in C_c^1(\mathbb{R} \times \mathbb{R}_+; \mathbb{R}).$$

Consider for example the case of a Riemann problem, that is to say the system (5.38) with the following initial condition

$$U_0 = \begin{cases} U_g \text{ for } x < 0, \\ U_d \text{ for } x > 0, \end{cases} \tag{5.39}$$

with $U_g, U_d \in D$. As in the scalar case ($p = 1$), we can show that a function of the following form,

$$U(x,t) = \begin{cases} U_g \text{ for } x < \sigma t, \\ U_d \text{ for } x > \sigma t, \end{cases} \quad \text{with } \sigma \in \mathbb{R},$$

is a weak solution if and only if the Rankine–Hugoniot relation is satisfied, that is to say if and only if

$$\sigma[U] = [F(U)]$$

(recall that $[U] = U_d - U_g$ designates the jump of U). The proof of this equivalence uses the proof of the same result in the case $p = 1$, for each of the p scalar equations of the system (independently of the other equations).

Let us now consider the case of the Riemann problem for a linear hyperbolic system, that is to say, for the system (5.38) with $F(U) = AU$, where A is a matrix diagonalisable in $\mathbb{R}$, and with an initial condition of the form (5.39). Let $\lambda_1, \ldots, \lambda_p$ be the eigenvalues (real) of A and $\varphi_1, \ldots, \varphi_p$ a basis (of $\mathbb{R}^p$) of associated eigenvectors. We decompose U_0 on the basis of the eigenvectors:

$$U_0 = \sum_{i=1}^{p} \alpha_i \varphi_i \text{ with } \alpha_i = \begin{cases} \alpha_{g,i} \text{ for } x < 0, \\ \alpha_{d,i} \text{ for } x > 0. \end{cases}$$

We can then show (see Problem 5.4) that the function U defined by

$$U(x,t) = \sum_{i=1}^{p} \alpha_i(x - \lambda_i t)\varphi_i$$

is the unique weak solution. Let us examine the structure of this solution: it is made up of p constant states, and we change state each time we cross a line $x = \lambda_i t$, since

$$\alpha_i(x - \lambda_i t) = \begin{cases} \alpha_{g,i} & \text{for } x < \lambda_i t, \\ \alpha_{d,i} & \text{for } x > \lambda_i t. \end{cases}$$

Let us return to the general case of the system (5.38). As in the scalar case, for a non-linear system, there is not in general uniqueness of a weak solution. We imitate the scalar case and introduce the notion of entropy solution.

Definition 5.40 (Entropy and entropy flux). Let $p > 1$, D a domain of $\mathbb{R}^p$ and $F \in C^1(D, \mathbb{R}^p)$. A function η from D to $\mathbb{R}$ is an entropy for the system (5.38) if

1. $\eta \in C^1(D, \mathbb{R})$ and η is convex;
2. there exists a function $\Phi \in C^1(D, \mathbb{R})$, called the entropy flux, such that, for all $U \in D$,

$$\nabla \Phi(U) = J_F(U)^t \, \nabla \eta(U),$$

which can also be written, component by component:

$$\partial_i \Phi(U) = \sum_{j=1}^{p} \partial_i F_j(U) \partial_j \eta(U).$$

Do entropies exist? The answer is obviously yes: just take η and Φ constant (which does not give much information about the solutions), or $\eta(U) = U_i$ and $\Phi(U) = F_i(U)$: in this case an entropy weak solution defined below is just a weak solution. These are so-called "trivial" entropies.

Are there any non-trivial ones? The answer is different depending on the value of p:

- $p = 1$ (scalar case). Any convex function η is an entropy, and the associated flux is a primitive of $\eta' F'$.
- $p = 2$. Non-trivial entropies exist (see, for example, [47]).
- $p \geq 3$. For any hyperbolic system, there is generally no entropy (other than the trivial entropies). However, many systems modelling physical phenomena possess an entropy, well known to the physicist (who generously explains it to the mathematician).

Definition 5.41 (Entropy weak solution of a hyperbolic system). Let $U_0 \in (L^\infty(\mathbb{R}))^p$ be a (vector) function from $\mathbb{R}$ to values in $D \subset \mathbb{R}^p$. We say that U is an entropy solution of the system (5.38) if $U \in L^\infty(\mathbb{R} \times \mathbb{R}_+; D)$ and satisfies

$$\iint_{\mathbb{R} \times \mathbb{R}_+} \eta(U(x,t)) \partial_t \varphi(x,t) \, dx \, dt + \iint_{\mathbb{R} \times \mathbb{R}_+} \Phi(U(x,t)) \partial_x \varphi(x,t) \, dx \, dt$$

$$+ \int_{\mathbb{R}} \eta(U_0(x)) \varphi(x,0) \, dx \geq 0, \ \forall \varphi \in C_c^1(\mathbb{R} \times \mathbb{R}_+; \mathbb{R}_+),$$

for (η, Φ) entropy and associated entropy flux.

As mentioned above, an entropy solution is necessarily a weak solution (taking η to be a linear function in the definition).

As in the case $p = 1$, it can be shown that a solution of the form

$$U(x, t) = \begin{cases} U_g \text{ for } x < \sigma t, \\ U_d \text{ for } x > \sigma t. \end{cases} \quad \text{with } U_g, U_d \in D \text{ and } \sigma \in \mathbb{R},$$

is an entropy solution if and only if, for every entropy η and associated flux Φ, the Rankine–Hugoniot conditions

$$\text{(i)} \qquad \sigma[U] = [F(U)],$$
$$\text{(ii)} \qquad \sigma[\eta(U)] \geq [\Phi(U)],$$

are satisfied (recall that $[V] = V_d - V_g$). Note that (ii)$\Rightarrow$(i). The proof of this equivalence is here also identical to that in the case $p = 1$; indeed, (ii) is a scalar equation (and (i) is equivalent to saying that U is a weak solution).

Do we have existence and uniqueness of the entropy solution? Kruzhkov's theorem (Theorem 5.29) allows us to answer affirmatively in the case $p = 1$. In the case $p > 1$ the situation is much more complex, in particular because a strictly hyperbolic system does not always admit a non-trivial entropy. Nevertheless, the Lax condition that was introduced in the scalar case (condition (5.18)) can be generalized to the case of a system. We use it below in the study of the Riemann problem (Section 5.3.3). Let us recall that in the case $p = 1$ and if F is strictly convex or strictly concave, this condition states that a weak discontinuous solution presents a shock (that is to say, an entropic discontinuity) if the characteristics from either side of a discontinuity curve meet this curve, see Theorem 5.20. Still in this case, the Lax condition is equivalent to the entropy condition, see Problem 5.3. This equivalence is still true for some systems. This is proved in Problem 5.18 for the Saint-Venant equations[10]. The proof is given in the case of a discontinuity line of the self-similar solution of the Riemann problem but generalises to the case of a regular discontinuity curve.

5.3.3 Resolution of the Riemann Problem

The resolution of the Riemann problem for a 1D scalar conservation law with a strictly convex or concave flux is the object of Problem 5.9.

Here we consider some hyperbolic systems described in Definition 5.38. The resolution of the Riemann problem for certain systems of this kind which arise in physics is interesting on the one hand because it allows us to understand the structure of entropy solutions, and on the other hand because some numerical schemes for the computation of an approximate solution of these systems are based on this resolution.

---

[10] Adhémar Jean Claude Barré de Saint-Venant (1797–1886), engineer and mathematician specialising in fluid mechanics.

5.3.3.1 Definitions

Recall that the Riemann problem reads

$$\partial_t U + \partial_x(F(U)) = 0, \ x \in \mathbb{R}, t \in \mathbb{R}_+, \tag{5.40a}$$

$$U(x,0) = \begin{cases} U_g \text{ for } x < 0, \\ U_d \text{ for } x > 0, \end{cases} \tag{5.40b}$$

with $F \in C^1(D, \mathbb{R}^P)$ and $U_g, U_d \in D$. The solution is known when $F(U) = AU$ (and $D = \mathbb{R}^P$, see the previous subsection). Let us seek a "self-similar" solution, that is to say, a solution of the form $U(x,t) = V(\frac{x}{t})$. Recall that such a solution defined by zone is a weak solution if and only if

1. it is a classical solution on each zone
2. it satisfies the Rankine–Hugoniot conditions at the passage of each discontinuity $x = \sigma t$ between two zones: $\sigma[U] = [F(U)]$.

To this notion of weak solution, an entropy condition must be added. If the system contains a non-trivial natural entropy (or entropies), the weak solution is entropic if and only if at each discontinuity $x = \sigma t$, it satisfies $\sigma[\eta(U)] \geq [\Phi(U)]$ for any entropy η and associated flux Φ. In practice, the Lax condition, which does not require the existence of an entropy, is also used, see Definition 5.51. The link between these two entropy conditions is studied in Problem 5.18 for the Saint-Venant equations.

Consider the following assumptions and notations:

$$\begin{cases} F \in C^2(D, \mathbb{R}^P). \\ \text{The problem (5.40) is strictly hyperbolic, and we denote by} \\ \lambda_1(U) < \ldots < \lambda_p(U) \text{ the eigenvalues of } J_F(U) \text{ and} \\ (\varphi_1(U), \ldots, \varphi_p(U)) \text{ a basis of } \mathbb{R}^P \text{ of associated eigenvectors.} \end{cases} \tag{5.41}$$

We admit that for all $i = 1, \ldots, p$, the eigenvalues satisfy $\lambda_i(U) \in C^1(D, \mathbb{R})$ and that it is possible to choose the function φ_i so that $\varphi_i \in C^1(D, \mathbb{R}^P)$.

Definition 5.42 (Genuinely non-linear field, linearly degenerate field)**.** Under the assumptions (5.41), we say that the i-th field associated with the non-linear function F is

- *genuinely non-linear* (GNL) if

$$\nabla \lambda_i(U) \cdot \varphi_i(U) \neq 0, \ \forall U \in D,$$

  and we then normalise $\varphi_i(U)$ so that $\nabla \lambda_i(U) \cdot \varphi_i(U) = 1$,
- *linearly degenerate* (LD) if

$$\nabla \lambda_i(U) \cdot \varphi_i(U) = 0, \ \forall U \in D.$$

(We also admit here that it is possible to choose $\varphi_i \in C^1(D, \mathbb{R}^P)$.)

For the study of the Riemann problem, we strongly use the fact that the system under study is strictly hyperbolic and that all fields are GNL or LD. Note that in the case of a linear system, the eigenvalues λ_i do not depend on U and all fields are linearly degenerate.

In the scalar case $p = 1$, there is only one eigenvalue $\lambda_1(U) = F'(U)$, we can choose $\varphi_1(U) = 1$, and

- the GNL case corresponds to the case where F'' never cancels out, in which case F is strictly convex or concave (and it is then this somewhat more general assumption that is important),
- the LD case corresponds to the case $F''(u) = 0$ for all u, in which case the function F is linear.

These are obviously not the only possible cases. An example is given by the Buckley[11]–Leverett[12] equation (Problem 5.12). A simpler example, with $p = 1$ and $D = \mathbb{R}$, consists of taking $F(s) = s^2 \mathrm{sgn}(s)$. The problem in this example is that $F'(0) = 0$ (the system is strictly hyperbolic but the field is neither GNL nor LD). As in the case of the Buckley–Leverett equation, the solution of the Riemann problem with ε non-zero as initial data on $]-\infty, 0[$ and $-\varepsilon$ on $]0, +\infty[$ is formed of a part "shock" and a part "rarefaction". It is also interesting to note that for $p > 2$, it is possible that all fields are GNL or LD without the system being strictly hyperbolic, see Problem 5.14.

An interesting example is the system of Euler equations for compressible fluid flows. For the complete system (conservation of mass, momentum and energy) with the equation of state of a perfect gas, it can be shown that there are two fields and one LD field. The case of barotropic Euler equations (the pressure depends only on ρ) is a bit simpler. For this system, we have $D = \mathbb{R}_+^\star \times \mathbb{R}$ and the system reads:

$$\partial_t \rho + \partial_x(\rho u) = 0, \tag{5.42a}$$

$$\partial_t(\rho u) + \partial_x(\rho u^2 + p) = 0, \tag{5.42b}$$

where p is related to ρ by a given function $\mathcal{P}$, that is increasing, convex and differentiable from $\mathbb{R}_+^\star$ to $\mathbb{R}_+^\star$; therefore $p = \mathcal{P}(\rho)$. Note that the Saint-Venant equations for shallow water flows fall into this framework, with ρ the water height, and $p = \alpha \rho^2$, with $\alpha > 0$ (see Problem 5.15).

Let us introduce the momentum $q = \rho u$; the system (5.42) is then written as:

$$\partial_t U + \partial_x(F(U)) = 0, \text{ with } U = \begin{bmatrix} \rho \\ q \end{bmatrix}, \ F(U) = \begin{bmatrix} q \\ \dfrac{q^2}{\rho} + \mathcal{P}(\rho) \end{bmatrix}.$$

---

[11] Stuart Edward Buckley (1908–1975), American petroleum engineer.

[12] Miles Leverett (1910–2001), American physicist, known for his work on porous media; he was also part of the Manhattan project.

Set $c = \sqrt{\mathcal{P}'(\rho)}$. The Jacobian matrix of F is then:

$$J_F(U) = \begin{bmatrix} 0 & 1 \\ -\frac{q^2}{\rho^2} + \mathcal{P}'(\rho) & \frac{2q}{\rho} \end{bmatrix} = \begin{bmatrix} 0 & 1 \\ -u^2 + c^2 & 2u \end{bmatrix},$$

whose characteristic polynomial is

$$P_{J_F(U)}(X) = X^2 - 2\frac{q}{\rho}X + \frac{q^2}{\rho^2} - \mathcal{P}'(\rho)$$
$$= X^2 - 2uX + u^2 - c^2.$$

The eigenvalues and (non normalised) associated eigenvectors are therefore

$$\lambda_1(U) = u - c, \ \lambda_2(U) = u + c, \ \varphi_1(U) = \begin{bmatrix} 1 \\ u - c \end{bmatrix}, \ \varphi_2(U) = \begin{bmatrix} 1 \\ u + c \end{bmatrix}.$$

Hence, since $\mathcal{P}' > 0$ and $\mathcal{P}'' \geq 0$,

$$\forall U \in D = \{(\rho, q)^t \ \rho > 0\}, \qquad
\begin{aligned}
\nabla \lambda_1(U) \cdot \varphi_1(U) &= -\frac{c}{\rho} - \frac{\mathcal{P}''(\rho)}{2c} < 0, \\
\nabla \lambda_2(U) \cdot \varphi_2(U) &= \frac{c}{\rho} + \frac{\mathcal{P}''(\rho)}{2c} > 0.
\end{aligned}$$

The two fields are therefore genuinely non-linear.

5.3.3.2 Study of a Decoupled System

The study of a strictly hyperbolic decoupled system whose fields are all either LD or GNL is easy and instructive. A decoupled system reads

$$\partial_t u_i + \partial_x(f_i(u_i(x,t))) = 0, \ x \in \mathbb{R}, t \in \mathbb{R}_+, i = 1, \ldots, p \tag{5.43a}$$
$$u_i(x,0) = u_i(0), \ x \in \mathbb{R}, \ i = 1, \ldots, p, \tag{5.43b}$$

where each component f_i of the function F only depends on the component u_i of the unknown. The domain D of the unknown vector U (whose components are $u_1, \ldots, u_p$) is here equal to $\mathbb{R}^p$. For $U = (u_1, \ldots, u_p)^t \in \mathbb{R}^p$ the eigenvalues of the Jacobian matrix of F at the point U (denoted by $J_F(U)$) are $f_i'(u_i)$, $i = 1, \ldots, p$. Since the system is strictly hyperbolic and $D = \mathbb{R}^p$, we therefore have, for all $i \neq j$ and all $u, v \in \mathbb{R}$, $f_i'(u) \neq f_j'(v)$. If necessary, by changing the order of the equations we can therefore assume $f_i'(u) < f_j'(v)$ for all $i \neq j$ and all $u \neq v$ (with $u, v \in \mathbb{R}$). Since all fields are either GNL or LD, each function f_i is either strictly convex, strictly concave, or linear.

Let us take for example the case $p = 2$ and f_1 and f_2 strictly convex. Consider the associated Riemann problem, that is

$$U(x,0) = \begin{cases} U_g \text{ for } x < 0, \\ U_d \text{ for } x > 0. \end{cases} \quad \text{, with} \quad \begin{cases} u_{g,1} < u_{d,1} \\ u_{g,2} < u_{d,2}. \end{cases}$$

We therefore have a scalar Riemann problem for each component of U. These two Riemann problems correspond to rarefaction waves. Denoting by u_1 and u_2 the components of a vector U of $\mathbb{R}^2$, the eigenvalues of the Jacobian matrix of the flux of this system are $f_1'(u_1)$ and $f_2'(u_2)$. Owing to the strict hyperbolicity of the system (cf. Definition 5.38), we have (if necessary by changing the order of the equations) that

$$f_1'(u_1) = \lambda_1(U) < \lambda_2(U) = f_2'(u_2), \text{ for all } u_1, u_2 \in \mathbb{R}. \tag{5.44}$$

In particular $f_1'(u_{d,1}) < f_2'(u_{g,2})$, and therefore the rarefaction zones (in the plane $\mathbb{R} \times \mathbb{R}_+^\star$) for the unknowns u_1 and u_2 are completely separated; the solution of the Riemann problem, in the plane $\mathbb{R} \times \mathbb{R}_+^\star$, appears as the superposition of the solutions of the two Riemann problems on u_1 and u_2.

Note that without the condition of strict hyperbolicity, the rarefaction waves are no longer separated. However, we can still construct the solution of the Riemann problem by taking advantage of the fact that the two equations of the system are decoupled. This will be less easy for systems where the equations are coupled. The other cases of relative positions between $u_{g,1}, u_{d,1}$ and $u_{g,2}, u_{d,2}$ are treated in a similar manner. For example, if $u_{g,1} > u_{d,1}$ and $u_{g,2} < u_{d,2}$, the solution (in the plane $\mathbb{R} \times \mathbb{R}_+^\star$) is formed of a shock, which we call the "1-shock", located on the line $x = \sigma t$, with $\sigma = \frac{f_1(u_{g,1}) - f_1(u_{d,1})}{u_{g,1} - u_{d,1}}$ (corresponding to the first equation), and a rarefaction, which we call the "2-rarefaction", located between the lines $x = f_2'(u_{g,2})t$ and $x = f_2'(u_{d,2})t$ (corresponding to the second equation). The 1-shock is not in the 2-rarefaction zone because $\sigma \in]f_1'(u_{d,1}), f_1'(u_{g,1})[$ and, thanks to (5.44), $f_1'(u_{g,1}) < f_2'(u_{g,2})$. Let us emphasize that here again, without the condition of strict hyperbolicity, the waves are not necessarily separated (see Problem 5.14).

5.3.3.3 Coupled Systems

We still consider the framework of hyperbolic systems (Definition 5.38) with the assumptions and notations 5.41. In the case of a non-decoupled system, the Rankine–Hugoniot relations are used to construct the discontinuous solutions called "shock waves", corresponding to the GNL fields, as in the scalar case. To determine the continuous solutions corresponding to the GNL fields, which are the "rarefaction waves", we use the Riemann invariants which are defined below. Finally, "contact waves", which correspond to the LD fields and are discontinuous, can be seen as the limit case of a rarefaction wave or a shock wave, so that their construction relies either on the Rankine–Hugoniot relations or on the Riemann invariants.

Riemann Invariants and Rarefaction Waves

Definition 5.43 (Riemann Invariant). Let $1 \leq i \leq p$. An i-Riemann invariant for the system (5.38a) is a non-constant mapping $r \in C^1(D, \mathbb{R})$ such that $\nabla r(U) \cdot \varphi_i(U) = 0$ for all $U \in D$.

The concept of Riemann invariant is only of interest for $p \geq 2$. Indeed for $p = 1$, $\varphi_1(U)$ is any non-zero real number and there is no Riemann invariant.

In the case of the decoupled system seen previously, we have $\lambda_i(U) = f_i'(u_i)$ and φ_i is collinear to the i-th vector of the canonical basis. Therefore all the mappings $U \mapsto u_j$, $j \neq i$, are i-Riemann invariants. (Recall that $u_1, \ldots, u_p$ are the components of the vector U of $\mathbb{R}^p$.) Thus for each field, that is, for each i, there are $(p-1)$ independent Riemann invariants (in the sense that these $(p-1)$ mappings are linearly independent). We can show that this situation generalises in the following way (see, for example, [47]): for all $U_0 \in D$, there exists a neighbourhood $\mathcal{V}$ of U_0 and $(p-1)$ i-Riemann invariants defined on $\mathcal{V}$ and linearly independent, which is equivalent to saying that their gradients are independent.

Let us calculate the Riemann invariants for the barotropic Euler equations (5.42). The eigenvalues and eigenvectors (not normalised) of the Jacobian $J_F(U)$ are

$$\lambda_1(U) = u - c, \ \lambda_2(U) = u + c, \ \varphi_1(U) = \begin{bmatrix} 1 \\ u - c \end{bmatrix}, \ \varphi_2(U) = \begin{bmatrix} 1 \\ u + c \end{bmatrix}.$$

Let us look for a 1-Riemann invariant in the form $r(U) = \frac{q}{\rho} + h(\rho)$. We then have

$$\nabla r(U) = \begin{bmatrix} -\frac{q}{\rho^2} + h'(\rho) \\ \frac{1}{\rho} \end{bmatrix},$$

and therefore

$$\nabla r(U) \cdot \varphi_1(U) = -\frac{q}{\rho^2} + h'(\rho) + \frac{1}{\rho}(\frac{q}{\rho} - c) = h'(\rho) - \frac{c}{\rho}.$$

The function r is therefore a 1-Riemann invariant if (and only if) $h'(\rho) = \frac{c}{\rho}$. It is therefore sufficient to take for h a primitive of the function $\rho \mapsto \frac{\sqrt{\mathcal{P}'(\rho)}}{\rho}$. In the case of the Saint-Venant system of equations (for which $\mathcal{P}(\rho) = \alpha \rho^2$, $\alpha > 0$), we have $c = \sqrt{2\alpha\rho}$, and a 1-Riemann invariant is $r_1(U) = \frac{q}{\rho} + 2c = u + 2c$, see also Problem 5.15. In the same way, we calculate a 2-Riemann invariant: $r_2(U) = \frac{q}{\rho} - h(\rho)$, where h is a primitive of the function $\rho \mapsto \frac{\sqrt{\mathcal{P}'(\rho)}}{\rho}$. In the case of the Saint-Venant equations, this 2-Riemann invariant is: $r_2(U) = u - 2c$.

Under regularity assumptions, for a system of p equations, a Riemann invariant common to $p - 1$ waves satisfies a particularly simple equation, as shown by the following proposition (see also, for example, [47]).

Proposition 5.44 (Evolution equation for a Riemann invariant). *Let $p > 1$; we consider a strictly hyperbolic system (Definition 5.38) of the form (5.38). Let $\lambda_i(U)$, $i = 1, \ldots, p$, be the p distinct real eigenvalues of the Jacobian matrix $J_F(U)$ of F at point $U \in D$. Let $i \in \{1, \ldots, p\}$; we suppose that there exists a mapping $r \in C^1(D, \mathbb{R})$ which is a j-Riemann invariant for all $j \neq i$. Then, for any function $U \in C^1(\mathbb{R} \times \mathbb{R}_+, D)$ satisfying (5.38a), we have*

$$\partial_t(r(U)) + \lambda_i(U)\partial_x(r(U)) = 0.$$

Proof: For all $V \in D$, let $\{\varphi_1(V), \ldots, \varphi_p(V)\}$ be a basis of $\mathbb{R}^p$ such that $J_F(V)\varphi_j(V) = \lambda_j\varphi_j(V)$ for all $j \in \{1, \ldots, n\}$. Since the eigenvalues of $J_F(V)^t$ are the same as those of $J_F(V)$, there also exists $\{\psi_1(V), \ldots, \psi_p(V)\}$ a basis of $\mathbb{R}^p$ such that $J_F(V)^t\psi_j(V) = \lambda_j\psi_j(V)$ for all $j \in \{1, \ldots, n\}$. Let $k \neq j$, since $\lambda_k(V) \neq \lambda_j(V)$, we have

$$\lambda_j(V)\psi_j(V) \cdot \varphi_k(V) = J_F(V)^t\psi_j(V) \cdot \varphi_k(V) = \psi_j(V) \cdot J_F(V)\varphi_k(V)$$
$$= \lambda_k(V)\psi_j(V) \cdot \varphi_k(V),$$

and therefore $(\lambda_j(V) - \lambda_k(V))\psi_j(V) \cdot \varphi_k(V) = 0$. This implies that $\psi_j(V) \cdot \varphi_k(V) = 0$. We also deduce, since $\{\varphi_1(V), \ldots, \varphi_p(V)\}$ and $\{\psi_1(V), \ldots, \psi_p(V)\}$ are bases of $\mathbb{R}^n$, that $\psi_j(V) \cdot \varphi_j(V) \neq 0$ for all $j \in \{1, \ldots, n\}$. We now use the regularity of r, and we decompose $\nabla r(V)$ in the basis $\{\psi_1(V), \ldots, \psi_p(V)\}$, $\nabla r(V) = \sum_{k=1}^n \alpha_k(V)\psi_k(V)$. Since r is a j-Riemann invariant for $j \neq i$, we have, for all $j \neq i$,

$$0 = \nabla r(V) \cdot \varphi_j(V) = \sum_{k=1}^n \alpha_k(V)\psi_k(V) \cdot \varphi_j(V) = \alpha_j(V)\psi_j(V) \cdot \varphi_j(V),$$

and therefore $\alpha_j(V) = 0$, which proves that $\nabla r(V) = \alpha_i(V)\psi_i(V)$. We finally use the regularity of the function U,

$$\partial_t r(U) = \nabla r(U) \cdot \partial_t U = \alpha_i(U)\psi_i(U)^t\partial_t U = -\alpha_i(U)\psi_i(U)^t J_F(U)\partial_x U$$
$$= -\alpha_i(U)(\partial_x U)^t J_F(U)^t\psi_i(U) = -\alpha_i(U)\lambda_i(U)(\partial_x U)^t\psi_i(U)$$
$$= -\lambda_i(U)(\partial_x U)^t(\alpha_i(U)\psi_i(U)) = -\lambda_i(U)\nabla r(U) \cdot \partial_x U = -\lambda_i(U)\partial_x r(U).$$

■

Proposition 5.44 essentially allows us to obtain a diagonal system with the functions $r_i(U)$ as unknowns (where r_i is a j-Riemann invariant for all $j \neq i$). Suppose for example $p = 2$ in Proposition 5.44, the mapping r_1 is a 1-Riemann invariant and the mapping r_2 is a 2-Riemann invariant. Proposition 5.44 gives, if U is regular,

$$\partial_t(r_1(U)) + \lambda_2(U)\partial_x(r_1(U)) = 0,$$
$$\partial_t(r_2(U)) + \lambda_1(U)\partial_x(r_2(U)) = 0.$$

If the mapping $U \mapsto R(U) = \begin{bmatrix} r_1(U) \\ r_2(U) \end{bmatrix}$ is a diffeomorphism (from D to its image), let s be its inverse mapping, so that $U = s(R)$. The functions $\bar{r}_1 : (x,t) \mapsto r_1(U(x,t))$ and $\bar{r}_2 : (x,t) \mapsto r_2(U(x,t))$ are then solutions of the following transport problem:

$$\partial_t(\bar{r}_1) + \lambda_2(s(\bar{R}))\partial_x(\bar{r}_1) = 0,$$
$$\partial_t(\bar{r}_2) + \lambda_1(s(\bar{R}))\partial_x(\bar{r}_2) = 0,$$

where $\bar{R}$ is the vector function whose components are $\bar{r}_1$ and $\bar{r}_2$. This allows us, for example, to obtain estimates on the Riemann invariants by considering them as solutions to this transport problem.

Thanks to the Riemann invariants, we now construct the rarefaction waves which are solutions of the Riemann problem that are self-similar and continuous on $\mathbb{R} \times \mathbb{R}_+^\star$, i.e. such that there exist $-\infty < a < b < +\infty$ with $U = U_g$ on $D_1 = \{(x,t), x \leq at\}$, $U(x,t) = V(\frac{x}{t})$ on $D_2 = \{(x,t), at < x < bt\}$, $U(x,t) = U_d$ on $D_3 = \{(x,t), x \geq bt\}$ and U is a classical solution in D_2. The continuity of U therefore imposes that $V(a) = U_g$ and $V(b) = U_d$. We fix a and b ($a < b$). We thus seek $V \in C([a,b]) \cap C^1(]a,b[)$ such that $V(a) = U_g$, $V(b) = U_d$, and U is a classical solution to (5.40a) in D_2 with $U(x,t) = V(\frac{x}{t})$. We have, for $(x,t) \in D_2$,

$$\partial_t U(x,t) = -\frac{x}{t^2}V'(\frac{x}{t}) \text{ and}$$

$$\partial_x(F(U(x,t))) = J_F(U(x,t))\partial_x(U(x,t)) = \frac{1}{t}J_F(V(\frac{x}{t}))V'(\frac{x}{t}).$$

Therefore,

$$\forall(x,t) \in D_2, \quad \partial_t U(x,t) + \partial_x(F(U(x,t))) = \frac{1}{t}\left(J_F(V(\frac{x}{t})) - \frac{x}{t}\right)V'(\frac{x}{t}).$$

If we limit ourselves to looking for a solution with $V' \neq 0$ on $]a,b[$, for $\partial_t U + \partial_x(F(U)) = 0$ in D_2, it is therefore necessary that, for all $\frac{x}{t} \in]a,b[$,

$\frac{x}{t}$ is an eigenvalue of $J_F(V(\frac{x}{t}))$, and $V'(\frac{x}{t})$ is a non-zero associated eigenvector.

Therefore, there exists an $i \in [\![1,p]\!]$ such that $\frac{x}{t} = \lambda_i(V(\frac{x}{t}))$. We are sure that i does not depend on $\frac{x}{t}$, because we have assumed V to be continuous and the system is strictly hyperbolic. Note that the existence and uniqueness of i is still valid if V' vanishes at a point between $]a,b[$ (still thanks to the fact that the system is strictly hyperbolic). On the other hand, if V' is zero over an entire interval $[c,d]$, $a < c < d < b$, i can be different on $]a,c[$ and on $]d,b[$. This situation occurs when the considered Riemann problem has a solution formed of two rarefaction waves (each rarefaction corresponding to a value of i). Finally, under the condition $V' \neq 0$ (or if V' only vanishes at isolated points), the function $V \in C([a,b]) \cap C^1(]a,b[)$ is suitable if and only if there exists an $i \in [\![1,p]\!]$ such that

$$\forall s \in]a, b[, \ \lambda_i(V(s)) = s \text{ and } V'(s) = \alpha(s)\varphi_i(V(s)), \text{ with } \alpha(s) \in \mathbb{R}, \qquad (5.45)$$

$$V(a) = U_g, \ V(b) = U_d. \qquad (5.46)$$

Note that the condition (5.46) and the fact that $\lambda_i(V(s)) = s$ (for $s \in]a, b[$ and therefore also $s \in [a, b]$) gives $\lambda_i(U_g) = a$ and $\lambda_i(U_d) = b$. The condition (5.45) implies that the field i is GNL. Indeed, since the mapping $U \mapsto \lambda_i(U)$ is differentiable and we are looking for V of class C^1 on $]a, b[$, the previous equalities give that

$$1 = \nabla\lambda_i(V(s)) \cdot V'(s) = \alpha(s)\nabla\lambda_i(V(s)) \cdot \varphi_i(V(s)), \ \forall s \in]a, b[. \qquad (5.47)$$

This proves that $\nabla\lambda_i(V(s)) \cdot \varphi_i(V(s)) \neq 0$ (and also $\alpha(s) \neq 0$) and therefore that the field i is GNL. Finally, by normalising $\varphi_i(U)$ by the condition $\nabla\lambda_i(U) \cdot \varphi_i(U) = 1$, the equality (5.47) gives $\alpha(s) = 1$ and $V'(s) = \varphi_i(V(s))$ for all $s \in]a, b[$. The necessary and sufficient condition on V is therefore that V is a solution to the following differential equation with initial and final condition as follows:

$$\begin{cases} V \in C([a, b], D) \cap C^1(]a, b[, D), \\ V'(s) = \varphi_i(V(s)), \text{ for all } s \in]a, b[, \\ V(a) = U_g, \ \lambda_i(U_g) = a, \ V(b) = U_d. \end{cases} \qquad (5.48)$$

We have just shown that this condition is indeed necessary. To see that it is sufficient, it is enough to note that if V is a solution to this differential equation, we indeed have $\lambda_i(V(s)) = s$ for all $s \in]a, b[$ because $\lambda_i(V(a)) = \lambda_i(U_g) = a$ and $\lambda_i(V(s))' = \nabla\lambda_i(V(s)) \cdot V'(s) = \nabla\lambda_i(V(s)) \cdot \varphi_i(V(s)) = 1$ for all $s \in]a, b[$. In particular, this gives $\lambda_i(U_d) = \lambda_i(V(b)) = b$.

Definition 5.45 (Rarefaction wave). An i-rarefaction wave of the system (5.40) is a solution to (5.40) such that

1. $U(x, t) = U_g$ if $x \leq \lambda_i(U_g)t$,
2. $U(x, t) = U_d$ if $x \geq \lambda_i(U_d)t$ (so we have $\lambda_i(U_d) > \lambda_i(U_g)$),
3. $U(x, t) = V(\frac{x}{t})$ if $\lambda_i(U_g)t < x < \lambda_i(U_d)t$, with $\frac{x}{t} = \lambda_i(V(\frac{x}{t}))$,
4. $V \in C([\lambda_i(U_g), \lambda_i(U_d)], \mathbb{R}), \ V(\lambda_i(U_g)) = U_g$ and $V(\lambda_i(U_d)) = U_d$.
5. $V \in C^1(]\lambda_i(U_g), \lambda_i(U_d)[, D), \ V'(\frac{x}{t}) = \varphi_i((V(\frac{x}{t}))$ for $\lambda_i(U_g)t < x < \lambda_i(U_d)t$.

We then say that U_d is connected to U_g by an i-rarefaction (note that this relation is not symmetric because $\lambda_i(U_g) < \lambda_i(U_d)$). Let $\Gamma_i(U_g)$ the set of U_d connected to U_g by an i-rarefaction. Recall that $\varphi_i(U)$ is normalised by the condition $\nabla\lambda_i(U) \cdot \varphi_i(U) = 1$.

The previous calculation therefore shows that U_d is connected to U_g by an i-rarefaction (i.e., $U_d \in \Gamma_i(U_g)$) if and only if $U_d \in \{V(s), \ s > a, \ V \text{ solution to (5.49), with } a = \lambda_i(U_g)\}$,

$$\begin{aligned} V'(s) &= \varphi_i(V(s)), s > a, \\ V(a) &= U_g. \end{aligned} \qquad (5.49)$$

Indeed, if U_d is connected to U_g by an i-rarefaction, the function V given by Definition 5.45 is a solution to (5.49) over the interval $[a, b]$ with $b = \lambda_i(U_d)$ and $U_d = V(\lambda_i(U_d)) = V(b) \in \Gamma_i(U_g)$. Conversely, if $U_d \in \{V(s), s > a, V$ a solution to (5.49), with $a = \lambda_i(U_g)\}$, there exists a b such that $U_d = V(b)$ with V a solution to (5.49) over the interval $[a, b]$ and therefore V is a solution to (5.48), which indeed shows that U_d is connected to U_g by an i-rarefaction.

Since the function φ_i is of class C^1, the Cauchy problem (5.49) admits a unique local solution. The set $\Gamma_i(U_g)$ is therefore the trajectory (in $\mathbb{R}^p$) of the solution of this differential equation. The vector $\varphi_i(U_g)$ is a tangent vector to this trajectory at the point U_g. This is summarised in the following theorem.

Theorem 5.46 (Rarefaction curve). *Let i be a GNL field and $U_g \in D$. Then $\Gamma_i(U_g)$ (defined in Definition 5.45) is a curve (in space $\mathbb{R}^p$) starting from U_g and the vector $\varphi_i(U_g)$ is a tangent vector to this trajectory at the point U_g.*

When solving the Riemann problem, we want to determine if there are rarefaction waves, and if a state U_d on the right is connected to a state U_g on the left by an rarefaction wave, that is to say, if $U_d \in \Gamma_i(U_g)$. With this aim, a fairly simple way is to involve the Riemann invariants. Suppose that U is an i-rarefaction wave. Using the notation of Definition 5.45, $U(x, t) = V(\frac{x}{t})$ for $a = \lambda_i(U_g) < \frac{x}{t} < \lambda_i(U_d) = b$, $V \in C^1(]a, b[) \cap C([a, b])$. Let r be an i-Riemann invariant, then $r(U_g) = r(U_d)$, since $r(U)$ is constant in the rarefaction zone: indeed, for all $\xi \in]a, b[$,

$$r(V(\xi))' = \nabla r(V(\xi)) \cdot V'(\xi) = 0 \text{ since } V'(\xi) = \varphi_i(V(\xi)).$$

This shows that

$$\Gamma_i(U_g) \subset \bar{\Gamma}_i(U_g) = \{U_d \in D : r(U_d) = r(U_g), \text{ for all Riemann invariant } r\}.$$

Since we generally have $p-1$ independent i-Riemann invariants, $\bar{\Gamma}_i(U_g)$ is a curve of $\mathbb{R}^p$ passing through U_d and U_g. On the other hand, if $U_d \in \Gamma_i(U_g)$, we necessarily have $\lambda_i(U_g) < \lambda_i(U_d)$. The set $\Gamma_i(U_g)$ therefore corresponds to "half" of the curve $\bar{\Gamma}_i(U_g)$. More precisely,

$$\Gamma_i(U_g) \subset \{U_d \in \bar{\Gamma}_i(U_g) \text{ such that } \lambda_i(U_g) < \lambda_i(U_d)\} = \tilde{\Gamma}_i(U_g).$$

In general, we have $\Gamma_i(U_g) = \tilde{\Gamma}_i(U_g)$ (if there are indeed $p-1$ independent i-Riemann invariants). Solving a Riemann problem therefore involves the construction of the curves $\Gamma_i(U_g)$. In practice, we construct the set $\bar{\Gamma}_i(U_g)$, and only keep the part such that $\lambda_i(U_g) < \lambda_i(U_d)$.

Let us start with the simple example of a decoupled system, with $p = 2$. In this case we have seen that a 1-Riemann invariant is u_2, and therefore the set $\bar{\Gamma}_i(U_g)$ is the line $u_2 = u_{g,2}$. If f_1 is strictly convex, the condition $\lambda_1(U_g) < \lambda_1(U_d)$ gives that $u_{g,1} < u_{d,1}$, which gives a half line. The set of points U_d connectable to U_g are the points such that $u_{d,2} = u_{g,2}$ and $u_{d,1} > u_{g,1}$. The solution U is then a function whose second component is constant (and equal to $u_{g,2}$) and whose first component corresponds to a rarefaction for a scalar equation.

Let us now consider the case of the barotropic Euler system (5.42). We have seen that a 1-Riemann invariant is $r_1(U) = u + h(\rho)$, where h is a primitive of $\rho \mapsto \frac{\sqrt{\mathcal{P}'(\rho)}}{\rho}$. For the shallow water equations $\mathcal{P}(\rho) = \alpha\rho^2$, $\alpha > 0$, we therefore have $r_1(U) = u + 2c$ (recall that $c = \sqrt{2\alpha\rho}$); we can draw the curve $\bar{\Gamma}_1(U_g) = \{U_d \in D : r(U_d) = r(U_g)$ for any 1-Riemann invariant $r\}$; in the variables ρ, u, by setting $\beta_g = u_g + 2\sqrt{2\alpha\rho_g}$, $r_1(U_d) = r_2(U_g)$ comes back to writing that $u + 2\sqrt{2\alpha\rho} = \beta_g$ i.e.

$$u = \beta_g - 2\sqrt{2\alpha}\sqrt{\rho}.$$

We then obtain the half curve $\tilde{\Gamma}_1(U_g)$ by imposing that $\lambda_1(U_g) < \lambda_1(U_d)$, that is to say, $u_g - c_g < u - c$ that is to say $\frac{q_g}{\rho_g} - \sqrt{2\alpha\rho_g} < \frac{q}{\rho} - \sqrt{2\alpha\rho}$.

Shock waves, contact discontinuities

We are still interested in the Riemann problem (5.40). As in the study of rarefaction waves, we fix U_g and seek a possible self-similar solution that connects U_g and U_d via a discontinuity. Specifically, we look for a solution where there exists a $\sigma \in \mathbb{R}$ such that the function U defined by $U(x, t) = U_g$ if $x < \sigma t$ and $U(x, t) = U_d$ if $x > \sigma t$ is a solution to (5.40). We have already established that for U to be a weak solution to (5.40) it is necessary and sufficient that

$$\sigma[U] = \sigma(U_d - U_g) = (F(U_d) - F(U_g)) = [F(U)]. \tag{5.50}$$

Equation (5.50) allows for multiple possible pairs (U_g, U_d), including those where there exists an eigenvalue λ_i of the Jacobian matrix satisfying $\lambda_i(U_g) < \lambda_i(U_d)$ with a certain i corresponding to a GNL field. In such cases, a rarefaction wave solution exists, however there is no uniqueness of the weak solution. This situation mirrors the scalar case, where an additional condition is needed to enforce uniqueness. However, in the general case of the systems considered here, an entropy function in the sense of (5.40) does not always exist. To address this, Lax [33] proposed the use of the Lax condition, originally introduced for scalar equations (see Theorem 5.20).

Definition 5.47 (Lax condition, system case). We place ourselves under the assumptions (5.41); let U be a weak solution to the problem (5.10). We say that U satisfies the Lax condition if there exists an $i \in 1, \ldots, p$ such that

$$\lambda_i(U_g) > \sigma > \lambda_i(U_d) \text{ if the field } i \text{ is GNL}, \tag{5.51a}$$
$$\lambda_i(U_g) = \sigma = \lambda_i(U_d) \text{ if the field } i \text{ is LD}. \tag{5.51b}$$

For a system like that of the shallow water equations, this Lax condition is equivalent to the entropy condition of the definition (5.41). This is proved in Problem 5.18.

By eliminating σ from one of the equations (5.50) and replacing it in the others, we obtain $(p-1)$ equations on U_d. The solutions of these $p-1$ equations generally give, at least in a neighbourhood of U_g, p curves in the $\mathbb{R}^d$ plane. (This is not proved here but an example is given in Problem 5.15.) We can then determine how these curves behave at point U_g. Let $V : [0, \bar{\varepsilon}[\to (\mathbb{R}^d)^\star$ be a mapping such that $\lim_{s \to 0} V(s) = 0$ and such that, for all $s \in [0, \bar{\varepsilon}[$, $U_d = U_g + V(s)$ is connectable to U_g by a discontinuity, that is, (5.50) is satisfied for a certain $\sigma = \sigma(s)$. Consider a sequence $(s_n)_{n \in \mathbb{N}}$ of $]0, \bar{\varepsilon}[$ such that $\lim_{n \to +\infty} s_n = 0$. We can assume that, up to a subsequence,

$$\lim_{n \to +\infty} \frac{V(s_n)}{\|V(s_n)\|} = \varphi,$$

with $\varphi \in \mathbb{R}^d$ (and $\|\varphi\| = 1$). We write the Rankine–Hugoniot relation (5.50) with $U_d = U_g + V(s_n)$; using the C^1 character of F, we therefore obtain

$$\sigma(s_n)V(s_n) = F(U_g + V(s_n)) - F(U_g)) = DF(U_g)(V(s_n)) + \|V(s_n)\|\,\varepsilon_n,$$

with $\lim_{n \to +\infty} \varepsilon_n = 0$, and therefore

$$\sigma(s_n)\frac{V(s_n)}{\|V(s_n)\|} = DF(U_g)\frac{V(s_n)}{\|V(s_n)\|} + \varepsilon_n.$$

This shows that the sequence $(\sigma(s_n))_{n \in \mathbb{N}}$ has a limit λ and that $\lambda\varphi = DF(U_g)\varphi$. Since $\varphi \neq 0$, there exists an $i \in \{1, \ldots, p\}$ such that $\lambda = \lambda_i(U_g)$ and φ is collinear with $\varphi_i(U_g)$ (and not null). The vectors $\varphi_1(U_g), \ldots, \varphi_p(U_g)$ are therefore tangent vectors (at the point U_g) to the p curves of vectors U_d connectable to U_g by a discontinuity. Let $\Gamma_i(U_g)$ be the curve associated with the field i (that is, the vector $\varphi_i(U_g)$ is tangent to this curve at the point U_g) and we assume that this field is GNL. The Lax condition (5.51a) then gives us $\lambda_i(U_g) > \sigma(s_n) > \lambda_i(U_d)$, at least for n large enough and for $\|U_g - U_d\|$ small enough (to ensure, thanks to the condition of strict hyperbolicity and the fact that $\sigma(s_n) \to \lambda_i(U_g)$, that this Lax condition cannot be satisfied for another value of i). Using the C^1 character of λ_i, this Lax condition then gives

$$\lambda_i(U_d) = \lambda_i(U_g) + \nabla\lambda_i(U_g) \cdot V(s_n) + \|V(s_n)\|\,\varepsilon_n < \sigma(s_n) < \lambda_i(U_g),$$

with $\lim_{n \to +\infty} \varepsilon_n = 0$, and therefore

$$\nabla\lambda_i(U_g) \cdot \frac{V(s_n)}{\|V(s_n)\|} + \varepsilon_n < 0.$$

Passing to the limit as $n \to +\infty$ we get $\nabla\lambda_i(U_g) \cdot \varphi \leq 0$, which proves that $\varphi = \alpha\varphi_i(U_g)$ with $\alpha < 0$ (because $\varphi_i(U_g)$ is normalised by $\nabla\lambda_i(U_g) \cdot \varphi_i(U_g) = 1$ and φ is non-null collinear with $\varphi_i(U_g)$). The curve $\Gamma_i(U_g)$ therefore has the same tangent at the point U_g as the curve $\Gamma_i(U_g)$ seen in the case of rarefaction waves, but these two curves depart from U_g in opposite directions.

In the case of the shallow water equations, $\mathcal{P}(\rho) = \alpha\rho^2$ with $\alpha = \frac{1}{2}$, Problem 5.15 gives the set of states U connectable to U_g by a 1-shock (resp. 2-shock). With the unknowns (h, u), it is the set of pairs (h, u) such that $h > h_g$ (resp. $h < h_g$) and $u = u_g - S$, $S = \sqrt{(h - h_g)(h^2 - h_g^2)/(2h_g h)}$. Given two states U_g and U_d, we can then construct, for the shallow water equations, all the states connectable to U_g by a 1-shock or a 1-rarefaction and all the states connectable to U_d by a 2-shock or a 2-rarefaction. We thus obtain two curves (one passing through U_g and the other through U_d). The intersection of these two curves gives a state, called the intermediate state, which allows us to construct the solution to the Riemann problem. This is formed by a 1-wave connecting U_g to the intermediate state followed by a 2-wave connecting the intermediate state to U_d. Figure 5.4 gives these curves with a particular choice of U_g and U_d.

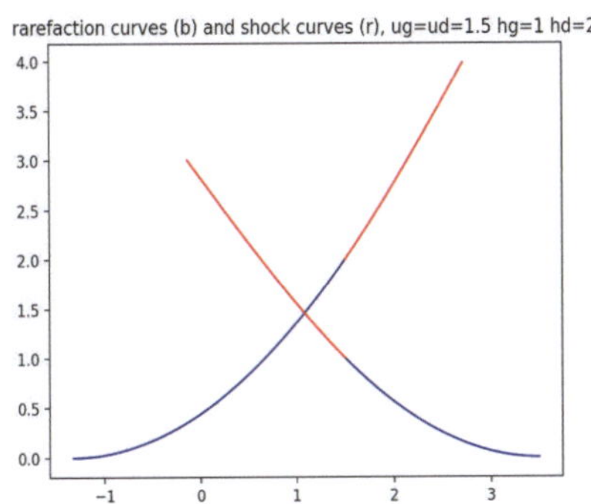

Fig. 5.4 The velocity u is on the x-axis and the height h on the y-axis. The intermediate state is given by the intersection of the red curve starting from U_g and the blue curve starting from U_d

From this study (shock waves, rarefaction waves and contact discontinuities), it is possible to prove Theorem 5.48, due to P.D. Lax (see [33, Theorem 5.4] for the initial statement and proof).

Theorem 5.48 (Solution of the Riemann problem, Lax). *Consider a strictly hyperbolic system, with GNL or LD fields, and $U_g \in D$. Then there exists an $\varepsilon > 0$ such that if $|U_g - U_d| < \varepsilon$, there is a solution to the Riemann problem (5.40), formed by at most $p + 1$ constant states connected by p waves (rarefactions, contacts, discontinuities or shocks) and satisfying the Lax condition.*

5.4 A Case study: The Shallow Water Equations

In this section, we consider the system of one-dimensional shallow water equations, which is the following system:

$$\partial_t h(x,t) + \partial_x(hu)(x,t) = 0, ; x \in \mathbb{R}, t \in \mathbb{R}_+, \tag{5.52a}$$

$$\partial_t(hu)(x,t) + \partial_x\left(hu^2 + \frac{g}{2}h^2\right)(x,t) = 0, ; x \in \mathbb{R}, t \in \mathbb{R}_+, \tag{5.52b}$$

where g is a given real number, $g > 0$. The system (5.52) is a simple model for the problem of shallow water flow. The unknown $h(x,t)$ represents the height of the water column at point x at time t. Here, we assume that the domain where the flow occurs has a flat bottom. The case of a non-flat bottom is discussed in Problems 5.16 and 5.17. The unknown $u(x,t)$ represents the velocity of this water column (located at point x at time t). The number g corresponds to the intensity of gravity. The function h takes its values in $\mathbb{R}_+^\star$, and the function u takes its values in $\mathbb{R}$.

We introduce two new unknowns:

- the momentum, $q : \mathbb{R} \times \mathbb{R}_+ \to \mathbb{R}$ defined by $q = hu$,
- the wave celerity, $c : \mathbb{R} \times \mathbb{R}_+ \to \mathbb{R}_+^\star$, defined by $c = \sqrt{gh}$.

We also define

$$U = \begin{bmatrix} h \\ q \end{bmatrix}, \ V = \begin{bmatrix} u \\ 2c \end{bmatrix}, \ p = \frac{gh^2}{2} \text{ and } D = \left\{ \begin{bmatrix} h \\ q \end{bmatrix} \in \mathbb{R}^2, h > 0 \right\}.$$

5.4.1 First Properties

By defining the function F from D to $\mathbb{R}^2$ as $F(U) = \begin{bmatrix} q \\ \dfrac{q^2}{h} + \dfrac{g}{2}h^2 \end{bmatrix}$, the system (5.52) also reads

$$\partial_t U(x,t) + \partial_x(F(U))(x,t) = 0, \ x \in \mathbb{R}, \ t \in \mathbb{R}_+. \tag{5.53}$$

The Jacobian matrix of F at point U is

$$DF(U) = \begin{bmatrix} 0 & 1 \\ -u^2 + gh & 2u \end{bmatrix}.$$

The characteristic polynomial of this matrix is $P_U(\lambda) = \lambda^2 - 2u\lambda + u^2 - gh = (u - \lambda)^2 - gh$. The eigenvalues of $DF(U)$ are therefore $\lambda_1(U) = u - c$ and $\lambda_2(U) = u + c$. They are real and distinct, and the system is therefore strictly hyperbolic. A basis of $\mathbb{R}^2$ formed by eigenvectors of this Jacobian matrix is then $\{\varphi_1(U), \varphi_2(U)\}$ with

$$\varphi_1(U) = \begin{bmatrix} 1 \\ u - c \end{bmatrix} \text{ and } \varphi_2(U) = \begin{bmatrix} 1 \\ u + c \end{bmatrix}.$$

Let us examine the nature of the two fields associated with the eigenvalues. Let $U \in D$, we have

$$\nabla \lambda_1(U) = \begin{bmatrix} -\dfrac{q}{h^2} + \dfrac{g}{2\sqrt{gh}} \\ \dfrac{1}{h} \end{bmatrix} = \begin{bmatrix} -\dfrac{u}{h} + \dfrac{g}{2c} \\ \dfrac{1}{h} \end{bmatrix},$$

which gives, with the previous notations:

$$\nabla \lambda_1(U) \cdot \varphi_1(U) = -\frac{u}{h} + \frac{g}{2c} + \frac{u - c}{h} = -\frac{g}{2c} \neq 0.$$

Since $\lambda_2(U) = u + c$, we have

$$\nabla \lambda_2(U) = \begin{bmatrix} -\dfrac{u}{h} - \dfrac{g}{2c} \\ \dfrac{1}{h} \end{bmatrix} \text{ and } \nabla \lambda_2(U) \cdot \varphi_1(U) = -\frac{u}{h} - \frac{g}{2c} + \frac{u + c}{h} = \frac{g}{2c} \neq 0.$$

The two fields are therefore GNL.

Let us compute the (non-trivial) Riemann invariants associated with each of the two fields of the system (recall that the Riemann invariants are given by Definition 5.43).

A 1-Riemann invariant is a function r_1 from D to $\mathbb{R}$ of class C^1 such that $\nabla r_1(U) \cdot \varphi_1(U) = 0$ for every U in D. If we look for r_1 in the form $r_1(U) = u + \psi(h)$, the condition on r_1 becomes

$$\begin{bmatrix} -\dfrac{q}{h^2} + \psi'(h) \\ \dfrac{1}{h} \end{bmatrix} \cdot \begin{bmatrix} 1 \\ u - c \end{bmatrix} = -\frac{u}{h} + \psi'(h) + \frac{u}{h} - \frac{c}{h} = \psi'(h) - \frac{c}{h} = 0,$$

One solution is to take $\psi(h) = 2c$, this gives indeed $\psi'(h) = \dfrac{g}{c} = \dfrac{c}{h}$. We therefore choose $r_1(U) = u + 2c$.

Similarly, a 2-Riemann invariant is $r_2(U) = u - 2c$.

Remark 5.49 (Equivalent form of the system for regular solutions). Let (h, u) be a regular solution of (5.52), and let $V = \begin{bmatrix} u \\ 2c \end{bmatrix}$. The equation (5.52b) gives

$$h\partial_t u + u\partial_t h + u\partial_x(hu) + hu\partial_x u + gh\partial_x h = 0.$$

Using (5.52a) and dividing by h (recall that $h > 0$),

$$\partial_t u + u\partial_x u + g\partial_x h = 0.$$

Since $\partial_x(2c) = \frac{c}{h}\partial_x h$ we have $g\partial_x h = \frac{gh}{c}\partial_x(2c) = c\partial_x(2c)$ and therefore

$$\partial_t u + u\partial_x u + c\partial_x(2c) = 0.$$

The equation (5.52a) gives

$$g\partial_t h + g(\partial_x h)u + gh\partial_x u = 0.$$

We have already seen that $g\partial_x h = c\partial_x(2c)$. Similarly $g\partial_t h = c\partial_t(2c)$. Thus

$$c\partial_t(2c) + cu\partial_x(2c) + c^2\partial_x u = 0,$$

which gives, by dividing by c,

$$\partial_t(2c) + u\partial_x(2c) + c\partial_x u = 0,$$

and therefore V is a solution to

$$\partial_t V(x,t) + B(V)\partial_x V(x,t) = 0, \ x \in \mathbb{R}, \ t \in \mathbb{R}_+, \ \text{avec } B(V) = \begin{bmatrix} u & c \\ c & u \end{bmatrix}. \qquad (5.54)$$

5.4.2 Entropy of the Shallow Water System

For all $U = \begin{bmatrix} h & q \end{bmatrix} \in D$, we define $\eta(U) = \frac{1}{2}hu^2 + p$ (recalling that $q = hu$ and $p = g\frac{h^2}{2}$).

Let us show that $\eta(U)$ is an entropy of the system, meaning that η is convex and that there exists a function Φ such that $\partial_t \eta(U) + \partial_x(\Phi(U)) = 0$ for any regular solution of (5.52) (and therefore of (5.53)).

Remark 5.50 (Entropy and Energy). For the Saint-Venant system (5.53), the quantity $\eta(U(x,t))$ represents the total energy of the water column located at point x at time t, i.e., the sum of the kinetic energy and the potential energy.

5.4.2.1 Entropy by an Algebraic Study of the System

Consider a regular solution to (5.52a)–(5.52b). Multiplying (5.52a) by gh and (5.52b) by u, we obtain

$$\partial_t\left(\frac{gh^2}{2}\right) + gh^2\partial_x u + u\partial_x\left(\frac{gh^2}{2}\right) = 0$$

$$h\partial_t\left(\frac{u^2}{2}\right) + u^2\partial_t h + u^2\partial_x(hu) + hu^2\partial_x u + u\partial_x p = 0.$$

We add these two equations. Using (5.52a) and $p = gh^2/2$, we obtain

$$\partial_t p + h\partial_t\left(\frac{u^2}{2}\right) + hu^2\partial_x u + 2p\partial_x u + 2u\partial_x p = 0.$$

Using again (5.52a), $h\partial_t\left(\frac{u^2}{2}\right) = \partial_t\left(h\frac{u^2}{2}\right) - \frac{u^2}{2}\partial_t h = \partial_t\left(h\frac{u^2}{2}\right) + \frac{u^2}{2}\partial_x(hu)$ and the previous equality gives

$$\partial_t\eta(U) + \frac{u^2}{2}\partial_x(hu) + hu^2\partial_x u + \partial_x(2pu) = 0.$$

Setting $\Phi(U) = (\frac{1}{2})hu^3 + 2pu$, the previous equality gives

$$\partial_t\eta(U) + \partial_x\Phi(U) = 0. \tag{5.55}$$

Another way to obtain this result (and we take this more general method in the sequel) is to notice that

$$\eta(U) = \frac{1}{2}hu^2 + p = \frac{q^2}{2h} + \frac{gh^2}{2}, \text{ so that } \nabla\eta(U) = \begin{bmatrix} -\frac{u^2}{2} + gh \\ u \end{bmatrix}.$$

We then multiply (5.52a) by $\partial_h\eta(U) = -\frac{u^2}{2} + gh$ and (5.52b) by $\partial_q\eta(U) = u$ and by similar manipulations to the previous ones we also obtain (5.55).

Note that $\partial_h\eta$ and $\partial_q\eta$ designate the partial derivatives of the function η, that is to say, of the function $U \mapsto \eta(U)$ (from D to $\mathbb{R}$) and therefore that $\partial_h\eta(U)$ and $\partial_q\eta(U)$ are these partial derivatives taken at the point U (belonging to D), while $\partial_t\eta(U)$ and $\partial_x\eta(U)$ designate the partial derivatives of the function $\eta \circ U$, that is to say, of the function $(x, t) \mapsto \eta(U(x, t))$ (from $\mathbb{R} \times \mathbb{R}_+^\star$ to $\mathbb{R}$).

It remains to verify that η is a convex function from D to $\mathbb{R}$, which is equivalent to showing that its Hessian matrix $H(U)$ is non-negative for all $U \in D$, or that $\xi^t H(U)\xi \geq 0$ for all $U \in D$ and $\xi \in \mathbb{R}^2$. Since

$$\nabla\eta(U) = \begin{bmatrix} -\dfrac{q^2}{2h^2} + gh \\ \dfrac{q}{h} \end{bmatrix},$$

we have

$$H(U) = \begin{bmatrix} \dfrac{u^2}{h} + g & -\dfrac{u}{h} \\ -\dfrac{u}{h} & \dfrac{1}{h} \end{bmatrix},$$

and therefore $\mathrm{tr}H(U) = \dfrac{u^2}{h} + g + \dfrac{1}{h} > 0$ and $\det H(U) = \dfrac{u^2}{h^2} + \dfrac{g}{h} - \dfrac{u^2}{h^2} = \dfrac{g}{h} > 0$. This shows that $H(U)$ is indeed non-negative for all $U \in D$.

5.4.2.2 Entropy by Passage to the Limit on Viscous Solutions

We add regularisation terms in the system (5.52a)-(5.52b), more precisely $-\varepsilon\partial_x^2 h$ in the first equation and $-\varepsilon\partial_{xq}^2 q$ in the second equation (with $\epsilon > 0$). Let h_ε and u_ε be the solutions of this new system (and $q_\varepsilon = h_\varepsilon u_\varepsilon$). Let us assume that these are regular functions, bounded in $L^\infty(\mathbb{R} \times \mathbb{R}_+)$ independently of ε, and that they converge in $L^1_{\mathrm{loc}}(\mathbb{R} \times \mathbb{R}_+)$, as $\varepsilon \to 0$ towards functions h and u respectively (with $h > 0$). We also assume that $\sqrt{\varepsilon}\partial_x h$ and $\sqrt{\varepsilon}\partial_x q$ are bounded in $L^2_{\mathrm{loc}}(\mathbb{R} \times \mathbb{R}_+)$. Let us then show that the pair (h, u) is a solution to (5.52a)–(5.52b) and satisfies, in the sense of the negativity of an element of $\mathcal{D}^\star(\mathbb{R} \times \mathbb{R}_+^\star)$ (Definition 1.5),

$$\partial_t \eta(U) + \partial_x \Phi(U) \leq 0.$$

Since h_ε and u_ε converge to h and u in $L^1_{\mathrm{loc}}(\mathbb{R} \times \mathbb{R}_+)$, the derivatives $\partial_x^2 h_\varepsilon$ and $\partial_x^2 u_\varepsilon$ converge to $\partial_x^2 h$ and $\partial_x^2 u$ at least in the sense of convergence in $\mathcal{D}^\star(\mathbb{R} \times \mathbb{R}_+^\star)$ and therefore $\varepsilon\partial_x^2 h_\varepsilon$ and $\varepsilon\partial_x^2 u_\varepsilon$ converge to 0 in the sense of convergence in $\mathcal{D}^\star(\mathbb{R}\times\mathbb{R}_+^\star)$. This proves that (h, u) is a weak solution to (5.52a)–(5.52b).

Let $U_\varepsilon = \begin{bmatrix} h_\varepsilon \\ q_\varepsilon \end{bmatrix}$. First note that $\eta(U_\varepsilon)$ and $\Phi(U_\varepsilon)$ converge in $L^1_{\mathrm{loc}}(\mathbb{R} \times \mathbb{R}_+)$, as $\varepsilon \to 0$, to $\eta(U)$ and $\Phi(U)$ respectively.

Multiplying (5.52a) by $\partial_h\eta(U_\varepsilon)$ and (5.52b) by $\partial_q\eta(U_\varepsilon)$, we obtain

$$\partial_t \eta(U_\varepsilon) + \partial_x \Phi(U_\varepsilon) + R_\varepsilon = 0, \tag{5.56}$$

with $R_\varepsilon = -\varepsilon\partial_h\eta(U_\varepsilon)\partial_x^2 h_\varepsilon - \varepsilon\partial_q\eta(U_\varepsilon)\partial_x^2 q_\varepsilon = S_\varepsilon + T_\varepsilon$, and

$$S_\varepsilon = -\sqrt{\varepsilon}\partial_x(\partial_h\eta(U_\varepsilon)\sqrt{\varepsilon}\partial_x h_\varepsilon) - \sqrt{\varepsilon}\partial_x(\partial_q\eta(U_\varepsilon)\sqrt{\varepsilon}\partial_x q_\varepsilon),$$
$$T_\varepsilon = \varepsilon\partial_x(\partial_h\eta(U_\varepsilon))\partial_x h_\varepsilon + \varepsilon\partial_x(\partial_q\eta(U_\varepsilon))\partial_x q_\varepsilon.$$

On the one hand, $S_\varepsilon \to 0$ in the sense of convergence in $\mathcal{D}^\star(\mathbb{R} \times \mathbb{R}_+)$ as $\varepsilon \to 0$ because $\partial_h\eta(U_\varepsilon))$ and $\partial_q\eta(U_\varepsilon))$ are bounded in $L^\infty(\mathbb{R} \times \mathbb{R}_+)$ and $\sqrt{\varepsilon}\partial_x h$ and $\sqrt{\varepsilon}\partial_x q$ are bounded in $L^1_{\mathrm{loc}}(\mathbb{R} \times \mathbb{R}_+)$.

On the other hand $T_\varepsilon = (\partial_x U_\varepsilon)^t H(U_\varepsilon)\partial_x U_\varepsilon$, where $H(U_\varepsilon)$ denotes the Hessian matrix of η at point U_ε. However, we showed previously that this Hessian matrix is non-negative, so we have $T_\varepsilon \geq 0$ and therefore, by taking the limit in (5.56) as $\varepsilon \to 0$, we find that in the sense of negativity in $\mathcal{D}^\star(\mathbb{R} \times \mathbb{R}_+^\star)$ (Definition 1.5), U satisfies

$$\partial_t \eta(U) + \partial_x \Phi(U) \leq 0.$$

5.4.3 The Riemann Problem

We are now interested in the Riemann problem, that is to say, the system (5.52a)–(5.52b) (equivalent to (5.53)) with the initial condition

$$h(x,0) = h_g, \ u(x,0) = u_g, \ x < 0, \tag{5.57}$$

$$h(x,0) = h_d, \ u(x,0) = u_d, \ x > 0, \tag{5.58}$$

where $h_g, h_d \in \mathbb{R}^{\star}_{+}$ and $u_d, u_g \in \mathbb{R}$ are given. Set $q_g = h_g u_g$, $q_d = h_d u_d$, $c_g = \sqrt{g h_g}$, $c_d = \sqrt{g h_d}$. Assume that $u_d - u_g < 2(c_g + c_d)$ (this condition is necessary for the problem (5.53), (5.57)–(5.58) to have a solution taking its values in D, this is the "no vacuum appearance" condition).

5.4.3.1 Solution with two Rarefaction Waves

We first assume that $2|c_g - c_d| \leq u_d - u_g$. We are going to construct the solution of the Riemann problem (5.53), (5.57)–(5.58). We are thus seeking a solution in the form of a 1-rarefaction wave and a 2-rarefaction wave separated by an intermediate state denoted by $(h_{\star}, u_{\star})$, with $h_{\star} > 0$. We look for the solution in the rarefaction zones with u and c as affine functions of $\frac{x}{t}$.

Let U_g, U_d and $U_{\star}$ be the constant states of this solution. We know that an i-Riemann invariant is constant in a i-rarefaction wave. Hence $u_g + 2c_g = u_{\star} + 2c_{\star}$ and $u_d - 2c_d = u_{\star} - 2c_{\star}$. This allows us to calculate $u_{\star}$, $c_{\star}$ and $h_{\star}$,

$$u_{\star} = \frac{u_g + u_d}{2} + c_g - c_d,$$

$$c_{\star} = \frac{u_g - u_d}{4} + \frac{c_g + c_d}{2}, \quad h_{\star} = \frac{c_{\star}^2}{g}.$$

We have indeed $c_{\star} > 0$ (and thus $h_{\star} > 0$) because $u_d - u_g < 2(c_g + c_d)$ (which is precisely the condition for non-appearance of the void).

To construct the solution, we know that the 1-rarefaction wave corresponds to the area $\{\lambda_1(U_g)t \leq x \leq \lambda_1(U_{\star})t\}$ and that the 2-rarefaction wave corresponds to the area $\{\lambda_2(U_{\star})t \leq x \leq \lambda_2(U_d)\}$. The construction of this solution formed of two rarefaction waves is therefore possible if and only if

$$\lambda_1(U_g) = u_g - c_g \leq u_{\star} - c_{\star} = \lambda_1(U_{\star}), \tag{5.59}$$

$$\lambda_2(U_{\star}) = u_{\star} + c_{\star} \leq u_d + c_d = \lambda_2(U_d). \tag{5.60}$$

The condition (5.59) is equivalent to $u_g - u_d \leq 2(c_g - c_d)$ and the condition (5.60) is equivalent to $u_g - u_d \leq 2(c_d - c_g)$. These two conditions are satisfied thanks to the condition $2|c_g - c_d| \leq u_d - u_g$.

The solution to the Riemann problem (5.53), (5.57)–(5.58) is therefore indeed made up of two rarefaction waves. It is constructed by distinguishing five zones D_i, $i = 1, \ldots, 5$; with

$$D_1 = \{x \leq (u_g - c_g)t\},$$

$$D_2 = \{(u_g - c_g)t < x < (u_{\star} - c_{\star})t\},$$

$$D_3 = \{(u_{\star} - c_{\star})t < x < (u_{\star} + c_{\star})t\},$$

$$D_4 = \{(u_\star + c_\star)t < x < (u_d + c_d)t\},$$
$$D_5 = \{(u_d + c_d)t < x\}.$$

- On D_1, we have $U = U_g$.
- On D_3, we have $U = U_\star$.
- On D_5, we have $U = U_d$.
- On D_2, we are dealing with a 1-rarefaction. We seek the solution in the form $U(x,t) = V(\frac{x}{t})$. Therefore, on the line $x = \alpha t$ with $(u_g - c_g) < \alpha < (u_\star - c_\star)$, the solution $U(x,t)$ is given by $\alpha = \frac{x}{t} = \lambda_1(V(\alpha)) = u - c$. In addition, writing the invariance of the 1-Riemann invariant, we obtain the solution by solving the system

$$u - c = \alpha,$$
$$u + 2c = u_g + 2c_g,$$

  which gives $3u = 2\alpha + u_g + 2c_g$, $3c = u_g + 2c_g - \alpha$.
- Finally, on D_4, we are dealing with a 2-rarefaction. Therefore, for $x = \alpha t$ with $(u_\star + c_\star) < \alpha < (u_d + c_d)$, the solution $U(x,t)$ is given by $\alpha = \frac{x}{t} = \lambda_2(V(\alpha))$, and by invariance of the 2-Riemann invariant, we obtain the solution by solving the system

$$u + c = \alpha,$$
$$u - 2c = u_d - 2c_d,$$

  which gives $3u = 2\alpha + u_d - 2c_d$, $3c = -u_d + 2c_d + \alpha$.
- Let $u_g \in \mathbb{R}$, h_g, $h_d \in \mathbb{R}_+^\star$ with $h_d > h_g$. For U to be a 1-shock, it is necessary and sufficient that $u_d = u_g - S$ (and $\sigma = [hu]/[h]$), that is to say,

$$u_d = u_g - \sqrt{\frac{g(h_g - h_d)(h_g^2 - h_d^2)}{2h_g h_d}}$$
$$= u_g - \sqrt{\frac{gh_g}{2}}\psi(\frac{h_d}{h_g}).$$

Of course, the reasoning is completely similar for a 2-shock. Let $u_g \in \mathbb{R}$, h_g, $h_d \in \mathbb{R}_+^\star$ with $h_d < h_g$. For U to be a 2-shock, it is necessary and sufficient that $u_d = u_g - S$ (and $\sigma = [hu]/[h]$), that is to say,

$$u_d = u_g - \sqrt{\frac{g(h_g - h_d)(h_g^2 - h_d^2)}{2h_g h_d}}$$
$$= u_g - \sqrt{\frac{gh_d}{2}}\psi(\frac{h_g}{h_d}).$$

5.4.3.2 Sufficient and Necessary Condition for a Shock

Here and in the following subsections, we assume that

$$2|c_g - c_d| > u_d - u_g, \text{ and we set } S = \sqrt{\frac{g(h_g - h_d)(h_g^2 - h_d^2)}{2h_g h_d}}. \tag{5.61}$$

If U is a weak solution then we have $h_g \neq h_d$ (indeed, if $h_g = h_d$, the Rankine–Hugoniot relation on the line $x = \sigma t$ gives, with the notations introduced earlier, $\sigma[h] = [hu]$ and therefore $u_g = u_d$ in contradiction with $2|c_g - c_d| > u_d - u_g$).

We start by looking for a sufficient and necessary condition for a shock, so we assume that there exists a $\sigma \in \mathbb{R}$ such that

$$U(x,t) = U_g \text{ if } x < \sigma t, \ U(x,t) = U_d \text{ if } x > \sigma t.$$

- Let us first show that U is a weak solution to (5.53), (5.57)–(5.58) if and only if $u_d = u_g \pm S$ and $\sigma(h_d - h_g) = (q_d - q_g)$.
  The function U is a weak solution if and only if the Rankine–Hugoniot conditions are satisfied on the line $x = \sigma t$, that is $\sigma[h] = [hu]$ and $\sigma[hu] = [hu^2 + p]$, which is equivalent to $[hu]^2 = [h][hu^2 + p]$ and $\sigma[h] = [hu]$. As

$$[hu]^2 = (h_d u_d - h_g u_g)^2 = h_d^2 u_d^2 + h_g^2 u_g^2 - 2h_g h_d u_g u_d \text{ and}$$
$$[h][hu^2 + p] = h_d^2 u_d^2 + h_g^2 u_g^2 - h_d h_g (u_g^2 + u_d^2) + [h][p],$$

  the function U is a weak solution if and only if $h_d h_g (u_g - u_d)^2 = [h][p]$ and $\sigma[h] = [hu]$. This indeed corresponds to $(u_g - u_d)^2 = S$ and $\sigma[h] = [hu]$.
  We now show that U is a 1-shock if and only if $u_d = u_g - S$ and $h_g < h_d$.

Necessary condition. Assume that U is a 1-shock. The Lax condition (5.51a) then gives, using $\sigma[h] = [hu]$,

$$u_g - c_g > \sigma = u_g + \frac{h_d(u_g - u_d)}{h_g - h_d} = u_d + \frac{h_g(u_g - u_d)}{h_g - h_d} > u_d - c_d. \tag{5.62}$$

The first inequality shows that $u_g - u_d$ and $h_g - h_d$ are non-zero and of opposite sign (since $c_g > 0$). Since $u_g - c_g > u_d - c_d$, we also have $u_g - u_d > c_g - c_d$. Since $c_g - c_d$ has the same sign as $h_g - h_d$, and therefore the opposite sign to that of $u_g - u_d$, we get $u_g - u_d > 0 > h_g - h_d$. This gives $u_d = u_g - S$ and $h_g < h_d$.

Sufficient condition. We now assume that $u_d = u_g - S$ and $h_g < h_d$ and we want to show that U is a 1-shock, that is, that (5.62) is satisfied. The first inequality is true if

$$gh_g < \frac{h_d^2(u_g - u_d)^2}{(h_g - h_d)^2} = \frac{h_d^2}{(h_g - h_d)^2}S^2$$

$$= \frac{h_d^2}{(h_g - h_d)^2}\frac{g(h_g - h_d)(h_g^2 - h_d^2)}{2h_g h_d} = g\frac{h_d}{h_g}\frac{h_g + h_d}{2}.$$

This is true because $h_g < h_d$. The second inequality is true if

$$gh_d > \frac{h_g^2(u_g - u_d)^2}{(h_g - h_d)^2} = \frac{h_g^2}{(h_g - h_d)^2}S^2$$

$$= \frac{h_g^2}{(h_g - h_d)^2}\frac{g(h_g - h_d)(h_g^2 - h_d^2)}{2h_g h_d} = g\frac{h_g}{h_d}\frac{h_g + h_d}{2}.$$

This is true because $h_d > h_g$.

Similarly, we show that U is a 2-shock if and only if $u_d = u_g - S$ and $h_g > h_d$. The Lax condition (which is (5.62) for the 1-shock) becomes

$$u_g + c_g > \sigma = u_g + \frac{h_d(u_g - u_d)}{h_g - h_d} = u_d + \frac{h_g(u_g - u_d)}{h_g - h_d} > u_d + c_d. \qquad (5.63)$$

The second inequality shows that $u_g - u_d$ and $h_g - h_d$ are non-zero and of the same sign (because $c_d > 0$). Since $u_g + c_g > u_d + c_d$, we also have $u_g - u_d > c_d - c_g$. Since $c_d - c_g$ has the same sign as $h_d - h_g$ and therefore the opposite sign of $u_g - u_d$, we get $u_g - u_d > 0 > h_d - h_g$. This gives $u_d = u_g - S$ and $h_g > h_d$. This shows that the condition $u_d = u_g - S$ and $h_g > h_d$ is necessary for a 2-shock. We now show that this condition is sufficient. We therefore assume that $u_d = u_g - S$ and $h_g > h_d$ and we want to show (5.63).

Showing the first inequality is equivalent to showing that

$$gh_g > \frac{h_d^2(u_g - u_d)^2}{(h_g - h_d)^2} = \frac{h_d^2}{(h_g - h_d)^2}S^2$$

$$= \frac{h_d^2}{(h_g - h_d)^2}\frac{g(h_g - h_d)(h_g^2 - h_d^2)}{2h_g h_d} = g\frac{h_d}{h_g}\frac{h_g + h_d}{2}.$$

This is true because $h_g > h_d$. Showing the second inequality is equivalent to showing that

$$gh_d < \frac{h_g^2(u_g - u_d)^2}{(h_g - h_d)^2} = \frac{h_g^2}{(h_g - h_d)^2}S^2$$

$$= \frac{h_g^2}{(h_g - h_d)^2}\frac{g(h_g - h_d)(h_g^2 - h_d^2)}{2h_g h_d} = g\frac{h_g}{h_d}\frac{h_g + h_d}{2}.$$

This is true because $h_d < h_g$.

- Let us now show that for any u_g, h_g and h_d, with $h_d > h_g$, there exists a unique u_d such that U is a 1-shock. Let $u_g \in \mathbb{R}$, $h_g, h_d \in \mathbb{R}_+^\star$ with $h_d > h_g$. We know that U is a 1-shock if and only if $u_d = u_g - S$ (and $\sigma = [hu]/[h]$), that is

$$u_d = u_g - \sqrt{\frac{g(h_g - h_d)(h_g^2 - h_d^2)}{2h_g h_d}}$$

$$= u_g - \sqrt{\frac{g h_g}{2}}\,\psi\left(\frac{h_d}{h_g}\right),$$

where ψ is the function defined by

$$x \in [1, +\infty[\,\mapsto\, \sqrt{\frac{(1 - 1/x)(x^2 - 1)}{x}}. \tag{5.64}$$

The approach is totally similar for a 2-shock. Let $u_g \in \mathbb{R}$, $h_g, h_d \in \mathbb{R}_+^\star$ with $h_d < h_g$. Then U is a 2-shock if and only if $u_d = u_g - S$ (and $\sigma = [hu]/[h]$), that is

$$u_d = u_g - \sqrt{\frac{g(h_g - h_d)(h_g^2 - h_d^2)}{2h_g h_d}}$$

$$= u_g - \sqrt{\frac{g h_d}{2}}\,\psi\left(\frac{h_g}{h_d}\right).$$

5.4.3.3 Solution formed by a 1-Rarefaction Wave and a 2-Shock

Still under the assumption (5.61), assuming here that $u_g - u_d < S$ and $h_d < h_g$, let us construct a solution to (5.53), (5.57)–(5.58) formed by a 1-rarefaction and a 2-shock connected by an intermediate state, denoted by $(h_\star, u_\star)$. We seek $(h_\star, u_\star)$ such that $u_\star + 2c_\star = u_g + 2c_g$, $u_\star = u_d + \sqrt{gh_d/2}\,\psi\left(\frac{h_\star}{h_d}\right)$, $h_\star > h_d$, with ψ the function defined by (5.64), and satisfying $u_g - c_g < u_\star - c_\star < \sigma$ where σ is the speed of the 2-shock.

Let U be the sought solution. In the zone $D_1 = \{x \le (u_g - c_g)t\}$, we have $U = U_g$. The zone $D_2 = \{(u_g - c_g)t < x < (u_\star - c_\star)t\}$ corresponds to the 1-rarefaction, and the solution can be calculated as it was done in subsection 5.4.3.1. The invariance of the 1-Riemann invariant in this zone gives $u_g + 2c_g = u_\star + 2c_\star$. In the zone $D_3 = \{(u_\star - c_\star)t \le x < \sigma t\}$, the solution is $U = U_\star$. In the zone $D_4 = \{x > \sigma t\}$, the solution is $U = U_d$ and the results of subsection 5.4.3.2 show that a necessary and sufficient condition for it to be a 2-shock is that $h_\star > h_d$, $u_d = u_\star - \sqrt{\frac{g h_d}{2}}\,\psi\left(\frac{h_\star}{h_d}\right)$ and $\sigma = (h_d u_d - h_\star u_\star)/(h_d - h_\star)$. In summary, the construction of a solution formed of a 1-rarefaction and a 2-shock connected by an intermediate state $(h_\star, u_\star)$ is possible if and only if

$$u_g - c_g < u_\star - c_\star, \tag{5.65}$$

$$u_g + 2c_g = u_\star + 2c_\star, \tag{5.66}$$

$$h_\star > h_d, \tag{5.67}$$

$$u_\star = u_d + \sqrt{\frac{gh_d}{2}} \psi\left(\frac{h_\star}{h_d}\right), \tag{5.68}$$

$$u_\star - c_\star < \sigma = \frac{h_d u_d - h_\star u_\star}{h_d - h_\star}. \tag{5.69}$$

For $h \in [h_d, h_g]$, we define $F(h) = u_d + \sqrt{\frac{gh_d}{2}} \psi\left(\frac{h}{h_d}\right) + 2\sqrt{gh}$. The function ψ is strictly increasing, as composed of the function $x \mapsto (1 - 1/x)(x^2 - 1)$ which is strictly increasing on $[1, +\infty[$ (as a product of strictly increasing positive functions) and of the function $x \mapsto \sqrt{x}$, also strictly increasing. The function F is therefore continuous, increasing, $F(h_d) = u_d + 2c_d$ and $F(h_g) = u_d + S + 2c_g > u_g + 2c_g$. Since $u_g + 2c_g > u_d + 2c_d$ (because $2(c_g - c_d) = 2|c_g - c_d| > u_d - u_g$), there exists a (unique) $h_\star \in]h_d, h_g[$ such that $F(h_\star) = u_g + 2c_g$.

We then define $u_\star = u_d + \sqrt{\frac{gh_d}{2}} \psi\left(\frac{h_\star}{h_d}\right)$ and the equations (5.66), (5.67) and (5.68) are indeed satisfied. Given (5.66), the inequality (5.65) is equivalent to $c_\star < c_g$, which is indeed true because $h_\star < h_g$. It remains to show (5.69). As

$$\sigma = \frac{h_\star u_\star - h_d u_d}{h_\star - h_d} = u_\star + \frac{h_d}{h_\star - h_d}(u_\star - u_d),$$

the condition (5.69) is satisfied because $c_\star > 0$, $h_\star > h_d$ and $u_\star > u_d$ (by (5.68)).

5.4.3.4 Solution formed by a 1-Shock and a 2-Rarefaction Wave

Still under the assumption (5.61), we now suppose that $u_g - u_d < S$ and $h_d > h_g$; we are going to show that the solution of the Riemann problem is then a 1-shock followed by a 2-rarefaction wave. The approach is very similar to the previous one. Let U be the sought solution and $(h_\star, u_\star)$ the intermediate state.

In the zone $D_1 = \{x \leq \sigma t\}$, we have $U = U_g$ and the study of subsection 5.4.3.2 shows that the necessary and sufficient condition for it to be a 1-shock is that $h_g < h_\star$,

$u_\star = u_g - \sqrt{\frac{gh_g}{2}} \psi\left(\frac{h_\star}{h_g}\right)$ and $\sigma = (h_g u_g - h_\star u_\star)/(h_g - h_\star)$.

In the zone $D_2 = \{\sigma t < x < (u_\star + c_\star)t\}$, the solution is $U = U_\star$. The zone $D_3 = \{(u_\star + c_\star)t < x < (u_d + c_d)t\}$ corresponds to the 2-rarefaction. The solution can be calculated as it was done in subsection 5.4.3.1 (zone D_4 of subsection 5.4.3.1). The invariance of the 2-Riemann invariants in this zone gives $u_d - 2c_d = u_\star - 2c_\star$. In the zone $D_4 = \{(u_d + c_d)t \leq x\}$, the solution is $U = U_d$.

In summary, the construction of a solution formed of a 1-shock and a 2-rarefaction connected by an intermediate state denoted by $(h_\star, u_\star)$ is possible if and only if

$$h_\star > h_g, \tag{5.70}$$

$$u_\star = u_g - \sqrt{\frac{gh_g}{2}}\,\psi\Big(\frac{h_\star}{h_g}\Big), \tag{5.71}$$

$$\sigma = \frac{h_g u_g - h_\star u_\star}{h_g - h_\star} < u_\star + c_\star, \tag{5.72}$$

$$u_\star + c_\star < u_d + c_d, \tag{5.73}$$

$$u_d - 2c_d = u_\star - 2c_\star, \tag{5.74}$$

For $h \in [h_g, h_d]$, we define $F(h) = u_g - \sqrt{\frac{gh_g}{2}}\,\psi\big(\frac{h}{h_g}\big) - 2\sqrt{gh}$. As we have already seen, the function ψ is increasing, and the function F is continuous, decreasing, $F(h_g) = u_g - 2c_g$ and $F(h_d) = u_g - S - 2c_d < u_d - 2c_d$. Since $u_d - 2c_d < u_g - 2c_g$ (because $2(c_d - c_g) = 2|c_g - c_d| > u_d - u_g$), there exists a (unique) $h_\star \in]h_g, h_d[$ such that $F(h_\star) = u_d - 2c_d$.

We then define $u_\star = u_g - \sqrt{\frac{gh_g}{2}}\,\psi\big(\frac{h_\star}{h_g}\big)$ and the equations (5.74), (5.70) and (5.71) are indeed satisfied. Given (5.74), the inequality (5.73) is equivalent to $c_\star < c_d$, which is indeed true because $h_\star < h_d$. It remains to show (5.72). Since

$$\sigma = \frac{h_\star u_\star - h_g u_g}{h_\star - h_g} = u_\star + \frac{h_g}{h_\star - h_g}(u_\star - u_g),$$

the condition (5.72) is satisfied because $c_\star > 0$, $h_\star > h_g$ and $u_\star < u_g$ (by (5.71)).

5.4.3.5 Solution formed by two Shocks

Still under the assumption (5.61), we now suppose that $u_g - u_d > S$, and we are going to find a solution of (5.53), (5.57)–(5.58) formed by a 1-shock and a 2-shock. For the solution U to be formed of two shocks separated by an intermediate state denoted by $(h_\star, u_\star)$, it is necessary and sufficient, according to the results of subsection 5.4.3.2, to have the following conditions

$$u_\star = u_g - \sqrt{\frac{gh_g}{2}}\,\psi\Big(\frac{h_\star}{h_g}\Big), \; h_g < h_\star, \tag{5.75}$$

$$u_\star = u_d + \sqrt{\frac{gh_d}{2}}\,\psi\Big(\frac{h_\star}{h_d}\Big), \; h_d < h_\star, \tag{5.76}$$

$$\frac{h_g u_g - h_\star u_\star}{h_g - h_\star} < \frac{h_d u_d - h_\star u_\star}{h_d - h_\star}. \tag{5.77}$$

The condition (5.77) indicates that the speed of the 1-shock is lower than the speed of the 2-shock.

Let $h_m = \min(h_d, h_g)$. For $h \geq h_m$, we define $G(h) = \sqrt{\frac{gh_d}{2}}\psi(\frac{h}{h_d}) + \sqrt{\frac{gh_g}{2}}\psi(\frac{h}{h_g})$. The conditions (5.75)–(5.76) give

$$u_g - u_d = G(h_\star).$$

The function G is continuous and increasing over the interval $[h_m, +\infty[$. Since $G(h_m) = S$, $u_g - u_d > S$ and $\lim_{h\to+\infty} G(h) = +\infty$, there exists a (unique) $h_\star \in \,]h_m, +\infty[$ such that $G(h_\star) = u_g - u_d$. We then define $u_\star$ with the conditions (5.75)–(5.76) (which are identical with this choice of $h_\star$). It remains to verify (5.77). This follows from $h_\star > \max(h_g, h_d)$, $u_g - u_\star > 0$ and $u_d - u_\star < 0$ because

$$\frac{h_g u_g - h_\star u_\star}{h_g - h_\star} = u_\star + h_g \frac{u_g - u_\star}{h_g - h_\star} < u_\star + h_d \frac{u_d - u_\star}{h_d - h_\star} = \frac{h_d u_d - h_\star u_\star}{h_d - h_\star}.$$

N.B. If $u_g - u_d = S$, we find a solution to (5.53), (5.57)–(5.58) containing only a 1-shock in the case $h_g < h_d$ and containing only a 2-shock in the case $h_g > h_d$.

5.5 Problem Set

Problem 5.1 (Maximum principle and positivity of regular solutions ($\star\star$)). *Solution on page 428.* Let $v \in C^1(\mathbb{R}, \mathbb{R})$ such that v' is bounded and let $u_0 \in C^1(\mathbb{R}, \mathbb{R})$. Consider the following two problems:

$$\begin{aligned}
&\partial_t u(x,t) + v\partial_x u(x,t) = 0, \ x \in \mathbb{R}, \ t \in \mathbb{R}_+^\star, \\
&u(x,0) = u_0(x), \ x \in \mathbb{R}.
\end{aligned} \tag{5.78}$$

$$\begin{aligned}
&\partial_t u(x,t) + \partial_x(vu)(x,t) = 0, \ x \in \mathbb{R}, \ t \in \mathbb{R}_+^\star, \\
&u(x,0) = u_0(x), \ x \in \mathbb{R}.
\end{aligned} \tag{5.79}$$

The problem (5.78) is a transport equation with initial data, and the problem (5.79) is a conservation equation with initial data. Note that if v is constant, the two problems are equivalent. The following results generalise to the case of a higher-dimensional space.

1. Let $A, B \in \mathbb{R}$ such that $A \leq u_0(x) \leq B$ for all $x \in \mathbb{R}$. Let $u \in C^1(\mathbb{R} \times \mathbb{R}_+, \mathbb{R})$ be a solution to (5.78). Show that $A \leq u(x,t) \leq B$ for all $x \in \mathbb{R}$ and $t \in \mathbb{R}_+$. Show (by giving an example) that this property can be false if u is a solution to (5.79). [Consider, for instance, for any $a \in \mathbb{R}$, the differential equation $x'(t) = v(x(t))$ with initial data $x(0) = a$.]
2. Assume that $u_0(x) \geq 0$ for all $x \in \mathbb{R}$. Let $u \in C^1(\mathbb{R} \times \mathbb{R}_+, \mathbb{R})$ be a solution to (5.79). Show that $u(x,t) \geq 0$ for all $x \in \mathbb{R}$ and $t \in \mathbb{R}_+$.

Problem 5.2 ($L^1(\mathbb{R}) \cap BV(\mathbb{R}) \subset L^\infty(\mathbb{R})$ ($\star$)). *Solution on page 429.* For $u \in L^1_{\text{loc}}(\mathbb{R})$ we recall (see Definition 5.27) that

$$|u|_{BV(\mathbb{R})} = \sup\{\int_\mathbb{R} u(x)\varphi'(x)\,dx, \varphi \in C_c^\infty(\mathbb{R}, \mathbb{R}), \|\varphi\|_{L^\infty(\Omega)} \le 1\},$$

and that $u \in BV(\mathbb{R})$ if $|u|_{BV(\mathbb{R})} < +\infty$. Let $u \in L^1(\mathbb{R}) \cap BV(\mathbb{R})$. Show that $u \in L^\infty(\mathbb{R})$ and

$$\|u\|_{L^\infty(\mathbb{R})} \le |u|_{BV(\mathbb{R})}.$$

Problem 5.3 (Lax Condition ($\star$)). *Solution on page 429.* Let f be a strictly convex function and $u_g, u_d \in \mathbb{R}$, $u_g \ne u_d$ and consider the Riemann problem

$$\partial_t u + \partial_x(f(u)) = 0 \text{ on } \mathbb{R} \times \mathbb{R}_+,$$

$$u(x,0) = \begin{cases} u_g \text{ if } x < 0, \\ u_d \text{ if } x > 0. \end{cases}$$

Consider a weak solution of the form

$$u(x,t) = \begin{cases} u_g \text{ if } x < \sigma t, \\ u_d \text{ if } x > \sigma t. \end{cases} \tag{5.80}$$

1. Express σ as a function of f, u_g, u_d.
2. Show that this solution is entropic if and only if the Lax condition $f'(u_g) > \sigma > f'(u_d)$ is satisfied.
3. We still assume that the solution is of the form (5.80), but we no longer assume that it is a weak solution. Show by a counterexample that in this case there is no longer equivalence between entropy solution and Lax condition.

Problem 5.4 (Linear hyperbolic system ($\star$)). *Solution on page 430.* Let $A \in M_n(\mathbb{R})$ and $u_0 \in L^\infty(\mathbb{R})^n$ and assume A is diagonalisable in $\mathbb{R}$; we seek a weak solution $u \in L^\infty(\mathbb{R} \times \mathbb{R}_+)^n$ of the following problem:

$$\partial_t u(x,t) + A\partial_x u(x,t) = 0, \ x \in \mathbb{R}, \ t \in]0, +\infty[, \tag{5.81a}$$

$$u(x,0) = u_0(x), \ x \in \mathbb{R}. \tag{5.81b}$$

Let $\{v_i, i \in \{1, \ldots, n\}\}$ be a basis of $\mathbb{R}^n$ formed of eigenvectors of A. We have $Av_i = \lambda_i v_i$ for all $i \in \{1, \ldots, n\}$ with $\lambda_i \in \mathbb{R}$. For $x \in \mathbb{R}$, we decompose $u_0(x)$ on the basis $\{v_i, i \in \{i, \ldots, n\}\}$. So we have $u_0 = \sum_{i=1}^n a_i v_i$ a.e. with $a_i \in L^\infty(\mathbb{R})$. Show that u defined a.e. by $u(x,t) = \sum_{i=1}^n a_i(x - \lambda_i t)v_i$ is a weak solution to (5.81).

Problem 5.5 (Uniqueness of the weak solution of the linear problem by duality ($\star\star$)). *Solution on page 431.* Let $c \in \mathbb{R}$ and $u_0 \in L^\infty(\mathbb{R})$. The purpose here is to prove the uniqueness of the weak solution u (in $L^\infty(\mathbb{R} \times \mathbb{R}_+)$) of the following transport problem:

$$\partial_t u(x,t) + c\partial_x u(x,t) = 0, \; x \in \mathbb{R}, \; t \in]0, +\infty[, \tag{5.82a}$$

$$u(x,0) = u_0(x), \; x \in \mathbb{R}. \tag{5.82b}$$

1. Show that it is sufficient to prove the uniqueness of the solution for $u_0 = 0$ a.e.

Hence from now on, we assume that u is a weak solution to (5.82) with $u_0 = 0$ a.e.

2. Let $\psi \in C_c(\mathbb{R} \times \mathbb{R})$. For $(x,t) \in \mathbb{R} \times \mathbb{R}_+$ let $\varphi(x,t) = -\int_t^{+\infty} \psi(x - c(t-s), s)\, ds$.

   a. Show that $\varphi \in C_c^1(\mathbb{R} \times [0, +\infty[)$ and that $\partial_t \varphi + c\partial_x \varphi = \psi$ in $\mathbb{R} \times [0, +\infty[$.
   b. Show that $\int_0^{+\infty} \int_{\mathbb{R}} u(x,t)\psi(x,t)\, dx\, dt = 0$.

3. Show that $u = 0$ a.e.
4. Deduce that the weak solution to the problem (5.82) is unique, as well as that of the system (5.81) of Problem 5.4.

Problem 5.6 (Linear equation with singular source term ($\star\star$)). *Solution on page 432.* Let $a, b, c, u_g, u_d \in \mathbb{R}$. With the real numbers b and c, we define the measure μ on the Borel sets of $\mathbb{R} \times \mathbb{R}_+$ by its action on $C_c^\infty(\mathbb{R} \times \mathbb{R}_+, \mathbb{R})$:

$$\int_{\mathbb{R} \times \mathbb{R}_+} \varphi\, d\mu = b \int_0^{+\infty} \varphi(ct, t)\, dt \text{ for all } \varphi \in C_c^\infty(\mathbb{R} \times \mathbb{R}_+, \mathbb{R}). \tag{5.83}$$

The measure μ is therefore supported by the half-line $\Gamma_c = \{(ct, t), t \in \mathbb{R}_+\}$. Define u_0 by $u_0(x) = u_g$ if $x < 0$ and $u_0(x) = u_d$ if $x > 0$. Consider the problem

$$\partial_t u + a\partial_x u = \mu \text{ in } \mathbb{R} \times \mathbb{R}_+, \tag{5.84}$$

$$u(\cdot, 0) = u_0 \text{ in } \mathbb{R}. \tag{5.85}$$

A weak solution to (5.84)–(5.85) is a function $u \in L^\infty(\mathbb{R} \times \mathbb{R}_+)$ such that

$$\int_{\mathbb{R} \times \mathbb{R}_+} u(\partial_t \varphi + a\partial_x \varphi)\, d(x,t) + \int_{\mathbb{R}} u_0(x)\varphi(x,0)\, dx = -b \int_0^{+\infty} \varphi(ct, t)\, dt$$

$$\text{for all } \varphi \in C_c^\infty(\mathbb{R} \times \mathbb{R}_+, \mathbb{R}). \tag{5.86}$$

1. *Uniqueness* Show that (5.84)–(5.85) has at most one weak solution. [One may refer to Problem 5.5.]
2. *Existence* Assume in this question that $c \neq a$. Show that (5.84)–(5.85) has a (unique) weak solution and construct it.
3. *Non-existence* Assume in this question that $c = a$ and $b \neq 0$. Show that (5.84)–(5.85) does not have a weak solution.

Problem 5.7 (Non-uniqueness of weak solutions ($\star$)). *Solution on page 434. Let u_g and u_d be real numbers such that $u_g < u_d$; we consider the Riemann problem*

$$\begin{cases} \partial_t u + \partial_x(u^2) = 0 \\ u(0,x) = \begin{cases} u_g & \text{if } x < 0 \\ u_d & \text{if } x > 0. \end{cases} \end{cases} \tag{5.87}$$

1. *Show that there exists a $\sigma \in \mathbb{R}$ such that if (for $t > 0$ and $x \in \mathbb{R}$)*

$$u(x,t) = \begin{cases} u_g & \text{if } x < \sigma t, \\ u_d & \text{if } x > \sigma t, \end{cases} \tag{5.88}$$

   then u is a weak solution to (5.87). Verify that u is not an entropy solution to (5.87).
2. *Show that u defined (on $\mathbb{R} \times \mathbb{R}_+$) by:*

$$\begin{cases} u(x,t) = u_g & \text{if } x < 2u_g t, \\ u(x,t) = \frac{x}{2t} & \text{if } 2u_g t \le x \le 2u_d t, \\ u(x,t) = u_d & \text{if } x > 2u_d t, \end{cases}$$

   is an entropy weak solution to (5.87).

Problem 5.8 (Discontinuity and entropy ($\star\star$)). *Solution on page 434.* The purpose here is to prove Proposition 5.16. Let $f \in C^1(\mathbb{R}, \mathbb{R})$ and $u_0 \in L^\infty(\mathbb{R})$. We are interested in the entropy solution of the Cauchy problem (5.1). Let $\sigma \in \mathbb{R}$, $D_1 = \{(x,t) \in \mathbb{R} \times \mathbb{R}_+^\star \ x < \sigma t\}$ and $D_2 = \{(x,t) \in \mathbb{R} \times \mathbb{R}_+^\star \ x > \sigma t\}$. Assume that $u_{|D_i} \in C^1(\bar{D}_i, \mathbb{R})$ ($i = 1, 2$) in the sense of Definition 1.28 and that u is a weak solution to (5.1). In particular, the Rankine–Hugoniot condition (5.7) is well satisfied (see Proposition 5.9); it reads

$$\sigma[u](\sigma t, t) = [f(u)](\sigma t, t) \text{ for all } t \in \mathbb{R}_+.$$

Show that u is an entropy solution to (5.1) if and only if the condition (5.16) is satisfied.

Problem 5.9 (Riemann problem ($\star\star$)). *Solution on page 436.* Let u_d and u_g be real numbers and f a function from $\mathbb{R}$ to $\mathbb{R}$ of class C^1.

1. Assume that f is strictly convex.

   a. Calculate the entropy solution of the Riemann problem (5.10) with data u_d and u_g, $u_g < u_d$. In this question, we assume that f is of class C^2 and $f''(s) > 0$ for all $s \in \mathbb{R}$.
   b. Calculate the entropy solution of the Riemann problem (5.10) with data u_d and u_g, $u_g > u_d$.

2. Assume that f is strictly concave. Can we reduce to the case of the first question to calculate the entropy solution of the Riemann problem (5.10) with data u_d and u_g?

Problem 5.10 (Construction of entropy weak solutions ($\star\star\star$)). *Solution on page 438.* Construct the entropy weak solution for the following Cauchy problem (5.9):

$$\partial_t u + \partial_x(u^2) = 0,$$
$$u_0(x) = u_0(x),$$

for the following initial conditions u_0

(a) $u_0(x) = \begin{cases} 1 & \text{if } x < 0, \\ 1 - x & \text{if } 0 < x < 1, \\ 0 & \text{if } x > 1. \end{cases}$

(b) $u_0(x) = \begin{cases} 1 & \text{if } x < 0, \\ 1 - x & \text{if } 0 < x < 1, \\ 1 & \text{if } x > 1. \end{cases}$

(c) $u_0(x) = \begin{cases} 0 & \text{if } x < 0 \\ x & \text{if } x \in [0, 1] \\ 0 & \text{if } > 1. \end{cases}$

(d) $u_0(x) = \begin{cases} 1 & \text{if } x < -1, \\ 0 & \text{if } -1 < x < 0, \\ 2 & \text{if } 0 < x < 1, \\ 0 & \text{if } x > 1. \end{cases}$

Problem 5.11 (Non-entropy solution ($\star\star$)). *Solution on page 443.* Consider the Burgers equation

$$\partial_t u(x, t) + \partial_x(u^2)(x, t) = 0, \ x \in \mathbb{R}, \ t \in \mathbb{R}_+, \tag{5.89}$$
$$u(x, 0) = 0, \ x \in \mathbb{R}. \tag{5.90}$$

Define $u : \mathbb{R} \times \mathbb{R}_+ \to \mathbb{R}$ by:

$$u(x, t) = \begin{cases} 0 & \text{for } x \in \mathbb{R}, \ t \in \mathbb{R}_+, \ x < -\sqrt{t}, \\ \dfrac{x}{2t} & \text{for } x \in \mathbb{R}, \ t \in \mathbb{R}_+^\star, \ -\sqrt{t} < x < \sqrt{t}, \\ 0 & \text{for } x \in \mathbb{R}, \ t \in \mathbb{R}_+, \ x > \sqrt{t}. \end{cases}$$

1. Show that $u^2 \in L^1_{\text{loc}}(\mathbb{R} \times \mathbb{R}_+)$ and $u \in L^1_{\text{loc}}(\mathbb{R} \times \mathbb{R}_+)$. (Recall that a function v from $\mathbb{R} \times \mathbb{R}_+$ to $\mathbb{R}$ belongs to $L^1_{\text{loc}}(\mathbb{R} \times \mathbb{R}_+)$ if $v\mathbf{1}_K \in L^1(\mathbb{R} \times \mathbb{R}_+)$ for all $K \subset \mathbb{R} \times \mathbb{R}_+ = \mathbb{R} \times [0, \infty[$, K compact.)
2. (Weak Solution (1)) Show that u satisfies:

$$\int_{\mathbb{R}_+} \int_{\mathbb{R}} (u(x, t)\partial_t \varphi(x, t) + u^2(x, t)\partial_x \varphi(x, t)) \, dx \, dt = 0,$$

$$\forall \varphi \in C_c^1(\mathbb{R} \times \mathbb{R}_+^\star, \mathbb{R}). \tag{5.91}$$

3. (Weak Solution (2)) Show that u is a weak solution to (5.89)–(5.90), i.e. that u satisfies:

$$\int_{\mathbb{R}_+} \int_{\mathbb{R}} (u(x, t)\partial_t(x, t) + u^2(x, t)\partial_x \varphi(x, t)) \, dx \, dt = 0,$$

$$\forall \varphi \in C_c^1(\mathbb{R} \times \mathbb{R}_+, \mathbb{R}). \tag{5.92}$$

4. (Entropy weak solution?) Let η be a convex function from $\mathbb{R}$ to $\mathbb{R}$, of class C^1. Define Φ from $\mathbb{R}$ to $\mathbb{R}$ by: $\Phi(s) = \int_0^s \eta'(\xi) 2\xi \, d\xi$, for all $s \in \mathbb{R}$. Study the sign of $\Phi(s) - s(\eta(s) - \eta(0))$ thanks to the convexity of η, and show that u satisfies:

$$\int_{\mathbb{R}_+} \int_{\mathbb{R}} (\eta(u)(x,t)\partial_t\varphi(x,t) + \Phi(u)(x,t)\partial_x\varphi(x,t)) \, dx \, dt \geq 0,$$

$$\forall \varphi \in C_c^1(\mathbb{R} \times \mathbb{R}_+^\star, \mathbb{R}_+). \quad (5.93)$$

5. Show that the function u is not the entropy weak solution of (5.89)–(5.90).

N.B. Question 4 shows that $\partial_t \eta(u) + \partial_x \Phi(u) \leq 0$ in the sense of derivatives by transposition in $\mathbb{R} \times \mathbb{R}_+^\star$. But we cannot show (5.93) (by adding $\int_{\mathbb{R}} \eta(0)\varphi(x,0) \, dx$ if $\eta(0) \neq 0$) for all $\varphi \in C_c^1(\mathbb{R} \times \mathbb{R}_+, \mathbb{R}_+)$, even when limiting ourselves to considering Krushkov's entropies. Indeed, $u(\cdot,t) \not\to 0$ in $L^1_{\text{loc}}(\mathbb{R})$ as $t \to 0$, and this convergence in $L^1_{\text{loc}}(\mathbb{R})$ would be true if u were an entropy weak solution of (5.89)–(5.90). We only have $u(\cdot,t) \to 0$ in $\mathcal{D}^\star(\mathbb{R})$ and this is what was used in question 3 to prove that u is a weak solution.

Problem 5.12 (Buckley–Leverett Equation ($\star\star\star$)). *Solution on page 445.* Consider the Riemann problem

$$\partial_t u + \partial_x(f(u)) = 0 \text{ on } \mathbb{R} \times \mathbb{R}_+, \quad (5.94)$$

$$u(x,0) = u_g \text{ for } x < 0, \ u(x,0) = u_d \text{ for } x > 0. \quad (5.95)$$

Assume here that the function $f \in C^2(\mathbb{R}, \mathbb{R})$ satisfies

(i) $f(0) = 0$, $f'(0) = f'(1) = 0$,
(ii) $\exists a \in\,]0,1[$, such that f is strictly convex over $]0,a[$, f is strictly concave over $]a,1[$.

Let $u_g = 1$ and $u_d = 0$.

1. Show that there is a unique point b in the interval $]a,1[$ such that $\frac{f(b)}{b} = f'(b)$. Then, show that there is a unique point c in $]0,b[$ such that $f'(c) = f'(b)$.

We keep this notation of points b and c in the following.

2. Let u be the function defined a.e. from $\mathbb{R} \times \mathbb{R}_+$ to $\mathbb{R}$ by:

$$\begin{cases} u(x,t) = 1 \text{ if } x \leq 0 \\ u(x,t) = \xi \text{ if } x = f'(\xi)t, \ b < \xi < 1 \\ u(x,t) = 0 \text{ if } x > f'(b)t \end{cases}$$

Show that u is the entropy weak solution to (5.10). [To show that the entropy condition is satisfied, one could start by noting that for any function η (from $\mathbb{R}$ to $\mathbb{R}$) of class C^1 and convex, we have $\int_0^b (f'(b) - f'(x))(\eta'(x) - \eta'(c)) \, dx \leq 0.$]

Thus, we can construct the entropy weak solution of the Riemann problem in the case relevant to petroleum engineering, specifically for the Buckley–Leverett equation, which reads (5.94) with

$$f(u) = \frac{u^2}{u^2 + \frac{(1-u)^2}{4}} \text{ and } u_g, u_d \in [0, 1].$$

To this end, we distinguish between the cases where the function f is convex-concave, purely convex, or purely concave between u_g and u_d, depending on the values of u_g and u_d.

Problem 5.13 (Landau Effect ($\star\star\star$)). *Solution on page 446.* Let f be a bounded Borel-measurable and periodic function from $\mathbb{R}$ to $\mathbb{R}$ (or, to simplify, a bounded, continuous and periodic function from $\mathbb{R}$ to $\mathbb{R}$). In this problem, we are interested in the limit as $t \to +\infty$ of the (weak) solution of the following problem:

$$\partial_t u(x, y, t) + y\partial_x u(x, y, t) = 0, \ x, y \in \mathbb{R}, \ t \in \mathbb{R}_+^\star, \tag{5.96a}$$

$$u(x, y, 0) = f(x), \ x, y \in \mathbb{R}. \tag{5.96b}$$

1. Give an explicit expression of the unique weak solution of (5.96) in terms of f.

In the following, let u denote this weak solution. Observe that u is continuous from $\mathbb{R}_+$ to $L_{\text{loc}}^p(\mathbb{R}^2)$ for all $p < +\infty$. Let m be the mean value of f over a period. Finally, for all $y \in \mathbb{R}$ and $r > 0$, let

$$F(y, r) = \frac{1}{2r} \int_{y-r}^{y+r} f(z) \, dz.$$

2. (Preliminary question) Show that $\lim_{r \to +\infty} F(y, r) = m$, uniformly with respect to $y \in \mathbb{R}$.
3. Let $a, b \in \mathbb{R}$ and $\delta > 0$. Show that

$$\lim_{t \to +\infty} \int_{b-\delta}^{b+\delta} \int_{a-\delta}^{a+\delta} u(x, y, t) \, dx \, dy = 4\delta^2 m.$$

   [Note that $\int_{b-\delta}^{b+\delta} f(x - yt) \, dy = 2\delta F(x - bt, \delta t)$.]
4. Show that $u(\cdot, \cdot, t) \to m$ $\star$-weakly in $L^\infty(\mathbb{R}^2)$, as $t \to +\infty$.

N.B.

1. By revisiting the previous proofs, it may be shown that the result of question 4 remains true if we replace in (5.96) $y\partial_x u$ by $a(y)\partial_x u$ where $a \in C^1(\mathbb{R}, \mathbb{R})$ and $a'(y) \neq 0$ for all $y \in \mathbb{R}$. (More generally, this result remains true under the weaker assumption that the set of points where a' vanishes is negligible.)
2. This problem shows that the oscillations of u at time 0 disappear as $t \to +\infty$. We therefore cannot retrieve with this limit at infinity the function u at time 0 while the knowledge of u at a time $t > 0$ allows us to retrieve u at time 0 (thanks to the equation (5.96a)). This is part of what is called the Landau[13] effect.

---

[13] Lev Davidovitch Landau (1908–1968), Soviet theoretical physicist, laureate of the 1962 Nobel Prize in Physics.

Problem 5.14 (Example of a non-strictly hyperbolic system ($\star$)). *Solution on page 450.* Let $p = 2$, $D = \mathbb{R}^2$. For $U = (u_1, u_2)^t$, we define $F = (f_1, f_2)^t$ by $f_1(U) = u_1^2$, $f_2(U) = (u_2 + 1)^2$.

1. Show that the system $\partial_t U + \partial_x(F(U)) = 0$ is hyperbolic but not strictly hyperbolic, and there exist two GNL fields associated with the function F.
2. We now consider the Riemann problem with initial data $U_g = (-1, -1)^t$ for $x < 0$ and $U_d = (0, -2)^t$ for $x > 0$. Show that the solution is then constituted by a shock wave (for the second equation) located within a rarefaction wave (for the first equation).

Problem 5.15 (The linearised Riemann problem, shallow water equations ($\star$)). *Solution on page 450.* We consider here the notations and assumptions of Section 5.4, in particular the assumption $u_d - u_g < 2(c_g + c_d)$, and we set $\overline{V} = \begin{bmatrix} \overline{u} \\ 2\overline{c} \end{bmatrix}$, $\overline{u} = (u_g + u_d)/2$ and $\overline{c} = (c_g + c_d)/2$.

We then replace in the Riemann problem (5.53), (5.57)–(5.58), the equation (5.53) by the following equation:

$$\partial_t V(x, t) + B(\overline{V})\partial_x V(x, t) = 0, \ x \in \mathbb{R}, \ t \in \mathbb{R}_+ \tag{5.97}$$

where B is defined by (5.54).

Construct the solution of the linearised Riemann problem (5.97), (5.57)–(5.58).

N.B.: We note that, thanks to the condition $u_d - u_g < 2(c_g + c_d)$, this new Riemann problem admits a solution with an intermediate state $(u_\star, 2c_\star)$ such that $c_\star > 0$, and therefore $c_\star = \sqrt{gh_\star}$ with a $h_\star > 0$. (This is not the case if $u_d - u_g \geq 2(c_g + c_d)$.) The fact that $h_\star > 0$ is important when we replace in the Godunov[14] scheme [29] the resolution of the Riemann problem by the resolution of this linearised Riemann problem (see e.g. [23, Chapter 5]).

Problem 5.16 (Entropy for the shallow water equations with non-flat bottom ($\star\star$)). *Solution on page 451.* In this problem, we are interested in the system of equations (in one space dimension) modelling a flow of water over a non-flat bottom. Let z be the regular function from $\mathbb{R}$ to $\mathbb{R}_+$ giving the level of the bottom and g the intensity of gravity. The considered system is then written as

$$\partial_t h(x, t) + \partial_x(hu)(x, t) = 0, \ x \in \mathbb{R}, \ t \in \mathbb{R}_+, \tag{5.98a}$$

$$\partial_t(hu)(x, t) + \partial_x(hu^2 + g\frac{h^2}{2})(x, t) + ghz'(x) = 0, \ x \in \mathbb{R}, \ t \in \mathbb{R}_+. \tag{5.98b}$$

Here, we adapt the studies conducted in Section 5.4.2 on the shallow water equations, now considering a non-flat bottom.

---

[14] Serguei Godunov (1929–2023), Russian mathematician, specialist in the numerical solution of partial differential equations.

1. (Entropy) For all $U = \begin{bmatrix} h \\ q \end{bmatrix} \in D$, we set $\eta(U) = \frac{1}{2}hu^2 + p + ghz$ (with $q = hu$ and $p = gh^2/2$). Show that $\eta(U)$ is an entropy of the system, i.e. that η is convex and that there exists a function Φ such that $\partial_t \eta(U) + \partial_x(\Phi(U)) = 0$ for any regular solution to (5.98).

   N.B. As in Section 5.4.2, the quantity $\eta(U(x,t))$ is the total energy of the water column located at point x at time t (that is to say, the sum of the kinetic energy and the potential energy).

2. (Limit of viscous solutions) We add regularisation terms in the system (5.98), that is to say, $-\varepsilon \partial_x^2 h$ for the first equation and $-\varepsilon \partial_x^2 q$ for the second equation (with $\varepsilon > 0$). Let h_ε and u_ε be the solutions of this new system (and $q_\varepsilon = h_\varepsilon u_\varepsilon$). Assume that these are regular bounded functions in $L^\infty(\mathbb{R} \times \mathbb{R}_+)$ independent of ε, and that they converge in $L^1_{\mathrm{loc}}(\mathbb{R} \times \mathbb{R}_+)$, as $\varepsilon \to 0$, towards functions h and u (with $h > 0$). We also assume that $\varepsilon \partial_x h$ and $\varepsilon \partial_x q$ are bounded in $L^2_{\mathrm{loc}}(\mathbb{R} \times \mathbb{R}_+)$. Show that the pair (h, u) is a solution to (5.98) and satisfies, in the sense of the negativity of an element of $\mathcal{D}^\star(\mathbb{R} \times \mathbb{R}_+^\star)$ (Definition 1.5),

$$\partial_t \eta(U) + \partial_x \Phi(U) \leq 0.$$

Problem 5.17 (Regular stationary solutions, shallow water equations ($\star\star\star\star$)). *Solution on page 451.* In this problem, we aim to construct regular stationary solutions to the one-dimensional system of equations (5.98), modelling a water flow over a non-flat bottom (see Problem 5.16). Let z be the regular function from $\mathbb{R}$ to $\mathbb{R}_+$ giving the bottom level and g the intensity of gravity. In the following, q denotes the function hu and ψ the function $\frac{u^2}{2} + gh + gz$. Hereafter, a stationary solution to (5.98) comprising a pair of C^1 class functions, denoted by (h, u), from $\mathbb{R}$ to $\mathbb{R}_+^\star \times \mathbb{R}$, will be called a "regular stationary solution" (note that $h(x) > 0$ for all $x \in \mathbb{R}$). We also assume that z is of class C^1.

1. Show that the pair (h, u) (from $\mathbb{R}$ to $\mathbb{R}_+^\star \times \mathbb{R}$) is a regular stationary solution to (5.98) if and only if the functions q and ψ are constant functions.

We are therefore given two non-negative numbers α and β and we are looking for a pair of functions (h, u) such that $q(x) = \alpha$ and $\psi(x) = \beta$ for all $x \in \mathbb{R}$. We let z_m be the maximum of the function z, and assume (of course) that $z_m < +\infty$.

2. (Lake at rest) In this question, we assume that $\alpha = 0$. Show that there is no regular stationary solution if $\beta \leq gz_m$ (recall that the function h must have positive values) and that the only regular stationary solution is given by $h(x) = \beta/g - z(x)$ (for all $x \in \mathbb{R}$) if $\beta > gz_m$.

In the rest of the problem, we assume $\alpha > 0$ and we set $\beta_m = gz_m + (3/2)(\alpha g)^{2/3}$.

3. Show that:

   a. If $\beta < \beta_m$, there is no regular stationary solution associated with the pair (α, β).

   b. If $\beta > \beta_m$, there are (exactly) two regular stationary solutions associated with the pair (α, β).

4. In this question, we assume that $\beta = \beta_m$. Show that:

   a. If z is a constant function, there is (exactly) one regular stationary solution associated with the pair (α, β).
   b. If $z(x) \neq z_m$ for all x (and therefore z is not constant), there are (exactly) two regular stationary solutions associated with the pair (α, β).

In the rest of the problem, we fix α and β with $\alpha > 0$ and $\beta > \beta_m$, and consider (h_i, u_i), $i = 1, 2$, the two pairs of regular stationary solutions.

5. Show that $h_1 - h_2$ has a constant sign. We can therefore assume $h_1(x) < h_2(x)$ for all x. Show that there exist regular functions φ_1 and φ_2 such that $h_i(x) = \varphi_i(z(x))$ for all x.

   Show that $h_2 + z$ is non-increasing when z is non-decreasing (and non-decreasing when z is non-increasing).

   Show that $h_1 + z$ is non-decreasing when z is non-increasing (and non-decreasing when z is non-increasing).

6. For $i = 1, 2$, give the sign of $u_i - \sqrt{gh_i}$ (recall that $\alpha > 0$ and $\beta > \beta_m$).

   N.B. If $u_i > \sqrt{gh_i}$, we say that the flow is supersonic. If $u_i < \sqrt{gh_i}$, we say that the flow is subsonic.

Problem 5.18 (Saint-Venant equations, entropy in the Lax sense ($\star\star\star$)). *Solution on page 454.* Consider once again the shallow water system (5.52), where g is a given real number, $g > 0$. Set $q = hu$.

Let $u_g, u_d \in \mathbb{R}$, $h_g, h_d \in \mathbb{R}^\star$ and $\sigma \in \mathbb{R}$. Set $q_g = h_g u_g$, $q_d = h_d u_d$. Define the function $U = \begin{bmatrix} h \\ q \end{bmatrix}$ from $\mathbb{R} \times \mathbb{R}_+$ to $\mathbb{R}_+^\star \times \mathbb{R}$ by

$$U(x, t) = \begin{bmatrix} h_g \\ q_g \end{bmatrix} \, if \, x < \sigma t, \text{ and } U(x, t) = \begin{bmatrix} h_d \\ q_d \end{bmatrix} \, if \, x > \sigma t.$$

Assume that U is a discontinuous weak solution to (5.52) with the initial condition $U_0 = \begin{bmatrix} h_g \\ q_g \end{bmatrix}$ if $x < 0$, $U_0 = \begin{bmatrix} h_d \\ q_d \end{bmatrix}$ if $x > 0$.

1. Show that U is an entropy weak solution in the sense of Definition 5.41 with the entropy
$$\eta(U) = hu^2/2 + gh^2/2$$

   if and only if $u_d < u_g$.

2. Using the tools of subsection 5.4.3.2, show that U is an entropy weak solution in the sense of Definition 5.41 with the entropy

$$\eta(U) = hu^2/2 + gh^2/2$$

   if and only if U satisfies the Lax condition 5.51.

5.6 Problem Set Solutions

Problem 5.1 (Maximum principle and positivity of regular solutions)

1. Let $a \in \mathbb{R}$. We consider the differential equation

$$x'(t) = v(x(t)), \tag{5.99}$$

$$x(0) = a. \tag{5.100}$$

Since v is a Lipschitz function from $\mathbb{R}$ to $\mathbb{R}$, this differential equation has a maximal solution $x_a \in C^1([0, +\infty[), \mathbb{R})$.

Let us then define $\varphi_a \in C^1(\mathbb{R}_+, \mathbb{R})$ by $\varphi_a(t) = u(x_a(t), t)$ so that, for all $t > 0$,

$$\varphi_a'(t) = \partial_x u(x_a(t), t)x_a'(t) + \partial_t u(x_a(t), t)$$
$$= \partial_t u(x_a(t), t) + v(x_a(t))\partial_x u(x_a(t), t) = 0.$$

The function φ_a is therefore constant and thus $\varphi_a(t) = \varphi_a(0) = u(a, 0) = u_0(a)$. This proves that $A \leq u(x_a(t), t) \leq B$ for all $t > 0$ and all $a \in \mathbb{R}$. It remains to show that $\{(x_a(t), t), t \in \mathbb{R}_+, a \in \mathbb{R}\} = \mathbb{R} \times \mathbb{R}_+$.

Let $t > 0$. The solution of the differential equation (5.99) depends continuously on the initial data; the mapping ψ_t defined by $\psi_t(a) = x_a(t)$ is therefore continuous from $\mathbb{R}$ to $\mathbb{R}$. On the other hand, $\lim_{a \to +\infty} x_a(t) = +\infty$ because $x_a(t) \geq a - \|v'\|_\infty t$, and $\lim_{a \to -\infty} x_a(t) = -\infty$ because $x_a(t) \leq a + \|v'\|_\infty t$. The intermediate value theorem then gives $\mathrm{Im}(\psi_t) = \mathbb{R}$. Hence $\{(x_a(t), t), t \in \mathbb{R}_+, a \in \mathbb{R}\} = \mathbb{R} \times \mathbb{R}_+$ and thus $A \leq u(x, t) \leq B$ for all $x \in \mathbb{R}$ and $t \in \mathbb{R}_+$.

This property is generally false if u is a solution to (5.79) as soon as $v' \neq 0$. Just consider for example: $u_0(x) = 1$ for all $x \in \mathbb{R}$ and take $A = B = 1$. The function $u(x, t) = 1$ for all $x \in \mathbb{R}$ and all $t \in \mathbb{R}_+$ is not a solution to (5.79) because $v' \neq 0$.

2. The equation (5.79) also reads

$$\partial_t u(x, t) + v(x)\partial_x u(x, t) + v'(x)u(x, t) = 0.$$

Let $a \in \mathbb{R}$; with the same functions x_a and φ_a as in the previous question, we now obtain

$$\varphi_a'(t) = \partial_x u(x_a(t), t)x_a'(t) + \partial_t u(x_a(t), t)$$
$$= \partial_t u(x_a(t), t) + v(x_a(t))\partial_x u(x_a(t), t)$$
$$= -v'(x_a(t))u(x_a(t), t) = -v'(x_a(t))\varphi_a(t).$$

Setting $g(t) = -v'(x_a(t))$ (so that $g \in C(\mathbb{R}_+, \mathbb{R})$), the function φ_a is therefore a solution of

$$\varphi_a'(t) = g(t)\varphi_a(t),$$
$$\varphi_a(0) = u_0(a).$$

Denoting by G the primitive of g that cancels at 0, we therefore have $\varphi_a(t) = u_0(a)e^{G(t)}$ for all $t \geq 0$. Since $u_0(a) \geq 0$, we get that $\varphi_a(t) \geq 0$ for all $t \geq 0$. We therefore finally have $0 \leq u(x_a(t), t)$ for all $t > 0$ and all $a \in \mathbb{R}$.

As in the previous question $\{(x_a(t), t), t \in \mathbb{R}_+, a \in \mathbb{R}\} = \mathbb{R}_+ \times \mathbb{R}$ and therefore $0 \leq u(x, t)$ for all $x \in \mathbb{R}$ and $t \in \mathbb{R}_+$.

Problem 5.2 ($L^1(\mathbb{R}) \cap BV(\mathbb{R}) \subset L^\infty(\mathbb{R})$)

Reminder of integration theory: Let $u \in L^1(\mathbb{R})$. We confuse u with one of its representatives (and therefore $u \in \mathcal{L}^1(\mathbb{R})$). We say that x is a Lebesgue point of u if

$$u(x) = \lim_{n \to +\infty} \frac{n}{2} \int_{-1/n}^{1/n} u(x + t) \, dt.$$

We can then show that almost every x in $\mathbb{R}$ is a Lebesgue point of u (see [26] problem 5.13).

Let x, y be Lebesgue points of u, $x < y$. For n such that $2/n < y - x$, we choose φ continuous and such that

$$\begin{cases} \varphi(z) = 0 \text{ if } z \leq x - \frac{1}{n} \text{ or } z \geq y + \frac{1}{n}, \\ \varphi(z) = 1 \text{ if } z \in [x + \frac{1}{n}, y - \frac{1}{n}] \\ \varphi \text{ affine on } [x - \frac{1}{n}, x + \frac{1}{n}] \text{ and } [y - \frac{1}{n}, y + \frac{1}{n}]. \end{cases}$$

We have $|\int u(z)\varphi'(z)\, dz| \leq |u|_{BV(\mathbb{R})}$. (With a regularisation, such a function φ is acceptable in the definition of $|u|_{BV(\mathbb{R})}$.) By letting $n \to +\infty$, we obtain $|u(x) - u(y)| \leq |u|_{BV(\mathbb{R})}$ and therefore

$$|u(x)| \leq |u|_{BV(\mathbb{R})} + |u(y)|,$$

for all Lebesgue points x, y. Since $u \in L^1(\mathbb{R})$, for any $\varepsilon > 0$, there exists a Lebesgue point y such that $|u(y)| \leq \varepsilon$ (otherwise $\int |u(y)|\, dy \geq \varepsilon(+\infty) = +\infty$). Hence $|u(x)| \leq |u|_{BV(\mathbb{R})}$ for all Lebesgue points x and therefore

$$\|u\|_{L^\infty(\mathbb{R})} \leq |u|_{BV(\mathbb{R})}.$$

Problem 5.3 (Lax Condition)

1. The Rankine–Hugoniot condition gives $\sigma = \dfrac{f(u_g) - f(u_d)}{u_g - u_d}$.

2. According to the Rankine–Hugoniot relation (and the mean value theorem), there exists a convex combination ξ of u_g and u_d such that $f'(\xi) = \sigma$. If the Lax condition is met, we therefore have in particular $f'(u_g) > f'(u_d)$. Since f is

strictly convex, f' is increasing, which implies that $u_g > u_d$; Proposition 5.18 then allows us to conclude that u is an entropy weak solution.

Conversely, suppose u is an entropy weak solution, then $u_g > u_d$ (Proposition 5.18) and therefore $\xi \in]u_d, u_g[$. Since f' is increasing, we indeed have $f'(u_g) > \sigma = f'(\xi) > f'(u_d)$.

3. Consider the Burgers equation $f(u) = u^2$, and the function

$$u(x,t) = \begin{cases} 1 & \text{if } x < t, \\ -1 & \text{if } x > t. \end{cases}$$

This function does indeed meet the Lax condition, since $f'(1) = 2$ and $f'(-1) = -2$, but it is not the entropy weak solution; it is not even a weak solution, since it does not meet the Rankine–Hugoniot relation. The entropy weak solution to this problem is the stationary function

$$u(x,t) = \begin{cases} 1 & \text{if } x < 0, \\ -1 & \text{if } x > 0. \end{cases}$$

Problem 5.4 (Linear hyperbolic system)

Let $u \in L^\infty(\mathbb{R} \times \mathbb{R}_+)^n$; let us decompose u on the basis of the eigenvectors $\{v_i, i \in \{1, \ldots, n\}\}$:

$$u(x,t) = \sum_{i=1}^{n} u_i(x,t)v_i,$$

and therefore

$$\partial_t u(x,t) = \sum_{i=1}^{n} \partial_t u_i(x,t)v_i \text{ and } A\partial_x u(x,t) = \sum_{i=1}^{n} \lambda_i \partial_x u_i(x,t)v_i.$$

Hence u is a weak solution to (5.81) if and only if, for all $i = 1, \ldots, n$, u_i is a weak solution of

$$\partial_t u_i(x,t) + \lambda_i \partial_x u_i(x,t) = 0, \; x \in \mathbb{R}, \; t \in]0, +\infty[, \tag{5.101a}$$

$$u_i(x,0) = a_i(x), \; x \in \mathbb{R}. \tag{5.101b}$$

Let us show that the function $u_i \in L^\infty(\mathbb{R} \times \mathbb{R}_+)$ defined by $u_i(x,t) = a_i(x - \lambda_i t)$ is a weak solution to (5.101). First, let us note that if a_i is a regular function, then u_i is a classical solution; it is therefore also a weak solution, and we are done. Now if a_i is only $L^\infty(\mathbb{R})$, we indeed have $u_i \in L^\infty(\mathbb{R} \times \mathbb{R}_+)$ and we still need to show that for any function $\varphi \in C_c^1(\mathbb{R} \times \mathbb{R}_+, \mathbb{R})$, the function u_i satisfies:

$$\int\int_{\mathbb{R} \times \mathbb{R}_+} [u_i(x,t)\partial_t \varphi(x,t) + \lambda_i u_i(x,t)\partial_x \varphi(x,t)] \, dx \, dt + \int_{\mathbb{R}} a_i(x)\varphi(x,0) \, dx = 0.$$

Let us set

$$X = \int\!\!\int_{\mathbb{R}\times\mathbb{R}_+} [u_i(x,t)\partial_t\varphi(x,t) + \lambda_i u_i(x,t)\partial_x\varphi(x,t)]\ dx\,dt.$$

Since $u_i(x,t) = a_i(x - \lambda_i t)$, we therefore have:

$$X = \int\!\!\int_{\mathbb{R}\times\mathbb{R}_+} [a_i(x - \lambda_i t)\partial_t\varphi(x,t) + \lambda_i a_i(x - \lambda_i t)\partial_x\varphi(x,t)]\ dx\,dt.$$

By applying the change of variables $y = x - \lambda_i t$ and using Fubini's theorem, we get:

$$X = \int_{\mathbb{R}} a_i(y) \int_{\mathbb{R}_+} [\partial_t\varphi(y + \lambda_i t, t) + \lambda_i \partial_x\varphi(y + \lambda_i t, t)]\ dt\,dy.$$

Let us then set
$$\psi_y(t) = \varphi(y + \lambda_i t, t).$$

Hence
$$X = \int_{\mathbb{R}} \left(a_i(y) \int_0^{+\infty} \psi_y'(t)\,dt \right) dy,$$

and as ψ is compactly supported on $[0, +\infty[$, we have

$$X = -\int_{\mathbb{R}} a_i(y)\psi_y(0)\,dy = -\int_{\mathbb{R}} a_i(y)\varphi(y,0)\,dy.$$

We have thus proved that the function u_i defined by $u_i(x,t) = a_i(x - \lambda_i t)$ is a weak solution of the equation (5.101). Hence the function $u : (x,t) \mapsto \sum_{i=1}^n a_i(x - \lambda_i t)v_i$ is a weak solution of the system (5.81).

Problem 5.5 (Uniqueness of the weak solution of the linear problem by duality)

1. Let u_1 and u_2 be weak solutions of (5.82), then $u_1 - u_2$ is a weak solution to (5.82) with $u_0 = 0$; it follows that to show that there is a unique solution to (5.82) is equivalent to showing that the null function is the unique weak solution to (5.82) with $u_0 = 0$.
2. a. For $x \in \mathbb{R}$ and $t \in \mathbb{R}_+$ the function $s \mapsto \psi(x - c(t - s), s)$ is continuous with compact support and therefore integrable. The function φ is thus well defined. The function φ has compact support in $\mathbb{R} \times [0, +\infty[$. Indeed, let $R > 0$ and $T > 0$ such that the support of ψ is included in $[-R, R] \times [-T, T]$. The support of ψ is then included in $[-S, S] \times [0, T]$ with $S = R + 2|c|T$. (Note that $\varphi(x, 0)$ can be non-zero.) Finally, we note that φ is the trace on $\mathbb{R} \times [0, +\infty[$ of a C^1 class function on $\mathbb{R}^2$. Indeed, $\varphi = \overline{\varphi}$ with $\overline{\varphi}(x,t) = -\int_t^{+\infty} \psi(x - c(t - s), s)\,ds$ for $x \in \mathbb{R}$ and $t \in \mathbb{R}$. The function $\overline{\varphi}$ is of class C^1 and the theorem of differentiation under the integral sign and the chain rule give

$$\partial_t \overline{\varphi}(x,t) = \psi(x,t) + \int_t^{+\infty} c\partial_x \psi(x - c(t-s), s)\, ds,$$

$$\partial_x \overline{\varphi}(x,t) = -\int_t^{+\infty} \partial_x \psi(x - c(t-s), s)\, ds.$$

Thus $\partial_t \varphi + c\partial_x \varphi = \psi$ in $\mathbb{R} \times [0, +\infty[$.

b. The function u is a weak solution with $u_0 = 0$ a.e. It therefore satisfies

$$\int \int_{\mathbb{R}\times\mathbb{R}_+} u(x,t)(\partial_t \varphi(x,t) + c\partial_x \varphi(x,t))dx\, dt = 0, \forall \varphi \in C_c^1(\mathbb{R} \times \mathbb{R}_+, \mathbb{R}).$$

We therefore deduce, by taking for φ the function defined with ψ, that

$$\int_0^{+\infty} \int_{\mathbb{R}} u(x,t)\psi(x,t)\, dx\, dt = 0.$$

3. The function ψ from question 2 can be arbitrarily chosen in $C_c^\infty(\mathbb{R}\times]0, +\infty[)$ (extending by 0 outside of $\mathbb{R}\times]0, +\infty[$). Question 1b then gives $u = 0$ a.e.
4. Question 3 proved the uniqueness of the weak solution of problem (5.82) (if u_1, u_2 are two solutions of problem (5.82), we then have $u_1 = u_2$ a.e. on $\mathbb{R}\times]0, +\infty[$). This proof also provides the uniqueness of the solution of the system (5.81) from Problem 5.4 because we show in the proof of Problem 5.4 that the system (5.81) is equivalent to n decoupled problems of the form (5.82).

Problem 5.6 (Linear equation with singular source term)

1. Let u, v be two weak solutions of (5.84)–(5.85). Set $w = u - v$. The function w is then a weak solution to (5.84)–(5.85) with $b = 0$ and $u_0 = 0$. Problem 5.5 then gives $w = 0$ a.e. This proves the uniqueness of the weak solution (5.84)–(5.85).
2. Assume $c > a$ (the case $c < a$ is similar). Set $D_1 = \{(x,t), t > 0, x < at\}$, $D_2 = \{(x,t), t > 0, at < x < ct\}$ and $D_3 = \{(x,t), t > 0, x > ct\}$. We seek the weak solution u of (5.84)–(5.85) in the form:

$$u = u_g \text{ in } D_1, \quad u = \overline{u} \text{ in } D_2, \quad u = u_d \text{ in } D_3, \tag{5.102}$$

with $\overline{u} \in \mathbb{R}$. We thus seek $\overline{u}$ so that u defined by (5.102) satisfies (5.86).
Let $\varphi \in C_c^\infty(\mathbb{R} \times \mathbb{R}_+, \mathbb{R})$ and let

$$E(\varphi) = \int_{\mathbb{R}\times\mathbb{R}_+} u(\partial_t \varphi + a\partial_x \varphi)\, d(x,t) + \int_{\mathbb{R}} u_0(x)\varphi(x,0)\, dx + b\int_0^{+\infty} \varphi(ct,t)\, dt.$$

We thus seek $\overline{u}$ (independent of φ) so that $E(\varphi) = 0$. Set $v = \begin{bmatrix} a \\ 1 \end{bmatrix}$. Let n_i be the outward normal unit vector to the domain D_i, div the divergence operator in the plane $\mathbb{R} \times \mathbb{R}_+$ and y the point (x,t); we then have

$$\int_{\mathbb{R}\times\mathbb{R}_+} u(\partial_t\varphi + a\partial_x\varphi)\, \mathrm{d}(x,t) = \int_{\mathbb{R}\times\mathbb{R}_+} u\,\mathrm{div}(v\varphi)\, \mathrm{d}(x,t)$$

$$= \sum_{i=1}^{3} \int_{\partial D_i} u(y)\varphi(y) v \cdot n_i(y)\, \mathrm{d}\gamma(y),$$

where γ denotes the Lebesgue measure 1-dimensional on ∂D_i (which is the edge of D_i) and the value of u in the integral over ∂D_i is taken from the side of D_i. Setting $\Gamma_a = \{(at, t),\, t \in \mathbb{R}_+\}$, we get that

$$\int_{\mathbb{R}\times\mathbb{R}_+} u(\partial_t\varphi + a\partial_x\varphi)\, \mathrm{d}(x,t) + \int_{\mathbb{R}} u_0(x)\varphi(x,0)\, \mathrm{d}x$$

$$= \int_{\Gamma_a} (u_g - \overline{u})\varphi(y) v \cdot n_1(y)\, \mathrm{d}\gamma(y) + \int_{\Gamma_c} (\overline{u} - u_d)\varphi(y) v \cdot n_2(y)\, \mathrm{d}\gamma(y).$$

We now note that

$$n_1 = (1/\sqrt{1 + a^2}) \begin{bmatrix} 1 \\ -a \end{bmatrix} \quad \text{and}$$

$$n_2 = (1/\sqrt{1 + c^2}) \begin{bmatrix} 1 \\ -c \end{bmatrix}.$$

Hence

$$E(\varphi) = \int_{\Gamma_c} (\overline{u} - u_d)\frac{1}{\sqrt{1 + c^2}}\varphi(ct, t)(a - c)\, \mathrm{d}\gamma(y) + b \int_0^{+\infty} \varphi(ct, t)\, \mathrm{d}t.$$

By parameterising Γ_c by t (which corresponds to a change of variable), the integration element $\mathrm{d}\gamma(y)$ becomes $\sqrt{1 + c^2}\, \mathrm{d}t$. We therefore obtain

$$E(\varphi) = \int_0^{+\infty} (\overline{u} - u_d)(a - c)\varphi(ct, t)\, \mathrm{d}t + b \int_0^{+\infty} \varphi(ct, t)\, \mathrm{d}t.$$

Taking $\overline{u}$ such that $\overline{u} = u_d + \frac{b}{c-a}$, we get that $E(\varphi) = 0$.

3. If $c = a$, we have, with the notations of the previous question, $D_2 = \emptyset$. The uniqueness reasoning of Problem 5.5 then allows us to show that $u = u_g$ on D_1 and $u = u_d$ on D_3. The reasoning of the previous question then gives, for all $\varphi \in C_c^\infty(\mathbb{R} \times \mathbb{R}_+, \mathbb{R})$,

$$E(\varphi) = b \int_0^{+\infty} \varphi(ct, t)\, \mathrm{d}t.$$

If $b \neq 0$, we get that there exists a $\varphi \in C_c^\infty(\mathbb{R} \times \mathbb{R}_+, \mathbb{R})$ such that $E(\varphi) \neq 0$. The problem (5.84)–(5.85) therefore has no weak solution.

Problem 5.7 (Non-uniqueness of weak solutions)

1. Let $\sigma \in \mathbb{R}$ and let u be defined a.e. on $\mathbb{R} \times \mathbb{R}_+$ by (5.88). Let $D_1 = \{(x,t) \in \mathbb{R} \times \mathbb{R}_+^\star, x < \sigma t\}$ and $D_2 = \{(x,t) \in \mathbb{R} \times \mathbb{R}_+^\star, x > \sigma t\}$. The function $u_{|D_i}$ satisfies $u_{|D_i} \in C^1(\overline{D}_i, \mathbb{R})$ (in the sense of Definition 1.28, i.e., restriction to D_i of a C^1 function on $\mathbb{R}^2$ and equal to u on D_i) and it is a classical solution in the domains D_i, $i = 1, 2$, and satisfies (a.e.) the initial condition. It is discontinuous on the line $x = \sigma t$. According to Proposition 5.9, it is a weak solution to (5.87) if and only if it satisfies the Rankine–Hugoniot relation on the line $x = \sigma t$. This relation reads
$$u_d^2 - u_g^2 = \sigma(u_d - u_g).$$
We therefore take $\sigma = (u_g + u_d)$. The function u is then a weak solution to (5.87). Since the function $s \mapsto s^2$ is strictly convex, according to Proposition 5.18 the function u is not an entropy weak solution because $u_g < u_d$.
2. Let $D_1 = \{(x,t) \in \mathbb{R} \times \mathbb{R}_+^\star, x < 2u_g t\}$, $D_2 = \{(x,t) \in \mathbb{R} \times \mathbb{R}_+^\star, 2u_g t 2u_d t\}$. The restriction $u_{|D_i}$ satisfies $u_{|D_i} \in C^1(\overline{D}_i, \mathbb{R})$ (still in the sense of Definition 1.28) and it is a classical solution in the domains D_i, $i = 1, 2, 3$. It satisfies (a.e.) the initial condition. It is continuous on the lines $x = 2u_g t$ and $x = 2u_d t$. Proposition 5.16 then gives that it is an entropy weak solution.

Problem 5.8 (Discontinuity and entropy)

According to Definition 5.11, u is an entropy weak solution if for any convex function η from $\mathbb{R}$ to $\mathbb{R}$, and for any associated flux function Φ defined by $\Phi(s) = \int_0^s f'(\tau)\eta'(\tau)\,d\tau$ (for $s \in \mathbb{R}$), we have

$$\int_{\mathbb{R}_+} \int_{\mathbb{R}} (\eta(u)\partial_t \varphi + \Phi(u)\partial_x \varphi)(x,t)\,dx\,dt + \int_{\mathbb{R}} \eta(u_0(x))\varphi(x,0)\,dx \geq 0,$$

$$\forall \varphi \in C_c^1(\mathbb{R} \times \mathbb{R}_+, \mathbb{R}_+).$$

As noted when defining the entropy solution, we can limit ourselves to considering η of class C^1.

To prove that this condition on u is equivalent to (5.16), we repeat the proof of Proposition 5.9. Let $\eta \in C^1(\mathbb{R}, \mathbb{R})$ and define Φ by $\Phi(s) = \int_0^s f'(\tau)\eta'(\tau)\,d\tau$. Let $\varphi \in C_c^1(\mathbb{R} \times \mathbb{R}_+, \mathbb{R}_+)$. Noting that thanks to Fubini's theorem we can integrate with respect to x then t or with respect to t then x, let

$$X_1 = \int_{\mathbb{R}} \int_{\mathbb{R}_+} \eta(u(x,t))\partial_t \varphi(x,t)\,dt\,dx,$$

$$X_2 = \int_{\mathbb{R}_+} \int_{\mathbb{R}} \Phi(u(x,t))\partial_x \varphi(x,t)\,dx\,dt.$$

We suppose for example $\sigma > 0$ (an analogous reasoning treats the cases $\sigma < 0$ and $\sigma = 0$). Since $u_{|D_i} \in C^1(\bar{D}_i, \mathbb{R})$ $(i = 1, 2)$, u is a classical solution to (5.1) in D_i, $i = 1, 2$, and we can integrate by parts in D_i $(i = 1, 2)$. We recall the following notations: $u_+(\sigma t, t) = \lim_{x \downarrow \sigma t} u(x, t)$, $u_-(\sigma t, t) = \lim_{x \uparrow \sigma t} u(x, t)$, $[u](\sigma t, t) = u_+(\sigma t, t) - u_-(\sigma t, t)$ and $[g(u)](\sigma t, t) = g(u_+(\sigma t, t)) - g(u_-(\sigma t, t))$ for $g = f, \eta$ or Φ. Since

$$
X_1 = \int_{\mathbb{R}_+} \left(\int_0^{\frac{x}{\sigma}} \eta(u(x,t)) \partial_t \varphi(x,t)\, dt \right) dx + \int_{\mathbb{R}_+} \left(\int_{\frac{x}{\sigma}}^{+\infty} \eta(u(x,t)) \partial_t \varphi(x,t)\, dt \right) dx
$$
$$
+ \int_{\mathbb{R}_-} \left(\int_{\mathbb{R}_+} \eta(u(x,t)) \partial_t \varphi(x,t)\, dt \right) dx,
$$

the integrations by parts give

$$
X_1 = \int_{\mathbb{R}_+} \eta(u_+(x, \tfrac{x}{\sigma})) \varphi(x, \tfrac{x}{\sigma})\, dx - \int_{\mathbb{R}_+} \eta(u(x,0)) \varphi(x,0)\, dx
$$
$$
- \int_{\mathbb{R}_+} \int_0^{\frac{x}{\sigma}} \eta'(u(x,t)) \partial_t u(x,t) \varphi(x,t)\, dt\, dx - \int_{\mathbb{R}_+} \eta(u_-(x, \tfrac{x}{\sigma})) \varphi(x, \tfrac{x}{\sigma})\, dx
$$
$$
- \int_{\mathbb{R}_+} \int_{\frac{x}{\sigma}}^{+\infty} \eta'(u(x,t)) \partial_t u(x,t) \varphi(x,t)\, dt\, dx - \int_{\mathbb{R}_-} \eta(u(x,0)) \varphi(x,0)\, dx
$$
$$
- \int_{\mathbb{R}_-} \int_{\mathbb{R}_+} \eta'(u(x,t)) \partial_t u(x,t) \varphi(x,t)\, dt\, dx.
$$

Therefore,

$$
X_1 = - \int_{\mathbb{R}} \eta(u(x,0)) \varphi(x,0)\, dx - \int_{D_1} \eta'(u(x,t)) \partial_t u(x,t) \varphi(x,t)\, d(x,t)
$$
$$
- \int_{D_2} \eta'(u(x,t)) \partial_t u(x,t) \varphi(x,t)\, d(x,t) + \int_{\mathbb{R}_+} [\eta(u)](x, \tfrac{x}{\sigma}) \varphi(x, \tfrac{x}{\sigma})\, dx.
$$

In the last integral, we perform the change of variables $t = \frac{x}{\sigma}$. We obtain

$$
X_1 = - \int_{\mathbb{R}} \eta(u(x,0)) \varphi(x,0)\, dx - \int_{D_1} \eta'(u(x,t)) \partial_t u(x,t) \varphi(x,t)\, d(x,t)
$$
$$
- \int_{D_2} \eta'(u(x,t)) \partial_t u(x,t) \varphi(x,t)\, d(x,t) + \sigma \int_{\mathbb{R}_+} [u](\sigma t, t) \varphi(\sigma t, t)\, dt.
$$

The calculation for X_2 is rather simpler

$$
X_2 = \int_{\mathbb{R}_+} \left(\int_{-\infty}^{\sigma t} \Phi(u(x,t)) \partial_x \varphi(x,t)\, dx \right) dt + \int_{\mathbb{R}_+} \left(\int_{\sigma t}^{+\infty} \Phi(u(x,t)) \partial_x \varphi(x,t)\, dx \right) dt.
$$

The integrations by parts give

$$X_2 = -\int_{D_1} \Phi'(u)\partial_x u(x,t)\varphi(x,t)\,\mathrm{d}(x,t) - \int_{D_2} \Phi'(u)\partial_x u(x,t))\varphi(x,t)\,\mathrm{d}(x,t)$$
$$- \int_{\mathbb{R}_+} [\Phi(u)](\sigma t,t)\varphi(\sigma t,t)\,\mathrm{d}t.$$

Since $\eta'(u)\partial_t u + \Phi'(u)\partial_x u = 0$ on D_1 and D_2, we therefore have:

$$X_1 + X_2 = -\int_{\mathbb{R}} \eta(u(x,0))\varphi(x,0)\,\mathrm{d}x$$
$$+ \int_{\mathbb{R}_+} (\sigma[\eta(u)](\sigma t,t) - [\Phi(u)](\sigma t,t))\varphi(\sigma t,t)\,\mathrm{d}t,$$

that is to say, for all $\varphi \in C_c^1(\mathbb{R} \times \mathbb{R}_+, \mathbb{R}_+)$,

$$\int_{\mathbb{R}_+} \int_{\mathbb{R}} (\eta(u)\partial_t\varphi + \Phi(u)\partial_x\varphi)(x,t)\,\mathrm{d}x\,\mathrm{d}t + \int_{\mathbb{R}} \eta(u_0(x))\varphi(x,0)\,\mathrm{d}x$$
$$= \int_{\mathbb{R}_+} (\sigma[\eta(u)](\sigma t,t) - [\Phi(u)](\sigma t,t))\varphi(\sigma t,t)\,\mathrm{d}t. \quad (5.103)$$

If condition (5.16) is satisfied, the second term of (5.103) is non-negative and u is therefore an entropy weak solution. Conversely, if u is an entropy weak solution, (5.103) gives

$$\int_{\mathbb{R}_+} (\sigma[\eta(u)](\sigma t,t) - [\Phi(u)](\sigma t,t))\varphi(\sigma t,t)\,\mathrm{d}t \geq 0$$

for any function $\varphi \in C_c^1(\mathbb{R} \times \mathbb{R}_+, \mathbb{R}_+)$. The function $t \mapsto \varphi(\sigma t,t)$ is an arbitrary function of $C_c^1(\mathbb{R}_+, \mathbb{R}_+)$. Since the function $t \mapsto \sigma[\eta(u)](\sigma t,t) - [\Phi(u)](\sigma t,t)$ is continuous, it follows that $\sigma[\eta(u)](\sigma t,t) - [\Phi(u)](\sigma t,t)) \geq 0$ for all $t \geq 0$.

Problem 5.9 (Riemann Problem)

1. a. Since f' is increasing, $f'(u_g) < f'(u_d)$ and therefore the characteristic curves originating from the axis $t = 0$ do not meet.
   We then define the zones D_i, $i = 1,2,3$, by

$$D_1 = \{(x,t) \in \mathbb{R} \times \mathbb{R}_+^\star, x < f'(u_g)t\},$$
$$D_2 = \{(x,t) \in \mathbb{R} \times \mathbb{R}_+^\star, f'(u_g)t < x < f'(u_d)t\},$$
$$D_3 = \{(x,t) \in \mathbb{R} \times \mathbb{R}_+^\star, x > f'(u_d)t\}.$$

   As suggested by the characteristic curves originating from the axis $t = 0$, we define u in D_1 by $u(x,t) = u_g$ and u in D_3 by $u(x,t) = u_d$. We then have, for $i = 1, 3$, $u_{|D_i} \in C^1(\overline{D}_i, \mathbb{R})$ (in the sense of Definition 1.28) and each restriction $u_{|D_i}$ is a classical solution (of the equation of the Riemann problem (5.10)) in the domain D_i. The function u also satisfies (a.e.) the initial condition of (5.10) (whatever the value of u in D_2).

We now look for u in D_2 in the form $u(x, t) = \phi(\frac{x}{t})$. If ϕ is of class C^1, for u to be a classical solution of the equation of the Riemann problem (5.10) it is necessary and sufficient that

$$\left(\frac{-x}{t^2} + f'(\phi(\frac{x}{t}))\frac{1}{t}\right)\phi'(\frac{x}{t}) = 0 \quad \text{for all} (x, t) \in D^2.$$

This therefore suggests taking u such that $f'(u(x, t)) = \frac{x}{t}$ for all $(x, t) \in D^2$. The function f' is increasing continuous from $\mathbb{R}$ onto its image $\text{Im}(f')$. It is therefore invertible and has a continuous inverse g; choose $u(x, t) = g(\frac{x}{t})$ for all $(x, t) \in D^2$. Since we have assumed that f is of class C^2 and that $f''(s) > 0$ for all $s \in \mathbb{R}$, the function g is of class C^1. To see this, it is enough to notice that, for $s \in \text{Im}(f')$ and $h \in \mathbb{R}^\star$ such that $s + h \in \text{Im}(f')$,

$$\frac{g(s + h) - g(s)}{h} = \frac{g(s + h) - g(s)}{f'(g(s + h)) - f'(g(s))}$$

$$= \frac{\varepsilon(h)}{f'(g(s) + \varepsilon(h)) - f'(g(s))},$$

with $\lim_{h \to 0} \varepsilon(h) = 0$ (because g is continuous). Thus g is differentiable at the point s and $g'(s) = \frac{1}{f''(g(s))}$. This indeed shows that g is of class C^1 (from $\text{Im}(f')$ to $\mathbb{R}$). Hence $u_{|D_2} \in C^1(\overline{D}_2, \mathbb{R})$ and is a classical solution of the Riemann problem equation (5.10) in the domain D_2.

Finally, we note that u is continuous on the lines $x = f'(u_g)t$ (which gives $u_g = g(\frac{x}{t})$) and $x = f'(u_d)t$ (which gives $u_d = g(\frac{x}{t})$). Proposition 5.16 then gives that it is the entropy weak solution to (5.10) and it is continuous from $\mathbb{R} \times \mathbb{R}_+$ to $\mathbb{R}$.

N.B. We assumed $f \in C^2(\mathbb{R})$ and $f''(s) > 0$ for all $s \in \mathbb{R}$ to obtain a classical solution in D_2. If f is only of class C^1, a regularisation of f allows us to show that the same choice of u in D_2, that is $u(x, t) = g(\frac{x}{t})$, gives a weak solution in D_2 and therefore also ultimately the entropy weak solution to (5.10).

b. Since f' is increasing, $f'(u_g) > f'(u_d)$ and therefore the characteristic curves originating from the axis $t = 0$ meet. We therefore look for a solution with a discontinuity.

Let $\sigma \in \mathbb{R}$. Define u a.e. on $\mathbb{R} \times \mathbb{R}_+$ by $u(x, t) = u_g$ if $x < \sigma t$ and $u(x, t) = u_d$ if $x > \sigma t$.

Let $D_1 = \{(x, t) \in \mathbb{R} \times \mathbb{R}_+^\star, x < \sigma t\}$ and $D_2 = \{(x, t) \in \mathbb{R} \times \mathbb{R}_+^\star, x > \sigma t\}$. The function $u_{|D_i}$ is in $C^1(\overline{D}_i, \mathbb{R})$ (that is, it is the restriction to D_i of a C^1 function on $\mathbb{R}^2$) and is a classical solution in the domains D_i, $i = 1, 2$. It satisfies (a.e.) the initial condition and is discontinuous on the line $x = \sigma t$. According to Proposition 5.9 it is a weak solution to (5.87) if and only if it satisfies the Rankine–Hugoniot relation on the line $x = \sigma t$. This relation reads

$$f(u_d) - f(u_g) = \sigma(u_d - u_g).$$

We therefore take $\sigma = \frac{f(u_d)-f(u_g)}{u_d-u_g}$. The function u is then a weak solution to (5.87).

Since the function f is strictly convex, according to Proposition 5.18 the function u is the entropy weak solution because $u_g > u_d$.

2. We can indeed reduce to the case of the first question.

With the function u from $\mathbb{R} \times \mathbb{R}_+$ to $\mathbb{R}$, we define the function v by $v(x, t) = u(-x, t)$. The function u is then an entropy weak solution of the Riemann problem (5.10) with data u_d and u_g if and only if the function v is then an entropy weak solution of the Riemann problem (5.10) with data u_d for $x < 0$ and u_g for $x > 0$ and with $-f$ instead of f. We are reduced to the case of the first question because $-f$ is strictly convex.

The Riemann problem (5.10) thus has a continuous solution if $u_d < u_g$ and a discontinuous solution if $u_d > u_g$.

Problem 5.10 (Construction of entropy weak solutions)

Initial condition (a) We start by constructing, for each x_0, the characteristic curve from x_0.

- For $x_0 < 0$, the characteristic curve is the half-line $\{(x_0 + 2t, t), t \geq 0\}$.
- For $0 \leq x_0 \leq 1$, the characteristic curve is the half-line $\{(x_0 + 2(1 - x_0)t, t), t \geq 0\}$, because $f'(u_0(x_0)) = 2(1 - x_0)$.
- For $x_0 > 1$, the characteristic curve is the half-line $\{(x_0, t), t \geq 0\}$.

For $0 < t < \frac{1}{2}$, the characteristic curves do not meet, as indicated in figure 5.5. The solution is therefore continuous for $0 < t < \frac{1}{2}$ and it is constant on each characteristic curve. For example, if $0 < t < \frac{1}{2}$ and $x = x_0 + 2(1 - x_0)t$ with $x_0 \in [0, 1]$, we have $u(x, t) = u_0(x_0) = 1 - x_0 = (1 - x)/(1 - 2t)$ because $x_0(1 - 2t) = x - 2t$. At $t = \frac{1}{2}$, the characteristic curves originating from the points x_0 of the interval $[0, 1]$ meet (at the point $x = 1$); a discontinuity appears and propagates at a speed consistent with the Rankine–Hugoniot relation. This allows us to construct the solution $u(x, t)$ for all $(x, t) \in \mathbb{R} \times \mathbb{R}_+$ in the following way. Set:

$$u(x, t) = 1 \text{ if } (x, t) \in D_1 = \{(x, t), 0 < t \leq \frac{1}{2}, x < 2t\} \cup \{(x, t), t > \frac{1}{2}, x < t + \frac{1}{2}\},$$

$$u(x, t) = \frac{1 - x}{1 - 2t} \text{ if } (x, t) \in D_2 = \{(x, t), 0 < t < \frac{1}{2}, 2t < x < 1\},$$

$$u(x, t) = 0 \text{ if } (x, t) \in D_3 = \{(x, t), 0 < t \leq \frac{1}{2}, 1 < x\} \cup \{(x, t), t > \frac{1}{2}, t + \frac{1}{2} < x\}.$$

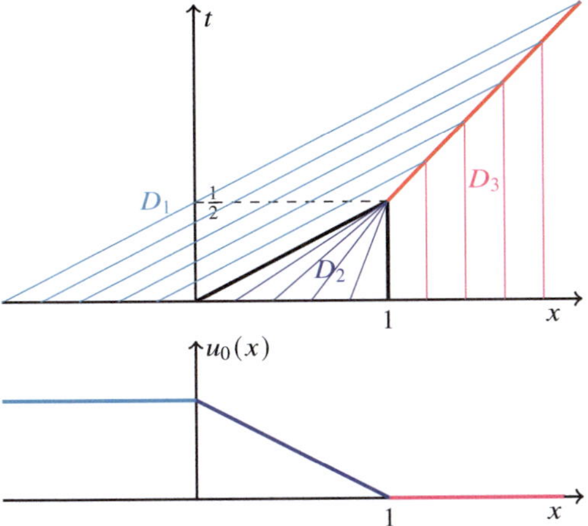

Fig. 5.5 Initial condition (a). At the top: Characteristic lines; in red the shock line, in black the C^1 discontinuity lines. – At the bottom: appearance of the initial condition u_0.

The function u is indeed a weak solution. In fact, it is an entropy weak solution (see Proposition 5.18).

Initial condition (b) Here again, we start by constructing, for each x_0, the characteristic curve from x_0.

- For $x_0 < 0$, the characteristic curve is the half-line $\{(x_0, t), t \geq 0\}$.
- For $0 \leq x_0 \leq 1$, the characteristic curve is the half-line $\{(x_0 + 2(1 - x_0)t, t), t \geq 0\}$, because $f'(u_0(x_0)) = 2(1 - x_0)$.
- For $x_0 > 1$, the characteristic curve is the half-line $\{(x_0 + 2t, t), t \geq 0\}$.

For $0 < t < \frac{1}{2}$, the characteristic curves do not meet, as indicated in figure 5.6. The solution is therefore continuous for $0 < t < \frac{1}{2}$ and it is constant on each characteristic curve.

As in the case of initial condition (a), we have, for example, if $0 < t < \frac{1}{2}$ and $x = x_0 + 2(1 - x_0)t$ with $x_0 \in [0, 1]$, $u(x, t) = u_0(x_0) = 1 - x_0 = (1 - x)/(1 - 2t)$ because $x_0(1 - 2t) = x - 2t$. The difference with initial condition (a) is that the solution now includes two rarefactions originating from points 0 and 1. In the rarefaction zone originating from point 0, we have $u(x, t) = \xi$ for $x = 2\xi t$ and $\xi \in [0, 1]$ (and therefore $u(x, t) = x/(2t)$). In the rarefaction zone originating from point 1, we have $u(x, t) = \xi$ for $x = 2\xi t + 1$ and $\xi \in [0, 1]$ (and therefore $u(x, t) = (x - 1)/(2t)$).

Then, at $t = \frac{1}{2}$, a shock appears. Using the Rankine–Hugoniot relation, it is shown (with the calculation of the solution on the characteristics of the equation, see below) that this shock propagates at speed 1. Owing to Proposition 5.18, the solution is indeed an entropy weak solution. In summary, this gives the following solution:

$$u(x,t) = \begin{cases} 0 & \text{on } D_1, \\ \frac{x}{2t} & \text{on } D_2, \\ \frac{1-x}{1-2t} & \text{on } D_3, \\ \frac{x-1}{2t} & \text{on } D_4, \\ 1 & \text{on } D_5, \end{cases}$$

with

$$D_1 = \{(x,t), 0 < t, x < 0\},$$

$$D_2 = \{(x,t), 0 < t \le \frac{1}{2}, 0 < x < 2t\} \cup \{(x,t), t > \frac{1}{2}, 0 < x < t + \frac{1}{2}\},$$

$$D_3 = \{(x,t), 0 < t < \frac{1}{2}, 2t < x < 1\},$$

$$D_4 = \{(x,t), 0 < t \le \frac{1}{2}, 1 < x < 1 + 2t\} \cup \{(x,t), t > \frac{1}{2}, t + \frac{1}{2} < x < 1 + 2t\},$$

$$D_5 = \{(x,t), 0 < t, 1 + 2t < x\}.$$

The function u is discontinuous on the set $\{(x,t), t > \frac{1}{2}, x = t + \frac{1}{2}\}$ (shock line, in red

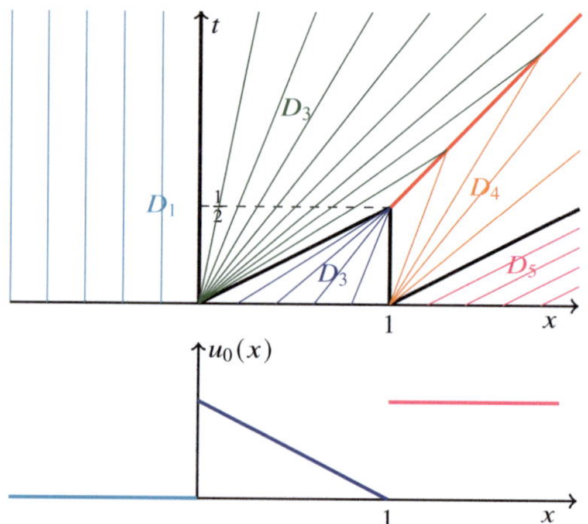

Fig. 5.6 Initial condition (b) – Top: characteristic lines; in red the shock line, in black the C^1 discontinuity lines – Bottom: the initial condition u_0.

on the figure). Let us check that the Rankine–Hugoniot relation is satisfied at every point of this set. Indeed, let $t > \frac{1}{2}$ and $x = t + \frac{1}{2}$. With the notations of Proposition 5.18, we have

$$u_-(x,t) = \frac{x}{2t} = \frac{t + \frac{1}{2}}{2t} = \frac{1}{2} + \frac{1}{4t} \qquad \text{(we use here } D_2\text{)},$$

$$u_+(x,t) = \frac{x-1}{2t} = \frac{t - \frac{1}{2}}{2t} = \frac{1}{2} - \frac{1}{4t} \qquad \text{(we use here } D_4\text{)}.$$

This gives $u_-(x,t) + u_+(x,t) = 1$ and the Rankine–Hugoniot relation is indeed satisfied. On the other hand, the constructed solution is indeed an entropy weak solution because $u_- > u_+$.

Initial condition (c) Given the initial condition, we can suspect that the entropy weak solution contains a discontinuity originating from the point $x = 1$. We therefore seek the solution in the form of a continuous function on the left and right of a discontinuity line denoted by L (in red in figure 5.7), defined by $L = \{(x,t), t > 0, x = \sigma(t)\}$, where σ is a C^1 non-decreasing function such that $\sigma(0) = 1$ and $\sigma'(t) < 2$ for all $t > 0$. Set (see figure 5.7):

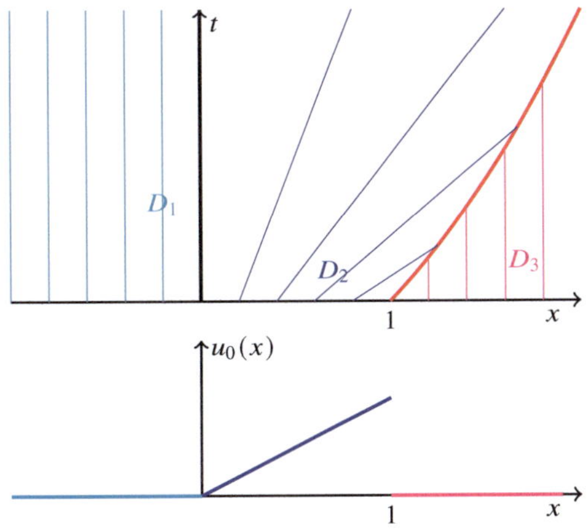

Fig. 5.7 Initial condition (c) – Top: Characteristic lines; in red the shock line, in black the C^1 discontinuity lines – Bottom: shape of the initial condition u_0.

$$D_1 = \{(x,t), 0 < t, x < 0\},$$
$$D_2 = \{(x,t), 0 < t, 0 < x < \sigma(t)\},$$
$$D_3 = \{(x,t), 0 < t, \sigma(t) < x\}.$$

We take $u(x,t) = 0$ if $(x,t) \in D_1$. In D_2, u is constructed using the characteristics, which gives $u(x,t) = x/(1 + 2t)$ if $(x,t) \in D_2$. Finally, we set $u(x,t) = 0$ if $(x,t) \in D_3$. For u to be a weak solution of the considered problem, it is enough to verify the Rankine–Hugoniot relation on L, that is (with the notations of Proposition 5.18) that

$$\sigma'(t) = u_-(x,t) + u_+(x,t) \text{ for all } (x,t) \in L.$$

Let $t > 0$ and $x = \sigma(t)$, we have $u_-(x,t) + u_+(x,t) = \sigma(t)/(1 + 2t)$. It is therefore sufficient that

$$\sigma'(t) = \sigma(t)/(1 + 2t) \text{ for all } t > 0,$$
$$\sigma(0) = 1.$$

The solution of this differential equation is $\sigma(t) = \sqrt{1 + 2t}$ (for all $t > 0$). With this choice of the function σ, the function u thus constructed is a weak solution to the problem considered. This function is even an entropy weak solution because $u_- > u_+$ on L (see Proposition 5.18).

$$\text{For } t > 0, \int_{\mathbb{R}} u(x,t)\,dx = \int_0^{\sqrt{1+2t}} \frac{x}{1+2t}\,dt = \frac{1}{2} = \int_{\mathbb{R}} u_0(x)\,dx.$$

Initial condition (d) A graphic of the entropy weak solution is given in figure 5.8. The discontinuity of u_0 at $x = -1$ begins to propagate at speed 1 (that is to say, on the line $x = -1 + t$), it is a shock wave; it separates the regions D_1 and D_2 in the figure. Note that, in accordance with the theory, the characteristics *enter* into the shock line. The discontinuity of u_0 at $x = 1$ begins to propagate at speed 2 (that is to say, on the line $x = 1 + 2t$), it is also a shock wave. The discontinuity of u_0 at $x = 0$ disappears, it gives a rarefaction wave (region D_3). In this rarefaction wave, we have $u(x,t) = x/(2t)$. Then, at $t = \frac{1}{2}$, the "head" of the rarefaction wave catches up

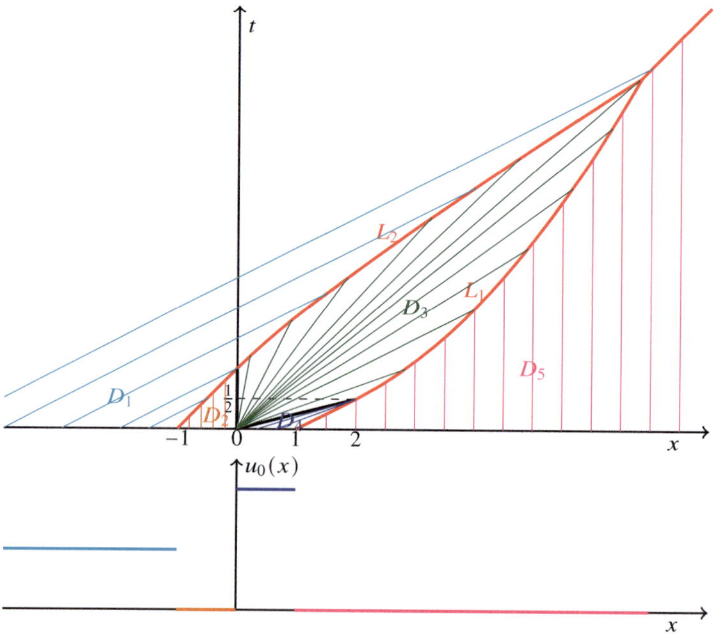

Fig. 5.8 Initial condition (d) – Top: characteristic lines; in red the shock line, in black the C^1 discontinuity lines – Bottom: the initial condition u_0.

with the right shock wave (at point $x = 2$) which then "slows down" and continues on a curve denoted by L_1, with $L_1 = \{(x,t),\ t > \frac{1}{2},\ x = \sigma_1(t)\}$. At $t = 1$, the left shock wave catches up with the "foot" of the rarefaction wave (at point $x = 0$). The shock wave "slows down" and continues on a curve that we denoted by L_2, with $L_2 = \{(x,t),\ t > 1,\ x = \sigma_1(t)\}$. The functions σ_1 and σ_2 are calculated using the Rankine–Hugoniot relations.

1. *Calculation of σ_1.* Let $(x,t) \in L_1$. The set L_1 separates the rarefaction wave D_2 from the area D_5 in which $u = 0$. The Rankine–Hugoniot relation yields

$$\sigma_1'(t) = u_-(x,t) + u_+(x,t) = \frac{x}{2t} = \frac{\sigma_1(t)}{2t}.$$

Since $\sigma_1(\frac{1}{2}) = 2$, the resolution of this differential equation gives $\sigma_1(t) = 2\sqrt{2t}$ for all $t > \frac{1}{2}$.

2. *Calculation of σ_2.* Let $(x,t) \in L_2$. The set L_2 separates the area D_1 in which $u = 1$ from the rarefaction wave D_2. So, with the Rankine–Hugoniot relation,

$$\sigma_2'(t) = u_-(x,t) + u_+(x,t) = 1 + \frac{x}{2t} = 1 + \frac{\sigma_2(t)}{2t}.$$

Since $\sigma_2(1) = 0$, the resolution of this differential equation gives $\sigma_2(t) = 2(t - \sqrt{t})$ for all $t > 1$.

The curves L_1 and L_2 meet at a t such that $1 + \frac{\sigma_2(t)}{2t} = \frac{\sigma_1(t)}{2t}$, i.e. $t = 3 + 2\sqrt{2}$, to give birth to a single discontinuity that propagates at speed 1, since this discontinuity separates the area D_1 in which $u = 1$ from the area D_5 in which $u = 0$.

The solution thus constructed is indeed entropic because on each discontinuity curve we have $u_g > u_d$, and the flux function $s \mapsto s^2$ of the Burgers equation is indeed convex.

Problem 5.11 (Non-entropy solution)

1. Since $L^1_{\text{loc}} \subset L^2_{\text{loc}}$, it is enough to show that $u^2 \in L^1_{\text{loc}}(\mathbb{R} \times \mathbb{R}_+)$. Let $T > 0$. For $0 < t < T$,

$$\int_{\mathbb{R}} u^2(x,t)\, dx = \int_{-\sqrt{t}}^{\sqrt{t}} \frac{x^2}{4t^2}\, dx = \frac{1}{6t^{\frac{1}{2}}},$$

and therefore

$$\int_0^T \left(\int_{\mathbb{R}} u^2(x,t)\, dx \right) dt = \int_0^T \frac{1}{6t^{\frac{1}{2}}}\, dt < +\infty,$$

which proves that $u^2 \in L^1_{\text{loc}}(\mathbb{R} \times \mathbb{R}_+)$.

2. A quick proof of this question consists of using Proposition 5.22 with, for all $\varepsilon > 0$, $\mathbb{R} \times [\varepsilon, +\infty[$ instead of $\mathbb{R} \times \mathbb{R}_+$. We give here another proof, closer to that given in Proposition 5.16.
Let $\varphi \in C^1_c(\mathbb{R} \times \mathbb{R}^\star_+, \mathbb{R})$. Using integration by parts, we obtain, for all $x \in \mathbb{R}$,

$$\int_0^\infty u(x,t)\partial_t \varphi(x,t)\, dt = \int_{x^2}^{+\infty} \frac{x}{2t^2}\varphi(x,t)\, dt - \frac{1}{2x}\varphi(x,x^2).$$

Noting that the two terms on the right are integrable (because $\varphi(x,t) = 0$ for t close to 0),

$$\int_{\mathbb{R}} \int_0^\infty u(x,t)\partial_t \varphi(x,t)\, dt\, dx$$

$$= \int_{\mathbb{R}} \int_{x^2}^{+\infty} \frac{x}{2t^2}\varphi(x,t)\, dt\, dx - \int_{\mathbb{R}} \frac{1}{2x}\varphi(x,x^2)\, dx. \quad (5.104)$$

Similarly, by integration by parts, we obtain, for all $t \in \mathbb{R}_+^\star$,

$$\int_{\mathbb{R}} u^2(x,t)\partial_x \varphi(x,t)\, dx = -\int_{-\sqrt{t}}^{\sqrt{t}} \frac{x}{2t^2}\, dx + \frac{1}{4t}\varphi(\sqrt{t},t) - \frac{1}{4t}\varphi(-\sqrt{t},t).$$

Here again the terms on the right are integrable (because $\varphi(x,t) = 0$ for t close to 0),

$$\int_0^{+\infty} \int_{\mathbb{R}} u^2(x,t)\partial_x \varphi(x,t)\, dx\, dt$$

$$= -\int_0^{+\infty} \int_{-\sqrt{t}}^{\sqrt{t}} \frac{x}{2t^2}\, dx\, dt + \int_0^{+\infty} \frac{1}{4t}(\varphi(\sqrt{t},t) - \varphi(-\sqrt{t},t))\, dt. \quad (5.105)$$

Fubini's theorem gives us

$$\int_{\mathbb{R}} \int_{x^2}^{+\infty} \frac{x}{2t^2}\varphi(x,t)\, dt\, dx = \int_0^{+\infty} \int_{-\sqrt{t}}^{\sqrt{t}} \frac{x}{2t^2}\, dx\, dt.$$

Then the change of variables $t = x^2$ gives

$$\int_0^{+\infty} \frac{1}{4t}\varphi(\sqrt{t},t)\, dt = \int_0^{+\infty} \frac{1}{4x^2}\varphi(x,x^2)2x\, dx = \int_0^{+\infty} \frac{1}{2x}\varphi(x,x^2)\, dx,$$

$$\int_0^{+\infty} \frac{1}{4t}\varphi(-\sqrt{t},t)\, dt = -\int_{-\infty}^0 \frac{1}{4x^2}\varphi(x,x^2)2x\, dx = -\int_{-\infty}^0 \frac{1}{2x}\varphi(x,x^2)\, dx.$$

By adding (5.104) and (5.105) we obtain (5.91).

3. We are given a function $\psi \in C^\infty(\mathbb{R},\mathbb{R})$ such that $\psi(t) = 0$ for $t \in] -\infty, 1]$ and $\psi(t) = 1$ for $t \in [2, +\infty[$. Then, for $n \in \mathbb{N}^\star$, we set $\psi_n(t) = \psi(nt)$.
   Let $\varphi \in C_c^1(\mathbb{R} \times \mathbb{R}_+, \mathbb{R})$. Since the function $(x,t) \mapsto \varphi(x,t)\psi_n(t)$ belongs to $\mathcal{D}(\mathbb{R} \times \mathbb{R}_+^\star)$, the previous question gives

$$\int_{\mathbb{R}_+} \int_{\mathbb{R}} (u(x,t)\partial_t(\varphi\psi_n)(x,t) + u^2(x,t)\partial_x(\varphi\psi_n)(x,t))\, dx\, dt = 0,$$

and therefore

$$\int_{\mathbb{R}_+} \int_{\mathbb{R}} u(x,t)\partial_t\varphi(x,t)\psi_n(t)\, dx\, dt + n\int_{\frac{1}{n}}^{2/n} \int_{\mathbb{R}} u(x,t)\varphi(x,t)\psi'(nt)\, dx\, dt$$

$$+ \int_{\mathbb{R}_+} \int_{\mathbb{R}} u^2(x,t)\partial_x\varphi(x,t)\psi_n(t)\, dx\, dt = 0. \quad (5.106)$$

Question 1 gives $u\partial_t\varphi$, $u^2\partial_x\varphi \in L^1(\mathbb{R}\times\mathbb{R}_+)$. Owing to the dominated convergence theorem,

$$\lim_{n\to+\infty}\int_{\mathbb{R}_+}\int_{\mathbb{R}} u(x,t)\partial_t\varphi(x,t)\psi_n(t)\,\mathrm{d}x\,\mathrm{d}t = \int_{\mathbb{R}_+}\int_{\mathbb{R}} u(x,t)\partial_t\varphi(x,t)\,\mathrm{d}x\,\mathrm{d}t,$$

$$\lim_{n\to+\infty}\int_{\mathbb{R}_+}\int_{\mathbb{R}} u^2(x,t)\partial_x\varphi(x,t)\psi_n(t)\,\mathrm{d}x\,\mathrm{d}t = \int_{\mathbb{R}_+}\int_{\mathbb{R}} u^2(x,t)\partial_x\varphi(x,t)\,\mathrm{d}x\,\mathrm{d}t.$$

Then,

$$n\int_{\frac{1}{n}}^{2/n}\int_{\mathbb{R}} u(x,t)\varphi(x,t)\psi'(nt)\,\mathrm{d}x\,\mathrm{d}t$$
$$= n\int_{\frac{1}{n}}^{2/n}\int_0^{\sqrt{t}} \frac{x}{2t}(\varphi(x,t)-\varphi(-x,t))\psi'(nt)\,\mathrm{d}x\,\mathrm{d}t.$$

Let M be an upper bound of $\psi'\partial_x\varphi$. We obtain

$$\left|n\int_{\frac{1}{n}}^{2/n}\int_{\mathbb{R}} u(x,t)\varphi(x,t)\psi'(nt)\,\mathrm{d}x\,\mathrm{d}t\right| \le nM\int_{\frac{1}{n}}^{2/n}\int_0^{\sqrt{t}} \frac{x^2}{t}\,\mathrm{d}x\,\mathrm{d}t$$
$$\le nM\int_{\frac{1}{n}}^{2/n}\sqrt{t}\,\mathrm{d}t \le M\sqrt{2/n}.$$

Passing to the limit as $n\to+\infty$ in (5.106), we get (5.92).

4. We can assume $\eta(0)=0$. This does not change the terms of (5.93). Set $\psi(s)=\Phi(s)-s\eta(s)$ so that $\psi'(s)=s\eta'(s)-\eta(s)$. The convexity of η then gives $\psi'(s)\ge 0$ for $s\ge 0$ and therefore $\psi(s)\ge 0$ for $s\ge 0$.

   To prove (5.93), we can then use Proposition 5.22 with, for all $\varepsilon>0$, $\mathbb{R}\times[\varepsilon,+\infty[$ instead of $\mathbb{R}\times\mathbb{R}_+$ or repeat the proof of question 2 by replacing u by $\eta(u)$ and u^2 by $\Phi(u)$.

5. The entropy weak solution of (5.89)–(5.90) is the function $u:(x,t)\in\mathbb{R}\times\mathbb{R}_+ \mapsto u(x,t)=0$. The function u is therefore not the entropy weak solution of (5.89)–(5.90).

Problem 5.12 (Buckley–Leverett Equation)

1. First note that f' is increasing on $[0,a]$ then decreasing on $[a,1]$. Hence $f'(x)>0$ and $f(x)>0$ for all $x\in]0,1]$.

   Since the function f is strictly convex on $]0,a[$ (and of class C^1 on $\mathbb{R}$),

$$0 > f(x) + (0-x)f'(x), \quad \text{for all } x\in]0,a],$$

   and therefore $f'(x) > f(x)/x$ for all $x\in]0,a]$. In particular $f'(a) > f(a)/a$.

For $x \in [a, 1]$, we set $h(x) = f(x) - xf'(x)$, so that $h(a) < 0$ and $h(1) = f(1) > 0$. Then, since $h'(x) = -xf''(x)$, the function h is increasing on $[a, 1]$. Therefore, there exists one and only one point $b \in]a, 1[$ such that $h(b) = 0$, i.e. $f'(b) = f(b)/b$.

Finally, since the function f' is increasing from 0 to $f'(a)$ on $[0, a]$ then decreasing from $f'(a)$ to 0 on $[a, 1]$, there exists a unique point $c \in]0, a[$ such that $f'(c) = f'(b)$ (because $b \in]a, 1[$).

2. We decompose the half-plane $\mathbb{R} \times \mathbb{R}_+$ into 3 zones,

$$D_1 = \{(x, t), \, t \geq 0, \, x < 0\},$$
$$D_2 = \{(x, t), \, t \geq 0, \, x = f'(\xi)t, \, b < \xi < 1\},$$
$$D_3 = \{(x, t), \, t \geq 0, \, x > f'(b)t\}.$$

In each of these three zones, the function u is a classical solution to (5.94) (and indeed satisfies (5.95)). The function u is continuous at the boundary between D_1 and D_2 (it is equal to 1, recall that $f'(1) = 0$). The function u is discontinuous at the boundary between D_2 and D_3. On the side of D_2, it equals b and it equals 0 in D_3. The boundary between D_2 and D_3 is the half-line of equation $x = f'(b)t$. Since $f'(b) = f(b)/b = (f(b) - f(0))/(b - 0)$, the Rankine–Hugoniot relation is indeed satisfied at the boundary between D_2 and D_3. This shows that u is a weak solution to (5.94)–(5.95).

It remains to verify the entropy condition at the boundary between D_2 and D_3. Let η be a function from $\mathbb{R}$ to $\mathbb{R}$ of class C^1 and convex. Since $(f'(b) - f'(x))(\eta'(x) - \eta'(c)) \leq 0$ for almost all $x \in [0, b]$ (this is obtained by distinguishing the cases $x < c$ and $x > c$ and using $f'(b) = f'(c)$), we indeed have $\int_0^b (f'(b) - f'(x))(\eta'(x) - \eta'(c)) \, dx \leq 0$.

Denoting the primitive of $\eta' f'$ by Φ, this gives

$$f'(b)(\eta(b) - \eta(0)) - f'(b)\eta'(c)b - (\Phi(b) - \Phi(0)) + f(b)\eta'(c) \leq 0.$$

Since $bf'(b) = f(b)$, we get $(\Phi(0) - \Phi(b)) \leq f'(b)(\eta(0) - \eta(b))$, which is indeed the entropy condition at the boundary between D_2 and D_3.

N.B. We could also have used the Oleinik condition, see Theorem 5.21.

Problem 5.13 (Landau Effect)

1. The first equation of (5.96) can be written $\partial_t u(x, y, t) + \partial_x (yu)(x, y, t) = 0$. The notion of weak solution for this equation is therefore perfectly defined.

Step 1, construction of a solution The problem (5.96) corresponds to a transport equation in the x direction, the speed of transport depending on the variable y (which can be seen here as a parameter). The beginning of Chapter 5 suggests the form of the weak solution. For $(x, y, t) \in \mathbb{R} \times \mathbb{R} \times \mathbb{R}_+$, we set

$$u(x, y, t) = f(x - yt).$$

The function u thus defined indeed belongs to $L^\infty(\mathbb{R} \times \mathbb{R} \times \mathbb{R}_+)$. We now show that u is a weak solution to (5.96). Let $\varphi \in C_c^1(\mathbb{R}^2 \times \mathbb{R}_+, \mathbb{R})$. Let us show that

$$
\int_{\mathbb{R}_+} \int_{\mathbb{R}} \int_{\mathbb{R}} u(x, y, t)(\partial_t \varphi(x, y, t) + y\partial_x \varphi(x, y, t))\, dx\, dy\, dt
$$
$$
= - \int_{\mathbb{R}} \int_{\mathbb{R}} f(x)\varphi(x, y, 0)\, dx\, dy, \quad (5.107)
$$

which shows that u is a weak solution to (5.96). Consider the left-hand side of (5.107), replacing $u(x, y, t)$ with $f(x - yt)$ in the integral. Owing to Fubini's theorem, for y and t fixed, we may perform the change of variables $x - yt = z$ in the integral with respect to x. We obtain

$$
\int_{\mathbb{R}_+} \int_{\mathbb{R}} \int_{\mathbb{R}} f(x - yt)(\partial_t \varphi(x, y, t) + y\partial_x \varphi(x, y, t))\, dx\, dy\, dt
$$
$$
= \int_{\mathbb{R}_+} \int_{\mathbb{R}} \int_{\mathbb{R}} f(z)(\partial_t \varphi(z + yt, y, t) + y\partial_x \varphi(z + yt, y, t))\, dz\, dy\, dt.
$$

(Note that $\partial_x \varphi$ (resp. $\partial_t \varphi$) denotes the derivative of φ with respect to its first (resp. third) variable.)

For $z, y \in \mathbb{R}$ and $t \in \mathbb{R}_+$, we define $\psi(z, y, t) = \varphi(z + yt, y, t)$, so that $\psi_t(z, y, t) = y\partial_x \varphi(z + yt, y, t) + \partial_t \varphi(z + yt, y, t)$. We thus obtain

$$
\int_{\mathbb{R}_+} \int_{\mathbb{R}} \int_{\mathbb{R}} f(x - yt)(\partial_t \varphi(x, y, t) + y\partial_x \varphi(x, y, t))\, dx\, dy\, dt
$$
$$
= \int_{\mathbb{R}_+} \int_{\mathbb{R}} \int_{\mathbb{R}} f(z)\psi_t(z, y, t)\, dz\, dy\, dt.
$$

We can now integrate the term on the right first with respect to t (thanks to Fubini's theorem). We obtain

$$
\int_{\mathbb{R}_+} \int_{\mathbb{R}} \int_{\mathbb{R}} f(x - yt)(\partial_t \varphi(x, y, t) + y\partial_x \varphi(x, y, t))\, dx\, dy\, dt
$$
$$
= - \int_{\mathbb{R}} \int_{\mathbb{R}} f(z)\psi(z, y, 0)\, dz\, dy,
$$

which gives

$$
\int_{\mathbb{R}_+} \int_{\mathbb{R}} \int_{\mathbb{R}} f(x - yt)(\partial_t \varphi(x, y, t) + y\partial_x \varphi(x, y, t))\, dx\, dy\, dt
$$
$$
= - \int_{\mathbb{R}} \int_{\mathbb{R}} f(z)\varphi(z, y, 0)\, dz\, dy.
$$

We have indeed shown (5.107): the function u is therefore a weak solution to (5.96).

Uniqueness of the weak solution to (5.96) Thanks to the linearity of the first equation of (5.96), it is sufficient to show that if u is a solution to (5.107) (for all $\varphi \in C_c^1(\mathbb{R}^2 \times \mathbb{R}_+, \mathbb{R})$) with $f = 0$ a.e., then $u = 0$ a.e. We therefore assume that u belongs to $L^\infty(\mathbb{R}^2 \times \mathbb{R}_+)$ and satisfies

$$\int_{\mathbb{R}_+} \int_{\mathbb{R}} \int_{\mathbb{R}} u(x, y, t)(\partial_t \varphi(x, y, t) + y\partial_x \varphi(x, y, t)) \, \mathrm{d}x \, \mathrm{d}y \, \mathrm{d}t = 0,$$

$$\text{for all } \varphi \in C_c^1(\mathbb{R}^2 \times \mathbb{R}_+, \mathbb{R}). \quad (5.108)$$

Let us show that $u = 0$ a.e. Let $\psi \in C_c^1(\mathbb{R}^2 \times \mathbb{R}_+, \mathbb{R})$. For x, $y \in \mathbb{R}$ and $t \in \mathbb{R}_+$, let

$$\varphi(x, y, t) = -\int_t^{+\infty} \psi(x - y(t - s), y, s) \, \mathrm{d}s.$$

Then $\varphi \in C_c^1(\mathbb{R}^2 \times \mathbb{R}_+, \mathbb{R})$ and for all x, $y \in \mathbb{R}$ and all $t > 0$, we have

$$\partial_t \varphi(x, y, t) + y\partial_x \varphi(x, y, t) = \psi(x, y, t) + y \int_t^{+\infty} \psi_x(x - y(t - s), y, s) \, \mathrm{d}s$$
$$-y \int_t^{+\infty} \psi_x(x - y(t - s), y, s) \, \mathrm{d}s$$

and so

$$\partial_t \varphi(x, y, t) + y\partial_x \varphi(x, y, t) = \psi(x, y, t).$$

Taking this function φ in (5.108) we obtain

$$\int_{\mathbb{R}_+} \int_{\mathbb{R}} \int_{\mathbb{R}} u(x, y, t)\psi(x, y, t) \, \mathrm{d}x \, \mathrm{d}y \, \mathrm{d}t = 0 \text{ for all } \psi \in C_c^1(\mathbb{R}^2 \times \mathbb{R}_+, \mathbb{R}).$$

Hence $u = 0$ a.e.

N.B. The method we have just used is a classical method to obtain the uniqueness of a problem by solving the adjoint problem (which is here $\partial_t \varphi + y\partial_x \varphi = \psi$ with $\varphi = 0$ as the "final" data.)

2. Let $T > 0$ such that $f(z + T) = f(z)$ for all $z \in \mathbb{R}$ (the function f is therefore of period T). Let $y \in \mathbb{R}$ and $r > T$. There exist p, $q \in \mathbb{Z}$ such that $(p - 1)T < y - r \leq pT < qT \leq y + r < (q + 1)T$. We then have

$$2rF(y, r) = \int_{y-r}^{pT} f(z) \, \mathrm{d}z + \int_{pT}^{qT} f(z) \, \mathrm{d}z + \int_{qT}^{y+r} f(z) \, \mathrm{d}z$$
$$= \int_{y-r}^{pT} f(z) \, \mathrm{d}z + (q - p)mT + \int_{qT}^{y+r} f(z) \, \mathrm{d}z.$$

This gives

$$F(y, r) = m + \left(\frac{(q - p)T}{2r} - 1\right)m + \frac{1}{2r} \int_{y-r}^{pT} f(z) \, \mathrm{d}z + \frac{1}{2r} \int_{qT}^{y+r} f(z) \, \mathrm{d}z.$$

Since $0 \leq 2r - (q - p)T \leq 2T$ and, with $M = \max\{|f(z)|, z \in \mathbb{R}\}$,

$$\left| \int_{y-r}^{pT} f(z)\, dz \right| \leq \int_{y-r}^{pT} |f(z)|\, dz \leq MT,$$

$$\left| \int_{qT}^{y+r} f(z)\, dz \right| \leq \int_{qT}^{y+r} |f(z)|\, dz \leq MT,$$

we therefore have

$$|F(y,r) - m| \leq \frac{(|m| + M)T}{r},$$

which indeed proves that $\lim_{r \to +\infty} F(y,r) = m$, uniformly with respect to $y \in \mathbb{R}$.

3. Let $x \in \mathbb{R}$ and $t > 0$. Using the change of variables $z = x - yt$, that is $y = (x - z)/t$, we obtain

$$\int_{b-\delta}^{b+\delta} f(x - yt)\, dy = \int_{x-bt-\delta t}^{x-bt+\delta t} \frac{f(z)}{t}\, dz = 2\delta F(x - bt, \delta t).$$

Now let $t > 0$. We know that $u(x, y, t) = f(x - yt)$, so we have

$$\int_{b-\delta}^{b+\delta} \int_{a-\delta}^{a+\delta} u(x, y, t)\, dx\, dy = \int_{b-\delta}^{b+\delta} \int_{a-\delta}^{a+\delta} f(x - yt)\, dx\, dy.$$

By Fubini's theorem, we therefore have

$$\int_{b-\delta}^{b+\delta} \int_{a-\delta}^{a+\delta} u(x, y, t)\, dx\, dy = \int_{a-\delta}^{a+\delta} \left(\int_{b-\delta}^{b+\delta} f(x - yt)\, dy \right) dx$$

$$= 2\delta \int_{a-\delta}^{a+\delta} F(x - bt, \delta t)\, dx.$$

The second question gives $\lim_{t \to +\infty} F(x - bt, \delta t) = m$, uniformly with respect to x, so we indeed have

$$\lim_{t \to +\infty} \int_{b-\delta}^{b+\delta} \int_{a-\delta}^{a+\delta} u(x, y, t)\, dx\, dy = 4\delta^2 m.$$

4. First note that (with $M = \max\{|f(z)|, z \in \mathbb{R}\}$)

$$\|u(\cdot, \cdot, t)\|_{L^\infty(\mathbb{R}^2)} \leq M \text{ for all } t > 0.$$

Set $C = \{\mathbb{1}_{]a-\delta,a+\delta[\times]b-\delta,d+\delta[}, \ a, b \in \mathbb{R}, \ \delta > 0\}$. The previous question shows that for all $\varphi \in C$ we have

$$\lim_{t \to +\infty} \int_{\mathbb{R}} \int_{\mathbb{R}} u(x, y, t)\varphi(x, y)\, dx\, dy = m \int_{\mathbb{R}} \int_{\mathbb{R}} \varphi(x, y)\, dx\, dy. \tag{5.109}$$

Let E be the vector space generated by C; by linearity of the integral, we then have (5.109) for all $\varphi \in E$. Now since $C_c(\mathbb{R}^2, \mathbb{R})$ is dense in $L^1(\mathbb{R}^2)$, any element of $C_c(\mathbb{R}^2, \mathbb{R})$ can be approximated as closely as desired for the norm on $L^1(\mathbb{R}^2)$ by an element of E, so that the vector space E is dense in $L^1(\mathbb{R}^2)$ (for the Lebesgue measure). Thanks to this density and to the $L^\infty(\mathbb{R}^2)$ bound on $u(\cdot, \cdot, t)$, we conclude that (5.109) is true for all $u \in L^1(\mathbb{R}^2)$. This indeed shows that $u(., ., t) \to m \star$-weakly in $L^\infty(\mathbb{R}^2)$, as $t \to +\infty$.

Problem 5.14 (Example of a non-strictly hyperbolic system)

1. The Jacobian $J_F(U)$ is a diagonal matrix and its eigenvalues are $f_1'(u_1) = 2u_1$ and $f_2'(u_2) = 2(u_2 + 1)$, which are real, so the system is indeed hyperbolic. These eigenvalues are equal for $u_1 = u_2 + 1$, which shows that the system is not strictly hyperbolic (even though we have $f_1'(u) < f_2'(u)$ for all $u \in \mathbb{R}$).
   Moreover, with the notations of Definition 5.42, we have $\nabla \lambda_i \cdot \phi_i = 2$ for $i = 1, 2$ and therefore the fields associated with F are GNL.
2. The system consists of two scalar equations whose non-linear flux is strictly convex. For the first equation, we have $u_{1,g} = -1 < u_{1,d} = 0$ and therefore the solution is a rarefaction, between the lines $x = f_1'(u_{1,g})t = -2t$ and $x = f_1'(u_{1,d})t = 0$. For the second equation, we have $u_{2,g} = -1 > u_{2,d} = -2$, we therefore have a shock, whose speed σ is given by the Rankine–Hugoniot relation:

$$\sigma = \frac{f_2(u_{2,d}) - f_2(u_{2,g})}{u_{2,d} - u_{2,g}} = \frac{1 - 0}{-2 + 1} = -1.$$

The shock is therefore in the middle of the rarefaction wave.

If, instead of $u_{2,g} = -1$, we take $u_{2,g} = 1$ (and still $u_{2,d} = -2$), the speed of the shock is then

$$\sigma = \frac{f_2(u_{2,d}) - f_2(u_{2,g})}{u_{2,d} - u_{2,g}} = \frac{1 - 4}{-2 - 1} = 1.$$

In this case, the shock is no longer in the rarefaction wave.

Problem 5.15 (The linearised Riemann problem, shallow water equations)

The eigenvalues of the matrix $B(\bar{V})$ are $\lambda_1 = \bar{u} - \bar{c}$ and $\lambda_2 = \bar{u} + \bar{c}$. A basis of $\mathbb{R}^2$ of associated eigenvectors is $\varphi_1 = \begin{bmatrix} 1 \\ -1 \end{bmatrix}$, $\varphi_2 = \begin{bmatrix} 1 \\ 1 \end{bmatrix}$. The decomposition of the vector $V = \begin{bmatrix} u \\ 2c \end{bmatrix}$ in the basis $\{\varphi_1, \varphi_2\}$ is

$$V = \frac{u - 2c}{2} \varphi_1 + \frac{u + 2c}{2} \varphi_2;$$

The solution to this new Riemann problem is therefore

$$
(u, c) = \begin{cases} (u_g, c_g) & \text{for } x < (\bar{u} - \bar{c})t, \\ (u_d, c_d) & \text{for } x > (\bar{u} + \bar{c})t \text{ and} \\ (u_\star, c_\star) & \text{for } (\bar{u} - \bar{c})t < x < (\bar{u} + \bar{c})t, \end{cases}
\qquad \text{with} \quad \begin{cases} u_\star + 2c_\star = u_g + 2c_g, \\ u_\star - 2c_\star = u_d - 2c_d, \end{cases}
$$

that is to say, $u_\star = \dfrac{u_g + u_d}{2} + (c_g - c_d)$, $c_\star = \dfrac{u_g - u_d}{4} + \dfrac{c_g + c_d}{2}$.

Problem 5.16 (Entropy for the shallow water equations with non-flat bottom)

1. (Entropy) Since $\eta(U) = \frac{1}{2}hu^2 + p + ghz = \frac{q^2}{2h} + g\frac{h^2}{2} + ghz$, we have

$$
\nabla \eta(U) = \begin{bmatrix} -\frac{u^2}{2} + gh + gz \\ u \end{bmatrix}.
$$

Multiplying (5.98a) by $-\frac{u^2}{2} + gh + gz$ and (5.98b) by u, we obtain

$$
-\left(\frac{u^2}{2}\right)\partial_t h + \partial_t\left(\frac{gh^2}{2}\right) + gz\partial_t h - \left(\frac{u^3}{2}\right)\partial_x h - h\frac{u^2}{2}\partial_x u + gh^2\partial_x u
$$
$$
+ u\partial_x\left(\frac{gh^2}{2}\right) + gz\partial_x(hu) = 0
$$

and

$$
h\partial_t\left(\frac{u^2}{2}\right) + u^2\partial_t h + u^3\partial_x h + u^2 h\partial_x u + hu^2\partial_x u + u\partial_x\left(\frac{gh^2}{2}\right) + ghuz' = 0.
$$

By adding these two equations, we obtain

$$
\partial_t \eta(U) + \partial_x \Phi(U) = 0,
$$

with $\Phi(U) = \frac{1}{2}hu^3 + guh^2 + ghuz = u(\frac{1}{2}hu^2 + 2p + ghz)$.
The proof that η is a convex function from D to $\mathbb{R}$ is identical to that proposed in Section 5.4.2, because the function η defined here is the same as that of Section 5.4.2 apart from the term ghz, which, being linear, is itself convex.
2. The added diffusion terms are the same as those of subsection 5.4.2.2 of Section 5.4.2. We can therefore answer this question in the same way.

Problem 5.17 (Regular stationary solutions for the shallow water equations)

1. Let (h, u) be a pair of C^1 class functions from $\mathbb{R}$ to $\mathbb{R}_+^\star \times \mathbb{R}$. The pair (h, u) is a regular stationary solution if and only if q is a constant function and

$$
u(x)q'(x) + h(x)u(x)u'(x) + gh(x)h'(x) + gh(x)z'(x) = 0, \text{ for all } x \in \mathbb{R}.
$$

Since q is a constant function and $h(x) > 0$ for all x, this last equation is equivalent to $\psi'(x) = 0$ for all x, that is to say, ψ is constant.

2. Let (h, u) be a regular stationary solution with $\alpha = 0$. Hence $u(x) = 0$ for all x (because $h(x) > 0$) and therefore $gh(x) = \beta - gz(x)$ for all x. This is only possible if $\beta > gz_m$ and the corresponding stationary solution is then given by $u(x) = 0$ and $h(x) = \beta/g - z(x)$ for all x.

N.B. If we allow h to take the value 0 and possibly be a non-regular function, we can construct, in the case where z is a non-constant function, an infinity of other stationary solutions with $u = 0$.

3. Let (h, u) be a regular stationary solution associated with the pair (α, β). Hence $u(x) = \alpha/h(x)$ for all x (recall that $h(x) > 0$). Since $\psi(x) = \beta$ for all x, we therefore have $\alpha^2/(2h^2) + gh + gz = \beta$, that is to say,

$$gh^3(x) + h^2(x)(gz(x) - \beta) + \alpha^2/2 = 0, \quad \text{for all } x \in \mathbb{R}. \tag{5.110}$$

Since $h(x) > 0$, equation (5.110) is impossible if $\beta \le gz(x)$. A first necessary condition (to have a regular stationary solution) is therefore $\beta > gz_m$ (which gives $\beta > gz(x)$ for all x). For $ga < \beta$, we define the polynomial P_a by $P_a(y) = gy^3 + y^2(ga - \beta) + \alpha^2/2$ so that (5.110) reads $P_{z(x)}(h(x)) = 0$.

Since $P_a'(y) = 3gy^2 + 2y(ga - \beta)$, the polynomial P_a has a (positive) local maximum at 0 and a local minimum at the point y_a given by $y_a = (2/(3g))(\beta - ga)$. The value of this local minimum is

$$P_a(y_a) = g(\frac{2}{3g})^3(\beta - ga)^3 - (\frac{2}{3g})^2(\beta - ga)^3 + \alpha^2/2$$

$$= -(\frac{4}{27g^2})(\beta - ga)^3 + \alpha^2/2.$$

Hence $P_a(y_a) > 0$ if $0 < (\beta - ga) < (3/2)(\alpha g)^{2/3}$ and $P_a(y_a) < 0$ if $(\beta - ga) > (3/2)(\alpha g)^{2/3}$. This explains the introduction of $\beta_m = gz_m + (3/2)(\alpha g)^{2/3}$.

a. Assume $\beta < \beta_m$. In this case, there are points x of $\mathbb{R}$ for which $\beta - gz(x) < (3/2)(\alpha g)^{2/3}$. For all these points, $P_{z(x)}(y) > 0$ for $y > 0$. We therefore cannot have $P_{z(x)}(h(x)) = 0$. This proves that there is no regular stationary solution associated with the pair (α, β).

b. Assume $\beta > \beta_m$. In this case, for all $a \le z_m$, the equation $P_a(y) = 0$ has two positive solutions denoted by $\varphi_1(a)$ and $\varphi_2(a)$ with $\varphi_1(a) < y_a = (2/(3g))(\beta - ga) < \varphi_2(a)$. For all $x \in \mathbb{R}$, we have $h(x) \in \{\varphi_1(z(x)), \varphi_2(z(x))\}$. Since h is continuous, we must therefore have $h(x) = \varphi_1(z(x))$ for all $x \in \mathbb{R}$ or $h(x) = \varphi_2(z(x))$ for all $x \in \mathbb{R}$.

   This suggests two stationary solutions (h_1, u_1) and (h_2, u_2) with, for $i = 1, 2$, $h_i(x) = \varphi_i(z(x))$ and $u_i(x) = \alpha/h_i(x)$. To show that these two solutions are indeed regular stationary solutions, it remains to verify that h_1 and h_2 are functions of class C^1. This is a consequence of the fact that φ_1 and φ_2 are functions of class C^1 in the neighbourhood of points a such that $(\beta - ga) >$

$(3/2)(\alpha g)^{2/3}$. Indeed, we set $F(a, y) = P_a(y)$ and we note that $\partial_y F(a, y) = 3gy^2 + 2y(ga - \beta) \neq 0$ for $y = \varphi_i(a)$ ($i = 1$ or 2, $(\beta - ga) > (3/2)(\alpha g)^{2/3}$). The implicit function theorem applied to the equation $F(a, y) = 0$ in the neighbourhood of points $(a, \varphi_i(a))$ then gives the C^1 character of the functions φ_i. Since we have assumed that z was of class C^1, we get that the functions h_i are of class C^1 (and therefore the functions u_i are as well).

Finally, as we reasoned from a necessary condition, there are indeed only two regular stationary solutions. Figure 5.9 gives an example of regular stationary solutions.

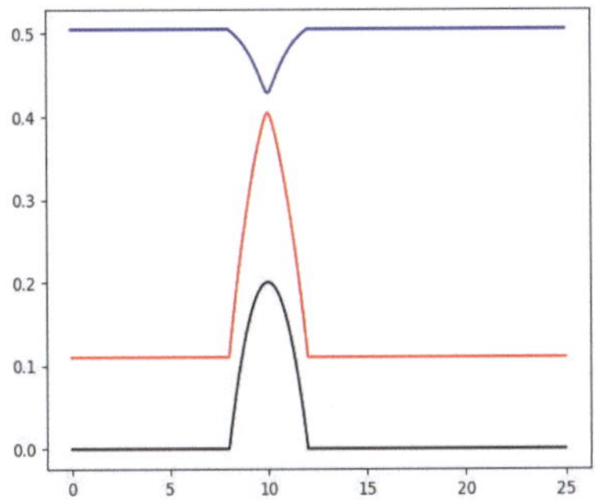

Fig. 5.9 Stationary solutions for $\beta = \beta_m + 0.001$ and $\alpha = 0.1$. Supercritical solution ($u > c$) in red and subcritical ($u < c$) in blue. The background is in black

4. As in question 3 and with the same notations, for all $a < z_m$, the equation $P_a(y)$ has two positive solutions denoted by $\varphi_1(a)$ and $\varphi_2(a)$ with $\varphi_1(a) < y_a = (2/(3g))(\beta - ga) < \varphi_2(a)$. But for $a = z_m$, we have $\varphi_1(a) = y_a = (2/(3g))(\beta - ga) = \varphi_2(a)$. The two solutions given in question 3 seem to still exist here and are indeed conflated if z is a constant function. The only remaining question is about the regularity of the functions h_i.

   a. If z is a constant function (thus $z = z_m$), there is a unique solution which is $h(x) = y_{z_m} = (2/(3g))(\beta_m - gz_m)$ for all x.
   b. If $z(x) \neq z_m$ for all x, there are exactly two regular stationary solutions associated with the pair (α, β). These are those calculated for the case $\beta > \beta_m$ (and the proof is identical).

If z is a non-constant function and there exists an x such that $z(x) = z_m$ (and still $\beta = \beta_m$) the situation is a bit more complex. The problem here is due to the fact that the functions φ_i are not differentiable at the point z_m. More precisely $\lim_{z \to z_m, z < z_m} |\varphi_i'(z)| = +\infty$. However, if the maximum of z is reached at a unique point denoted by x_m and $z''(x_m) < 0$, it can be shown that there are (exactly) two regular stationary solutions associated with the pair (α, β), obtained by taking

i different depending on whether $x < x_m$ and $x > x_m$. This case is not detailed here.

5. The first part of this question was resolved in question 3b. With the notations of question 3b, we have indeed, for $i = 1, 2$, $h_i(x) = \varphi_i(z(x))$ for all $x \in \mathbb{R}$ and the functions φ_i are regular. Finally, we have $h_1 < h_2$.

   For the second part of the question we note that $h_i(x) + z(x) = \psi_i(z(x))$ with $\psi_i(a) = \varphi_i(a) + a$ (note that $\psi_i(a)$ is defined, like $\varphi_i(a)$, for all a such that $ga < \beta - (3/2)(\alpha g)^{2/3}$). Since $h_i'(x) + z'(x) = \psi_i'(z(x))z'(x)$, the second part of the question is a consequence of the fact that $\psi_2'(a) < 0$ and $\psi_1'(a) > 0$ for all $a \leq z_m$. To prove this fact, let $a \leq z_m$. We have $ga < \beta - (3/2)(\alpha g)^{2/3}$ and

$$g\varphi_i^3(a) + \varphi_i^2(a)(ga - \beta) + \alpha^2/2 = 0.$$

   This first equality gives, in particular, $g\varphi_i(a) + (ga - \beta) < 0$. On the other hand, by differentiating this equation with respect to a, we obtain

$$\varphi_i'(a)\varphi_i(a)(3g\varphi_i(a) + 2(ga - \beta)) = -g\varphi_i^2(a). \tag{5.111}$$

   For $i = 2$, we use $g\varphi_2(a) + ga - \beta < 0$ and $y_a = (2/(3g))(\beta - ga) < g\varphi_2(a)$. This gives $0 < (3g\varphi_2(a) + 2(ga - \beta)) < g\varphi_2(a)$. Thus, with (5.111), $\varphi_2'(a) < -1$ and therefore $\psi_2'(a) < 0$.

   For $i = 1$, we use $y_a = (2/(3g))(\beta - ga) > \varphi_1(a)$, this gives $3g\varphi_1(a) + 2(ga - \beta) < 0$. Thus, with (5.111), $\varphi_1'(a) > 0$ and therefore $\psi_1'(a) > 1 > 0$.

6. We use the notations from the solutions of the previous questions. For $i = 1$ or 2, $u_i^2 = \alpha^2/h_i^2$, which gives with (5.110)

$$2gh_i(x) + 2(gz(x) - \beta) + u_i^2(x) = 0, \text{ for all } x \in \mathbb{R},$$

and therefore

$$3gh_i(x) + 2(gz(x) - \beta) = gh_i(x) - u_i^2(x), \text{ for all } x \in \mathbb{R}. \tag{5.112}$$

   Recall that the choice of φ_1 and φ_2 is such that $\varphi_1(a) < (2/(3g))(\beta - ga) < \varphi_2(a)$ for all $a \leq z_m$. For $i = 2$, this gives, with $a = z(x)$, $u_2^2(x) < gh_2(x)$ and therefore $0 < u_2(x) < \sqrt{gh_2(x)}$ for all $x \in \mathbb{R}$. The flow here is subsonic (recall that $\sqrt{gh}$ is, for this system, the "speed of sound"). For $i = 1$, this gives, with $a = z(x)$, $u_1^2(x) > gh_1(x)$ and therefore $u_1(x) > \sqrt{gh_1(x)}$ for all $x \in \mathbb{R}$. The flow here is supersonic.

Problem 5.18 (Shallow water equations, entropy in the Lax sense)

1. The function U is a weak solution if and only if the Rankine–Hugoniot conditions are satisfied on the line $x = \sigma t$, that is, with the usual notations, $\sigma[h] = [hu]$ and $\sigma[hu] = [hu^2 + p]$. First note that $h_d \neq h_g$, otherwise $\sigma[h] = [hu]$ gives $u_g = u_d$ in contradiction with the fact that U is discontinuous. We then also

have $u_g \neq u_d$ because $u_g = u_d$ gives $\sigma = u_g$ and $\sigma[hu] = [hu^2 + p]$ then gives $[p] = 0$ and therefore $h_g = h_d$, in contradiction with the fact that U is discontinuous.

Assume, for the sake of readability, that $g = 1$ and set $v_g = \bar{v}$, $v_d = v$ for $v = h, u$ and p ($p = h^2/2$). The Rankine–Hugoniot conditions read

$$\sigma(h - \bar{h}) = (hu - \bar{h}\bar{u}),$$
$$\sigma(hu - \bar{h}\bar{u}) = (hu^2 + p - \bar{h}\bar{u}^2 - \bar{p}).$$

We multiply the first equation by $(h + \bar{h})/2 - (u + \bar{u})^2/8$, the second by $(u + \bar{u})/2$ and we add. We obtain

$$\sigma[p + \frac{1}{2}hu^2] - \sigma\frac{[h]}{8}(u - \bar{u})^2 = [\frac{1}{2}hu^3 + 2pu] - \frac{[hu]}{8}(u - \bar{u})^2 - \frac{[u]}{4}[h]^2,$$

and therefore, as $\sigma[h] = [hu]$, with $\eta(U) = \frac{1}{2}hu^2 + p$ and $\Phi(U) = \frac{1}{2}hu^3 + 2pu$,

$$\sigma[\eta(U)] = [\Phi(U)] - \frac{[u]}{4}[h]^2.$$

This proves that it is an entropy weak solution in the sense of Definition 5.41 with the entropy $\eta(U) = hu^2/2 + gh^2/2$ if and only if $[u] \leq 0$, that is to say $u_d < u_g$ (since we already know that $u_g \neq u_d$).

2. If U is an entropy weak solution in the sense of Definition 5.41 with the entropy $\eta(U) = hu^2/2 + gh^2/2$, the previous question shows that $u_d < u_g$, that is, with the notations of subsection 5.4.3.2, that $u_d = u_g - S$ (we have $S > 0$). The result of this latter subsection then gives that U is a 1-shock if $h_g < h_d$ and a 2-shock if $h_d < h_g$. The function U therefore satisfies the Lax condition.

Conversely, if U is a weak solution that satisfies the Lax condition, we necessarily have $u_d < u_g$, this is the first step of subsection 5.4.3.2 (and we then showed that it is a 1-shock if $h_g < h_d$ and a 2-shock if $h_g > h_d$). The previous reasoning then gives us

$$\sigma[\eta(U)] = [\Phi(U)] - \frac{[u]}{4}[h]^2 > [\Phi(U)],$$

which proves that U is an entropy weak solution in the sense of Definition 5.41 with the entropy $\eta(U) = hu^2/2 + gh^2/2$.

References

1. R. A. Adams and J. J. F. Fournier. *Sobolev Spaces*, volume 140 of *Pure and Applied Mathematics (Amsterdam)*. Elsevier/Academic Press, Amsterdam, second edition, 2003. 1, 5, 11, 12

2. G. Allaire. *Analyse numérique et optimisation*. Éditions de l'École Polytechnique, 2012. 196

3. H. W. Alt and S. Luckhaus. Quasilinear elliptic-parabolic differential equations. *Math. Z.*, 183(3):311–341, 1983. 320

4. W. Arendt, I. Chalendar, and Eymard. Lions' representation theorem and applications. *J. Math. Anal. Appl.*, 522(2), 2023. 280

5. M. Artola. Sur une classe de problèmes paraboliques quasi-linéaires. *Boll. Un. Mat. Ital. B (6)*, 5(1):51–70, 1986. 186

6. J.-P. Aubin. Analyse mathématique - un théorème de compacité. *C. R. A. S.*, 256(24):5042–5044, 1963. 296, 301

7. C. Bardos, A. LeRoux, and J. Nédélec. First order quasilinear equations with boundary conditions. *Comm. Partial Differential Equations*, 9:1017–1034, 1979. 374, 376

8. P. Bénilan, L. Boccardo, T. Gallouët, R. Gariepy, M. Pierre, and J. L. Vázquez. An L^1-theory of existence and uniqueness of solutions of nonlinear elliptic equations. *Ann. Scuola Norm. Sup. Pisa Cl. Sci. (4)*, 22(2):241–273, 1995. URL: http://www.numdam.org/item?id=ASNSP_1995_4_22_2_241_0. 257

9. S. Benzoni-Gavage and D. Serre. *Multidimensional Hyperbolic Partial Differential Equations*. Oxford Mathematical Monographs. The Clarendon Press, Oxford University Press, Oxford, 2007. First-order systems and applications. 389

10. L. Boccardo and T. Gallouët. Nonlinear elliptic and parabolic equations involving measure data. *J. Funct. Anal.*, 87(1):149–169, 1989. doi:10.1016/0022-1236(89)90005-0. 255

11. F. Boyer and P. Fabrie. *Éléments d'analyse pour l'étude de quelques modèles d'écoulements de fluides visqueux incompressibles*, volume 52 of *Mathématiques & Applications (Berlin) [Mathematics & Applications]*. Springer-Verlag, Berlin, 2006. 81, 314

12. H. Brezis. *Analyse fonctionnelle*. Masson, 1985. vi

13. H. Brezis. *Functional Analysis, Sobolev Spaces and Partial Differential Equations*. Universitext. Springer, New York, 2011. vi, 1, 8, 10, 13, 253

14. L. E. J. Brouwer. Über Abbildung von Mannigfaltigkeiten. *Math. Ann.*, 71(4):598, 1912. doi:10.1007/BF01456812. 173

15. T. Cazenave and A. Haraux. *Introduction aux problèmes d'évolution semi-linéaires*, volume 1 of *Mathématiques & Applications (Paris) [Mathematics and Applications]*. Ellipses, Paris, 1990. 253

16. G. A. Chechkin and A. Y. Goritsky. S.N. Kruzhkov's lectures on first-order quasilinear PDEs. In E. Emmrich and P. Wittbold, editors, *De Gruyter Proceedings in Mathematics*, Analytical and Numerical Aspects of Partial Differential Equations, pages pp. 1–68. De Gruyter, July

2009. Traduit du russe par B. Andreianov. URL: https://hal.archives-ouvertes.fr/hal-00363287, doi:10.1515/9783110212105.1. 368

17. K. Deimling. *Nonlinear Functional Analysis*. Springer-Verlag, Berlin, 1985. 173

18. F. Demengel and G. Demengel. *Traces of Functions on Sobolev Spaces*, pages 113–177. Springer London, London, 2012. doi:10.1007/978-1-4471-2807-6_3. 89

19. J. Droniou. Intégration et Espaces de Sobolev à Valeurs Vectorielles. Working paper or preprint, Apr. 2001. URL: https://hal.archives-ouvertes.fr/hal-01382368. 258, 260, 263

20. J. Droniou. Quelques Résultats sur les Espaces de Sobolev. Working paper or preprint, Apr. 2001. URL: https://hal.archives-ouvertes.fr/hal-01382370. 11, 13, 14, 81, 111, 127

21. J. Droniou, R. Eymard, T. Gallouët, C. Guichard, and R. Herbin. *The Gradient Discretization Method*, volume 82 of *Mathématiques & Applications (Paris) [Mathematics and Applications]*. Springer, 2018. 305

22. L. Evans. *Partial Differential Equations, second edition*. Graduate studies in Mathematics, vol 19, AMS, 2010. 13, 67, 80, 81, 245, 249, 250

23. R. Eymard, T. Gallouët, and R. Herbin. Finite volume methods. In *Handbook of Numerical Analysis, Vol. VII*, Handb. Numer. Anal., VII, pages 713–1020. North-Holland, Amsterdam, 2000. doi:10.1086/phos.67.4.188705. 320, 378, 385, 388, 389, 425

24. R. Eymard, T. Gallouët, and R. Herbin. Existence and uniqueness of the entropy solution to a nonlinear hyperbolic equation. *Chin. Ann. of Math.*, 16B(1):1–14, 1995. 368, 384

25. T. Gallouët. Analyse fonctionnelle. Lecture, Nov. 2021. URL: https://cel.hal.science/hal-04026364. vi, 10, 76, 261

26. T. Gallouët and R. Herbin. *Mesure, intégration, probabilités*. Ellipses, 2013. URL: https://hal.science/hal-01283567. vi, 1, 3, 4, 7, 8, 9, 15, 22, 31, 32, 41, 44, 47, 65, 70, 133, 135, 142, 145, 167, 168, 169, 192, 201, 202, 214, 229, 230, 236, 237, 258, 259, 429

27. T. Gallouët, R. Herbin, and J.-C. Latché. Lax–Wendroff consistency of finite volume schemes for systems of non linear conservation laws: extension to staggered schemes. *SeMA*, 2021. doi:10.1007/s40324-021-00263-0. 387

28. E. Godlewski and P.-A. Raviart. *Numerical Approximation of Hyperbolic Systems of Conservation Laws*, volume 118 of *Applied Mathematical Sciences*. Springer-Verlag, New York, second edition, [2021] ©2021. doi:10.1007/978-1-0716-1344-3. 389

29. S. K. Godunov. A difference scheme for numerical solution of discontinuous solution of hydrodynamic equations. *Mat. Sbornik*, 47:271–306, 1969. 425

30. P. Grisvard. *Elliptic Problems in Nonsmooth Domains*, volume 69 of *Classics in Applied Mathematics*. Society for Industrial and Applied Mathematics (SIAM), Philadelphia, PA, 2011. doi:10.1137/1.9781611972030.ch1. 25, 90

31. O. Guibé, A. Mokrane, Y. Tahraoui, and G. Vallet. Lewy-Stampacchia's inequality for a pseudomonotone parabolic problem. *Adv. Nonlinear Anal.*, 9(1):591–612, 2020. doi:10.1515/anona-2020-0015. 314

32. O. Kavian. *Introduction à la théorie des points critiques et applications aux problèmes elliptiques*, volume 13 of *Mathématiques & Applications (Berlin) [Mathematics & Applications]*. Springer-Verlag, Paris, 1993. 173

33. P. D. Lax. *Hyperbolic Systems of Conservation Laws and the Mathematical Theory of Shock Waves*, pages 1–48. Society for Industrial and Applied Mathematics, 1987. URL: https://epubs.siam.org/doi/abs/10.1137/1.9781611970562.ch1, arXiv:https://epubs.siam.org/doi/pdf/10.1137/1.9781611970562.ch1, doi:10.1137/1.9781611970562.ch1. 372, 403, 405

34. H. Le Dret. *Équations aux dérivées partielles elliptiques non linéaires*, volume 72 of *Mathématiques & Applications (Berlin) [Mathematics & Applications]*. Springer, Heidelberg, 2013. doi:10.1007/978-3-642-36175-3. 203

35. J. Leray. Sur le mouvement d'un liquide visqueux emplissant l'espace. *Acta Math.*, 63(1):193–248, 1934. doi:10.1007/BF02547354. vi, 2

36. J. Leray and L. Jacques-Louis. Quelques résultats de Višik sur les problèmes elliptiques non linéaires par les méthodes de minty-browder. *Bulletin de la S. M. F.*, 93:97–107, 1965. URL: http://www.numdam.org/item?id=BSMF_1965__93__97_0. 190

37. J. Leray and J. Schauder. Topologie et équations fonctionnelles. *Ann. Sci. École Norm. Sup. (3)*, 51:45–78, 1934. URL: http://www.numdam.org/item?id=ASENS_1934_3_51__45_0. 173, 174

38. J.-L. Lions. *Quelques méthodes de résolution des problèmes aux limites non linéaires*. Dunod, Paris; Gauthier-Villars, Paris, 1969. 303

39. J.-L. Lions and E. Magenes. *Non-Homogeneous Boundary Value Problems and Applications. Vol. I*, volume Band 181 of *Die Grundlehren der mathematischen Wissenschaften*. Springer-Verlag, New York-Heidelberg, 1972. Translated from the French by P. Kenneth. 90

40. N. G. Meyers. An L^p estimate for the gradient of solutions of second order elliptic divergence equations. *Ann. Scuola Norm. Sup. Pisa Cl. Sci. (3)*, 17:189–206, 1963. 83, 256

41. L. Nirenberg. On elliptic partial differential equations. *Annali della Scuola Normale Superiore di Pisa - Scienze Fisiche e Matematiche*, Ser. 3, 13(2):115–162, 1959. URL: http://www.numdam.org/item/ASNSP_1959_3_13_2_115_0/. 80

42. O. A. Oleĭnik. On the uniqueness of the generalized solution of the Cauchy problem for a nonlinear system of equations occurring in mechanics. *Uspehi Mat. Nauk (N.S.)*, 12(6(78)):169–176, 1957. 372

43. F. Otto. *Ein Randwertproblem f¨ur skalare Erhaltnungssätze (German)*. PhD thesis, Universität Bonn, 1993. 374

44. F. Otto. Initial-boundary value problem for a scalar conservation law. *C. R. Acad. Sci. Paris Sér. I Math.*, 8:729–734, 1996. 375, 386

45. A. Prignet. Remarks on existence and uniqueness of solutions of elliptic problems with right-hand side measures. *Rend. Mat. Appl. (7)*, 15(3):321–337, 1995. 257

46. L. Schwartz. *Théorie des distributions*, volume IX-X of *Publications de l'Institut de Mathématique de l'Université de Strasbourg*. Hermann, Paris, 1966. Nouvelle édition, entiérement corrigée, refondue et augmentée. 2

47. D. Serre. *Systems of Conservation Laws 1: Hyperbolicity, Entropies, Shock Waves*. Cambridge University Press, 1999. Translated by I. N. Sneddon. doi:10.1017/CBO9780511612374. 389, 392, 398

48. J. Serrin. Pathological solutions of elliptic differential equations. *Ann. Scuola Norm. Sup. Pisa Cl. Sci. (3)*, 18:385–387, 1964. 257

49. J. Simon. Compact sets in the space $L^p(0, T; B)$. *Ann. Mat. Pura Appl. (4)*, 146:65–96, 1987. doi:10.1007/BF01762360. 296, 301

50. J. Smoller. *Shock Waves and Reaction-Diffusion Equations*, volume 258 of *Grundlehren der Mathematischen Wissenschaften [Fundamental Principles of Mathematical Sciences]*. Springer-Verlag, New York, 1994. 249

51. L. Tartar. *An Introduction to Sobolev Spaces and Interpolation Spaces*, volume 3 of *Lecture Notes of the Unione Matematica Italiana*. Springer, Berlin; UMI, Bologna, 2007. 5

52. G. Teschl. *Topics in Linear and Nonlinear Functional Analysis*, volume to appear of *Graduate Studies in Mathematics*. American Mathematical Society (AMS), Providence, Rhode Island, 2021,https://www.mat.univie.ac.at/ gerald/ftp/book-fa/fa.pdf. 173

53. J. Vovelle. Convergence of finite volume monotone schemes for scalar conservation laws on bounded domains. *Numer. Math.*, 90(3):563–596, 2002. doi:10.1007/s002110100307. 387

Abbreviations

v.s.   vector space
n.v.s. normed vector space
v.s.s. vector subspace
a.e.   almost everywhere
s.t.   such that

BC   Boundary Conditions
PDE Partial Differential Equations
LD   Linearly Degenerate (Definition 5.42)
GNL Genuinely Non-Linear (Definition 5.42)

Notations

Id   the identity map of a set to itself
div   divergence

E a normed real vector space.
E' the topological dual of a normed real vector space (see Definition 1.12)
Ω open set of $\mathbb{R}^N$

$C^k(\Omega)$ set of functions from Ω to $\mathbb{R}$, k times continuously differentiable
$D(\mathcal{A})$ domain of an operator $\mathcal{A}$.
$\mathcal{D}(\Omega)$ set of functions from Ω to $\mathbb{R}$, of class C^∞ and with compact support in Ω
$\mathcal{D}^\star(\Omega)$ the algebraic dual of $\mathcal{D}(\Omega)$, see Definition 1.3
$H^1(\Omega)$ set of (classes of) L^2 functions with weak L^2 derivative, Definition 1.8
$H_0^1(\Omega)$ closure of $\mathcal{D}(\Omega)$ in $H^1(\Omega)$
$H^m(\Omega)$ Sobolev space, Definition 1.8
$H_{\mathrm{div}}(\Omega)$ set of (classes of) functions of $L^2(\Omega)$ with (weak) divergence in $L^2(\Omega)$

$\mathcal{L}^p$	set of measurable functions with integrable p-th power
L^p	set of (classes of) measurable functions with integrable p-th power
L^p_{loc}	set of (classes of) measurable functions with locally integrable p-th power (see Definition 1.1)
$\mathrm{Lip}_{\mathrm{loc}}(\mathbb{R}, \mathbb{R})$	space of locally Lipschitz functions
$\mathcal{L}(E, F)$	set of continuous linear maps from E to F
$\mathcal{M}_n(\mathbb{R})$	set of real square matrices of order n.
$\mathrm{Vect}\{e_p, p = 1, \ldots n\}$	vector space generated by the family $\{e_p, p = 1, \ldots n\}$.
$W^{m,p}(\Omega)$	Sobolev space, Definition 1.8
$[a, b]$	closed interval of $\mathbb{R}$ with lower bound a and upper bound b
$]a, b[$	open interval of $\mathbb{R}$ with lower bound a and upper bound b
$[\![a, b]\!]$	$[a, b]$ if $a \leq b$, $[b, a]$ otherwise
$|x|$	Euclidean norm of $x \in \mathbb{R}^N$
$x \cdot y$	scalar product of x and y in $\mathbb{R}^N$
$(u \mid v)_E$	inner product of u and v in a Hilbert space E
$\langle T, \varphi \rangle_{F,E}$	action of the element T of the dual (topological or algebraic) F of E on an element φ of E
$\mathbb{N}$	set of natural numbers
$\mathbb{Z}$	set of integers
$\mathbb{R}$	set of real numbers
$\mathbb{R}_+$	set of non-negative real numbers, that is $\{x \in \mathbb{R},\ x \geq 0\}$.
$\mathbb{R}^\star$	set of non-zero real numbers
$d\gamma$	measure on the boundary of an open set
γ	trace on the boundary of an open set
λ	Lebesgue measure on the Borel sets of $\mathbb{R}$ (or a subset of $\mathbb{R}$)
λ_N	Lebesgue measure on the Borel sets of $\mathbb{R}^N$ (or a subset of $\mathbb{R}^N$)
$\mathrm{sgn}(s)$	sign of s (1 if $s \geq 0$, -1 otherwise)
$\mathbb{1}_A$	characteristic function of a set A ($\mathbb{1}_A(x) = 1$ if $x \in A$, 0 otherwise)
$[V]$	jump of V (across a discontinuity)
$D_i \varphi$	weak derivative of the function φ with respect to its i-th space variable
$d\varphi$	differential of the function φ
$\partial_i \varphi$	partial derivative of the function φ with respect to its i-th space variable
$\partial_t \varphi$	partial (weak) derivative of the function φ of space and time with respect to its time variable
$\rho(T)$	set of regular values of the operator T (resolvent set)
$\sigma(T)$	spectrum: set of singular values of the operator T
$\mathcal{VP}(T)$	set of *eigenvalues* of the operator T

Index